Advances in Intelligent Systems and Computing

Volume 1306

The series "Advances in Intelligent Systems and Computing" contains publications on theory, applications, and design methods of Intelligent Systems and Intelligent Computing. Virtually all disciplines such as engineering, natural sciences, computer and information science, ICT, economics, business, e-commerce, environment, healthcare, life science are covered. The list of topics spans all the areas of modern intelligent systems and computing such as: computational intelligence, soft computing including neural networks, fuzzy systems, evolutionary computing and the fusion of these paradigms, social intelligence, ambient intelligence, computational neuroscience, artificial life, virtual worlds and society, cognitive science and systems, Perception and Vision, DNA and immune based systems, self-organizing and adaptive systems, e-Learning and teaching, human-centered and human-centric computing, recommender systems, intelligent control, robotics and mechatronics including human-machine teaming, knowledge-based paradigms, learning paradigms, machine ethics, intelligent data analysis, knowledge management, intelligent agents, intelligent decision making and support, intelligent network security, trust management, interactive entertainment, Web intelligence and multimedia.

The publications within "Advances in Intelligent Systems and Computing" are primarily proceedings of important conferences, symposia and congresses. They cover significant recent developments in the field, both of a foundational and applicable character. An important characteristic feature of the series is the short publication time and world-wide distribution. This permits a rapid and broad dissemination of research results.

Indexed by SCOPUS, DBLP, EI Compendex, INSPEC, WTI Frankfurt eG, zbMATH, Japanese Science and Technology Agency (JST), SCImago.

All books published in the series are submitted for consideration in Web of Science.

More information about this series at http://www.springer.com/series/11156

Rafik A. Aliev · Janusz Kacprzyk ·
Witold Pedrycz · Mo Jamshidi ·
Mustafa Babanli · Fahreddin M. Sadikoglu
Editors

14th International Conference on Theory and Application of Fuzzy Systems and Soft Computing – ICAFS-2020

 Springer

Editors
Rafik A. Aliev
Department of Control Systems
Azerbaijan State Oil and Industry University
Baku, Azerbaijan

Janusz Kacprzyk
Systems Research Institute
Polish Academy of Sciences
Warsaw, Poland

Witold Pedrycz
Department of Electrical
and Computer Engineering
University of Alberta
Edmonton, AB, Canada

Mo Jamshidi
Department of Electrical
and Computer Engineering
University of Texas at San Antonio
San Antonio, TX, USA

Mustafa Babanli
Azerbaijan State Oil and Industry University
Baku, Azerbaijan

Fahreddin M. Sadikoglu
Department of Mechatronics
Near East University
Mersin, Turkey

ISSN 2194-5357   ISSN 2194-5365   (electronic)
Advances in Intelligent Systems and Computing
ISBN 978-3-030-64057-6   ISBN 978-3-030-64058-3 (eBook)
https://doi.org/10.1007/978-3-030-64058-3

This Springer imprint is published by the registered company Springer Nature Switzerland AG
The registered company address is: Gewerbestrasse 11, 6330 Cham, Switzerland

Preface

The Fourteenth International Conference on Application of Fuzzy Systems, Soft Computing and Artificial Intelligence Tools(ICAFS-2020) is the premier international conference organized by Azerbaijan Association of "Zadeh's Legacy and Artificial Intelligence" (Azerbaijan), Azerbaijan State Oil and Industry University (Azerbaijan), University of Texas, San Antonio (USA), University of Alberta (Alberta Canada), University of Toronto (Toronto, Ontario, Canada), University of California, Berkeley (Berkeley, USA), Polish Academy of Sciences, System Research Institute(Poland), Near East University (North Cyprus).

This volume presents an edited selection of the presentations from the Fourteenth International Conference on Application of Fuzzy Systems, Soft Computing and Artificial Intelligence Tools (ICAFS-2020) which was held in Budva, Montenegro, August 27–28, 2020. ICAFS-2020 is held as a meeting for the communication of research on application of fuzzy logic, uncertain computation, Z-information processing, neurofuzzy approaches and different constituent methodologies of soft computing applied in economics, business, industry, education, medicine, earth sciences and other fields. The conference provided an opportunity to present and discuss state-of-the-art research in this expanding domain.

This volume will be a useful guide for academics, practitioners and graduates in fuzzy logic and soft computing. It will allow for increasing of interest in development and applying of soft computing and artificial intelligence methods in various real-life fields.

August 2020

Rafik Aliev
Chairman of ICAFS-2020

Organization

Chairman

R. A. Aliev   Azerbaijan State Oil and Industry University, Azerbaijan

Co-chairmen and Guest Editors

J. Kacprzyk   Systems Research Institute Polish Academy of Sciences, Poland
M. Jamshidi   University of Texas at San Antonio, USA
W. Pedrycz   University of Alberta, Canada
M. B. Babanli   Azerbaijan State Oil and Industry University, Azerbaijan
F. S. Sadikoglu   Near East University, North Cyprus

International Program Committee

A. Averkin   Plekhanov Russian University of Economics, Russia
A. Musayev   The Azerbaijan University, Azerbaijan
B. Fazlollahi   Georgia State University, USA
C. Kahraman   Istanbul Technical University, Turkey
D. Dubois   Université Paul Sabatier, IRIT, France
D. Enke   Missouri University of Science and Technology, USA
D. Kumar Jana   Haldia Institute of Technology, India
E. Babaei   University of Tabriz, Iran
F. Aminzadeh   University of Houston, USA
H. Berenji   Intelligent Inference Systems Corporation, USA
H. Hamdan   Université Paris-Sud, Université Paris-Saclay, France, UK

H. Prade	Université Paul Sabatier, IRIT, France
H. Roth	University of Siegen, Germany
I. Batyrshin	National Polytechnic Institute, Mexico
I. G. Akperov	Southern University (IMBL), Russia
I. Perfilieva	University of Ostrava, Czech Republic
K. Atanassov	Institute of Biophysics and Biomedical Engineering, Bulgaria
K. Bonfig	University of Siegen, Germany
K. Takahashi	Hitachi Research Laboratory, Japan
M. Gupta	University of Saskatchewan, Canada
M. Nikravesh	University of California, Berkeley, USA
N. Allahverdi	Karatay Univeristy, Turkey
N. Yusupbekov	Tashkent State Technical University, Uzbekistan
O. Huseynov	Azerbaijan State Oil and Industry University, Azerbaijan
O. Kaynak	Bogazici University, Turkey
P. Moog	University of Siegen, Germany
G. Imanov	Institute of Control Systems of the Azerbaijan National Academy of Sciences, Azerbaijan
R. Gurbanov	Azerbaijan State Oil and Industry University, Azerbaijan
R. R. Aliev	Eastern Mediterranean University, North Cyprus
R. Yager	Iona College, USA
S. Ulyanov	Dubna State University, Russia
T. Allahviranloo	Bahçeşehir University, Turkey
T. Fukuda	Nagoya University, Japan
T. Takagi	Meiji University, Japan
V. Kreinovich	University of Texas at El Paso, USA
V. Loia	University of Salerno, Italy
V. Niskanen	University of Helsinki, Finland
V. Novak	University of Ostrava, Czech Republic
V. B. Tarassov	Bauman Moscow State Technical University, Russia

Organizing Committee

Chairman

| U. Eberhardt | University of Siegen, Germany |

Co-chairmen

| L. Gardashova | Azerbaijan State Oil and Industry University, Azerbaijan |
| T. Abdullayev | Odlar Yurdu University, Azerbaijan |

Members

N. Adilova	Azerbaijan State Oil and Industry University, Azerbaijan
A. Alizadeh	Azerbaijan State Oil and Industry University, Azerbaijan
B. Guirimov	State Oil Company of Azerbaijan Republic, SOCAR, Azerbaijan
A. Guliyev	Azerbaijan Tourism and Management University, Azerbaijan
R. Rzayev	Institute of Control Systems of the Azerbaijan National Academy of Sciences, Azerbaijan
M. Elamin	Near East University, North Cyprus
M. Salahli	Çanakkale Onsekiz Mart University, Turkey
E. Tuncel	Near East University, North Cyprus
G. Sadikoglu	Near East University, North Cyprus
S. Uzelaltinbulat	Near East University, North Cyprus
Ş. Akdağ	Near East University, North Cyprus

Conference Organizing Secretariat

Azadlig Ave. 20, AZ 1010 Baku, Azerbaijan
Phone: +99 412 493 45 38, Fax: +99 412 598 45 09
E-mail: latsham@yandex.ru, adilovanigarr@gmail.com

Keynote Speaker Abstracts

Cognitive Biases in Soft Human Consistent Decision Making and Optimization Models

Janusz Kacprzyk ⓘ

Polish Academy of Sciences Member, Academia Europaea Member, European Academy of Sciences and Arts Foreign Member, Bulgarian Academy of Sciences Foreign Member, Spanish Royal Academy of Economic and Financial Sciences (RACEF) Foreign member, Finnish Society of Sciences and Letters Systems Research Institute, Warsaw, Poland

kacprzyk@ibspan.waw.pl

Abstract. We deal with a broadly perceived human-centric decision-making problem in which a best (sufficiently, good, reasonable, satisfactory,...) option is to be found from among feasible ones. To increase the trustworthiness of the solution process itself and a solution obtained, we postulate that the process should be transparent and explainable to the human decision maker and analyst, and—to be more effective and efficient—should proceed in a human-in-the-loop like context. To increase the implement ability, we advocate models and solutions that reflect some human-specific characteristics. In particular, in the models developed, we reflect some of Kahneman and Tversky's cognitive biases.

The cognitive biases are some diversion from what traditional models, based usually on the utility maximization, postulate. People make judgments or decisions in the ways that are systematically different from what the traditional economic models say but this does not necessarily lead to suboptimal outcomes.

We consider the following main classes cognitive biases within which we just mention a few: (1) decision making, belief and behavioral biases (e.g., the bandwagon effect, i.e., to do what a majority thinks), (2) social biases (e.g., status quo bias, i.e., a tendency to defend and bolster the status quo), (3) memory errors and biases (e.g., consistency bias, i.e., the present resembles the past), etc.

We use elements of some of these cognitive biases to develop a new class of decision making and optimization models, notably fuzzy. In particular, we use the status quo and bandwagon effect biases to extend our fuzz group decision making and consensus reaching models. Moreover, we show the use of the status quo bias to obtain a realistic and easier implementable fuzzy LP models of regional agriculture. We mention some remedies, i.e., debasing or cognitive bias mitigation, but also show a positive role of including cognitive biases.

Z-relation Equations

R. A. Aliev Aliev[ID]

Joint MBA Program, Georgia State University, USA, Azerbaijan State Oil and Industry University, Azerbaijan, 20 Azadlig Ave., AZ1010 Baku, Azerbaijan

raliev@asoa.edu.az

Abstract. Since 1965, fuzzy set theory and its extensions has played an important role in developing effective tools in such fields as decision making, fuzzy control, data mining, forecasting, image analysis, etc. Fuzzy relation equations were main tool for solving optimization problems in a lot of theoretical and practical tasks. Unfortunately, these approaches did not take into account reliability of existing information. As a formal construct to deal with fuzziness and partial reliability of information, a Z-number concept was introduced by Zadeh. Nowadays, a series of works on arithmetic of Z-numbers, decision making under Z-number-based information, Z-differential equations, etc. are suggested. No investigations exist on Z-relation equations and their applications. In this study, we introduce a definition of Z-relation, and some operations over Z-relations. On this basis, Z-relation equation is formulated and some results on its solvability are proposed.

Advances in Rule-Based Architectures: A Study in the Design of Granular and Scalable Interpretable Models

Witold Pedrycz◉

Department of Electrical and Computer Engineering, University of Alberta, Edmonton AB T6R 2V4 Canada

wpedrycz@ualberta.ca

Abstract. Rule-based models have enjoyed a great deal of interest in the previous decades being regarded as a fundamental vehicle of knowledge representation and serving as a computational environment for an array of knowledge-based systems and their applications. They have gained popularity in fuzzy modeling realized as a collection of "if-then" statements endowed with information granules and formalized as fuzzy sets. The recent developments in so-called explainable AI (XAI) also link to rule-based architectures. The inherent modularity of the model adheres to the general way of problem solving through structuring a given complex problem into a collection of subproblems.

In spite of the well-established position, there are two challenging issues of rule-based modeling. Albeit their origin is very different, they can be solved in a unified manner. The first one is about the scalability of rule-based models, which arises when being faced with high-dimensional data. We demonstrate that some fundamental limitations such as a concentration effect hamper a design of good quality rules. To alleviate the problem, we consider a decomposition of data space (input variables) and building a slew of low-dimensional (or even one dimensional) models. Finally, the obtained results have to be combined through some aggregation mechanism. The second issue arises when building a global model on a basis of a series of multiview models (viz. models built on a basis of locally available input variables).

In the two categories of problems identified above, an interest is to aggregate the results delivered by the collection of the models. We investigate a way of aggregation realized with the use of the principle of justifiable granularity—one of the fundamentals of granular computing. We advocate that the diversity of local models gives rise to the granular format (viz. information granule of type-1) of the results of the aggregation. Furthermore, we highlight the phenomenon of elevation of type of information granules: if the local models produce results that are information granules of type-1, the ensuing aggregation yields the results that become information granules of type-2.

Contents

Fuzzy Clique Set Determination Method as an Example of Fuzzy Temporal Graph Invariant

Alexander Bozhenyuk[1]([⊠]) [iD], Vitalii Bozheniuk[2] [iD],
Janusz Kacprzyk[3] [iD], and Margarita Knyazeva[1] [iD]

[1] Southern Federal University, Nekrasovskiy Str., 44, 347928 Taganrog, Russia
avb002@yandex.ru, margarita.knyazeva@gmail.com
[2] RWTH Aachen University, Templergraben 55, 52056 Aachen, Germany
square.nabla@gmail.com
[3] Systems Research Institute Polish Academy of Sciences, Newelska 6,
01-447 Warsaw, Poland
janusz.kacprzyk@ibspan.waw.pl

Abstract. The paper investigates the problem of the invariant determination in graphs that have fuzzy-estimated parameters and temporal characteristics. The basic idea is to find fuzzy clique set in such graphs. Fuzzy temporal graph capture the notion of temporal characteristics and uncertainty while processing some operations on it. In this paper the idea of temporality in such graph models is treated in the way that adjacency of the vertices may change over time periods. Cliques normally refer to subgraphs in a graph such that vertices in each subgraph are pairwise adjacent. Adjacency may be uncertain due to some features of the network and may vary in time. In the problem of maximum clique determination the idea is to search the clique with most vertices within a graph. So the fuzzy adjacency here can be interpreted in terms of immediate likelihood of vertex to capture or to share whatever is flowing though the graph network. The idea of maximum clique subset for the fuzzy graph with fuzzy-estimated characteristics is presented in this paper. A method for determination of all maximum clique sets is proposed, as well as a fuzzy clique set is determined. The illustrative examples are given as well.

Keywords: Fuzzy temporal graph · Subgraph · Fuzzy clique set · Clique degree · Logical variable · Absorption rule

1 Introduction to Temporal Graph Modelling

Graphs are normally used by different users and analysts for modelling spatial knowledge with a certain structure. Besides their theoretical value as NP-hard problem solution methodology, the invariants in graphs (such as maximum clique problem) have different practical applications in community search in social networks, anomaly detection in security science, bioinformatics and in other models with predefined structure of its elements [1–3]. In some cases the relations between the elements can be partially predefined or uncertain to some extent. Fuzzy graphs can be used as a model

© The Author(s), under exclusive license to Springer Nature Switzerland AG 2021
R. A. Aliev et al. (Eds.): ICAFS 2020, AISC 1306, pp. 1–9, 2021.
https://doi.org/10.1007/978-3-030-64058-3_1

of such uncertain systems [4–6]. Graphs can be static and dynamic. In static information representation models [7] the relations, interdependencies, connections or
certain specific characteristics between the elements of graph cannot be changed within
the time period. For example, resource availability, capacity of the road and other
important characteristics of the network. Another case is when the relations or other
features of the graph model may vary over time in a discrete (or continuous) time.
These graphs capture the idea of temporality or dynamicity [8]. No doubt there are
different approaches to graph classification according to their practical application and
the way graph model can handle uncertainty: temporal or dynamic graphs, computational graphs, Markov chains and Petri nets [9–14]. Various applications of these
graphs ranging from social networks search, scheduling problems, transportation and
routing problems, geoinformation systems to biological networks under uncertainty
have grown steadily.

Cliques are widely used to represent dense communities in complex networks and
the maximum clique problem is known to be an effective tool to analyze the structure
of the graph model. Cliques are example of invariants of the graph. Let's consider a
graph invariant as a common property which is usually preserved by isomorphism.
Then a graph invariant is a property (it can be numerical or Boolean) which must be the
same for any two isomorphic graphs. A graph feature or graph invariant is a property of
the graph that depends only on the abstract structure, but not on the way the graph can
be represented: the chromatic number, the degree of a graph, the number of connected
components of a graph or bridges and others. There are the number of papers devoted
to the problem of the maximum clique determination in massive graphs, in large space
graphs and uncertain probabilistic graphs [15–17].

This article focuses on the problem of maximum clique determination in the graph
with a certain structure: relations between the elements (vertices) of a graph may vary
in time and have uncertain characteristics, for example capacity of the road when
traffic-jam, while the elements themselves have crisp characteristics and preserve
unchanged during time, for example service stations, gas stations, buildings etc. [18,
19].

2   Introduction to Fuzzy Click Set

The concept of a fuzzy temporal graph was introduced in the paper [20] as a triple
$\tilde{G} = (X, \{\tilde{\Gamma}_t\}, T)$, where X is a set of vertices ($|X| = n$); T is a natural number, which
specifies the discrete time; $\{\tilde{\Gamma}_t\}$ is a family of correspondences, which fuzzy displays
the vertices X into itself at times $t \in T$.

Example 1. Let's consider an example of a temporal fuzzy graph $\tilde{G}$ which is shown in
Fig. 1. Here the vertices $X = \{x, y, z, v\}$, the time $T = \{1, 2, 3\}$.

So, temporal fuzzy graph $\tilde{G}$ can be represented as the union of T fuzzy subgraphs:

$$\tilde{G} = \bigcup_{t \in T} \tilde{G}^{(t)}.$$

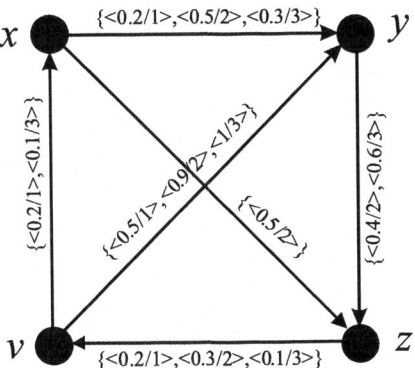

Fig. 1. Example of fuzzy temporal graph

Here $\tilde{U}^{(t)} = \{ <\mu_t(x,y), (x,y)> \,|\, (x,y) \in X^2 \}$ is the set of fuzzy edges with the membership function $\mu_t : X^2 \to [0,1]$, which is considered at time $t \in T$.

The concept of a maximal clique of a fuzzy graph $\tilde{G}^{(t)} = (X, \tilde{U}^{(t)})$ was introduced in the paper [21] as a subset of vertices $Y \subset X$ with clique degree

$$\delta_Y = \min_{\forall x \in Y} \min_{\substack{\forall y \in Y \\ x \neq y}} \left(\mu_t(x,y) \vee \mu_t(y,x) \right)$$

if the following condition is satisfied: $(\forall Y' \supset Y)[\delta_{Y'} < \delta_Y]$.

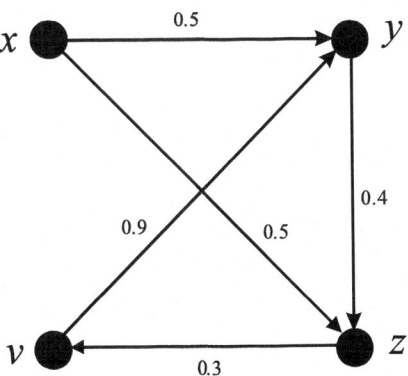

Fig. 2. Fuzzy subgraph $\tilde{G}^{(2)}$

Example 2. Let $t = 2$, then the fuzzy subgraph $\tilde{G}^{(2)}$ of fuzzy temporal graph $\tilde{G}$ is shown in Fig. 2:

In this fuzzy subgraph, the subset of the vertices $Y_1 = \{y\ v\}$ is the maximum clique with clique degree $\delta_{Y_1} = 0.9$. The subset $Y_2 = \{z,\ v\}$ is not such, since there is a subset of $Y_3 = \{y, z, v\}$ whose clique degree $\delta_{Y_3} = \delta_{Y_2} = 0.3$ and $Y_2 \ \square \ Y_3$.

Let us denote by $\delta_i^{\max}$ the greatest degree among all maximum cliques with the number of vertices equal to i in the fuzzy subgraph $\tilde{G}^{(t)}$. Then the fuzzy set $\tilde{K}_t = \{ <\delta_1^{\max}/1> , <\delta_2^{\max}/2> , \ldots, <\delta_n^{\max}/n> \}$ is the fuzzy set of cliques of the sub-graph $\tilde{G}^{(t)}$ [21].

We introduce the concept of a fuzzy clique set of a temporal fuzzy graph as the intersection of the clique sets of its subgraphs:

Definition 1. A set

$$\tilde{K} = \bigcap_{t \in T} \tilde{K}_t \tag{1}$$

is called fuzzy clique set of the temporal fuzzy graph $\tilde{G}$.

A fuzzy set of clicks sets the highest clique degree δ_i for a given number of vertices $i \in \overline{1, n}$ at any time $t \in T$.

Property 1. Let's $\tilde{K} = \{ <\delta_1/1> , <\delta_2/2> , \ldots, <\delta_n/n> \}$. Then the following proposition is true:

$$1 = \delta_1 \geq \delta_2 \geq \ldots \geq \delta_n \geq 0.$$

In other words, the greater the number of vertices in a subgraph, the less the minimum degree of connectivity (membership function) between any two vertices of this subgraph at any time.

3   Finding Fuzzy Clique Set

Let's consider an approach to finding a fuzzy set of cliques of a fuzzy temporal graph. This approach is a generalization of the approach that was proposed for finding the invariants of fuzzy and intuitive graphs [22, 23].

Suppose that $Y^{(t)} \subseteq X$ is a clique of the subgraph $\tilde{G}_t$ with clique degree δ. Then the following statement must be fulfilled:

$$(\forall x, y \in X)[x \notin Y^{(t)} \vee y \notin Y^{(t)} \vee \mu_t(x, y) \geq \delta \vee \mu_t(x, y) \geq \delta]. \tag{2}$$

For each vertex $x \in X$, we assign a Boolean variable:

$$p_x = \begin{cases} 1, & \text{if } x \in Y^{(t)}; \\ 0, & \text{if } x \notin Y^{(t)}. \end{cases} \tag{3}$$

For inequality $\mu_t(x, y) \geq \delta$, we introduce into consideration the fuzzy variable $r_{xy} = \mu_t(x, y)$. Then, the expression (2), taking into account the expression (3), can be represented as the formula (4):

$$\Phi_t = \bigwedge_{\forall x, y \in X} (\bar{p}_x \vee \bar{p}_y \vee r_{xy} \vee r_{yx}). \tag{4}$$

Let us simplify expression (4) using the absorption rules (5):

$$\begin{aligned} \bar{p}_x \vee \bar{p}_x \wedge p_y &= \bar{p}_x, \\ \xi' \wedge \bar{p}_x \vee \xi'' \wedge \bar{p}_x \wedge \bar{p}_y &= \xi' \wedge \bar{p}_x, \text{ if } \xi' \geq \xi''. \end{aligned} \tag{5}$$

Here, $0 \leq \xi', \xi'' \leq 1$.
Thus, expression (4) using rules (5) can be converted to expression (6):

$$\Phi_t = \bigvee_{i=\overline{1,l}} (\bar{p}_{x_i} \wedge \bar{p}_{y_i} \wedge \ldots \wedge \bar{p}_{z_i} \wedge \delta_i). \tag{6}$$

Property 2. Each bracket in expression (6) defines the maximum clique as follows: those vertices of the subgraph $\tilde{G}_t$ whose Boolean variables are absent in this bracket form a maximum clique $Y^{(t)} \subseteq X$ with degree δ_i. The number l in expression (6) determines the number of maximum clicks of the considered subgraph.

Based on this property, the following approach can be proposed for defining a fuzzy click set in a fuzzy temporal graph:

– For all subgraphs $\tilde{G}_t$, $t \in T$ of fuzzy temporal graph $\tilde{G}$, we define families of maximal cliques $\{Y^{(t)}\}$ with calculated clique degrees;
– Using the obtained families of maximal cliques, we determine the fuzzy click set $\tilde{K}_t$ of subgraph $\tilde{G}_t$;
– According to expression (1), we define a fuzzy click set $\tilde{K}$ of the graph $\tilde{G}$.

4   Calculation Example of Fuzzy Clique Set

Let's consider an example of calculating the fuzzy clique set $\tilde{K}$ of the graph shown in Fig. 1. First, we find the maximum cliques of subgraphs $\tilde{G}_t$, $t = \{1, 2, 3\}$, the adjacency matrices of which have the form:

$$R_1 = \begin{array}{c} \\ x \\ y \\ z \\ v \end{array} \begin{array}{|cccc|} x & y & z & v \\ \hline 0 & 0.2 & 0 & 0 \\ 0 & 0 & 0 & 0 \\ 0 & 0 & 0 & 0.2 \\ 0.2 & 0.5 & 0 & 0 \end{array}, \quad R_2 = \begin{array}{c} \\ x \\ y \\ z \\ v \end{array} \begin{array}{|cccc|} x & y & z & v \\ \hline 0 & 0.5 & 0 & 0 \\ 0 & 0 & 0.4 & 0 \\ 0 & 0 & 0 & 0.3 \\ 0 & 0.9 & 0 & 0 \end{array},$$

$$R_3 = \begin{array}{c} \\ x \\ y \\ z \\ v \end{array} \begin{array}{|cccc|} x & y & z & v \\ \hline 0 & 0.3 & 0 & 0 \\ 0 & 0 & 0.6 & 0 \\ 0 & 0 & 0 & 0.1 \\ 0 & 1.0 & 0 & 0 \end{array}.$$

According to matrix R_1, expression (3) will look like:

$$\Phi_1 = (\bar{p}_x \vee \bar{p}_y \vee 0.2 \vee 0) \wedge (\bar{p}_x \vee \bar{p}_z \vee 0) \wedge (\bar{p}_x \vee \bar{p}_v \vee 0 \vee 0.2) \wedge (\bar{p}_y \vee \bar{p}_z \vee 0)$$
$$\wedge (\bar{p}_y \vee \bar{p}_v \vee 0.5 \vee 0) \wedge (\bar{p}_z \vee \bar{p}_v \vee 0.2 \vee 0).$$

Using rules (4), we get:

$$\Phi_1 = 1\bar{p}_x\bar{p}_y\bar{p}_v \vee 1\bar{p}_x\bar{p}_z\bar{p}_v \vee 1\bar{p}_x\bar{p}_y\bar{p}_z \vee 1\bar{p}_y\bar{p}_z\bar{p}_v \vee 0.5\bar{p}_x\bar{p}_z \vee 0.2\bar{p}_x\bar{p}_v \vee 0.2\bar{p}_z.$$

The resulting expression shows that the subgraph $\tilde{G}_1$ has 7 maximum cliques which are given in Table 1:

Table 1. Maximum cliques with clique degrees.

Maximum clique	Clique degree
$Y_1 = \{x\}$, $Y_2 = \{y\}$, $Y_3 = \{z\}$, $Y_4 = \{v\}$	1.0
$Y_5 = \{y, v\}$	0.5
$Y_6 = \{y, z\}$, $Y_7 = \{x, y, v\}$	0.2

Thus, fuzzy subgraph $\tilde{G}_1$ has the fuzzy set of cliques:

$$\tilde{K}_1 = \{<1/1>, <0.5/2>, <0.2/3>, <0/4>\}.$$

Carrying out similar calculations for the adjacency matrices R_2 and R_3, we obtain:

$$\tilde{K}_2 = \{<1/1>, <0.9/2>, <0.3/3>, <0/4>\},$$
$$\tilde{K}_3 = \{<1/1>, <0.6/2>, <0.1/3>, <0/4>\}.$$

From here we get:

$$\tilde{K} = \tilde{K}_1 \cap \tilde{K}_2 \cap \tilde{K}_3 = \{<1/1>, <0.5/2>, <0.1/3>, <0/4>\}.$$

The resulting fuzzy set of cliques means, in particular, that in the considered graph at any time there is a subgraph with 2 vertices with the clique degree at least 0.5 and there is no subgraph with 3 vertices with the clique degree greater than 0.1.

5 Results

The paper discuses fuzzy temporal graph discrete model, where vertices are represented as a number of crisp objects and the edges reflect fuzzy relations between objects within discrete time moments. The notion of fuzzy clique set with a certain degree of adjacency is introduced. The idea of fuzzy clique set that determines the greatest clique degree for the temporal graph with uncertain characteristics on the edges is presented. The formal algorithm for maximal fuzzy clique sets determination is suggested. The Maghout's method is modified to cope with fuzzy variables on temporal graphs. And finally a numerical example with 4 vertices is considered to illustrate the approach. The result of the example shows that a fuzzy temporal graph at discrete time moment t may have a subset of two vertices with the clique degree 0.5, and a subset of three vertices with the clique degree 0.1. This may help to model dynamical features and characteristics of fuzzy graph according to its invariant.

6 Conclusion

In this paper the idea of fuzzy clique set as an invariant for the uncertain temporal graph was presented. In the example of the temporal graph given in the paper, the relations between elements in graph (or the degree of connectivity between the vertices) may vary in time. Invariant analytics allows researchers to investigate special characteristics of any given graph which may be fuzzy as well. A fuzzy clique set allows to find the greatest clique degree of the graph for a given number of vertices at any discrete moment. The method fuzzy clique set determination is considered in the paper and the illustrative numerical example is given as well.

Acknowledgments. The reported study was funded by RFBR according to the research project N 20-01-00197.

References

1. Ore, O.: Theory of graphs. Amer. Math. Soc. Colloq. Publ. Providence (1962)
2. Kaufmann, A.: Introduction a la theorie des sous-ensemles flous. Masson, Paris (1977)
3. Christofides, N.: Graph Theory. An Algorithmic Approach. Academic Press, London (1976)

4. Rosenfeld, A.: Fuzzy graphs. In: Zadeh, L.A., Fu, K.S., Shimura, M. (eds.) Fuzzy Sets and Their Applications, pp. 77–95. Academic Press, New York (1975). https://doi.org/10.1016/B978-0-12-775260-0.50008-6

5. Monderson, J., Nair, P.: Fuzzy Graphs and Fuzzy Hypergraphs. Physica-Verl, Heidelberg (2000)

6. Mordeson, J.N., Peng, C.S.: Operations on fuzzy graphs. Inf. Sci. **79**, 159–170 (1994)

7. Kostakos, V.: Temporal graphs. In: Proceedings of Physica A: Statistical Mechanics and its Applications, vol. 388, no. 6, pp. 1007–1023. Elsevier (2008). https://doi.org/10.1016/j.physa.2008.11.021

8. Brézillon, P., Pasquier, L., Pomerol, J.-C.: Reasoning with contextual graphs. Eur. J. Oper. Res. **136**(2), 290–298 (2002). https://doi.org/10.1016/S0377-2217(01)00116-3

9. Barzilay, R., Elhadad, N., McKeown, K.: Inferring strategies for sentence ordering in multidocument news summarization. J. Artif. Intell. Res. **17**, 35–55 (2002). https://doi.org/10.1613/jair.991

10. Bramsen, P.J.: Doing Time: Inducing Temporal Graphs. Technical report, Massachusetts Institute of Technology (2006)

11. Baldan, P., Corradini, A., Konig, B.: Verifying finite-state graph grammars: an unfolding-based approach. Lecture Notes in Computer Science, vol. 3170, pp. 83–98 (2004). https://doi.org/10.1007/978-3-540-28644-8_6

12. Baldan, P., Corradini, A., Konig, B., Konig, B.: Verifying a behavioural logic for graph transformation systems. Electron. Notes Theor. Comput. Sci. **104**, 5–24 (2004). https://doi.org/10.1016/j.entcs.2004.08.018

13. Collberg, C., Kobourov, S., Nagra, J., Pitts, J., Wampler, K.: A system for graph-based visualization of the evolution of software. In: Proceedings of ACM Symposium on Software Visualization (SoftVis 2003), San Diego, CA, USA, pp. 77–86 (2003). https://doi.org/10.1145/774841.774844

14. Dittmann, F., Bobda, C.: Temporal graph placement on mesh-based coarse grain reconfigurable systems using the spectral method. In: IFIP Advances in Information and Communication Technology, vol. 184, pp. 301–310. Springer (2005). https://doi.org/10.1007/11523277_29

15. Lu, C., Yu, J.X., Wei, H., Zhang, Y.: Finding the maximum clique in massive graphs. Proc. VLDB Endow. **10**(11), 1538–1549 (2017). https://doi.org/10.14778/3137628.3137660

16. Cheng, J., Ke, Y., Fu, A.W.-C., Yu, J.X., Zhu, L.: Finding maximal cliques in massive networks. ACM Trans. Database Syst. **36**(4), 21 (2011). https://doi.org/10.1145/2043652.2043654

17. Li, R.-H., Dai, Q., Wang, G., Ming, Z., Qin, L., Yu, J.X.: Improved algorithms for maximal clique search in uncertain networks. In: Proceedings of IEEE 35th International Conference on Data Engineering (ICDE), Macau, China, pp. 1178–1189 (2019). https://doi.org/10.1109/icde.2019.00108

18. Bozhenyuk, A., Belyakov, S., Knyazeva, M., Rozenberg, I.: Searching method of fuzzy internal stable set as fuzzy temporal graph invariant. In: Communications in Computer and Information Science, vol. 583, pp. 501–510 (2018). https://doi.org/10.1007/978-3-319-91473-2_43

19. Bozhenyuk, A., Belyakov, S., Rozenberg, I.: Coloring method of fuzzy temporal graph with the greatest separation degree. In: Advances in Intelligent Systems and Computing, vol. 450, pp. 331–338 (2016). https://doi.org/10.1007/978-3-319-33609-1_30

20. Bozhenyuk, A., Belyakov, S., Knyazeva, M.: Modeling objects and processes in gis by fuzzy temporal graphs. In: Studies in Fuzziness and Soft Computing, vol. 393, pp. 277–286. (2020). https://doi.org/10.1007/978-3-030-47124-8_22

21. Bershtein, L., Bozhenyuk, A., Knyazeva, M.: Definition of cliques fuzzy set and estimation of fuzzy graphs isomorphism. Procedia Comput. Sci. **77**, 3–10 (2015). https://doi.org/10.1016/j.procs.2015.12.353
22. Bershtein, L., Bozhenuk, A.: Maghout method for determination of fuzzy independent, dominating vertex sets and fuzzy graph kernels. Int. J. Gener. Syst. **30**(1), 45–52 (2001). https://doi.org/10.1080/03081070108960697
23. Bozhenyuk, A., Belyakov, S., Kacprzyk, J., Knyazeva, M.: The method of finding the base set of intuitionistic fuzzy graph. In: Advances in Intelligent Systems and Computing, vol. 1197, pp. 18–25 (2020). https://doi.org/10.1007/978-3-030-51156-2_3

Z-set Based Approach to Control System Design

Rafik A. Aliev(✉) and Latafat A. Gardashova

Azerbaijan State Oil and Industry University,
Azadlig 35, Nasimi, Baku, Azerbaijan
raliev@asoa.edu.az, l.qardashova@asoiu.edu.az

Abstract. In real life problems If-Then rules antecedent and consequent parts are characterized with the bimodal information. The bimodal information covers probabilistic and fuzzy uncertainty (not separately) in existing literature. There are fuzzy reasoning approach and probabilistic approximate reasoning approach that deal separately with fuzzy uncertainty and probabilistic uncertainty. There is the need to develop new approach to approximate reasoning with If-...-Then rules which are based on bimodal information. In this paper, we suggest new approach for approximate reasoning with Z-set-based rules. Numerical example with 7 Z-set-based rules for approximate reasoning are considered.

Keywords: Z-set · Bimodal information · Fuzzy implication · Probabilistic implication · Z-set based If-Then rules

1 Introduction

Reasoning is the one area of artificial intelligence which depend on the human judgment level and has been a dynamic research field since the 1950s.

In the scientific literature reasoning is characterized by the action of extracting a conclusion from basic properties using an existing methodology. If uncertainty in the given problem is characterized by random effect, then knowledge and reasoning is represented using probability theory. If uncertainty depends on inaccuracy of information, then fuzzy logic reasoning is effective approach. In approximate reasoning a possible inaccurate conclusion is inferred from a collection of inexact premises. First work about approximate reasoning is given in [1] by Zadeh.

Many existing methods of fuzzy reasoning are based on fuzzy relation and Zadeh's Compositional Rule of Inference (CRI). The detailed comparison of them is discussed in [2–4]. In scientific literature there are fuzzy reasoning methods which do not use Zadeh's CRI and use degree of similarity. In [5] fuzzy production rules, fuzzy reasoning methods and six similarity-based fuzzy reasoning methods are compared and analyzed.

IF - THEN rules along with the reasoning mechanism are the kernel of a fuzzy model. RBR is an ideal method for solving problems in which exist few rules [6]. The problem-solving complication is proportional to the number of rules which using to match the pattern of data. Moreover, RBR lacks the ability to learn due to the difficulty of getting new expertise in new rules [7].

R. A. Aliev et al. (Eds.): ICAFS 2020, AISC 1306, pp. 10–21, 2021.
https://doi.org/10.1007/978-3-030-64058-3_2

A comparison of fuzzy reasoning methods is discussed in [8–11]. These methods based on implication operators. Zadeh's alternative approach [12] is about management of uncertainty. This approach subsumes probability theory and predicate logic and makes it possible to deal with different types of uncertainty.

The probability theory provides no answer to the problem of combining two probability measures on the same space, because evidently it is not a problem of mathematics, but judgment. Probabilistic implication-based reasoning is discussed in [13]. The presented implications are based on conditional copulas and is a bridge between fuzzy logic and probability theory.

Modeling combination of probabilistic and fuzzy uncertainties is devoted in a series of works, consisting on fuzzy belief networks, fuzzy belief rule-based systems, fuzzy and probabilistic information combination and etc. [14–21]. In this point of view the Z-number concept is a kernel of combination of fuzzy and probabilistic uncertainties. Today Z-numbers and Z-sets are effective tools for representation of imperfect information. Z-valued t-norm and t-conorm operators-based aggregation is discussed in [22] and these operators are applied to aggregating expert's opinions described with Z-numbers in real life problems. Z-Distance Based IF-THEN Rules are discussed in [23]. In this study, authors for the first time applied the concept of distance of Z-numbers to the approximate reasoning with Z-number based IF-THEN rules. In [24] authors proposed a new approach developed to study approximate reasoning with Z-rules on a basis of linear interpolation.

Z-number includes fuzzy and probabilistic uncertainty. The main disagreement or difficulty is necessity accomplishing both two uncertainties' reasoning In [25], a novel Z-network model and its associated reasoning algorithm are proposed to overcome the difficulty.

Authors [23–25] used the Z-number and did not use the Z-set. In this paper we suggest a new approach for approximate reasoning with Z-set based rules. The proposed approach has been implemented using of the MATLAB toolbox and the table processor.

The rest of the paper is organized as follows. Section 2 gives the preliminaries. A statement of the problem is given in Sect. 3. Section 4 considers the proposed approach to reasoning with Z-rules. In Sect. 5, a numerical example is illustrated. Finally, Sect. 6 concludes the paper.

2   Preliminaries

Definition 1 [24]. Z IF-THEN Rules. If X_1 is $Z_{X_1,n} = (A_{X_1,n}, B_{X_1,n})$ and, ..., and X_m is $Z_{X_m,n} = (A_{X_m,n}, B_{X_m,n})$ then Y is $Z_Y = (A_{Y,n}, B_{Y,n})$.
$Z_{X_m,n} = (A_{X_m,n}, B_{X_m,n})$ is the Z-sets based inputs and $Z_Y = (A_{Y,n}, B_{Y,n})$ is Z sets-based outputs of the rues.

Definition 2 [24]. Z-sets. Let X be a space of objects and $x \in X$. A Z-set Z in X is defined by membership function $A(x) \in [0, 1]$, probability distributions P_X associated with X, membership function $B(P(x)) \in [0, 1]$. Where $A(x)$ denotes that element x belongs to A with membership degree $A(x)$, $B(P(x))$ denotes that probability measure value belongs to B with degree $B(P(x))$, P_X is the underlying probability density of X. There is restriction

$$P_X = \int_X A(x)p(x)dx$$

Taking to accounting that the degree to which P_X satisfies Z-set $\{Prob\}_P(A(x), \forall x \in X)$ is B we can get $G_P = B(\int_X A(x)p(x)dx)$.

So Z-set in X can be expressed as triple $Z = (A, B, G)$, where A is fuzzy subset of X, B is fuzzy set (restriction) on probability measure of A, G is set of probability distribution over the space P.

$P(A)$, induced by a set of distributions G:

$$G = \left\{ p_Z(x) : \int_X p_Z(x)dx = 1, \int_X p_Z(x)A(x)dx \text{ is } B, x \in X \right\} \tag{1}$$

3   Statement of the Problem

Given n Z-set based rule base and current input to find output.

Bimodal information based IF-Then rules is given below:

If X_1 is $Z_{X_1,1} = (A_{X_1,1}, B_{X_1,1})$ and ,..., and X_m is $Z_{X_m,1} = (A_{X_m,1}, B_{X_m,1})$ then Y is $Z_Y = (A_{Y,1}, B_{Y,1})$

If X_1 is $Z_{X_1,2} = (A_{X_1,2}, B_{X_1,2})$ and ,..., and X_m is $Z_{X_m,2} = (A_{X_m,2}, B_{X_m,2})$ then Y is $Z_Y = (A_{Y,2}, B_{Y,2})$

...

If X_1 is $Z_{X_1,n} = (A_{X_1,n}, B_{X_1,n})$ and ,..., and X_m is $Z_{X_m,n} = (A_{X_m,n}, B_{X_m,n})$ then Y is $Z_Y = (A_{Y,n}, B_{Y,n})$

and a current observation

X_1 is $Z'_{X_1} = (A'_{X_1}, B'_{X_1})$ and ,..., and X_m is $Z'_{X_m} = (A'_{X_m}, B'_{X_m})$, find the Z-value of Y. Here m is the number of Z-valued input variables and n is the number of rules.

4   Solution Method

Using fuzzy implications such as Zadeh, Mamdani, Aliev etc. fuzzy implications to calculate A''. This step is to define for each rule a fuzzy relation matrix R_s, $s = 1 \ldots 7$ in a two dimensional space consisting of the error E_i, and the control action U_j:

$$R_s \subseteq \tilde{E}_i \times \tilde{U}_j$$

The membership function $\mu_s(u, e)$ of the correspondent relation R_s has the form according as Ali-3 implication [8]

$$\mu_s(u,e) = \begin{cases} 1, & \mu_{E_i}(e) \le \mu_{U_j}(u) \\ \dfrac{\mu_{U_j}(u)}{\mu_{E_i}(e) + (1 - \mu_{U_j}(u))}, & \mu_{E_i}(e) > \mu_{U_j}(u) \end{cases}$$

$$\text{i,j,s} = \overline{1,n} \ (n = 7)$$

To define a composed fuzzy relation matrix R.

$R = \cup \, R_S, s = 1, \ldots, n$ according as logical connective of Ali 3 logic-conjunction:

The membership function of the output defined using the compositional rule of inference,

$$\mu_U(u) = \max_u \min[\mu_E(e^c_{ic}), \mu_R(u,e)]$$

Next step to calculate B:

Using points of support B over antecedents of the all rules obtain set of probability distributions G.

Distributions are aggregating by using probability aggregation method weighted sum:

$$P_G(p) = \sum_{i-1}^{n} w_i P_i(p).$$

Using points of support B over consequents of the all rules are obtained set of probability distributions G.

For every rule using the distribution of the input and output calculate a copula the following form

$$p_{ii}p_{oj} \le p_{ij} \le \min(p_{ii}, p_{oj})$$
$$\sum p_{ij} = 1$$

Obtained set of probability distributions over all rules are aggregating by using aggregation method.

We have A part of Z number, aggregated probability vector(over inputs of the rule) and aggregated probability vector over obtained distributions for all rules.

New input is given as Z number. Obtain set of probability distributions using points of support B over new input. We are getting probability vector.

Calculate distance between two vector (new input based probability vector and aggregated probability vector over inputs of the rule) by using Hellinger distance of two discrete probability distribution vector:

$$ZHD(P_{Z_1}, P_{Z_2}) = \min_{p_{zi} \in G} \left(\frac{1}{\sqrt{2}} \left(\sum_{j=1}^{k} (p_{z1_j}^{1/2} - p_{z2_j}^{1/2}) \right)^{1/2} \right)$$

Calculate similarity measure: $SM = 1/(1 + ZHD)$.

We have A, similarity measure and probability distributions (P_i). Define support points of output B:

$$Bi = \sum_{j=1}^{n} \mu_{jout} \cdot SM \cdot P_{ji}$$

To calculate the membership value of output B: Select min of all point membership values (input and output) which make P_i.

Assume, that $I_{p_{B_i}}, i = \overline{1,n}$ are number of points on universe of B in Z number over antecedents of the rules.

$O_{P_{B_l}}, l = \overline{1,n}$ number of points on universum of B in Z number over consequents of the rules.

$R_k, k = \overline{1,m}$ number of the rules, $j = \overline{1,n^2}$ number of points of the new B in obtained Z number.

$$\mu_{B_j} = \min((\underset{R_1}{\Lambda}(\mu_{I_{P_{B_j}}}, \mu_{O_{P_{B_j}}}), \ldots, (\underset{R_m}{\Lambda}(\mu_{I_{P_{B_j}}}, \mu_{O_{P_{B_j}}})))$$

For instance, $\mu_{B_1} = \min((\underset{R_1}{\Lambda}(\mu_{I_{P_{B_1}}}, \mu_{O_{P_{B_1}}}), \ldots \ldots, (\underset{R_m}{\Lambda}(\mu_{I_{P_{B_1}}}, \mu_{O_{P_{B_1}}})))$.

5 Numerical Example

Let's assume that these following 7 rules are involved in Z-valued control system [26].

1. If the error e is [negative big (NB), small sure] THEN the control action u is [negative big (NB), small sure].
2. If the error e is [negative medium (NM), sure] THEN the control action u is [negative medium (NM), sure].
3. If the error e is [negative small (NS), very sure] THEN the control action u is [negative small (NS), very sure].
4. If the error e is [zero (ZE), medium] THEN the control action u is [zero, medium].
5. If the error e is [positive small (PS), sure] THEN the control action u is [positive small (PS), medium].
6. If the error e is [positive medium (PM), very sure] THEN the control action u is [positive medium (PM), sure].
7. If the error e is [positive big (PB), medium] THEN the control action u is [positive big (PB), sure].

Errors for Z-valued control system are defined with e, control actions are described as u. Through rules membership function can be computed as follows:

small sure $= 0.3/0.7 + 1/0.73 + 0.2/0.75;$ *medium* $= 0.2/0.5 + 1/0.6 + 0.4/0.8;$
sure $= 0.1/0.7 + 1/0.8 + 0.3/0.9;$ *very sure* $= 0.2/0.8 + 1/0.9 + 0.2/1$

Rule 1

$$\mu_{E1}(e) = 1.00/ - 10 + 0.73/ - 7 + 0.34/ - 3 + 0.20/0 + 0.13/3 + 0.08/7 + 0.06/10$$
$$\mu_{U1}(u) = 1.00/ - 1 + 0.92/ - 0.7 + 0.67/ - 0.3 + 0.50/0 + 0.37/0.3 + 0.26/0.7 + 0.20/1$$

...

Rule 7

$$\mu_{E7}(e) = 0.06/ - 10 + 0.08/ - 7 + 0.13/ - 3 + 0.20/0 + 0.34/3 + 0.73/7 + 1.00/10$$
$$\mu_{U7}(u) = 0.20/ - 1 + 0.26/ - 0.7 + 0.37/ - 0.3 + 0.50/0 + 0.67/0.3 + 0.92/0.7 + 1.00/1$$

For given 7 rules and their two properties (errors and control actions) membership functions were evaluated. Problem is to perform Z inference.

New input: $(\mu_E(e^c_{ic}) = 0.11/ - 10 + 0.17/ - 7 + 0.34/ - 3 + 0.61/0 + 0.96/3 + 0.73/7 + 0.41/10, sure)$

Relations created by using ALI-III implication for above given 7 rules. Intersection of relations is realized by using logical connective of Ali 3 and determined the following form relation $(R = \cup R_S, s = 1, \ldots, n)$, n = 7. In Table 1 is given $\mu_R(u, e)$-is the value of membership function on composed fuzzy relation.

Table 1. Composed fuzzy relation

$R = \cup R_S, s = 1, \ldots, n$						
1	0.92	0.67	0	0	0	0
1	1	0.734631	0	0	0	0
1	1	1	1	1	0	0
1	1	1	1	1	1	1
1	1	1	1	1	1	1
1	1	1	1	1	1	1
1	1	1	1	1	1	1

The membership function of the control action defined using the compositional rule of inference and obtained

$$\mu_{Eout}(e) = 0.17/ - 1 + 0.61/ - 0.7 + 0.96/ - 0.3 + 0.96/0 + 0.96/0.3 + 0.96/0.7 + 0.73/1$$

Calculation of probability distribution for all antecedents and consequents of the rules. For example, probability distributions on antecedent and consequent of the first rules obtained according to (1) is given in Table 2.

Table 2. Probability distributions on antecedent and consequent

Distributions (on input of first rule)			Distributions (on output of first rule)		
0.7	0.73	0.75	0.7	0.73	0.75
0.448868	0.506573	0.552072	0.254563558	0.346735	0.388148
0.259223	0.246233	0.231103	0.186579117	0.233073	0.321775
1.29E−01	0.096136	0.057315	0.178358714	0.12748	0.016603
0.043783	0.010553	0	0.166239062	0.061789	0
0.018731	0	0	0.140509473	0.026961	0
0.034664	0.022479	0.008288	0.073751076	0.029392	0
0.065264	0.118026	0.151222	0	0.17457	0.273474

Aggregation of probability distribution of antecedent of 7 rules using probability aggregation method- a weighted sum is determined as follow (Table 3):

Table 3. Aggregation of probability distribution

p1	p2	p3
0.360862588	0.281393	0.276036123
0.312332067	0.423765	0.543411345
0.152748362	0.170579	0.028657468
0.043531749	0.006059	0
0.012607359	0	0
0.035833254	0.01124	0.004143985
0.082084621	0.106965	0.147751078

Calculation of Copula. For every rule, using the distribution of the input and output calculate a copula and are getting for first rule over one point (for example, points (0.7, 0.73) are given in Table 4)).

Table 4. Copula over points (0.7, 0.73)

Copula						
0.114	0.084	0.080	0.075	0.063	0.033	0.000
0.066	0.048	0.046	0.043	0.036	0.019	0.000
0.033	0.024	0.023	0.022	0.018	0.010	0.000
0.011	0.008	0.008	0.007	0.006	0.003	0.000
0.005	0.003	0.003	0.003	0.003	0.001	0.000
0.009	0.006	0.006	0.006	0.005	0.003	0.000
0.017	0.012	0.012	0.011	0.009	0.005	0.000

Probabilistic reasoning. Using copula and probability implicationis defined the probability distributions for all points. By using (0.7, 0.73) is determined the following form distribution over first rule (Table 5)

$$P = (0.346735094, 0, 233073, 0.12748, 0.061789, 0.026961, 0.029392, 0.17457)$$

Table 5. Probability distributions (I rule)

Points	Distributions						
(0.7, 0.7)	p11=	0.254564	0.186579	0.178359	0.166239	0.140509	0.073751	0
(0.7, 0.73)	p12=	0.346735	0.233073	0.12748	0.061788793	0.026961	0.029392	0.17457
(0.7, 0.75)	p13=	0.388148	0.321775	0.016603	0	0	0	0.273474
(0.73, 0.7)	p14=	0.254564	0.186579	0.178359	0.166239062	0.140509	0.073751	0.00E+00
(0.73, 0.73)	p15=	3.47E−01	0.233073	1.27E−01	6.18E−02	2.70E−02	2.94E−02	0.17457
(0.73, 0.75)	p16=	3.88E−01	0.321775	1.66E−02	0.00E+00	0.00E+00	0.00E+00	0.273474
(0,75, 0.7)	p17=	0.254564	0.186579	0.178359	0.166239062	0.140509	0.073751	0
(0.75, 0.73)	p18=	3.47E−01	0.233073	0.12748	0.061789	0.026961	0.029392	0.17457
(0.75, 0.75)	p19=	0.388148	0.321775	0.016603	0	0	0	0.273474

Table 6. Probability distributions vector

Probability distributions						
0.357542	0.127325	0.113683	0.121433	0.117446	0.079467	0.083104
0.188826	0.274558	0.207415	0.129151	0.074473	0.038292	0.087285
0.209533	0.318909	0.151977	0.098256	0.060992	0.023596	0.136737
0.357542	0.127325	0.113683	0.121433	0.117447	0.079467	0.083104
0.188826	0.274558	0.207415	0.129151	0.074473	0.038292	0.087285
0.194074	0.160887	0.496272	0	0	0	0.148767
0.357542	0.127325	0.113683	0.121433	0.117447	0.079467	0.083104
0.188826	0.274558	0.207415	0.129151	0.074473	0.038292	0.087285
0.194074	0.160887	0.496272	0	0	0	0.148767

For 7 rules was determined 63 distributions. Obtained set of probability distributions over all rules are aggregating by using aggregation method. Aggregation of set of probability distributions is determined the probability vector as following form (Table 6):

A part of the Z-set is defined as follow:

$\mu_{Eout}(e) = 0.17/-1 + 0.61/-0.7 + 0.96/-0.3 + 0.96/0 + 0.96/0.3 + 0.96/0.7 + 0.73/1$

Calculate similarity measurement between current vector of probability distributions with aggregated input vector of probability distributions. We have A part of Z number, aggregated probability vector (over inputs of the rule (Table 6)) and aggregated probabilty vector over obtained distributions for all rules.

New input is given as Z number. Obtain set of probability distributions using points of support B over new input (Table 7). We are getting probability vector.

Table 7. New input distribution

Points and distributions		
0.7	0.8	0.9
0	0	0
0.042575	1E−06	1E−06
0.116257	0.038604	0
0.199474	0.274913	0.154821
0.334469	0.513237	0.819908
0.264072	0.173247	0.025272
0.043152	0	0

Calculate distance between two vector (new input based probability vector and aggregated probability vector over inputs of the rule). Obtained value of the distant:

$$ZHD(P_{Z_1}, P_{Z_2}) = 0.263482$$

Result of the calculation similarity measurement between current vector of probability distributions with aggregated input vector of probability distributions

$$SM = 0.787751$$

Calculate probability distributions in consequent part using similarity degree. We have A, similarity measure and probability distributions.

To calculate the membership value of output B, select min of all point membership values (input and output) which make $P_i, i = \overline{1,9}$.

For example

$$P_1 = \left(\begin{array}{l} \left(\frac{0.3}{0.7}, \frac{0.3}{0.7}\right), \left(\frac{0.1}{0.7}, \frac{0.1}{0.7}\right), \left(\frac{0.2}{0.8}, \frac{0.2}{0.8}\right), \left(\frac{0.2}{0.5}, \frac{0.2}{0.5}\right), \left(\frac{1}{0.7}, \frac{0.2}{0.5}\right), \\ \left(\frac{0.2}{0.8}, \frac{1}{0.7}\right), \left(\frac{0.2}{0.5}, \frac{1}{0.7}\right) \end{array} \right)$$

$\ldots$

$$P_9 = \left(\begin{array}{l} \left(\frac{0.2}{0.75}, \frac{0.2}{0.75}\right), \left(\frac{0.3}{0.9}, \frac{0.3}{0.9}\right), \left(\frac{0.2}{1}, \frac{0.2}{1}\right), \left(\frac{0.4}{0.8}, \frac{0.4}{0.8}\right), \left(\frac{0.3}{0.9}, \frac{0.4}{0.8}\right), \\ \left(\frac{0.2}{1}, \frac{0.3}{0.9}\right), \left(\frac{0.4}{0.8}, \frac{0.3}{0.9}\right) \end{array} \right)$$

Obtained result:

$$A = 0.17/-1 + 0.61/-0.7 + 0.96/-0.3 + 0.96/0 + 0.96/0.3 + 0.96/0.7 + 0.73/1$$

$$\begin{aligned}B = {} & 0.1/0.579987076 + 0.1/0.582751463 + 0.1/0.582751662 \\ & + 0.2/0.599465802 + 0.2/0.599466255 + 1/0,599466383 \\ & + 0.1/0.638780318 + 0.2/0.640317693 + 0.2/0.640318266\end{aligned}$$

Graphical representation of the obtained result is given below (Fig. 1):

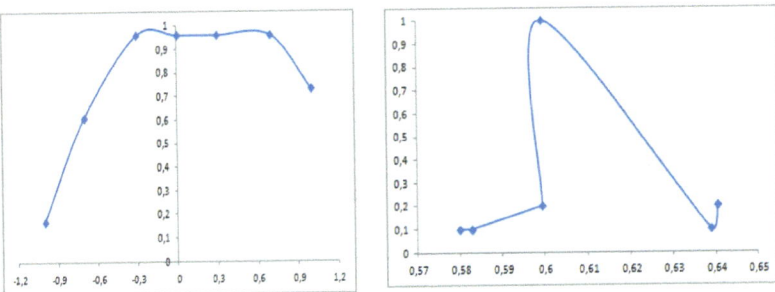

Fig. 1. The obtained result

6   Conclusion

In this paper, we develop a new approach to approximate reasoning using If-Then rules with Z-set based antecedents and consequents. An approach for reasoning with Z-rules is detailed in this paper. Advantage of this method is characterized by ability to operate with combination of fuzzy and probabilistic uncertainties. Separately using fuzzy and probability reasoning did not able to keep basic properties of information. This reasoning method based on probabilistic implication, a distance, similarity measure, fuzzy and probability aggregations. The proposed approach is able to deal with real-world bimodal information and the approach is characterized by an ability to compute both the inaccurate value of a variable and reliability. The proposed approach can be applied on different areas of the real life, in creating decision making and decision support systems.

References

1. Zadeh, L.A.: Fuzzy logic and approximate reasoning. Synthese **30**, 407–428 (1975)
2. Mizumoto, M., Zimmermann, H.-J.: Comparison of fuzzy reasoning methods. Fuzzy Sets Syst. **8**, 253–283 (1982)
3. Pal, S.K., Mandal, D.P.: Fuzzy logic and approximate reasoning: an overview. IETE J. Res. **37**(5), 548–560 (1991). https://doi.org/10.1080/03772063.1991.11437008

4. Nakanishi, H., Turksen, I.B., Sugeno, M.: A review and comparison of six reasoning methods. Fuzzy Sets Syst. **57**, 257–294 (1993)
5. Yeung, D.S., Tsang, E.C.C.: A comparative study on similarity-based fuzzy reasoning methods. Trans. Syst. Man Cybern. Part B: Cybern. **27**(2), 216–227 (1997)
6. Lim, T.P., Husain, W., Zakaria, N.: Recommender system for personalized wellness therapy. Int. J. Adv. Comput. Sci. Appl. **4**, 54–60 (2013)
7. Prentzas, J., Hatzilygeroudis, I.: Categorizing approaches combining rule-based and case-based reasoning. Expert Syst. **24**, 97–122 (2007)
8. Aliev, R.A., Aliev, R.R.: Soft Computing and its Application. World Scientific, Singapore (2001)
9. Aliev, R.A., Tserkovny, A.: Systemic approach to fuzzy logic formalization for approximate reasoning. Inform. Sci. **181**(6), 1045–1059 (2011)
10. Saner, T., Gardashova, L.A., Allahverdiyev, R.A., Eyupoglu, S.Z.: Analysis of the job satisfaction index problem by using fuzzy inference. Procedia Comput. Sci. **102**, 45–50 (2017)
11. Aliyarov, Y.R., Qardashova, L.A., Ahmadov, Sh.A.: Fuzzy expert system for rectal cancer. In: Advances in Intelligent Systems and Computing, vol. 896, pp. 160–167. Springer (2018)
12. Zadeh, L.A.: Outline of a new approach to the analysis of complex systems and decision processes. IEEE Trans. Syst. Man Cyberns. **SMC-3**, 28–44 (1973)
13. Grzegorzewski, P.: Probabilistic implications. Fuzzy Sets Syst. **226**, 53–66 (2013). https://doi.org/10.1016/j.fss.2013.01.003
14. Touazi, F., Cayrol, C., Dubois, D.: Possibilistic reasoning with partially ordered beliefs. J. Appl. Logic **13**(4), 770–798 (2015)
15. Dubois, D., Durrieu, C., Prade, H., Rico, A., Ferro, Y.: Extracting decision rules from qualitative data using Sugeno integral: a case-study. In: 13th European Conference on Symbolic and Quantitative Approaches to Reasoning with Uncertainty, pp. 14–24. Springer, Switzerland (2015)
16. Couso, I., Dubois, D.: Statistical reasoning with set-valued information: Ontic vs. epistemic views. Int. J. Approx. Reason. **55**(7), 1502–1518 (2014)
17. Jiao, L., Pana, Q., Denœux, T., Liang, Y., Feng, X.: Belief rule-based classification system: Extension of FRBCS in belief functions framework. Inform. Sci. **309**, 26–49 (2015)
18. Couso, I., Dubois, D.: A perspective on the extension of stochastic orderings to fuzzy random variables. In: 16th World Congress of the International Fuzzy Systems Association (IFSA), Gijon, pp. 1486–1492 (2015)
19. Zhou, S., Chen, Q., Wang, X.: Fuzzy deep belief networks for semi-supervised sentiment classification. Neurocomputing **131**, 312–322 (2014)
20. Dubois, D., Prade, H.: Possibility theory and its applications: where do we stand? In: Kacprzyk, J., Pedrycz, W. (eds.) Springer Handbook of Computational Intelligence, pp. 31–60. Springer, Heidelberg (2015)
21. Dubois, D., Liu, W., Ma, J., Prade, H.: The basic principles of uncertain information fusion. An organized review of merging rules in different representation frameworks. Inform. Fusion **32**, 12–39 (2016)
22. Aliev, R.A., Huseynov, O.H., Aliyeva, K.R.: Z-valued t-norm and t-conorm operators-based aggregation of partially reliable information. Procedia Comput. Sci. **102**, 12–17 (2016). https://doi.org/10.1016/j.procs.2016.09.363
23. Aliev, R.A., Huseynov, O.H., Zulfugarova, R.X.: Z-distance based IF-THEN Rules. Sci. World J. **2016**, 1–9(online) (2016). https://doi.org/10.1155/2016/1673537
24. Aliev, R.A., Pedrycz, W., Huseynov, O.H., Eyupoglu, S.Z.: Approximate reasoning on a basis of Z-number valued If-Then rules. IEEE Trans. Fuzzy Syst. **25**(6), 1589–1600 (2017)

25. Jiang, W., Cao, Y., Deng, X.: A novel Z-network model based on bayesian network and Z-number. IEEE Trans. Fuzzy Syst. **28**(8), 1585–1599 (2020). https://doi.org/10.1109/TFUZZ.2019.2918999
26. Aliev, R., Mamedova, G., Aliev, R.: Fuzzy Set Theory and its Application. Iran, Tabriz (1993)

Logics of Estimates for Fuzzy Statements

Gerald S. Plesniewicz[1]([✉]) and Valery B. Tarassov[2]

[1] National Research University MPEI, 14 Krasnokazarmennaya St.,
Moscow, Russia
salve777@mail.ru
[2] Bauman Moscow State Technical University, 52nd Baumanskaya St.,
Moscow, Russia
Vbulbov@yahoo.com

Abstract. Let **Z** be Zadeh's fuzzy propositional logic, i.e., a fuzzy proposi-
tional logic in which $\sim x$, $x \wedge y$, $x \vee y$ and $x \rightarrow y$ are interpret as $1 - x$, $\min\{x, y\}$, $\max\{x, y\}$ and $\max\{1 - x, y\}$ (correspondingly). *Estimates* for a **Z** sentence
φ are expressions of the form $\varphi \geq a$ and $\varphi \leq a$ where a is a number from the
unit interval [0, 1]. Let **E-Z** denote the set of all estimates for **Z** sentences, and
BE-Z be the set of Boolean combinations of the estimates. One can consider
BE-Z as a metalogic for the (object) logic **Z** if to interpret naturally estimates
and Boolean combinations of estimates by expanding valuations for atoms
(propositional variables). We call **BE-Z** the *logic of estimates* (for **Z** sentences).
For the logic of estimates, we define some inference system **S(Z)** consisting in
inference rules and acting in accordance with the method of analytic tableaux. **S
(Z)** is a sound and complete inference system. We also consider, by an example,
how use the inference system **S(Z)** to evaluate queries to fuzzy ontologies
written in the logic **BE-Z**.

Keywords: Fuzzy logics · Estimates for fuzzy statements · Analytic tableaux ·
Inference systems · Fuzzy ontologies · Query answering over fuzzy ontologies

1 Introduction. Main Definitions

Let **Z** be Zadeh's fuzzy propositional logic, i.e. a fuzzy propositional logic in which
$\sim x$, $x \wedge y$, $x \vee y$ and $x \rightarrow y$ are interpreted as (respectively) $1 - x$, $\min\{x, y\}$, $\max\{x, y\}$
and $\max\{1 - x, y\}$. *Estimates* for a **Z** sentence φ are expressions of the form
$\varphi \geq a$ and $\varphi \leq a$, where a is a number from the unit interval [0, 1]. Let **E-Z** denote
the set of all estimates for **Z** sentences, and let **BE-Z** be the set of Boolean combi-
nations of the estimates (i.e.it is the set such that: (*i*) **E-Z** $\subseteq$ **BE-Z**; (*ii*) $\sim \lambda$, $\lambda \wedge \mu$, $\lambda \vee$
μ and $\lambda \rightarrow \mu$ belong to **BE-Z**, if λ and μ belong to **BE-Z**).

Let A to be the set of atoms (i.e. propositional variables) in Z. A fuzzy valuation of
atoms is a function "•" from A to [0.1]. This function extends naturally to E-Z and then
to BE-Z, such that:

(i) "λ" $\in$ {0 (false), 1 (true)} for each $\lambda \in$ **BE-Z**;
(ii) "$\varphi \geq a$" = 1 if and only if "φ" $\geq a$";
(iii) "$\varphi \leq a$" = 1 if and only if "φ" $\leq a$, "$\varphi \leq a$" = 1;

R. A. Aliev et al. (Eds.): ICAFS 2020, AISC 1306, pp. 22–29, 2021.
https://doi.org/10.1007/978-3-030-64058-3_3

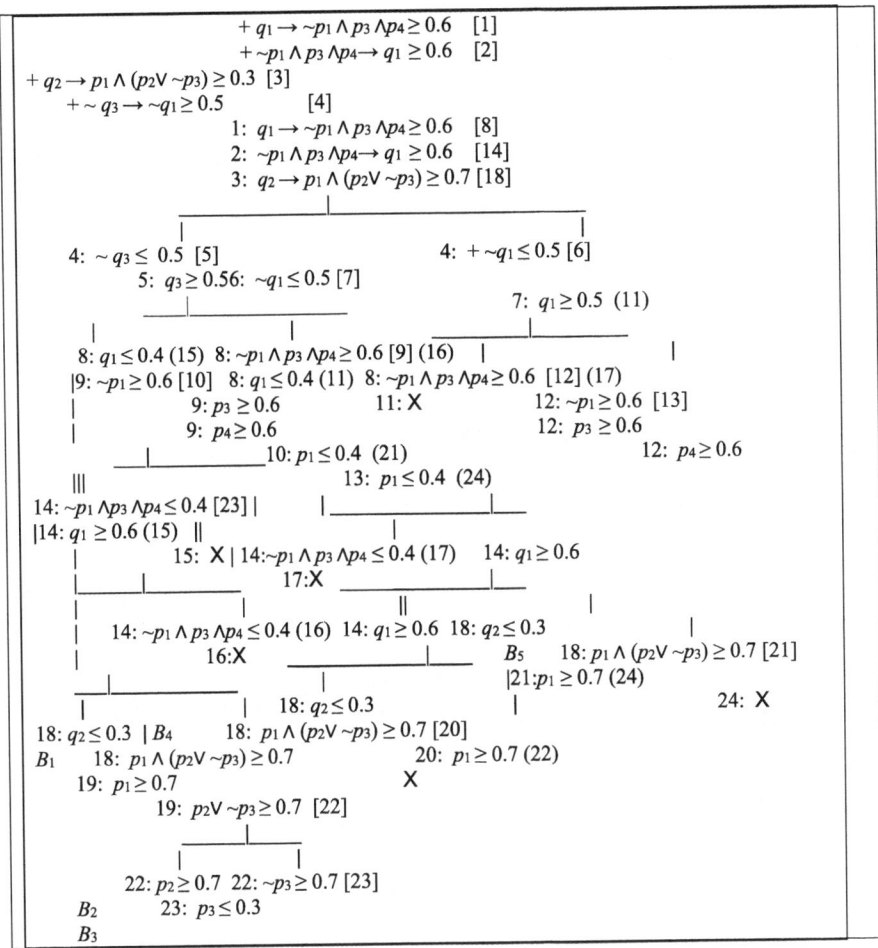

Fig. 1. Inference tree for the ontology from Example 2.

(iv) "$\sim\lambda$" = $\sim$"λ",   "$\lambda \wedge \mu$" = "λ" $\wedge$ "μ",   "$\lambda \vee \mu$" = "λ" $\vee$ "μ",   "$\lambda \rightarrow \mu$" = $\sim$"λ" $\vee$ "μ".

As any logic, the logic of estimates **BE-Z** induces the relation '|=' of (semantic) *logical consequence*. Let O be an *ontology* written in the logic **BE-Z**, i.e. a finite set of sentences, and φ be any sentence of the logic. Then $O \models \varphi$ if and only if there is no interpretation "•" such that φ is false ("φ" = 0) and all sentences $\psi \in O$ are true ("ψ" = 1). One can consider **BE-Z** as a metalogic for the (object) logic **Z**.

Earlier, the logic of estimates has been considered by J. Chen and S. Kundu [2, 4]. They have built a sound and complete inference system for this logic. Their system is based on generalized resolution. In this paper we propose a sound and complete inference system S(**Z**) whose rules act in accordance with analytic tableaux techniques [1, 3].

Let us remind that an inference system is *complete* if the logical consequence relation '⊨' implies the inference relation '⊢', i.e., φ is inferred from O by the rules of the inference system when O ⊨ φ. An inference system is *sound* if '⊢' implies '⊨'.

An important motive for introducing logics of estimates was that, in practical applications of fuzzy logics, experts have difficulties in determining the exact values of fuzziness for statements, while in the logic of estimates we can use only bounds for truth values and can manipulate the bounds.

One of applications of the logic of estimates is related to query answering. Estimates of the forms $p \geq r$ and $p \leq r$ ($p \in \textbf{At}$, $r \in [0, 1]$) are called *elementary*. A finite set O of non-elementary estimates from **BE-Z** is called *ontology* (written in **BE-Z**). Elementary estimates are considered as *facts*. A finite set F of facts is called *fact base*. We will consider *queries* to an ontology with a fact bases $O \cup F$. We write a query to $O \cup F$ in the form? (max$x$, min$y$) – $x \leq$ φ $\leq y$, where φ is a propositional formula, φ $\in \textbf{Z}$. The query reads as follows: "Find the maximum value of the variable x and the minimum value of the variable y such that $O \cup F$ implies the estimates φ $\geq x$ and φ $\leq y$. Thus, the *answer* to this query is the pair($x = a$, $y = b$) such that $O \cup F \models$ φ $\geq a$ and $O \cup F \models$ φ $\geq b$ but $O \cup F \not\models$ φ $\geq a'$ and $O \cup F \not\models$ φ b' for each $a' > a$ and each $b' < b$.

We will show, by example, how to evaluate queries to ontologies with fact bases written in the logic **BE-Z** of estimates for Zadeh's fuzzy propositional logic.

2   Equivalences in the Logic BE-Z

The following lemma contains basic equivalences of the logic **BE-Z**.

Lemma 1. Let φ and ψ are sentences of the logic and a, $b \in [0, 1]$. The following equivalences hold:

$$\sim \varphi \geq a \equiv \varphi \leq 1 - a, \tag{1}$$

$$\sim \varphi \leq a \equiv \varphi \geq 1 - a,$$

$$\varphi \wedge \psi \geq a \equiv (\varphi \geq a) \wedge (\psi \geq a),$$

$$\varphi \wedge \psi \leq a \equiv (\varphi \geq a) \vee (\psi \geq a),$$

$$\varphi \vee \psi \geq a \equiv (\varphi \geq a) \vee (\psi \geq a),$$

$$\varphi \vee \psi \leq a \equiv (\varphi \leq a) \wedge (\psi \leq a), \tag{2}$$

$$(\varphi \geq a) \wedge (\varphi \geq b) \equiv \varphi \geq \max\{a, b\}, \tag{3}$$

$$(\varphi \leq a) \wedge (\varphi \leq b) \equiv \varphi \leq \min\{a, b\}.$$

Here are the proofs of the equivalences (1), (2) and (3):

- "$\sim \varphi \geq a$" $= 1 \Longleftrightarrow$ "$\sim \varphi$" $\geq a \Longleftrightarrow 1 -$ "φ" $\geq a \Longleftrightarrow$ "φ" $\leq 1 - a \Longleftrightarrow$ "φ $1 - a$" $= 1$. Hence $\sim \varphi \geq a \equiv \varphi \leq 1 - a$.
- "$\varphi \vee \psi \leq a$" $= 1 \Longleftrightarrow$ "$\varphi \vee \psi$" $\leq a \Longleftrightarrow$ "φ" $\vee$ "ψ" $\leq a \Longleftrightarrow$ "φ" $\leq a$ and "ψ" $\leq a$ (since $\max\{x, y\} \leq a \Longleftrightarrow x \leq a$ and $y \leq b$) $\Longleftrightarrow$ "$\varphi \leq a$" $= 1$ and "$\psi \leq b$" $= 1 \Longleftrightarrow$ "$\varphi \leq a$" $\wedge$ "$\psi \leq b$" $= 1 \Longleftrightarrow$ "$\varphi \geq a$" $\wedge$ "$\psi \geq b$" $\Longleftrightarrow$ "$(\varphi \geq a) \wedge (\psi \geq)$" $= 1$.
- "$(\varphi \geq a) \wedge (\varphi \geq b)$" $= 1 \Longleftrightarrow$ "$\varphi \geq a$" $= 1$ and   "$\varphi \geq b$" $= 1 \Longleftrightarrow$ "φ" $\geq a$ and "φ" $\geq b \Longleftrightarrow$ "φ" $\geq \max\{a, b\} \Longleftrightarrow$ "$\varphi \geq \max\{a, b\}$" $= 1$.

3   Method of Analytic Tableaux for the Logic BE-Z

The Tables 1, 2, 3 and 4 show the inference rules in the style of analytic tableaux for the logic **BE-Z**. We denote by **S(Z)** the inference system of these rules.

Table 1. Inference rules for propositional connectives.

Rule number	Antecedent	Consequents
1	$+ \sim \varphi$	$- \varphi$
2	$\sim \varphi$	$+ \varphi$
3	$+ \varphi \wedge \psi$	$+ \varphi$ and $+ \psi$
4	$- \varphi \wedge \psi$	$- \varphi$ or $- \psi$
5	$+ \varphi \vee \psi$	$+ \varphi$ or $+ \psi$
6	$- \varphi \vee \psi$	$- \varphi$ and $- \psi$
7	$+ \varphi \rightarrow \psi$	$- \varphi$ or $+ \psi$
8	$- \varphi \rightarrow \psi$	$+ \varphi$ and $- \psi$
9	$+ \lambda$	λ
10	$- \lambda$	$\sim \lambda$

φ and ψ are **BE-F(L)** sentences, λ is a **E-F(L)** sentence

In Table 1 the rules contain sentences with the '+' or '−' signs. The sign '+' means that the sentence is true, and '−' means that it false. The set of rules in Table 1 is the standard inference system in the style of analytic tableaux for propositional logic with signed sentences [1, 4], if we consider φ and ψ as formulas of classic propositional logic.

When applying the method of analytical tables, the inferences are presented in the form of trees, the vertices of which are logical sentences, and the edges are determined from the applied inference rules.

Example 1. Consider an example of applying the inference system **S(Z)** to the ontology.

Table 2. Inference rules for estimates.

Rule number	Antecedent	Consequents
1	$\sim\varphi \geq a$	$\varphi \leq \mathbf{n}(a)$
2	$\sim\varphi \leq a$	$\varphi \geq \mathbf{n}(a)$
3	$\varphi \wedge \psi \geq a$	$\varphi \geq a$ and $\psi \geq a$
4	$\varphi \wedge \psi \leq a$	$\varphi \leq a$ or $\psi \leq a$
5	$\varphi \vee \psi \geq a$	$\varphi \geq a$ or $\psi \geq a$
6	$\varphi \vee \psi \leq a$	$\varphi \leq a$ and $\psi \leq a$
7	$\varphi \rightarrow \psi \geq a$	$\varphi \leq 1 - a$ or $\psi \geq a$
8	$\varphi \rightarrow \psi \leq a$	$\varphi \geq 1 - a$ and $\psi \leq a$

Table 3. Inference rules for estimates.

Rule number	Antecedent	Consequents
1	$\sim(\sim\varphi \geq a)$	$\sim(\varphi \leq 1 - a)$
2	$\sim(\sim\varphi \leq a)$	$\sim(\varphi \geq 1 - a)$
3	$\sim(\varphi \wedge \psi \geq a)$	$\sim(\varphi \geq a)$ or $\sim(\psi \geq a)$
4	$\sim(\varphi \wedge \psi \leq a)$	$\sim(\varphi \leq a)$ and $\sim(\psi \leq a)$
5	$\sim(\varphi \vee \psi \geq a)$	$\sim(\varphi \geq a)$ and $\sim(\psi \geq a)$
6	$\sim(\varphi \vee \psi \leq a)$	$\sim(\varphi \leq a)$ or $\sim(\psi \leq a)$
7	$\sim(\varphi \rightarrow \psi \geq a)$	$\sim(\varphi \leq 1 - a)$ and $\sim(\psi \geq a)$
8	$\sim(\varphi \rightarrow \psi \leq a)$	$\sim(\varphi \geq 1 - a)$ or $\sim(\psi \leq a)$

$O = \{q_1 \rightarrow \sim p_1 \wedge p_3 \wedge p_4 \geq 0.6, \ \sim p_1 \wedge p_3 \wedge p_4 \rightarrow q_1 \geq 0.6, q_2 \rightarrow p_1 \wedge (p_2 \vee \sim p_3) \geq 0.3,$
$q_3 \rightarrow (\sim q_1 \leq 0.5)\}.$

Table 4. Binary inference rules.

Rule number	Antecedent	Consequents
1	$+\varphi$	$-\varphi X$
2	λ	$\sim\lambda\ X$
3	$\varphi \geq a$	$\varphi \leq b, \ \sim(a \leq b)X$
4	$\varphi \geq a$	$\sim(\varphi \geq b), b \leq aX$
5	$\varphi \leq a$	$\sim(\varphi \geq b), a \leq bX$
6	$\varphi \leq a$	$\varphi \leq b\ \varphi \leq \min\{a, b\}$
7	$\varphi \geq a$	$\varphi \geq b\ \varphi \geq \max\{a, b\}$

Here it is supposed that φ and ψ are arbitrary **BE-Z** sentences.

We begin to build the inference tree with an initial branch been the sequence of the signed by '+' sentences of O (see Fig. 1).The first vertex of the tree is the first sentence of O marked with a '+' and with the label '[1]'.The sign '+' indicates that this sentence is assumed to be true, and the label '[1]' indicates that in step 1 there was applied some inference rule to the sentence (in this case, the rule 9 of Table 1). The result of applying this rule is a vertex with a sentence without sign '+' and with a left label '1:', which indicates that this vertex was attached to the current tree in step 1. The vertex '1: $q_1 \rightarrow \sim p_1 \wedge p_3 \wedge p_4 \geq 0.6$ [8]' contains the following information: the estimate $q_1 \rightarrow \sim p_1 \wedge p_3 \wedge p_4 \geq 0.6$ was obtained in step 1, and some inference rule was applied to this estimate.

In constructing this tree, we used standard tactics for the sequence of inference rules applicable at current step [4]. In particular, with this choice, we gave priority to rules that have no alternative in consequents. There are 5 open branches in the complete tree in Fig. 1. Other 5 branches are closed. In the terminology of analytic tableaux, a branch of the inference tree is called closed if it contains a contrary pair of formulas, and therefore, the branch is inconsistent set. Thus, if all branch of a complete inference tree for an ontology are closed then the ontology is inconsistent.

In general, let us build a complete inference tree for a given consistent ontology O. Then the tree has open branches.

Lemma 2. Let O be a consistent ontology in **BE-Z** and B_1, B_2, ..., B_m are all open branches of an inference tree built for O. For each B_i, let us write out all the estimates from B_i, and let C_i be the conjunct made up of these estimates. Also, let $O^{\wedge}$ be the conjunct made up all sentences from O. Then $O^{\wedge} \equiv C_1 \vee C_1 \vee ... \vee C_m$.

Theorem. The inference system S(**Z**)for the Zadeh's logic **Z** is sound and complete.

The soundness of these systems follows from Lemma 1 (see Sect. 2). The proof of the completeness is obtained using Lemma 2.

Query Answering over Ontologies in Logic of Estimates BE-Z.

Let us consider, how to evaluate queries to ontologies written in the logic **BE-Z**.

Example 2. Consider a problem of medical diagnosis in a case when there are three diseases q1, q2, q3 and 4 symptoms p1, p2, p3, p4. Suppose the following knowledge defines the relationship between these diseases and symptoms:

(a) The disease q_1 with a confidence degree ≥ 0.6 is uniquely determined by the presence of symptoms p_3 and $p_{\overline{4}}$ and by the absence of symptom p_1;

(b) At a confidence degree ≥ 0.3, if the disease q_2 occurs then symptom p_1 is observed, and also there is a symptom p_2 or there is no symptom p_3;

(c) At a confidence degree ≥ 0.5, if the disease $q_3$does not occurs, then disease $q_3$also does not occur.

Suppose symptom p_1 is observed with estimate $p_1 \geq 0.6$. Consider query.

(d) Find the best lower and upper estimates for disease q_1.

In the logic of **BE-Z**, this knowledge is represented by the following sentences:

$$q_1 \rightarrow \sim p_1 \wedge p_3 \wedge p_4 \geq 0.6 \text{ and } \sim p_1 \wedge p_3 \wedge p_4 \rightarrow q_2 \geq 0.6 \text{ for (a)}; \quad (4)$$

$$q_2 \rightarrow p_1 \wedge (p_2 \vee \sim p_3) \geq 0.3 \text{ for (b)}; \quad (5)$$

$$\sim q_3 \rightarrow \sim q_1 \geq 0.5 \text{ for (c)}. \quad (6)$$

So, we have the ontology O from Example 1. We also represent the query (d) as Q: ? (max x, min y) $- x \leq q_1 \leq y$.

The assumption that symptom p_1 is observed with estimate $p_1 \geq 0.6$ represents as the fact base $\mathbf{F} = \{p_1 \geq 0.6\}$.

Figure 1 shows the inference tree built for the ontology $\mathbf{O}$. As we see, the tree has 5 open branches from which the following conjuncts are extracted:

$$C_1 = (q_3 \geq 0.5) \wedge (q_1 \leq 0.4) \wedge (q_2 \leq 0.7),$$
$$C_2 = (q_3 \geq 0.5) \wedge (q_1 \leq 0.4) \wedge (q_2 \leq 0.7) \wedge (p_1 \geq 0.7) \wedge (p_2 \geq 0.7),$$
$$C_3 = (q_3 \geq 0.5) \wedge (q_1 \leq 0.4) \wedge (q_2 \leq 0.7) \wedge (p_1 \geq 0.7) \wedge (p_3 \leq 0.3),$$
$$C_4 = (q_3 \geq 0.5) \wedge (p_3 \geq 0.6) \wedge (p_4 \geq 0.6) \wedge (p_1 \leq 0.4) \wedge (q_1 \geq 0.6),$$
$$C_5 = (p_3 \geq 0.6) \wedge (p_4 \geq 0.6) \wedge (p_1 \leq 0.4) \wedge (q_1 \geq 0.6) \wedge (q_2 \leq 0.3).$$

The set $\mathbf{O} \cup \mathbf{F} \cup \{\sim(q_1 \geq x)\}$ of sentences is inconsistent if and only if $\mathbf{O} \cup \mathbf{F} \models q_1 \geq x$. Due to Lemma 2, this set is inconsistent if and only if $C_1' \vee C_2' \vee C_3' \vee C_4' \vee C_5' \equiv 0$, where $C_i' \vee C_i \wedge (p_1 \geq 0.6) \wedge \sim (q_3 \geq x) = C_i \wedge (p_1 \geq 0.6) \wedge (q_3 < x)$. It is easy to see that $C_i' = 0$ ($i = 1, 2, 3$) for $x \leq 0.5$ since C_i' contains the contrary pair $(q_3 \geq 0.5, q_3 < x)$ if $x \leq 0.5$. Therefore, 0.5 is the maximum of the variable x.

When finding the minimum of the variable y, we take $C_i'' = C_i \wedge (p_1 \geq 0.6) \wedge \sim (q_3 \leq y) = C_i \wedge (p_1 \geq 0.6) \wedge (q_3 > y)$. It is clear that $C_i'' = 0$ only if $y = 1$. Therefore, 1 is the minimum value of the variable y.

Thus, $(x = 0.5, y = 1)$ is the answer to the query Q.

4   Conclusion

We have introduced the logic **BE-Z** of estimates for fuzzy statements represented by sentences of Zadeh's fuzzy propositional logic **Z**. For the logic **BE-Z**, we built the system **S(Z)** of inference rules acting on ontologies with fact bases in accordance with analytic tableaux techniques. **S(Z)** is a sound and complete inference system for the logic of estimates **BE-F(L)**.

Acknowledgment. This work was supported by Russian Foundation for Basic Research (projects 18-29-03088, 20-07-00615 and 20-07-00770).

References

1. Agostino, M., Gabbay, D., Hahnle, R., Possega, J.: Handbook of Tableaux Methods. Springer, Berlin (2001). https://doi.org/10.1007/978-94-017-1754-0
2. Chen, J., Kundu, S.: A sound and complete fuzzy logic system using Zadeh's implication operator. In: Ras, Z.W., Maciek, M. (eds.) Proceedings of the 9th International Symposium on Methodologies for Intelligent Systems. Lecture Notes in Artificial Intelligence, vol. 1079, pp. 233–242. Springer, Berlin (1996). https://doi.org/10.1007/3-540-61286-6_148

3. Fitting, M.: First-Order Logic and Automated Theorem Proving. Springer, Berlin (1996). https://doi.org/10.1007/978-1-4612-2360-3
4. Kundu, S.: An improved method for fuzzy-inferencing using Zadeh's implication operator. In: Proceedings of IJCAL Workshop on Fuzzy Logic in AI, pp. 117–125 (1995)

Construction of Consistent Z-Preferences in Decision Making for a Foreign Market Selection

Rafig R. Aliyev$^{(\boxtimes)}$

Research Laboratory of Intelligent Control and Decision-Making Systems
in Industry and Economics, Azerbaijan State Oil, and Industry University,
Azadlyg Avenue, 20, AZ1010 Baku, Azerbaijan
`rafig.aliyev@hotmail.com`

Abstract. The objective of this paper is to help a case company in its initial foreign market selection (IFMS) decision. Therefore, paper attempted to design a decision model business for internationalization process. In order to validate and test created model Multi Attribute Decision Making (MADM) methods was evaluated. The notion of consistency is used to estimate the quality of preference knowledge and its stability for reliable evaluation of decision alternatives in MADM. In the existing MADM models, there is a set of strict consistency conditions and requirements are not achievable in the real situations when decision maker has limited rationality and partially reliable preferences. In this study, we use an approach to deriving consistency-driven preference degrees for such kind of situations. A preference degree is described by a Z-number to reflect imprecision and partial reliability of preference knowledge in MADM for a foreign market selection.

Keywords: Z-number · Z-number-valued preference · Decision making · Pairwise comparison matrix · Consistency

1 Introduction

On one hand, rising globalized environment and intensity of competition, escalated dependency of firms for internationalized business activities in order to secure their survival in the world market, on the other hand comprehensive improvement of technological innovations, reduced barriers and favorable government incentives encouraged companies to internationally expand and maximize their sustainable growth.

Foreign Market Entry (FME) is essential managerial and strategical decision that have been researched widely under various critical concepts. Existing literatures of internationalization structured around Entry Mode Selection (EMS), International Market Selection (IMS), timing of entry, motivations behind expansion, marketing strategies and many more [1]. Although, it was argued that these concepts are interlocked and integrated with each other, this relation is more practical and applicable in sequential process of market entry strategies [2]. That means when identifying the FME strategy, firstly it is expected to clarify the target market/location and then the

© The Author(s), under exclusive license to Springer Nature Switzerland AG 2021
R. A. Aliev et al. (Eds.): ICAFS 2020, AISC 1306, pp. 30–37, 2021.
https://doi.org/10.1007/978-3-030-64058-3_4

appropriate entry mode, market plan and others [3, 4]. However, eclectic paradigm is one of the examples that concentrates on choice of entry mode and selects the market accordingly as single decision process [5]. This illustrates the lack of paid attention to decision of IMS.

Efficiency and validity of results of decision making mainly is related to consistency and stability of decision maker's preferences.

A decision maker's (DM's) preference may be formally described by a pairwise comparison matrix (PCM) (aij), where an entry a_{ij} denotes a degree to which an *i-th* alternative (criterion) is preferred to *j*-th one [6]. Natural conditions used for a_{ij} are $a_{ii} = 1$ and $a_{ji} = 1/a_{ij}$ (reciprocity), $\forall i,j = 1, \ldots, n$. Traditionally, consistency of (a_{ij}) is based on multiplicative transitivity condition (though different constructs are also used):

$$a_{ij}a_{jk} = a_{ik}, \forall i,j,k.$$

This implies that degree of preference a_{ik} is equal to the product of preferences degrees staying at all possible ways from i to k through j.

An important area of study is devoted to inconsistency reduction, also referred to as construction of consistency-driven PCM. In [7] a sensitivity analysis is used to uncover PCM entries most influential for Saaty's index. In [8] it is proposed a more general framework based on lp distance (p = 2 is the Euclidean distance case). A goal programming approach is used to derive a consistent PCM closest to a given inconsistent one.

Being imprecise and partially consistent, real-world preferences are also partially reliable. The reasons are restricted competence of DM's, complexity of alternatives, imperfect decision-relevant information, psychological biases etc. Up to day, no works have been proposed on consistency of partially reliable preferences.

In this paper we deal with construction of consistency-driven partially reliable preferences described by PCM with Z-number-valued entries in MADM for a foreign market selection.

The paper is structured as follows. In Sect. 2 necessary definitions are provided such as Z-number, Z-number-valued PCM etc. In Sect. 3, a statement of problem of construction of a consistent Z-number-valued PCM that describes partially reliable preference for MADM for a foreign market selection is formulated. The solution of the MADM problem is given. Section 4 concludes.

2 Preliminaries

Definition 1. Continuous Z-number **[9, 10].** A continuous Z-number is an ordered pair $Z = (A, B)$ where A is a continuous fuzzy number playing a role of a fuzzy constraint on values that a random variable X may take:

$$X \text{ is } A,$$

and B is a continuous fuzzy number with a membership function $\mu_B : [0, 1] \rightarrow [0, 1]$, playing a role of a fuzzy constraint on the probability measure of A:

$$P(A) = \int_R \mu_A(x)p(x)dx \text{ is } B.$$

Definition 2. A Z-number-valued PCM. A Z-number-valued PCM (Z_{ij}) is a square matrix of Z-numbers:

$$(Z_{ij} = (A_{ij}, B_{ij})) = \begin{pmatrix} Z_{11} = (A_{11}, B_{11}) & \cdots & Z_{1n} = (A_{1n}, B_{1n}) \\ \cdot & \cdots & \cdot \\ Z_{n1} = (A_{n1}, B_{n1}) & \cdots & Z_{nn} = (A_{nn}, B_{nn}) \end{pmatrix}.$$

A Z-number $Z_{ij} = (A_{ij}, B_{ij})$, $i, j = 1, \ldots, n$ describes partially reliable information on degree of preference for *i-th* alternative (criterion) against *j-th* one.

Definition 3. An inconsistency index for Z-number-valued PCM. An inconsistency index K for Z-number-valued PCM (Z_{ij}) is defined as follows:

$$K((Z_{ij})) = \max_{i<j<k} \min \left\{ D\left(Z(1), \left(\frac{z_{ik}}{z_{ij}z_{jk}}\right)\right) D\left(Z(1), \left(\frac{z_{ij}z_{jk}}{z_{ik}}\right)\right) \right\},$$

where the components of Z-number $Z(1) = (A, B)$, are fuzzy singletons $A = 1$ and $B = 1$.

3   Statement of the Problem

Let us consider extraction of a consistent Z-number-valued matrix to describe preferences over multiple criteria in a foreign market selection problem. We will deal with three criteria that describe a series of economical and institutional characteristics: *Institutional Proximity* (C_1) (government performance and economic freedom issues), *Economic Proximity* (C_2) (socioeconomic issues) and *Social and Cultural Proximity* (C_3) (cultural characteristics). Due to complexity of the considered criteria, a DM's preferences may be characterized by fuzziness and partial reliability. In view of this, we use partially reliable preference degrees of the Saaty scale to represent comparative importance of criteria (Table 1):

This information can be formalized by a 3×3 matrix of Z-numbers with TFNs-based components:

Table 1. Z-number-valued preference knowledge about criteria importance

	C_1	C_2	C_3
C_1	(equally important, very sure)	(slightly less important, sure)	(between slightly and more important, sure)
C_2	(slightly more important, sure)	(equally important, very sure)	(absolutely more important, almost sure)
C_3	(between slightly and less important, sure)	(absolutely less important, almost sure)	(equally important, very sure)

$$
\begin{pmatrix}
Z_{11} = ((0.93, 0.95, 0.97), & Z_{12} = ((0.327, 0.333, 0.34), & Z_{13} = ((3.92, 4, 4.08), \\
(0.95, 0.98, 1)) & (0.7, 0.8, 0.9)) & (0.7, 0.8, 0.9)) \\
Z_{21} = ((2.94, 3, 3.06), & Z_{22} = ((0.93, 0.95, 0.97), & Z_{23} = ((8.82, 9, 9.02), \\
(0.7, 0.8, 0.9)) & (0.95, 0.98, 1)) & (0.6, 0.7, 0.8)) \\
Z_{31} = ((0.245, 0.25, 0.255), & Z_{32} = ((0.1108, 0.111, 0.113), & Z_{33} = ((0.93, 0.95, 0.97), \\
(0.7, 0.8, 0.9)) & (0.6, 0.7, 0.8)) & (0.95, 0.98, 1))
\end{pmatrix}
$$

Table 2. The codebook for A parts of Z-number

Linguistic term	Fuzzy number
Absolutely less important	$(0.1108, 0.111, 0.113)$
Slightly less important	$(0.327, 0.333, 0.34)$
Between slightly and less important	$(0.245, 0.25, 0.255)$
Equally important	$(0.93, 0.95, 0.97)$
Slightly more important	$(2.94, 3, 3.06)$
Absolutely more important	$(8.82, 9, 9.02)$

Table 3. The codebook for B parts of Z-number

Linguistic term	Fuzzy number
Sure	$(0.7, 0.8, 0.9)$
Very sure	$(0.95, 0.98, 1)$
Almost sure	$(0.6, 0.7, 0.8)$

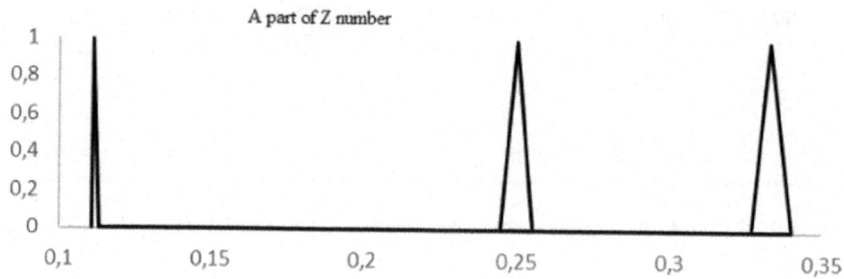

Fig. 1. Triangles for the first 3 linguistic terms (absolutely less important, slightly less important, between slightly and less important)

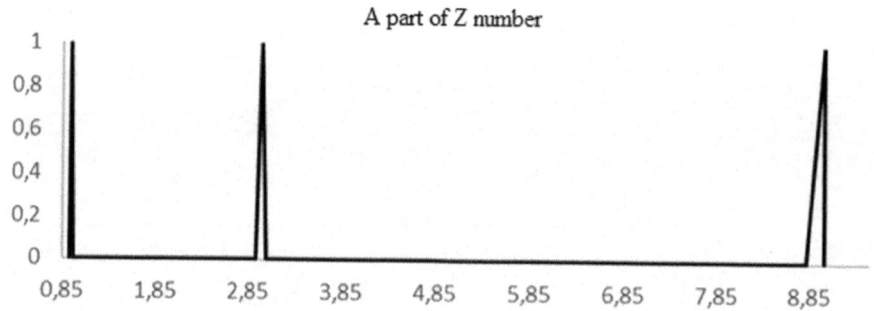

Fig. 2. Triangles for the last 3 linguistic terms (equally important, elightly more important, absolutely more important)

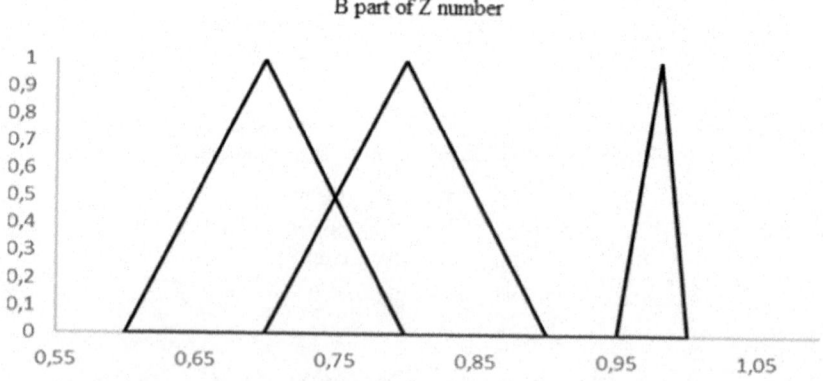

Fig. 3. Triangles for B parts

Codebooks for A and B parts of Z-number are given in Table 2, 3 and their graphical representations are shown in Fig. 1, 2 and 3.
The problem is stated as follows.

Objective function:

$$\sum_{i=1}^{3} \sum_{j=1}^{3} D(Z'_{ij}, Z_{ij}) \rightarrow \min$$

For solution of the problem reciprocity, transitivity and non-negativity conditions are given below. Differential evaluation optimization method is used.

Reciprocity conditions:

$$Z'12 * Z'21 = Z\_1,$$
$$Z'13 * Z'31 = Z\_1,$$
$$Z'23 * Z'32 = Z\_1,$$

where $Z\_1 = ((1, 1, 1), (0.8, 0.9, 1))$

Transitivity conditions:

$$Z'11 * Z'12 = Z'11,$$
$$Z'11 * Z'13 = Z'13,$$

$$Z'12 * Z'22 = Z'12,$$
$$Z'12 * Z'23 = Z'13,$$
$$Z'13 * Z'32 = Z'12,$$
$$Z'13 * Z'33 = Z'13,$$

$$Z'21 * Z'11 = Z'21,$$
$$Z'21 * Z'13 = Z'23,$$
$$Z'22 * Z'21 = Z'21,$$
$$Z'22 * Z'23 = Z'23,$$
$$Z'23 * Z'31 = Z'21,$$
$$Z'23 * Z'33 = Z'23,$$

$$Z'31 * Z'11 = Z'31,$$
$$Z'31 * Z'12 = Z'32,$$
$$Z'32 * Z'21 = Z'31,$$
$$Z'32 * Z'22 = Z'32,$$
$$Z'33 * Z'31 = Z'31,$$
$$Z'33 * Z'32 = Z'32.$$

Non-negativity conditions:

$$Z'11 \geq Z\_0,$$
$$Z'12 \geq Z\_0,$$
$$Z'13 \geq Z\_0,$$

$$Z'21 \geq Z\_0,$$
$$Z'22 \geq Z\_0,$$
$$Z'23 \geq Z\_0,$$

$$Z'31 \geq Z\_0,$$
$$Z'32 \geq Z\_0,$$
$$Z'33 \geq Z\_0,$$

where $Z\_0 = ((0,0,0),(0.8,0.9,1))$

Optimal Z-number-valued matrix $(Z'_{ij}) = MP(V'_{best})$ is retrieved:

$$
\begin{pmatrix}
\begin{array}{lll}
Z_{11} = ((1.0008, 1.0028, 1.0028), & Z_{12} = ((0.4029, 0.4029, 0.402935), & Z_{13} = ((3.65, 3.651, 3.6512), \\
(0.9996, 1, 1)) & (0.72, 0.996, 0.996)) & (0.997, 0.9973, 0.9973)) \\
Z_{21} = ((2.4827, 2.483, 2.483), & Z_{22} = ((0.995452, 1, 1), & Z_{23} = ((9.06, 9.06, 9.064), \\
(0.78, 0.78, 0.991447)) & (0.95, 0.98, 1)) & (0.9947, 0.9947, 0.995)) \\
Z_{31} = ((0.278, 0.278, 0.278), & Z_{32} = ((0.110, 0.1104, 0.1104), & Z_{33} = ((0.999, 1.003, 1.003), \\
(0.0078, 0.008, 0.50)) & (0.499, 0.999, 0.999)) & (0.999, 1, 1))
\end{array}
\end{pmatrix}
$$

At the final step, we have to verify whether the value of K for the obtained (Z'_{ij}) exceeds a predefined threshold $\theta_K = 0.1$. The computed value of K is $K((Z'_{ij})) = 0.003$ which does not exceed θ_K. Thus, the obtained matrix can be considered as consistent.

4   Conclusion

Real-world preferences are characterized by imprecise and partially reliable information. In this paper, we propose an approach to construct a consistency-driven Z-number-valued PCM characterized by imprecision and partial reliability of a DM's preferences in a foreign market selection. The approach is based on formalism of Z-numbers. An inconsistency reduction of Z-number-valued PCM is formulated as an optimization problem with Z-number valued decision variables. A DE optimization-based solution method is proposed.

References

1. Surdu, I., Mellahi, K.: Theoretical foundations of equity based foreign market entry decisions: a review of the literature and recommendations for future research. Int. Bus. Rev. **25**(5), 1169–1184 (2016). https://doi.org/10.1016/j.ibusrev.2016.03.001
2. Hart, S.J., Webb, J.R., Jones, M.V.: Export marketing research and the effect of export experience in industrial SMEs. Int. Mark. Rev. **11**(6), 4–22 (1994). https://doi.org/10.1108/02651339410072980
3. Root, F.R.: Entry Strategies for International Markets. Revised and Expanded. Wiley, San Francisco (1998)
4. Górecka, D., Szałucka, M.: Country market selection in international expansion using multicriteria decision aiding methods. Multiple Criteria Decis. Making **8**, 31–55 (2013)
5. Dunning, J.H.: The eclectic paradigm of international production: a restatement and some possible extensions. J. Int. Bus. Stud. **19**, 1–31 (1988)
6. Brunelli, M.: A survey of inconsistency indices for pairwise comparisons. Int. J. Gen. Syst. Sep. **47**(8), 751–771 (2018). https://doi.org/10.1080/03081079.2018.1523156
7. Aliev, R.A., Perdycz, W., Huseynov, O.H.: Hukuhara difference of Z-numbers. Inform. Sci. **466**, 13–24 (2018)
8. Dopazo, E., González-Pachón, J.: Consistency-driven approximation of a pairwise comparison matrix. Kybernetika **5**, 561–568 (2003)
9. Aliev, R.A., Huseynov, O.H., Aliyev, R.R., Alizadeh, A.V.: The Arithmetic of Z-Numbers: Theory and Applications. World Scientific, Singapore (2015)
10. Aliev, R.A., Huseynov, O.H., Zeinalova, L.M.: The arithmetic of continuous Z-numbers. Inform. Sci. **373**, 441–460 (2016)

Fuzzy Decision Method Based on Zadeh's Data Aggregation Approach

Mustafa B. Babanli[1($\boxtimes$)] and Jale M. Babanli[2]

[1] Azerbaijan State Oil and Industry University,
Azadlyg Avenue, 20, AZ1010 Baku, Azerbaijan
mustafababanli@yahoo.com
[2] "Eazi START" Startup and Innovation Center of Azerbaijan State Oil
and Industry University, Azadlyg Avenue, 20, AZ1010 Baku, Azerbaijan
babanlijale@gmail.com

Abstract. Decision making often takes place in an environment of uncertainty and imprecision of decision relevant information. There are many effective methods of decision making with uncertain information. However, these methods are too complex to be applied to problems with a huge number of alternatives and criteria. In this work, a new effective fuzzy decision-making approach based on Zadeh's idea for aggregation of data is suggested. A problem of university units ranking is considered to prove feasibility and validity of the proposed approach.

Keywords: Fuzzy set · Fuzzy number · Similarity measure

1 Introduction

Decision making problems underlie human activity in various fields, such as economics, business, system analysis etc. Decision making theory is a large field of various approaches. Determination of an optimal decision is related to a series of factors that are difficult to be quantified. Indeed, actual decision making takes place under imprecise, uncertain, and incomplete information. Under such conditions, classical logic-based approach are often not suitable.

Real-world problems of decision making are characterized by high complexity. In particular, a decision maker (DM) often finds it difficult to describe perception and preference under uncertainty. In order to handle linguistic information, Zadeh introduced the concepts of fuzzy set (FS) [1]. FSs were successfully used for decision analysis, and this lead to emergence of a series of extensions, such as intuitionistic fuzzy sets (IFSs) and hesitant fuzzy sets (HFSs). IFSs are used to describe uncertainty of non-membership degree [2], and HFSs are used to describe hesitant information [3]. Furthermore, extended linguistic sets have been proposed [4–7] to adopt application of IFSs and HFSs under qualitative information.

Fuzzy logic [1, 8] is widely applied in various fields. The most preferred methods in the decision-making field are fuzzy AHP, ANP, ELECTRE, TOPSIS and other methods. However, these methods are too complex for practical application when decision environment is characterized with a huge number of alternatives and criteria.

© The Author(s), under exclusive license to Springer Nature Switzerland AG 2021
R. A. Aliev et al. (Eds.): ICAFS 2020, AISC 1306, pp. 38–46, 2021.
https://doi.org/10.1007/978-3-030-64058-3_5

In this paper, an effective method for fuzzy multicriteria decision making based on Zadeh idea of data aggregation [9] is used for ranking university units.

The rest of the paper is organized as follows. Section 2 includes definitions of the main concepts used in the paper. Section 3 describes the statement of problem. Section 4 is devoted to the proposed methodology of fuzzy decision making. An application of the approach to faculties ranking in university is given in Sect. 5.

2 Preliminaries

Definition 1. A fuzzy number. Let X be a universe of discourse. The fuzzy set C on X, whose membership function is the mapping of $\mu_C : \mathcal{R} \to [0, 1]$, is a continuous fuzzy number if it fulfils the following conditions:

1. C is a normal fuzzy set;
2. C is a convex fuzzy set;
3. α-cut C^α is a closed interval for any $\alpha \in [0, 1]$;
4. The support $\text{supp}(C)$ is bounded.

Definition 2. Aggregation of fuzzy numbers. Let $C_1, \ldots, C_m$ be fuzzy numbers. An arithmetic mean-based aggregation of fuzzy numbers, C is defined as follows:

$$C = \frac{\sum_{i=1}^m C_i}{m}$$

Definition 3. Distance between triangular fuzzy numbers. Let $C_1 = (c_{11}, c_{12}, c_{13})$, $C_2 = (c_{21}, c_{22}, c_{23})$ be two triangular fuzzy numbers. A distance between C_1 and C_2 is defined as

$$d(C_1, C_2) = |P(C_1) - P(C_2)|,$$

where $P(C_1) = \frac{c_{11} + 4c_{12} + c_{13}}{6}$, $P(C_2) = \frac{c_{21} + 4c_{22} + c_{23}}{6}$.

3 Statement of the Problem

This paper addresses the issue of ranking multiattribute alternatives by using group decision approach. A set of alternatives $A = \{a_1, \ldots, a_m\}$, and criteria vector $F = (f_1, \ldots, f_n)$ are given. Preference values of evaluated alternatives are based on knowledge of an expert team. This means that each member of the expert group evaluates every alternative concerning the criteria provided (Table 1).

Table 1. Criteria values of alternatives

	f_1	...	f_j	...	f_n
a_1	f_{11}^k	...	f_{1j}^k	...	f_{1n}^k
a_i	f_{1i}^k	...	f_{ij}^k	...	f_{in}^k
...	...	...	...	...	...
a_m	f_{m1}^k	...	f_{mj}^k	...	f_{mn}^k

Here f_{ij}^k is the k-th expert's evaluation of the i-th alternative w.r.t. the j-th criterion. Criteria are grouped by importance degrees. The considered problem is to determine an optimal alternative. In other words, a^* (optimal alternative) is to be found as

$$\text{Agg}(a^*) = \max_{a \in A} \text{Agg}(a), \tag{1}$$

where Agg is an aggregated general index of an alternative.

4 Solution Method

The considered issue is addressed through the application of Group Decision Making Method. The method implementation consists of the following steps:

1. Establish an expert team for criteria evaluation of alternatives.
2. Each expert evaluates each a_i alternative based on criteria vector $F = (f_1, \ldots, f_n)$, so forms a vector $\vec{f}_i^k = (f_{i1}^k, \ldots, f_{ij}^k, \ldots, f_{in}^k)^T$.
3. For each a_i, vectors $\vec{f}_i^k$, $k = 1, \ldots, K$ are aggregated to a single vector $\bar{f}_i = (f_{i1}, \ldots, f_{ij}, \ldots, f_{in})^T$ as:

$$\bar{f}_i = \frac{\sum\limits_{k=1}^{K} f_i^k}{K}. \tag{2}$$

That is, each component of $\bar{f}_i = (f_{i1}, \ldots, f_{ij}, \ldots, f_{in})^T$, f_{ij} is calculated as:

$$\bar{f}_{ij} = \frac{\sum\limits_{k=1}^{K} f_{ij}^k}{K}$$

4. It is calculated the arithmetic mean for every importance subgroup of criteria for each alternative: $\varphi_l(a_i), l \in \{1, \ldots, L\}$, L is the number of subgroups.

5. Weighted average of values $\varphi_l(a_i)$ is computed to yield a general aggregated index of an alternative:

$$Agg(a_i) = IG_1\varphi_1(a_i) + \dots + IG_1\varphi_1(a_i) + \dots + IG_L\varphi_L(a_i). \qquad (3)$$

Here $IG_i, i = 1, \dots L$ is the coefficient reflecting the importance of groups.

6. Alternatives $a_i, i = 1, \dots, n$ are ranked on basis of their indices. $Agg(a_i)$ is compared on basis of the distance to the fuzzy number Q, which represents the highest linguistic term of the scale of estimation:

$$a_1 \succ a_2 \text{ iff } d(Agg(a_1), Q) < d(Agg(a_2), Q).$$

Here d is the distance between triangular fuzzy numbers (Definition 3). Thus, an alternative which has a close distance to the fuzzy number Q is considered as superior.

5 An Application

Let's look at the problem of ranking of the faculties at university. Alternatives are faculties denoted $a_1, a_2, a_3, a_4, a_5, a_6, a_7$. Criteria: f_1 - the quality of teaching subjects; f_2 - providing the educational process with visual aids; f_3 - providing the educational process with modern audiences; f_4 - acquisition of practical knowledge by students through practical tasks, production experience and group project training; f_5 - theoretical and practical knowledge of teachers; f_6 - internet access for auditoriums; f_7 - facility created for students; f_8 - involvement of students in creative research; f_9 - general assessment of ethical norms; f_{10} - number and quality of scientific results of scientists; f_{11} - level of material and technical equipment; f_{12} - level of student-centered teaching.

The problem was solved on the basis of the solution method (Sect. 4) as follows. At first, a group of 5 experts was formed.

Each expert evaluated each faculty through questionnaires. Criteria vectors $\bar{f}_i$ of the expert group are calculated for the alternatives based on (2). The linguistic values of the criteria used in the questionnaires are described as fuzzy numbers (Fig. 1).

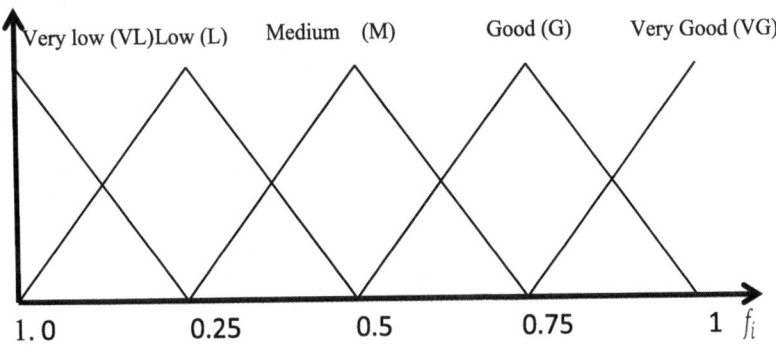

Fig. 1. Codebook of linguistic terms

Then, by using the formula (2), the vector $\bar{f}_i$ is found. For example, the calculated values for a_1 are as follows:

$$f_1 = \frac{M+G+G+G+G}{5} = \frac{(0.25\ 0.5\ 0.75)+(0.5\ 0.75\ 1)+(0.5\ 0.75\ 1)+(0.5\ 0.75\ 1)+(0.5\ 0.75\ 1)}{5}$$
$$= \frac{(2.25\ 3.5\ 4.75)}{5} = (0.45\ 0.7\ 0.95),$$

$$f_2 = \frac{M+G+G+M+G}{5} = \frac{(0.25\ 0.5\ 0.75)+(0.5\ 0.75\ 1)+(0.5\ 0.75\ 1)+(0.25\ 0.5\ 0.75)+(0.5\ 0.75\ 1)}{5}$$
$$= \frac{(2\ 3.25\ 4.5)}{5} = (0.4\ 0.65\ 0.9),$$

$$f_3 = \frac{G+G+G+VG+G}{5} = \frac{(0.5\ 0.75\ 1)+(0.5\ 0.75\ 1)+(0.5\ 0.75\ 1)+(0.75\ 1\ 1)+(0.5\ 0.75\ 1)}{5}$$
$$= \frac{(2.75\ 4\ 5)}{5} = (0.55\ 0.8\ 1),$$

$$f_4 = \frac{M+M+G+G+G}{5} = \frac{(0.25\ 0.5\ 0.75)+(0.25\ 0.5\ 0.75)+(0.5\ 0.75\ 1)+(0.5\ 0.75\ 1)+(0.5\ 0.75\ 1)}{5}$$
$$= \frac{(2\ 3.25\ 4.5)}{5} = (0.4\ 0.65\ 0.9),$$

$$f_5 = \frac{M+M+G+G+G}{5} = \frac{(0.25\ 0.5\ 0.75)+(0.25\ 0.5\ 0.75)+(0.5\ 0.75\ 1)+(0.5\ 0.75\ 1)+(0.5\ 0.75\ 1)}{5}$$
$$= \frac{(2\ 3.25\ 4.5)}{5} = (0.4\ 0.65\ 0.9),$$

$$f_6 = \frac{VG+VG+G+VG+M}{5} = \frac{(0.75\ 1\ 1)+(0.75\ 1\ 1)+(0.5\ 0.75\ 1)+(0.75\ 1\ 1)+(0.25\ 0.5\ 0.75)}{5}$$
$$= \frac{(3\ 4.25\ 4.75)}{5} = (0.6\ 0.85\ 0.95),$$

$$f_7 = \frac{G+G+G+G+G}{5} = \frac{(0.5\ 0.75\ 1)+(0.5\ 0.75\ 1)+(0.5\ 0.75\ 1)+(0.5\ 0.75\ 1)+(0.5\ 0.75\ 1)}{5}$$
$$= \frac{(2.5\ 3.75\ 5)}{5} = (0.5\ 0.75\ 1),$$

$$f_8 = \frac{L+L+M+G+M}{5} = \frac{(0\ 0.25\ 0.5)+(0\ 0.25\ 0.5)+(0.25\ 0.5\ 0.75)+(0.5\ 0.75\ 1)+(0.25\ 0.5\ 0.75)}{5}$$
$$= \frac{(1\ 2.25\ 3.5)}{5} = (0.2\ 0.45\ 0.7),$$

$$f_9 = \frac{M+G+G+G+G}{5} = \frac{(0.25\ 0.5\ 0.75)+(0.5\ 0.75\ 1)+(0.5\ 0.75\ 1)+(0.5\ 0.75\ 1)+(0.5\ 0.75\ 1)}{5}$$
$$= \frac{(2.25\ 3.5\ 4.75)}{5} = (0.45\ 0.7\ 0.95),$$

$$f_{10} = \frac{M+M+G+M+M}{5} = \frac{(0.25\ 0.5\ 0.75) + (0.5\ 0.75\ 1) + (0.5\ 0.75\ 1) + (0.5\ 0.75\ 1) + (0.5\ 0.75\ 1)}{5}$$

$$= \frac{(1.5\ 2.5\ 4)}{5} = (0.3\ \ 0.55\ \ 0.8),$$

$$f_{11} = \frac{L+M+G+M+G}{5} = \frac{(0\ 0.25\ 0.75) + (0.25\ 0.5\ 0.75) + (0.5\ 0.75\ 1) + (0.25\ 0.5\ 0.75) + (0.5\ 0.75\ 1)}{5}$$

$$= \frac{(1.5\ 2.75\ 4)}{5} = (0.3\ \ 0.55\ \ 0.8),$$

$$f_{12} = \frac{VL+VL+VL+VL+VL}{5} = \frac{(0\ 0\ 0.25) + (0\ 0\ 0.25) + (0\ 0\ 0.25) + (0\ 0\ 0.25) + (0\ 0\ 0.25)}{5}$$

$$= \frac{(0\ 0\ 1.25)}{5} = (0\ \ 0\ \ 0.25).$$

The values of $\bar{f}_i$ for the other alternatives are found analogously.

Further, for each faculty, it is needed to calculate $\varphi_1(a_i)$, $\varphi_2(a_i)$, $\varphi_3(a_i)$ values. The importance subgroups of criteria are shown in Table 2.

Table 2. Importance Subgroups of Criteria

Importance rates	Criteria
High importance, (0.5 0.6 0.7)	f_1
	f_4
	f_5
	f_{10}
Medium importance, (0.2 0.3 0.4)	f_7
	f_8
	f_{11}
	f_{12}
Low importance, (0 0.1 0.2)	f_2
	f_3
	f_6
	f_9

So, $\varphi_1(a_i)$, $\varphi_2(a_i)$, $\varphi_3(a_i)$ are found as follows:

$$\varphi_1(a_i) = \frac{f_{i1} + f_{i4} + f_{i5} + f_{i10}}{4},$$

$$\varphi_2(a_i) = \frac{f_{i7} + f_{i8} + f_{i11} + f_{i12}}{4},$$

$$\varphi_3(a_i) = \frac{f_{i2} + f_{i3} + f_{i6} + f_{i9}}{4}.$$

For example, we obtained for a_1:

$$\varphi_1(a_1) = \frac{(0.45\ 0.7\ 0.95) + (0.4\ 0.65\ 0.9) + (0.4\ 0.65\ 0.9) + (0.3\ 0.65\ 0.8)}{4}$$

$$= (0.3875\ \ 0.6375\ \ 0.8875),$$

$$\varphi_2(a_1) = \frac{(0.5\ 0.75\ 1) + (0.2\ 0.45\ 0.7) + (0.3\ 0.55\ 0.8) + (0\ 0\ 0.25)}{4}$$

$$= (0.25\ \ 0.4375\ \ 0.6875),$$

$$\varphi_3(a_1) = \frac{(0.45\ 0.7\ 0.95) + (0.55\ 0.8\ 1) + (0.6\ 0.85\ 0.95) + (0.45\ 0.7\ 0.95)}{4}$$

$$= (0.5125\ \ 0.7625\ \ 0.9625).$$

The values φ for other faculties have been calculated analogously.

Finally, weighted average of values $\varphi_l(a_i)$ are calculated for each alternative by using (3):

$\mathrm{Agg}(a_1) = (0.5\ 0.6\ 0.7)(0.39\ 0.64\ 0.89) + (0.2\ 0.3\ 0.4)(0.25\ 0.438\ 0.688) + (0\ 0.1\ 0.2)(0.51\ 0.76\ 0.96)$
$\quad = (0.24\ \ 0.59\ \ 1),$

$\mathrm{Agg}(a_2) = (0.5\ 0.6\ 0.7)(0.24\ 0.48\ 0.73) + (0.2\ 0.3\ 0.4)(0.19\ 0.35\ 0.6) + (0\ 0.10.2)(0.45\ 0.7\ 0.9)$
$\quad = (0.16\ \ 0.46\ \ 0.93),$

$\mathrm{Agg}(a_3) = (0.5\ 0.6\ 0.7)(0.34\ 0.59\ 0.83) + (0.2\ 0.3\ 0.4)(0.26\ 0.45\ 0.7) + (0\ 0.1\ 0.2)(0.41\ 0.66\ 0.86)$
$\quad = (0.22\ \ 0.55\ \ 1),$

$\mathrm{Agg}(a_4) = (0.5\ 0.6\ 0.7)(0.21\ 0.44\ 0.69) + (0.2\ 0.3\ 0.4)(0.15\ 0.33\ 0.58) + (0\ 0.1\ 0.2)(0.41\ 0.66\ 0.86)$
$\quad = (0.14\ \ 0.43\ \ 0.89),$

$\mathrm{Agg}(a_5) = (0.5\ 0.6\ 0.7)(0.3\ 0.54\ 0.79) + (0.2\ 0.3\ 0.4)(0.23\ 0.41\ 0.66) + (0\ 0.1\ 0.2)(0.51\ 0.76\ 0.95)$
$\quad = (0.2\ \ 0.52\ \ 1),$

$\mathrm{Agg}(a_6) = (0.5\ 0.6\ 0.7)(0.28\ 0.5\ 0.73) + (0.2\ 0.3\ 0.4)(0.28\ 0.45\ 0.69) + (0\ 0.1\ 0.2)(0.48\ 0.73\ 0.91)$
$\quad = (0.19\ \ 0.51\ \ 0.97),$

$\mathrm{Agg}(a_7) = (0.5\ 0.6\ 0.7)(0.33\ 0.63\ 0.76) + (0.2\ 0.3\ 0.4)(0.23\ 0.12\ 0.24) + (0\ 0.1\ 0.2)(0.41\ 0.66\ 0.86)$
$\quad = (0.21\ \ 0.56\ \ 0.95).$

The aggregated values of alternatives are ranked by measuring the distance (Definition 3) to the highest term, "very good" (Fig. 1). For example, for a_1, the distance of its aggregated values to "very good" term is calculated as follows:

$$d(Agg(a_1), very\ good) = \left| \frac{0.24 + 4 * 0.59 + 1}{6} - \frac{0.75 + 4 * 1 + 1}{6} \right| = |0.62 - 0.96|$$
$$= 0.34$$

The obtained results are given in Table 3.

Table 3. The results of ranking of the faculties

Faculty	$d(Agg(a_i), very\ good)$
a_1	0.342917
a_3	0.380625
a_7	0.391067
a_5	0.409792
a_6	0.427083
a_2	0.471042
a_4	0.504167

As smaller distance value is better, the ranking results are $a_1 \succ a_3 \succ a_7 \succ a_5 \succ a_6 \succ a_2 \succ a_4$.

6 Conclusion

In this paper we use an approach to decision making problem characterized by fuzzy information. The approach is based on Zadeh's idea to deal with a large number of alternative and criteria. A complex decision problem of ranking university faculties on the basis of 12 criteria is used to illustrate applicability of the approach. Information on criteria evaluations and criteria importance is characterized by fuzziness. The obtained results show computational efficiency of the proposed approach.

References

1. Zadeh, L.A.: Fuzzy sets. Inf. Control **8**(3), 338–356 (1965). https://doi.org/10.1016/S0019-9958(65)90241-X
2. Atanassov, K.T.: Intuitionistic fuzzy sets. Fuzzy Sets Syst. **20**(1), 87–96 (1986). https://doi.org/10.1016/S0165-0114(86)80034-3
3. Torra, V., Narukawa, Y.: On hesitant fuzzy sets and decision. In: 18th IEEE International Conference on Fuzzy Systems, pp. 1378–1382. IEEE Press (2009). https://doi.org/10.1109/FUZZY.2009.5276884
4. Peng, H.G., Wang, J.Q.: Cloud decision model for selecting sustainable energy crop based on linguistic intuitionistic information. Int. J. Syst. Sci. **48**(15), 3316–3333 (2017)

5. Liu, P.D., Zhang, X.H.: A novel picture fuzzy linguistic aggregation operator and its application to group decision-making. Cogn. Comput. **10**(2), 242–259 (2018). https://doi.org/10.1007/s12559-017-9523-z

6. Ye, J.: Multiple attribute decision-making methods based on the expected value and the similarity measure of hesitant neutrosophic linguistic numbers. Cogn. Comput. **10**, 454–463 (2018). https://doi.org/10.1007/s12559-017-9535-8

7. Wang, J., Wang, J.Q., Tian, Z.P., Zhao, D.Y.: A multi-hesitant fuzzy linguistic multicriteria decision-making approach for logistics outsourcing with incomplete weight information. Int. Trans. Oper. Res. **25**(3), 831–856 (2018). https://doi.org/10.1111/itor.12448

8. Aliev, R., Tserkovny, A.: Systemic approach to fuzzy logic formalization for approximate reasoning. Inf. Sci. **181**(6), 1045–1059 (2011)

9. Zadeh, L.A.: A Very Simple Formula for Aggregation and Multicriteria Optimization. Computer Science Division, Department of Electrical and Computer Sciences, University of California, Berkeley, CA 94720-1776, USA, pp. 961–962 (2016)

Fuzzy Soft Sets and Image Processing Application

Jiří Močkoř$^{(\boxtimes)}$ and Petr Hurtik

Institute for Research and Applications of Fuzzy Modeling,
University of Ostrava, 30. dubna 22, 701 03 Ostrava 1, Czech Republic
{Jiri.Mockor,Petr.Hurtik}@osu.cz
http://irafm.osu.cz/

Abstract. A monad defined by a powerset object of fuzzy soft sets
is constructed. Using this monad, a monadic fuzzy soft relation in a
fuzzy soft set and an approximation operator defined by this relation are
introduced. The results are used in selective color segmentation problem.

Keywords: Fuzzy soft sets · Fuzzy soft relation · Approximation
of fuzzy soft sets · Color segmentation algorithm

1 Introduction

Soft set and fuzzy soft set theories introduced by Molodtsov [1] and Maji et al
[2] are natural generalizations of a fuzzy set theory that have many applications
in data analysis, texture classification, optimizations, and many others, [3–7], for
example. Since the beginning of these theories, many publications have appeared
dealing with the theoretical properties of fuzzy soft sets and their use, see, e.g.,
[4,8,9]. From this short list of publications it follows that the theory of fuzzy
soft sets is of interest for its application possibilities to many specialists in the
field of fuzzy sets. In general, it seems that the issue of fuzzy soft sets is more
elaborated in the field of applications, while theoretical results concerning fuzzy
soft sets are less common. Few publications deal with the theory of fuzzy soft
sets to the extent that is currently common in the field of classical fuzzy sets,
including the use of tools from, for example, the theory of categories.

One of the key theoretical tool in the theory of fuzzy sets that allows the
application of standard algebraic and topological techniques in fuzzy sets is a
theory of monads which represents a basic tool in constructions of many fuzzy
structures.

In our paper, we show that the theory of fuzzy soft set, similar to classical
fuzzy set theory, can be based on the theory of monads in appropriate categories.
This allows us to use in fuzzy soft set structures methods, which are commonly
used in the fuzzy set theory, such as approximation operators based on various
relations. We will show that with the use of the mentioned monadic constructions in fuzzy soft sets, we can also introduce these approximation operators

R. A. Aliev et al. (Eds.): ICAFS 2020, AISC 1306, pp. 47–54, 2021.
https://doi.org/10.1007/978-3-030-64058-3_6

in these structures. We show how approximation operators defined by monadic relations can be used in a selective color segmentation in image processing as an application of approximation operators in fuzzy soft sets.

The proofs of the presented results will be published elsewhere.

2   Methods

The main theoretical tools that we will use in the article relate to monads in categories and subsequently to relations between two objects in the category, defined by the monad. For the convenience of readers, we recall the basic definition of a monad in a category (see [10–12]).

Definition 1. *Let* **K** *be a category. Then* $\mathcal{F} = (F, \Diamond, \xi)$ *is called a monad in* **K**, *if*

1. $T : \mathbf{K} \to \mathbf{K}$ *is a mapping defined on objects of* **K**,
2. $\xi = \{\xi_S | \xi_S : X \to T(S) \text{ is a } \mathbf{K} - \text{morphism}, S \in \mathbf{K}\}$,
3. *If* $p : X \to T(Y)$ *and* $q : Y \to T(Z)$ *are* **K**-*morphisms, then* $q \Diamond p : X \to T(Z)$ *is a* **K**-*morphism, called Kleisli composition of* p *and* q *and this composition is associative,*
4. *If* $h : X \to T(Y)$ *and* $f : X \to Y$ *and* $g : Y \to T(Z)$ *are* **K**-*morphisms, then* $\xi_Y \Diamond h = h$ *and* $g \Diamond (\eta_Y . f) = g . f$

The basic example of a monad in the fuzzy set theory is the powerset monad of L-valued fuzzy sets presented in the following example, where $L = (L, \vee, \wedge, \otimes, \to, 0_L, 1_L)$ is a complete residuated lattice.

Example 1. The monad $\mathcal{Z} = (Z, \Diamond, \chi)$ in the category **Set** is such that

1. The object function Z from **Set** to **Set** is defined by $Z(A) = L^A$, for arbitrary $A \in \mathbf{Set}$,
2. For each set $A \in \mathbf{Set}$, $\chi_A : A \to L^A$ is a characteristic map $\chi_{\{x\}}^A : A \to L$ of a an element x in A,
3. For each $p : A \to L^B$ and $q : B \to L^C$ in **Set**, $q \Diamond p : A \to L^C$ is defined by

$$x \in A, z \in C, \quad (q \Diamond p)(x)(z) = \bigvee_{y \in B} p(x)(y) \otimes q(y)(z).$$

Using a monad in a category, we can introduce the concept of a relation defined by this monad. Such relations are related to morphisms in Kleisli categories, which, in contrast to classical mappings mostly used in the theory of fuzzy structures, allows the use of various relations as morphisms between structures, see, e.g., [10,13].

Definition 2 ([10]). *Let* **K** *be a category and let* $\mathcal{F} = (F, \Diamond, \xi)$ *be a monad in* **K**. *An* $\mathcal{F}$-*relation* R *from a* **K**-*object* X *to a* **K**-*object* Y *is a morphism* $R : X \to F(X)$ *in the category* **K** *and it is denoted by* $R : X \rightsquigarrow Y$. *The composition of* $\mathcal{F}$-*relations is defined by* $\Diamond$.

Although the concept of a relation defined by a monad is very general, it actually comprises the full range of relations standardly used in the fuzzy set theory. A typical example of a relation defined by a monad is a classical L-valued fuzzy relation in a set, as follows from the following example.

Example 2. Let $\mathbf{K} = \mathbf{Set}$ and let $\mathcal{Z}$ be the monad from Example 2.1. There exists a bijection φ between the set of all $\mathcal{Z}$-relations from X to Y and the set of all L-valued fuzzy relations from X to Y, such that for arbitrary $R : X \rightsquigarrow Y$ and $S : Y \rightsquigarrow V$ we have

$$\varphi(S \lozenge R) = \varphi(S) \star \varphi(R)$$

where $\star$ is a composition of L-valued fuzzy relations.   □

Each $\mathcal{F}$-relation $R : X \rightsquigarrow Y$ can define a $\mathbf{K}$-morphism $R^\uparrow : F(X) \to F(Y)$, such that

$$R^\uparrow = 1_{F(X)} \lozenge R,$$

which is called an *upper approximation*. This name is motivated by the example of $\mathcal{Z}$-relations $R : X \rightsquigarrow Y$ (i.e., classical fuzzy relations $R : X \times Y \to L$), where $R^\uparrow$ is a standard upper approximation operator defined by a fuzzy relation R.
    The main entity we will deal with is the *fuzzy soft set*, introduced in [2]:

Definition 3 ([2]). *Let U be a fixed set of objects to be evaluated and let X be a set (= the set of all possible criteria). A pair (A,t) is called an L-valued fuzzy soft set in a set X, if $A \subseteq X$ and $t : A \to L^U$.*

The intuitive interpretation of a fuzzy soft set (A, t) can be described as follows. A subset $A \subseteq X$ represents the criteria from the set X that are currently used for evaluation, while for the criterion $x \in A$, the L-valued fuzzy set $t(x) \in L^U$ represents evaluation how the objects from U correspond to this criterion x.

3   Results: Monad and Monadic Relations in Fuzzy Soft Sets

In this section, we will focus on the existence of a monad in the category $\mathbf{Set}$ of sets, which, analogously as the monad $\mathcal{Z}$ for classical fuzzy sets, would represent the powerset object of all fuzzy soft sets. This monad will then allows to define the concept of a monadic relation in the category $\mathbf{Set}$ and subsequently, to define approximation operators using monadic relations.
    We start with brief description of the construction of the monad $\mathcal{M}$ in the category $\mathbf{Set}$, representing powerset objects of all fuzzy soft sets in a given set X. As we mentioned before, L is supposed to be a complete residuated lattice.

Construction of $\mathcal{M}$: Let $X \in \mathbf{Set}$. By $M(X)$ we denote the set of all L-valued fuzzy soft sets in X, i.e.,

$$M(X) = \{(A, t) : \emptyset \neq A \subseteq X, t : A \to L^U\}.$$

We use the following notation. If $p : X \to M(Y)$ is a mapping, for $a \in X$ we set $p(a) = (Y_a^p, p_a) \in M(Y)$, where $\emptyset \neq Y_a^p \subseteq Y$ and $p_a : Y_a^p \to L^U$.

The structure $\mathcal{M} = (M, \square, \xi)$ in the category **Set** is defined by

1. The object function is $M : \mathbf{Set} \to \mathbf{Set}$,
2. For each $X \in \mathbf{Set}$, $\xi_X : X \to M(X)$ is defined by $\xi_X(x) = (\{x\}, \xi_x)$, $\xi_x : \{x\} \to L^U$, $\xi_x(x)(u) = 1_L$ for arbitrary $u \in U$,
3. For arbitrary mappings $p : X \to M(Y)$ and $q : Y \to M(Z)$, the composition map $q \square p : X \to M(Z)$ is defined by

$$a \in X, \quad q \square p(a) = (Z_a^{q \square p}, (q \square p)_a) \in M(Z),$$

$$Z_a^{q \square p} = \bigcup_{b \in Y_a^p} Z_b^q, \quad (q \square p)_a : Z_a^{q \square p} \to L^U,$$

$$c \in Z_a^{q \square p}, u \in U, \quad (q \square p)_a(c)(u) = \bigvee_{\{b \in Y_a^p : b \in Z_b^q\}} p_a(b)(u) \otimes q_b(c)(u),$$

Then the following theorem holds, proof of which will be published elsewhere.

Theorem 1. $\mathcal{M} = (M, \square, \xi)$ *is a monad in the category* **Set**, *which represents a powerset structure of fuzzy soft sets.*

Using the monad $\mathcal{M}$, we can now use $\mathcal{M}$-relations from Definition 2.2. However, since the specific interpretation, an $\mathcal{M}$-relation is relatively confusing and must be obtained using the tools given in the construction of the monad $\mathcal{M}$, we give an explicit definition of $\mathcal{M}$-relation, which, as follows from the following theorem, is equivalent to the original $\mathcal{M}$-relation.

Definition 4. *Let X be an set. An L-valued fuzzy soft relation in X is an L-valued fuzzy soft set (G, V) in the set $X \times X$ with naturally defined composition $\oplus$ of L-fuzzy soft relations.*

The following theorem shows that L-fuzzy soft relations in X are equivalent to $\mathcal{M}$-relations.

Theorem 2. *Let $\mathcal{M} = (M, \square, \xi)$ be the monad in the category* **Set** *from Theorem 3.1. For a set X we denote*

$$Rel(X) = \{(G, V) | (G, V) \text{ is an } L\text{-valued fuzzy soft relation in } X\},$$
$$Rel_{\mathcal{M}}(X) = \{R | R : X \rightsquigarrow X \text{ is a } \mathcal{M} - relation\}.$$

1. *There exists a bijection map*

$$\Phi : Rel(X) \to Rel_{\mathcal{M}}(X).$$

2. *For arbitrary $(G, V), (H, W) \in Rel(X)$ the following equality holds*

$$\Phi((H, W) \oplus (G, V)) = \Phi(H, W) \square \Phi(G, V).$$

Proposition 1. *Let (G, V) be an L-valued fuzzy soft relation in X and let (A, s) be an L-valued fuzzy soft set in X. The upper approximation of (A, s) defined by the $\mathcal{M}$-relation $\Phi(G, V)$ is a fuzzy soft set $\Phi(G, V)^{\uparrow}(A, s)$, such that*

$$\Phi(G, V)^{\uparrow}(A, s) = (G[A], s^{\uparrow}), \quad where$$

$$G[A] = \{y \in X : \exists x \in A, (x, y) \in G\},$$

$$s^{\uparrow} : G[A] \to L^{U},$$

$$y \in G[A], u \in U, \quad s^{\uparrow}(y)(u) = \bigvee_{\{x \in A : (x, y) \in G\}} s(x)(u) \otimes V(x, y)(u).$$

4 Application: Fuzzy Soft Sets and Color Segmentation in Color Images

The problem of color segmentation in color images is interesting from several points of view. The first one is that there is no automatic tool to realize the technique perfectly, so the graphic designers are forced to create such images manually, which may be time-consuming. The second reason is the uncertainty of a term *color* that can be handled naturally by a fuzzy approach. Both of these reasons led us to try to use the fuzzy soft sets theory, which focuses on assessing the suitability of the elements (in our case, pixels in color images) concerning several criteria.

Let (U, d) represent a finite metric space of pixels of a color image with a metric d. For each pixel $u \in U$, a value $S(u) \in E$ representing the color in a pixel u is defined. This value can be represented in different ways, one of possible representations of the color value $S(u)$ of a pixel $u \in U$ can be given by the so-called HSV representation as a vector $S(u) = [h, s, v]$.

Let $E_S = S(U) = \{e \in E : \exists u \in U, S(u) = e\} \subseteq E$, then we can consider a $[0, 1]$-valued fuzzy soft set (E_S, s) in the space E, where $s : E_S \to [0, 1]^{U}$ and $s(e)(u)$ is a membership value which quantifies the grade of membership of the pixel $u \in U$ to the color segment represented by a color $e \in E_S$. Obviously, this membership value should be greater, the more pixels $v \in U$ with the color $S(u)$, or a color similar to $S(u)$ exist in the immediate vicinity of the pixel u. Using some simplification, we can formally describe this situation as follows.

On the set E of colors we can define a fuzzy similarity relation $\sigma : E \times E \to [0, 1]$, such that $\sigma(e, e')$ represents a similarity degree of two colors $e, e' \in E$. Analogously, using the metric space (U, d) of pixels, we can define another fuzzy similarity relation $\rho : U \times U \to [0, 1]$ such that $\rho(u, v)$ expresses the fact that the pixels u and v are "similar", i.e. they are close to each other. Then the grade of membership $s(e)(u)$ of the pixel $u \in U$ to the color segment represented by the color e can be described by the value

$$e \in E_S, u \in U, \quad s(e)(u) = \frac{\sum_{v \in U} \rho(u, v) . \sigma(S(v), e)}{\sum_{v \in U} \rho(u, v)},$$

which defines how true it is that in the vicinity of the pixel u there are rather enough pixels with a color similar to $\mathbf{e}$. Therefore, from that point of view, the color segmentation of a color image S is represented by the fuzzy soft set $(E_S, s) \in M(E)$.

The fuzzy soft relation (H, R) in E, i.e., a fuzzy soft set in the space $E \times E$ can be defined as follows. Since σ is a fuzzy similarity relation, for a fixed level of importance $\alpha \in (0, 1]$ the binary relation

$$H = \sigma_\alpha = \{(\mathbf{e}, \mathbf{e}') \in E \times E : \sigma(\mathbf{e}, \mathbf{e}') \geq \alpha\},$$

is an equivalence relation in E, representing the α-level relation of E. The map $R : H \to L^U$ is defined by

$$R(\mathbf{e}, \mathbf{e}')(u) = \sigma(S(u), \mathbf{e}) \leftrightarrow \sigma(S(u), \mathbf{e}'),$$

where $\leftrightarrow$ is the bi-residuum operation in the Lukasiewicz lattice $[0, 1]$. Then, $R(\mathbf{e}, \mathbf{e}')(u)$ expresses the grade how the color $S(u)$ of a pixel $u \in U$ is similar to both colors $\mathbf{e}, \mathbf{e}'$.

The upper approximation of a fuzzy soft set (E_S, s) which expresses the color segmentation of an image S, defined by a fuzzy soft relation (H, R) is

$$\Phi(H, R)^\uparrow (E_S, s) = (H[E_S], s^\uparrow),$$

where $H[E_S] = \{\mathbf{e} \in E : \exists \mathbf{e}' \in E_S, (\mathbf{e}', \mathbf{e}) \in H\}$ is the set of all colors which are equivalent to colors from E_S in the equivalence relation H and

$$s^\uparrow : H[E_S] \to L^U,$$

$$\mathbf{e} \in H[E_S], u \in U, \quad s^\uparrow(\mathbf{e})(u) = \bigvee_{\{\mathbf{f} \in E_S : (\mathbf{f}, \mathbf{e}) \in H\}} s(\mathbf{f})(u) \otimes R(\mathbf{f}, \mathbf{e})(u).$$

Therefore, $(H[E_S], s^\uparrow)$ is a fuzzy soft set which represents an extended color segmentation of an image S.

Using the theoretical approach presented in both examples we implemented an illustrative version of the application with GUI for Windows and made the build with source codes available on our webpage[1]. It should be mentioned that according to the examples, it is necessary to define two distance functions, ρ and σ for spatial and intensity distance, respectively. In our application we fixed them to the form of:

$$\rho(u, v) = 1/2^{|u-v|},$$

$$\sigma(S(v), \mathbf{e}) = |S(v) - \mathbf{e}|.$$

The spatial distance function decreases fast, and for a *distant enough* points, it tends to be near zero, so we propose to realize the computation only for points close enough to each other. That is controlled in our application by δ radius in δ-neighborhood, which is set by a user.

---

[1] http://graphicwg.irafm.osu.cz/storage/selective-color.zip.

For the visual example, see Figs. 1 and 2, where we show the performance on two images downloaded from the pixabay.com website. For demonstration purposes, the figures also include a comparison with the fuzzy selection tool available through GIMP software and GrabCut algorithm [14]. The demonstration shows that our proposed method is very competitive and can be used for the task of selective color segmentation.

Fig. 1. The performance on Apple image. From left: Original image and the crops of GrabCut, Gimp fuzzy selection, and our method. GrabCut and our approach segmented the apple precisely, Gimp missed a part of it.

Fig. 2. The performance on Shoe image. From left: Original image and cuts of GrabCut, Gimp fuzzy selection, our method. GrabCut segmented shoe together with background. Gimp fuzzy selection was unable to include the small apart parts of the shoe. Our method segmented the image correctly.

5   Conclusion

In this paper, we developed a theory of upper approximations of fuzzy soft sets by fuzzy soft relations, which are based on the theory of monads in categories. We used the fact that the powerset structures in fuzzy soft sets define a monad, and using this monad, we introduced the fuzzy soft relations. Using this general method, we introduced the upper approximation of fuzzy soft sets defined by these fuzzy soft relations. In this paper, we also investigated the possible use of both fuzzy soft sets and upper approximations of fuzzy soft sets defined by fuzzy soft relations in color segmentation. We have shown that both of these theoretical tools can be naturally used in selective color segmentation problem.

Acknowledgment. This research partially supported by the ERDF/ESF project CZ.02.1.01/0.0/0.0/17-049/0008414.

Compliance with Ethical Standards. Authors Jiří Močkoř and Petr Hurtik declare that they have no conflict of interest. This article does not contain any studies with human participants or animals performed by any of the authors.

References

1. Molodtsov, D.: Soft set theory - first results. Comput. Math. Appli. **37**, 19–31 (1999)
2. Maji, K., Biswas, R., Roy, A.: Fuzzy soft sets. J. Fuzzy Math. **9**(3), 589–602 (2001)
3. Zou, Y., Xiao, Z.: Data analysis approaches of soft sets under incomplete information. Knowl.-Based Syst. **21**(8), 941–945 (2008)
4. Xiao, Z., Gong, K., Zou, Y.: A combined forecasting approach based on fuzzy soft sets. J. Comput. Appl. Math. **228**(1), 326–333 (2009)
5. Pei, D., Miao, D.: From soft sets to information systems. In: 2005 IEEE international conference on granular computing, vol. 2, 617–621. IEEE (2005)
6. Mushrif, M.M., Sengupta, S., Ray, A.K.: Texture classification using a novel, soft-set theory based classification algorithm. In: Asian Conference on Computer Vision, pp. 246–254. Springer (2006)
7. Kovkov, D., Kolbanov, V., Molodtsov, D.: Soft sets theory-based optimization. J. Comput. Syst. Sci. Int. **46**(6), 872–880 (2007)
8. Roy, A.R., Maji, P.: A fuzzy soft set theoretic approach to decision making problems. J. Comput. Appl. Math. **203**(2), 412–418 (2007)
9. Kharal, A., Ahmad, B.: Mappings on fuzzy soft classes. In: Advances in Fuzzy Systems 2009 (2009)
10. Manes, E.G.: Algebraic Theories. Graduate Text in Mathematics, vol. 26. Springer, Heidelberg (1976)
11. Höhle, U.: Partially ordered monads. In: Many Valued Topology and its Applications, pp. 29–54. Springer (2001)
12. Eklund, P., Galán, M.Á.: Partially ordered monads and rough sets. In: Transactions on Rough Sets VIII, pp. 53–74. Springer (2008)
13. Manes, E.: Book review "fuzzy sets and systems, theory and applications". Bull. (New Seri.) Am. Math. Soc. **7**(3) (1982)
14. Rother, C., Kolmogorov, V., Blake, A.: GrabCut: interactive foreground extraction using iterated graph cuts. In: ACM transactions on graphics (TOG), vol. 23, pp. 309–314. ACM (2004)

Quality Criteria of Fuzzy IF-THEN Rules and Their Calculations

Nigar E. Adilova[(⊠)] [ID]

Joint MBA Program, Azerbaijan State Oil and Industry University,
20 Azadlig Avenue, Baku AZ1010, Azerbaijan
nigar.adilova@asoiu.edu.az

Abstract. Fuzzy IF-THEN rules are usually taken as specific characterization of dependencies among objects. The conception of IF-Then rules contributes all researchers and practitioners to equip conditional statements. It can directly relate with fuzzy inference methods and provides to create a comprehensive logical theory. This paper encompasses the quality criteria of fuzzy If-Then rules based on their calculations. Fuzzy If-Then rules supply the solution of decision-making problem. Numerical examples for the quality of fuzzy If-Then rules are described for their 5 criteria.

Keywords: Fuzzy number · Fuzzy If-Then rules · Interpretability measure · Complexity measure · Coverage degree · Inconsistency of fuzzy rules · Accuracy measure

1 Introduction

A system of fuzzy IF-THEN rules is considered as a knowledge-based system where inference is made based on the rules. The generation of fuzzy If-Then rules in each fuzzy subspace facilitated decision-making problems. Accordingly, the theory of fuzzy If-Then rules has been originated by many authors [1–5]. To identify the quality criteria of fuzzy If-Then rules it is expedient the usage of five different criteria: Complexity, Coverage, Partition, Inconsistency and Accuracy indices. Ordinarily, the first three variables specify the interpretability index of fuzzy rule-based systems (FRBSs). Therefore, they are included in the calculation of interpretability index based on the Nauck index suggested by Nauck [6]. Nonetheless, some researchers took a close interest in the investigation and development of coverage degree.

Particularly, coverage degree evaluates qualitative parameter in fuzzy sets. The coverage parameter assesses the degree for instances of an outcome for If-Then rules. The degree of coverage allows for a more accurate read of the results of analyses [7].

Mostly, interpretability is characterized like the readability or transparency and measured for the number of rules, the number of antecedents, and the number of linguistic terms used in the rule base. This necessary index for FRBSs is acknowledged as the advantage of fuzzy systems and it plays a main role in fuzzy modeling. There have been many researches related to the interpretability issue in recent years [8–10].

Moreover, it is important to know, whether our preferences are sufficiently coherent, in other words, whether our preferences are consistent or inconsistent [11].

© The Author(s), under exclusive license to Springer Nature Switzerland AG 2021
R. A. Aliev et al. (Eds.): ICAFS 2020, AISC 1306, pp. 55–62, 2021.
https://doi.org/10.1007/978-3-030-64058-3_7

The issue of consistency in fuzzy Analytic Hierarchy Process (AHP) has been developed by many authors after introduced by Saaty T.L. Simultaneously, various methods also have been discussed to resolve inconsistency resolution [12–18]. Inconsistency is an important issue that directly reveals anomalies in rule-based systems by detection. In [19] authors focused on the extraction of easily interpretable knowledge from large amount of data measured by using consistency of fuzzy systems. As an important matter for fuzzy rule bases in [13] authors noticed a systematic approach to understand the inconsistency problem. The other contradictory parameter for FRBS is considered accuracy measure and researches on accuracy problem are still scars.

This paper investigates the computational approach to complexity, coverage, partition, accuracy and inconsistency measures. Author demonstrates the use of criteria in an example.

The paper is organized as follows: Sect. 2 introduces some preliminary information about criteria of fuzzy If-Then rules. Section 3 describes computational results of given criteria. Finally, Sect. 4 is devoted to conclusion.

2 Preliminaries

Definition 1. Fuzzy number [2]: A fuzzy number is a set A on R which has the following properties: a) A is a normal fuzzy set; b) A is a convex fuzzy set; c) $\alpha-$ cut of A, A^α is a closed interval for every $\alpha \in (0, 1]$; d) the support of A, A^{+0} is bounded.

Definition 2. Fuzzy If-Then Rules [3, 4, 20]:   Fuzzy if-then rule statements are commonly used to define the conditional statements that possess fuzzy logic.
A single fuzzy if-then rule is shown in this form:

$$\text{If x is } A \text{ then y is } B$$

where both A and B are linguistic values defined by fuzzy sets.
Multi-input multi-output fuzzy system is described as follows

$$\text{If } X_1 \text{ is } A_1 \text{ and } X_2 \text{ is } A_2 \text{ and } \ldots \text{ and } X_n \text{ is } A_n$$

$$\text{then } Y_1 \text{ is } B_1 \text{ and } Y_2 \text{ is } B_2 \text{ and } \ldots \text{ and } Y_m \text{ is } B_m$$

where A_i and B_i are information granules.

Criteria of Fuzzy If-Then Rules

Complexity Measure [6, 8, 9]: Interpretability is computed as the product of complexity (com), the degree of coverage (cov) and partition measure $(\overline{part})$.

Individually, complexity is measured by the total number of rules, the total number of premises, and the total number of labels defined by input.

$$comp = m \left/ \sum_{k=1}^{r} n_k \right.$$

(1)

where m is the number of classes (outputs), r is the number of rules and n_k is the number of variables used in the k-th rule.

Coverage Degree [6, 8, 9]: The degree of coverage indicates the degree for instances of an outcome for If-Then rules. cov is the average normalized coverage degree on cov_i

$$cov_i = \frac{\int_{X_i} \overline{h}_i(x)dx}{N_i} \quad \text{where } \overline{h}_i(x) = \begin{cases} h_i(x) = \sum_{k=1}^{p_i} \mu_i^{(k)}(x), & \text{if } 0 \leq h_i(x) \leq 1 \\ \frac{p_i - h_i(x)}{p_i - 1}, & \text{otherwise} \end{cases} \quad (2)$$

where X_i is the domain of the ith variable and this domain is partitioned by p_i fuzzy sets and with $N_i = \int_{X_i} dx$ for continuous domains or with $N_i = |X|$ for discrete domains.

Partition Complexity Measure [6, 8, 9]: Partition complexity measure (part) - stands for the partition index which is computed as the inverse of the number of MFs minus one for each input variable.

$$part_i = 1/(p_i - 1) \quad (3)$$

p_i is the number of MFs in the ith input variable.

Consistency and Inconsistency Indexes [13, 14, 18]: The importance of consistency is that, where the proposed criteria is compared with others and the most correct and alternative one is chosen among them. To calculate the consistency of two arbitrary fuzzy rules, definitions of Similarity of Rule Premise (SRP) and Similarity of Rule Consequent (SRC) are given. For instance two fuzzy rules R_1 and R_k are given as below:

$$R_1 : \text{ If } x_1 \text{ is } A_{i1}(x_1) \text{ and } x_2 \text{ is } A_{i2}(x_2) \text{ and } \dots x_n \text{ is } A_{in}(x_n), \text{ then } y \text{ is } B_i(y)$$

$$R_k : \text{ If } x_1 \text{ is } A_{k1}(x_1) \text{ and } x_2 \text{ is } A_{k2}(x_2) \text{ and } \dots x_n \text{ is } A_{kn}(x_n), \text{ then } y \text{ is } B_k(y)$$

Consistency index (*Cons*) is calculated as follows for two given rules $R(i)$ and $R(k)$:

$$Cons(R(i), R(k)) = \exp\left\{-\frac{\left(\frac{SRP(i,k)}{SRC(i,k)} - 1.0\right)^2}{\left(\frac{1}{SRP(i,k)}\right)^2}\right\}$$

where *SRP* is the similarity of rule premises and *SRC* is the similarity of rule consequents, and they are calculated as:

$$SRP(i, k) = \min_{j=1}^{a} S(A_{ij}, A_{kj}), \quad SRC(i, k) = S(B_i, B_k)$$

where S is the fuzzy similarity measure. The S measure is the fuzzy relation that expresses the degree to which fuzzy sets A and B are equal and is defined as follows:

$$S(A, B) = \frac{|A \cap B|}{|A \cup B|} = \frac{|A \cap B|}{|A| + |B| - |A \cap B|},$$

where intersection ($\cap$) and union ($\cup$) are defined by a proper couple of t-norm and t-conorm and $|\cdot|$ is the cardinality of the resulting fuzzy set. This value ranges from 0 to 1.

Inconsistency has been defined in terms of heights of intersecting fuzzy sets by L. A.Zadeh. Inconsistent rules can be obtained when the control actions of experts are directly and naively represented. Two fuzzy rules are inconsistent if they have the same if-part, but different then-parts. Two or more rules are said to be explicitly inconsistent if they have identical antecedents ("if" part) but have different consequents ("then" part). The set of rules are said to be implicitly inconsistent if two or more rules can be modified to be the explicitly inconsistent rules by appending the logical-and of arbitrary conditions to the antecedents.

During construction of rule-base collision such inconsistency is applied. If antecedents overlap, the consequents come to be different from given two rules, it means that these two rules are inconsistent. For more interpretation, these two following rules will be inconsistent:

If X_1 is A_1 and X_2 is A_2 THEN Y is C

If X_1 is A_1 and X_2 is A_2 THEN Y is D

The degree of inconsistency of a given RB (f_{incons}) is calculated as:

$$f_{incons} = \sum_{i=1}^{N} Incons(i),$$

where $Incons(i)$ is the degree of inconsistency for the ith. It is defined as:

$$Incons(i) = \sum_{\substack{1 \leq k \leq N \\ k \neq i}} [1.0 - Cons(R^1(i), R^1(k))] + \sum_{\substack{1 \leq l \leq L \\ i=1,2,...N}} [1.0 - Cons(R^1(i), R^2(l))],$$

$$(4)$$

where R^1 and R^2 denote the RB generated from data and the RB extracted from prior knowledge (since the authors defined this index considering the possibility of including rules provided by experts) and N and L are the rule numbers of R^1 and R^2, respectively.

Accuracy Measure **[8]:** Accuracy is measured by the number of correctly classified training patterns. It is deviation between responses of fuzzy system with all rules FS_{all} and fuzzy system FS with some of the rules (RMSE):

$$J = \sqrt{\frac{1}{n}\sum_{i=1}^{n}(y_i - \widehat{y}_i)^2} \qquad (5)$$

where n is number of samples (current input vectors), $\widehat{y}_i = FS_{all}(x_i)$ is response of fuzzy system with all rules, $y_i = FS(x_i)$, x_i is the i-th sample (current input vector).

3   The Calculation of the Criteria of Fuzzy If-Then Rules

Let's presume that these following 27 rules are involved in fuzzy rule-based system (Table 1).

Table 1. Fuzzy rule-based system [21]

Rules	Inputs			Output
№	X_1	X_2	X_3	Y
1	H	L	L	S
2	H	L	M	M
3	H	L	H	H
4	H	M	L	M
5	H	M	M	M
...	...	...	...	...
23	L	M	M	H
24	L	M	H	M
25	L	H	L	S
26	L	H	M	M
27	L	H	H	H

Each fuzzy rule is represented by its 3 antecedents (X_1, X_2, X_3) and a consequent (Y).

In accordance with three inputs (X_1, X_2, X_3), the membership functions are:

for X_1 $Less = (0; 6; 10)$ $Medium = (7; 11; 25)$ $High = (20; 27; 50)$

for X_2 $Less = (0; 7; 10)$ $Medium = (7; 19; 25)$ $High = (20; 26; 50)$

for X_3 $Less = (1000; 1200; 1500)$ $Medium = (400; 900; 1200)$ $High = (50; 400; 500)$

for Y $Less = (0; 7; 10)$ $Medium = (8; 24; 30)$ $High = (25; 30; 60)$

Graphical representations of membership functions for 3 inputs and an output variables for original case are described in Fig. 1 (respectively (a, b, c, d)).

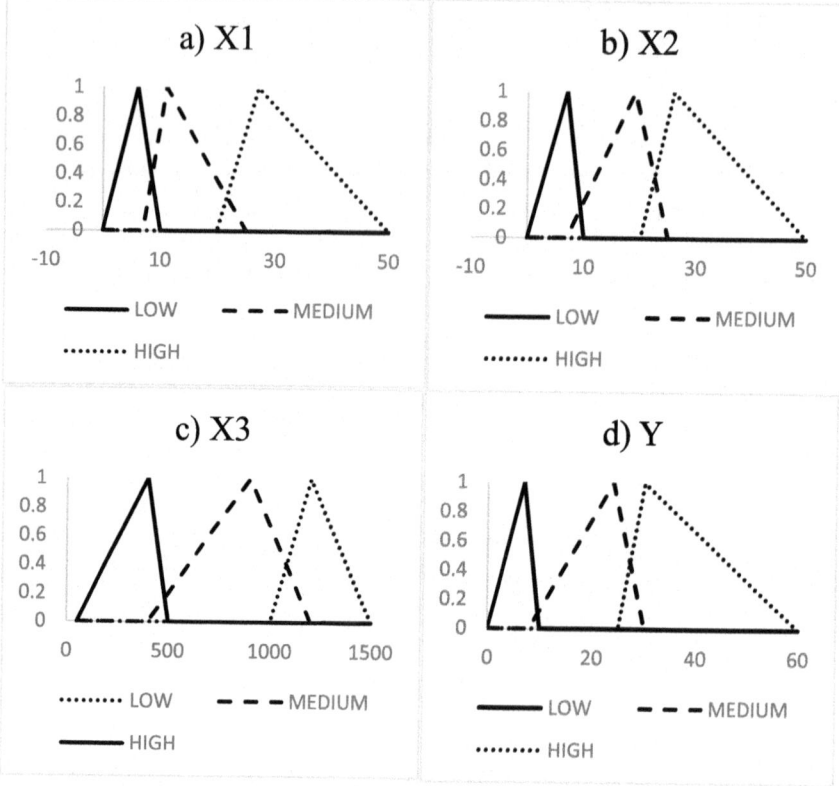

Fig. 1. (a, b, c, d) Membership functions for original case

The problem is to compute the measures for FRBS based on shown criteria. Corresponding to (1)–(5) formulas quality criteria is obtained for original case and the second case (Table 2). The second case demonstrates the results based on randomly decreased rules.

Table 2. Computational results for quality criteria

Quality criteria	Original case	Second case
The number of rules	27	22
Complexity measure	0,037	0,045
Coverage degree	0,588	0,592
Partition measure	0,5	0,5
Inconsistency measure	0,0172	0,0184
Accuracy measure (RMSE)	0,089	8,824

4   Conclusion

In this paper, the author investigated the quality criteria and their calculation methods for fuzzy If-Then rules. The computational results show that, the number of rules affects the change of the values for quality criteria. To achieve the best values for the criteria it is needed to solve multi-criterial optimization problem.

References

1. Zadeh, L.A.: Calculus of fuzzy If-Then rules and its applications. Appl. Artif. Intelli. **1708**, 426–430 (1992)
2. Aliev, R.A.: Fundamentals of the Fuzzy Logic-Based Generalized Theory of Decisions. Springer, Heidelberg (2013)
3. Aliev, R.A., Pedrycz, W.: Fundamentals of a fuzzy-logic-based generalized theory of stability. IEEE Trans. Syst. Man Cyber. Part B (Cybern.) **39**(4), 971–988 (2009)
4. Aliev, R., Tserkovny, A.: Systemic approach to fuzzy logic formalization for approximate reasoning. Inform. Sci. **181**(6), 1045–1059 (2011)
5. Ishibuchi, H., Nakashima, T., Nii, M.: Fuzzy If-Then rules for pattern classification. Fuzzy If-Then Rules Comput. Intell. **553**, 267–295 (2000)
6. Nauck, D.D.: Measuring interpretability in rule-based classification systems. In: FUZZ-IEEE 2003, St. Louis, Missouri, USA, vol. 1, pp. 196–201 (2003)
7. Charles, R.: Set relations in social research: evaluating their consistency and coverage. Politic. Anal. **14**, 291–310 (2006)
8. Gacto, M.J., Alcala, R., Herrera, F.: Interpretability of linguistic fuzzy rule-based systems: an overview of interpretability measure. Inform. Sci. **181**, 4340–4360 (2011)
9. Razak, T.R., Garibaldi, J.M., Wagner, C., Pourabdollah, A., Soria D.: Interpretability indices for hierarchical fuzzy systems. In: Proceedings IEEE International Fuzzy System Conference (2017)
10. Alonso, J., Magdalena, L., Gonzalez-Rodriguez, G.: Looking for a good fuzzy system interpretability index: an experimental approach. Int. J. Approx. Reason. **51**, 115–134 (2009)
11. Zadeh, L.A.: Fuzzy Sets and Applications. Wiley, New York (1987). Selected papers by L. Zadeh, Yager, R.R. et al. (eds.)
12. Saaty, T.L.: The Analytic Hierarchy Process. McGraw-Hill International, New York (1980)
13. Roychowdhury Sh., Wang B.-H.: Measuring inconsistency in fuzzy rules. In: IEEE International Conference on Fuzzy System, vol. 2 (1998)
14. Yu, W., Bin, Z.: Design of fuzzy logic controller with inconsistent rule base. J. Intel. Fuzzy Syst. **2**(2), 147–159 (1994)
15. Gottwald, S.: On some types of consistency consideration in fuzzy logic. Fuzzy Neuro Syst. **5**, 366–373 (1997)
16. González, A., Pérez, R.: Completeness and consistency conditions for learning fuzzy rules. Fuzzy Sets Syst. **96**(1), 37–51 (1998)
17. Gottwald, S., Novák, V.: An approach toward consistency degrees of fuzzy theories. Int. J-Gener. Syst. **29**(4), 499–510 (2000)
18. Adilova, N.E.: Consistency of fuzzy If-Then rules for control system. In: 10th International Conference on Theory and Application of Soft Computing, Computing with Words and Perceptions - ICSCCW-2019, pp. 137–142 (2019)

19. Jin, Y., Seelen, W., Sendhoff, B.: An approach to rule-based knowledge extraction. In: Proceedings of IEEE International Conference on Fuzzy System (1998)
20. Aliev, R.A., Aliev, F., Babaev, M.: Fuzzy Process Control and Knowledge Engineering in Petrochemical and Robotic Manufacturing. Verlag, Germany (1991)
21. Mohanaselvi, S., Shanpriya, B.: Application of fuzzy logic to control traffic signals. In: AIP Conference Proceedings, vol. 2112, p. 020045 (2019). https://doi.org/10.1063/1.5112230

Multiple Z-Regression with Fuzzy Coefficients

Olga M. Poleshchuk$^{(\boxtimes)}$ (iD)

Moscow Bauman State Technical University, Moscow, Russia
olga.m.pol@yandex.ru

Abstract. A multidimensional regression model under Z-information with fuzzy coefficients has developed in the paper. The initial data are Z-numbers, the first and second components of which are values of linguistic variables with the properties of completeness and orthogonality. In order to construct the regression model, aggregation indicators for Z-numbers are determined in the form of weighted segments of the product of both components. The optimization function is defined as the sum of the squares of the distances between the aggregating segments for model Z-numbers and initial Z-numbers and between the aggregating segments of model first components of Z-numbers and initial first components of Z-numbers. To determine the parameters of unknown coefficients, the minimum function is found. Reliability recognition of model output information is based on a comparative analysis of the weighted segments for model output Z-number and the weighted segments of Z-numbers, the first components of which are model first components of output Z-number, and the second components are formalization of the linguistic values of reliability.

Keywords: Z-number · Z-information · Regression model · Linguistic variable · Weighted segment

1 Introduction

The result of prediction substantially depends on the reliability of the information on the basis of which this decision is made. Therefore, it is difficult to overestimate the contribution that was made by Professor Lotfi Zadeh, who gave the notion of Z-numbers and opened up new opportunities for obtaining stable and adequate final results in prediction and decision-making problems [1].

Given the fuzziness of the real world and, accordingly, the fuzziness of its estimates, Professor Zadeh combined in the Z-number the fuzzy value of a variable and a fuzzy estimate of the reliability of this value. Information that contains Z -numbers has been named Z-information. The concept of Z-number was the logical result of many years of research. It was an invaluable contribution to the theoretical and practical components of real-world modeling, taking into account the different types of uncertainty and the characteristics of human thinking.

Over the past 2011, research related to the Z-information processing has been and remains relevant. Papers from this area contained theoretical foundations and practical applications. These papers are devoted to the formalization of expert group information, decision-making and prediction.

R. A. Aliev et al. (Eds.): ICAFS 2020, AISC 1306, pp. 63–70, 2021.
https://doi.org/10.1007/978-3-030-64058-3_8

In [2], arithmetic operations on Z-numbers were defined. The author [3] used Z-numbers for reasoning. Operations on Z-numbers, using their transformation into ordinary fuzzy numbers was applied in [4]. The authors [5] suggested to operate on Z-numbers by transforming Z-numbers into ordinary fuzzy numbers, all operations with which are well developed [6]. The authors [7] used components of the Z-number in conjunction with the weighted first component to determine the distance between the Z-numbers. The weights of all three components are taken equal. It is proposed to weigh the first component by defuzzifying the second component according to the well-known the center of gravity method [8]. Currently, there are different models of Z-numbers ranking. The authors [7] proposed ranking the Z-numbers by the first component and comparing the second components when the first components are equal. In [9], the distance between Z-numbers is determined on the basis of weighted segments, which are aggregating segments for Z-numbers, determined using all their α - cuts.

A significant contribution to the development of the Z-information processing has been made by Professor Rafik R. Aliyev and his team [10–17]. The authors [10, 11], developed operations on discrete and continuous Z-numbers. In [12], a model was developed for aggregating the opinions of an expert group. In [13], Z-numbers were used for solution a decision -making problem. In [14], the authors suggested ranking Z-numbers with the help of the expected utility function. In [15], the distance between Z-numbers is determined using the Jaccard similarity measure.

A logical question that arises in the processing of data obtained during the evaluation of various characteristics is the question of the relationship of these characteristics and the output prediction of data. This issue is also relevant in the processing of Z-information. However, at the moment there is the only approach of the authors [16, 17] to the construction of a regression model under Z-information, which is carried out on the basis of operations with fuzzy numbers and probability distributions, followed by the use of the Jaccart measure to formulate the optimization problem.

Therefore, the urgent problem is the further development of regression analysis under Z-information. Development of a Z-regression with fuzzy coefficients is the purpose of this paper.

Section 2 of the paper gives the basic concepts and definitions. Section 3 proposes a regression model under Z-information. Section 4 gives conclusions.

2   Basic Concepts and Definitions

An ordered pair $Z = (\tilde{A}, \tilde{R})$ of fuzzy numbers was named a Z-number, where $\tilde{A}$ is a fuzzy value of a variable X $(\mu_A(x) : X \rightarrow [0, 1])$ and $\tilde{R}$ is a fuzzy estimate of reliability $\tilde{A}$ $(\mu_R(x) : [0, 1] \rightarrow [0, 1])$ [1].

According to the [18] five $\{X, T(X), U, V, S\}$ was named a linguistic variable, where X - is a name; $T(X) = \{X_l, l = \overline{1, m}\}$ - a term-set; V - is a syntactical rule; S - is a semantic rule; U - a universal set.

Let the membership functions $\mu_l(x), l = \overline{1, m}$ of linguistic variable satisfy the following conditions [19–22]: 1. Generally $\hat{U}_l = \{x \in U : \mu_l(x) = 1\}$ are intervals.

2. $\mu_l(x), l = \overline{1,m}$ has a non-decreasing property to the left of $\widehat{U}_l$ and has a non-increasing property to the right of $\widehat{U}_l$. 3. $\mu_l(x), l = \overline{1,m}$ can have no more two first type discontinuity points. 4. $\forall x \in U \sum\limits_{l=1}^{m} \mu_l(x) = 1$.

A linguistic variable with properties 1–4 has been named Full Orthogonal Semantic Scope [20].

Consider a trapezoidal number $\tilde{A}$ $(\mu_A(x) = (a_1, a_2, a_L, a_R))$, then α-cut of $\tilde{A}$ is A_α such that:

$$A_\alpha = \{x \in R : \mu_A(x) \geq \alpha\} = \left[A_\alpha^1, A_\alpha^2\right] = [a - (1-\alpha)a_L, a + (1-\alpha)a_R], \ \alpha \in [0,1].$$

In [23], for a triangular number $\tilde{A}$ $(\mu_A(x) = (a, a_L, a_R))$ the definition of weighted point Θ is given:

$$\Theta = \frac{\int\limits_0^1 \left(\frac{A_\alpha^1 + A_\alpha^2}{2}\right) 2\alpha d\alpha}{\int\limits_0^1 2\alpha d\alpha} = a + \frac{1}{6}(a_R - a_L). \tag{1}$$

However, the authors of [24] were not satisfied with the fact that different fuzzy numbers with this definition can have the same weighted point, that is, individual informational features of different fuzzy numbers may be lost. Therefore in [24], the concept of a weighted segment $[\Theta_1, \Theta_2]$ for a number $\tilde{A}$ with membership function $\mu_A(x) = (a_1, a_2, a_L, a_R)$ was introduced, which is obtained as a union of weighted points of all the triangular numbers belonging to the number $\tilde{A}$:

$$\Theta_1 = \int\limits_0^1 \frac{2a_1 - (1-\alpha)a_L}{2} 2\alpha d\alpha = a_1 - \frac{1}{6}a_L,$$

$$\Theta_2 = \int\limits_0^1 \frac{2a_2 + (1-\alpha)a_R}{2} 2\alpha d\alpha = a_2 + \frac{1}{6}a_R. \tag{2}$$

In [25], based on weighted segments, arithmetic operations for fuzzy numbers are defined.

If $\tilde{A}, \tilde{C}$ with the membership functions $\mu_A(x) = (a_1, a_2, a_L, a_R), \mu_C(x) = (c_1, c_2, c_L, c_R)$ are trapezoidal fuzzy numbers, $[A_1, A_2], [C_1, C_2]$ are their weighted segments, then the weighted segment of the sum $A + C$ has the form $[A_1 + C_1, A_2 + C_2]$.

If $\tilde{A}, \tilde{C}$ with the membership functions $\mu_A(x) = (a_1, a_2, a_L, a_R), \mu_C(x) = (c_1, c_2, c_L, c_R)$ are non-negative trapezoidal fuzzy numbers $(a_1 - a_L \geq 0, c_1 - c_L \geq 0)$, then the weighted segment $\left[\theta_{AC}^1, \theta_{AC}^2\right]$ of the product $A \times C$ has the form $\theta_{AC}^1 = c_1\left(a_1 - \frac{1}{6}a_L\right) - c_L\left(\frac{1}{6}a_1 - \frac{1}{12}a_L\right), \theta_{AC}^2 = c_2\left(a_2 + \frac{1}{6}a_R\right) + c_R\left(\frac{1}{6}a_2 + \frac{1}{12}a_R\right).$

3 Problem Formulation and Solution

Let $Z_k^v = \left(\tilde{A}_k^v, \tilde{B}_k^v\right), k = \overline{1,m}, v = \overline{1,n}$ and $Z^v = \left(\tilde{A}^v, \tilde{B}^v\right), v = \overline{1,n}$ are the input and output information accordingly. $\tilde{A}_k^v, k = \overline{1,m}, v = \overline{1,n}$ $\left(\mu_{A_k^v} = \left(a_{1k}^v, a_{2k}^v, a_{Lk}^v, a_{Rk}^v\right)\right)$, $\tilde{A}^v, v = \overline{1,n}$ $\left(\mu_{A^v} = \left(a_1^v, a_2^v, a_L^v, a_R^v\right)\right)$ formalize the terms of Full Orthogonal Semantic Scopes A_k and A accordingly, $\tilde{B}_k^v, \tilde{B}^v, k = \overline{1,m}, v = \overline{1,n}$ $\left(\mu_{B_k^v} = \left(b_{1k}^v, b_{2k}^v, b_{Lk}^v, b_{Rk}^v\right)\right.$, $\mu_{B^v} = \left(b_1^v, b_2^v, b_L^v, b_R^v\right)$, $k = \overline{1,m}, v = \overline{1,n}\right)$ formalize the terms of Full Orthogonal Semantic Scope B - "Reliability". In [26, 27], methods for constructing of Full Orthogonal Semantic Scopes are described in detail.

Let construct a regression model in the form:

$$Z = \tilde{c}_1 Z_1 + \tilde{c}_2 Z_2 + \ldots + \tilde{c}_m Z_m, \tag{3}$$

where $\tilde{c}_k, k = \overline{1,m}$ are fuzzy numbers with membership functions $\mu_{c_k} = \left(c^k, c_L^k, c_R^k\right)$, $k = \overline{1,m}$.

Let assume that all the linguistic variables have the same universal set $[0, 1]$, and therefore the fuzzy numbers that formalize their terms are non-negative. If for some linguistic variables the universal sets are not a segment $[0, 1]$, then these sets can be normalized to the segment $[0, 1]$.

We extend the definition of the weighted segment given in [24] and define the weighted segment $\left[\theta_Z^1, \theta_Z^2\right]$ for $Z = (\tilde{A}, \tilde{B})$, $\mu_A = \left(a_1, a_2, a_L, a_R\right)$, $\mu_B = \left(b_1, b_2, b_L, b_R\right)$ as the weighted segment of the product of its components: $\theta_Z^1 = b_1\left(a_1 - \frac{1}{6}a_L\right) - b_L\left(\frac{1}{6}a_1 - \frac{1}{12}a_L\right)$, $\theta_Z^2 = b_2\left(a_2 + \frac{1}{6}a_R\right) + b_R\left(\frac{1}{6}a_2 + \frac{1}{12}a_R\right)$.

Then for model output $\tilde{c}_1 Z_1^v + \tilde{c}_2 Z_2^v + \ldots + \tilde{c}_m Z_m^v, v = \overline{1,n}$, $Z_k^v = \left(\tilde{A}_k^v, \tilde{B}_k^v\right), k = \overline{1,m}, v = \overline{1,n}$, $\mu_{A_k^v} = \left(a_{1k}^v, a_{2k}^v, a_{Lk}^v, a_{Rk}^v\right)$, $\mu_{B_k^v} = \left(b_{1k}^v, b_{2k}^v, b_{Lk}^v, b_{Rk}^v\right)$, $k = \overline{1,m}, v = \overline{1,n}$, we get the weighted segment $\left[\theta_{\mathcal{M}}^{1v}, \theta_{\mathcal{M}}^{2v}\right]$ in the following form [28]:

$$\theta_{\mathcal{M}}^{1v} = \sum_{k=1}^m \left(c^k\left[b_{p_k k}^v a_{p_k k}^v + \frac{(-1)^{p_k}}{6}b_{p_k k}^v a_{Mp_k k}^v + \frac{(-1)^{p_k}}{6}b_{Mp_k k}^v a_{p_k k}^v + \frac{1}{12}b_{Mp_k k}^v a_{Mp_k k}^v\right]\right)$$
$$-\sum_{k=1}^m \left(c_L^k\left[\frac{1}{6}b_{p_k k}^v a_{p_k k}^v + \frac{(-1)^{p_k}}{12}b_{p_k k}^v a_{Mp_k k}^v + \frac{(-1)^{p_k}}{12}b_{Mp_k k}^v a_{p_k k}^i + \frac{1}{20}b_{Mp_j}^v a_{Mp_k k}^v\right]\right),$$

$$\theta_{\mathcal{M}}^{2v} = \sum_{k=1}^m \left(c^k\left[b_{q_k k}^v a_{q_k k}^v + \frac{(-1)^{q_k}}{6}b_{q_k k}^v a_{Mq_k k}^v + \frac{(-1)^{q_k}}{6}b_{Mq_k k}^v a_{q_k k}^v + \frac{1}{12}b_{Mq_k k}^v a_{Mq_k k}^v\right]\right)$$
$$+\sum_{k=1}^m \left(c_R^k\left[\frac{1}{6}b_{q_k k}^v a_{q_k k}^v + \frac{(-1)^{q_k}}{12}b_{q_k k}^v a_{Mq_k k}^v + \frac{(-1)^{q_k}}{12}b_{Mq_k k}^v a_{q_k k}^i + \frac{1}{20}b_{Mq_j}^v a_{Mq_k k}^v\right]\right)$$

Let define weighted segment $\left[\vartheta_{\mathcal{M}}^{1v}, \vartheta_{\mathcal{M}}^{2v}\right]$ for $\tilde{c}_1 \tilde{A}_1^v + \tilde{c}_2 \tilde{A}_2^v + \ldots + \tilde{c}_m \tilde{A}_m^v, v = \overline{1,n}$ and weighted segment $\left[\vartheta^{1v}, \vartheta^{2v}\right]$ for $\tilde{A}^v, v = \overline{1,n}$:

$$\vartheta_{\mathcal{M}}^{1v} = \sum_{k=1}^{m} \left(c^k \left[a_{p_k k}^v + (-1)^{p_k} \frac{1}{6} a_{M_{p_k k}}^v \right] - c_L^k \left[\frac{1}{6} a_{p_k k}^v + (-1)^{p_k} \frac{1}{12} a_{M_{p_k k}}^v \right] \right),$$

$$\vartheta_{\mathcal{M}}^{2v} = \sum_{k=1}^{m} \left(c^k \left[a_{q_k k}^v + (-1)^{q_k} \frac{1}{6} a_{M_{q_k k}}^v \right] + c_R^k \left[\frac{1}{6} a_{q_k k}^v + (-1)^{q_k} \frac{1}{12} a_{M_{q_k k}}^v \right] \right).$$

$$\vartheta^{1v} = a_1^v - \frac{1}{6} a_L^v, \quad \vartheta^{2v} = a_2^v + \frac{1}{6} a_R^v.$$

$$p_k = \begin{cases} 1, & c^k - c_L^k \geq 0 \\ 2, & c^k + c_R^k < 0 \end{cases}, \quad M_{p_k} = \begin{cases} L, & p_k = 1 \\ R, & p_k = 2 \end{cases}, \quad q_k = \begin{cases} 2, & c^k - c_L^k \geq 0 \\ 1, & c^k + c_R^k < 0 \end{cases}, \quad M_{q_k}$$
$$= \begin{cases} L, & q_k = 1 \\ R, & q_k = 2 \end{cases}.$$

We define the weighted segment $[\theta^{1v}, \theta^{2v}]$ of the initial output data in the following form: $\theta^{1v} = \sum_{k=1}^{m} \left(b_1^v \left(a_1^v - \frac{1}{6} a_L^v \right) - b_L^v \left(\frac{1}{6} a_1^v - \frac{1}{12} a_L^v \right) \right)$, $\theta^{2v} = \sum_{k=1}^{m} \left(b_2^v \left(a_2^v + \frac{1}{6} a_R^v \right) + b_R^v \left(\frac{1}{6} a_2^v + \frac{1}{12} a_R^v \right) \right)$.

The optimization problem for finding the unknown coefficients $c_k, k = \overline{1, m}$ of the regression model we define as follows:

$$F(c_1, \ldots, c_m) = \sum_{v=1}^{n} \left[\left(\theta_{\mathcal{M}}^{1v} - \theta^{1v} \right)^2 + \left(\theta_{\mathcal{M}}^{2v} - \theta^{2v} \right)^2 + \left(\vartheta_{\mathcal{M}}^{1v} - \vartheta^{1v} \right)^2 + \left(\vartheta_{\mathcal{M}}^{2v} - \vartheta^{2v} \right)^2 \right]$$
$$\rightarrow \min.$$

Function $F(c_1, \ldots, c_m)$ is piecewise differentiable since $\theta_{\mathcal{M}}^{1v}, \theta_{\mathcal{M}}^{2v}, \vartheta_{\mathcal{M}}^{1v}, \vartheta_{\mathcal{M}}^{2v}$ are piecewise linear functions, then the optimization problem is solved according to the method [29].

It is easy to define the first component of model output Z-number. The question is how to determine the reliability (the second component) of this number? Define the weighted segment $\theta_{\mathcal{M}}^{1v}, \theta_{\mathcal{M}}^{2v}$ of the model value $\tilde{c}_1 Z_1^v + \tilde{c}_2 Z_2^v + \ldots + \tilde{c}_m Z_m^v, v = \overline{1, n}$ and weighted segments $[\delta_l^{1v}, \delta_l^{2v}]$ of the products $\tilde{B}^l \times (\tilde{c}_1 \tilde{A}_1^v + \tilde{c}_2 \tilde{A}_2^v + \ldots + \tilde{c}_m \tilde{A}_m^v)$, $l = \overline{1, q}$, where $\tilde{B}^l, l = \overline{1, q}$ are formalizations of the q values of a linguistic variable with name B - "Reliability". Let define $\rho^2(v, l) = \left(\theta_{\mathcal{M}}^{1v} - \delta_l^{1v} \right)^2 + \left(\theta_{\mathcal{M}}^{2v} - \delta_l^{2v} \right)^2$. If $\rho^2(v, p) = \min_l \rho^2(v, l)$, then the reliability of the model value is fuzzy number $\tilde{B}^v$ and accordingly p-th term of linguistic variable "Reliability".

4   Conclusions

Regression analysis is given the opportunity to identify the relationship between initial data and to predict output data. Therefore, this analysis is a necessary tool for solving many practical problems, and theoretical studies related to regression analysis are always relevant.

With the advent of the ability to process information, taking into account its reliability, the question arose of developing regression models in the conditions of Z-information.

The initial data of the developed regression are Z-numbers, whose components are formalizations of semantic scopes terms with the properties of completeness and orthogonality. The coefficients of Z-regression are triangular numbers. Regression models with fuzzy coefficients are not currently developed.

One of the problems in constructing regression models under the initial Z-information is the difficulty of directly operating Z-numbers. Therefore, in the paper aggregation segment for Z-number based on α-cuts its components has been determined. This segment has been called weighted segment and has been used for operating with Z-numbers, determining the distance between Z-numbers, constructing an optimization problem and recognizing the output model results.

The optimization problem is solved by requiring a minimum of the sum of the squares of the distances between the aggregating segments for the model and initial data and between the aggregating segments for the first components of the model and initial data.

In order to recognize the reliability of the model value we find the distances between segments for the model value and segments of Z-numbers, whose first components are first component of model data and the second components are values of linguistic variable "Reliability". The reliability of the model value is determined by the linguistic value that corresponds to the minimum distance.

Thus, this paper makes it possible to construct the relationships between input and output data and predict the output data under Z-information. This provides new opportunities for solving tasks in problem areas with the active participation of experts.

References

1. Zadeh, L.A.: A Note on Z-numbers. Inform. Sci. **14**(181), 2923–2932 (2011). https://doi.org/10.1016/j.ins.2011.02.022
2. Zadeh, L.A.: Methods and systems for applications with z-numbers United States patent. Patent No.: US 8,311,973 B1 (2012)
3. Yager, R.R.: On Z-valuations using Zadeh's Z-numbers. Int. J. Intell. Syst. **3**(27), 259–278 (2012). https://doi.org/10.1002/int.21521
4. Kang, B., Wei, D., Li, Y., Deng, Y.: Decision making using Z-numbers under uncertain environment. J. Inform. Comput. Sci. **7**(8), 2807–2814 (2012)
5. Kang, B., Wei, D., Li, Y., Deng, Y.: A method of converting Z-number to classical fuzzy number. J. Inform. Comput. Sci. **9**(3), 703–709 (2012)

6. Dutta, P., Boruah, H., Ali, T.: Fuzzy arithmetic with and without α-cut method: a comparative study. IJLTC **1**(2), 99–107 (2011)
7. Wang, F., Mao, J.: Approach to multicriteria group decision making with Z-numbers based on TOPSIS and power aggregation operators. Mat. Probl. Eng. 1–18 (2019). https://doi.org/10.1155/2019/3014387
8. Yager, R.R., Filev, D.P.: On the issue of defuzzification and selection based on a fuzzy set. Fuzzy Set. Syst. **5**(3), 255–272 (1993). https://doi.org/10.1016/0165-0114(93)90252-d
9. Poleshchuk, O.M.: Novel approach to multicriteria decision making under Z-information. In: Proceedings of the International Russian Automation Conference (RusAutoCon-2019), p. 8867607 (2019). https://doi.org/10.1109/rusautocon.2019.8867607
10. Aliev, R.A., Alizadeh, A.V., Huseynov, O.H.: The arithmetic of discrete Z-numbers. Inform. Sci. **1**(290), 134–155 (2015). https://doi.org/10.1016/j.ins.2014.08.024
11. Aliev, R.A., Huseynov, O.H., Zeinalova, L.M.: The arithmetic of continuous Z-numbers. Inform. Sci. **373**, 441–460 (2016). https://doi.org/10.1016/j.ins.2016.08.078
12. Aliev, R.K., Huseynov, O.H., Aliyeva, K.R.: Aggregation of an expert group opinion under Z-information. In: Proceedings of the Eighth International Conference on Soft Computing, Computing with Words and Perceptions in System Analysis, Decision and Control (ICSCW-2015), pp. 115–124 (2015)
13. Aliev, R.A., Zeinalova, L. M.: Decision-making under Z-information. In: Human-Centric Decision-Making Models for Social Sciences, pp. 233–252 (2013)
14. Aliyev, R.R., Talal Mraizid, D.A., Huseynov, O.H.: Expected utility based decision making under Z-information and its application. Comput. Intell. Neurosci. **3**, 364512 (2015). https://doi.org/10.1155/2015/364512
15. Aliyev, R.R.: Similarity based multi-attribute decision making under Z-information. In: Proceedings of the Eighth International Conference on Soft Computing, Computing with Words and Perceptions in System Analysis, Decision and Control (ICSCW-2015), pp. 33–39 (2015)
16. Sadikoglu, F., Huseynov, O., Memmedova, K.: Z-Regression analysis in psychological and educational researches. Proc. Comput. Sci. **102**, 385–389 (2016). https://doi.org/10.1016/j.procs.2016.09.416
17. Zeinalova, L.M., Huseynov, O.H., Sharghi, P.: A Z-number valued regression model and its application. Intell. Autom. Soft Comput. **24**, 187–192 (2017). https://doi.org/10.1080/10798587.2017.1327551
18. Zadeh, L.A.: The Concept of a linguistic variable and its application to approximate reasoning. Inform. Sci. **8**, 199–249 (1975). https://doi.org/10.1016/0020-0255(75)90036-5
19. Ryjov, A.P.: The concept of a full orthogonal semantic scope and the measuring of semantic uncertainty. In: Proceedings of the Fifth International Conference Information Processing and Management of Uncertainty in Knowledge-Based Systems (IPMU-1994), pp. 33–34 (1994)
20. Poleshchuk, O.M., Komarov, E.G., Darwish, A.: Comparative analysis of expert criteria on the basis of complete orthogonal semantic spaces. In: Proceedings of the 19th International Conference on Soft Computing and Measurements (SCM-2016), p. 7519784 (2016). https://doi.org/10.1109/scm.2016.7519784
21. Darwish, A., Poleshchuk, O.: New models for monitoring and clustering of the state of plant species based on sematic spaces. J. Intell. Fuzzy Syst. **3**(26), 1089–1094 (2014). https://doi.org/10.3233/ifs-120702
22. Poleshchuk, O., Komarov, E.: The determination of rating points of objects and groups of objects with qualitative characteristics. In: Proceedings of the Annual Conference of the North American Fuzzy Information Processing Society – NAFIPS 2009, p. 5156416 (2009). https://doi.org/10.1109/nafips.2009.5156416

23. Chang, Y.-H.: Hybrid fuzzy least- squares regression analysis and its reliability measures. Fuzzy Set. Syst. **119**, 225–246 (2001). https://doi.org/10.1016/s0165-0114(99)00092-5
24. Domrachev, V.G., Poleshuk, O.M.: A regression model for fuzzy initial data. Automat. Rem. Control **64**(11), 1715–1724 (2003). https://doi.org/10.1023/a:1027322111898
25. Poleshuk, O.M., Komarov, E.G.: Multiple hybrid regression for fuzzy observed data. In: Proceedings of the Annual Conference of the North American Fuzzy Information Processing Society (NAFIPS-2008), p. 4531224 (2008). https://doi.org/10.1109/nafips.2008.4531224
26. Poleshchuk, O.M.: Creation of linguistic scales for expert evaluation of parameters of complex objects based on semantic scopes. In: Proceedings of the International Russian Automation Conference (RusAutoCon – 2018), p. 8501686 (2018). https://doi.org/10.1109/rusautocon.2018.8501686
27. Poleshchuk, O.M.: Formalization, prediction and recognition of expert evaluations of telemetric data of artificial satellites based on type-II fuzzy sets. In: Studies in Computational Intelligence, vol. 836, pp. 39–64 (2020). https://doi.org/10.1007/978-3-030-20212-5_3
28. Poleshchuk, O., Komarov, E.: A fuzzy linear regression model for interval type-2 fuzzy sets. In: Proceedings of the Annual Conference of the North American Fuzzy Information Processing Society (NAFIPS-2012), p. 6290970 (2012). https://doi.org/10.1109/nafips.2012.6290970
29. Coleman, T.F., Li, Y.: A reflective newton method for minimizing a quadratic function subject to bounds on some of the variables. SIAM J. Optim. **6**, 1040–1058 (1996). https://doi.org/10.1137/s1052623494240456

Solution of the Retail Marketing Problem of Rational Choice of the Location of Trade Enterprises Using the Method of Hierarchy Analysis and Fuzzy Set Theory

Ahmed Veliyev[1], Tarlan Abdullayev[2(✉)] (iD), Ramiz Alekperov[3], and Vuqar Salahli[4]

[1] Odlar Yurdu Univercity, Baku AZ1072, Azerbaijan
oyu-asp@mail.ru
[2] Scientific Affairs, Odlar Yurdu Univercity,
Baku AZ1072, Azerbaijan
oyu-asp@mail.ru
[3] Informathion Technology and Programming, Azerimed LLC,
Baku AZ1147, Azerbaijan
ramizalekper@gmail.com
[4] Department of Computer Engineering, Odlar Yurdu Univercity,
Baku AZ1072, Azerbaijan
vsalahli@ekin.com

Abstract. The principles of solving the multicriteria problem of retail marketing for the rational choice of the location of trade enterprises (alternatives) based on the classical method of analysis of Saaty hierarchies, using the theory of fuzzy sets, are considered. The choice of the best alternative is carried out on the basis of the arguments of the maxima (abbreviated arg max), as the arguments, which are evaluated based on the operation of intersection of fuzzy sets of criteria. This approach is very practical from the point of view of using the experience and intuition of marketing workers and requires minimal costs to solve similar problems.

Keywords: Commercial establishments · Retail marketing · Multi-criteria decision-making · Saaty hierarchy analysis · Fuzzy logic · Fuzzy inference sets

1 Introduction

The success of a retailer depends mainly on the amount of capital invested, the good choice of business and its geographical location, which are the marketing objectives of the retailer. Choosing a location for a retail outlet is one of the most important decisions a retailer must make. Choosing the right location for retail outlet determines the potential number of customers and turnover. In this regard, one of the main factors that determine the effectiveness of retail trade enterprises is their rational geographical location. The solution to this problem requires consideration of appropriate approaches, principles, and factors that are formulated, taking into account the types of product

© The Author(s), under exclusive license to Springer Nature Switzerland AG 2021
R. A. Aliev et al. (Eds.): ICAFS 2020, AISC 1306, pp. 71–79, 2021.
https://doi.org/10.1007/978-3-030-64058-3_9

groups. Usually, goods are divided into three main groups [1–4]: consumer goods - cheap, regularly purchased goods that are included in the obligatory consumer basket (drinks, milk and dairy products, bread and bakery products, sweets, etc.) are bought, as a rule, out of habit and sold in numerous outlets; large expensive items (electronic household appliances, cars, etc.); luxury goods - materials from gold, silver, jewelry etc.

Logically, outlets selling expensive goods and luxury goods are usually located in central areas, where customers are more likely to travel than for everyday goods or medicines. From the point of view of the location of retail trade enterprises, two approaches are distinguished [2, 5]: 1. trade in "flows" - gravitation towards the central areas of the city; 2. Trade type "traffic" - the location of the largest trade enterprises in the outskirts.

There are also three zones: 1. zones of increased commercial activity - the city center; areas of the main city highways; areas of transport hubs (intersections of central streets) and areas of railway stations; areas of enterprises and markets, etc.; 2. zones of medium commercial activity - areas adjacent to the center and having good transport links with it; areas of secondary city highways; areas of main significance. 3. commercial risk zones - outskirts of the city.

The following basic methods and approaches are known for solving this problem [6, 7]:

1. Financial analysis method;
2. Checklist method;
3. Analog approach;
4. Regression analysis;
5. The method of using the Huff gravity model;
6. Revealed preference.

The analysis of the considered approaches shows that they are mainly designed to determine the location of the retail enterprise, considering the forecast of turnover. Usually they do not consider the factors of uncertainty, incompleteness, and difficult formalization of selection criteria. It needs to note also that due to the importance of location for a retail enterprise and the presence of methodological difficulties in forecasting turnover, it becomes necessary to use special methods to solve this problem.

In practice, in general, the choice of the location of trade enterprises is done using experience and intuition, in other words, by the judgments of experts in this field. In most cases, they evaluate alternatives by simple points [8] for each of the decision criteria. For example, work [9] provides an example of evaluating options for the location of trade/catering establishments, representing a system for evaluating outlets (using 8 criteria), which is shown in Table 1. The values in the table have been slightly modified to show the difficulty of rational choice from many alternatives if their total score values are the same.

Table 1. Examples for assessing the location options of commercial enterprises

№	Criteria (K)	Evaluation (Q – satisfactory)				
		Norms	Cretier's prioritet (cp)	Alternative options (A)		
				a_1	a_2	a_3
K_2	Relative purchasing power for a resident	For a resident 200	2	100	66.67	75
K_3	Number of customers in the impact zone	8000	3	66.67	79.17	95.83
K_1	The flow of passers-by within 1 h	500	1	62.33	83.33	50
K_4	The width of the sales area of the object	min 800 kV.m	4	79.17	83.33	91.67
K_5	Area of sales outlets	min 40 kV.m	5	100	100	83.33
K_6	Number of stops	min 10 places	5	75	91.67	100
K_7	Possibility of delivery	–	6	96	83.33	90
K_8	Public transport	3 min away	7	100	91.67	93.34
	Sums for comparison			679.17	679.17	679.17

As can be seen from Table 1, managers as alternatives selected 3 potential locations of the sales point of sale, which are evaluated on a point system, taking into account 8 criteria. Analysis of the criteria shows that they are vague and difficult to formalize (for example, the criterion - Possibility of delivery). Evaluation of alternatives according to these criteria are subjective and fuzzy. The best alternative is determined according to the maximum point system. The question remains open if there are alternatives with the same number of points. The solution to similar problems is based on the application of the theory of fuzzy sets [11–15], which allows considering the knowledge and experience of specialists, as well as factors of uncertainty and difficult - formalizability. The analysis of the above-mentioned methods for solving the problem of choosing the potential location of a shopping facility and the factors influencing this selection process makes it possible to substantiate that this issue is a matter of rational choice of alternatives in multi-criteria conditions and is implemented under various socio-economic factors.

In our case, for the rational choice of the location of the outlet, we applied an approach based on the choice of a feasible solution - an alternative that best meets the criteria and goals of decision makers. The solution to this multicriteria problem is based on the use of the hierarchy analysis method (Saaty method) [10] and the theory of fuzzy sets [11–14].

2 Mathematical Formulation of the Solution to a Multicriteria Problem

Let's say there are many alternatives (locations of outlets) $A = \{a_1, a_2, \ldots, a_m\}$ (where m is total number of alternatives) and many criteria $K = \{K_1, \ldots, K_n\}$ where n is total number of criteria), by which these alternatives are evaluated according to the stated goal. The assessment of alternatives for each criterion is represented by a fuzzy set $\tilde{K}$, which is represented as follows:

$$\tilde{K}_i = \left\{ \frac{\mu_{K_i}(a_1)}{a_1}, \ldots, \frac{\mu_{K_i}(a_m)}{a_m} \right\}, i \in n, \tag{1}$$

where $\mu_{K_i}(a_j), i \in n, j \in m$ is the membership function of a fuzzy set, which determines the degree of membership of the alternative a_j from the point of view of the criterion K_i to the concept of the best - rational decision.

Suppose that each of the criteria has a different degree of weight (weighting factors) $\alpha_i, (i \in 1, n)$ which are calculated by the following formula

$$\alpha_i = n \cdot \omega_i, i \in n \tag{2}$$

ω_i is the Saaty importance coefficient [10], defined as follows:

1. For each criterion from the set $K_i \in K, i \in n$, their Saaty importance coefficients are determined $\omega_i, i \in n$, using expert knowledge. To start with an expert (managers, marketing workers, etc.) according to the following logic - "How many times the criterion K_i exceeds criterion K_j?" and Table 2, pairwise comparison of the criteria $K = \{K_1, \ldots, K_n\}$ relative to some goal, G is the degree of influence on the choice of the best solution, and the comparison results are recorded in the survey matrix:

$$
\begin{array}{c}
 \quad (G) \quad K_1 \ \ldots \ K_n \\
\begin{array}{cc}
K_1 \\
\\
K_n
\end{array}
\begin{pmatrix}
g_{11} & \cdots & g_{1n} \\
\cdot & \cdots & \cdot \\
g_{n1} & \cdots & g_{nn}
\end{pmatrix}
\end{array}
$$

When filling in the matrix, the following relations are observed: $g_{ii} = 1; \ g_{ij} = 1/g_{ji}, \ i \neq j$.

2. To determine the degree of consistency of the constructed survey matrix, the criteria importance coefficients are calculated as follows:

$$g_i = \sum_{j=1}^{n} g_{ij}, i \in n \tag{3}$$

$$\omega_i = \frac{g_i}{\sum\limits_{j=1}^{n} g_j} \tag{4}$$

Table 2. Empirical Saaty scale

Meaning g_{ij}		Value g_{ij}
K_i Equally significant with	K_j	1
K_i Weakly superior	K_j	3
K_i Surpasses	K_j	5
K_i Far surpasses	K_j	7
K_i Absolutely superior	K_j	9
K_i The situation when you need it	K_j	2, 4, 6, 8

ω_i – also interpreted as the degree of compatibility of the criteria K_i with the set goal $G - \mu_G(K_i)$. Further, to determine the degree of consistency of the constructed matrix, the consistency index is calculated by the following formula (CI):

$$CI = \frac{\lambda - n}{\sigma \cdot (n - 1)} \tag{5}$$

where n is number of criteria considered.

$$\begin{pmatrix} g_{11} & \cdots & g_{1n} \\ \cdots & \cdots & \cdots \\ a_{n1} & \cdots & a_{nn} \end{pmatrix} \cdot \begin{pmatrix} \omega_1 \\ \cdots \\ \omega_n \end{pmatrix} = \begin{pmatrix} y_1 \\ \cdots \\ y_n \end{pmatrix} \tag{6}$$

$$\lambda \approx \frac{1}{n} \cdot \left(\frac{y_1}{\omega_1} + \ldots + \frac{y_n}{\omega_n} \right) \tag{7}$$

σ is a random index depending on the number of compared criteria (n), the value of which is taken from Table 3.

Table 3. Definitions of value σ

N	3	4	5	6	7	8	9	10	11	12	13	14
Σ	0,58	0,90	1,12	1,24	1,32	1,41	1,45	1,49	1,51	1,48	1,56	1,57

In the case of a consistent survey, the inequality must be fulfilled $CI \leq 0.20$. This will mean that the polling procedure has been completed successfully. If this inequality is not met, the expert is polled again, or the correctness of the task is checked. The rule for choosing the best alternative is defined as an intersection.

$$P = K_1^{\alpha_1} \cap \ldots \cap K_n^{\alpha_n} \tag{8}$$

An alternative is considered rational, which is determined according to the formula:

$$a^* = \arg \max_{i=1,\dots,\,m} \mu_P(a_j), j \in m \tag{9}$$

where for everyone ι - that criterion $K_i (i \in n)$, $\mu_P(a_i)$ is calculated as follows:

$$\mu_P(a_j) = \min\left\{ \frac{\mu_{K_1}^{\alpha_1}(a_j)}{a_j}, \dots, \frac{\mu_{K_i}^{\alpha_i}(a_j)}{a_j} \right\}, i \in n, j \in m \tag{10}$$

3 Solving the Problem of Choosing the Location of the Outlet

As can be seen from Table 1, it is necessary to choose the most rational solution from among the proposed three alternative (a_1, a_2, a_3) options for the location of the outlet, taking into account three factors (criteria): K_1- human flow through the object for 1 h; K_2 - number of customers in the effect zone, K_3- related buying possibility for a resident,K_1- wide range of sale of the object, K_2 - area of shopping sales, K_3 - number of stations, K_1 - delivery opportunity, K_2 - public transport. After interviewing experts, according to Table 2, the following survey matrix was obtained (Table 4), which reflects the opinion of experts about the preference of the criteria. Table 4 also presents the results of calculating the coefficients of the importance of Saaty-ω_i and the weighting factors for each of the criteria. Multiplying the matrices G and W (6), we determine the matrix $Y(Y = G * W)$, the results of which are shown in Table 4 (matrix Y) Using this matrix and (5)–(7), we calculate the consistency index (CI):

$$\lambda \approx \frac{1}{8} \cdot \left(\frac{3.24}{0.23} + \frac{2.83}{0.21} + \frac{1.73}{0.18} + \frac{1.22}{0.14} + \frac{0.74}{0.08} + \frac{0.66}{0.08} + \frac{0.28}{0.05} + \frac{0.14}{0.01} \right) = 9.90 \tag{11}$$

According to Table 3, we select values for 8 criteria for $\sigma = 1.41$.

$$CI = \frac{\lambda - n}{\sigma \cdot (n-1)} = \frac{9.90 - 8}{1.41 \cdot 7} = 0.1928 \leq 0.20 \tag{12}$$

Table 4. Questionnaire matrix with the results of calculating the values of the weight coefficients.

	Matrix G									Matrices		
	K_1	K_2	K_3	K_4	K_5	K_6	K_7	K_8		W	α_i	Y
K_1	1	2	3	3	5	7	9	9	39	0.23	1.86	3.24
K_2	0.5	1	3	3	5	7	7	9	35.5	0.21	1.7	2.83
K_3	0.33	0.33	1	3	3	4	5	9	30.67	0.18	1.47	1.73
K_4	0.33	0.33	0.33	1	3	3	5	9	24	0.14	1.15	1.22
K_5	0.2	0.2	0.25	0.33	1	3	3	5	12.98	0.08	0.62	0.74
K_6	0.14	0.14	0.25	0.33	0.33	3	3	5	14.2	0.08	0.68	0.66
K_7	0.11	0.14	0.11	0.14	0.33	0.2	1	7	9.04	0.05	0.43	0.28
K_8	0.11	0.11	0.11	0.11	0.2	0.2	0.14	1	1.99	0.01	0.09	0.14
Total (g)									167.38			

As can be seen from formula 10, the concordance index conditions are satisfied, which indicates the success of the survey. On the other hand, as can be seen from the obtained values, the coefficients of the importance of the criteria are consistent with the priority of the criteria given in Table 1. Further, using the linguistic concept "the advisability of choosing an alternative taking into account the considered criterion" for each criterion using the Gauss membership function [11–14] (Fig. 1).

$$\mu(u) = \exp\left(-(u - 10)^2/\sigma_k^2\right) \ (k = 1 \div 8) \text{ (where values } \sigma_k \text{ is determined in}$$

accordance with the priority of the criteria, for our case $\sigma_k = 20 + cp$) fuzzy sets are constructed, which are presented in Table 5. Using the values of Table 5, we construct the membership function of the fuzzy set P in accordance with the above formulas formulas 8 and 10, as well as using the values of the weight coefficients α_i (Table 4).

Table 5. Values of membership functions

	K1	K2	K3	K4	K5	K6	K7	K8
a1	0.0400	1.0000	0.1225	0.4703	1.0000	0.3679	0.9766	1.0000
a2	0.5325	0.1007	0.4403	0.6173	1.0000	0.8949	0.6629	0.9092
a3	0.0035	0.2749	0.9677	0.8865	0.6411	1.0000	0.8625	0.9410

$$\mu_P(a_1) = \min(0.0400^{1.86}, 1^{1.70}, 0.1225^{1.47}, 0.4703^{1.15}, 1^{0.62}, 0.3679^{0.68},$$
$$0.9766^{0.43}, 1^{0.09}) = 0.00248;$$

$$\mu_P(a_2) = \min(0.5325^{1.86}, 0.1007^{1.70}, 0.4403^{1.47}, 0.6173^{1.15}, 1^{0.62}, 0.8949^{0.68},$$
$$0.6629^{0.43}, 0.9092^{0.09}) = 0.01635;$$

$$\mu_P(a_3) = \min(0.0035^{1.86}, 0.2749^{1.70}, 0.9677^{1.47}, 0.8865^{1.15}, 0.6411^{0.62}, 1^{0.68},$$
$$0.8625^{0.43}, 0.9410^{0.09}) = 0.00002;$$

Then, according to formula 9, the choice of the alternative

$$a^* = \arg\max_{i=1,\dots,3} \mu_P(x_i) = \arg\max(0.00248; 0.01635; 0.00002) = a_2$$

can be considered rational. If the obtained values are expanded in ascending order: $a_1 \rightarrow 0.00002 < a_3 \rightarrow 0.00248 < a_2 \rightarrow 0.01635$, it's obvious that $a^* = a_2$, and therefore the most preferred is the second alternative - the location of the outlet.

4 Conclusion

Analysis of the results obtained shows that the alternative is practically not suitable as an alternative. But the gap between the second and third alternatives is very large, which depends on the degree of importance of the decision criteria. The solution of this problem by using the method of numerical (point) estimates using the theory of fuzzy logic, described in [16], also showed the rationality of choosing an alternative a_2 as the best solution. The analysis of the results obtained makes it possible to substantiate the pragmatism and effectiveness of applying the theory of fuzzy sets for solving similar multicriteria decision-making problems, with a large number of fuzzy and difficult-to-formalize criteria. The considered method of solving the multicriteria problem of retail marketing on the rational choice of the location of trade enterprises (alternatives) is based on the classical method of analysis of Saaty hierarchies, using the theory of fuzzy sets. The choice of the best alternative is done based on the arguments of the maxima (abbreviated arg max), as the arguments, which are evaluated by the operation of intersection of fuzzy sets of criteria.

This approach is very practical from the point of view of using the experience and intuition of marketing workers and requires minimal costs to solve such problems, and in the future, a software system will be developed in order to provide the possibility of its application for solving similar problems.

References

1. http://www.elitarium.ru/torgovoe-predpriyatie-roznichnaya-torgovlya-rynok-magazin-gorod-territoriya-mestoraspolozhenie-planirovanie/. Accessed 15 July 2020
2. https://www.cfin.ru/press/marketing/2001-4/08.shtml. Accessed 15 July 2020
3. Kiselev, V. M.: Methodological apparatus of marketing communications at points of sale: olfactory merchandising. Retail Marketing (in Russian) FBK-PRESS, Moskva (2004)
4. Brizhasheva, O.V.: Trade Marketing (in Russian). UlGTU, Ulyanovsk (2007)
5. https://studfile.net/preview/5183229/page:3/. Accessed 15 July 2020
6. Hiduke, G., Ryan, J.D.: Small Business: An Entrepreneur's Business Plan. 9th edn. South-Western College Pub. (2013)
7. https://dis.ru/library/520/25648/. Accessed 15 July 2020
8. Bezborodova T.M., Dyuzheva M.B.: Trade enterprises management. Tutorial (in Russian) Omsk (2011)
9. https://www.marketing.spb.ru/read/m11/index.htm. Accessed 26 Mar 2020
10. Saaty, T.L.: The Analytical Hierarchy Process, vol. 1. Springer, Heidelberg (1989). 265 p

11. Zadeh, L.A.: Fuzzy logic, neural networks, and soft computing. Commun. ACM **37**, 77–84 (1994)
12. Aliev, R.A.: Fundamentals of the Fuzzy Logic-Based Generalized Theory of Decisions. Springer, Heidelberg (2013)
13. Zimmermann, H.-J.: Fuzzy Set Theory and its Applications, 4th edn. Kluwer Academic Publishers, Dordrecht (2001)
14. Piegat, A.: Fuzzy Modeling and Control, vol. 1, 742 p.. Physica; Softcover reprint of hardcover (2010)
15. Oglu, A.R.B., Kizi, I.I.T.: A method for forecasting the demand for pharmaceutical products in a distributed pharmacy network based on an integrated approach using fuzzy logic and neural networks. In: Kahraman, C., Cevik Onar, S., Oztaysi, B., Sari, I., Cebi, S., Tolga, A. (eds.) Intelligent and Fuzzy Techniques: Smart and Innovative Solutions. INFUS 2020. Advances in Intelligent Systems and Computing, vol. 1197, pp. 998–1007. Springer, Cham (2021). https://doi.org/10.1007/978-3-030-51156-2_116
16. Oglu, A.R.B., Oglu, S.V.M.: Estimation of potential locations of trade objects on the basis of fuzzy set theory. In: Kahraman, C., Cevik Onar, S., Oztaysi, B., Sari, I.U., Cebi, S., Tolga, A.C. (eds.) Intelligent and Fuzzy Techniques: Smart and Innovative Solutions. INFUS 2020. Advances in Intelligent Systems and Computing, vol. 1197, pp. 228–237. Springer, Cham (2020). https://doi.org/10.1007/978-3-030-51156-2_28

Multi-criterial Optimization Problem for Fuzzy If-Then Rules

O. H. Huseynov[1(✉)] [iD] and Nigar E. Adilova[2] [iD]

[1] Research Laboratory of Intelligent Control and Decision Making Systems in Industry and Economics, Azerbaijan State Oil and Industry University, Baku, Azerbaijan
oleg_huseynov@yahoo.com
[2] Joint MBA Program, Azerbaijan State Oil and Industry University, 20 Azadlig Avenue, Baku AZ1010, Azerbaijan
nigar.adilova@asoiu.edu.az

Abstract. Fuzzy IF-Then rules as universal approximator very frequently are used for linguistic modelling of real-world complex system. This type of models is characterized mainly two contradictory requirements: Interpretability and accuracy. Unfortunately, researchers in existing works usually focused on the accuracy of the models without paying attention to their interpretability. However, very important problem in developing of fuzzy If-Then rule base is to achieve the best trade-off between interpretability and accuracy. In this paper, we will consider the problem of investigation of trade-off between complexity and semantic-based interpretability and accuracy of fuzzy If-Then knowledge base. The proposed approach is related with differential evolutionary optimization-based multicriteria decision-making problem. Numerical example related to the multi-criterial optimization problem is given.

Keywords: Fuzzy If-Then rules · Interpretability measure · Accuracy-interpretability trade-off

1 Introduction

Fuzzy logic system can provide the solution of substantial problems in decision support system and knowledge extraction with the assistance of interpretability matter. To improve interpretability index for fuzzy logic systems some researchers took a close interest in the development of Nauck and Fuzzy indices. The usage of these indexes linguistic fuzzy rule base system can be more interpretable.

Nauck index being the most commonly used was proposed by D. Nauck for measuring interpretability in rule-based classification system. In addition to discussion some aspects of interpretability fuzzy systems he introduced an interpretability index in the form of the product of complexity, partition and coverage variables [1].

A new method for inducing a set of interpretabile rules from data was presented in [2]. For the importance of interpretability authors [3] proposed an overview to obtain more interpretable linguistic fuzzy rule-based systems. Then [4] introduced the

R. A. Aliev et al. (Eds.): ICAFS 2020, AISC 1306, pp. 80–88, 2021.
https://doi.org/10.1007/978-3-030-64058-3_10

extensions to the Nauck and fuzzy interpretability indices for Hierarchial Fuzzy Systems (HFSs).

Interpretability is a capability to express the behavior of the real system in an understandable way. However, the accuracy is other contradictory capability to faithfully represent the real system. Therefore, authors considered accuracy-interpretability trade-off in [5]. They concentrated on the balance of both accuracy and interpretability indices for modeling problems. A little later the approach to the issue of developing fuzzy systems have been proposed by [6]. In their related work they suggested interpretability-accuracy trade-off in Evolutionary multi-objective fuzzy systems. The trade-off between the size of fuzzy rule-based systems and their classification performance was described by illustrating fuzzy partition of each input variable [7]. Obviously, to define a good trade-off is extremely difficult and requires many applications. Expressing more precisely, interpretability and accuracy are two important features of fuzzy systems which are conflicting in their nature. These two conflicting goals are often involved in the design of fuzzy rule-based systems. Some rearchers tried to simultaneously perform the accuracy maximization and the complexity maximization in order to design fuzzy rule based systems with high accuracy and high interpretability. Whereas, it is impossible to optimize these two objectives. The accuracy-interpretability trade-off in fuzzy rule-based systems classifiers using a multiobjective fuzzy genetic-based machine mearning algorithm has been discussed [8, 9]. Looking for a good trade-off between interpretability and accuracy indicesis one of the most difficult and challenging tasks.

In [10] authors proposed a solution approach to multi-criterial optimization problem with conflicting criterion, such as, cardinality and specificity in fuzzy If-Then rules.

A number of approaches have been proposed for finding a fuzzy rule-based systems with a good accuracy-interpretability trade-off. However, identifying the trade-off between interpretability and accuracy indexes of fuzzy systems still remains an open problem [11].

In this paper we propose to concentrate on the trade-off among multi criteria for fuzzy rule-based systems. The problem is to find the multi-criterial optimization for fuzzy If-Then rules. Hence, to deal with this problem fuzzy TOPSIS method was considered for calculating alternative preferences and weights of criteria.

The paper is structured as follows: Sect. 2 describes the preliminary information on multi criterial optimization problem. Section 3 includes the statement of the problem. The solution of the problem is applied in Sect. 4. Section 5 contains some concluding remarks.

2 Preliminaries

*Definition 1. **Fuzzy number:*** A fuzzy number is a set A on R which possesses the following properties: a) A is a normal fuzzy set; b A is a convex fuzzy set; c) $\alpha-$ cut of A, A^{α} is a closed interval for every $\alpha \in (0, 1]$; d) the support of A, A^{+0} is bounded [12].

Definition 2. Fuzzy If-Then rules [13, 14]: Fuzzy if-then rule statements are used to define the conditional statements that possess fuzzy logic. Multi-input multi-output fuzzy system is described as follows

If X_1 is A_1 and X_2 is A_2 and ... and X_n is A_n

then Y_1 is B_1 and Y_2 is B_2 and ... and Y_m is B_m

where A_i and B_i are information granules.

Definition 3. Complexity measure: *comp* - represents the complexity of a classifier measured as the number of classes divided by the total number of premises:

$$comp = m \Big/ \sum_{k=1}^{r} n_k$$

where m is the number of classes (outputs), r is the number of rules and n_k is the number of variables used in the k-th rule.

Definition 4. Coverage degree: cov is the average normalized coverage degree on cov_i

$$cov_i = \frac{\int_{X_i} \overline{h_i}(x)dx}{N_i} \quad \text{where } \overline{h_i}(x) = \begin{cases} h_i(x) = \sum_{k=1}^{p_i} \mu_i^{(k)}(x), & \text{if } 0 \le h_i(x) \le 1 \\ \frac{p_i - h_i(x)}{p_i - 1}, & \text{otherwise} \end{cases}$$

where X_i is the domain of the ith variable and this domain is partitioned by p_i fuzzy sets and with $N_i = \int_{X_i} dx$ for continuous domains or with $N_i = |X|$ for discrete domains.

Definition 5. Partition complexity measure: Partition complexity measure (part) - stands for the partition index which is computed as the inverse of the number of MFs minus one for each input variable.

$$part_i = 1/(p_i - 1)$$

p_i is the number of MFs in the ith input variable.

Definition 6. Consistency and Inconsistency of fuzzy rules: *Cons* is the consistency index calculated as follows for two given rules $R(i)$ and $R(k)$:

$$Cons(R(i), R(k)) = \exp\left\{ -\frac{\left(\frac{SRP(i,k)}{SRC(i,k)} - 1.0\right)^2}{\left(\frac{1}{SRP(i,k)}\right)^2} \right\}$$

where *SRP* is the similarity of rule premises and *SRC* is the similarity of rule consequents. The degree of inconsistency of a given RB (f_{incons}) is calculated as:

$$f_{incons} = \sum_{i=1}^{N} Incons(i),$$

where $Incons(i)$ is the degree of inconsistency for the ith. It is defined as:

$$Incons(i) = \sum_{\substack{1 \leq k \leq N \\ k \neq i}} [1.0 - Cons(R^1(i), R^1(k))] + \sum_{\substack{1 \leq l \leq L \\ i=1,2,\ldots,N}} [1.0 - Cons(R^1(i), R^2(l))],$$

where R^1 and R^2 denote the RB generated from data and the RB extracted from prior knowledge.

Definition 7. Accuracy measure: Accuracy is deviation between responses of fuzzy system with all rules FS_{all} and fuzzy system FS with some of the rules (RMSE):

$$RMSE = \sqrt{\frac{1}{P} \sum_{p=1}^{P} (y_p - \hat{y}_p)^2}$$

where p is number of samples (current input vectors), $\hat{y}_p = FS_{all}(x_p)$ is response of fuzzy system with all rules, $y_p = FS(x_p)$, x_p is the p-th sample (current input vector).

3 Statement of the Problem and Solution Approach

We need to obtain the trade-off between the following criteria of FB:

1. Interpretability
   1.1. Complexity measure
   1.2. Coverage degree
   1.3. Partition measure
   1.4. Inconsistency of If-Then rules
2. Accuracy

The trade-off is obtained by varying the fuzzy terms of input variables and excluding some rules from FB. Excluding rules leads to improvement of interpretability.

Decision variables:

1) fuzzy terms of input variables described by TFNs $A_i^j = (a_{i\,l}^j, a_{i\,m}^j, a_{i\,r}^j)$.
2) A binary vector (Rule 1, Rule 2, ..., Rule R) describing collection of the used rules. The components of the vector take values 0 or 1. For example, if Rule 2 is 1 then rule 2 is included into FB, otherwise not. The FB with all rules is (1, 1..., 1)
3) A binary vector (Term 1, Term 2, ..., Term Q) describing collection of fuzzy terms of each input variables. The components of the vector takes values 0 or 1. The first 3 components correspond to terms Less, Medium, High of the 1st input variable, the second 3 components to those of 2nd one, the last 3 to those of 3rd one. For example, if Term 1 is 1 then Less term of 1st input is used in partition, otherwise not (in this case all the rules with this term are excluded).

Inconsistency index among rule sets is not desirable in rule-based system. Consequently, it must approach to minimum value. Due to the accuracy-interpretability trade-off relation, the accuracy maximization often leads to the deterioration in the interpretability of fuzzy rule-based systems. This means that fuzzy rule-based system is an accurate and complicated system. When we try to improve the interpretability index for FRBS we must take into account to maximize 3 variables (complexity, coverage and partition). Moreover, following general restrictions for fuzzy number have to consider achieving the good trade-off. The problem is formulated as follows:

$$comp \rightarrow \max, \ \text{cov} \rightarrow \max, \ part_j \rightarrow \max, \ f_{incons} = \sum_{Rule\#=1}^{R} Incons(Rule\#) \rightarrow \min,$$

$$RMSE = \sqrt{\frac{1}{P}\sum_{p=1}^{P}(y_p - \hat{y}_p)^2} \rightarrow \min$$

$$(1)$$

Constraints are given below (in notations below upper index $j = 1, ..., N$ denotes input variable, lower index $i = 1, ..., M$; denotes fuzzy term. For example, a_{2l}^j denotes lower bound of TFN describing the first term (Less) of the 1st input variable (arriving vehicle).

1. Constraints on Fuzzy Terms:

1) Sequence law, i = 1, ..., M

$$a_{i,l}^j \leq a_{(i+1),l}^j, \ l \leq a_{(i+2),l}^j; \ a_{i,m}^j \leq a_{(i+1),m}^j, \ m \leq a_{(i+2),m}^j; \ a_{i,r}^j \leq a_{(i+1),r}^j, \ r \leq a_{(i+2),r}^j$$

$$(2)$$

2) Overlapping law

$$a_{i+1,l}^j \leq a_{i,r}^j$$

$$(3)$$

3) Length of fuzzy number

$$a_{i\,m}^j - a_{i\,l}^j \geq d_{i\,l}^j, \ a_{i\,r}^j - a_{i\,m}^j \geq d_{i\,r}^j$$

$$(4)$$

2. Constraints on Collection of the Rules:

The number of the components of (Rule 1, Rule 2, ..., Rule R) equal to 1 is equal or higher than K (minimum 2/3 of all rules may be used K = 2/3R):

$$Rule\ 1 + Rule\ 2 + ... + Rule\ R \geq K$$

$$(5)$$

3. Constraints on Collection of the Terms:

The number of the components of (Term 1, Term 2, ..., Term Q) equal to 1 is equal or higher than N*S (minimum S terms are used for each input):

$$Term\,1 + Term\,2 + \ldots + Term\,Q \geq N * S \tag{6}$$

The considered optimization problem is a complex problem that includes objective functions (1) and constraints (2)–(6). For solving this problem, we propose using differential evolution method [15] to obtain Pareto-Optimal solutions. Further, the best solution may be found among the Pareto-Optimal ones by applying a multi-attribute decision-making method.

4 Application

Let us consider solving problem (1)–(6) for FB for traffic control with 3 inputs (X_1, X_2, X_3), and output (Y) [16]. The FB consists of the following 27 rules (Table 1).
The membership functions for input (X_1, X_2, X_3), and output Y variables are:
for X_1 *Less* $= (0; 6; 10)$ *Medium* $= (7; 11; 25)$ *High* $= (20; 27; 50)$
for X_2 *Less* $= (0; 7; 10)$ *Medium* $= (7; 19; 25)$ *High* $= (20; 26; 50)$
for X_3 *Less* $= (1000; 1200; 1500)$ *Medium* $= (400; 900; 1200)$ *High* $= (50; 400; 500)$
for Y *Less* $= (0; 7; 10)$ *Medium* $= (8; 24; 30)$ *High* $= (25; 30; 60)$

Table 1. Fuzzy rule-based system [16]

Rules	Inputs			Output
№	X_1	X_2	X_3	Y
1	H	L	L	S
2	H	L	M	M
3	H	L	H	H
4	H	M	L	M
5	H	M	M	M
...	...	...	...	...
23	L	M	M	H
24	L	M	H	M
25	L	H	L	S
26	L	H	M	M
27	L	H	H	H

The optimization problem for this FRBS is formulated as follows:

1. Constraints on Fuzzy Terms

Sequence law, i = 1, ..., 3

$$a_{i,\,l}^{j} \le a_{(i+1),\,l}^{j},\ {}_{l} \le a_{(i+2),\,l}^{j},\ {}_{l};\ a_{i,\,m}^{j} \le a_{(i+1),\,m}^{j},\ {}_{m} \le a_{(i+2),\,m}^{j},\ {}_{m};\ a_{i,\,r}^{j} \le a_{(i+1),\,r}^{j},\ {}_{r} \le a_{(i+2),\,r}^{j},\ {}_{r}$$

Overlapping law

$$a_{2,\,l}^{1} \le a_{1,\,r}^{1};\ a_{3,\,l}^{1} \le a_{2,\,r}^{1};\ a_{2,\,l}^{2} \le a_{1,\,r}^{2};\ a_{3,\,l}^{2} \le a_{2,\,r}^{2};\ a_{2,\,l}^{3} \le a_{1,\,r}^{3};\ a_{3,\,l}^{3} \le a_{2,\,r}^{3}.$$

Length of fuzzy number

$$a_{1\,m}^{1} - a_{1\,l}^{1} \ge 3;\ a_{1\,r}^{1} - a_{1\,m}^{1} \ge 2;\ a_{1\,m}^{2} - a_{1\,l}^{2} \ge 2;\ a_{1\,r}^{2} - a_{1\,m}^{2} \ge 7;\ a_{1\,m}^{3}$$
$$- a_{1\,l}^{3} \ge 4;\ a_{3\,r}^{1} - a_{3\,m}^{1} \ge 10;$$

$$a_{2\,m}^{1} - a_{2\,l}^{1} \ge 3;\ a_{2\,r}^{1} - a_{2\,m}^{1} \ge 1;\ a_{2\,m}^{2} - a_{2\,l}^{2} \ge 6;\ a_{2\,r}^{2} - a_{2\,m}^{2} \ge 3;\ a_{2\,m}^{3}$$
$$- a_{2\,l}^{3} \ge 3;\ a_{2\,r}^{3} - a_{2\,m}^{3} \ge 12;$$

$$a_{3\,m}^{1} - a_{3\,l}^{1} \ge 100;\ a_{3\,r}^{1} - a_{3\,m}^{1} \ge 150;\ a_{3\,m}^{2} - a_{3\,l}^{2} \ge 250;\ a_{3\,r}^{2}$$
$$- a_{3\,m}^{2} \ge 150;\ a_{3\,m}^{3} - a_{3\,l}^{3} \ge 120;\ a_{3\,r}^{3} - a_{3\,m}^{3} \ge 50;$$

2. Constraints on Collection of the Rules:

The number of the components of (Rule 1, Rule 2, ..., Rule 27) equal to 1 is equal or higher than 18 (minimum 2/3 of all rules are used):

$$Rule\ 1 + Rule\ 2 + \ldots + Rule\ 27 \ge 18$$

3. Constraints on Collection of the Terms:

The number of the components of (Term 1, Term 2, ..., Term 9) equal to 1 is equal or higher than 6 (minimum 2 terms are used for each input):

$$Term\ 1 + Term\ 2 + \ldots + Term\ 9 \ge 6$$

As we mentioned in Sect. 3 we will obtain several Pareto-Optimal solutions of the considered problem. To develop 3 criteria (complexity, partition, and coverage) in FRBS the number of rules should be decreased. The Pareto-Optimal solutions which are several FRBS with 22 rules each are shown in Table 2.

Table 2. Multi-criteria values for FRBS.

	RMSE	Complexity	Partition	Coverage	Inconsistency
1st case	8,704	0,045	0,5	0,588	0,0153
2nd case	6,578	0,045	0,5	0,584	0,0178
3rd case	8,824	0,045	0,5	0,592	0,0184
...	...	...	...	...	...
nth case	34,92	0,045	0,5	0,617	0,017

To find the best case among them, we use TOPSIS method. For this purpose the criteria values are normalized (Table 3).

Table 3. Normalized multi-criteria values for FRBS.

	RMSE	Complexity	Partition	Coverage	Inconsistency
1st case	0,231281767	0,5	0,5	0,493795	0,44568547
2nd case	0,17478992	0,5	0,5	0,490436	0,5185099
3rd case	0,234470394	0,5	0,5	0,497154	0,53598776
...	...	...	...	...	...
nth case	0,927890544	0,5	0,5	0,518149	0,49520608

The best case will be found as the case should with the shortest distance (Euclidean distance) from the ideal solution. The ideal solution is given below (Table 4):

Table 4. Values for ideal solution.

	RMSE	Complexity	Partition	Coverage	Inconsistency
Ideal solution	0,17478992	0,5	0,5	0,518149	0,44568547

The best solution is 2nd case as it has the minimal distance to ideal one.

5 Conclusion

In this paper we examined a method for the solution of optimization problem with the respect to 5 criteria of fuzzy rule-based system. Fuzzy TOPSIS method was used in the choosing of the best case for 5 criteria.

References

1. Nauck, D.D.: Measuring interpretability in rule-based classification systems. In: FUZZ-IEEE 2003, St. Louis, Missouri, USA, vol. 1, pp. 196–201 (2003)
2. Guillauma, S., Charnomordic, B.: A new method for inducing a set of interpretable fuzzy partitions and fuzzy inference systems from data. In: Interpretability Issues in Fuzzy Modeling, pp. 148–175. Springer, Berlin (2003)
3. Gacto, M.J., Alcala, R., Herrera, F.: Interpretability of linguistic fuzzy rule-based systems: An overview of interpretability measure. Inform. Sci. **181**, 4340–4360 (2011)
4. Razak, T.R., Garibaldi, J.M., Wagner, C., Pourabdollah, A., Soria, D.: Interpretability indices for hierarchical fuzzy systems. In: Proceedings of IEEE International Fuzzy System Conference (2017)
5. Galende, M., Sainz, G.I., Fuente, M.J.: Accuracy-interpretability trade-off for precise fuzzy modeling using simple indices. Application to Industrial plants. In: Proceeding of the 18th World Congress, Milano, pp. 12656–12661 (2011)
6. Shukla, P.K., Tripathi, S.P.: A review on the interpretability-Accuracy trade-off in evolutionary multi-objective fuzzy systems (EMOFS). Inform. (Switzerland) **3**(4), 256–277 (2012)
7. Ishibuchi, H., Yamamoto, T.: Trade-off between the number of fuzzy rules and their classification performance. In: Accuracy Improvements in Linguistic Fuzzy Modeling. Studies in Fuzziness and Soft Computing, vol. 129, pp. 72–99 (2003)
8. Ishibuchi, H., Nojima, Y.: Analysis of interpretability-accuracy tradeoff of fuzzy systems by multiobjective fuzzy genetics-based machine learning. Int. J. Approx. Reason. **44**, 4–31 (2007)
9. Alonso, J.M., Magdalena, L., Cordon, O.: Embedding HILK in a three-objective evolutionary algorithm with the aim of modeling highly interpretable fuzzy rule-based classifiers. In: 4th International Workshop on Genetic and Evolutionary Fuzzy Systems, Spain, pp. 15–20 (2010)
10. Aliev, R.A., Huseynov, O.H., Adilova, N.E.: Multi-criterial optimization of information granules in fuzzy IF-THEN rules. In: 10th World Conference Intelligent Systems for Industrial Automation, WCIS-2018, Tashkent, Uzbekistan, pp. 52–55 (2018)
11. Adilova, N.E.: Quality criteria of fuzzy If-Then rules. In: ICAFS-2020 (2020)
12. Aliev, R.A.: Fundamentals of the Fuzzy Logic-Based Generalized Theory of Decisions. Springer, Heidelberg (2013)
13. Aliev, R.A., Pedrycz, W.: Fundamentals of a fuzzy-logic-based generalized theory of stability. IEEE Trans. Syst. Man Cyber. Part B (Cybern.) **39**(4), 971–988 (2009)
14. Aliev, R., Tserkovny, A.: Systemic approach to fuzzy logic formalization for approximate reasoning. Inform. Sci. **181**(6), 1045–1059 (2011)
15. Storn, R., Price, K.: Differential evolution – a simple and efficient heuristic for global optimization over continuous spaces. J. Glob. Optim. **11**, 341–359 (1997). https://doi.org/10.1023/A:1008202821328
16. Mohanaselvi, S., Shanpriya, B.: Application of fuzzy logic to control traffic signals. In: AIP Conference Proceedings, vol. 2112, p. 020045 (2019). https://doi.org/10.1063/1.5112230

Evaluation of the Management and Prevention of Covid-19 Pandemic in Most Infected Countries via Fuzzy PROMETHEE Approach

Omid Mirzaei[✉] , Gülsüm Aşıksoy , and Ayse Gunay Kibarer

Department of Biomedical Engineering, Near East University,
via Mersin 10, Nicosia, North Cyprus, Turkey
{omid.mirzaei,gulsum.asiksoy,
aysegunay.kibarer}@neu.edu.tr

Abstract. Nowadays, the whole world is deadly involved with Coronavirus disease. The superpowers are placed at the top of the list of countries affected by this deathly virus. The main objective of this work is to compare totally fifteen countries with the highest number of populations infected with the disease to identify the countries which keep their healthcare superior to their economic aspects. In this research, fuzzy. PROMETHEE method was employed for the evaluation of management and prevention of this deadly disease caused by coronavirus (covid-19) in most infected countries taking into consideration several parameters such as total cases, total deaths, total recovered, active cases, critical and serious cases, total test and population. According to the outcome of the results, Germany was found to appear as the first country in the PRO-METHEE Rainbow which was capable of managing to prevent and control its population against covid-19 infection, caring about healthcare more than the economy or military power, compared to that of other countries under study.

Keywords: Covid-19 · Evaluation · Fuzzy PROMETHEE · Pandemic · Decision making

1 Introduction

During the year 2019, December 19, cases of a serious disease were reported by the World Health Organization (WHO) which mostly resulted in severe pneumonia and death. This infectious disease was first observed in Wuhan, the capital of Hubei, China [1]. The number of cases rose quickly soon afterwards, spreading across China and all over the world [2].

Currently, as the epidemic has developed in many different countries, a huge amount of cases has been recorded in other nations from all continents except for Antarctica. In USA, Italy, and Spain outside China, the velocity of new cases has overcome the rate of the disease in China. WHO named the disease as COVID-19 in February 2020 due to the first reported date (Corona Virus December - 19) [3]. The virus is known to be a severe acute respiratory syndrome coronavirus 2 (SARS-CoV-2), previously described as 2019-nCoV (the novel coronavirus) [4].

R. A. Aliev et al. (Eds.): ICAFS 2020, AISC 1306, pp. 89–94, 2021.
https://doi.org/10.1007/978-3-030-64058-3_11

Governments worldwide are busy working hard to establish counter measures to fight against possible devastating effects. Health organizations are coordinating information flows, giving directives and rules to best reduce the effect of the threat caused by COVID-19. On the other hand, researchers around the world are working tirelessly at the same time. Treatment procedures are enormously progressing together with the information reported for transmission mechanisms. Novel test kits are being prepared with regard to the clinical range of disease for early prevention. Several instabilities still exist with the expectation of the period when the pandemic will reach its peak point [5].

The initial cases of the coronavirus disease were related with direct subjection concerning Huanan Seafood Wholesale Market of Wuhan, suspected to occur from animals to humans. However, following events showed the possibility of that the virus being transmitted from human-to-human as well, the highest spread being due to people with characteristics of that disease. Even the people showing no sign of those symptoms were capable of transmitting COVID-19. Thus, isolation was accepted as the best way to fight this pandemic [5]. Data provided by WHO, reported 3,679,499 cases of COVID-19, including 254,199 deaths. Until now, pandemic exists in 215 Countries. By June 17, 2020, more than 8,256,725 cases have been reported exceeding 445,958 deaths [6].

A promising drug has been proposed by research collaborators for inhibiting the pandemic infecting cells through Spike (S) proteins of the COVID-19 virus which binds to the human angiotensin converting enzyme (ACE2) found in the lungs. The Karolinska Institutet researchers in Sweden and the University of British Columbia (UBC) in Canada, proved that COVID-19 could be prevented from entering human cells by the addition of a genetically modified variant of ACE2 (hrsACE2) which is dependent on the dose affecting the viral growth of the virus reducing it by 1000 to 5000 in cell cultures [7].

Nowadays, most developed and populous countries are affected by this pandemic virus. Most of the countries with the highest number of infected viruses are the world's most developed countries, including the United States, Russia, and France etc.

The aim of this study is to investigate countries with regard to their military and economic power as well, related to controlling their population and health struggle with Covid-19 virus and its resulting fatal effects.

2 Methodology

During the year 1965, fuzzy sets as an elongation of the classic set hypothesis were assigned by open sets launched by Zade [8]. PROMETHEE method was developed in 1985. PROMETHEE may be a straightforward concept, which is far too easier to utilize than any other multi-criteria investigation methods that are right now accessible for implementation. In issues with a limited number of options, PROMETHEE is fabulous for ranking with respect to multiple and complex criteria [9]. PROMETHEE needs two sorts of data: (a) The relative significance (weights) of the considered criteria, and (b) data with respect to the preferences of the decision-makers taking interest in participating with the problem under study [10].

This method is used in so many fields (engineering, education, etc.) for the evaluation of complex systems that have fuzzy parameters and produce effective comparison results. During recent years, researchers have started to employ this method largely in the field of medicine and healthcare as well.

During the year 2018, Maisaini et al. used this method to evaluate lung cancer treatment techniques. The details of this research are as follows: Fuzzy PROMETHEE was utilized to analyze the therapy techniques on the basis of variables like the dose of radiation, treatment cost and period, survival rate, secondary impacts and the expenses of the healing center. Consequently, surgery technique proved to exhibit the best performance compared to other methods for the treatment of lung cancer concerning the selected criteria such as importance and weight [11]. This research may be described as "the state of the art" being related to a healthcare application of fuzzy PROMETHEE approach for the topic under study.

Another research about the health field by using fuzzy PROMETHEE was achieved in 2019 by Özsahin et al., who used this method for the evaluation of treatment techniques in lung cancer. Fuzzy PROMETHEE was utilized to analyze and compare the strategies employed in certain pancreatic cancer treatments involving "surgery", "chemotherapy", "immunotherapy" and many other novel techniques ending up with the surgery method being more superior to others. Treatment and equipment cost, survival probability and outcoming in favorable results faced with, following surgery were investigated. The outcome of this research provides a valuable information for alternative treatments of pancreatic cancer cases [12].

Mghandi and Aroozbahani in 2019, used this method for management of pre-crisis in Tehran. In that research, the Fuzzy PROMETHEE II method was employed for the positioning of Tehran City's s water supply risk management scenarios counting anticipation and preparedness in pre-crisis conditions, with the thought of experts' suppositions in Tehran Area Water and Wastewater Company. The proposed decision-making model can offer assistance choice makers to prioritize drinking water supply scenarios under crisis conditions [13].

As can be seen in Table 1, the importance of parameters was determined with the Fuzzy language scale.

Table 1. The linguistic scale of importance

Linguistic scale for evaluation	Triangular fuzzy scale
Very high (VH)	(0.75, 1, 1)
High (H)	(0.50, 0.75, 1)
Moderate(M)	(0.25, 0.50, 0.75)
Low (L)	(0, 0.25, 0.50)
Very low (VL)	(0, 0, 0.25)

In this application, 15 countries being most infected with the COVID-19 virus were selected by using World meters. Seven criteria were involved such as population, total cases, total deaths, total recovered, active cases, serious critical, and total tests taken from [6]. Table 2 exhibits all the data and the application tools used for this analysis.

Table 2. Infected Countries

Country	Total cases	Total deaths	Total recovered	Active cases	Critical and serious cases	Total test	Population
USA	VH	VH	VH	VH	VH	VH	VH
Brazil	VH	M	VH	H	H	VL	H
Russia	H	L	H	M	M	VH	H
Spain	H	M	NA	NA	L	N	VL
UK	H	H	NA	NA	L	H	L
India	H	L	M	M	H	M	VH
Italy	H	M	H	L	VL	M	L
Peru	M	L	L	M	L	L	VL
Germany	M	L	H	VL	L	M	M
Turkey	M	VL	M	L	L	L	M
Iran	M	L	M	VL	M	L	M
France	M	M	L	L	L	L	L
Chile	L	VL	L	VL	L	VL	VL
Mexico	L	L	L	VL	VL	VL	H
Saudi Arabia	VL	VL	L	VL	L	VL	VL

3 Result and Discussion

According to the results, Germany follows the best route to fight against the coronavirus disease and is ranked as second in Mexico, however, Brazil is in the 15th position following USA indicating the weakness of the country in the policy of significant prevention against this disease. Table 3 shows these results.

Table 3. Ranking of the infected countries

Rank	Action	Phi	Phi+	Phi-
1	Germany	0,3274	0,4940	0,1667
2	Mexico	0,3036	0,5238	0,2202
3	Turkey	0,2202	0,4702	0,2500
4	Saudi Arabia	0,2143	0,4643	0,2500
5	Chile	0,1786	0,4405	0,2619
6	Iran	0,0893	0,4107	0,3214
7	Italy	0,0774	0,4286	0,3512
8	Russia	0,0060	0,3869	0,3810
9	India	−0,1310	0,3155	0,4464
10	Peru	−0,1310	0,2619	0,3929

(continued)

Table 3. (*continued*)

Rank	Action	Phi	Phi+	Phi-
11	France	−0,1369	0,2738	0,4107
12	UK	−0,1429	0,1905	0,3333
13	Spain	−0,1548	0,1548	0,3095
14	USA	−0,3452	0,2857	0,6310
15	Brazil	−0,3750	0,2381	0,6131

The advantages and disadvantages of those parameters which affect the success of the struggle with the virus are shown in Fig. 1.

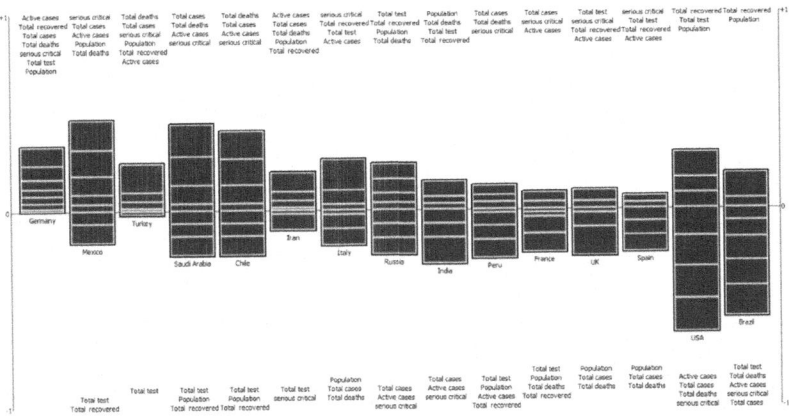

Fig. 1. Evolution results

4 Conclusion

Utilizing Fuzzy PROMETHEE approach, it was possible to end up with trustable decisions obtained from the fuzzy input data. In this study, we showed that countries which are concerned mainly with military, economic, or political power have not paid much attention to healthcare, highly encountering with this pandemic. Consequently, they have not been successful in dealing with prevention, patient recovery and mortality of this deadly virus. This research implies that the proposed method is a powerful approach for the solution of healthcare problems involving difficulty in decision making and may be defined as "the state of the art" in its own field.

References

1. Di Gennaro, F., Pizzol, D., Marotta, C., Antunes, M., Racalbuto, V., Veronese, N., Smith, L.: Coronavirus diseases (COVID-19) current status and future perspectives: a narrative review. Int. J. Environ. Res. Pub. He. (2020). https://doi.org/10.3390/ijerph17082690
2. Ahn, D.G., Shin, H.J., Kim, M.H., Lee, S., Kim, H.S., Myoung, J., Kim, S.J.: Current status of epidemiology, diagnosis, therapeutics, and vaccines for novel coronavirus disease 2019 (COVID-19). J. Microbiol. Biotechn. **30**(3), 313–324 (2020). https://doi.org/10.4014/jmb.2003.03011
3. Özdemir, Ö.: Coronavirus disease 2019 (COVID-19): diagnosis and management (narrative review). Erciyes. Med. J. **42**(3) (2020). https://doi.org/10.14744/etd.2020.99836
4. World Health Organization. https://www.who.int/dg/speeches/detail/who-director-general-s-remarks-at-the-media-briefing-on-2019-ncov-on-11-February-2020. Accessed 02 Nov 2020
5. Cascella, M., Rajnik, M., Cuomo, A., Dulebohn, S. C., Di Napoli, R.: Features, evaluation and treatment coronavirus (COVID-19). In: Statpearls. StatPearls Publishing (2020)
6. Worldometer. https://www.worldometers.info/coronavirus. Accessed 18 Sept 2020
7. Drugtargetreview. https://www.drugtargetreview.com/news/59290/decoy-ace2-receptors-could-be-promising-covid-19-infection-preventing-drug. Accessed 03 Apr 2020
8. Badiru, A.B., Cheung, J.Y.: Fuzzy Engineering expert systems with Neural Network Applications. 1st Edn. 320 p. Wiley-Interscience (2002)
9. Tuzkaya, G., Gulsun, B., Kahraman, C., Ozgen, D.: An integrated fuzzy multi-criteria decision making methodology for material handling equipment selection problem and an application. Expert Syst. Appl. **37**, 2853–2863 (2010). https://doi.org/10.1016/j.eswa.2009.09.004
10. Gul, M., Celik, E., Gumus, A.T., Guneri, A.F.: A fuzzy logic based PROMETHEE method for material selection problems. Beni-Suef Univ. J. Basic Appl. Sci. (2017). https://doi.org/10.1016/j.bjbas.2017.07.002
11. Maisaini, M., Uzun, B., Ozsahin, I., Uzun, D. Evaluating Lung Cancer Treatment Techniques Using Fuzzy PROMETHEE Approach. In International Conference on Theory and Applications of Fuzzy Systems and Soft Computing, ICAFS-2018, Springer, Cham, 209–215 (2018)
12. Ozsahin, I., Ozsahin, D.U., Nyakuwanikwa, K., Simbanegavi, T.W.: Fuzzy PROMETHEE for ranking pancreatic cancer treatment techniques. In: 2019 IEEE Advanced Science, Engineering and Technology Conference, pp. 1–5 (2019). https://doi.org/10.1109/icaset.2019.8714554
13. Ghandi, M., Roozbahani, A.: Risk management of drinking water supply in critical conditions using fuzzy PROMETHEE V technique. Water Resour. Manag. **34**(2), 595–615 (2020). https://doi.org/10.1007/s11269-019-02460-z

Fuzzy Dynamic Programming-Based Multi-stage Transportation Problem

Shamil A. Ahmadov[1,2]([⊠]) [iD]

[1] French-Azerbaijani University, 183 Nizami street, Baku, Azerbaijan
shamilahmadov@yandex.ru
[2] Azerbaijan State Oil and Industry University,
Azadlig 35, Nasimi, Baku, Azerbaijan

Abstract. In this study we are analyzing the fuzzy dynamic programming based multi-stage transportation problem. The purpose of this paper is to evolve a method relating to fuzzy dynamic programming for solving the fuzzy least cost route problem. In this work the costs of transportation are represented as a generalized trapezoidal fuzzy number. Backward recursive equation characterized by fuzziness is used to determine the optimal solution of the given problem. The analyzed approach is explained by a numerical example. The special characteristic of the used method is the uncertainty in the least cost route models, and it is effectively eliminated by fuzzy dynamic programming.

Keywords: Multi-stage transportation problem · Fuzzy dynamic programming · Generalized trapezoidal fuzzy number · Recursive equation · Fuzzy least cost route problem

1 Introduction

Dynamic programming (DP) is a significant optimization procedure using multi-stage decision process. DP as an optimization tool is adapted to operations in which decision-making can be divided into separate stages.

Dynamic programming method was developed by R. Bellman in the 50th year and has found applications in different fields. The method of dynamic programming is based on the principle of optimality formulated by Bellman. As a practical method of optimization, the method of dynamic programming became possible only with the use of modern computer technology [1, 2]. For the first time, this method solved the problems of optimal inventory management, after this the class expanded significantly. Many scientists contributed their research work in the field of fuzzy dynamic programming. The many research papers, the applications in the field of fuzzy dynamic programming are detailed by Aliev, Kacprzyk [3–7].

Multi-stage multi-objective transportation problem under uncertainty environment is discussed in [8]. In this study a two-stage transportation model is evolved with fuzzy numbers. The practical implementation of this model is realized by using Mamdani fuzzy inference and genetic algorithm.

© The Author(s), under exclusive license to Springer Nature Switzerland AG 2021
R. A. Aliev et al. (Eds.): ICAFS 2020, AISC 1306, pp. 95–101, 2021.
https://doi.org/10.1007/978-3-030-64058-3_12

In [9], the author has offered a model as an extension of Bellman and et al. model [10]. Solving a fuzzy optimal subdivision problem using least cost route problem is discussed in [11].

In [12], the authors offered an approach for explaining the fuzzy shortest path problem. They have applied the fuzzy weights and a ranking approach for comparing the fuzzy verge.

Authors of [13] involved an approach based on dynamic programming to determine the path in fuzzy environment. The proposed sorting approach can eliminate the generation of the Pareto paths, due to there exist an extreme number of possible Pareto shortest paths. In given structures, it will be hard for a decision maker to choose a specific path.

In [14] authors modified the Dijkstra's approach to find the solution of shortest path problem with fuzzy data.

Authors of the paper [15] proposed an approach for the solution of fuzzy transportation problems by using triangular fuzzy number.

Obtaining an optimal solution for a multi objective two stage fuzzy transportation problem is discussed in [16, 17].

The basic motivation of this work is to introduce a new type of fuzzy least cost route problem which will be effective for the present real-life problems.

Its specific characteristic can be summarized substantially in three main points: 1) define the problem by using the fuzzy decision variables and fuzzy objective function; 2) set the transportation function as a function of the fuzzy stage variable and fuzzy decision variable: 3) obtaining the optimal decision by using recursive equation.

The structure of the paper is organized as follows. Section 2 gives the preliminaries. A statement of the problem and its solution is given in Sect. 3. Finally, Sect. 4 concludes the paper.

2 Preliminaries

Definition 1 Fuzzy Set [18].

A fuzzy set is a class of objects with a continuum of grades of membership. Such a set is characterized by a membership (characteristic) function which assigns to each object a grade of membership ranging between zero and one.

Definition 2 Fuzzy Forward Recursive Equation [19].

$$\tilde{f}_n^*(s_n) = \underset{x_n}{Min}\{C_{S_n x_n} + \tilde{f}_{n-1}^*(x_n)\}$$

where x_n is the decision variable that determine the destination and s_n is the state variables describes, for instance, a specific city any stage, $C_{S_n x_n}$ is the cost associated with the state variable s_n and x_n for the current nth stage.

Definition 3 Generalized Fuzzy Number [11].

A generalized fuzzy number $\tilde{A} = (a_1, a_2, a_3, a_4; w)$, where w is the height of the fuzzy number, $0 < w \leq 1$ is called a generalized trapezoidal fuzzy number if its membership function is given by

$$\mu_{\tilde{A}}(x) = \begin{cases} 0, & x \leq a_1 \\ w\frac{x - a_1}{a_2 - a_1}, & a_1 \leq x \leq a_2 \\ w\frac{a_4 - x}{a_4 - a_3}, & a_3 \leq x \leq a_4 \\ 0, & x \geq a_4 \end{cases}$$

Definition 4 Ranking Method [11].

Let $\tilde{A} = (m, \alpha, \beta, w)$ then the ranking function of $\tilde{A}$ is defined as $R(\tilde{A}) = w\left[m + \frac{\beta - \alpha}{4}\right]$.

3   Least Cost Route Problem

Suppose an enterprise (company) has to transport its products from city A_1 to city A_n. The value of transportation cost is represented along the lines which relate the nodes. Every node represents every city.

Our aim is to define the optimal (least cost) route connecting the cities A_1 and A_n.. The basic constituent of the least cost route problem are stages, states and decision alternative at each stage. $\tilde{f}_n^*(s_n)$ is the optimal or minimal path when the salesman is in state with n more stages to go for attaining the final stage. For one stage return forward recursive equation is expressed by $\tilde{f}_1 = \tilde{c}_1(s_1, x_1)$ and the optimal value of $\tilde{f}$ under the state variable s_1

$$\tilde{f}_1^*(s_1) = \underset{x_1}{Min}\{\tilde{c}_1(s_1, x_1)\}$$

The range of x_1 is defined by s_1 and s_1 depends on what has happened in second stage. Then return function is represented as

$$\tilde{f}_2(s_2, x_2) = c_{S_2 x_2} x_n + \tilde{f}_1^*(s_1)\}$$

For n stage return forward recursive equation is expressed by

$$\tilde{f}_n^*(s_n) = \underset{x_n}{Min}\{C_{S_n x_n} + \tilde{f}_{n-1}^*(x_n)\}$$

3.1   Numerical Example

Graphical representation of the problem is given in Fig. 1, where $A_1, \ldots, A_{11}$ are cities, the value of transportation cost is represented along the lines which connect the nodes, as $\tilde{C}_1, \tilde{C}_2, \ldots \tilde{C}_{19}$. It is required to define the optimal least cost route connecting the cities A_1 to A_{11}.

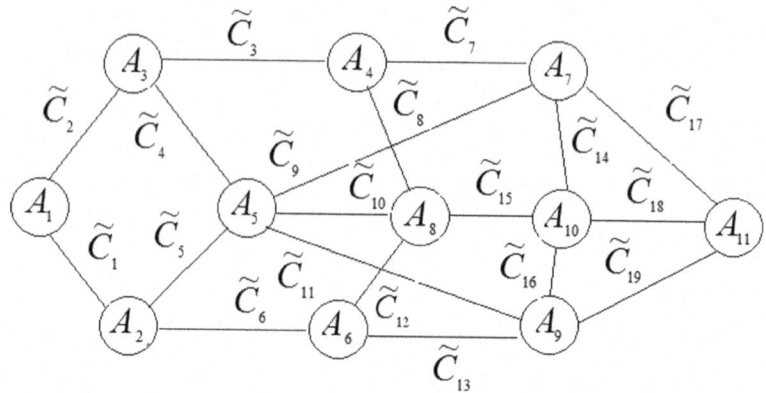

Fig. 1. Fuzzy least cost route problem

Consider the solution process of the given problem. Values of cost variables are defined as

$$\tilde{C}_1 = (4, 5, 6, 7; 0.1); \quad \tilde{C}_2 = (6, 7, 8, 9; 0.3); \quad \tilde{C}_3 = (4, 5, 6, 7; 0.2);$$

$$\tilde{C}_4 = (3, 4, 5, 6; 0.1); \quad \tilde{C}_5 = (2, 3, 4, 5; 0.1); \quad \tilde{C}_6 = (6, 7, 8, 9; 0.3);$$

$$\tilde{C}_7 = (3, 4, 5, 6; 0.1); \quad \tilde{C}_8 = (6, 7, 8, 9; 0.3); \quad \tilde{C}_9 = (5, 6, 7, 8; 0.2);$$

$$\tilde{C}_{10} = (7, 8, 9; 10; 0.3); \quad \tilde{C}_{11} = (5, 6, 7, 8; 0.2); \quad \tilde{C}_{12} = (8, 9, 10, 11; 0.3);$$

$$\tilde{C}_{13} = (10, 11, 12; 13; 0.3); \quad \tilde{C}_{14} = (7, 8, 9, 10; 0.3); \quad \tilde{C}_{15} = (2, 3, 4, 5; 0.1);$$

$$\tilde{C}_{16} = (6, 7, 8, 9; 0.3); \quad \tilde{C}_{17} = (2, 3, 4, 5; 0.1); \quad \tilde{C}_{18} = (1, 2, 3, 4; 0.3);$$

$$\tilde{C}_{19} = (10, 11, 12; 13; 0.2).$$

Solution of the problem depend on 5 stages. The result of each stage will consist of choosing a path out of several alternatives. Starting with node A_{11} at stage $5\tilde{f}_5$ has 3 values $\tilde{C}_{17}$, $\tilde{C}_{18}$, $\tilde{C}_{19}$. In solution process, operations over GtrFN are based on parametric representations of data. Parametric form $(m, \alpha, \beta; w)$ of given data is given below:

$$\tilde{C}_1 = (5.5, 1, 1; 0.1); \quad \tilde{C}_2 = (7.5, 1, 1; 0.3); \quad \tilde{C}_3 = (5.5, 1, 1; 0.2);$$

$$\tilde{C}_4 = (4.51, 1; 0.1); \quad \tilde{C}_5 = (3.5, 1, 1; 0.1); \quad \tilde{C}_6 = (7.5, 1, 1; 0.3);$$

$$\tilde{C}_7 = (4.5, 1, 1; 0.1); \quad \tilde{C}_8 = (7.5, 1, 1; 0.3); \quad \tilde{C}_9 = (6.5, 1, 1; 0.2);$$

$\tilde{C}_{10} = (8.5, 1, 1; 0.3); \quad \tilde{C}_{11} = (6.5, 1, 1; 0.2); \quad \tilde{C}_{12} = (9.5, 1, 1; 0.3);$

$\tilde{C}_{13} = (11.5, 1, 1; 0.3); \quad \tilde{C}_{14} = (8.5, 1, 1; 0.3); \quad \tilde{C}_{15} = (3.5, 1, 1; 0.1);$

$\tilde{C}_{16} = (7.5, 1, 1; 0.3); \quad \tilde{C}_{17} = (3.5, 1, 1; 0.1); \quad \tilde{C}_{18} = (2.5, 1, 1; 0.3);$

$\tilde{C}_{19} = (11.5, 1, 1; 0.2).$

Table 1. Stage 5 $(\tilde{f}_1 = \tilde{c}_1(s_1, x_1))$

s_1	x_1	$\tilde{f}_1(s_1, x_1)$	$\tilde{f}_1^*(s_1)$	x_1
	A_7	(3.5, 1, 1; 0.1)	(3.5, 1, 1; 0.1)	A_{11}
	A_{10}	(2.5, 1, 1; 0.3)	(2.5, 1, 1; 0.3)	A_{11}
	A_9	(11.5, 1, 1; 0.2)	(11.5, 1, 1; 0.2)	A_{11}

Table 2. Stage 4 $(\tilde{f}_2(s_2, x_2) = \tilde{c}_{s_2, x_2} + \tilde{f}_1^*(s_1))$, States s_2

x_2	A_7	A_9	A_{10}	$\tilde{f}_2^*(s_2)$	x_2
A_7			(11, 2, 2; 0.3)	(11, 2, 2; 0.3)	A_{10}
A_8			(6, 2, 2; 0.1)	(6, 2, 2; 0.1)	A_{10}
A_9			(10, 2, 2; 0.3)	(10, 2, 2; 0.3)	A_{10}

Table 3. Stage 3 $(\tilde{f}_3(s_3, x_3) = \tilde{c}_{s_3, x_3} + \tilde{f}_2^*(s_2))$, States s_3

x_3	A_7	A_8	A_9	$\tilde{f}_3^*(s_3)$	x_3
A_4	(14, 3, 3; 0.1)	(13.5, 3, 3; 0.3)		(13.5, 3, 3; 0.3)	A_8
A_5	(17.5, 3, 3; 0.1)	(14.5, 3, 3; 0.1)	(18.5, 3, 3; 0.2)	(14.5, 3, 3; 0.1)	A_8
A_6		(15.5, 3, 3; 0.1)	(21.5, 3, 3; 0.2)	(15.5, 3, 3; 0.1)	A_8

Table 4. Stage 2 $(\tilde{f}_4(s_4, x_4) = \tilde{c}_{s_4, x_4} + \tilde{f}_3^*(s_3))$, States s_4

x_4	A_4	A_5	A_6	$\tilde{f}_4^*(s_4)$	x_4
A_3	(19, 4, 4; 0.2)	(19, 4, 4; 0.1)		(19, 4, 4; 0.1)	A_4
A_2		(18, 4, 4; 0.1)	(23, 4, 4; 0.2)	(18, 4, 4; 0.1)	A_5

Table 5. Stage 1 $(\tilde{f}_5(s_5, x_5) = \tilde{c}_{s_5,x_5} + \tilde{f}_4^*(s_4))$, States s_5

x_5	A_2	A_3	$\tilde{f}_5^*(s_5)$	x_5
A_1	(23.5, 5, 5; 0.1)	(26.5, 5, 5; 0.1)	(23.5, 5, 5; 0.1)	A_2

By using backward path from A_{11} to A_1, in stage 5 the fuzzy costs between node A_{11} and nodes A_7, A_9, A_{10} are defined and obtained optimal fuzzy routs are given in Tables 1, 2, 3, 4, 5.

In the last stage the minimum cost of the route connecting $A_1, \ldots, A_{11}$ is defined as $A_{11} \rightarrow A_{10} \rightarrow A_8 \rightarrow A_5 \rightarrow A_2 \rightarrow A_1$. The minimum cost is obtained as $\tilde{C}_1 + \tilde{C}_5 + \tilde{C}_{10} + \tilde{C}_{15} + \tilde{C}_{18} = (23.5, 5, 5; 0.1)$.

Results of ranking by using equation of the GtrFN (Definition 4), obtained minimum crisp cost is 2.35.

4   Conclusion

This paper introduces a dynamic programming model for fuzzy least cost route problem. The costs of transportation are represented as generalized trapezoidal fuzzy numbers. In this study, the fuzzy backward recursive equations are formed to achieve an optimal solution of the problem. Here we present an implementation of algorithmic tool for solving this problem. The proposed algorithm can be applied to real-life problems in transportation and other fields.

References

1. Aliev, R.A., Mamedova, G., Aliev, R.R.: Fuzzy Set Theory and its Application. Iran, Tabriz (1993)
2. Liao, X., Wang, J., Ma, L.: An algorithmic approach for finding the fuzzy constrained shortest path in a fuzzy graph. Complex Intell. Syst. 1–11 (2020). http://doi.org/10.1007/s40747-020-00143-6
3. Aliev, R.A., Aliev, R.R.: Soft computing and its application. World Scientific, New Jersey, London, Singapore, Hong Kong (2001
4. Kacprzyk, J.: Fuzzy dynamic programming-basic issues. In: Delgado, M., e t al. (eds.) Fuzzy Optimization: Recent Advances, pp. 321–331. Physica, Heidelberg (1994)
5. Kacprzyk, J.: Multistage Fuzzy Control: A Prescriptive Approach. Wiley, New York (1997)
6. Kacprzyk, J., Esogbue, A.O.: Fuzzy dynamic programming: main developments and applications. Fuzzy Sets Syst. **81**(1), 31–45 (1996). https://doi.org/10.1016/0165-0114(95)00239-1
7. Ahmadov, Sh.A., Gardashova L.A.: Fuzzy dynamic programming approach to multistage control of flash evaporator system. In: Advances in Intelligent Systems and Computing, vol. 1095, pp. 101–105. Springer (2019). https://doi.org/10.1007/978-3-030-35249-3_12
8. Mishra, M., Panda, D.: Multi-stage multi-objective transportation problem under uncertainty environment. Int. J. Recent Tec. En. **8**(3), 4056–4060 (2019). https://doi.org/10.35940/ijrt.C5370.098319

9. Baldwin, J.F., Pilswoth, B.W.: Dynamic programming for fuzzy systems with fuzzy environment. J. Math. Anal. Appl. **85**, 1–23 (1982)
10. Bellman, R.E., Zadeh, L.A.: Decision-making in a fuzzy environment. Manag. Sci. **17**(4), 141–164 (1970). https://doi.org/10.1287/mnsc.17.4.B141
11. Nagalakshmi, T., Uthra, G.: An approach of finding an optimal solution for a fuzzy least cost route problem by using generalized trapezoidal fuzzy numbers. Adv. Fuzzy Math. **12**(3), 737–745 (2017)
12. Hernandes, F., Lamata, M.T., Verdegay, J.L., Yamakami, A.: The shortest path problem on networks with fuzzy parameters. Fuzzy Set Syst. **158**(14), 1561–1570 (2007). https://doi.org/10.1016/j.fss.2007.02.022
13. Mahdavi, I., Nourifar, R., Heidarzade, A., Amiri, N.M.: A dynamic programming approach for finding shortest chains in a fuzzy network. Appl. Soft Comput. **9**(2), 503–511 (2009)
14. Deng, Y., Chen, Y., Zhang, Y., Mahadevan, S.: Fuzzy Dijkstra algorithm for shortest path problem under uncertain environment. Appl. Soft Comput. **12**(3), 1231–1237 (2012)
15. Gowthami, R. Prabakaran, K.: Solution of multi objective transportation problem under fuzzy environment. J. Phys. Conf. Ser. 1–11 (2019). https://doi.org/10.1088/1742-6596/1377/1/012038
16. Sudhakar, V.J.: Solving the multi-objective two stage fuzzy transportation problem by zero suffix method. J. Math. Res. **2**(4), 135–140 (2010)
17. Muruganandam, S., Srinivasan, R.: Optimal solution for multi-objective two stage fuzzy transportation problem. Asian J. Res. Soc. Sci. Hum. **6**(5), 744–752 (2016). https://doi.org/10.5958/2249-7315.2016.00149.0
18. Zadeh, L.A.: Fuzzy sets. Inform. Control **8**(3), 338–353 (1965)
19. Mohanaselvi, S., Mondel, S.S.: A fuzzy dynamic programming approach to fuzzy least cost route problem. J. Phys: Conf. Ser. **1377**, 1–8 (2019). https://doi.org/10.1088/1742-6596/1377/1/012042

Comparative Assessment of the Regional Freight Transportation by Method of Fuzzy Linear Regression

Taras Bogachev$^{(\boxtimes)}$ (ID), Tamara Alekseychik (ID),
Anatoly Chuvenkov (ID), and Svetlana Batygova (ID)

Rostov State University of Economics, Bolshaya Sadovaya street 69,
344002 Rostov-on-Don, Russia
bogachev73@yandex.ru, alekseychik48@mail.ru,
chuvenkovaf@mail.ru, batygova@yandex.ru

Abstract. The paper assesses the state of road freight transport in Russia by the method of fuzzy linear regression, depending on three factors: the density of paved public roads, gross regional product per capita and freight tariff indices. The corresponding algorithm is implemented in the Python programming language for regions, each of which is included in one of the six federal regions of Russia: Central, Volga, North-West, South, Ural and North Caucasus, which play an important role in the development of the Russian economy. Fuzzy linear regression coefficients are fuzzy symmetric triangular numbers, for finding which the corresponding linear programming problems were solved. The verification of the constructed models carried out using the control sample confirmed their adequacy. Also, an analysis of factors affecting the studied indicator in each of the selected regions was carried out. The constructed models can be used to predict the indicator of the volume of cargo transportation by road, namely, to build a fuzzy forecast for this indicator, indicating the modal value of the volume of cargo transportation, as well as pessimistic and optimistic forecasts of this indicator. The proposed technique allows you to build not only a fuzzy forecast of the analyzed indicator, but also to rank the factors according to the degree of their influence on this indicator.

Keywords: Fuzzy set theory · Analysis of transportation systems · Fuzzy linear regression model

1 Introduction

The progressive development of the transport industry is a necessary condition for the functioning of any economic system [1]. With the help of transport binds together all the regions, while providing interaction between the economic areas and the various production sectors. Due to high maneuverability, a special role is played by road transport, which carries out mixed transport, servicing industries, agriculture, trade and construction. This type of transport is used for both intraregional and interregional transport of goods and passengers. The contribution of freight transport to the development of various countries, economies is considered in [2–6].

In Russia, due to its geographical features, road transport is developed unevenly, the main roads are concentrated in the European part, and beyond the Urals, due to climatic conditions, the density of roads decreases. More and more attention has been paid to the development of road transport in recent years, but the effectiveness of its use must be considered taking into account many factors: the technical and operational conditions of the relevant sections of roads, the distance of transportation, the type of cargo, its delivery time and mass, the structure of cargo turnover, etc.

The most developed transport system is in the Central, North-Western (except for the European North), Southern, and Volga Federal regions, and the least developed is in the far Eastern and Siberian regions. The availability of paved roads is the most important indicator of transport development in the country, but unfortunately in Russia the quality of the road network is still low, there are local problems in the material and technical condition of the industry and the development of transport infrastructure. All this proves the relevance of the research topic.

For these reasons, the influence of various factors on the development of transport systems in Russia's regions, in particular, the analysis of indicators of road freight transport, is an urgent task in transport research usually use econometric methods [7, 8], methods of operations research, (for example, economic and mathematical models of the distribution of financial resources), optimization (including multi-criteria [9, 10]) models. However, due to the large number of statistical data, often incomplete and distorted, it is advisable to use fuzzy modeling methods [11, 12]. This paper offers an example of applying one of these methods to assess the state of the transport industry in the region.

2 Purpose of the Study

When analyzing various economic processes, the data used usually contains uncertainty. Fuzzy modeling makes it possible to take into account the uncertainty not only in the analyzed data, but also the uncertainty associated with factors not accounted for in the model.

In this regard, to assess the state of road freight transport in Russia, it is proposed to apply the fuzzy modeling method [13–17] for six Federal regions of Russia: Central, Volga, North-Western, Southern, Ural and North-Caucasian, which play an important role in the development of the Russian economy. This method combines the advantages of the fuzzy logic approach and the classical linear optimization. To illustrate the proposed methodology for assessment of road freight transport of Russia in the given work in each of these regions were selected from regions in the Central Federal region–Yaroslavlregion, in the Volga–TatarstanRepublic, in the North-Western –Novgorod region, in the Southern– Rostov region, in the Ural– Tyumen region, in the North Caucasus– Stavropol region.

The following factors were selected for the analysis: the density of paved public roads, the gross regional product per capita, and indices of freight rates, for which the following designations were introduced:

Y – carriage of goods by road transport organizations of all activities (million tons);
X_1 – density of public roads with hard surface (km of roads per 10000 km^2 of territory);
X_2 – gross regional product per capita (thousand rubles);
X_3 – tariff indices for freight traffic (December to December of the previous year, in percent).

3 Description of a Fuzzy Linear Regression Model

The construction of fuzzy linear regression is usually based on the concept of triangular fuzzy numbers A, which are given by a triple of numbers a_m, a and a_M.

$$a_m \leq a \leq a_M \tag{1}$$

Under the condition of strict inequality (1), A has the following membership function:

$$\mu_A(x) = \begin{cases} 0, & x \notin [a_m; a_M] \ ; \\ \dfrac{x - a_m}{a - a_m}, & a_m \leq x \leq a; \\ \dfrac{a_M - x}{a_M - a}, & a \leq x \leq a_M. \end{cases}$$

The number $A = <a_m, a, a_M>$ is called symmetric if the condition $a - a_m = a_M - a = r$ is satisfied, where $r \geq 0$. Then the problem of fuzzy linear regression is formulated as follows. For given k observations of the dependent variable y_j on n factors $x_i = (x_{i1}, x_{i2}, \ldots, x_{ik})^T$, where $i = 1, \ldots, n$, one needs to find fuzzy coefficients $A_0, A_1, \ldots, A_n$ such that for a fuzzy set $Y_j = A_0 + A_1 x_{1j} + \ldots + A_n x_{nj}$ its membership function $\mu_j(Y_j)$ satisfies the condition $\mu_j(Y_j) \geq h$, where h is the given reliability threshold, and the uncertainty associated with these coefficients would be minimal.

Let the model parameters be the numbers $A_i = <a_i - r_i, a_i, a_i + r_i>$, where $a_i \in R$ and $r_i \geq 0$. In this case, Y_j is also a triangular fuzzy number. In accordance with [12, 13, 17] to find a_i and r_i, we get the linear programming problem using the values of the indicators from [18] for 1996–2013:

$$f = 18a_0 + \sum_{j=1}^{18}\sum_{i=1}^{3} r_i x_{ij} \rightarrow min \tag{2}$$

$$\begin{cases} y_j \geq a_0 + \displaystyle\sum_{i=1}^{3} a_i x_{ij} - (1-h)\left(r_0 + \displaystyle\sum_{i=1}^{3} r_i x_{ij}\right), \\ y_j \leq a_0 + \displaystyle\sum_{i=1}^{3} a_i x_{ij} + (1-h)\left(r_0 + \displaystyle\sum_{i=1}^{3} r_i x_{ij}\right), \\ \qquad\qquad r_i \geq 0, \ j = 1..18. \end{cases}$$

Solving this problem for different values of the reliability threshold h of a fuzzy linear model, we obtain the target function in the form of fuzzy triangular numbers $Y = <Y_d, Y_m, Y_u>$, where Y_m– model the value of the Y, Y_d – constraint on the left of the Y, Y_u – constraint on the right of the Y.

The implementation of the solution of problem (2) for each region is carried out in Python. The values of the indicators of the 2014–2017 years are used to check the built model for the quality of the forecast.

4 Application of a Fuzzy Linear Regression Model for the Explore of Goods Carriage by Road Transport in the Region

4.1 Rostov Region

The equation of fuzzy linear regression with $h = 0.4$ has the form:

$$Y = \langle 123.691;\ 142.47;\ 161.249 \rangle - \langle 0.422552;\ 0.2785;\ 0.13452 \rangle X_1 - 0.00014\, X_2 - 0.0223 X_3$$

The fuzzy parameters of the regression model are the free term and the parameter at the indicator X_1, and the parameters at the indicators X_2 and X_3 are not fuzzy numbers, and the indicator X_1 has a greater weight among the factors studied in the regression. It is necessary to note a rather large influence on the degree of fuzziness of the exponent Y of the free term, which may be due, for example, to factors not taken into account in the model or to the uncertainty in the data being analyzed. The parameter at X_2 (gross regional product per capita) is quite small, almost zero, which means there is no effect of this indicator on the volume of freight traffic.

Let us construct a graph of fuzzy linear regression, which for this region reflects a fuzzy forecast of the volume of cargo transported by road (see Fig. 1). Y_m reflects the modal value of the indicator Y, Y_d is the pessimistic forecast of this indicator, Y_u is the optimistic forecast, Y_c is the real value of the indicator Y.

Evaluation of the results of the study of the influence of factors X_1, X_2 and X_3 on the volume of road transport of goods in the studied region was carried out using a control sample for the period 2014–2017. From Table 1 it is seen that the constructed equation of fuzzy linear regression for the Rostov region is correct and can be used for forecasting.

Let us briefly present the solution of problem (2) for the remaining regions considered in the work. Evaluation of the results of a study of the influence of factors X_1,

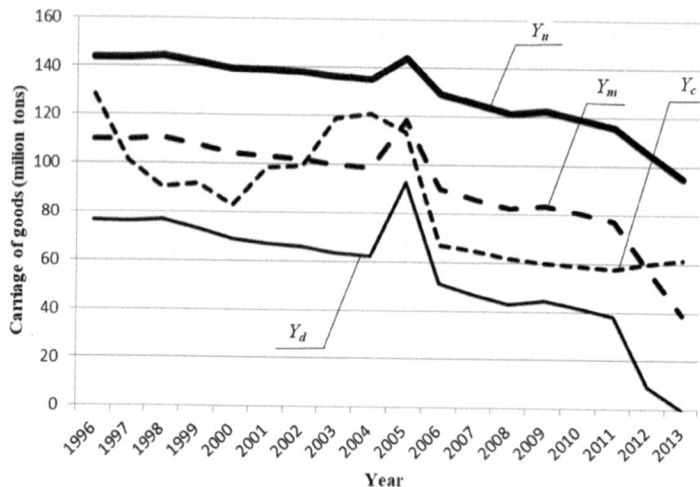

Fig. 1. Fuzzy linear regression graph for indicator of carriage of goods by road transport organizations of all activities (million tons) in the Rostov region, $h = 0.4$

Table 1. Evaluation of results

Year	Y controlling	Left border of fuzziness interval	Right border of fuzziness interval	Indicator Y controlling
2014	60.3	−20.6999	91.76603	+
2015	60.0	−27.0918	85.66225	+
2016	60.4	−30.0337	83.00842	+
2017	54.6	−32.907	80.42321	+

X_2 and X_3 on Y, carried out using the corresponding control samples, also showed the correctness of the constructed models.

4.2   Stavropol Region

The equation of regression with $h = 0.4$ has the form:

$$Y = 60.7637 - 0.007X_1 - 0.00017X_2 + \langle -0.195; -0.026; 0.143 \rangle X_3$$

In this equation, the fuzzy parameter of the model is the parameter with the indicator X_3 (freight tariff index), and it also has a greater weight in the resulting regression compared to other factors. The parameters at X_1, X_2, as well as the free term, are not fuzzy, and the factors X_1, X_2 to a lesser extent affect the indicator Y, especially X_2 (gross regional product per capita).

4.3   Tatarstan Republic

The equation of regression with $h = 0.4$ has the form:

$$Y = 122.18 - 0.147X_1 - \langle 0.002; 0.104; 0.207 \rangle X_3$$

In this equation, the fuzzy parameter of the model is also the parameter with indicator X_3 (freight tariff index), the parameter with indicator X_1 (density of public roads with hard surface) and the free term are not fuzzy values, the parameter with variable X_2 is zero.

4.4   Novgorod Region

The equation of regression with $h = 0.4$ has the form:

$$Y = 43.693 - \langle 0.284; 0.248; 0.213 \rangle X_1 - 0.1328X_3$$

Note that, as in Tatarstan Republic, the parameter with the variable X_2 is zero. But for this region, the parameter with the indicator X_3 is not fuzzy, and the parameter with the indicator X_1 is fuzzy, and it is the indicator X_1 that has the greatest weight in the resulting regression. The fuzzy forecast for Y has more blurred boundaries than, for example, in Tatarstan Republic.

4.5   Yaroslavl Region

The equation of regression with $h = 0.4$ has the form:

$$Y = -6.806 + 0.177X_1 + \langle -0.012; 0.071; 0.155 \rangle X_3$$

For this region, the parameter with the variable X_3 is fuzzy, the parameter with the variable X_1 is not fuzzy, the free term is rather small compared with other regression equations. It should be noted that factor X_1 has a slightly greater weight in the constructed regression than factor X_3. The fuzzy forecast for indicator Y has more blurred boundaries than, for example, in Tatarstan Republic.

4.6   Tyumen Region

The equation of regression with $h = 0.4$ has the form:

$$Y = 92.053 + \langle 12.26; 21.732; 31.203 \rangle X_1 - 0.2278X_3$$

In this region, the fuzzy parameter with the variable X_1 has a rather strong influence on the volume of cargo transportation than the factor X_3, and the parameter with X_2 is zero. The fuzzy forecast for indicator Y has less blurry boundaries than, for example, in the Stavropol region.

5 Conclusion

An assessment of the state of the volume of freight traffic by road transport in the Rostov, Yaroslavl, Novgorod, Tyumen and Stavropol regions, Tatarstan Republic, carried out by the method of fuzzy modeling, depending on the density of public roads with hard surface, gross regional product per capita and the index of tariffs for freight transportation showed that the gross regional product per capita in all the studied regions does not have a significant effect on the volume of freight traffic. The assessment of the results of the study of the influence of these factors on the volume of cargo transportation is also confirmed with the help of the control sample for the period 2014–2017.

It should be noted that only for the Rostov region and the Stavropol Territory, the fuzzy regression equations contain all three factors, and for the Rostov region the fuzzy parameter in the regression equation is the parameter at the density of public roads with hard surface, which has a greater weight among the factors under study. For the Stavropol Territory, the fuzzy parameter in the regression equation is the parameter at the freight rate index and it also has a greater weight compared to other factors.

For the Yaroslavl Region and the Republic of Tatarstan, a fuzzy parameter in the regression equations is the parameter for the freight rates index, but this factor has less weight in the regression equation compared to the density of public roads with hard surface. For the Novgorod and Tyumen regions, the fuzzy parameter in the regression equations is the parameter at the indicator density of public roads with hard surface, and it also has a greater weight in the obtained regression equations.

It should be noted that the constructed models can be used to predict the indicator of the carriage of goods by road transport, namely, to build a fuzzy forecast in the form of an interval of its values, indicating the modal value, as well as pessimistic and optimistic forecasts of this indicator.

The result of the research is the proposed methodology that can be used not only to assess the state of the volume of freight transport by road, but also to analyze any economic system.

Based on the correlation analysis, among the set of economic indicators, factors are selected that affect the economic indicator under study.

- For each studied region, the task of constructing a fuzzy regression using Python is solved.
- The results of the study are evaluated using the control sample for the constructed model, and the correspondence of the actual values of the indicator under study is compared with the interval of its fuzzy forecast.
- Based on the results of the study, strategies, or main directions of development of the economic process under study are developed.

In further studies, it is planned to carry out a more thorough selection of factors affecting the state of road freight transport in the regions of Russia, as well as for further use of fuzzy linear regression, apply the linearization method.

References

1. Tight, M.R., Delle Site, P., Meyer-Ruhle, O.: Decoupling transport from economic growth: towards transport sustainability in Europe. Eur. J. Transp. Infrast. **4**(4), 381–404 (2004)
2. Tapio, P.: Towards a theory of decoupling: degrees of decoupling in the EU and the case of road traffic in Finland between 1970 and 2001. Transp. Policy **12**(2), 137–151 (2005)
3. McKinnon, A.C.: Decoupling of road freight transport and economic growth trends in the UK: an exploratory analysis. Transp. Rev. **27**(1), 37–64 (2007)
4. Alises, A., Vassallo, J.M., Guzman, A.F.: Road freight transport decoupling: a comparative, analysis between the United Kingdom and Spain. Transp. Policy **32**, 186–193 (2014)
5. Engström, R.: The roads' role in the freight transport. Transp. Res. Procedia **142016**, 1443–1452 (2016)
6. Park, M., Hahn, J.-S.: Regional freight demand estimation using korean commodity flow survey data. Transp. Res. Procedia **112015**, 504–514 (2017)
7. Stefanis, V., Botzoris, G.: An aggregate econometric model for the forecast of rail passenger demand. Inform. IT Today **2**(1), 12–19 (2014)
8. Profillidis, V., Botzoris, G.: A comparative analysis of performances of econometric, fuzzy and time-series models for the forecast of transport demand. In: Proceedings of IEEE Int. Fuzzy System Conference, pp. 1–6, London (2007)
9. Kopytov, E., Abramov, D.: Multiple-criteria analysis and choice of transportation alternatives in multimodal freight transport system. Transp. Telecommun. **13**(2), 148–158 (2012)
10. Chislov, O.N., Zadorozhniy, V.M., Bogachev, T.V., Kravets, A.S., Egorova, I.N., Bogachev, V.A.: Time parameters optimization of the export grain traffic in the port railway transport technology system. Adv. Intel. Syst. Comput. **1091**, 126–137 (2020)
11. Teodorovic, D.: Fuzzy sets theory applications in traffic and transportation. Eur. J. Oper. Res. **74**(3), 379–390 (1994)
12. Alekseychik, T., Bogachev, T., Bogachev, V.: Comparative assessment of the transport systems of the regions using fuzzy modeling. Adv. Intel. Syst. Comput. **896**, 651–658 (2019)
13. Tanaka, H., Uejima, S., Asai, K.: Linear regression analysis with fuzzy model. IEEE T. Syst. Man Cybern. **12**(6), 903–907 (1982)
14. Tanaka, H.: Fuzzy data analysis by possibilistic linear models. Fuzzy Sets Syst. **24**(3), 363–375 (1987)
15. Kim, K.J., Moskowitz, H., Koksalan, M.: Fuzzy versus statistical linear regression. Eur. J. Oper. Res. **92**(2), 417–434 (1996)
16. Hojati, M., Bector, C.R., Smimou, K.: A simple method for computation of fuzzy linearregression. Eur. J. Oper. Res. **1**, 172–184 (2005)
17. Bogachev, T., Alekseychik, T., Pushkar, O.: Analysis of indicators of the state of regional freight traffic by method of fuzzy linear regression. Adv. Intel. Syst. Comput. **1095**, 632–638 (2019)
18. The ROSSTAT: Region of Russia. Socio-economic indicators. Statistical compendium. https://gks.ru/bgd/regl/b19_14p/Main.htm. Accessed 21 Apr 2020

Construction of Device for Fuzzy Chaos Signal Generation

K. M. Babanli$^{(\boxtimes)}$ (iD)

Istanbul University Cerrahpasa, 34320 Avcılar, Istanbul, Turkey
kenanbabanli@gmail.com

Abstract. Construction of device for fuzzy chaos signal generation is valuable problem in the field of communication and practical application. Problems of communication can be modeled by different types of differential equations. In this paper is used fuzzy differential equation for modelling the chaotic dynamics of an electronic circuit. Our aim of this study is to design an electronic circuit for generation of chaotic fuzzy signals.

Keywords: Fuzzy number · Fuzzy differential equation · Electronic circuit · Chaotic signals

1 Introduction

Series of work about chaotic secure communications exist in scientific literature.

Chaotic masking secure communication is discussed in [1, 2]. Authors of paper improved masking secure communication scheme. Design and Hardware Implementation of a New Chaotic Secure Communication Technique is presented in [3]. Later adding electronic components the chaotic secure communication circuit was implemented new device [4, 5]. This device structure has implemented by using Lorens system.

Chaotic Duffing oscillators and frequency estimation based scheme is discussed in [6]. Possibility of this scheme is limited by transmission of binary-coded messages.

In scientific literature is discussed the three main types of chaotic secure communication. They are chaotic masking, chaotic shift keying [7] and chaotic modulation [8]. Advantage of the chaotic modulation method is using to hide chaotic signal spectrum information allowing for a wide spectrum of feature and it is sensitive to parameter variation.

In [9] proposed method able to establish secure communication based on an improved Lorenz chaotic optimization circuit. In this case generation fuzzy chaotic signals is important problem. The goal of this paper is to construction of device for fuzzy chaotic signal generation.

This paper has six sections. The preliminaries are given in section second. The third section is discussed statement of the problem. Section 4 describes the solution of the problem and experimental results. Conclusion is proposed in Sect. 5.

R. A. Aliev et al. (Eds.): ICAFS 2020, AISC 1306, pp. 110–117, 2021.
https://doi.org/10.1007/978-3-030-64058-3_14

2 Preliminaries

Definition 1 [1]. A fuzzy number is a fuzzy set A on R which possesses the following properties: a) A is a normal fuzzy set; b) A is a convex fuzzy set; c) α-cut of A, A^α is a closed interval for every $\alpha \in (0, 1]$; d) the support of A, A^{+0} is bounded.

Definition 2. Lorenz System [10]. Chaotic Lorenz system is described by the following differential equations

$$
\begin{aligned}
x_1 &= a(x_2 - x_1), \\
x_2 &= cx_1 - x_1x_3 - x_2, \\
x_3 &= x_1x_2 - bx_3,
\end{aligned}
$$

where a, b, c are the parameters. The behavior of this system under some variables of a, b and c is c
   haotic

Definition 3 Fractal Dimension. Fractal dimension dF coincides with the Lyapunov's dimension dL calculated as:

$$
dL = j + \sum_{i=1}^{j} \lambda_i / |\lambda_{j+1}|
$$

where $\lambda_1 \geq \lambda_2 \geq \ldots \geq \lambda_n$, and value of j is found from the requirements:

$$
\sum_{i=1}^{j} \lambda_i \geq 0, \sum_{i=1}^{j+1} \lambda_i < 0
$$

3 Statement of the Problem

Our aim is to construct of device for fuzzy chaotic signal generation by using Lorenz system. We consider the following FDEs as a model of the considered system:

$$
\begin{aligned}
x_1 &= a(x_2 - x_1), \\
x_2 &= cx_1 - x_1x_3 - x_2, \\
x_3 &= x_1x_2 - bx_3,
\end{aligned}
\tag{1}
$$

where the initial condition is decribed by fuzzy triangle number.

The problem is to find parameters and fuzzy initial conditions such that the solution of (1) will be represented as a fuzzy chaotic signal and construct of device for fuzzy chaotic signal generation.

The structure of electronic circuit Lorenz system is given in Fig. 1.

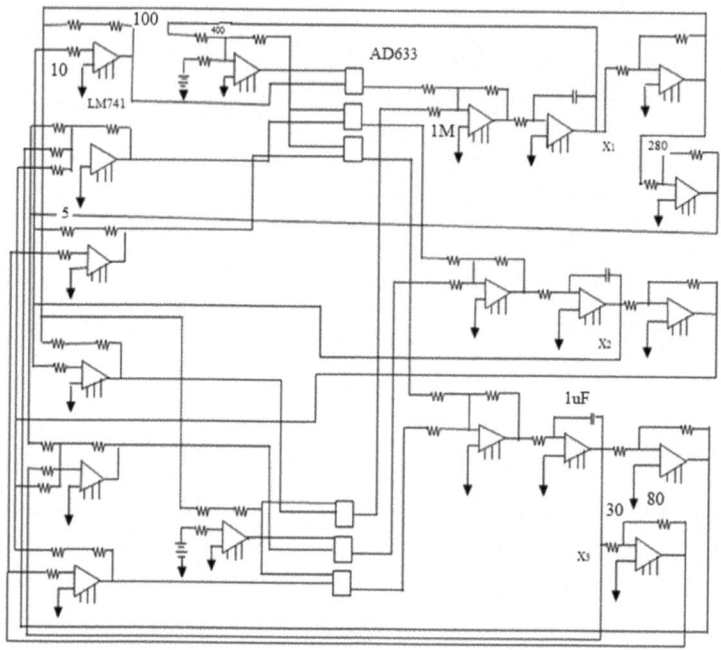

Fig. 1. The structure of electronic circuit Lorenz system (Experimental improvement circuit schematic).

For construction the device is used the following equipments: Op-amp:

LM741–19 pieces, Analog Multiplier: AD633–6 pieces, Resistors: 5K – 4 pieces, 10K – 15 pieces, 30K – 1 piece, 80K – 1 piece, 100K – 21 pieces, 280K – 1 piece, 400K – 2 pieces, 1M – 3 pieces, Capacitors: 1uF (Ceramic) – 9 pieces.

4   Solution of the Problem

4.1   Signal Analysis

Assume that following parameters and initial conditions are used for system (1):

Assume that $a = 10$, $b = 8/3$ and $c = 28$. Initial conditions: $x(0) = -0.1$, $y(0) = 0.2$, $z(0) = 0.30$. With this parameters Lorenz system has a chaotic behavior. For the Lorenz system described by these parameters, the Lyapunov exponents values are as follows.

$LE1 = 0.9022$, $LE2 = 0.0003$, $LE3 = -14.5691$.

When the initial condition is decribed by fuzzy triangle number

$(x10L, x10M, x10R) = (-0.2, -0.1, 0.1)$; $(x20L, x20M, x20R) = (0.1, 0.2, 0.3)$; $(x30L, x30M, x30R) = (0.2, 0.3, 0.4)$

Then $xR0 = [0.10; 0.30; 0.40]$; $x0 = [-0.10; 0.20; 0.30]$; $xL0 = [-0.20; 0.10; 0.20]$.

The components of Lorentz's fuzzy trajectories, considered within the initial conditions is analyzed (Figs. 2, 3, 4, 5).

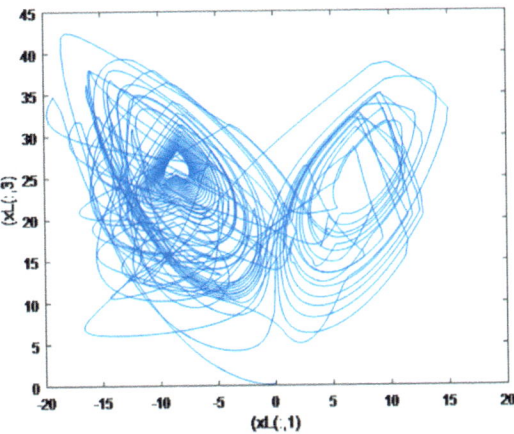

Fig. 2. Phase portrait of left dilation x1L(t) and x3L(t)

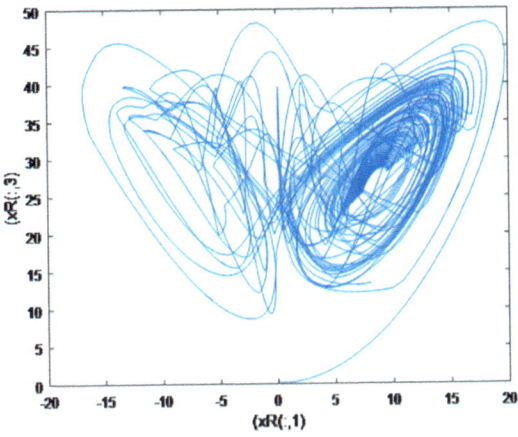

Fig. 3. Phase portrait of left dilation x1R(t) and x3R(t)

Chaotic systems are dynamical systems that are highly sensitive to initial conditions and values of parameters. This sensitivity is popularly known as the butterfly effect. The chaos phenomenon was first observed in weather models by Lorenz [10, 11]. The Lyapunov exponent is a measure of the divergence of phase points that are initially very close and can be used to quantify chaotic systems. A positive maximal Lyapunov exponent and phase space compactness are usually taken as defining conditions for a chaotic system.

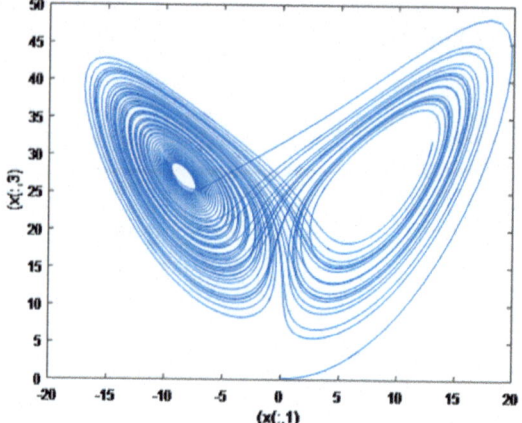

Fig. 4. Phase portrait of left dilation x1(t) and x3(t)

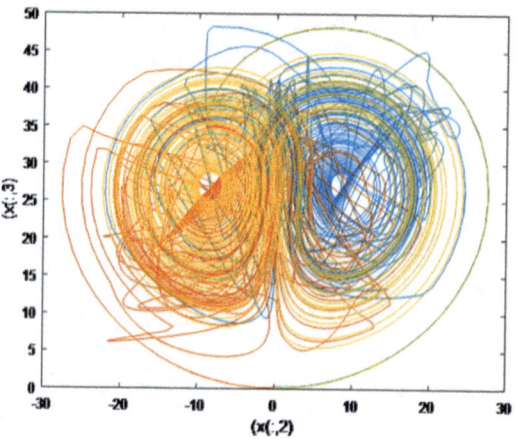

Fig. 5. Phase portrait of fuzzy x2(t) and x3(t)

For special case

$a = 5, b = \frac{8}{3}, c = 28$ Lyapunov exponents values are as follows:

LE1 = −0.082430, LE2 = −0.168676, LE3 = −8.415582.

Note that,

$b = \frac{8}{3}$, $c = 28$ and $a \in (6, 18)$ or $a = 10$, $c = 28$ and $b \in (\frac{1}{55}, \frac{11}{3})$ or when $a = 10$, $b = \frac{8}{3}$ and $c > 24$ the behaviour of the Lorenz system is chaotic.

By using values of parameters-k, λ_1, λ_2, λ_3 is calculated Lyapunov dimension.

In this case using above obtained values of parameters, for example k = 2, $\lambda_1 = 0.9022$; $\lambda_2 = 0.0003$ and $\lambda_3 = -14.5691$ is defined the following results:

$$DL = k + \frac{\lambda_1 + \lambda_2 + \ldots \lambda_k}{|\lambda_k + 1|}, \ where$$

$$k = \max\{i : \lambda 1 + \lambda 2 + \ldots + \lambda i > 0\}$$

$$k = \max\{i : \lambda 1 + \lambda 2 > 0\} = 2$$

$$DL = k + \frac{\lambda_1 + \lambda_2}{|\lambda_3} = 2, + \frac{0.9022 + 0.00031}{1 - 14.5691} = 2.061946$$

4.2   Constuction of Device for Fuzzy Chaos Signal Generation

Using above given equipments has constructed the following device. Experimental circuit board photo is given Fig. 6.

Fig. 6. Experimental circuit board photo

5   Conclusion

To improve security in communication systems, the existing work was analyzed and the use of fuzzy chaotic signals was proved. Lorenz system were used to create an electronic circuit that gives chaotic signals, created a system of differential equations with initial conditions and parameters which solution is based on chaotic behavior. Lorenz system, characterized by fuzzy differential equations and having chaotic behavior based solution, were analyzed in detail. Computer simulation for the analysis of chaotic signals was carried out, Lyapunov's dimension and fractal dimension were determined. Sensitivity analysis was performed and determined the interval values of the system parameters that give chaotic behaviors of electronic circuit. An electronic chip has been created, on basis of suggested research, which generates chaotic signals. Real signals generated constructed device by points is described in Table 1 and comparison modeled and generated signals is given Table 2:

Table 1. Real signals generated constructed device (by points)

Points	Time	Volt(x1)	Time	Volt(x2)	Time	Volt(x3)
1	−3E−08	0,00E+00	−3,00E−08	−1,09E+01	−3,00E−08	−1,18E+01
2	−3E−08	1,98E+01	−2,99E−08	−1,09E+01	−2,99E−08	−1,18E+01
...	...	...	...	...	...	...
200	−1E−08	1,99E+01	−1,01E−08	−1,10E+01	−1,01E−08	−1,18E+01
201	−1E−08	1,99E+01	−1,00E−08	−1,11E+01	−1,00E−08	−1,18E+01
...	...	...	...	...	...	...
598	2,97E−08	1,96E+01	2,97E−08	−1,09E+01	2,97E−08	−1,19E+01
599	2,98E−08	1,96E+01	2,98E−08	−1,09E+01	2,98E−08	−1,19E+01
600	2,99E−08	1,96E+01	2,99E−08	−1,09E+01	2,99E−08	−1,19E+01

Table 2. Comparison modeled and generated signals

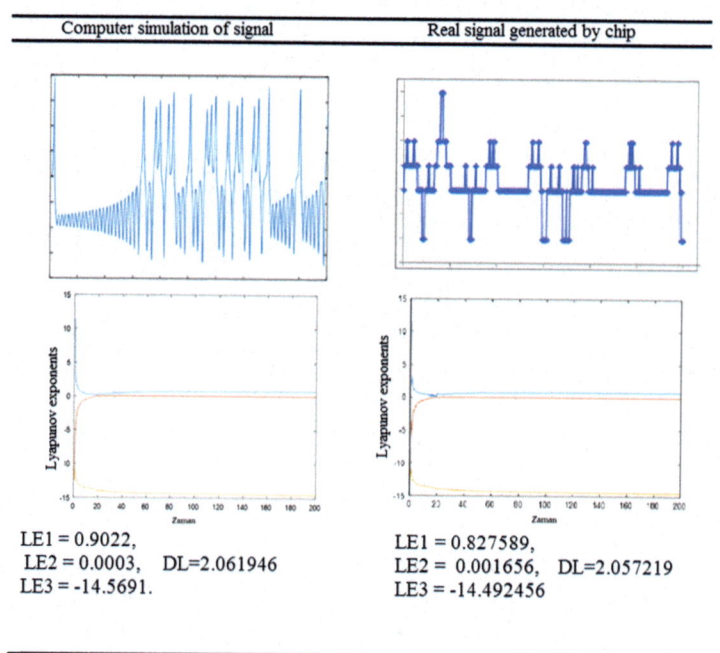

Computer simulation of signal	Real signal generated by chip
LE1 = 0.9022, LE2 = 0.0003, DL=2.061946 LE3 = -14.5691.	LE1 = 0.827589, LE2 = 0.001656, DL=2.057219 LE3 = -14.492456

It is established that the results obtained by computer simulation are close to the real results given by an electronic chip.

References

1. Aliev, R.A., Aliev, R.R.: Soft Computing and its Application. World Scientific, Singapore (2001)
2. Milanović, V., Zaghloul, M.E.: Improved masking algorithm for chaotic communications systems. Electron. Lett. **32**(1), 11–12 (1996)
3. Li, X., Yan-Jun, L., Zhang, Y.-F., Zhang, X.-G., Gupta, P.: Design and hardware implementation of a new chaotic secure communication technique. PLoS ONE **11**(8), 1–19 (2016). https://doi.org/10.1371/journal.pone.0158348
4. Cuomo, K.M., Oppenheim, A.V., Strogatz, S.H.: Synchronization of Lorenz-based chaotic circuits with applications to communications. Circ. Syst. II: Anal. Dig. Signal process. IEEE Trans. **40**(10), 626–633 (1993)
5. Yu-Min, L., Yu-Hong. Z., Jin-Quan Y.: Circuit implementation of secure communication system based on improved chaotic masking algorithm. J. Circ. Syst. **14**(1), 116–118 (2009)
6. Zapateiro, M., Vidal, Y., Acho, L.: A secure communication scheme based on chaotic Duffing oscillators and frequency estimation for the transmission of binary-coded messages. Commun. Nonlinear Sci. Numer. Simul. **19**(4), 991–1003 (2014)
7. Galias, Z., Maggio, G.M.: Quadrature chaos-shift keying: theory and performance analysis. Circ. Syst. I: Fund. Theory Appl. IEEE Trans. **48**(12), 1510–1519 (2001)
8. Li, S., Alvarez, G., Chen, G.: Breaking a chaos-based secure communication scheme designed by an improved modulation method. Chaos Solitons Fractals **25**(1), 109–120 (2005)
9. Filali, R.L., Benrejeb, M., Borne, P.: On observer-based secure communication design using discrete-time hyperchaotic systems. Commun. Nonlinear Sci. Numer. Simul. **19**(5), 1424–1432 (2014)
10. Lorenz, E.N.: Deterministic non-periodic flows. J. Atmos. Sci. **20**(2), 130–141 (1963)
11. Li, S.-Y., Yang, C.-H., Lin, C.-T., Ko, L.-W., Chiu, T.-T.: Chaotic motions in real fuzzy electronic circuits. Abstr. Appl. Anal. **2013**, 1–14 (2013)

Monadic Automata with Relational Morphisms

Jiří Močkoř[(✉)]

Centre of Excellence IT4Innovations, Institute for Research and Applications of Fuzzy
Modeling, University of Ostrava, 30. dubna 22, 701 03 Ostrava 1, Czech Republic
jiri.mockor@osu.cz

Abstract. Categories of monadic automata and monadic automata
with inpu-output objects are introduced where morphisms between
automata are relations defined by monads. These categories with relational morphisms include most known automata, including deterministic, nondeterministic, and fuzzy automata, where morphisms between
automata are relations between sets or fuzzy relations. Relationships
between relational morphisms of input-output monadic automata and
languages accepted by these automata are investigated.

1 Introduction

The notion of a fuzzy automaton and a language accepted by this automaton
were investigated by many authors and many special examples of these notions
contain well-known types of fuzzy or classical automata ([1,2,8–10,13], for example). During the development of this theory, the theory of categories was actively
used, among other things, which made it possible to generalize the description
of many constructions in this theory which enabled the further development of
this theory [2,9,16].

A common feature of all these results was that the relationships between individual automata and languages are described using various *mappings* between
sets. However, current developments in fuzzy set theory are increasingly focused
on the use of *fuzzy relations* instead of classical mappings. We therefore consider
it appropriate to apply this trend in the theory of monadic automata and languages and thus extend this theory in accordance with current developments in
the field of fuzzy structures.

2 Methods

The principal tool in this paper is a monad in a category. Application of monadic
tools in automata theory was for the first time used in [1,2] and elaborated in
many other papers, including our last paper [12]. In the present paper we want
to extend the monadic automata theory and to define categories of monadic
automata with monadic relations as morphisms. We use the so-called partially
ordered monads, which were introduced and elaborated in [3–6,14]. The following
definition of this structure is modified from definitions in [6] and [15].

R. A. Aliev et al. (Eds.): ICAFS 2020, AISC 1306, pp. 118–125, 2021.
https://doi.org/10.1007/978-3-030-64058-3_15

Definition 1. *Let* **K** *be a category with a forgetful functor to the category* **Set** *and with a monad* $\mathcal{F} = (F, \Diamond, \xi)$. *Then* $\mathcal{F}$ *is called a partially pre-ordered monad in* **K**, *if*

1. $F : \mathbf{K} \to \mathbf{K}$ *is a mapping defined on objects of* **K**,
2. $\xi = \{\xi_X | \xi_X : X \to F(X) \text{ is a } \mathbf{K} - morphism, X \in \mathbf{K}\}$,
3. *If* $f : X \to F(Y)$ *and* $g : Y \to F(Z)$ *are* **K**-*morphisms, then* $g \Diamond f : X \to F(Z)$ *is a* **K**-*morphism, called Kleisli composition of* f *and* g *and this composition is associative,*
4. *If* $f : X \to F(Y)$ *is a* **K**-*morphism, then* $\xi_Y \Diamond f = f$,
5. *If* $f : X \to Y$ *and* $g : Y \to F(Z)$ *are* **K**-*morphisms, then* $g \Diamond (\eta_Y . f) = g.f$.
6. *For arbitrary* **K**-*object* X, *on the set* $F(X)$ *is defined a partially order relation* $\leq_{F(X)}$,
7. *For arbitrary* **K**-*objects* X, Y, *on the set* $Mor_\mathbf{K}(X, F(Y))$ *of* **K**-*morphisms from* X *to* $F(Y)$, *a partially pre-order relation* $\preceq_{X,Y}$ *is defined such that for arbitrary* **K**-*morphisms* $h, k : X \to F(Y)$,

$$h \preceq_{X,Y} k \Leftrightarrow \forall x \in X, \quad h(x) \leq_{F(Y)} k(x).$$

If the domain X *and the codomain* $F(Y)$ *of the morfisms* h *and* k *are obvious, for simplicity, instead of* $\preceq_{X,Y}$ *we use only* $\preceq$.
8. *If* $f, f' : X \to F(Y), g, g' : Y \to F(Z)$ *are* **K**-*morphisms, the following implications hold*

$$g \preceq g' \Rightarrow g \Diamond f \preceq g' \Diamond f,$$
$$f \preceq f' \Rightarrow g \Diamond f \preceq g \Diamond f'.$$

If the forgetful functor is faithful, then $\preceq$ is an order relation and $\mathcal{F}$ will be called a *partially ordered monad*.

A typical examples of partially ordered monads are the following:

Example 1. The partially ordered monad monad $\mathcal{P} = (P, \Diamond, \eta)$ in **Set** is such that

1. For $X \in \mathbf{Set}$, $P(X) = (2^X, \subseteq)$,
2. $\eta_X : X \to P(X)$ is such that $\eta_X(x) = \{x\}$, $x \in X$,
3. For each $f : X \to P(Y), g : Y \to P(Z)$ we have

$$x \in X, \quad (g \Diamond f)(x) = \bigcup_{y \in f(x)} g(y).$$

Example 2. Let $L = (L, \leq, \otimes)$ be a quantale. The partially ordered monad $\mathcal{Z} = (Z, \Diamond, \chi)$ in **Set** is such that

1. $Z : \mathbf{Set} \to \mathbf{Set}$, $X \in \mathbf{Set}$, $Z(X) = (L^X, \leq)$, where $\leq$ is a point-wise ordering,
2. For $X \in \mathbf{Set}$, $\chi_X : X \to L^X$, $\chi_X(x)(y) = 1_L$, if $x = y$ and $\chi_X(x)(y) = 0_L$, otherwise, for arbitrary $x, y \in X$,
3. For $f : X \to L^Y$, $g : Y \to L^Z$ in **Set**,

$$x \in X, z \in Z, \quad (g \Diamond f)(x)(z) = \bigvee_{y \in Y} f(x)(y) \otimes g(y)(z).$$

3   Results: Monadic Automata with Relational Morphisms

In the theory of automata monads in categories were for the first time used in [1]. In this paper, we focus on categories of monadic automata, where the morphisms between automata are monadic relations, generalizing morphisms in given categories. These results significantly expand the existing results on automata, presented e.g. in [12]. The proofs and more details will be published elsewhere.

Monadic relations were defined by E. G. Manes in [11] in the following way.

Definition 2. *For a monad $\mathcal{F} = (F, \Diamond, \xi)$ in a category $\mathbf{K}$ and $\mathbf{K}$-objects X, Y, a $\mathcal{F}$-relation $R : X \rightsquigarrow Y$ is a $\mathbf{K}$-morphism $R : X \rightarrow F(Y)$. The composition of $\mathcal{F}$-relations $R : X \rightsquigarrow Y$ and $S : Y \rightsquigarrow Z$ is defined by $S \Diamond R : X \rightsquigarrow Z$.*

Using the notion of a monadic relation we can introduce the motion of a category of monadic automata with monadic relations as morphisms.

Definition 3. *For a partially pre-ordered monad $\mathcal{F} = (F, \Diamond, \xi)$ in $\mathbf{K}$, $\mathbf{RAut_K}[\mathcal{F}]$ is the category such that*

1. *Objects of $\mathbf{RAut_K}[\mathcal{F}]$ are $\mathcal{F}$-automata $(U, (M, *), d)$ in $\mathbf{K}$, where*
   (a) *$U \in \mathbf{K}$,*
   (b) *$(M, *)$ is a monoid,*
   (c) *$d : (M, *) \rightarrow (Mor_{\mathbf{K}}(U, F(U)), \Diamond)$ is a monoid homomorphism,*
2. *Relational morphisms $(g, \beta) : (U, (M, *), d) \dashrightarrow (V, (N, \times), h)$ are such that*
   (a) *g is a $\mathcal{F}$-relation $U \rightsquigarrow V$,*
   (b) *$\beta \subseteq M \times N$ is such that $(m, n) \in \beta, (m', n') \in \beta \Rightarrow (m * m', n \times n') \in \beta$ holds (it is abbreviated by $\beta : (M, *) \multimap (N, \times)$),*
   (c) *for each $m \in M, n \in N$, $(m, n) \in \beta$, we have*

$$h(n) \Diamond g \succeq g \Diamond d(m),$$

   where $\succeq = \succeq_{U, F(V)}$ is defined in Definition 2.
3. *Composition $\odot$ of $\mathcal{F}$-morphisms $(g, \beta) : (U, (M, \star_M), d) \dashrightarrow (V, (N, \star_N), h)$ and $(k, \gamma) : (V, (N, \star_N), h) \dashrightarrow (W, (O, \star_O), t)$ is defined by*

$$(k, \gamma) \odot (g, \beta) = (k \Diamond g, \gamma \circ \beta) : (U, (M, \star_M), d) \dashrightarrow (W, (O, \star_O), t),$$

   where $\gamma \circ \beta$ is a standard composition of binary relations.

Morphisms in $\mathbf{RAut_K}[\mathcal{F}]$ are called $\mathcal{F}$-morphisms or relational morphisms only.

Categories of non-deterministic, deterministic or fuzzy automata are frequently used, where morphisms are defined by mappings. In the following examples we introduce relational versions of these categories, using explicit definitions of relational morphisms between corresponding types of automata.

Example 3. The relational category $\mathbf{RDA}$ of deterministic automata is defined by

1. Object are deterministic automata $(U, (M, *_M), d)$, where U is a set, $(M, *_M)$ is a monoid and $d : U \times M \to U$ satisfies

$$s \in U, m, n \in M, \quad d(s, m *_M n) = d(d(s, m), n), \quad d(s, 1_M) = s.$$

2. A pair $(g, \beta) : (U, (M, \star_M), d) \to (V, (N, \star_N), h)$ is a relational morphism in **RDA** if
   (a) $g \subseteq U \times V$ is a relation,
   (b) $\beta : (M, *_M) \multimap (N, *_N)$,
   (c) For arbitrary $s \in U, m \in M$,

$$(m, n) \in \beta, (s, r) \in g \Rightarrow g^{\to}(d(s, m)) \subseteq \{h(r, n)\}$$

holds, where $g^{\to}(x) = \{t \in R : (x, t) \in g\}$.

Analogously can be introduced the relational version of a category of non-deterministic automata.

Example 4. The relational category **RNDA** of non-deterministic automata is defined by

1. Objects are non-deterministic automata $(U, (M, *_M), d)$, where U is a set, $(M, *_M)$ is a monoid and $d : U \times M \to P(U)$ is such that
   (a) $d(s, 1_M) = \{s\}$,
   (b) $d(s, m * n) = \bigcup_{t \in d(s,m)} d(t, n) \subseteq U$.
2. $(g, \beta) : (U, M, d) \to (V, N, h)$ is a relational morphism in **RNDA**, if
   (a) $g \subseteq U \times V$ is a relation, such that for arbitrary $s \in U$ there exists $r \in V$ such that $(s, r) \in g$,
   (b) $\beta : (M, *) \multimap (N, \times)$,
   (c) For arbitrary $s \in U, m \in M$ the following implication holds:

$$(s, r) \in g, (m, n) \in \beta \Longrightarrow g^{\Rightarrow}(d(s, m)) \subseteq h(r, n),$$

where $g^{\Rightarrow} : P(U) \to P(V)$ is defined by $g^{\Rightarrow}(X) = \{r \in V : \exists a \in X, (a, r) \in g\}$.

We introduce the notion of a relational version of a category of L-valued fuzzy automata, where $L = (L, \leq, \otimes)$ is a quantale.

Example 5. The relational category **RFuz** of fuzzy automata is defined by

1. Object are L-valued automata $(U, (M, *_M), d)$, where U is a set, $(M, *_M)$ is a monoid and $d : U \times M \times U \to L$ satisfies

$$d(s, 1_M, r) = \begin{cases} 1_L, & s = r \\ 0_L, & \text{otherwise} \end{cases}, \quad d(s, m *_M n, t) = \bigvee_{x \in U} d(s, m, x) \otimes d(x, n, t).$$

2. $(g, \beta) : (U, M, d) \to (V, N, h)$ is a relational morphism in **RFuz** if
   (a) $g : U \times V \to L$ is an L-valued fuzzy relation such that for arbitrary $s \in U$ there exists $r \in V$ such that $g(s, r) = 1_L$,

(b) $\beta : (M, *) \multimap (N, \times)$,

(c) For arbitrary $s \in U, r \in V$ and $(m, n) \in \beta$,

$$g^{\rightarrow}(d(s, m, -)) \le h(r, n, -) \otimes g(s, r),$$

holds, where $g^{\rightarrow} : L^U \to L^V$ is defined by

$$u \in L^U, r \in V, \quad g^{\rightarrow}(u)(r) = \bigvee_{x \in U} u(x) \otimes g(x, t).$$

We show an example of a relational morphism between automata from the category $\mathbf{RAut_K}[\mathcal{F}]$.

Example 6. Let $\mathcal{Z} = (Z, \Diamond, \chi)$ be the monad from Example 2 and let $\mathcal{A} = (S, (M, \star), d)$ be an object in the category $\mathbf{RAut_{Set}}[\mathcal{Z}]$, such that the transition $\mathcal{Z}$-relation $d(m) : S \to Z(S)$ is extensional with respect to a given fuzzy equivalence relation $q : S \times S \to L$, i.e, for arbitrary $m \in M$,

$$\forall s, t, t' \in S, \quad d(m)(s)(t) \otimes q(t, t') \le d(m)(s)(t').$$

Then

$$(\overline{q}, 1_M) : \mathcal{A} \dashrightarrow \mathcal{A}$$

is a relational morphism where $\overline{q} : S \to Z(S)$ is defined by $\overline{q}(s)(s') = q(s, s') \in L$. It follows from the inequality

$$\overline{q} \Diamond d(m)(s)(t) = \bigvee_{x \in S} d(m)(s)(x) \otimes \overline{q}(x)(t) \le d(m)(s)(t) \le$$

$$\bigvee_{x \in S} \overline{q}(s)(x) \otimes d(m)(x)(t) = d(m) \Diamond \overline{q}(s, t).$$

Although automata and morphisms between these automata from categories presented in Examples 3–5 differ greatly from one example to another, in reality they are only special examples of automata and morphisms from a category $\mathbf{RAut_K}[\mathcal{F}]$ for appropriate monad $\mathcal{F}$ in $\mathbf{K}$. In fact, the following theorem holds.

Theorem 1. *The following diagram of embedding functors commutes*

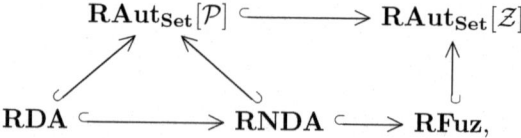

where $\mathcal{P}, \mathcal{Z}$ are monads from Examples 2.1 and 2.2, respectively.

Since its inception, the theory of automata has been closely linked to the theory of languages, especially in the field of acceptance of a given language by an automaton. In connection with the use of category theory in the theory of automata, this relationship between languages and automata has extended to the field of automata defined by categories. Our goal is to extend this relationship to categories of monadic automata, where morphisms between automata are monadic relations. This significantly generalizes the previous results describing the relationships between automata and languages.

Definition 4. *Let* $\mathcal{F} = (F, \Diamond, \xi)$ *be a partially pre-ordered monad in* **K**. *Let* $J(H, respectively)$ *be an* **K**-*object, called an input (output) object, respectively.*

1. $(U, (M, \star), d, i, q)$ *is an input-output* $(J, H, \mathcal{F})$-*automaton in* **K**, *if* $(U, (M, \star), d)$ *is an* **RAut$_K$**$[\mathcal{F}]$-*object and* $i : J \to U$ *and* $q : U \to F(H)$ *are* **K**-*morphisms,*
2. $(f, \alpha) : (U, (M, \star), d, i, q) \dashrightarrow (V, (N, \circledast), h, j, p)$ *is a relational morphism of* $(J, H, \mathcal{F})$-*automata if* (f, α) *is* **RAut$_K$**$[\mathcal{F}]$-*morphism of automata and*

$$\xi_V \cdot j \succeq f.i, \quad p \Diamond f \succeq q.$$

The relationship between the input-output automaton in the category and the language defined above the input alphabet M of this automaton, which is accepted by this automaton was first specified in [7]. In order for such a language $\mathcal{L}$ to be accepted by an automaton with input J and output H, $\mathcal{L}$ must be defined as a mapping $M \to Mor_K(J, H)$. In [12] we extended this definition to monadic automata with input and output objects as follows.

Definition 5. *Let* $\mathcal{F} = (F, \Diamond, \xi)$ *be a partially pre-ordered monad in* **K** *and let* $\mathcal{R} = (R, (N, \star), d, i, q)$ *be an input-output* $(J, H, \mathcal{F})$-*automaton in* **K**.

1. $(J, H, \mathcal{F})$-*language defined over* N *is a mapping* $\mathcal{L} : N \to Mor_K(J, F(H))$.
2. $\mathcal{L}$ *is accepted by* $\mathcal{R}$, *if for all* $n \in N$, $\mathcal{L}(n) = (q \Diamond d(n)).i$.

The following relationship between relational morphisms of input-output $\mathcal{F}$-automata where $\mathcal{F}$ is a partially pre-ordered monad in **K** and languages accepted by these automata represents a generalization of one of principal result between automata and languages.

Theorem 2. $\mathcal{F} = (F, \Diamond, \xi)$ *be a partially pre-ordered monad in* **K** *and let* $\mathcal{R} = (U, (A, \star), d, i, q)$ *and* $\mathcal{S} = (V, (B, \times), h, j, p)$ *be input-output* $(J, H, \mathcal{F})$-*automata in* **K**. *Let* $\mathcal{L}_\mathcal{R}$ *and* $\mathcal{L}_\mathcal{S}$ *be languages defined over* A *and* B, *respectively, accepted by these automata. If* $(g, \beta) : \mathcal{R} \dashrightarrow \mathcal{S}$ *is a relational morphism from Definition 3.3, for arbitrary* $(a, b) \in \beta$ *we have*

$$\mathcal{L}_\mathcal{R}(a) \preceq \mathcal{L}_\mathcal{S}(b).$$

4   Conclusions

Our goal was to continue building the theory of automata in general categories, where the structures of automata are based on monads in these categories. This general approach has great potential for further development of the theory of automata, because, among other things, it allows to generate a wide range of examples of such automata. This is due, among other things, to the fact that monads in categories can be constructed from any pair of adjoint functors between two categories, which creates a virtually unlimited supply of examples. Examples of this type include classic powerset monads for sets or fuzzy sets, or more general, monads of various filter structures. For each monad in a category it is then possible to define an automaton whose states are represented by an object in this category.

Instead of standard mappings as morphisms we choose monadic relations in categories of automata defined by partially pre-ordered monads which significantly generalizes the existing results for automata in categories. We also extended these results to identify relationships between partially pre-ordered monads couples and relational morphisms of languages accepted by these monads automata. The obtained results will be possible to use, for example, in defining completely new types of automata, based on previously unused partially pre-ordered monads, including the possibility to examine the relationships between these new types of automata and the languages accepted by these automata.

Funding. This work was partly supported from ERDF/ESF project CZ.02.1.01/0.0/0.0/17-049/0008414.

References

1. Arbib, M., Manes, E.: Fuzzy morphisms in automata theory. Lect. Notes Comp. Sci. **25**, 80–86 (1975)
2. Arbib, M., Manes, E.: Fuzzy machines in a category. Bull. Austra. Math. Soc. **13**, 169–210 (1975)
3. Eklund, P., Galán, M.A., Monads, P.O., Sets, R., Peters J.F., Skowron A. (eds) Transactions on Rough Sets VIII. LNCS, vol 5084. Springer, Heidelberg (2008)
4. Gähler, W.: General topology - the monadic case, examples, applications. Acta Math. Hungar. **88**, 279–290 (2000)
5. Gähler, W.: Extension structures and completions in topology and algebra. Seminarberichte aus dem Fachbereich Mathematik, Band 70, FernUniversität, Hagen (2001)
6. Höhle, U.: Partially Ordered Monads. In: Many Valued Topology and Its Applications. Kluwer Academic Publishers, Dordrecht (2001)
7. Kawahara, Y.: Automata in categories and regular languages, memoirs of the faculty of science. Kyushu Univ. Ser. A **38**, 47–59 (1984)
8. Li, Z.H., Li, Y.: The relationships among several types of fuzzy automata. Inform. Sci. **176**(15), 2208–2226 (2006)
9. Li, Y.: A categorical approach to lattice-valued fuzzy automata. Fuzzy Sets Syst. **157**, 855–864 (2006)
10. Li, Y., Pedrycz, W.: The equivalence between fuzzy Mealy and fuzzy Moore machines. Soft. Comput. **10**, 953–959 (2006)
11. Manes, E.G.: Algebraic Theories. Springer, Berlin (1976)
12. Močkoř, J.: Monads and a common framework for fuzzy type automata. Int. J. Gen. Syst. **48**(4), 406–442 (2019)
13. Mordeson, J., Malik, D.S.: Fuzzy Automata and Languages: Theory and Applications. Chapman and Hall/CRC, New York (2002)
14. Rodabaugh, S.E.: Relationship of algebraic theories to powerset theories and fuzzy topological theories for lattice-valued mathematics. Int. J. Math. Math. Sci. **2007**, 1–71 (2007)

15. Gavin, S.: On the monadic nature of categories of ordered sets. Cahiers de Topologie et Géométrie Différentielle Catégoriques **52**, 163–187 (2011)
16. Xing, H., Qiu, D.: Automata theory based on complete residuated lattice-valued logic: a categorical approach. Fuzzy Sets Syst. **160**(16), 2416–2428 (2009)

Dental Disease Detection Using Fuzzy Logic

L. A. Gardashova[1]($\boxtimes$)(iD) and B. F. Aliyev[2](iD)

[1] Azerbaijan State Oil and Industry University, Azadliq Avenue 20, Baku, Azerbaijan
l.qardashova@asoiu.edu.az
[2] Azerbaijan Medical University, Samad Vurgun 167, Baku, Azerbaijan
dr.bahadur1234@mail.ru

Abstract. Diagnosis of dental diseases is a labor-intensive process for dentists with inaccurate information. Diagnostic depends on uncertainty information and using artificial intelligence is required by dentists. Dental diseases detection needs expert opinion and information in order to improve their perception. Accurate detection can define an approach which capable of processing observed data. This work offers a method based on fuzzy logic and neural network. It can operate with any type of data such as imprecise, inaccurate or noisy information and Fuzzy Logic Systems is understandable. Reasoning by Fuzzy logic is quite simple and it provides efficient solution to difficult problems in all fields of life, such as stomatology.

Keywords: Dental disease · Fuzzy logic · Neural network · Plaque · Inflamed gums · Gingivitis disease · Pulpits disease · Membership function · Neuro-fuzzy inference system

1 Introduction

The high intensity of dental caries and the widespread of pulpitis and periodontitis among the population, errors in the diagnosis of these diseases, post-treatment complications, the necessity of a new approach to the diagnosis, treatment and prevention of major dental diseases are indicated. Improving the effectiveness of diagnosis and treatment of caries, pulpitis and periodontitis are basic problems of dental treatment [1, 2].

Currently, there are several types of dental diseases. Dental disease can be classified, such as periodontitis, pulpitis, genguitis and progressive periodontitis. These diseases have similar symptoms [3] such as plaque, pain, inflamed gums, bleeding gums, red gums, breath odor, swollen gums, wobbly teeth. But the difference between these diseases lies in the severity of each symptom.

Pulpitis is one of the inflamed disease that involves dental pulp caused by a bacterial infection in dental caries.

In Gingivitis gum disease inflammation can be seen with bone changing which leads periodontitis.

Periodontitis is tissue inflammation of the teeth, and mostly the main reason is gingivitis keperiodontium expansion inflammation. Many researchs are devoted to basic characteristics of the dental disease.

Studies [4] have shown that Adaptive Neuro Fuzzy Inference System (ANFIS) is a very powerful tool in medical diagnosis and have produced satisfactory results.

Authors of [5] proposed a Convolutional Neural Network for diagnosing periodontal disease. The network was trained using multi-layer perception and a backward propagation learning algorithm.

A fuzzy rule based system for dental diagnosis from x-ray is discussed in [6]. It based on fuzzy clustering method and Mamdani inference model was used to generate the outcome.

A Fuzzy Expert System Design for Diagnosis of Periodontal Dental Disease is discussed in [7]. The aims of the study are to design a Fuzzy Expert System (FES) for diagnosis of periodontal dental disease by using a computer program. It helps dentists for investigation, diagnosis and treatment of the disease.

Our previous work [8] is devoted to problems of clinic, diagnostics, treatment and preventive maintenance of stomatology diseases. Generalizing the data of the world literature regarding this problem and our own clinical experience, we developed diagnostic expert system for determination of types of a pulpitis on the basis of environment ESPLAN [9]. In this paper there are examples of the use of intellectual diagnostic system of the pulpitis, which are adequate to the modern requirements to decision-making systems.

Despite of the facts there are a series of works devoted to the definition of dental disease types, there are unresolved problems. These problems lead to difficulties in achieving the desired degree of accuracy of diagnosis. In this work we use fuzzy inference system and clustering method, because it is more flexible in the system design and it is used in Multiple Input and Single Output systems.

The rest of the paper is the following form. Section 1 is devoted to the review related on the solution of dental disease problem. Section 2 contains preliminaries. Section 3 is devoted to processing of knowledge and experimental researches for diagnostics of a dental diseases. Section 4 is conclusion.

2   Preliminaries

Definition 1 [9] Fuzzy Number. Afuzzy number is a fuzzy set A on R which possesses the following properties: a) A is a normal fuzzy set; b) A is a convex fuzzy set; c) α-cut of A, A^{α} is a closed interval for every $\alpha \in (0, 1]$; d) the support of A, A^{+0} is bounded.

A fuzzy number $\tilde{A} = (a_1, a_2, a_3, a_4)$, is called a trapezoidal fuzzy number if its membership function is given by

$$\mu_{\tilde{A}}(x) = \begin{cases} 0, & x \le a_1 \\ \frac{x-a_1}{a_2-a_1}, & a_1 \le x \le a_2 \\ \frac{a_4-x}{a_4-a_3}, & a_3 \le x \le a_4 \\ 0, & x \ge a_4 \end{cases}$$

Definition 2 Subtractive Clustering [10, 11]. The subtractive clustering approach considers every data is a potential cluster center that calculates a level of the likelihood such as every data point should define the center of cluster, based on the density of surrounding data points. Assume that m dimensions, n data point $(x_1, x_2, \ldots, x_n)$ and every data point is potential cluster center, the density function D_i of data point at x_i is represented by:

$$D_i = \sum_{i=1}^{n} e^{\left(\frac{\|x_i - x_j\|^2}{\left(\frac{r_a}{2} \right)} \right)}$$

where r_a is a positive data. The highest potential data point is covered by more data points. A radius defines a neighbor area then the data points, which exceed r_a, have no influence on the density of data point.

It is possible to choose a data point with the highest potential and select the first cluster center if density of each data point is calculated beforehand.

Assuming that x_{c1} is selected and D_{c1} is its density, the density of each data point can be corrected by:

$$D_i = D_i - D_{c1} e^{\left(\frac{\|x_i - x_j\|^2}{\left(\frac{r_b}{2} \right)} \right)}$$

The density function of data point which is close to the first cluster center is reduced, so these data points cannot become the next cluster center. r_b defines an neighbor area. As a rule constant $r_b > r_a$. For avoiding the overlapping of cluster centers near to other (s) is given by [9]:

$$r_b = 1.5 \cdot r_a$$

Correcting the density function of data points is possible to find the next cluster center. Looping process goes until finding all cluster centers. The subtractive clustering is used to find the number of clusters over given data. The fuzzy model is being created by using these clusters.

3 Statement of the Problem

In this paper, the aim is to define dental disease detection by using Fuzzy logic, sub clustering method and neural-network.

Problems on dental disease detection was examined through observations and interviews with experts. It is based on disease symptoms which becomes input variables criterion affecting output parameters.

This study used 80 patients data and their symptoms is given in Table 1. The criteria used are disease symptoms as indicators affecting output variables (out1). This study used 8 input criteria are as follows:

1. Plaque (in1)
2. Inflamed gums (in2)
3. Pain (in3)
4. Red gums (in4)
5. Swollen gums (in5)
6. Easily bleeding gums (in6)
7. Breath odor (in7)
8. Wobbly teeth (in8)

Value of the variable out1 are: 1-pulpitis, 2- Gingvitis, 3-Periodontitis, 4-Advanced Periodontitis. The effectiveness of treatment in all groups of patients with caries, pulpitis and periodontitis is assessed by means of objective examination methods, collecting anamnesis, identification of relapses and complications.

For the decision of these problems the effective architecture of fuzzy recurrent neural networks with 8 inputs, and one output has been developed. The network is trained on the basis of error back propagation method. Such neural network is characterized by smaller computing complexity and good ability to training.

The dataset used in this study comprises of 80 diagnosed data and 2/3 of it was used for training, and 1/3 data for test. The fragment of training data is given in Table 1.

Table 1. Fragment of training data

N	in1	in2	in3	in4	in5	in6	in7	in8	out1
1	20	10	10	25	15	15	25	0	1
2	20	70	70	70	10	10	10	0	1
3	45	45	45	45	50	55	10	10	2
4	75	70	70	60	50	50	35	35	4
5	30	25	25	25	20	20	10	10	1
...	...	...	...	...	...	...	...	...	...
51	45	45	45	45	45	45	0	0	2
52	70	70	70	60	50	50	35	35	4
53	65	50	50	40	45	45	40	0	4
54	45	45	45	45	45	45	45	35	3

Fragment of the extracted rules using subclustering method is described in Fig. 1.

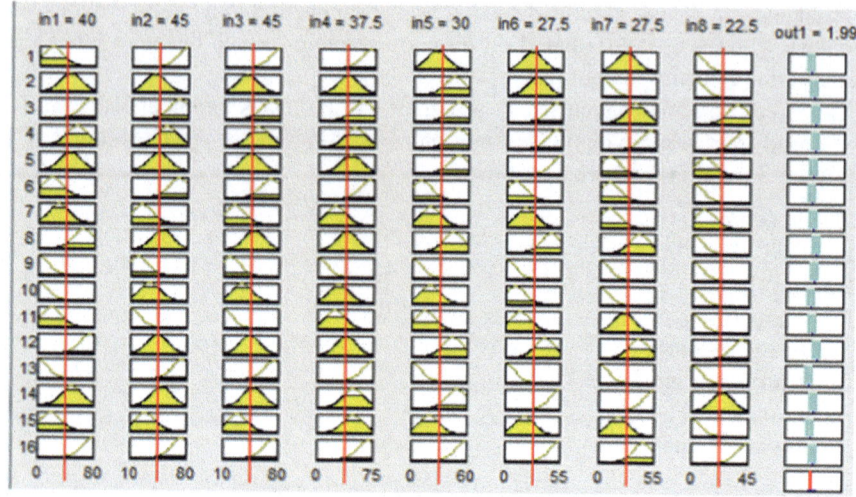

Fig. 1. Fragment of the extracted rules

Fragment of testing data for subclustering method and fuzzy inference based method is described in Fig. 2:

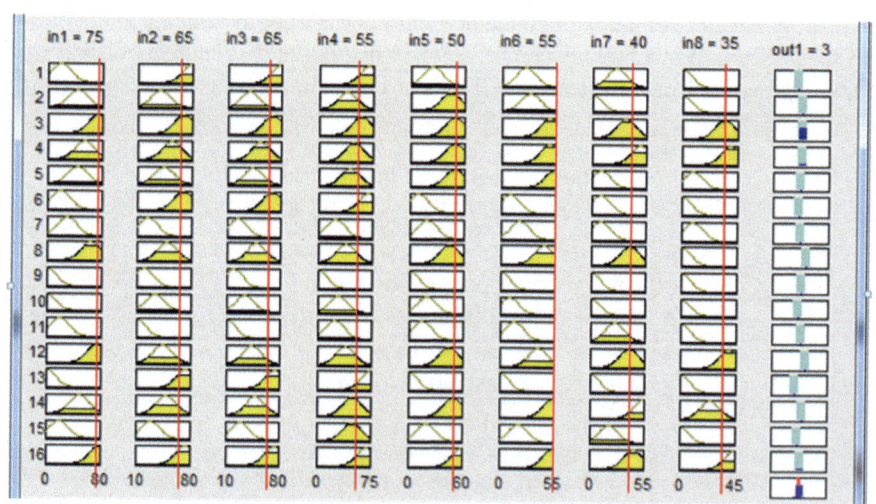

Fig. 2. Fragment of testing data for subclustering method

For finding the type of disease it is possible to use rules which are demonstrated in Figure. If we use initial condition such that in1 = 75, in2 = 65, in3 = 65, in4 = 55, in5 = 50, in6 = 55, in7 = 40, in8 = 35 then we define disease is pulpits or output of the system is out1 = 3.

4   Conclusion

In this paper, it is analyzed Fuzzy logic based dental disease problem. Subclustering method is used for rule extraction. The dataset used in training the network was collected from various medical laboratories. The dataset comprises of 80 data instances and 8 symptoms attribute which was collected using feature selection. The network was trained using a backward propagation learning algorithm. The system was able to accurately diagnose dental disease.

All calculations are made in MATLAB environment. Fragment of obtained results are given below:

Test 1:
IF Plaque is LOW and inflamed gums is SEVERE and pain is SEVERE and red gums is MODERATE and swollen gums LOW and Gum bleeds easily is LOW and bread odor is LOW wobbly teeth is LOW **THEN** detected **Pulpitis Disease.**

Test 2:
IF Plaque is LOW and inflamed gums is MODERATE and pain is MODERATE and red gums is LOW and swollen gums LOW and Gum bleeds easily is LOW and bread odor is LOW wobbly teeth is LOW **THEN** detected **Gingivitis Disease.**

Clinical diagnosis of dental diseases are quite difficult and most dentists' diagnosis might be subjective. Artificial intelligence has shown commendable results in medical diagnosis. This study utilized Fuzzy logic toolbox and sub clustering method for diagnosing dental disease. In order to differentiate disease and to define diagnosis for the patients we used observed data. 27 dataset is tested and result defined according to the practical data.

References

1. Aliev, R.A., Aliyev, B.F., Gardashova, L.A., Huseynov, O.H.: Selection of an optimal treatment method for acute periodontitis disease. J. Med. Syst. **36**, 639–646 (2012). https://doi.org/10.1007/s10916-010-9528-6
2. Aliev, B.F., Gardashova, L.A.: Selection of an optimal treatment method for acute pulpitis disease. Procedia Comput. Sci. **120**, 539–546 (2017). https://doi.org/10.1016/j.procs.2017.11.276
3. Parewe, A.M.A.K., Mahmudy, W.F.: Dental disease identification using fuzzy inference system. J. Environ. Eng. Sustain. Technol. **3**(1), 33–419 (2016)
4. Papantonopoulos, G., Takahashi, K., Bountis, T., Loos, B.G.: Artificial neural networks for the diagnosis of aggressive periodontitis trained by immunologic parameters. PLoS ONE **9**(3), e89757 (2014). https://doi.org/10.1371/journal.pone.0089757
5. Shehnaz, W., Bhardwaj, A.: Convolutional neural network for periodontal disease. Int. J. Comput. Sci. Inform. Technol. Secur. **7**(2), 2249–9555 (2017)
6. Tran, M.T., Nguyen, D.T., Pham, D.T., LeHoang, S.: Dental diagnosis from X-Ray images using fuzzy rule-based systems. Intl. J. Fuzzy Syst. Appl. **6**(1), 1–16 (2017)
7. Novruz, A., Tevfik, A.: A fuzzy expert system design for diagnosis of periodontal dental disease. IEEE Trans. Fuzzy Syst. **7546**, 1–5 (2011)

8. Abdullayev, T.S., L.A., Gardashova, Aliev, B.F., Aliev, A.G., Ismailov, B.I.: Fuzzy expert system ESPLAN and its application in business, medicine and technics. In: Seventh International Conference on Application of Fuzzy Systems and Soft Computing, Germany, pp. 205–215 (2006)
9. Aliev, R.A., Aliev, R.R.: Soft Computing and its Application. World Scientific, London (2001)
10. Yager, R., Filev, D.: Generation of fuzzy rules by mountain clustering. J. Intell. Fuzzy Syst. 2(3), 209–219 (1994)
11. Collazo-Cuevas, J.I., Aceves-Fernandez, M.A., Gorrostieta-Hurtado, E., Pedraza-Ortega, J. C., Sotomayor-Olmedo, A., Delgado-Rosas, M.: Comparison between Fuzzy C-means clustering and fuzzy clustering subtractive in urban air pollution. In: 20th International Conference on Electronics Communications and Computers (CONIELECOMP), Cholula, pp. 174–179 (2010). https://doi.org/10.1109/conielecomp.2010.5440772

Evaluation of the Effectiveness of Integration Processes in Production Enterprises Based on the Fuzzy Logic Model

Rahib Imamguluyev ⓘ, Tural Suleymanli$^{(\boxtimes)}$, and Niyar Umarova

Odlar Yurdu University, AZ1072 Baku, Azerbaijan
rahib.aydinoglu@gmail.com,
suleymanli.tural@gmail.com, niyar6l@list.ru

Abstract. Integration should be seen as a strategic alliance of enterprises that allows participants to strengthen their competitive position and gain financial and other advantages of business development. If we look at integration as a specific project, a set of organizational and economically specialized measures, it can become a more purposeful and efficient process. This specialization is very important, given that integrated companies do not stop their current activities in the process of change. In the process of integration, both costs and results are considered, regardless of the measurement methods. Thus, in the process of integration, the scale of business changes, it is necessary to find a balanced compatibility of relative and absolute indicators when creating a system of evaluation indicators. The advantage of the fuzzy logic approach is that their time limits are defined, which allows the time intervals of the efficiency calculation to be accurately determined. Considering the integration process as an object, it is expedient to assess the effectiveness of this object on the basis of fuzzy logic, both at the stage of decision-making on its realization and at the stage of evaluation of the obtained results. In addition, the assessment should be based on both the position of each participant and the synergistic effect obtained. Thus, different degrees of efficiency in the evaluation process were considered on the basis of fuzzy logic, and calculations were performed in the Fuzzy section of the Matlab program.

Keywords: Fuzzy logic · Economy · Integration processes · Integration effectiveness · Matlab

1 Introduction

The mechanism of realization of the integration process includes a set of objects, subjects, processes, and it is this set that forms its cost and result part. When considering options for project implementation, it is necessary to focus on maximum economic efficiency when choosing the mechanism of implementation [1].

Systematic dynamics must occur in the integration process and as a result, when an additional, synergistic effect is provided for the whole system due to the coordinated change of local loops. This is a controlled change in the parameters of the subsystem of the whole system as a result of the interaction of the dynamic processes. It is necessary

© The Author(s), under exclusive license to Springer Nature Switzerland AG 2021
R. A. Aliev et al. (Eds.): ICAFS 2020, AISC 1306, pp. 133–139, 2021.
https://doi.org/10.1007/978-3-030-64058-3_17

to determine these dynamics, to measure them quantitatively, and on this basis to determine the effectiveness and efficiency of integration on the basis of fuzzy logic.

Integration processes that have the ability to interact with a positive synergistic outcome:

- complementary integration processes as a result of the joint realization of a large number of individual results.
- interdependent integration processes, in which only joint activity can yield positive results, and individual activity can only lead to negative results.

There are several types of integration synergies, including the following.

- Functional (improvement of the management system is provided by the reduction of conventional fixed costs as the business grows).
- Purposeful (manifested in the increase of financial and other opportunities for the development, production and sale of products by minimizing the cost of interaction with the external environment).
- Complex (allows you to achieve additional results by improving the supply of resources and conditions of sale of the product).
- Conglomerate (consists of distributing risks and reducing their impact).

It is also important to note the missed benefits in assessing the effectiveness of integration projects. Such cases occur when the enterprises involved in the integration process are not able to work at full capacity [2–4].

At the stage of economic justification of integration processes, its potential effectiveness will be determined on the basis of fuzzy logic, and the system of indicators used to evaluate investment projects, i.e. net income, net discount income, profitability index can be used. The methods for determining these indicators are well known, but given the specifics of the projects we are considering, it is important to pay attention to: what should be the reporting period of the project and how to choose the rate of return to determine the current value of cash flows before and after the merger?

It is advisable to take the reporting period equal to the duration of the integration project. In turn, the duration of the project depends not only on the implementation of organizational, economic and other measures provided for in the project, but also on the institutionalization of the new structure, the new business process model, the adaptation of staff, suppliers and consumers. (one year or more, depending on the scale).

Efficiency can be measured both as a system of expenditure and final indicators, and with an integrated indicator. The latter should reflect the essence of the integration project, its main purpose [5, 6].

The integration of manufacturing enterprises can be attributed to the most important strategic decision of each of the participants. In this case, many believe that market capitalization can be considered as an integrated indicator that reflects the purpose and scale of changes in the integration process. This approach is quite common in countries with developed market economies. In such an approach, the synergistic effect can be defined as the excess of the capitalization of a joint venture over the total of its capitalization prior to the merger of the participating companies. However, there are certain methodological problems in the implementation of this approach. In particular, it is necessary to take into account the organizational and legal form of project participants. Determining the market value of merging companies with open joint stock companies can be used to determine the market price of shares for calculations. The merits of a combined company may not be visible in the share price, or they may be delayed to some extent.

The methodology for calculating market value also depends on the profitability of the merging companies. For strategic purposes, low-profit and completely successful companies that combine losses, depending on resources, production or sales, can unite. In such cases, it is necessary to model the cash flow for options before and after the merger, choosing the right approach and the method of profit normalization for loss-making companies [7].

In line with the value-based approach to performance appraisal, they note the positive and negative dynamics of market capitalization change criteria. Positive system dynamics correspond to the common interests of the participants.

Thus, the fuzzy approach is appropriate here, as the positive and negative effectiveness of merging companies is not known in advance.

2   Fuzzy Logic

The concept of Fuzzy Logic was first proposed by Lotfi A. Zadeh in an article published in 1965. However, it was used as a more frequently expressed term in the 1970s. Fuzzy logic applications, also known as hazy logic, are computer-aided artificial intelligence systems that reason in a way to imitate human behavior and the functioning of nature.

Can process data and turn it into meaningful information. Can create control and decision-making processes by using meaningful information with modeling styles close to human thinking. However, when it is enriched with algorithms, it can process data and control or decision-making processes at the same time.

Fuzzy logic principles do not work like classical logic principles. Fuzzy logic principles do not have completely correct or completely false values. Instead, uncertain options such as false or near-true partially true and partially false are considered within the variable range [6, 8].

3   Evaluating the Effectiveness of Integration Based on Fuzzy Logic

Research has shown that the development of scientifically based methods for assessing the effectiveness of integration requires a more in-depth study of the economic nature of the "effectiveness" category (including the characteristics of the relationship between agro-industrial entities). Thus, from the point of view of the essence and economic content of the problem of efficiency, based on the study of the works of local and foreign scientists dedicated to this problem, we have identified the following [6, 7, 9].

1. There is no terminological unity in the use of the term "efficiency" in the analysis of production and economic systems. As a rule, the term "efficiency" is used as a synonym for efficiency, optimality, and sustainability Table 1.

Table 1. Modern approaches to the interpretation of the concept of "effectiveness"

Categories	Contents
1. Economic approach	
Efficiency, effectiveness	General parameter of creating a production system
2. Production-technological approach	
Optimality and balance of the system	Internal characteristics of the resource potential, the proportions between its components
3. Systematic approach, complex approach	
Flexibility, durability, stability	The state of the system, its ability to resist external influences
Synergetics	The result of cooperative interactions of subsystems that change the quality of the system in the process of self-organization
4. Commercial approach	
Competitiveness	Ability to maintain stable positions in the competition
Importance	The level of meeting the needs of consumers
Innovation	Ability to use modern ICT

2. In some cases, this concept is considered unilaterally. In a process approach, performance evaluation is limited to comparing the characteristics of "inputs" and "outputs". From the point of view of the system approach, efficiency reflects the individual characteristics of the quality of the system.

Thus, the substantiation of the qualitative aspect of integration effectiveness can be built on the basis of the categories of "efficiency" and "optimality" on the basis of fuzzy logic. The main characteristics of these approaches are systematized and given in Table 2.

Table 2. Degrees of integration effectiveness

Efficiency	Optimality
Very very little	Very very little
Very little	Very little
Little	Little
Medium	Medium
Much	Much
More	More

Based on the above-mentioned approaches to the effectiveness category, a concrete report was made on the basis of fuzzy logic of its individual elements.

4 Practical Application of Fuzzy Technologies

The first step in applying fuzzy logic to a system is to define the inputs and outputs of the system. Given the effectiveness of integration, the most important expectation is to take into account the efficiency and optimality levels of efficiency. To meet these expectations, the efficiency parameters that make up the results of the fuzzy logic model are important. The inputs and outputs of the fuzzy logic model are defined as shown (see Fig. 1, 2). Membership function numbers, names, lower and upper limits of all parameters are determined by the effect of input and output parameters on the problem to be modeled. (see Fig. 1) shows the membership functions of the input parameters and (see Fig. 2) shows the lower and upper limits of the membership functions of the output parameters [6, 8–10].

Once the membership functions and the lower and upper limits of the parameters required to build the model have been determined, 31 rules (see Fig. 3) have been established to establish the necessary relationships between the parameters affecting the system.

The value of the efficiency factor is classified into one of six fuzzy sets: very-very little, very little, little, medium, much, and more.

This value is between [0–100]. When the process is complete, a list is created and can be used as a model.

Each language variable has six fuzzy values with triangular or trapezoidal membership functions:

- For input variables – (see Fig. 1): VVL - very very little; VL - very little; L - little; ME - medium; MU - much; MO - more.
- For output variables – (see Fig. 2): VB - very bad; B - bad; A - average; G - good; VG - very good.

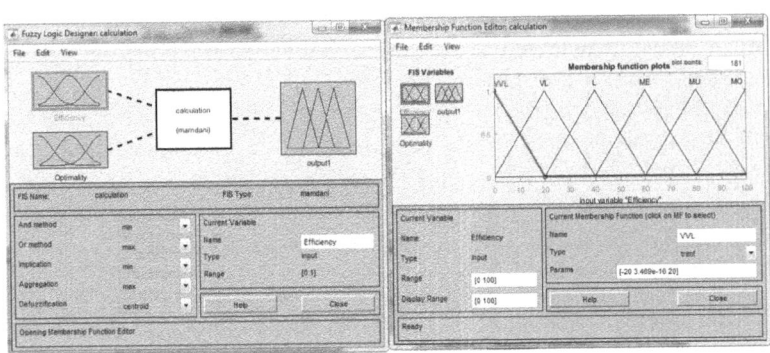

Fig. 1. Input variables

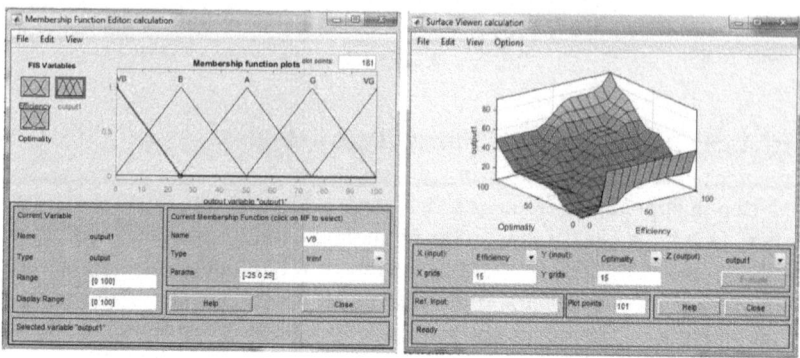

Fig. 2. Output variables.

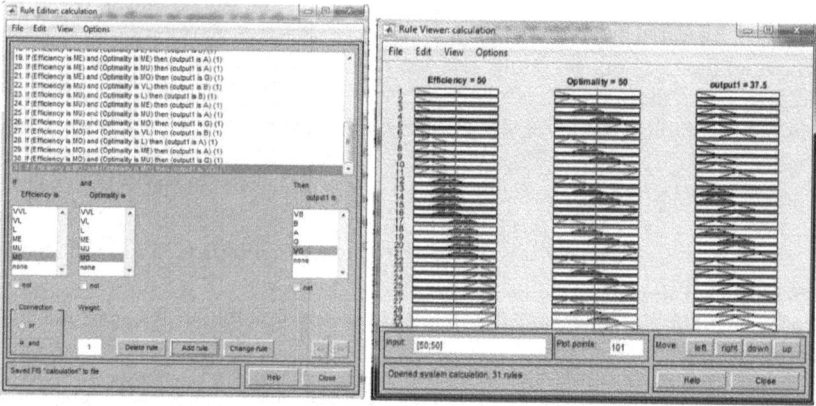

Fig. 3. Rules window

5   Conclusion

The evaluation of the alternatives in Table 1 in the MATLAB\Fuzzy Change Systems note described above allowed the following results to be obtained: Alternatively, the value of the Integration Effectiveness was set to very very little, very little, little, medium, much, more instead of less and more.

The results are more rhetorical, showing the effectiveness of integration based on fuzzy complementary theory. On the other hand, given other simple but vague and difficult formation criteria, it is useful to use fuzzy set theory when determining effectiveness with a simple objective method based on a perfect integration effect and decisions based on them are reasonable and economically feasible.

References

1. Gubarech, M.A., Masilkina, E.M.: Promotion and positioning in marketing (in Russian). Success. Bus. Strategy, Moscow, Russia **9**(2), 222–223 (2011)
2. Grant, R.: Modern strategic analysis (in Russian), Piter, MBA Classics, 672 p. (2018)
3. Burov, M.P.: State Regulation of the National Economy: Modern Paradigms and Mechanisms for the Development of Russian Regions (in Russian), 342 p. Dashkov and K. (2018)
4. Samigulin, E.V.: Economic law and parameters of the retail product market (in Russian), AUCA Academic Review, pp. 117–125 (2008)
5. Petrova, A.V.: Analysis of the product market and assortment (in Russian), Yekaterinburg, p. 67 (2016)
6. Aliev, R.A., Fazlollahi, B., Aliev, R.R.: Soft Computing and its Applications in Business and Economics, 462 p. Springer, Heidelberg (2004)
7. Novikova, I.S.: Analysis of forecasting methods of trade turnover (in Russian). www.c: Premier/Downloads/ЭПМ
8. Zadeh, L.A.: Fuzzy logic, neural networks, and soft computing. Commun. ACM **37**, 77–84 (1994)
9. Beale, M.H., Hagan, M.T., Demuth, H.B.: Neural Network Toolbox. User's Guide. Math Works, Inc., Natick (2014)
10. Vasilios, N.K.: Matlab: A Fundamental Tool for Scientific Computing and Engineering Applications, vol. 1, 533 p. InTeOp (2012)

Fuzzy Approach for Evaluation of Student's Performance

J. M. Babanli[✉]

Azerbaijan State University of Oil and Industry, Azadlig Avenue, 20,
Baku AZ1010, Azerbaijan
babanlijale@gmail.com

Abstract. A series of works on fuzzy logic application in educational systems exist. Student's performance is a basic attribute in this realm and it depends on level of education. Student's performance evaluation includes analysis of skills and ability which are characterized by uncertainty. Always student's results are evaluated by using linguistic terms instead of numerical values. This paper uses fuzzy logic-based approach to find a student with the highest performance. The purpose of the paper is to make university give full consideration of the grades of students to define their performance.

Keywords: Fuzzy set · Fuzzy number · Student's performance

1 Introduction

Student evaluation is a significant part of the educational process. An effective system for evaluating of students is a basis for realizing goals of education. Student evaluation can come in many forms and depends on terms of response. Each university produces its own evaluations, which are administered across the subject range. Basic components of evaluating student performance are external evaluation and student self-evaluation. External evaluation is how teachers evaluate the student's performance. Student self-evaluation is how the students evaluate their own performance.

Nowadays, a lot of approaches for students' performance evaluation based on fuzzy inference exist. Paper [1] is devoted to the problem of student performance evaluation based on Data mining method. The method was applied to perform Criterion-Referenced Evaluation and Norm-Referenced Evaluation. In [2] they proposed a model of fuzzy expert system to evaluate overall performance of teachers. In [3], a new fuzzy inference system for evaluating the learning progress is proposed. For evaluation, they consider three factors of questions, such as difficulty, importance, and complexity. A student performance evaluation system based on fuzzy logic is discussed in [4]. In [5, 6], fuzzy logic is used for evaluation of academic performance of students. A fuzzy decision support system model for evaluation of student's performance is given in [7]. The authors of [8] proposed reasoning system based on fuzzy logic for performance evaluation of students.

All these are related to application of fuzzy logic [9] in education problems, but important issue is development of descriptive evaluation-based fuzzy approach. In this

R. A. Aliev et al. (Eds.): ICAFS 2020, AISC 1306, pp. 140–147, 2021.
https://doi.org/10.1007/978-3-030-64058-3_18

paper, an effective method for fuzzy multicriteria decision making based on Zadeh idea of data aggregation is used for student performance evaluation.

The paper is structured as follows. The preliminaries are given in Sect. 2. The statement of problem is given in Sect. 3. Section 4 describes the solution method. Numerical example is discussed in Sect. 5. Section 6 is conclusion.

2 Preliminaries

Definition 1. A fuzzy number. Let X be a universe of discourse. The fuzzy set C on X, whose membership function is the mapping of $\mu_C : \mathcal{R} \rightarrow [0, 1]$, is a continuous fuzzy number if it fulfils the following conditions:

1. C is a normal fuzzy set;
2. C is a convex fuzzy set;
3. α-cut C^α is a closed interval for any $\alpha \in [0, 1]$;
4. The support $\text{supp}(C)$ is bounded.

Definition 2. Aggregation of fuzzy numbers. Let $C_1, \ldots, C_m$ be fuzzy numbers. An arithmetic mean-based aggregation of fuzzy numbers, C is defined as follows:

$$C = \frac{\sum_{i=1}^{m} C_i}{m}.$$

Definition 3. Distance between triangular fuzzy numbers. Let $C_1 = (c_{11}, c_{12}, c_{13})$, $C_2 = (c_{21}, c_{22}, c_{23})$ be triangular fuzzy numbers. A distance between C_1 and C_2 is defined as

$$d(C_1, C_2) = |P(C_1) - P(C_2)|,$$

where $P(C_1) = \frac{c_{11} + 4c_{12} + c_{13}}{6}$, $P(C_2) = \frac{c_{21} + 4c_{22} + c_{23}}{6}$.

3 Statement of the Problem

This paper addresses the issue of ranking multiattribute alternatives by using group decision approach. A set of alternatives $A = \{a_1, \ldots, a_m\}$ and criteria vector $F = (f_1, \ldots, f_n)$ are given. Preference values of evaluated alternatives are based on knowledge of an expert team. This implies that each member of an expert group evaluates every alternative concerning the criteria provided (Table 1).

Table 1. Criteria values of alternatives

	f_1	$\cdots$	f_j	$\cdots$	f_n
a_1	f_{11}^k	$\cdots$	f_{1j}^k	$\cdots$	f_{1n}^k
a_i	f_{1i}^k	$\cdots$	f_{ij}^k	$\cdots$	f_{in}^k
$\cdots$	$\cdots$	$\cdots$	$\cdots$	$\cdots$	$\cdots$
a_m	f_{m1}^k	$\cdots$	f_{mj}^k	$\cdots$	f_{mn}^k

Here f_{ij}^k is the k-th expert's evaluation of the i-th alternative w.r.t. the j-th criterion. Criteria are grouped by importance degrees. The considered problem is to determine an optimal alternative. In other words, a^* (optimal alternative) is to be found as

$$\text{Agg}(a^*) = \max_{a \in A} \text{Agg}(a), \tag{1}$$

where Agg is an aggregated general index of an alternative.

4 Solution Method

The considered issue is addressed through the application of Group Decision Making Method. The method implementation consists of the following steps:

1. Establish an expert team for criteria evaluation of alternatives.
2. Each expert evaluates each a_i alternative based on criteria vector $F = (f_1, \ldots, f_n)$, so forms a vector $\overline{f}_i^k = (f_{i1}^k, \ldots, f_{ij}^k, \ldots, f_{in}^k)^T$.
3. For each a_i, vectors $\overline{f}_i^k$, $k = 1, \ldots, K$ are aggregated to a single vector $\overline{f}_i = (f_{i1}, \ldots, f_{ij}, \ldots, f_{in})^T$ as:

$$\overline{f}_i = \frac{\sum\limits_{k=1}^{K} f_i^k}{K}. \tag{2}$$

That is, each component of $\overline{f}_i = (f_{i1}, \ldots, f_{ij}, \ldots, f_{in})^T$, f_{ij} is calculated as:

$$\overline{f}_{ij} = \frac{\sum\limits_{k=1}^{K} f_{ij}^k}{K}$$

4. It is calculated the arithmetic mean for every importance subgroup of criteria for each alternative: $\varphi_l(a_i), l \in \{1, \ldots, L\}$, L is the number of subgroups.
5. Weighted average of values $\varphi_l(a_i)$ is computed to yield a general aggregated index of an alternative:

$$Agg(a_i) = IG_1 \phi_1(a_i) + \dots + IG_l \phi_l(a_i) + \dots + IG_L \phi_L(a_i). \qquad (3)$$

Here $IG_i, i = 1, \dots L$ is the coefficient reflecting the importance of groups.

6. Alternatives a_i, $i = 1, \dots, n$ are ranked on basis of their indices. $Agg(a_i)$ is compared on basis of the distance to the fuzzy number Q, which represents the highest linguistic term of the scale of estimation:

$$a_1 \succ a_2 \text{ iff } d(Agg(a_1), Q) < d(Agg(a_2), Q)$$

Here d is the distance between triangular fuzzy numbers (Definition 3). Thus, an alternative which has a close distance to the fuzzy number Q is considered as superior.

5 Application

Let us consider a problem of students performance evaluation w.r.t. to courses. Students are denoted $a_1, \dots a_{24}$. Criteria are performance w.r.t. courses: f_1- Descriptive statistics; f_2- Systems modeling and simulation; f_3- Fundamentals of industrial engineering; f_4- Machine elements; f_5- Dynamics; f_6- Investment and Financing; f_7- Microeconomics; f_8- International production and Investment management; f_9- Production; f_{10}- Communication management; f_{11}- Marketing; f_{12}- Metrology laboratory; f_{13}- Risk and finance. The problem was solved on the basis of the solution approach given in Sect. 4 as follows.

At first, a group of 5 experts was formed.

Each expert evaluated each student w.r.t. the criteria. Criteria vectors $\bar{f}_i$ of the expert group are calculated for the alternatives based on (2). The linguistic values of the criteria used in the questionnaires are described as fuzzy numbers (Fig. 1).

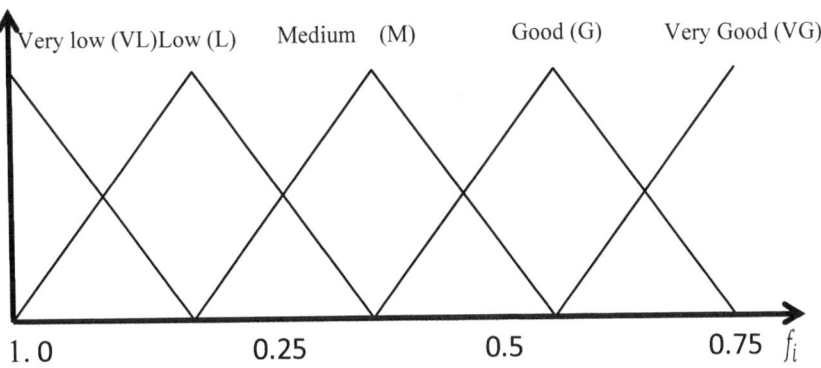

Fig. 1. The codebook of linguistic terms

Then, by using the formula (2), the vector $\bar{f}_i$ is found. For example, the calculated values for a_1 are as follows:

$$f_1 = (0.6\ 0.8\ 1),\ f_8 = (0.8\ 1\ 1),\ f_2 = (0.8\ 1\ 1),\ f_9 = (0.8\ 1\ 1),$$
$$f_3 = (0.8\ 1\ 1),\ f_{10} = (0.8\ 1\ 1),\ f_4 = (0.8\ 1\ 1),\ f_{11} = (0.8\ 1\ 1),$$
$$f_5 = (0.8\ 1\ 1),\ f_{12} = (0.8\ 1\ 1),\ f_6 = (0.8\ 1\ 1),\ f_{13} = (0.8\ 1\ 1),$$
$$f_7 = (0.8\ 1\ 1).$$

The values of $\bar{f}_i$ for the other alternatives are found analogously. Further, for each student, it is needed to calculate $\varphi_1(a_i)$, $\varphi_2(a_i)$, $\varphi_3(a_i)$ values. The importance subgroups of criteria are shown in Table 2.

Table 2. Importance subgroups of criteria

Importance rates	Criteria
High importance, (0.5 0.6 0.7)	f3
	f4
	f5
Medium importance, (0.2 0.3 0.4)	f2
	f8
	f9
	f11
	f12
Low importance, (0 0.1 0.2)	f1
	f6
	f7
	f10
	f15

So, $\varphi_1(a_i)$, $\varphi_2(a_i)$, $\varphi_3(a_i)$ are found as follows:

$$\phi_1(a_i) = \left(\frac{f_3 + f_4 + f_5}{3}\right),$$

$$\phi_2(a_i) = \left(\frac{f_2 + f_8 + f_9 + f_{11} + f_{12}}{5}\right),$$

$$\phi_3(a_i) = \left(\frac{f_1 + f_6 + f_7 + f_{10} + f_{13}}{5}\right).$$

The results are shown below:

$$\phi_1(a_1) = (0.53\ 1\ 1),$$

$$\phi_2(a_1) = (0.8\ 1\ 1),$$

$$\phi_3(a_1) = (0.76\ 0.96\ 1),$$

$$\cdots\cdots$$

$$\phi_1(a_{24}) = (0.33\ 0.53\ 0.73),$$

$$\phi_2(a_{24}) = (0.16\ 0.36\ 0.56),$$

$$\phi_3(a_{24}) = (0\ 0.12\ 0.32).$$

Next, the general aggregated index for each student is calculated using (3). The obtained results:

$$Agg(a_1) = (0.56\ 0.996\ 1.3),$$

$$Agg(a_2) = (0.361\ 0.694\ 1.147),$$

$$\cdots\cdots\cdots\cdots$$

$$Agg(a_{23}) = (0.28\ 0.588\ 1),$$

$$Agg(a_{24}) = (0.197\ 0.438\ 0.799).$$

Finally, the aggregated values of alternatives are ranked by measuring the distance (Definition 3) to the highest term, "very good" (Fig. 1). For example, for a_1, the distance of its aggregated values to "very good" term is calculated as follows:

$$d(\text{Agg}(a_1), \text{very } good) = \left| \frac{0.24 + 4 * 0.59 + 1}{6} - \frac{0.75 + 4 * 1 + 1}{6} \right| = |0.62 - 0.96|$$
$$= 0.34.$$

The obtained results are given in Table 3.

Table 3. The results of ranking of students

Student	$d(Agg(a_i), very\,good)$	Student	$d(Agg(a_i), very\,good)$
a_5	0.009	a_7	0.2143
a_{19}	0.013	a_2	0.2443
a_{11}	0.0157	a_{17}	0.2725
a_{15}	0.0368	a_{21}	0.2578
a_{11}	0.0591	a_8	0.2883
a_{18}	0.0701	a_{23}	0.3583
a_{16}	0.0768	a_{20}	0.3663
a_{22}	0.0806	a_{24}	0.5003
a_{12}	0.0968	a_6	0.5363
a_4	0.1071	a_9	0.5483
a_3	0.2016	a_{13}	0.5643
a_{10}	0.2123	a_{14}	0.6083

The smaller distance has a higher rank. Thus, $a_5 \succ a_{19} \succ a_1 \succ \ldots \succ a_{14}$.

6   Conclusion

In this paper we consider a complex problem of ranking students. For solving this problem, we rely on Zadeh's idea to deal with a large number of alternatives and criteria. The considered problem is characterized by 13 criteria. Information on criteria evaluation and importance is characterized by fuzziness. The method suggested in our previous paper is used for solving this problem. The obtained results show efficiency of the proposed approach.

References

1. Rasmani, K.A., Shen, Q.: Data-driven fuzzy rule generation and its application for student academic performance evaluation. Appl. Intell. **25**, 305–319 (2006). https://doi.org/10.1007/s10489-006-0109-9
2. Jyothi, G., Parvathi, Ch., Srinivas, P., Althaf Rahaman, Sk.: Fuzzy expert model for evaluation of faculty performance in technical educational institutions. Int. J. Eng. Res. Appl. **4**(5, Version 7), 41–50 (2014)
3. Goodarzi, M.H., Amiri, A.: Evaluating students' learning progress by using fuzzy inference system. In: Sixth International Conference on Fuzzy Systems and Knowledge Discovery, pp. 561–565. IEEE Press, New York (2009). https://doi.org/10.1109/fskd.2009.313
4. Annabestani, M., Rowhanimanesh, A., Mizani, A., Rezaei, A.: Descriptive evaluation of students using fuzzy approximate reasoning. https://arxiv.org/ftp/arxiv/papers/1905/1905.02549.pdf
5. Yadav, R.S., Singh, V.P.: Modelling academic performance evaluation using soft-computing techniques: a fuzzy logic approach. Int. J. Comput. Sci. Eng. **3**(2), 676–686 (2011)

6. Yadav, R.S., Soni, A.K., Pal, S.: A study of academic performance evaluation using fuzzy logic techniques. In: IEEE International Conference on Computing for sustainable global development (INDIACom), INSPEC Accession Number: 14382897. IEEE Press, New York (2014). https://doi.org/10.1109/indiacom.2014.6828010
7. Yildiz, Z., Baba, A.F.: Evaluation of student performance in laboratory applications using fuzzy decision support system model. In: IEEE Global Engineering Education Conference (EDUCON), pp. 1023–1027. IEEE Press (2014). https://doi.org/10.1109/educon.2014.6826230
8. Kharola, A., Kunwar, S., Choudhury, G.B.: Students performance evaluation: a fuzzy logic reasoning approach. PM World J. 4(9), 1–11 (2015). https://pmworldlibrary.net/wp-content/uploads/2015/09/pmwj38-Sep2015-Kharola-Kunwar-Choudhury-students-performance-evaluation-Featured-Paper.pdf
9. Aliev, R., Tserkovny, A.: Systemic approach to fuzzy logic formalization for approximate reasoning. Inf. Sci. 181(6), 1045–1059 (2011)

The Man Who Changed the Scientific World: To the Centenary of the Birth of Lotfi Zadeh

Rafik A. Aliev[1(✉)] and Valery B. Tarassov[2]

[1] Azerbaijan State Oil and Industry University, Baku AZ1010, Azerbaijan
raliev@asoa.edu.az
[2] Bauman Moscow State Technical University, Moscow 105005, Russia

Abstract. The paper is dedicated to the memory of Lotfi Zadeh, the founder of fuzzy logic. His ideas and theories brought about a considerable shift of modern scientific paradigm, related to the development of «human-centric» mathematical theories with applications. Some pages of Zadeh's extraordinary biography are discussed. The main components of his scientific contribution, including fuzzy sets and linguistic variables, fuzzy logic and approximate reasoning, possibility theory and soft computing, information granulation and computing with words, are reviewed. An attempt to trace the chronology of Zadeh's scientific works is made.

Keywords: Fuzzy set · Fuzzy logic · Approximate reasoning · Possibility theory · Soft computing · Computing with words · Information granulation · Generalized uncertainty theory

1 Introduction

On February 4, 2021 we will celebrate the centenary of the birth of Prof. Lotfi Zadeh (Fig. 1), a great scientist of our time, the founder of a number of major scientific trends in the modern theory of control, applied mathematics, electrical engineering, computer science, artificial intelligence, the Father of Fuzzy Logic. He was part of a cohort of very few pioneering scientists who generate new, original scientific ideas and form the basic scientific paradigms.

Professor L. A. Zadeh was the creator of fuzzy set theory and linguistic variables, fuzzy logic and approximate reasoning, possibility theory and soft computing, information granulation and generalized uncertainty theory, z-numbers and generalized constraints, etc. His ideas and theories not only opened a new era in the development of science free from restrictions of narrow disciplines to enable interdisciplinary synergies. They highly contributed to the emergence of new information and cognitive technologies, brought about the arrival of effective industrial applications, such as fuzzy controllers and computers, fuzzy chips and networks, fuzzy recognition and clustering systems, and so on [1].

The objective of this paper consists in recalling some important biographical details and reviewing basic contributions by the Father of Fuzzy Logic by decades of his scientific activity.

© The Author(s), under exclusive license to Springer Nature Switzerland AG 2021
R. A. Aliev et al. (Eds.): ICAFS 2020, AISC 1306, pp. 148–164, 2021.
https://doi.org/10.1007/978-3-030-64058-3_19

Fig. 1. Professor L. A. Zadeh

2   Biography Pages

Lotfi A. Zadeh was born in Novkhany, Baku Region, Soviet Azerbaijan (now, Republic of Azerbaijan) on February 4, 1921 as Lotfi Aliasker-zadeh [1]. His father, Rahim Aliaskerzade, an Iranian Azerbaijani from Ardabil, was a journalist, the foreign correspondent for the newspaper Iran in Baku, and his Russian Jewish mother, Feiga (Fanya) Korenman, from Odessa, studied medicine and became a pediatrician.

Lotfi attended elementary school in Baku for three years, but in 1931, when he was ten years old, the family moved to Tehran in Iran, his father homeland. Young Lotfi was enrolled in American Alborz College in Tehran, which was a Presbyterian missionary school, where he was educated for the next eight years.

After graduating from Alborz college, Lotfi Zadeh passed the exams to the University of Tehran and placed third in the entire country. In 1942, he graduated from the University of Tehran with a degree in electrical engineering. Then, despite a rather high income, L. Zadeh decided to leave behind a comfortable life in Tehran and immigrate to the United States to fulfill his dream of a career in the academic world.

Lotfi left Tehran early in 1944 traveling to the United States by air and sea. He arrived in New York in July 1944 and moved to Cambridge after spending the summer months working at the International Electronics Corporation.

In 1944 Lotfi Zadeh entered the Massachusetts Institute of Technology as a graduate student and received his MS degree in electrical engineering from MIT in February 1946. Then he decided to move to Columbia University, where he got the position of instructor in electrical engineering. After spending three years as instructor, he obtained his PhD degree in 1949 under the supervision of Professor J. Ragazzini. The thesis was concerned with the frequency analysis of time-varying networks.

In 1950's Zadeh's scientific interests shifted from classical electrical engineering to systems analysis and information science [2]. Already in 1950 he published a significant paper «An extension of Wiener's theory of prediction», co-authored with Professor Ragazzini. This work found application in designing finite-memory filters; today it is considered classical. In 1952 Zadeh again together with Ragazzini proposed the z-transform method for discrete systems. Nowadays this method is also viewed as classical one; it is widely used in digital signal processing. In 1953 he developed a new approach to non-linear filtration and constructed a hierarchy of non-linear systems, which was based on the Volterra-Wiener presentation. Thus, the fundamentals of optimal non-linear processors to detect useful signals in noise were formulated.

Also in 1950's L.A. Zadeh became very interested in probability theory and its application to decision analysis [2]. He met H. Robbins, a brilliant mathematician, and R. Bellman, the father of dynamic programming, who later became his close friends. In 1956–1957 he was a visiting member of the Institute for Advanced Study in Princeton, New Jersey. There he was inspired by a course of logic taught by S. Kleene.

In 1954 Lotfi Zadeh was promoted to the rank of Associate Professor and he received a full professor rank in 1957.

Zadeh taught for ten years at the Columbia University. In January 1959 Professor J. Whinnery, the Chair of Electrical Engineering Department at the University of California, proposed him to move to Berkeley. There were pros and cons. After weighting it, in July 1959 the 38-year-old Lotfi Zadeh with his family started a long journey by car from New York to Berkeley.

Professor Lotfi Zadeh joined the Department of Electrical Engineering at the University of California, Berkeley, in 1959 and served as its chairman from 1963 to 1968. In 1963 he published an important book [3], co-authored by Prof. Desoer, where a new state-based approach to linear system theory was described. This book has survived four editions. Its ideas and results were the sources of various modern approaches in systems analysis and automatic control. Nowadays the state space approach is widely used in system engineering ranging from industrial robots to space guidance control.

Thus, in mid-sixties, Professor L. A. Zadeh had already become a leading scientist in the field of systems theory, automatic control theory, and their applications. Nevertheless, the innovation spirit proper to Lotfi Zadeh did not allow him to rest on his laurels. He made an unexpected and sharp turn over of scientific interests from «honorable strict science» to non-classical «vague science». It was really a risky step that put in question his further scientific career.

3   Fuzzy Sets and Other Early Papers in 1960's

In 1965 44-year-old Prof. L. A. Zadeh published in Information and Control a main scientific work of his life – the pioneering paper «Fuzzy Sets» [4]. This work is of great historical significance. It opened a new scientific area that induced a powerful resonance all over the world and generated an enormous flow of publications. This flow is not exhausted up to now.

A basic Zadeh's idea was simple: a real human reasoning based on natural language cannot be adequately modeled in the framework of classical mathematical methods. The introduction of *fuzzy set* – a class with vague boundaries, described by membership function, provided a suitable basis for developing more flexible approach to reasoning, decision-making and modeling of complex humanistic systems. The behavior of such systems is characterized rather by linguistic variables than ordinary numeric variables.

In the above mentioned paper «Fuzzy Sets» L. Zadeh defined the concept of fuzzy set and its ordinary level sets, suggested various ways of specifying intersection and union operations, introduced pseudo-complementation operation, and new unary operations – concentration and dilatation. Fuzzy relations, their composition and projections were defined. The extension principle was formulated, and mappings of fuzzy sets were considered. Fuzzy restrictions and translations rules for fuzzy propositions were proposed. It is worth noticing that already in [4] fuzzy sets with fuzzy membership functions called fuzzy sets of type 2 were introduced as early counterparts of fuzzy granular structures.

The development of fuzzy models for complex systems to bridge the gap between classical logic and intuition, creation of innovative formal approaches allowing adaptation of strict mathematics to real human ways of everyday thinking and communication – there were novel keynote scientific problems formulated by Lotfi Zadeh.

The first two papers on fuzzy sets by Lotfi Zadeh, «Fuzzy Sets» [4] and «Shadows of Fuzzy Sets» [5] appeared almost at the same time in the USA and Soviet Union. The first Zadeh's translator into Russian V.L. Stefanuk confirmed that Lotfi himself selected adequate words for Russian translation and wished that his new ideas on fuzzy sets and fuzzy logics were known both in the West and in the East.

In 1966, together with R. Bellman and R. Kalaba, L. Zadeh published a paper [6] on the use of fuzzy sets in abstraction and pattern classification in prestigious Journal of Mathematical Analysis and Applications.

In late 1960's Lotfi Zadeh published such papers as «Probability Measures of Fuzzy Events» [7], «Fuzzy Algorithms» [8], «Note on Fuzzy Languages» [9]. He was always interested in probability theory and searched for natural ways of its extension. In [7] Zadeh introduced the notion of a fuzzy event. Usually an event is seen as a precisely specified collection of points in the sample space. By contrast, in everyday experience one frequently encounters situations in which an «event» is ill-defined or fuzzy. Zadeh cited as examples of ill-defined events: «It is a warm day», «x is approximately equal to 5», «In twenty tosses of a coin there are several more heads than tails». These expressions are fuzzy because of the meaning imprecision of the underlined words. He generalized the mathematical expressions for mean, variance and entropy in probability theory in case of fuzzy events. In his opinion, there are many concepts and results in probability theory, information science and related fields which admit of such generalization.

A new conceptual framework for decision-making under fuzziness was proposed in the paper «Decision-Making in a Fuzzy Environment» [10]. Here a fuzzy decision is obtained as a convolution of fuzzy goals and fuzzy constraints. The most important feature of this framework is its symmetry with respect to goals and constraints – a symmetry that erases the differences between them and makes it possible the specification of fuzzy goals and constraints in the same set of alternatives.

This confluence principle was detailed by considering three cases: intersection, product and convex combination of fuzzy goals and constraints. It was also shown that the case where the goals and the constraints were defined as fuzzy sets in different spaces could be easily reduced to the previous case as they would be defined in the same space. Furthermore, the authors illustrated the new decision-making framework by examples of multi-stage decision processes, stochastic systems in a fuzzy environment and systems with implicitly defined termination time.

This paper was considered as a manifesto of fuzzy revolution in decision-making. It was a desktop guide for several generations of both decision-making theorists and practitioners.

4   Basic Publications of 1970's: Fuzzy Relations, Fuzzy Semantics, Linguistic Variables, Fuzzy Logics

Prof. Lotfi A. Zadeh was always deeply interested in the problems of natural and artificial languages that stimulated his studies on semantics. His main semantic question was «Can the fuzziness of meaning be treated quantitatively, at least in principle? In the paper «Quantitative Fuzzy Semantics» [11] he gave an affirmative answer to this question. In the section «Meaning» of this paper he formulated the basics: «We consider two spaces: a universe of discourse U and a set of terms T, which play the roles of names of subsets of U. Let the generic elements of T and be denoted by x and y, respectively. Then the meaning $M(x)$ of a term x is given by a fuzzy subset of U characterized by a membership function $\mu(y|x)$ which is conditioned on x. For instance, if we take a color palette, then the meaning of «red» $M(red)$, is a fuzzy subset of U». In the following section «Language", Zadeh defined a language L as a fuzzy binary relation in $T \times U$ that is expressed by the membership function μ_L: $T \times U \rightarrow [0,1]$.

Another semantic-oriented papers which appeared in 1972, concerned the concept of linguistic hedge. A basic idea suggested in [12] was that linguistic hedges such as «very», «more», «more or less», «much», «essentially», «slightly» etc. may be modeled by unary nonlinear operations which act on the fuzzy set representing the meaning of its operand.

In «Similarity Relations and Fuzzy Orderings» [13] two basic kinds of fuzzy relations were defined. The degrees of similarity and preference were introduced. A fuzzy similarity generalizes the crisp equivalence; it is defined as reflexive, symmetric and transitive fuzzy relation. A fuzzy ordering is a transitive fuzzy relation. In particular, a fuzzy partial ordering is a fuzzy ordering which is anti-symmetric and reflexive. At last, fuzzy linear ordering meets the extended condition of linearity: for any two alternatives x, y either x is preferred to y with a degree $\mu > 0$ or inversely y is preferred to x with a degree $\mu > 0$. Various properties of fuzzy similarity and fuzzy ordering relations were investigated and, as an illustration, an extended version of Szpilrajn's theorem was proved.

Among Zadeh's works in 1970's, four seminal papers are of special concern: «Outline of a New Approach to the Analysis of Complex Systems and Decision Processes» [14], «The Concept of Linguistic Variable and Its Application to Approximate Reasoning» [15], as well as «Fuzzy Logic and Approximate Reasoning» [16] and «Fuzzy Sets as a Basis for a Theory of Possibility» [17].

In [14] Zadeh's *Principle of Incompatibility* was formulated: «As the complexity of a system increases, our ability to make precise and yet significant statements about its behavior diminishes until a threshold is reached beyond which precision and significance (or relevance) become almost mutually exclusive characteristics». And further: «the key elements in human thinking are not numbers, but labels of fuzzy sets, that is, classes of objects in which the transition from membership to membership is gradual rather than abrupt. Indeed, the pervasiveness of fuzziness in human thought processes suggests that much of the logic behind human reasoning is not the traditional two-valued or even multi-valued logic, but the logic with fuzzy truths, fuzzy connectives, and fuzzy rules of inference».

Three main features of the proposed new approach were noticed: 1) use of linguistic variables instead of or in addition to numerical variables; 2) expression of simple relations between fuzzy variables by conditional statements; 3) characterization of complex relations by fuzzy algorithms.

In particular, if x and y are values of linguistic variables, the conditional statements describing the dependence of y on x can be written in the form: "If x is small then y is very large", "If x is not small and not large then y is not very large", and so on. A fuzzy algorithm [8] is an ordered sequence of instructions (like a computer program) in which some of the instructions may contain labels of fuzzy sets, e.g. "Reduce x slightly if y is large", "If x is small then stop; otherwise increase x by 2".

Besides, a compositional rule of inference was proposed and the notion of «Computation of the Meaning of Values for a Linguistic Variable» was introduced. Fuzzy relational and behavior algorithms, in particular, algorithm Behavior, algorithm Oval, algorithm Intersection, algorithm Obstacle and others were constructed.

«The Outline of a New Approach...» was really a keynote paper. It served as a foundation of fuzzy control: on its basis E.Mamdani developed the first fuzzy controller [18].

Another important paper [15] contained a basic definition of linguistic variables: «by a linguistic variable we mean a variable whose values are words or sentences in a natural or artificial language». More specifically, a linguistic variable LV is characterized by a quintuple $\langle L, T(L), U, G, M \rangle$, where L is the name of the variable, $T(L)$ is the term-set of L, that is, the collection of its linguistic values; U is a universe of discourse, G is a syntactic rule which generates the terms in $T(L)$ and M is a semantic rule which associates with each linguistic value X its meaning $M(X)$. Here $M(X)$ denotes a fuzzy subset of U. The meaning of a linguistic value X is characterized by a compatibility function $c: U \to [0,1]$, which associates with each u in U its compatibility with X.

In this paper, the examples of term-sets were specified for *Age, Appearance, Truth,* etc. «The specification of *Truth* as a linguistic variable with values such as *true, very true, completely true, not very true, untrue*, etc., leads to what is called *fuzzy logic*. By providing a basis for approximate reasoning, that is, a mode of reasoning which is neither exact nor very inexact, such logic may offer a more realistic framework for human reasoning than the traditional two-valued logic». Basic logical connectives for fuzzy logic were specified. An example of approximate Modus Ponens rule was given. It was also shown that probability values, too, can be seen as terms of linguistic variable such as *likely, very likely, unlikely*, etc.

Zadeh's preprints on linguistic variables were translated into Russian and published in the form of monograph «The Concept of a Linguistic Variable and Its Application to Approximate Reasoning» [19]. It was the first (and for a long time the only) book by L. Zadeh.

The paper entitled «Fuzzy Logic and Approximate Reasoning» [16] was published in 1975; it was the first Zadeh's big article with reflection on fuzziness in logic (the short paper of 1974 [20] can be mentioned only in a historical retrospective).

The term «fuzzy logic» is used in this paper to describe an imprecise logical system, in which the truth-values are fuzzy subsets of the unit interval with linguistic labels such as *true, false, not true, very true, more or less true, rather false, very false,* etc. Linguistic truth values are not allowed in traditional logical systems, but are routinely used by humans in everyday discourse. The truth-value set is assumed to be generated by a context-free grammar, with a semantic rule providing a means of computing the meaning of each linguistic truth-value as a fuzzy subset of [0,1]. Since it is not closed under the operations of negation, conjunction, disjunction and implication, the result of an operation on truth-values requires, in general, a linguistic approximation. As a consequence, the truth tables and the rules of inference in fuzzy logic are inexact and depend on the meaning associated with the primary truth-value *true* as well as the modifiers *very, quite, more or less.*

In [16] L. Zadeh summarized: «Perhaps the simplest way of characterizing fuzzy logic is to say that it is the logic of approximate reasoning. As such, it is a logic whose distinguishing features are: (i) fuzzy truth-values expressed in linguistic terms with modifiers; (ii) imprecise truth tables; and (iii) rules of inference whose validity is approximate rather than exact.

In these respects, fuzzy logic differs significantly from standard logical systems ranging from the classical Aristotelian logic to inductive logics and many-valued logic with set-valued truth-values».

In the paper «Local and Fuzzy Logics» [21] the authors emphasized that «Fuzzy logic is local, i.e. both the truth values and their conjunctions such as "AND", "OR" and "IF-THEN" have variable rather than fixed meanings. This is the reason why fuzzy logic can be viewed as a local logic. Hence, the inference process has a semantic character rather than a syntactic one: in FL, the conclusion depends on the meaning assigned to the fuzzy sets that appear in the set.

Consequently, fuzzy logic is the result of a double weakening of the basic laws of classical logic. On the one hand, the principle of bivalence and the law of excluded middle are rejected, that gives rise to a multi-valued logic and, finally, to a membership function that allows us to interpret the predicates. On the other hand, the variability of the meaning related both to truth values and connectives, makes the logical inference imprecise.

A detailed lecture «A Theory of Approximate Reasoning» [22] was delivered by L. Zadeh at the International Workshop on AI in Repino near Leningrad in 1977 (Fig. 2). It already contained the concepts of fuzzy restriction and fuzzy constraint propagation.

Fig. 2. The International Workshop on Artificial Intelligence in Repino near Leningrad (April 18–24, 1977). From right to left: L. Zadeh is participating in a discussion, J. McCarthy, the computer scientist known as the father of AI, V. I. Varshavsky, the Soviet classic in the field of collective behavior of automata, D. A. Pospelov, the founder of AI in Soviet Union.

Later on, in 1994 L. Zadeh already noticed that «The term fuzzy logic is actually used in two different senses. In a narrow sense, fuzzy logic is a logical system which is an extension of multi-valued logic serving as logic of approximate reasoning. But in a wider sense, fuzzy logic is more or less synonymous with the theory of fuzzy sets» [23].

Nowadays, a broad concept of fuzzy logic includes fuzzy sets and linguistic variables (specifically, linguistic truth values), fuzzy relations and approximate reasoning, fuzzy rules and fuzzy constraints, test-score (experience-based) semantics and generalized uncertainty theory, etc. It encompasses a variety of soft formal methods and tools for fuzzy control, pattern recognition, natural language processing, and so on. In some sense, it implements the engineering approach to logical modeling.

A seminal paper «Fuzzy Sets as a Basis for a Theory of Possibility» [17] was published in 1978 in the first issue of Fuzzy Sets and Systems. Here the main Zadeh's thesis was as follows: «When our main concern is with the meaning of information rather than with its measure, the proper framework for information analysis is possibilistic in nature than probabilistic one. What is needed for such an analysis is not probability theory but an analogous and yet different. This theory might be called the theory of possibility. The mathematical apparatus of the theory of fuzzy sets provides a natural basis for the theory of possibility, playing a role which is similar to that of measure theory in relation to the theory of probability».

In [17] Zadeh introduced the concept of possibility distribution function via membership function of fuzzy set and considered it as an interpretation of fuzzy restriction. The possibility measure can be easily constructed by possibility distribution with using max-normalization. To a large extent, possibility theory is comparable to probability theory because it is based on set-functions. However, the possibility measure is a modification of probability measure: the two first axioms of classical measure–boundary condition and monotonousness axiom – are preserved, but the additivity axiom is replaced by max-axiom (the either-or condition). To differ from the probability measure, the possibility measure has its dual called the necessity measure; it was introduced by D. Dubois and H. Prade [24] by taking instead of the additivity axiom the min-axiom (the both-and condition). The possibility-probability consistency principle was proposed to explain a weak connection between possibility and probability. In particular, Zadeh spoke about the primacy of possibility: what is probable must preliminarily be possible. At the same time, following Dempster-Shafer approach, imprecise probabilities are introduced by taking possibility as upper estimate and necessity as lower estimate of probability measure.

Marginal possibility distributions and conditional possibility distributions were also studied in [17]. Moreover, the possibility distributions of composite and qualified propositions were introduced. Conditional translation rules of type I and type II were proposed.

In brief, possibility theory is an uncertainty theory aimed at handling of incomplete information. It is not additive measure and makes sense on ordinal structures.

Possibility theory is one of the most promising off-springs of fuzzy sets that helps bridge the gap between artificial intelligence and statistics. It clarifies the role of fuzzy sets in uncertainty management and explains why probability degrees, viewed as frequency or betting rates, can be used to derive membership functions.

The concepts of possibility theory were successfully applied in PRUF (Possibilistic Relational Universal Fuzzy) – a meaning representation language for natural languages [25]. In addition to approximate reasoning, PRUF can be employed as a language for the representation of imprecise knowledge and as a means of precisiation of fuzzy propositions expressed in a natural language.

According to B. Turksen [26], by the late 1970's, Lotfi Zadeh and his followers essentially developed the foundation of applied fuzzy mathematics.

5 How to Make McCarthy's Dream Come True: Fuzzy Approaches to the Modeling of Common-Sense Knowledge

The father of Artificial Intelligence, John McCarthy throughout all his life believed that a long-term goal of AI is human-level AI (see [27]). In his opinion, it could be based on the formalizing of commonsense knowledge and reasoning in mathematical logic. Since 1980's, many Zadeh's papers were dedicated to applications of fuzzy sets, fuzzy logic, possibility theory in artificial intelligence. In [28] it was stressed that «management of uncertainty is an intrinsically important issue in the design of expert systems because much of the information in the knowledge base of a typical expert system is imprecise, incomplete or not totally reliable. Fuzzy logic subsumes both predicate logic and probability theory, and makes it possible to deal with different types of uncertainty within a single conceptual framework. In fuzzy logic, the deduction of a conclusion from a set of premises is reduced, in general, to the solution of a nonlinear program through the application of projection and extension principles. This approach to deduction leads to various basic syllogisms which may be used as rules of combination of evidence in expert systems. Among them are the intersection/product syllogism, the generalized modus ponens, the consequent conjunction syllogism, and the major-premise reversibility rule».

In [29] this fuzzy syllogistic topic was continued: «A fuzzy syllogism in fuzzy logic is defined as an inference schema in which the major premise, the minor premise, and the conclusion are propositions containing fuzzy quantifiers. A basic fuzzy syllogism in fuzzy logic is the intersection/product syllogism. Furthermore, it is noticed that syllogistic reasoning in fuzzy logic provides a basis for reasoning with dispositions, that is, with propositions that are preponderantly but not necessarily always true. It is also shown that the concept of dispositionality is closely related to the notion of usuality and serves as a gateway to what might be called a theory of usuality, a theory that may eventually provide a computational framework for commonsense reasoning».

The fundamentals of commonsense knowledge were discussed in [30], and the theory of disposition was outlined in [31, 32]. The basic idea was that «commonsense knowledge may be viewed as a collection of dispositions, that is, propositions with implied fuzzy quantifiers». Typical examples of dispositions are: *Icy roads are slippery. Tall men are not very agile. What is rare is expensive, etc.* It is understood that, upon restoration of fuzzy quantifiers, a disposition is converted into a proposition with explicit fuzzy quantifiers, e.g., *Tall men are not very agile → Most tall men are not very agile.*

Since traditional logical systems do not provide methods for representing the meaning of propositions containing fuzzy quantifiers, such systems are unsuitable for dealing with commonsense knowledge. An appropriate computational framework for dealing with commonsense knowledge is provided by fuzzy logic, which is the logic underlying approximate reasoning. A summary of the basic concepts and techniques underlying the application of fuzzy logic to knowledge representation was given in [32].

6 New Trends in 1990's: Soft Computing, Information Granulation, Computing with Words

In 1990's Professor L.A. Zadeh perceived a new inspiration and burst of creative energy that was resulted in opening innovative scientific areas – Soft Computing and Information Granulation. In 1994 two pioneering papers on Soft Computing [33, 34] appeared. This Zadeh's initiative is closely related to the emergence of hybrid systems in computer science and AI.

In biology, hybridization is considered as the most powerful form of integration, when in one organism the various hereditary features are merged. By analogy, a hybrid system in Computer Science includes two or more heterogeneous subsystems, integrated by a shared goal or joint actions, although these subsystems can have both different nature and specification languages. In brief, hybrid computer systems use two or more various computer technologies.

According to Zadeh [33, 34], the basis of soft computing is that unlike the traditional hard computing, soft computing is aimed at an accommodation with pervasive imprecision of real world. The guiding principle of soft computing is to exploit the tolerance for imprecision and uncertainty to achieve tractability, robustness, low solution cost and better rapport with reality. The role model for soft computing is the human mind.

Soft computing is not a single methodology. Moreover, it is not a simple collection of methodologies, but their partnership. The principal partners in this juncture are fuzzy logic, neurocomputing, genetic algorithms, probabilistic reasoning and chaos theory. Here fuzzy logic is mainly concerned with imprecision and approximate reasoning, probabilistic reasoning – with uncertainty and propagation of beliefs, neural network – with learning, genetic algorithm – with search and optimization, and chaos theory – with nonlinear dynamics. In essence, fuzzy logic, neurocomputing, genetic algorithms, etc. are complementary and synergistic rather than competitive technologies. For this reason, it is worth using them in combination.

At the turn of the XXth and XXIst centuries such initiatives of the Father of Fuzzy Logic as an Information Granulation Theory and a non-traditional Granular Mathematics program seem to be of primary concern. In 1979 his work appeared entitled «Fuzzy Sets and Information Granularity», where information granules were introduced [35]. For some time this work remained imperceptible. The situation changed in 1997 when L. Zadeh formulated fundamentals of the Theory of Fuzzy Information Granulation (TFIG) in his seminal paper «Toward a Theory of Fuzzy Information Granulation and its Centrality in Human Reasoning and Fuzzy Logic» [36].

The term «granule» is originated from Latin word *granum* that means grain, to denote a small particle in the real world. In [36] L. Zadeh specified *granule* as «a collection of objects which are drawn together by indistinguishability, similarity, proximity or functionality». There are various classifications of granules: crisp and fuzzy granules, information and knowledge granules, time and space granules, etc.

Specifically, the following Zadeh's granulation principle was formulated in 2000 in a qualitative form [37]: «To exploit the tolerance for imprecision, employ the coarsest level of granulation, which is consistent to allowable level of imprecision». Later on W. Pedrycz [38] suggested the following condition: find the compromise by two criteria – the size of granule and the number of granules, i.e. data specificity and data coverage.

Granulation is a basic property of human cognition. According to L. Zadeh, there are three basic concepts that underlie human cognition: granulation, organization and causation. Informally, granulation involves decomposition of whole into parts; organization involves integration of parts into whole; and causation involves association of causes with effects.

In [36] L. A. Zadeh points out that «the TFIG is inspired by the ways in which humans granulate information and reason with it. However, the basics of TFIG and its methodology are mathematical in nature. The point of departure in TFIG is the concept of a generalized constraint. A granule is characterized by a generalized constraint which defines it. The principal types of granules are: possibilistic, veristic and probabilistic».

In Zadeh's opinion, «the fuzzy information granulation may be viewed as a mode of generalization which may be applied to any concept, method or theory. Related to fuzzy granule, there are the following principal modes of generalization:

1. *Fuzzification* (f-generalization). In this mode of generalization a crisp set is replaced by a fuzzy set.
2. *Granulation* (g-generalization). In this case, a set is partitioned into granules.
3. *Randomization* (r-generalization). In this case, a variable is replaced by a random variable.
4. *Usualization* (u-generalization). In this case, a proposition expressed as X *is* A is replaced with: usually $(X$ *is* $A)$».

The TFIG enables a natural foundation for Computing with Words (CW) [39]. The keystone of CW is the idea that natural language words and word combinations are the labels of fuzzy granules. Moreover, in computing with words, the meaning of a proposition is related to a represented fuzzy constraint.

In CW the initial and terminal data sets, IDS and TDS, are given in the form of propositions expressed in a natural language. These propositions are translated, respectively, into antecedent and consequent constraints. Here consequent constraints are derived from antecedent constraints by employing the rules of constraint propagation. The main constraint propagation rule is the generalized extension principle.

From the viewpoint of Prof. L. A. Zadeh [39], there are two major imperatives for computing with words. At first, CW is a necessity when the available information is too imprecise to justify the use of numbers. At second, computing with words is expedient when there is a tolerance for imprecision, uncertainty and partial truth that can be exploited to achieve tractability, robustness, low solution cost and better rapport with reality.

7   2000–2010's: From Computing with Perceptions and GTU to Z-Numbers and Fuzzy Logic Geometry

Even in 2000's after his 80th anniversary Professor L. A. Zadeh continued to work very actively (Figs. 3, 4). He extended CW to computing with perceptions [40, 41], developed the theory of generalized constraints [39–43] and the generalized uncertainty theory [44].

Manipulation of perceptions plays a key role in human recognition, decision and execution processes. To compute with perceptions, we can use the generalized constraint language (GCL). In this language, the meaning of a proposition is expressed as a generalized constraint X *isr* R, where X is the constrained variable, R is the constraining relation and *isr* is a variable copula in which *r* is a variable whose value defines the way in which R constrains *X*. Generalized constraints may be qualified, combined and propagated. The set of all generalized constraints together with rules governing qualification,combination and propagation procedures constitutes the GCL.

In [42] the foundations of perception-based theory of probabilistic reasoning with imprecise probabilities are considered, and in [43] the transition from imprecise to granular probabilities is discussed.

A principal premise of Generalized Theory of Uncertainty (GTU) [44] is that, fundamentally, information is a generalized constraint on the values which a variable is allowed to take. The principal tools which GTU draws from fuzzy logic include precisiated natural language [45] and protoform theory.

In 2011 L. A. Zadeh introduced the concept of a Z-number that relates to the issue of reliability of information [46]. A Z-number is defined as an ordered pair of fuzzy numbers $Z = (A, B)$, where the first component A is a restriction (constraint) on the values which a real-valued uncertain variable X is allowed, and the second component B is a measure of reliability (certainty) of the first component. An example of Z-number is: (about 45min, very sure). Basic operations over Z-numbers are considered in [47].

In a series of papers on extended fuzzy logics [48–51] such new concepts as fuzzy validity, restriction-centered theory of truth, and, finally, an extended fuzzy logic based on fuzzy geometry (fuzzily defined points, lines, circles, etc.) and fuzzy transforms are introduced.

Fig. 3. Lotfi Zadeh and Rafik Aliev, the co-authors of the book «Fuzzy Logic Theory and Applications: Part I and Part II»

At the World Congress of the International Fuzzy Systems Association in June 2007, Cancun, Mexico, 86-year old L.A.Zadeh gave the hour-long plenary talk «Fuzzy Logic as the Logic of Natural Languages».

Fig. 4. Lotfi Zadeh and vice-president of Russian Association for Fuzzy Systems and Soft Computing Valery Tarassov at the IFSA-2007 World Congress in Cancun, June 19, 2007.

The 6th World Conference on Soft Computing dedicated to 95th anniversary of L.A, Zadeh took place in Berkeley in May 22–25, 2016. It was the last international event with Lotfi Zadeh.

8   Conclusion

The scientific power of the theory is largely determined by the possibilities of its further evolution and extension, the resonance it causes in the scientific community. Fuzzy set theory has not been the only model introduced to deal with imprecise and uncertain information. During the past 55 year a lot of new models have been proposed to mathematically tackle incomplete information. Some of them are extensions of fuzzy set theory and others use different paths.

The following nonstandard set models were created after Zadeh's paper on fuzzy sets [4]: Type-2 fuzzy set and L-fuzzy set, Flou set and L-flou set, Interval-valued fuzzy set and Intuitionistic fuzzy set, Level fuzzy set and Twofold fuzzy set, Random set and Probabilistic set, Grey set and Soft set, Toll set and Bipolar fuzzy set, Vague set and Shadow set, Rough set, Fuzzy rough set and Rough fuzzy set theory, etc. (see [52, 53]).

The role of Lotfi Zadeh in the modern world is not limited only to scientific achievements that opened the age of innovative fuzzy technologies. His extraordinary destiny, his own graded membership to different nations and cultures, comfortably carried through the whole life, that had directed his international activities, made a valuable contribution to the formation of planetary scientific community of XXI century. It accelerated the emergence of a new synergistic scientific vision that supposes a symbiosis of eastern and western traditions.

Acknowledgment. The work is supported by Russian Foundation for Basic Research, the project No 20-07-00770.

References

1. Zadeh, L.A.: Wikipedia
2. Zadeh, L.A.: My life and work: a retrospective view applied and computational mathematics. Spec. Issue Fuzzy Set Theory Appl. **10**(1), 4–9 (2011)
3. Zadeh, L.A., Desoer, Ch.: Linear System Theory: The State Space Approach. McGraw-Hill Book Company Inc., New York (1963)
4. Zadeh, L.A.: Fuzzy sets. Inf. Control **8**, 338–353 (1965)
5. Zadeh, L.A.: Shadows of fuzzy sets, Russian translation by V. Stefanuk. Inf. Trans. Prob. **2**(1), 37–44 (1966). in Russian
6. Bellman, R.E., Kalaba, R., Zadeh, L.A.: Abstraction and pattern ckassification. J. Math. Anal. Appl. **13**, 1–7 (1966). Reprinted in: Fuzzy Models for Pattern Recognition. Bezdek, J. C., Pal, S.K. (eds.) IEEE Press, New York, pp. 231–235 (1992)
7. Zadeh, L.A.: Probability measures of fuzzy events. J. Math. Anal. Appl. **23**, 421–427 (1968)
8. Zadeh, L.A.: Fuzzy algorithms. Inf. Control **12**, 94–102 (1968)
9. Lee, E.T., Zadeh, L.A.: Note on fuzzy languages. Inf. Sci. **1**(4), 421–434 (1969)
10. Bellman, R.E., Zadeh, L.A.: Decision-making in a fuzzy environment. Manag. Sci. **17**(4), 141–164 (1970)
11. Zadeh, L.A.: Quantitative fuzzy semantics. Inf. Sci. **3**(2), 159–176 (1971)
12. Zadeh, L.A.: A fuzzy-set-theoretic interpretation of linguistic hedges. J. Cybern. **2**, 4–34 (1972)
13. Zadeh, L.A.: Similarity relations and fuzzy orderings. Inf. Sci. **3**(2), 177–200 (1971)
14. Zadeh, L.A.: Outline of a new approach to the analysis of complex systems and decision processes. IEEE Trans. Syst. Man Cybern. **3**(1), 28–44 (1973)
15. Zadeh, L.A.: The concept of linguistic variable and its application to approximate reasoning. Parts 1 and 2. Inf. Sci. **8**(199–249), 301–357 (1975)
16. Zadeh, L.A.: Fuzzy logic and approximate reasoning. Synthese **30**, 407–428 (1975)
17. Zadeh, L.A.: Fuzzy sets as a basis for a theory of possibility. Fuzzy Sets Syst. **1**(1), 3–28 (1978)
18. Mamdani, E.H., Assilian, S.: An experiment in linguistic synthesis with a fuzzy logic controller. Int. J. Man Mach. Stud. **7**(1), 1–13 (1975)
19. Zadeh, L.A.: The Concept of Linguistic Variable and its Application to Approximate Reasoning. Mir, Moscow (1976)
20. Zadeh, L.A.: Fuzzy logic and its application to approximate reasoning. In: Proceedings of 1974 IFIP Congress on Information Processing. North-Holland, Amsterdam, vol. 3, pp. 591–594 (1974)
21. Bellman, R.E., Zadeh, L.A.: Local and fuzzy logics. In: Dunn, J.M., Epstein, G. (eds.) Modern Uses of Multiple-Valued Logics, pp. 105–116. D. Reidel, Dordrecht (1977)
22. Zadeh, L.A.: Theory of Approximate Reasoning. Machine Intelligence. In: Hayes, J., Michie, D., Mikulich, L.I. (eds.) pp. 149–194. Halstead Press, New York (1979)

23. Zadeh, L.A.: Fuzzy logic and soft computing: issues, contentions and perspectives. In: Proceedings of the 3rd International Conference on Fuzzy Logic, Neural Nets and Soft Computing, Iizuka, Japan, March 2004, vol. 1–2, pp. 77–84 (1994).

24. Dubois, D., Prade, H.: Possibility Theory: An Approach to Computerized Processing of Uncertainty. Plenum Press, New York (1988)

25. Zadeh, L.A.: PRUF: a meaning representation language for natural languages. Int. J. Man Mach. Stud. **10**, 395–460 (1978)

26. Turksen, B.: Scientific and philosophical contribution of L. A. Zadeh. In: Proceedings of the 1st International Conference on Soft Computing and Computing with Words in System Analysis, Decision and Control, pp. 17–23. B Quadrat Verlag, Kaufering (2001)

27. McCarthy, J.: The future of AI – a manifesto. AI Mag. **26**(4), 39 (2005)

28. Zadeh, L.A.: The role of fuzzy logic in the management of uncertainty in expert systems. Fuzzy Sets Syst. **11**, 199–227 (1983)

29. Zadeh, L.A.: Syllogistic reasoning in fuzzy logic and its application to usuality and reasoning with disposition. IEEE Trans. Syst. Man Cybern. **15**, 754–763 (1985)

30. Zadeh, L.A.: A Theory of Common-Sense Knowledge. In: Skala, H.J., Termini, S., Trillas, E. (eds.) Aspects of Vagueness, pp. 257–295. D.Reidel, Dordrecht (1984)

31. Zadeh, L.A.: A computational theory of dispositions. Int. J. Intell. Syst. **2**, 39–63 (1987)

32. Zadeh, L.A.: Knowledge representation in fuzzy logic. IEEE Trans. Knowl. Data Eng. **1**(1), 89–100 (1989)

33. Zadeh, L.A.: Fuzzy logic, neural network and soft computing. Commun. ACM **37**(3), 77–84 (1994)

34. Zadeh, L.A.: Soft computing and fuzzy logic. IEEE Softw. **11**(6), 48–58 (1994)

35. Zadeh, L.A.: Fuzzy sets and information granularity. In: Gupta, M.M., Ragade, R.K., Yager, R.R. (eds.) Advances in Fuzzy Sets Theory and Applications, North-Holland, Amsterdam, pp. 3–20 (1979)

36. Zadeh, L.A.: Toward a theory of fuzzy information granulation and its centrality in human reasoning and fuzzy logic. Fuzzy Sets Syst. **90**, 111–127 (1997)

37. Zadeh, L.A.: Toward a logic of perception based on fuzzy logic. In: Novak, V., Perfilieva, I. (eds.) Discovering the World with Fuzzy Logic, pp. 4–28. Physica-Verlag, Heidelberg (2000)

38. Pedrycz, W.: Granular Computing. Analysis and Design of Intelligent Systems. CRC Press, Boca Raton (2013)

39. Zadeh, L.A.: Fuzzy logic = computing with words. IEEE Trans. Fuzzy Syst. **4**, 103–111 (1996)

40. Zadeh, L.A.: From computing with numbers to computing with words – from manipulation of measurements to manipulation of perceptions. In: Wang, P.P. (ed.) Computing with Words, pp. 35–68. Wiley and Sons, New York (2001)

41. Zadeh, L.A.: A new direction in ai: toward a computational theory of perceptions. AI Mag. **22**(1), 73–84 (2001)

42. Zadeh, L.A.: Toward a perception-based theory of probabilistic reasoning with imprecise probabilities. J. Stat. Plan. Infer. **105**, 233–264 (2002)

43. Zadeh, L.A.: From imprecise to granular probabilities. Fuzzy Sets Syst. **154**, 370–374 (2005)

44. Zadeh, L.A.: Toward a generalized theory of uncertainty (GTU) – an outline. Inf. Sci. **172**, 1–40 (2005)

45. Zadeh, L.A.: Precisiated natural language (PNL). AI Mag. **25**(3), 74–91 (2004)

46. Zadeh, L.A.: A note on Z-numbers. Inf. Sci. **181**(2011), 2923–2932 (2011)

47. Aliev, R.A., Huseynov, O.H., Alyev, R.R.: The Arithmetic of Z-Numbers. World Scientific, Singapore (2015)

48. Zadeh, L.A.: Fuzzy logic – a new direction. the concept of f-validity. in: new dimensions in fuzzy logic and related technologies. In: Proceedings of the 5th EUSFLAT Conference, vol. 1, pp. 29 (2007)
49. Zadeh, L.A.: Toward extended fuzzy logic – a first step. Fuzzy Sets Syst. **160**, 3175–3181 (2009)
50. Zadeh, L.A.: Toward a restriction-centered theory of truth and meaning. Inf. Sci. **248**, 1–4 (2013)
51. Zadeh, L.A.: Fuzzy logic: a personal perspective. Fuzzy Sets Syst. **281**, 4–20 (2015)
52. Kerre, E.: The impact of fuzzy set theory on contemporary mathematics. Int. J. Appl. Comput. Math. **10**(1), 20–34 (2011)
53. Tarassov, V.B.: Information granulation by cognitive agents and non-standard fuzzy sets. In: Aliev, R.A., Bonfig, K.W., Kreinovich, V., Turksen, I.B. (eds.) Proceedings of the Sixth International Conference on Soft Computing, Computing with Words and Perceptions in System Analysis, Decision and Control, pp. 59–74. b-Quadrat Verlag, Kaufering (2011)

Combined Heat and Power Economic Emission Dispatch Applying Exchange Market Algorithm with Fuzzy Satisfying Techniques

Fahreddin Sadikoglu[1]([✉]) [iD] and Ebrahim Babaei[2] [iD]

[1] Engineering Faculty, Near East University, Mersin 10, 99138 Nicosia,
North Cyprus, Turkey
fahreddin.sadikoglu@neu.edu.tr
[2] Faculty of Electrical and Computer Engineering, University of Tabriz,
Tabriz, Iran
e-babaei@tabrizu.ac.ir

Abstract. Increasing fuel cost and environmental issues force power producers to use high efficiency methods with low pollutant gas emission for power generation as in combined heat and power (CHP) generators. Non-linearity, non-convexity and complication of CHP generators turn combined heat and power economic emission dispatch (CHPEED) setback to an intricate maximization procedure. Because of inability of conventional methods in solving CHPEED problem, heuristic and evolutionary algorithms are needed to solve. The CHPEED problem is a multi-objective maximization setback where production cost and emission are the two competing objective functions. In this paper, the challenges of multi-objective CHPEED is regarded as a single objective optimization problem by utilizing weighting coefficients. Application of exchange market algorithm (EMA) is necessitated to resolve this setback and also find compromise solutions. In view of selecting the optimal accommodating solution, fuzzy satisfying method is applied.

Keywords: Combined heat and power · Economic emission dispatch · Exchange market algorithm · Fuzzy decision making · Valve-point effects

1 Introduction

In recent past years, energy efficiency is of critical importance. The energy efficiency in conventional thermal units is less than 60%. While in combined heat and power (CHP) units, both heat and power are produced from fuel source and efficiency is approximately 90% due to the application of wasted heat. Emission and production costs are decreased in CHP units about 13–18% and 10–40%, respectively [1, 2].

Co-generation has attracted attention of many researchers due to reduction in cost and emission and increasing heat demand in the industry. As a result, the importance of CHP economic dispatch (CHPED) problem grows significantly. CHPED is to identify the appropriate output of individual unit so as to minimize the total value and also satisfy the constraints. The complicated parts in the CHPED are twin reliance on power/heat generation and steam valve admission effects. Thus, the CHPED becomes a non-convex,

R. A. Aliev et al. (Eds.): ICAFS 2020, AISC 1306, pp. 165–173, 2021.
https://doi.org/10.1007/978-3-030-64058-3_20

non-linear and non-smooth setback that conventional methods are unable to find optimum solution [3]. In order to solve such complicated problem, different heuristic algorithms are proposed. A better particle swarm maximization-based method named selective particle swarm optimization (SPSO) was proposed in [4]. In [5] genetic algorithm (GA) was used to resolve challenges of CHPED. Firefly algorithm (FA) [6], cuckoo algorithm optimization (COA) [7], group search optimization (GSO) [8], time varying acceleration coefficients particle swarm optimization (TVAC-PSO) [3] and ant colony search algorithm (ACSA) [9] are the other methods applied to overcome the CHPED setback. So as to better operations of primary group search optimization (GSO) algorithm, adaptive scrounger and ranger procedures were utilized and a new modified group search optimization (MGSO) method was presented in [10]. A hybrid optimization method with the aid of civilized swarm optimization (CSO) being a comprehensive search method and Powell's pattern search (PPS) method as local search method has been presented in [11] to solve the CHPED problem. A thorough review of heuristic algorithms for CHPED is presented in [12].

In all of above-mentioned researches environmental aspects of dispatch problem is not considered. In recent years, due to environmental issues and global concern about pollutant gases emission, various techniques were proposed to resolve CHP economic emission dispatch problem (CHPEED). A multi-objective advanced technique with the aid of strength Pareto evolutionary algorithm (SPEA) and fuzzy set theory has been proposed in [13]. Multi-objective line-up competition algorithm (MLCA) with the aid of fuzzy decision-making process are used to provide Pareto-optimal solutions and choose the best compromise solution, respectively, in [14]. Nondominated classification genetic procedure II is implemented in the CHPEED setback in [15]. All mentioned methods used to solve the CHPEED problem have deterministic solutions with high computational time. So, there is a need to find a method that is able to find improved solutions in relation to cost and emission in less computational time.

Exchange market algorithm (EMA) is a newly proposed method for deciphering optimization problems like economic load dispatch [16, 17], reactive power dispatch [18], optimal economic and emission dispatch [19] and reliable economic dispatch of microgrids [20]. EMA is influenced by selling and buying shares in the stock market in balanced and unbalanced market modes [21]. Economic dispatch of CHP systems using EMA is surveyed in [17] while pollutant gases emission is not considered in this reference. With regard to importance of environmental issues and global trends to reduce pollutant gases emission, in this paper, EMA is applied to reduce the cost and emission in CHP units by considering the limitations. It is a complex task to locate a balanced solution involving cost of fuel and emission. Hence, multi-objective setback is solved as a single objective problem by applying weighting coefficients for competing objective operations and fuzzy satisfying method is applied in choosing suitable compromise solution in this paper. Comparing results with other methods shows capability and supremacy of the proposed method.

2 Formulation of the Problem

In this study a system comprising of heat-only units, power-only units, and co-generation units that produce power and heat simultaneously has been considered. The heat-power operational attainable portion of a CHP unit has been depicted in Fig. 1. The twin reliance on heat and power is shown in this figure. Also, non-convexity is taken into account in workable operation domain of this sample CHP units.

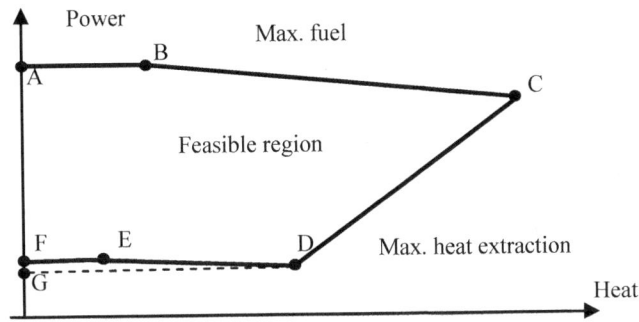

Fig. 1. Heat-power viable operational domain for a co-generation unit.

2.1 Objective Function

Reducing the cost and emission are the objectives of the CHPEED problem while power and heat requirements and system operational challenges are met. Objective function is mentioned in (1):

$$\min\ C_T = w \times F_1 + (1 - w) \times F_2 \tag{1}$$

where F_1 and F_2 are the two objective functions related to production cost of power and amount of pollutant gas emission, accordingly. w is weighting coefficient in the interval [0, 1].

2.2 Production Cost Functions

The cost of power production can be stated using (2):

$$F_1(\$/h) = \sum_{i=1}^{N_p} C_i(P_i^p) + \sum_{j=1}^{N_c} C_j(P_j^c, H_j^c) + \sum_{k=1}^{N_h} C_k(H_k^h) \tag{2}$$

where C is the cost function for each unit. N_P is the power-only unit count, N_c co-generation unit count and N_h is the heat-only unit count. i, j and k are the parameters for mentioned entities. Cost functions of the power only units, cogeneration and heat only units are shown in (3–5) respectively.

$$C_i(P_i^p) = \alpha_i(P_i^p)^2 + \beta_i P_i^p + \gamma_i (\$/h) \tag{3}$$

$$C_j(P_j^c, H_j^c) = a_j(P_j^c)^2 + b_j P_j^c + c_j + d_j(H_j^c)^2 + e_j H_j^c + f_j P_j^c H_j^c (\$/h) \tag{4}$$

$$C_k(H_k^h) = a_k(H_k^h)^2 + b_k H_k^h + c_k(\$/h) \tag{5}$$

where $C_i(P_i^p)$ represents the cost of ith power-only unit to produce P_i^p MW. α_i, β_i and γ_i are the cost coefficients of power-only units. $C_j(P_j^c, H_j^c)$ is the cost function, a_j, b_j, c_j, d_j, e_j and f_j are the cost coefficients of jth co-generation unit while producing P_j^c MW and H_j^c MWth. $C_k(P_k^h)$ is the cost function and a_k, b_k and c_k are the coefficients of kth heat-only entity whilst providing H^h MWth heat.

Quadratic cost algorithms for traditional thermal units as in (3) have been used widely in the literature [22, 23]. In order to consider steam valve admission effects in the production cost, a sinusoidal quantity is included in the quadratic cost function. Cost function of power-only units considering valve point reactions is shown in (6):

$$C_i(P_i^p) = \alpha_i(P_i^p)^2 + \beta_i P_i^p + \gamma_i + |\lambda_i \sin(\rho_i(P_i^{pmin} - P_i^p))| (\$/h) \tag{6}$$

2.3   Emission Functions

The pollutant gas emission can be defined as:

$$F_2 \text{ (kg)} = \sum_{i=1}^{N_p} E_i(P_i^p) + \sum_{j=1}^{N_c} E_j(P_j^c, H_j^c) + \sum_{k=1}^{N_h} E_k(H_k^h) \tag{7}$$

where E is known as emission function for each unit. Emission functions for thermal, CHP and heat-only units are shown in (8–10), respectively:

$$E_i(P_i^p) = \alpha_{ei}(P_i^p)^2 + \beta_{ei} P_i^p + \gamma_{ei} + \lambda_{ei} \times exp(\rho_{ei} \times P_i^p) \text{ (kg)} \tag{8}$$

$$E_j(P_j^c, H_j^c) = a_{ej} \times P_j^c (\text{kg}) \tag{9}$$

$$E_k(H_k^h) = a_{ek} \times H_k^h (\text{kg}) \tag{10}$$

where $E_i(P_i^p)$ is the emission of ith conventional thermal unit while producing P_i^p MW. α_{ei}, β_{ei}, γ_{ei}, λ_{ei} and ρ_{ei} are emission coefficients of ith conventional thermal unit. $E_j(P_j^c, H_j^c)$ is the amount of emission in kg and a_{ej} is the emission coefficient for jth co-generation unit. $E_k(H_k^h)$ and a_{ek} are the emission amount and emission coefficient for kth heat-only unit, respectively.

2.4   Constraints

The summation of power produced by power-only and CHP units should satisfy electrical power demand and system power loss as depicted in (11). The heat demand and produced heat equality constraint is shown in (12):

$$\sum_{i=1}^{N_p} P_i^p + \sum_{j=1}^{N_c} P_j^c = P_d \tag{11}$$

$$\sum_{j=1}^{N_c} H_j^c + \sum_{k=1}^{N_h} H_k^h = H_d \tag{12}$$

where P_d is the electrical power demand and H_d is the complete heat requirement of the system. The constraints related to upper and lower limits of generating units are as follows:

$$P_i^{p\,\min} \leq P_i^p \leq P_i^{p\,\max} \quad i = 1, 2, 3, \ldots, N_p \tag{13}$$

$$P_j^{c\,\min}(H_j^c) \leq P_j^c \leq P_j^{c\,\max}(H_j^c) \quad j = 1, 2, 3, \ldots, N_c \tag{14}$$

$$H_j^{c\,\min}(P_j^c) \leq H_j^c \leq H_j^{c\,\max}(P_j^c) \quad j = 1, 2, 3, \ldots, N_c \tag{15}$$

$$H_k^{h\,\min} \leq H_k^h \leq H_k^{h\,\max} \quad k = 1, 2, 3, \ldots, N_h \tag{16}$$

where $P_i^{p\,\max}$ and $P_i^{p\,\min}$ are the maximum and minimum power generation limits of power-only units. $P_j^{c\,\max}(H_j^c)$ and $P_j^{c\,\min}(H_j^c)$ are the maximum and minimum power generation limits of the CHP units. $H_j^{c\,\max}$ and $H_j^{c\,\min}(P_j^c)$ are the maximum and minimum heat generation limits of CHP units. $H_k^{h\,\max}$ and $H_k^{h\,\min}$ are the maximum and minimum heat generation bounds of the heat-only units.

3   Proposed Method

The proposed methodology for solving the CHPEED problem consists of three steps. First, different weighting coefficients are assigned, normalized objective functions using (17) are combined using weighting coefficients and single objective problem is obtained. In next step, the single objective optimization difficulty is solved using EMA and a group of accommodating solutions is achieved. In the last step, fuzzy productive technique is applied to select optimal accommodating solution. The flowchart of mentioned method is presented in Fig. 2.

$$F_j^{normalized} = \frac{F_j - F_j^{\min}}{F_j^{\max} - F_j^{\min}} \quad j = 1, 2 \tag{17}$$

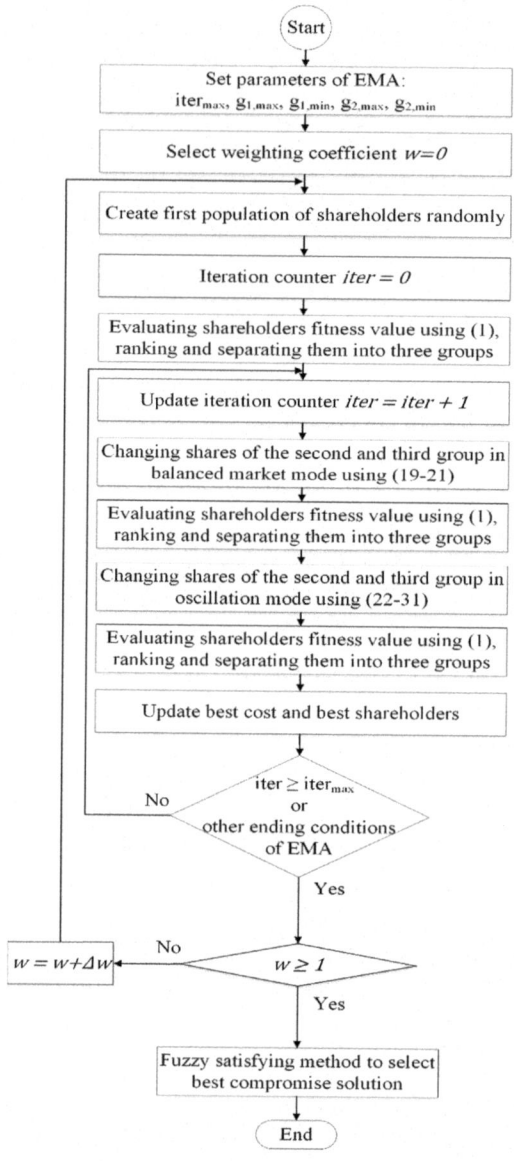

Fig. 2. Flowchart of preferred technique for solving CHPEED setback

3.1 Exchange Market Algorithm

EMA is an optimization procedure which relies on purchasing and selling shares by shareholders in exchange market. In this method, there are two market situations comprising balanced mode and oscillation mode. In the balanced mode, algorithm is used in locating the local optimal points while the oscillation mode algorithm tries to find the unknown global optimal points. Stockholders in each mode are divided into three groups. Members of the first group are top stockholders and also are the optimal answers to the optimization problem. In the second and third groups, level of risk increases as the fitness value decreases [17, 21].

3.2 Fuzzy Satisfying Method

So as to find best compromise solution amidst every compromise solution, a satisfying fuzzy method is applied. In relation to [24], fuzzy membership function for ith objective function in kth point is computed utilizing (18):

$$\mu_i^k = \begin{cases} 1 & f_i \leq f_i^{\min} \\ \frac{f_i^{\max} - f_i}{f_i^{\max} - f_i^{\min}} & f_i^{\min} < f_i < f_i^{\max} \\ 0 & f_i \geq f_i^{\max} \end{cases} \quad (18)$$

where minimum ith objective function is $f_i^{\min}$ whiles maximum ith objective function is $f_i^{\max}$. Normalized membership function for kth solution is computed utilizing (19):

$$\mu_{norm}^k = \frac{\sum_{i=1}^{N} \mu_i^k}{\sum_{k=1}^{P} \sum_{i=1}^{N} \mu_i^k} \quad (19)$$

3.3 Test System

In this system there are 5 generation units which are three co-generation units and one unit of conventional thermal and e heat-only accordingly. Cost and emission functions of power-only unit are included from [15]. The power and heat requirements for this test system are 300 MW and 150 MW, accordingly.

Compromise solutions set is shown in Table 1.

Table 1. Comparison results obtained from EMA and other methods

	Economic dispatch		Emission dispatch		Economic emission dispatch		
	RCGA [15]	EMA	RCGA [15]	EMA	NSGA-II [15]	SPEA 2 [15]	EMA
P_1 (MW)	134.9904	135	39.2000	35	93.9044	96.4846	92.5913
P_2 (MW)	49.9525	40.0751	125.8000	125.8000	72.8298	71.1705	70.4532
P_3 (MW)	25.0827	20	45	34.200	43.3448	44.5018	31.9555
P_4 (MW)	89.9744	104.9249	90	105	89.9210	87.8431	105
H_2 (MWth)	73.5089	73.9673	32.3998	135.6000	84.9250	84.7660	76.5496
H_3 (MWth)	35.8519	36.2419	55	20.4776	22.6032	10.2186	33.4989
H_4 (MWth)	1.2916	0	24	2.9560	2.6268	17.9054	0
H_5 (MWth)	39.3476	39.7908	37	0	39.8449	37.1100	39.9516
Cost ($)	13776.14	13673.2982	17048.75	17234.5494	15008.70	14964.30	14903.9397
Emission (kg)	12.0647	12.0561	1.4460	1.1748	6.0563	6.3667	5.8857
CPU times (s)	18.2134	3.1027	20.5417	3.3792	7.3627	50.0416	3.8409

4   Conclusion

In this paper EMA is utilized to resolve challenges of CHPEED and find compromise solutions set. Fuzzy satisfying technique is used to choose the optimal accommodating solution. Valve-point effects and CHP units' constraints are considered to solve multi-objective problem for minimizing emission and cost simultaneously. Comparing obtained results from EMA with previous studies illustrates that EMA has better performance than the other mentioned techniques in relation to finding optimal solutions in a better CPU time. Results show that total save in generation cost and emission for a year are equal to 917700.228$ and 1494.456 kg compared to NSGA-II method also 528756.228$ and 4213.56 kg compared to SPEA 2. These results demonstrate that greater benefits over a long period of time can be achieved by using mentioned method to resolve challenges of CHP economic emission dispatch. As a future study, due to the high performance of the EMA, this algorithm can be used to solve other power system problems.

References

1. Karki, S., Kulkarni, M., Mann, M.D., Salehfar, H.: Efficiency improvements through combined heat and power for on-site distributed generation technologies. Cogener. Distrib. Gener. J. 22, 19–34 (2007)
2. Bahmani-Firouzi, B., Farjah, E., Seifi, A.: A new algorithm for combined heat and power dynamic economic dispatch considering valve-point effects. Energy 52, 320–332 (2013)
3. Mohammadi-Ivatloo, B., Moradi-Dalvand, M., Rabiee, A.: Combined heat and power economic dispatch problem solution using particle swarm optimization with time varying acceleration coefficients. Electr. Power Syst. Res. 95, 9–18 (2013)
4. Ramesh, V., Jayabarathi, T., Shrivastava, N., Baska, A.: A novel selective particle swarm optimization approach for combined heat and power economic dispatch. Electr. Power Compon. Syst. 37, 1231–1240 (2009)
5. Song, Y., Xuan, Q.: Combined heat and power economic dispatch using genetic algorithm based penalty function method. Electr. Mach. Power Syst. 26, 363–372 (1998)
6. Yazdani, A., Jayabarathi, T., Ramesh, V., Raghunathan, T.: Combined heat and power economic dispatch problem using firefly algorithm. Front. Energy 7, 133 (2013)
7. Mehdinejad, M., Mohammadi-Ivatloo, B., Dadashzadeh-Bonab, B.: Energy production cost minimization in a combined heat and power generation systems using cuckoo optimization algorithm. Energy Efficiency 10, 81–96 (2017)
8. Group search optimization for combined heat and power economic dispatch. Int. J. Electr. Power Energy Syst. 78, 138–147 (2016)
9. Song, Y., Chou, C., Stonham, T.: Combined heat and power economic dispatch by improved ant colony search algorithm. Electr. Power Syst. Res. 52, 115–121 (1999)
10. Davoodi, E., Zare, K., Babaei, E.: A GSO-based algorithm for combined heat and power dispatch problem with modified scrounger and ranger operators. Appl. Thermal Eng. 120, 36–48 (2017)
11. Narang, N., Sharma, E., Dhillon, J.: Combined heat and power economic dispatch using integrated civilized swarm optimization and Powell's pattern search method. Appl. Soft Comput. 52, 190–202 (2017)

12. Nazari-Heris, M., Mohammadi-Ivatloo, B., Gharehpetian, G.: A comprehensive review of heuristic optimization algorithms for optimal combined heat and power dispatch from economic and environmental perspectives. Renew. Sustain. Energy Rev. **81**, 2128–2143 (2017)
13. Abido, M.A.: Environmental/economic power dispatch using multiobjective evolutionary algorithms. IEEE Trans. Power Syst. **18**, 1529–1537 (2003)
14. Shi, B., Yan, L.-X., Wu, W.: Multi-objective optimization for combined heat and power economic dispatch with power transmission loss and emission reduction. Energy **56**, 135–143 (2013)
15. Basu, M.: Combined heat and power economic emission dispatch using nondominated sorting genetic algorithm-II. Int. J. Electr. Power Energy Syst. **53**, 135–141 (2013)
16. Ghorbani, N., Babaei, E.: Exchange market algorithm for economic load dispatch,". Int. J. Electr. Power Energy Syst. **75**, 19–27 (2016)
17. Ghorbani, N.: Combined heat and power economic dispatch using exchange market algo-rithm. Int. J. Electr. Power Energy Syst. **82**, 58–66 (2016)
18. Rajan, A., Malakar, T.: Exchange market algorithm based optimum reactive power di   n. Appl. Soft Comput. **43**, 320–336 (2016)
19. Rajan, A., Malakar, T.: Optimum economic and emission dispatch using exc   e market algorithm. Int. J. Electr. Power Energy Syst. **82**, 545–560 (2016)
20. Pourghasem, P., Sohrabi, F., Mohammadi-Ivatloo, B., Abapour, M.: Reliable economic dispatch of microgrids by exchange market algorithm. In: Smart Grid Conference (SGC), pp. 1–5 (2017)
21. Ghorbani, N., Babaei, E.: Exchange market algorithm. Appl. Soft Comput. **19**, 177–187 (2014)
22. Khorram, E., Jaberipour, M.: Harmony search algorithm for solving combined heat and power economic dispatch problems. Energy Convers. Manag. **52**, 1550–1554
23. Wang, L., Singh, C.: Stochastic combined heat and power dispatch based on multi-objective particle swarm optimization. Int. J. Electr. Power Energy Syst. **30**, 226–234 (2008)
24. Arul, R., Velusami, S., Ravi, G.: A new algorithm for combined dynamic economic emission dispatch with security constraints. Energy **79**, 496–511 (2015)

Intuitionistic Fuzzy Assessment of Aggregated Quality of Life Index

G. Imanov[1(✉)] and A. Aliyev[2]

[1] Control Systems Institute of the Azerbaijan National Academy of Sciences,
Baku AZ1141, Azerbaijan
korkmazi2000@gmail.com
[2] Research and Training Center for Labor and Social Problems,
Baku AZ1000, Azerbaijan
msc.aaliyev@gmail.com

Abstract. In this work Intuitionistic fuzzy hybrid weighted average approach were proposed to define aggregated quality of life (QOL) index. Evaluation of QOL has recently become one of the study frontiers of multiple criteria decision making (MCDM) research. In this study, to evaluate the QOL, the weights of data and aggragation - associated weights of criteria are estimated with the application of intuitionistic fuzzy linguistic analysis. Then, intuitionistic fuzzy hybrid weighted averaging (IFHWA) operator is employed to aggregate influence of life objects criteria on QOL. Paper consist of four parts. In introduction we describe the nessicity of incorporation of criteria weights besides data weights. The Subsect. 1.1 of the article provides an estimation of weights of data that converted into intuitionistic fuzzy values; Next section presents the estimation of aggragation - associated weights of criteria; The last section presents estimation of the QOL as aggragated intuitionistic fuzzy value with the application of IFHWA operator and its interpretation.

Keywords: Quality of life · Intuitionistic fuzzy hybrid weighted average · Intuitionistic fuzzy preference relation · Aggregated life quality index

1 Introduction

QOL research has almost a centurial history. As science fields expand the frontiers of their research, new research prospects are emerging. Development and application of fuzzy sets and fuzzy logic theory added a lot to the research of the QOL.

The work has been carried out in this regard [1, 2] are of great importance, in which they applied MCDM and intuitionistic fuzzy aggregation techniques in the assessment of life satisfaction and QOL.

As an extension of fuzzy sets, intuitionistic fuzzy set theory (IFS) [3] opened new horizons to explore. An IFS is constituted of three parameters, which are membership function (MF), non-membership function (NMF) and hesitation margin, while fuzzy set is outlined by only membership function. This area of science is a more effective way to deal with uncertainty of the decison makers (DMs) who might not be capable to exactly state their satisfaction (or membership) level for alternatives, due to that (1) the

DMs have not definite and satisfactory information about the problem; (2) the DMs are not capable to distinguish clearly the priority of the alternatives [4].

In this paper the novatory intention is to incorporate criteria weights besides data weights in the assessment of QOL with the application of an IFHWA operator.

Within the scope of decision making with IFSs, aggregation operators play a very crucial role since they can be used to compound multi-dimensional estimation values provided as intuitionistic fuzzy values (IFVs) into a generic value. All intuitionistic fuzzy weighted averaging (IFWA) operators can be applied to fuse intuitionistic fuzzy values into an aggregated IFV, and different operators have various properties and application extent. Notice that, the IFWA operators take into account the weights of the intuitionistic fuzzy arguments but leave out the importance degrees of the ordered positions of the criteria. The IFHWA operators address this problem and consider both the weights of given arguments alongside with their ordered positions [5].

1.1   Obtaining the Weights of Arguments

The Economist Intelligence Unit has developed a new "QOL" index based on a unique methodology that associates the results of subjective life satisfaction surveys to the objective determinants of QOL accross countries [6]. In many countries, inequality is fleetly increasing, inflating the wealth of the theoretical average individual as represented by GDP per capita, while in reality most individuals in these countries see almost no increase in wealth [7].

The EIU identified the following nine factors that best predict the QOL:

Material Wellbeing (MW) - Measured by GDP per capita;

Health (H) - Measured by life expectancy at birth;

Political Stability and Security (PS) - Measured by political and security ratings developed by the EIU;

Political Freedom (PF) - Measured by the Freedom of the World Index;

Family Life (FL) - Measured by the divorce rate;

Community Life (CL) - Measured through church attendance or union membership;

Climate and Geography (CG) - Measured by latitude to distinguish between warm and cold climates;

Job Security (JS) - Measured by the unemployment rate;

Gender Equality (GE) - Measured by the ratio between female to male average earnings.

In order to convert crisp data into intuitionistic fuzzy numbers (IFNs) max-min normalization method is utilized. In the next step, the normalised QOL indicators are converted into IFNs using the intuitionistic fuzzy triangular functions *iftrif* [11].

Table 1. QOL indicators over life objects (domains) in Azerbaijan

Indicators	2014	2015	2016	2017	2018
Material wellbeing	6268.0	5706.6	6269.6	7226.0	8126.2
Health	74.2	75.2	75.2	75.4	75.8
Political stability and security	−0.56	−0.73	−0.80	−0.75	−0.70
Political freedom	6.0	6.0	6.5	6.5	6.5
Family life	1.3	1.3	1.4	1.5	1.5
Community life	1.0	1.0	0.99	1	1
Climate and geography	25.6	22.1	22.3	22.9	24.8
Job security	4.9	5.0	5.0	5.0	4.9
Gender equality	1.95	1.96	2.13	2	1.86

Sources: [8–10].

Intuitionistic fuzzy triangular MF and NMF of A come in the following structure:

$$\mu_A(x) = \begin{cases} 0; \\ \left(\frac{x-a}{b-a}\right); \\ \left(\frac{c-x}{c-b}\right); \\ 0; \end{cases} \tag{1}$$

$$v_A(x) = \begin{cases} 1-\epsilon; & x \leq a \\ 1-\left(\frac{x-a}{b-a}\right); & a < x \leq b \\ 1-\left(\frac{c-x}{c-b}\right); & b \leq x < c \\ 1-\epsilon; & x \geq c \end{cases} \tag{2}$$

In order to apply the *iftrif*, for the sake of simplicity isosceles triangles were used, where hesitation margin $\epsilon = 0.1$ is taken. Afterwards of normalization, the data converted into IFNs are given in the Table 2.

Table 2. QOL indicators over life objects (domains) in Azerbaijan

Indicators	2014	2015	2016	2017	2018
Material wellbeing	(0.36,0.54)	(0,0.90)	(0.37,0.53)	(0.64,0.26)	(0,0.90)
Health	(0,0.90)	(0.65,0.25)	(0.65,0.25)	(0.40.0.50)	(0,0.90)
Political stability and security	(0,0.90)	(0.48,0.42)	(0,0.90)	(0.32,0.58)	(0.73,0.17)
Political freedom	(0,0.90)	(0,0.90)	(0,0.90)	(0,0.90)	(0,0.90)
Family life	(0,0.90)	(0,0.90)	(0.90,0.10)	(0,0.90)	(0,0.90)
Community life	(0,0.90)	(0,0.90)	(0,0.90)	(0,0.90)	(0,0.90)
Climate and geography	(0,0.90)	(0,0.90)	(0.01,0.89)	(0.36,0.54)	(0.36,0.54)
Job security	(0,0.90)	(0,0.90)	(0,0.90)	(0,0.90)	(0,0.90)
Gender equality	(0.57,0.33)	(0.66,0.24)	(0,0.90)	(0.86,0.04)	(0,0.90)

Weights of the QOL indicators are estimated as the weights of decision makers as proposed by Boran et al. [12]. This area of science is a more effective way to deal with uncertainity of the decison makers (DMs) who might not be capable to exactly state their satisfaction (or membership) level for alternatives, due to that (1) the DMs have not definite and satisfactory information about the problem; (2) the DMs are not capable to distinguish clearly the priority of the alternatives [4].

Let $D_k = [\mu_k, v_k, \pi_k]$ be an IFN for rating of k-th decision maker. So the weight of k-th DM can be formulated as follows:

$$\lambda_k = \frac{\left(\mu_k + \pi_k\left(\frac{\mu_k}{v_k}\right)\right)}{\sum_{k=1}^{l}\left(\mu_k + \pi_k\left(\frac{\mu_k}{v_k}\right)\right)} \tag{3}$$

and $\sum_{k=1}^{l} \lambda_k = 1$.

The estimation results of the weights of QOL indicators are provided in Table 3.

Table 3. Weights of QOL indicators

Indicators	2014	2015	2016	2017	2018
Material wellbeing	0.37	0	0.19	0.17	0
Health	0	0.37	0.40	0.10	0
Political stability and security	0	0.24	0	0.07	0.73
Political freedom	0	0	0	0	0
Family life	0	0	0.40	0	0
Community life	0	0	0	0	0
Climate and geography	0	0	0.01	0.08	0.27
Job security	0	0	0	0	0
Gender equality	0.63	0.39	0	0.58	0

1.2   Obtaining the Aggregation - Associated Weights of Criteria

The weights of life objects from Economist.com survey, for the World 2005 [6] give a preliminary view about the relation of life objects (domains) and can be theoretical background in building intuitionistic fuzzy preference relation (IFPR) matrix.

Fuzzy preference degree within 0.1–0.9 scale was modified by incorporating their meanings in works of Gong et al. [13]. This scale can be a basis for composing intuitionistic fuzzy preference scale. The intuitionistic fuzzy preference of life objects (domains) in linguistic terms with their IFN counterparts have been developed, that is illustrated in Table 4.

Table 4. Linguistic terms for classification the preference of criteria

Linguistic terms	IF preference numbers
Exactly equal (EE)	(050, 0.50)
Slightly preferred (SP)	(0.60, 0.30)
Definitely preferred (DP)	(0.70, 0.20)
Strongly preferred (STP)	(0.80, 0.10)
Extremely preferred (EP)	(0.90, 0.10)

Let $(\mathcal{R})^*$ be an **IFPR**, then the following conditions must hold [14]:

$$r_{ij} = (0.50, 0.50), \mu_{ij} = v_{ji}, v_{ij} = \mu_{ji}, \pi_{ij} = \pi_{ji}, \tag{4}$$

$$\mu_{ij} + v_{ij} + \pi_{ij} = 1$$

The last condition provides the additive consistency of an IFPR. The additive consistency is inappropriate in modeling consistency, so all attention has to be focused on multiplicative consistency of the IFPR [15]. In the intuitionistic fuzzy analytic hierarchy process, with the objective to get a reasonable solution, before assessment of the priorities of the criteria, we need to check whether the IFPR is consistent or not. Consistency is a weighty problem in preference relations and the lack of consistency may give rise to wrong results [16].

Referring to Table 4, and considering additive consistency condition the following IFPR matrix $(\mathcal{R})^*$ is composed:

	MW	H	PS	PF	FL	CL	GF	JS	GE
MW	(0.50,0.50)	(0.50,0.50)	(0.60,0.30)	(0.60,0.30)	(0.70,0.20)	(0.70,0.20)	(0.80,0.10)	(0.80,0.10)	(0.90,0.10)
H	(0.50,0.50	(0.50,0.50)	(0.60,0.30)	(0.60,0.30)	(0.70,0.20)	(0.70,0.20)	(0.80,0.10)	(0.80,0.10)	(0.90,0.10)
PS	(0.30,0.60)	(0.30,0.60)	(0.50,0.50)	(0.50,0.50)	(0.60,0.30)	(0.60,0.30)	(0.70,0.20)	(0.70,0.20)	(0.80,0.10)
PF	(0.30,0.60)	(0.30,0.60)	(0.50,0.50)	(0.50,0.50)	(0.60,0.30)	(0.60,0.30)	(0.70,0.20)	(0.70,0.20)	(0.80,0.10)
PS	(0.20,0.70)	(0.20,0.70)	(0.30,0.60)	(0.30,0.60)	(0.50,0.50)	(0.50,0.50)	(0.60,0.30)	(0.60,0.30)	(0.70,0.20)
FL	(0.20,0.70)	(0.20,0.70)	(0.30,0.60)	(0.30,0.60)	(0.50,0.50)	(0.50,0.50)	(0.60,0.30)	(0.60,0.30)	(0.70,0.20)
CL	(0.10,0.80)	(0.10,0.80)	(0.20,0.70)	(0.20,0.70)	(0.30,0.60)	(0.30,0.60)	(0.50,0.50)	(0.50,0.50)	(0.60,0.30)
JS	(0.10,0.80)	(0.10,0.80)	(0.20,0.70)	(0.20,0.70)	(0.30,0.60)	(0.30,0.60)	(0.50,0.50)	(0.50,0.50)	(0.60,0.30)
GE	(0.10,0.90)	(0.10,0.90)	(0.10,0.80)	(0.10,0.80)	(0.20,0.70)	(0.20,0.70)	(0.30,0.60)	(0.30,0.60)	(0.50,0.50)

According to Genç et al. [17, 18] the following way is proposed to transform the given matrix into the multiplicative consistent IFPR matrix: where, $(r_{ij})^* = (\mu_{ij}, v_{ij})^*$ $(i = 1, 2, \ldots, n; j = 1, 2, \ldots, n)$ and satisfies the following condition:

$$\left(\mu_{ij}\right)^* = \max\left\{\mu_{ij}, \max_p\left\{\frac{\mu_{ip}\mu_{pj}}{\mu_{ip}\mu_{pj} + (1 - \mu_{ip})(1 - \mu_{pj})}\right\}\right\} \tag{5}$$

$$\left(v_{ij}\right)^* = \max\left\{v_{ij}, \max_p\left\{\frac{v_{ip}v_{pj}}{v_{ip}v_{pj} + (1 - v_{ip})(1 - v_{pj})}\right\}\right\} \tag{6}$$

where, the elements of $(\mathcal{R})^*$ matrix $\left(\mu_{ij}\right)^*$ and $\left(v_{ij}\right)^*$, are the membership degree and the non-membership degree of the alternative x_i over x_j, respectively.

Applying formulas (5) and (6) the following multiplicative consistent IFPR matrix $(\mathcal{R})^*$ is composed:

	MW	H	PS	PF	FL	CL	CG	JS	GE
MW	(0.50,0.50)	(0.50,0.50)	(0.60,0.31)	(0.60,0.31)	(0.70,0.21)	(0.70,0.21)	(0.80,0.14)	(0.80,0.14)	(0.90,0.10)
H	(0.50,0.50)	(0.50,0.50)	(0.60,0.31)	(0.60,0.31)	(0.70,0.21)	(0.70,0.21)	(0.80,0.14)	(0.80,0.10)	(0.90,0.10)
PS	(0.31,0.60)	(0.31,0.60)	(0.50,0.50)	(0.50,0.50)	(0.60,0.30)	(0.60,0.30)	(0.70,0.20)	(0.70,0.20)	(0.80,0.14)
PF	(0.31,0.60)	(0.31,0.60)	(0.50,0.50)	(0.50,0.50)	(0.60,0.30)	(0.60,0.30)	(0.70,0.20)	(0.70,0.20)	(0.80,0.14)
FL	(0.21,0.70)	(0.21,0.70)	(0.30,0.60)	(0.30,0.60)	(0.50,0.50)	(0.50,0.50)	(0.60,0.30)	(0.60,0.30)	(0.70,0.21)
CL	(0.21,0.70)	(0.21,0.70)	(0.30,0.60)	(0.30,0.60)	(0.50,0.50)	(0.50,0.50)	(0.60,0.30)	(0.60,0.30)	(0.70,0.21)
CG	(0.14,0.80)	(0.14,0.80)	(0.20,0.70)	(0.20,0.70)	(0.30,0.60)	(0.30,0.60)	(0.50,0.50)	(0.50,0.50)	(0.60,0.31)
JS	(0.14,0.80)	(0.14,0.80)	(0.20,0.70)	(0.20,0.70)	(0.30,0.60)	(0.30,0.60)	(0.50,0.50)	(0.50,0.50)	(0.60,0.31)
GE	(0.10,0.90)	(0.10,0.90)	(0.14,0.80)	(0.14,0.80)	(0.21,0.70)	(0.21,0.70)	(0.31,0.60)	(0.31,0.60)	(0.50,0.50)

After obtaining multiplicative consistent aggregated IFPR matrix, the priority vector of criteria $w = (w_1, w_2, \ldots, w_n)^T$ can be estimated with the following equation proposed by Genç et al. [17, 18]:

Consequently, the priority vector of criteria $w = (w_1, w_2, \ldots, w_n)^T$ can be obtained as follows [17, 19]:

$$
w_j = \left[w_j^L, w_j^U\right] = \left(\frac{1}{\sum_{j=1}^{n}\left(\frac{\left(1-\bar{\mu}_{ij}^*\right)}{\bar{\mu}_{ij}^*}\right)}, \frac{1}{\sum_{i=1}^{n}\left(\frac{\bar{v}_{ij}^*}{\left(1-\bar{v}_{ij}^*\right)}\right)} \right) \tag{7}
$$

$$ w_1 = w_2 = [0.208, 0.259], \qquad w_7 = w_8 = [0.036, 0.055], $$

$$ w_3 = w_4 = [0.112, 0.153], \qquad w_9 = [0.023, 0.029] $$

$$ w_5 = w_6 = [0.063, 0.093], $$

Following the way proposed by Boran et al. [20] in the next step, optimal weight vector of criteria $w^* = (w^*, w^*, \ldots, w^*)$ is estimated so that all closeness coefficients of criteria could be as big as possible. For this purpose, the following multi-objective optimization model is applied:

$$
\text{Maximize} \quad \sum_{j=1}^{n}\sum_{i=1}^{m} w_j \frac{d\left(r_{ij}, \alpha_j^-\right)}{d\left(r_{ij}, \alpha_j^+\right) + d\left(r_{ij}, \alpha_j^-\right)} \tag{8}
$$

$$
\text{Subject to} \quad \begin{aligned} w_1^L &\leq w_1 \leq w_1^U \\ w_2^L &\leq w_2 \leq w_2^U \\ &\vdots \\ w_n^L &\leq w_n \leq w_n^U \end{aligned} \tag{9}
$$

$$\sum_{j=1}^{n} w_j = 1 \tag{10}$$

$$w_j \geq 0 \text{ for } j = 1, 2, 3, \ldots n.$$

Solving the linear programming model by utilizing "Matlab Optimization Toolbox", the optimal weights were obtained as follows:

$w^* = \left(w_1^*, w_2^*, w_3^*, w_4^*, w_5^*, w_6^*, w_7^*, w_8^*, w_9^*\right) = (0.208,\ 0.208,\ 0.147,\ 0.112,\ 0.093,\ 0.093,\ 0.055,\ 0.055,\ 0.029)$.

Estimation of Aggregated Quality of Life Index by Application of IFHWA Operator

For a series of IFVs $\alpha_j = \left(\mu_{\alpha_j}, v_{\alpha_j}\right)$ ($j = 1, 2, \ldots, n$), the aggregated value obtained by the application of the IFHWA operator [5] is also an IFN, and

$$IFHWA_{\lambda,\omega} = \left[1 - \prod_{k=1}^{l}\left(1 - \mu_{\alpha_j}\right)^{\frac{\omega_{\varepsilon(j)}\lambda_j}{\sum_{j=1}^{n}\omega_{\varepsilon(j)}\lambda_j}}, \prod_{k=1}^{l}\left(v_{\alpha_j}\right)^{\frac{\omega_{\varepsilon(j)}\lambda_j}{\sum_{j=1}^{n}\omega_{\varepsilon(j)}\lambda_j}}\right] \tag{11}$$

Where, $IFHWA_{\lambda,\omega} = \left(\mu_{A_i}(x_j), v_{A_i}(x_j), i = 1, 2, \ldots, m; j = 1, 2, \ldots, n.\right)$ and $\omega = (\omega_1, \omega_2, \ldots, \omega_n)^T$ is an associated weighting vector with $\omega_j \in [0, 1]$ and $\sum_{j=1}^{n} \omega_j = 1$, $\varepsilon : \{1, 2, \ldots, n\} \rightarrow \{1, 2, \ldots, n\}$ is the permutation such that α_j is the $\varepsilon(j)$th largest element of the series of IFVs $\alpha_j (j = 1, 2, \ldots, n)$, and $\lambda = (\lambda_1, \lambda_2, \ldots, \lambda_n)^T$ is the weighting vector of the IFVs $\alpha_j (j = 1, 2, \ldots, n)$, with $\lambda_j \in [0, 1]$ and $\sum_{j=1}^{n} \lambda_j = 1$ (Table 5).

Table 5. Intuitionistic fuzzy hybrid weighted average results

Indicators	2014	2015	2016	2017	2018
$IFHWA_{\lambda,\omega}$	(0.41,0.49)	(0.61,0.29)	(0.70,0.24)	(0.62,0.24)	(0.70,0.20)

The results as aggregated intuitionistic values are not quite enough to comprehend and deduce reasonable conclusions. To this end, interval value intuitionistic fuzzy scale (IVIFS) that could reflect humanistic perception for rating the QOL has been developed, referring the method used by Kahraman et al. [21].

According to Table 6, the IFVs of $IFHWA_{\lambda,\omega}$ falling into the corresponding interval value intuitionistic fuzzy scale have been confronted with linguistic terms, which represent Aggregated Quality of Life Index for each year:

AQLI(2014) = Medium Low, **AQLI(2017)** = Medium High, **AQLI(2015)** = Medium High, **AQLI(2018)** = High, **AQLI(2016)** = High.

Table 6. Linguistic terms and their matching IVIFS

Indicators	Membership and non-membership values
Absolutely high (AH)	([0.85,0.90], [0.00,0.10])
Very high (VH)	([0.75,0.85], [0.10,0.15])
High (H)	([0.65,0.75], [0.15,0.25])
Medium high (MG)	([0.55,0.65], [0.25,0.35])
Medium (M)	([0.45,0.55], [0.35,0.45])
Medium low (ML)	([0.35,0.45], [0.45,0.55])
Low (L)	([0.25,0.35], [0.55,0.65])
Very low (VL)	([0.15,0.25], [0.65,0.75])
Absolutely low (AL)	([0.10,0.15], [0.75,0.85])

2 Conclusions

In this paper, we have estimated QOL as a problem of MCDM with the application of IFS. The notion of IFS can handle both vagueness and ambiguity (non-specifity) type of uncertainties. Concept of IFS is suitable to convert crisp data into intuitive fuzzy data then aggregate and deduce a result. For this purpose, we used IFHWA operator, which fuses IFVs into an overall IFV considering argument weights and aggregation – association weights of criteria. The obtained data are compatible with intuitionistic linguistic values that replicate real human decisions. This approach can be useful in elaborating sustainable macroeconomic development policies by improving the QOL across life objects (domains) and reducing poverty.

References

1. Imanov, G.J.: Assessment Level of Humanism in National Sustainable Development. Desafios De La Nueva Sciedad Sobrecompleja: Humanizmo, Transhumanizmo, Dataismo Y Otros Ismos, pp. 119–133 (2019)
2. Imanov, G.J., Bayramov, V.I.: Fuzzy approach to assessment of the national life satisfaction index. Neyro-necitki texnologii modelyuvannya v ekonomisi **4**, 46–61 (2015)
3. Atanassov, K.T.: Intuitionistic fuzzy sets. Fuzzy Sets Syst. **20**, 87–96 (1986)
4. Herrera-Viedma, E., Chiclana, F., Herrera, F., Alonso, S.: Group decision-making model with incomplete fuzzy preference relations based on additive consistency. IEEE Trans. Syst. Man Cybern. Part B-Cybern. **37**(1), 176–189 (2007)
5. Huchang, L., Xu, Z.S.: Intuitionistic fuzzy hybrid weighted aggregation operators. Int. J. Intell. Syst. **29**, 971–993 (2014). https://doi.org/10.1002/int.21672
6. The Economist intelligence Unit's quality-of-life index. The World In 2005.https://www.economist.com/media/pdf/qualityoflife.pdf
7. Stiglitz, J.E., Amartya, S., Fitoussi, J.-P.: Mismeasuring Our Lives: Why GDP Doesn't Add Up: The Report. New York: New (2010)
8. Azerbaijan State Statistical Commitee. https://www.stat.gov.az
9. https://en.wikipedia.org/wiki/Freedom_in_the_World
10. https://www.theglobaleconomy.com/Azerbaijan/wb_political_stability/

11. Radhika, C., Parvathi, R.: Intuitionistic fuzzification functions. Global J. Pure Appl. Math. **12**(2), 1211–1227 (2016)
12. Boran, F.E., Genc, S., Kurt, M., Akay, D.: A multi criteria intuitionistic fuzzy group decision making for supplier selection with TOPSIS method. Expert Syst. Appl. **36**, 11363–11368 (2009). https://doi.org/10.1016/j.eswa.2009.03.039
13. Gong, Z.W., Lin, Y., Yao, T.: Uncertain fuzzy preference relations and their applications. (2012). https://doi.org/10.1007/978-3-642-28448-9
14. Gong, Z.W., Li, L.S., Zhou, F.X.: Goal programming approaches to obtain the priority vectors from the intuitionistic fuzzy preference relations. Comput. Ind. Eng. **57**, 1187–1193 (2009). https://doi.org/10.1016/j.cie.2009.05.007
15. Xu, Z.S., Huchang, L.: Intuitionistic fuzzy analytic hierarchy process. IEEE Trans. Fuzzy Syst. **22**(4) (2014). https://doi.org/10.1109/TFUZZ.2013.2272585
16. Huchang, L., Xu, Z.S.: Priorities of intuitionistic fuzzy preference relation based on multiplicative consistency. IEEE Trans. Fuzzy Syst. **22**(6), 1669–1681 (2014). https://doi.org/10.1109/TFUZZ.2014.2302495
17. Genç, S., Boran, F.E., Akay, D.: Some approaches on estimating criteria weights from intuitionistic fuzzy preference relations under group decision making. J. Multiple-Valued Logic Soft Comput. inpress (2010)
18. Genç, S., Boran, F.E., Akay, D., Xu, Z.S.: Interval multiplicative transitivity for consistency, missing values and priority weights of interval fuzzy preference relations. Inf. Sci. **180**, 4877–4891 (2010)
19. Boran, F.E.: An integrated intuitionistic fuzzy multi criteria decision making method for facility location selection. Math. Comput. Appl. **16**(2), 487–496 (2011)
20. Boran, F.E., Genc, S., Akay, D.: A method for solving multi criteria intuitionistic fuzzy decision making problems. In: 1st International Fuzzy Systems Symposium, pp. 47–51. Ankara (2019)
21. Kahraman, C., Öztayşi, B., Onar, S.Ç.: An integrated intuitionistic fuzzy AHP and TOPSİS approach to evaluation of outsource manufacturers. J. Intell. Syst. **29**(1), 283–297 (2020). https://doi.org/10.1515/jisys-2017-0363

Review of Place Recognition Approaches: Traditional and Deep Learning Methods

Mohammed Abdulghani Taha[1(✉)] ⓘ, Melike Şah[2] ⓘ,
and Cem Direkoğlu[3] ⓘ

[1] Department of Computer Engineering, Tishik International University,
Erbil, Iraq
mohammed.abdulghani@tishik.edu.iq
[2] Department of Computer Engineering, Near East University, via Mersin 10,
Nicosia, North Cyprus, Turkey
melike.sah@neu.edu.tr
[3] Department of Electrical and Electronics Engineering, Middle East Technical
University - Northern Cyprus Campus, via Mersin 10, Kalkanli, Guzelyurt,
North Cyprus, Turkey
cemdir@metu.edu.tr

Abstract. Place recognition is an important topic in many vision-based applications. For this purpose, many traditional and artificial intelligence-based methods have been proposed. Despite years of expertise accrued in this area, it still remains a challenging problem due to the various ways in which the appearance of places in the real world can take place. Recently, rapid improvements have been made in image processing and recognition using deep neural networks. In particular, convolutional neural frameworks (CNN) are widely used in object detection, image classification, as well as in place recognition. The advantage of CNN-based place recognition is that CNN methods can automatically learn image patterns using sample images without any pre-processing and can handle appearance variations better than traditional methods. In this paper, we review and discuss traditional and deep learning-based place recognition algorithms, as well as existing datasets that can be used for performance measurement.

Keywords: Deep learning · Place recognition · Sparse feature · Text-based localization · Topological maps · Convolutional neural networks

1 Introduction

Place recognition has attracted considerable attention in the field of computer vision and the robotic communities in order to improve long-term autonomous navigation and autonomy [1]. There are variety of applications of place recognition such as autonomous driving, robotic navigation through virtual reality, retrieval of images and geo-location of archival images (i.e. determining the location of the given image).

Visual Place Recognition (VPR) is one of the elemental structures in a range of computer vision applications and has produced critical intrigue for an impressive period [1]. The objective is to accurately identify the places according to the visual place information. However, place appearances may change during the day/night

R. A. Aliev et al. (Eds.): ICAFS 2020, AISC 1306, pp. 183–191, 2021.
https://doi.org/10.1007/978-3-030-64058-3_22

(because of the lighting variances), according to the climate (spring, summer, fall, winter) and because of other weather and human factors [2]. Therefore, VPR methods have to tackle wide variety of problems. For example, a comparable environment may be totally different day and night due to changing lighting conditions. Humans in seconds can recognize places even if it is day/night, or summer/winter. However, computer-based approaches have to solve these problems, such as by using territorial image descriptions and photo coordination action plans. Recently, advanced VPR systems heavily rely on deep learning methods [3–5] to tackle these difficult circumstances. Like [4], the most of the deep learning based methods do not directly use high-level image information. On the other hand, VPR is also applied for automated road navigation [6]. Because of the scene compositions found in nature, sequence-based methods and topological maps are also widely for VPR for automated driving.

The purpose of this paper is to review and compare traditional and deep learning-based place recognition approaches. In particular, we discuss algorithms, datasets and performances of these methods. There are few review papers in this domain [1, 7]. [1] mainly focuses how VPR approaches handle appearance changes in the environment. [7] also focuses on appearance changes and how VPR approaches handle heterogeneous data. In this review paper, we discuss traditional methods as well as explain the details of the recent deep learning methods for VPR. Especially, recent deep learning methods provide very promising results and improve the state-of-the-art.

2   Existing Works

Visual place recognition can generally be categorized into narrow fields. In this section, we will summarize different place recognition methods that use traditional techniques (such as sparse feature, sequence-based, text-based, image retrieval techniques and topological maps), and deep learning models.

2.1   Sparse Feature-Based Place Recognition

Many place recognition applications use SIFT [8], SURF [9], HOG [10] features, Bag of Words, probabilistic models, and other features like 3D-scene sparse-visual feature maps [11]. These strategies rely on poor vitality; uncertain when faced with changes in lighting and complex scenes.

2.2   Sequence-Based Place Recognition

In [6], a sequence-based method is proposed for VPR that can handle varying weather conditions, seasonal differences as well as lighting changes. In this method, a robot map is generated by using multiple image sequences at various points across time. SeqSLAM is an extension of the popular Simultaneous Localization and Mapping (SLAM) algorithm for robot navigation. SeqSLAM gives the best matching candidate location in a sequence of images (Fig. 1a). SeqSLAM is a popular and one of the most important efforts to address the problems that arose with the hardening of changes in knowledge, management of spatial and transitory data. [12] uses multi-sequence maps for effective place recognition. They localize a vehicle by creating a map using Google street view images.

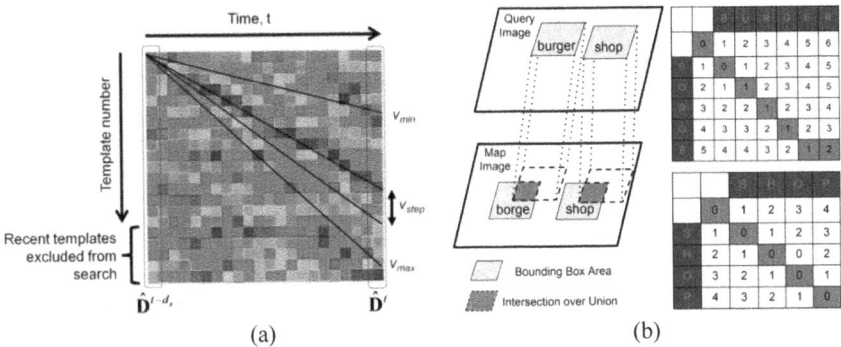

(a)   (b)

Fig. 1. Sample traditional methods: (a) Sequence-based VPR; searching for coherent matching sequences [6], (b) Place recognition using TextPlace; Levenshtein distance measurement between two pairs of query and map strings [13]

2.3 Text-Based Localization and Place Recognition

There are different works that use text-based localization for place recognition. Traffic signs, highway lines, notes, and store signs regularly contain a wide range of prohibitive attractions, which can be viewed as markers. [13] introduces an algorithm called TextPlace for recognizing places using screen text in images. Screen texts contain rich high-level information that is also invariant to lighting and illumination changes. [13] uses screen texts and topological localization for place recognition as shown in Fig. 1b. [14] applies a global localization approach by utilizing multiple texts observed on a map. [15] finds the GPS location of road signs for place recognition. Recent approaches combine text detection and deep learning for place recognition such as the work of [16].

2.4 Pure Image Retrieval

Image matching is a major element of many computer vision tasks, such as object or scene recognition, resolution of various 3D structures, stereo correspondence and motion tracking [17, 18]. [19] utilizes virtual line descriptor and a semi-automatic matching technique. [20] uses a different peer vocabulary tree to obtain a reasonable visual image of locations in a city-size dataset. FAB-MAP 2.0 [21] also used a review glossary to show visually distinctive evidence of in a 1000 km trajectory, which is a probabilistic method.

2.5 Topological Maps

Topological technologies attempt to reduce the number of models that combine information from different regions and the availability of associations between them.

The highlights and the outward appearance of the neighbor can be used to make these models such as in the works of TextPlace [6] and FAB-MAP [21].

2.6   Deep Learning-Based Methods

Recently, deep learning-based methods are applied for VPR [1, 22–24]. [22] optimizes the extraction of features and similarity metric is used to train a model especially for the task of place recognition in order to manage the varying appearances across time. A convolutional neural highlight extraction strategy is being constructed near the SAES metric [22], which improves the ability to recognize comparisons between positive and negative sets. Frame efficiency is exceptionally moved forward using SAES measurements as shown in Fig. 2.

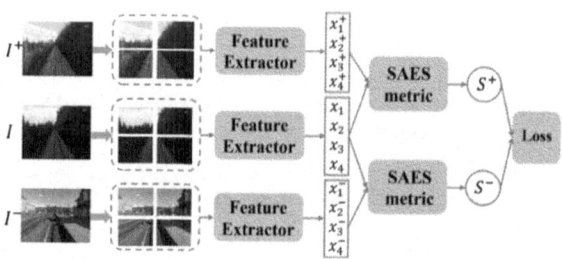

Fig. 2.   SAES based descriptor place recognition [22]

[23] also applies CNN; a pre-trained network model, VGG16-places365, is used to automatically learn image descriptors and optimizing them through some pooling, fusion and binarization operations, then presenting the resemblance result of place recognition with the place sequence Hamming distance (Fig. 3). In [4], they train two CNN models on a massive scale for the place-recognition task and use a multi-scale encoding technique to create condition-and viewpoint-invariant characteristics.

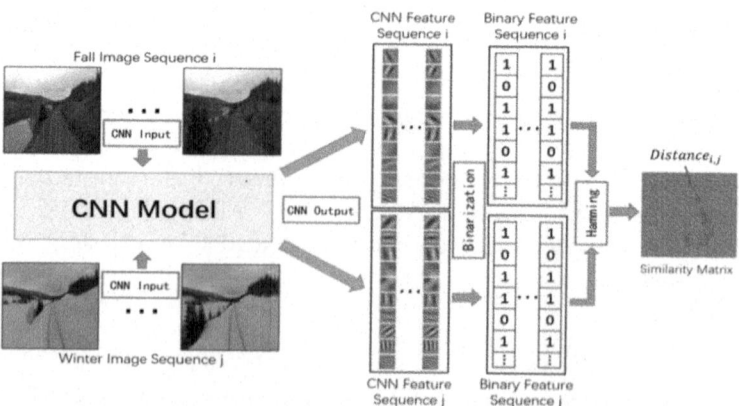

Fig. 3.   CNN based place recognition [23]

In [24], a CNN model is developed that incorporates image data attributes and localization opportunity signals through using indoor scene datasets and simulating the indoor position probability situation. A feature fusion model is developed that is composed mainly of the Inception V3 feature extraction module, and the feature fusion and selection module (Fig. 4). The image feature module based on the Inception V3 is achieved by adjusting with standard scene images. The function fusion and decision modules consist of a fully connected layer and final prediction, and the parameters are obtained by means of the simulative localization and picture training set.

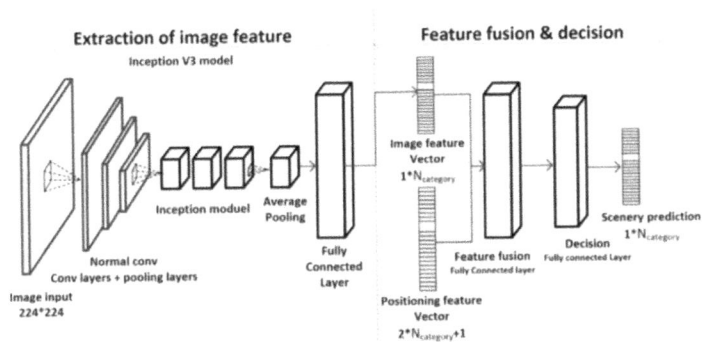

Fig. 4. Fusion-based Place recognition [24]

Another CNN-based deep learning-based method is NetVLAD [5] shown in the Fig. 5. The key element of this architecture is NetVLAD, which is a simplified layer of the image representation "Vector of Locally Aggregated Descriptors" (VLAD). VLAD is widely used in image retrieval. The layer can be conveniently inserted into any CNN architecture, which is ideal for backpropagation processing [5]. They establish a training technique, based on the new poorly supervised failure in the ranking, to learn end-to-end design parameters from images depicting the same locations throughout time retrieved from Google Street View Time Machine [5]. As a result, the approach determine the location of a given question image easily and accurately.

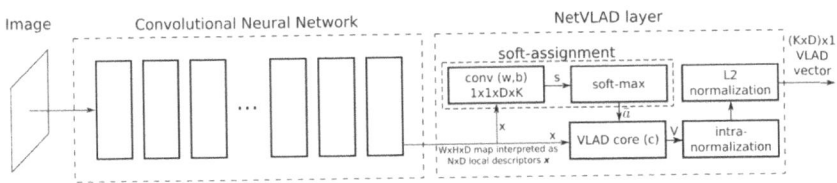

Fig. 5. CNN architecture with the NetVLAD layer [5]

Another approach is [25] that combine results of multiple image processing methods, such as sum of absolute differences, histogram of oriented gradients (HOG) and two deep learning features into one Hidden Markov Model. This method is called Multi-Process Fusion (Fig. 6) that automatically selects the best image processing technique for the environment. In this way, strengths of different image processing methods are combined using an automated weighting scheme [25].

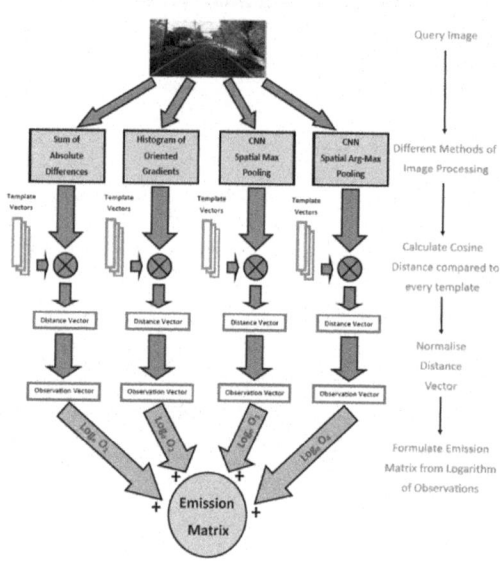

Fig. 6. Multi-Process Fusion: Selecting the best image processing methods [25]

3   Place Recognition Datasets

There are two main types of datasets: image-based and motion-based structure datasets. For image-based datasets, the Google Maps Street View, the IM2GPS dataset, the YFCC100M dataset, the San Francisco Landmark dataset, and the Alps100K dataset are used. For motion-based, the Rome16K dataset, the Dubrovnik6K dataset, the Quad dataset, the Landmark 3D dataset, and the Cambridge Emblem dataset are used. On the other hand, for VPR evaluations, in the recent publications, the following datasets are used for evaluations.

The Nordland dataset is captured by a moving train in Norway across four different seasons [26]. It consists of four videos of the same preparation in different seasons over 728 km long journey in Norway. St Lucia dataset is recorded in an area in the Saint Lucia suburb of Brisbane with a front camera mounted on a car [27]. Oxford RobotCar dataset is captured over a year across different periods of day, seasons and routes in the city of Oxford [28]. It contain 2.5 km route. SYNTHIA dataset is a publicly available datasets that contains synthetic images of urban scenes [29].

4 Discussions

The reviewed papers compare their results on different datasets and using different accuracy measurements. For example, some of the papers use Precision-Recall curves and others use F-measure scores. Therefore, it is not possible to combine the results of all these methods. Instead, we summary the most promising VPR methods: From the traditional methods, SeqSLAM [6], FAB-MAP [21] are popular. Among the deep learning methods, SEAS [22], CNN [23], NVLAD [5] and Multi-Process Fusion [25] proved that they perform better than traditional methods in terms of accuracy based on their published results. Particularly, it is reported that deep-learning based methods can handle difficult circumstances such as lighting changes, illumination, day/night, and season changes better than the traditional methods.

5 Conclusions

In this paper, we review both traditional and deep learning-based methods for visual place recognition (VPR). These VPR approaches can be divided into different categories, such as sparse feature-based, sequence-based, text-based, image retrieval-based, topological maps (can be combined with others), and deep learning-based. Traditional methods use features like SIFT, SURF, HOG, sequence information and topological maps/metrics. Recently, deep learning based VPR approaches improve the performances of the traditional methods. In future, with the availability of large image datasets for CNN training, the robustness of the CNN methods can be improved even further. On the other hand, rather than training only images with CNN methods, combined image and image features (such as HOG) can be trained with CNN approaches. It is evident that deep learning will lead the future of place recognition.

References

1. Lowry, S., Sunderhauf, N., Newman, P., Leonard, J.J., Cox, D., Corke, P., Milford, M.J.: Visual place recognition: a survey. IEEE T. Robot. **32**(1), 1–19 (2016). https://doi.org/10.1109/TRO.2015.2496823
2. Zhou, B., Lapedriza, A., Khosla, A., Oliva, A., Torralba, A.: Places: a 10 million image database for scene recognition. IEEE TPAMI **40**, 1452–1464 (2017). https://doi.org/10.1109/TPAMI.2017.2723009
3. Crandall, D.J., Li, Y., Lee, S., Huttenlocher, D.P.: Recognizing landmarks in large-scale social image collections. In: Zamir, A., Hakeem, A., Van Gool, L., Shah, M., Szeliski, R. (eds.) Large-Scale Visual Geo-Localization. Advance Computer Vision Pattern Recognition. Springer, Cham. https://doi.org/10.1007/978-3-319-25781-5_7
4. Chen, Z., Jacobson, A., Sunderhauf, N., Upcroft, B., Liu, L., Shen, C., Reid, I., Milford, M.: Deep learning features at scale for visual place recognition. In: Proceedings of IEEE International Conference Robotics Automation, pp. 3223–3230 (2017). https://doi.org/10.1109/ICRA.2017.7989366
5. Arandjelović, R., Gronat, P., Torii, A., Pajdla, T., Sivic, J.: NetVLAD: CNN architecture for weakly supervised place recognition. IEEE TPAMI **40**(6), 1437–1451 (2018)

6. Milford, M.J., Wyeth, G.F.: SeqSLAM: Visual route-based navigation for sunny summer days and stormy winter nights. IEEE ICRA, pp. 1643–1649 (2012). https://doi.org/10.1109/ICRA.2012.6224623

7. Piasco, N., Sidib, D., Demonceaux, C., Gouet-Brunet, V.: A survey on visual-based localization: on the benefit of heterogeneous data. Pattern Recogn. **74**, 90–109 (2018). https://doi.org/10.1016/j.patcog.2017.09.013

8. Lowe, D.G.: Object recognition from local scale-invariant features. IEEE ICCV **2**, 1150–1157 (1999). https://doi.org/10.1109/ICCV.1999.790410

9. Bay, H., Ess, A., Tuytelaars, T., van Gool, L.: Speeded-up robust features (SURF). Comput. Vis. Image Und. **110**(3), 346–359 (2008). https://doi.org/10.1016/j.cviu.2007.09.014

10. Dalal, N., Triggs, B.: Histograms of oriented gradients for human detection. In: IEEE CVPR, pp. 886–893 (2005). https://doi.org/10.1109/CVPR.2005.177

11. Cieslewski, T., Stumm, E., Gawel, A., Bosse, M., Lynen, S., Siegwart, R.: Point cloud descriptors for place recognition using sparse visual information. In: IEEE ICRA, pp. 4830–4836 (2016). https://doi.org/10.1109/ICRA.2016.7487687

12. Vysotska, O., Stachniss, C.: Effective visual place recognition using multi-sequence maps. IEEE Robot Autom Let **4**(2), 1730–1736 (2019). https://doi.org/10.1109/LRA.2019.2897160

13. Hong, Z., Petillot, Y., Lane, D., Miao, Y., Wang, S.: TextPlace: visual place recognition and topological localization through reading scene texts. In: IEEE/CVF ICCV, pp. 2861–2870 (2019). https://doi.org/10.1109/ICCV.2019.00295

14. Radwan, N., Tipaldi, G.D., Spinello, L., Burgard, W.: Do you see the bakery? Leveraging georeferenced texts for global localization in public maps. In: IEEE ICRA, pp. 4837–4842 (2016). https://doi.org/10.1109/ICRA.2016.7487688

15. Ranganathan, A., Ilstrup, D., Wu, T.: Lightweight localization for vehicles using road markings. In: IEEE/RSJ IROS, pp. 921–927 (2013). https://doi.org/10.1109/IROS.2013.6696460

16. Liao, M., Shi, B., Bai, X.: Textboxes++: a single-shot oriented scene text detector. IEEE T Image Process **27**(8), 3676–3690 (2018). https://doi.org/10.1109/TIP.2018.2825107

17. Lowe, D.G.: Distinctive image features from scale-invariant keypoints. IJCV 91–110 (2004). https://doi.org/10.1023/B:VISI.0000029664.99615.94

18. Tang, J., Acton, S.: An image retrieval algorithm using multiple query images. IEEE ISSPA 193–196 (2003). https://doi.org/10.1109/ISSPA.2003.1224673

19. Liu, Z., Marlet, R.: Virtual line descriptor and semi-local matching method for reliable feature correspondence. BMVC 16 (2012). https://doi.org/10.5244/C.26.16

20. Schindler, G., Brown, M., Szeliski, R.: City-scale location recognition. In: IEEE CVPR, pp. 1–7 (2007). https://doi.org/10.1109/CVPR.2007.383150

21. Cummins, M., Newman, P.: Appearance-only SLAM at large scale with FAB-MAP 2.0. Int. J. Robot. Res. **30**(9), 1100–1123 (2011). https://doi.org/10.1177/0278364910385483

22. Zhao, C., Ding, R., Key, H.L.: End-to-end visual place recognition based on deep metric learning and self-adaptively enhanced similarity metric. In: IEEE ICIP, pp. 275–279 (2019). https://doi.org/10.1109/ICIP.2019.8802931

23. Zhu, J., Ai, Y., Tian, B., Cao, D., Scherer, S.: Visual place recognition in long-term and large-scale environment based on CNN feature. In: IEEE IVS, pp. 1679–1685 (2018). https://doi.org/10.1109/IVS.2018.8500686

24. Guo, W., Wu, R., Chen, Y., Zhu, X.: Deep learning scene recognition method based on localization enhancement. Sensors (2018). https://doi.org/10.3390/s18103376

25. Hausler, S., Jacobson, A., Milford, M.: Multi-process fusion: visual place recognition using multiple image processing methods. IEEE Robot Autom. Let. **4**(2), 1924–1931 (2019). https://doi.org/10.1109/LRA.2019.2898427

26. Nordland Dataset. https://nrkbeta.no/2013/01/15/nordlandsbanen-minute-by-minute-season-by-season/
27. St Lucia data. https://wiki.qut.edu.au/display/cyphy/StCLuciaCMultipleCTimesCofCDay
28. Vi, W., Maddern, G., Pascoe, C., Linegar, Newman, P.: 1 year, 1000 km: the oxford robotcar dataset. Int. J. Robot. Res. **36**(1), 3–15 (2017). https://doi.org/10.1177/0278364916679498
29. Synthia dataset. https://synthia-dataset.net/

Fuzzy Reasoning in Music

Javanshir Guliyev[1] 🔘 and Konul Memmedova[2](✉) 🔘

[1] Department of Acting, Near East University, Nicosia, North Cyprus
javanshir55@gmail.com
[2] Department of Psychological Counselling and Guidance, Near East University,
Nicosia, North Cyprus
konul.memmedova@neu.edu.tr

Abstract. Music composition has been recently used by the many musicians and scientists on artificial intelligence. One of main problem here is how to evaluate synthesized music by a computer. This article presents concepts how fuzzy logic may be applied to music analysis and composition. Main idea is that ambiguity in the music is maintained not by classical set, with use of fuzzy set approach.

Keywords: Fuzzy sets · Fuzzy number · Music composition · Computer music · ESPLAN

1 Introduction

Music generation includes different aspects: melody generation, harmony, rhythm analysis etc. Music composition methods also can be divided into note-based and signal-based methods. Musicians are interested what kind of computer music is preferred by them. How to estimate the quality of the generated by computer music? Does computer compose new music as famous musician?

Research evidence in the area of application of Fuzzy Logic to music composition is very scarce. The conceptual use of Fuzzy Logic in music composition is presented in [1]. The author provides a brief overview of state – of – the –art of semiotics of music in automated composition systems. [2] uses Fuzzy Logic as the tool to advance computer music argument. The author proposes an automated Fuzzy Logic system which could analyze each piece of music for its patterning. In [3] arhythmic compositions through the control of several sound synthesis parameters is considered. In [4] author has developed a chord creating mechanism. In [5] it is developed a Fuzzy Logic module Music program. The paper [6] outlines the rationale for using fuzzy logic and granular computing, to emulate compositional decision-making processes. Authors present research related to the identification of the significant compositional elements and their connective grammar.

This paper describes some of the core concepts of fuzzy logic and demonstrates how they may be applied to problems common in music analysis and composition.

The paper is structured as follows. Section 2 includes basic concepts used in the paper. In Sect. 3 we give representation of music elements as fuzzy sets. Section 4 is dedicated to illustrating the solution method of the problem. Section 5 offers concluding remarks.

© The Author(s), under exclusive license to Springer Nature Switzerland AG 2021
R. A. Aliev et al. (Eds.): ICAFS 2020, AISC 1306, pp. 192–197, 2021.
https://doi.org/10.1007/978-3-030-64058-3_23

2 Preliminaries

Definition 1. Fuzzy number: A fuzzy number is a set A on R which possesses the following properties: a) A is a normal fuzzy set; b A is a convex fuzzy set; c) $\alpha-$ cut of A, A^α is a closed interval for every $\alpha \in (0, 1]$; d) the support of A, A^{+0} is bounded [7].

Definition 2. Approximate reasoning [8]: Efficiency of inference engine considerably depends on the knowledge base internal organization. That is why ESPLAN realizes paradigm of "network of production rules" similar to semantic network. Here the nodes are rules and vertexes are objects. Inference mechanism act as follows. First some objects take some values (initial data). Then, all production rules, containing each of these objects in antecedent, are chosen from the knowledge base. For these rules the truth degree is computed (in other words, the system estimates the truth degree of the fact that current values of objects correspond to values fixed antecedents). If the truth degree exceeds some threshold then imperatives from consequent are executed. At that time the same objects as well as a new one takes new values and the process continues till work area contains "active" objects ("active" object means untested one).

$$IF\ x_1 = a_1^j\ AND\ x_2 = a_2^j\ AND\ldots THEN\ y_1 = b_1^j\ AND\ y_2 = b_2^j\ AND\ldots$$
$$IF\ldots THEN\ Y_1 = AVRG(y_1)\ AND\ Y_2 = AVRG(y_2)\ AND\ldots$$

ESPLAN has a built-in function AVRG which calculates the average value. This function simplifies the organization of compositional approximate reasoning with possibility measures.

Definition 3. Trapezoidal fuzzy number (TFN): TFN has a corresponding membership function which is built using parametric LR-representation (Fig. 1).

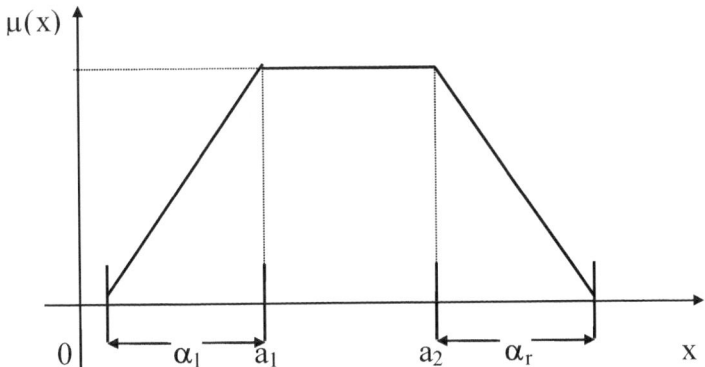

Fig. 1. Membership function

So, each membership function is defined by 4 parameters: α_l - left deviation, a_1 – left peak, a_2 – right peak, α_r – right deviation, i.e. $\mu_a(x) = (\alpha_l, a_1, a_2, \alpha_r)$.

So we have an analytic form of membership function:

$$\mu_a(x) = \begin{cases} 1 - \frac{a_1 - x}{\alpha_l} & \text{if } a_1 - \alpha_l \leq x \leq a_1 \\ 1 & \text{if } a_1 \leq x \leq a_2 \\ 1 - \frac{x - a_2}{\alpha_r} & \text{if } a_2 \leq x \leq a_2 + \alpha_r \\ 0 & \text{in other cases} \end{cases}$$

Definition 4. ESPLAN [9]: This shell consists of the following steps:

Step 1. Calculation the truth degree of the given rule as:

$$d_{jk} = Poss(\delta_k / a_{jk}) \cdot cf_k$$
$$\tau_j = \min(d_{jk})$$

Step 2. Evaluated w_i objects has intend linguistic value defined as (δ_i, cf_i). where δ_i is linguistic value, $cf_k \in]0, 100]$ is confidence degree of the value δ_i. δ_k - linguistic value of the rule object, a_{jk} - current linguistic value (j is index of the rule, k is index of relation) value(for example, A_ir)

Step 3. Calculation $D_j = (\min_j d_{jk}) * CF_j / 100$, where CF is the confidence degree of the rule.

Step 4. Designer of the rule determine the firing level (π) and $D_j \geq \pi$ is checked. If the antecedent holds true, then the consequent of rule is calculated.

Step 5. Each evaluated w_i objects has S_i value: $w_i, (\delta_i^1, cf_i^1), \ldots, \ldots, (\delta_i^{S_i}, cf_i^{S_i}) S_i$ is the number of the rules in fuzzy inference process.

Step 6. The mean value is defined as follows:

$$\bar{v}_i = \frac{\sum_{n=1}^{S_i} \delta_i^n \cdot cf_i^n}{\sum_{n=1}^{S_i} cf_i^n}$$

3   Representation of Music Elements

In classical music system pitches are represented as numbers, typically by the MIDI code [10]. The main drawback of this system is that it does not represent nearness of B (pc = 11) to C (pc = 0). As it is shown in [10] a more adequate representation is the pitch sets. Pitch sets can represent chords. In existing literature, a pitch set has twelve members, which are either 0 or 1. Unfortunately, such type of representation can not deal with ambiguity inherent in musical systems.

A more flexible system of representation in musical environment is fuzzy sets [11]. Fuzzy pitch sets can represent chords more flexible and adequate. Chord inversion construction is multidimensional and depends on the quality of sound desired, the avoidance of parallel fifth, fingering difficulties, etc. [10]. This relationship can be represented by IF-THEN rules [10]:

- If root position keeps common tones, then root position.
- If first inversion keeps common tones, then first inversion.
- If second inversion keeps common tones, then second inversion.
- If last position was root, then first inversion or second inversion.   (1)
- If there have been too many firsts in a row, then root or second.
- If there have been too many seconds in a row, then root or first.

In IF-THEN rules (1) predicate part has crisp and fuzzy values of musical factors. For example, "too many first inversion" may be describes as trapezoidal fuzzy number given in Fig. 2.

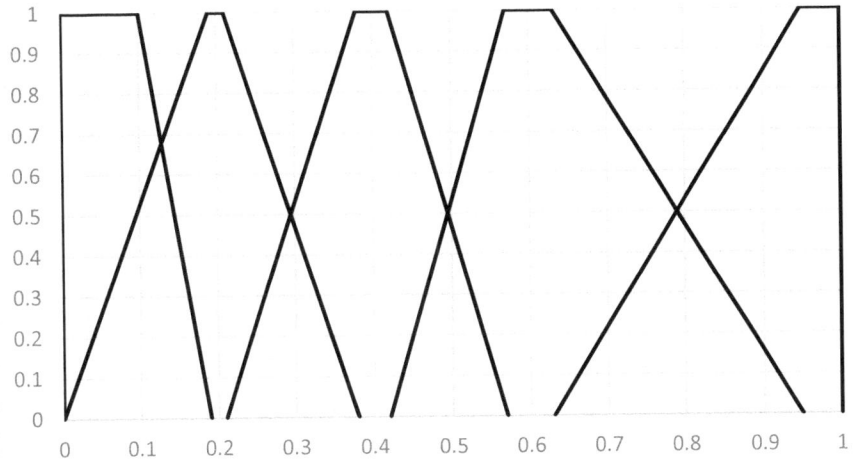

Fig. 2. Trapezoidal fuzzy number for "too many first inversion"

Approximate reasoning on based on these rules gives the final answer that can be used to rotate the chord to the desired inversion.

4   Solution of the Problem

In general, many musical concepts are vague category. For example, for attribute "loudness", musicians usually define *mf* as a softer than *f*, and louder than *mp*, which is louder than *p*.

Codebook of fuzzy sets for linguistic term, "loudness" can be expressed as it shown in Fig. 3.

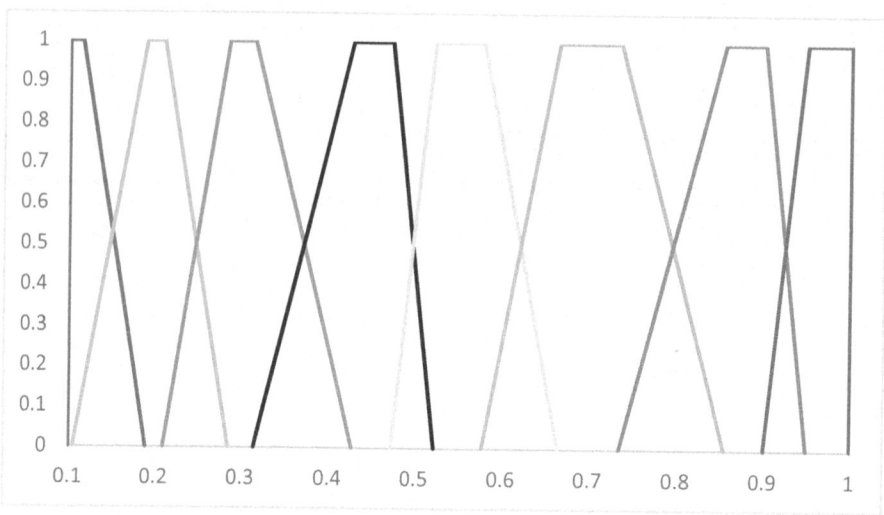

Fig. 3. Trapezoidal fuzzy number for "loudness"

For reasoning outputs on use of IF-THEN rules we use fuzzy expert system ESPLAN [9]. For simulation musical fragments given in Figs. 4, 5 we represented.

Fig. 4. Musical fragment from "Xaric Segah"

Fig. 5. Musical fragment from "Bayatı Şiraz"

5   Conclusion

In this paper we have considered application of fuzzy logic concept to the production of music elements. IF-THEN rules are used to produce chord inversions in music syntheses. For musical knowledge representation and approximate reasoning fuzzy shell ESPLAN was used.

References

1. Whalley, I.: towards a closed system automated composition engine: linking 'Kansei' and musical language recombinicity. In: ICMC2002, Goteborg, Sweden, pp. 200—203 (2002)
2. Milicevic, M.: Positive emotion learning through music listening. http://myweb.lmu.edu/mmilicevic/NEWpers/_PAPERS/papers.html
3. Cadiz, R.F.: Compositional control of computer music by fuzzy logic. Ph.D. thesis, Northwestern University (2006)
4. Elsea, P.: Fuzzy logic and musical decisions. ftp://arts.ucsc.edu/pub/ems/FUZZY/Fuzzy_logic_and_Music.pdf
5. Sorensen, A., Brown, A.: Music: music composition in Java. https://explodingart.com/jmusic/
6. Suiter, W.: The promise of fuzzy logic in generalized music composition. In: Cultural Computing. In: Proceedings of 2nd IFIP TC 14; Entertainment Computing Symposium, Brisbane, Australia, Germany, pp. 118–128. Springer, Heidelberg (2010)
7. Aliev, R.A.: Fundamentals of the Fuzzy Logic-Based Generalized Theory of Decisions. Springer, Heidelberg (2013)
8. Aliev, R.A., Aliev, R.R.: Soft Computing and Its Applications. Singapore, New Jersey (2001)
9. Aliev, R.A.: Uncertain Computation-Based Decision Theory. World Scientific Publishing, Singapore (2018)
10. Elsea, P.: Fuzzy logic and musical decisions, 30 p. University of California, Santa Cruz (1995)
11. Zadeh, L.A., Aliev, R.A.: Fuzzy Logic Theory and Its Applications: Part I and Part II. World Scientific, Singapore, New Jersey (2018)

Integrative Jurisprudence from Fuzzy Logic Perspective

Mehman A. Damirli[✉][iD]

Faculty of Law, University of Kyrenia, Kyrenia, Mersin 10, Cyprus, Turkey
mdamirli@yahoo.com

Abstract. The main purpose of this article is to find out the logical bases of integrative jurisprudence. For the purposes of this research, integrative jurisprudence is defined as the entire legal science with a systematic application of an integrative approach, i.e. it is not just restricted to separates legal disciplines or understanding of law. The author proceeds from the assumption that fuzzy logic can be utilized to provide an appropriate logical-ontological and logical-methodological ground of integrative jurisprudence. The justification proceeds in two main stages. This discussion leads into revealing of the cognitive possibilities of fuzzy logic for methodological enrichment of integrative jurisprudence. The fuzzy logic dialogical "both-and" approach suggested by the author will enable providing expanded, comprehensive understanding of legal reality, finding a way out from the polar opposition between competing conceptions of law and approaches to it, and opening up creative ways for legal scientists and practitioners.

Keywords: Integrative jurisprudence · Law · Fuzzy set · Fuzzy logic · Fuzzy logic dialogical "both-and" approach

1 Introduction

The term "integrative jurisprudence" was first proposed by the American scholar Jerome Hall in 1964 [1]. Meanwhile, the very first idea of a comprehensive, pluralistic perception of law originated in Ukrainian and Russian legal science dating back to in the second half of the 19th century, and the first projects to conceptualize these ideas were carried out at the beginning of the 20th century by B.A. Kistyakovsky in his pluralistic conception of law [2] and by A.S. Yashchenko in his synthetic theory of law [3]. Representatives of Western legal science J. Hall and H. Berman, although they preferred the use of the term "integrative jurisprudence", also talked about the synthesis of the basic conceptions (theories) of law in their opinion (Hall about the synthesis of the philosophy of natural law, positivism and sociology of law [1], Berman – legal positivism, theory of natural law and the historical school of law [4]). In this regard, modern versions of the integrative conception of law do not differ either.

It is noteworthy that for all the scientists mentioned above, the common thing is the rejection of a one-sided approach to law, without rejecting classical theories as erroneous, the desire to overcome dualism and to synthesize some theoretically significant points worked out by competing concepts of law. However, some scholars are calling for the unification of existing theories of law, which have different bases. Moreover, this is done (especially by scientists in the post-Soviet space) in order to search for an exhaustive, holistic, single definition of law, reflecting its complex social nature. This approach inevitably leads in most cases to their eclecticism. Obviously, it is for this reason that the various versions of the integrative concept of law so far presented have aroused fair criticism. Thus, transcend of dualism and fostering new approaches are most important task of contemporary jurisprudence.

It is well-known that in the base of any science there are some fundamental axiomatic propositions that are not justified within the framework of own subject-matter and require going beyond it. Today, the inadequacy of the approaches and models used in legal science, built on the basis of classical binary logic, is beyond doubt. It seems that such a basis can be a fuzzy logic that allows overcoming of the shortcomings of the binary approach and opening up of new perspectives for a comprehensive understanding of law. Moreover, as the analysis shows, authors of integrative conceptions of law sometimes consciously or unconsciously used some concepts and provisions that are now known as characteristic of multivalent, fuzzy logic. In addition, over the past decades, scientists have done certain work on the application of the apparatus of fuzzy logic in the field of law and on the definition of some fuzzy characteristics of law, thereby preparing the ground for solving both the problems of understanding of law and many other conceptual and practical problems in this sphere.

In this paper, it is attempted to provide a logical-ontological and logical-methodological justification of integrative jurisprudence based on the provisions of the theory of fuzzy logic. It should be noted right away that here the concept of "jurisprudence" used in the expression "integrative jurisprudence" is understood as the entire legal science, i.e. it is not restricted to its separate disciplines (while Hall and Berman limited it to general theory of law and philosophy of law, respectively), or just understanding of law, though understanding of law is one of its main questions (if not the main one), because "behind any question of what the law is in each particular case, there is a fundamental question: 'What is the law as such?'" [5].

The rest of the paper is organized as follows. Section 2 includes necessary terms and definitions on fuzzy logic and methodological instruments used in the research. The following two sections, being crucial for this research, are devoted to the logical-ontological (Sect. 3) and logical-methodological (Sect. 4) bases of the integrative jurisprudence in the light of fuzzy logic. Conclusions are recapped in Sect. 5.

2   Preliminaries

2.1   Fuzzy Set

Fuzzy set "is a class of objects with a continuum of grades membership. Such set is characterized by a membership (characteristic) function which assigns to each object a grade of membership ranging between zero and one" [6, 7]. In other words, fuzzy set is a class of objects that takes any value in the interval [0,1], not just 0 or 1 (where 0 denotes "full non-membership", 1 – "full membership"). The values for partial, intermediate membership (or "belonging") are calibrated in accordance with the defined criteria.

2.2   Fuzzy Logic

As the founder of fuzzy logic L. Zadeh himself noted, the term fuzzy logic is used in two different senses. In the narrow sense, fuzzy logic is a logical system that is an extension of multivalued logic (although in basic operations it is very different both in spirit and in content than multivalued logic) and "denotes a set of inaccurate, approximate methods for solving problems of weakly structured systems management in the field of Humanities, robust control, biology, and medicine" [8]. In the broad sense of the word, which prevails today, fuzzy logic is equivalent to the theory of fuzzy sets. Thus, fuzzy logic, understood in a narrow sense, is a section of fuzzy logic in a broad sense [8]. Classical logic operates with only two concepts: true and false, excluding any intermediate values, while fuzzy (multi-valued) logic allows assigning any (infinite set) truth value in the interval [0, 1] and suggests "tolerance for imprecision, uncertainty and partial truth to achieve... Better agreement with reality" [9]. In the fuzzy logic "the truth-values are fuzzy subsets of the unit interval with linguistic labels such as true, false, not true, very true, quite true, not very true and not very false, etc." [6, 7] An important characteristic of fuzzy logic is that any theory T can be fuzzified and, therefore, generalized by replacing the notion of a crisp set in T with a notion of a fuzzy set. Benefits from fuzzification are greater generality and better fit of the model to reality [8].

2.3   Fuzzy Logic "Both-And" Approach

As it is known, dualistic dichotomous thinking, based on binary (two-valued) logic, proceeds from the "either-or" approach, which prefers one side of a dualism and excludes the other (A and not-A), and ultimately, such a monistic essentially view leads to fails to account for the complexity of the studied phenomena and "amputate our understanding" [10].

Fuzzy logic, going beyond the limiting effects of dualistic "either-or" approach on understandings of reality, instead of this offers as an alternative model of thinking a nondualistic, pluralistic and multivalence "both-and" approach. Its essence lies in the recognition of the possibility of not only coexistence of two or more seemingly opposing perspectives, also intermediate opportunities between them, and in giving value to traditionally excluded from analyses aspects.

2.4   Dialogical Approach

If the dualistic "either-or" approach assumes a monological discourse, the fuzzy logic "both-and" approach requires a dialogical discourse, which itself, in substance, is based on a pluralistic, multivalent logical thinking. The dialogical approach accepts different, even opposing, views simultaneously, and enables to create a more detailed picture of the studied phenomena that better reflect their complexity, fuzziness, and paradoxes. It places emphasis on the interconnections of conceptions, fosters an appreciation of how the possibilities and limitations of these conceptions are most apparent from opposing views. As it is noted in [11], the aim of this approach is not to arrive at a finite point of understanding, conclusions, but to strive toward harmony and agreements between different social concepts.

The indicated close connections between the two above-described approaches (fuzzy logic "both-and" and dialogical) allow us to propose their combination – fuzzy logic dialogical "both-and" approach. Applying this approach will allow to open up ways for overcoming shortcomings in existing approaches to law by blending polar opposing positions and, ultimately, to arrive at an expanded view of the law in general and separate legal phenomena in particular.

3   Fuzzy Logic and Logical-Ontological Basis of Integrative Jurisprudence

As it is known that ontology is characterized by fundamental assumptions about the nature of reality. Consequently, the logical-ontological basis concerns logical assumptions about the nature of reality. With regard to legal reality in the form of a question, it can be formulated as follows: "What are logical assumptions about the nature of legal reality?" In this particular case, we are interested in fuzzy logical-ontological assumptions about the nature of reality by the studied jurisprudence.

L. Zadeh, at the beginning of the construction of fuzzy logic theory, noted humanistic systems among the spheres of application of this theory. By a humanistic system Zadeh means "a system of which behaviour is strongly influenced by human judgement, perception or emotions, and the examples of humanistic systems are: economic systems, political systems, legal systems, educational systems, etc. A single individual and his thought processes may also be viewed as a humanistic system." [9] Zadeh, proceeding from the principle of incompatibility, claims that high precision is incompatible with high complexity of the system [9].

According to fuzzy logic, objective characteristics of reality are fuzziness, vagueness, uncertainty, imprecision, ambiguity, abstractness and approximateness. When applied to legal reality, this can be revealed as follows. To begin with, in the literature it is clearly demonstrated that ontological complexity [12] and legal pluralism is a fact [13, 14] and that the law (legal reality) is a complex multidimensional system. This is brilliantly illustrated by S. Maksimov's conception of legal reality. According to this conception, legal reality as a complex multi-level system, as the whole set of all legal phenomena is structured from such relatively autonomous levels (forms of the being of law) as: a) the world of ideas (the idea of law); b) the world of signs forms (legal norms and laws); c) the world of social interactions (legal life) [15]. It is noteworthy that fuzziness is characteristic for each of these levels/forms (the contributions of some researchers [16–25] are demonstrations of this). However, these levels, as well as their particular elements, differ on the extent of fuzziness. In particular, a high level of fuzziness (super-fuzziness) is especially a characteristic of legal ideas and principles, soft rules of law – soft law (the high extent of fuzziness of the latter allowed O. Perez to call it "fuzzy law" [25]), judicial language, reasoning and decision-making. The fact is that not only the above-mentioned forms of being of law, but also the boundaries between them are imprecise, vague, and uncertain, since they are only relatively autonomous aspects of the being of law and closely interconnected between themselves.

Moreover, it should be taken into account that law (legal reality) is an integral part of social reality, where it has close ties with its other components: economy, politics, religion, culture, and so on. Moreover, the boundaries of these spheres are blurred. This was well demonstrated by W. Menski in relation to the relationship between law and religion. In particular, he noted that "erecting walls of separation around 'law' and 'religion' is not possible because everything is interlinked and fuzzy, these negotiable boundaries will simply be crossed in many ways, no matter how many obstacles and barbed wire regulatory systems of various kinds we may seek to invent" [26].

It should be added that both the establishment and implementation of the norms of positive law, the realization of legal reasoning and judicial decision-making, and the cognition of legal reality are the product of a human thought activity. This activity is also objectively characterized by indistinctness and imprecision, which is partly demonstrated in [27–29].

Thus, the foregoing allows to argue that defined objectively, fuzziness are a property of legal reality themselves – legal reality is a fuzzy set, and it is inevitable due to the following main origins: (1) uncertainty from nature of legal phenomena themselves, (2) blurriness from lack of firm boundaries (a) between levels (forms) of being of law, also (b) between them and other components of social reality, and (3) imprecision from nature of human legal thinking and cognition of legal reality.

4 Fuzzy Logic and Logical-Methodological Basis of Integrative Jurisprudence

Methodology assumes the nature of ways and means of studying phenomena. Hence, the logical-methodological basis concerns logical assumptions about ways and means of studying phenomena. From the standpoint of fuzzy logic, here we are talking about the fact that fuzziness are a property of researchers' beliefs about law (legal reality), speaking in the language of fuzzy logic, law (legal reality) X is fuzziness to researcher Y [30].

The accepting of fuzziness of a law leads to the recognition also that the logical and methodological foundations of legal cognition should be based on the fuzzy logic, because the nature of the subject-matter of cognition determines the features of the methodology to be used here, the productivity of the methodological tools themselves depends on their adequacy to the subject [31]. Moreover, according to the theory of fuzzy logic, the very process of human thinking is a fuzzy set; essentially all human concepts are fuzzy. In the literature, it is rightly notes that "all social concepts can be usefully" fuzzified, that is, "formulated as fuzzy sets" [32]. However, human consciousness always strives for accurate scientific knowledge: "Mankind likes to think in terms of extreme opposites. It is given to formulating its beliefs in terms of Either-Ors, between which it recognizes no intermediate possibilities" [33]. The problem is that many core concepts in social, including legal, "research are best understood as graded sets" [32]. Instances in the field of legal knowledge are such dichotomies as truth versus falsehood, objectivity versus subjectivity, rationalism versus irrationalism, real versus ideal, existing versus due, factual versus critical, etc. Very such dualistic approach divides fundamental philosophical assumptions underpinning main two competing legal conceptions – legal positivism and natural law theory (its classical versions), and this results in a one-sided monistic interpretations of the nature of law.

Now it is obvious that it is necessary to transcend the dualism based on binary logic and to encourage an alternative model of thinking - fuzzy logic and approaches based on it. One of such, in particular, is the fuzzy logic dialogical "both-and" approach proposed here. Integrative, multivalent research with using this approach gives opportunities for overcoming dualism and its impacts on existing approaches to law. As a model of analysis it allows to emphasize the interconnections within the law and between it and other social phenomena, and to give value to the traditionally excluded aspects of these interconnections. This approach fosters an appreciation of how the possibilities and limitations of legal conceptions are most apparent from polar opposing positions. Thus, the given approach facilitates interplay between of legal conceptions and, consequently, contributes to explaining the complexity of the legal reality. It also suggests an infinite number of possibilities to solving the legal problems.

Thus, whereas the utilize of a single conception can ensure a valuable but restricted view, integrative, multivalent research with using fuzzy logic dialogical "both-and" approach and others methodological tools of fuzzy logic can create multi-faceted pictures that reflect the complexity and fuzziness of phenomena in legal research – pictures that reveal different but interwoven aspects of these phenomena.

Fuzzy logic requires "to abandon the high standards of rigor and precision" for the sake of greater correspondence to reality, "and become more tolerant of approaches which are approximate in nature" [9]. Applied to integrative jurisprudence, this also means that we must abandon the assignment to the law a fixed, unchanging, particular definition as such and focus on agreement existing conceptions of law, instead of only two extremes. All the more so that it is "the fact that the bodies of jurisprudential knowledge designated 'legal positivism,' 'natural law philosophy' and 'legal sociology' are not mutually exclusive. There is nonetheless a significant degree of common interest and common product" [1], because they reveal from different points of view a unified phenomenon "law". From this point of view, attention is drawn to the contribution of the E. Adams and T. Spaak who, by "fuzzifying the natural law – legal positivist debate", highlight that "fuzzy logic is a significant new tool which may be used to bridge the gap between the bivalent world views espoused by natural law theorists and legal positivists" [34]. Authors proceed from the fact that for dualistic theorists "truth is either 0 or 1, legal positivism or natural law theory. For fuzzy theorists, truth is somewhere in the continuum of gray values between 0 and 1" [34]. In search of agreement between the two classical theories, the authors applying the apparatus of fuzzy logic, and formulating and combining fuzzy rules, create a fuzzy system, which ultimately allows them to indicate areas of significant common ground and helps build consensus between natural law theorist and positivist, where none existed previously [34]. It remains for me to add that such schemes, indicating the unique cognitive possibilities of fuzzy logic, will work and have productive results, if not count the elaboration of a fixed definition of law as the only task of research.

5   Conclusion

In this research, the logical-ontological and logical-methodological bases of integrative jurisprudence were ascertained and the following results have been obtained.

It was found out that the fuzziness is a property of legal reality themselves – legal reality inherently is a fuzzy, and it is inevitable due to the following main origins: (1) uncertainty from nature of legal phenomena themselves, (2) blurriness from lack of firm boundaries (a) between levels (forms) of being of law, also (b) between law and other components of social reality, and (3) imprecision from nature of human legal thinking and cognition of legal reality. The fuzzy nature of law, in turn, stipulates the logical and methodological foundations of integrative jurisprudence, putting in the foreground the methodological provisions, tools and procedures of fuzzy logic. In this regard, in the represented article was demonstrated the role of the approach proposed here (fuzzy logic dialogical "both-and" approach) in overcoming dualistic thinking and elicitation points of contact (common ground) and reaching consensus in the understanding of law and elaboration of many others legal issues.

Today fuzzy logic is a fast-growing and promising interdisciplinary field with many unique possibilities. Integrative jurisprudence, using these possibilities of fuzzy logic and thereby enriching own methodological arsenal, I believe, will become a new face of legal science at the present stage of its development.

Justification of logical-ontological and logical-methodological bases of integrative jurisprudence in the light of fuzzy logic is the first step toward its formation. The rationale set out above ensures a starting point, but more work needs to be done. For example, it is necessary to demonstrate the capabilities of fuzzy logic in the synthesis of various concepts, as well as for solving other important issues of jurisprudence.

References

1. Hall, J.: From legal theory to integrative jurisprudence. Univ. Cincinnati Law Rev. **33**(2), 153–205 (1964)
2. Kistyakovsky, B.A.: Social sciences and law. Essays on the Methodology of the Social Sciences and the General Theory of Law. Moscow (1916). (in Russian)
3. Yashchenko, A.S.: Theory of Federalism: the Experience of the Synthetic Theory of Law and State. Yuriev (1912). (in Russian)
4. Berman, H.J.: Toward an integrative jurisprudence: politics, morality history. California Law Rev. **76**(4), 779–801 (1988)
5. Maksymov, S.: What is law? Key points of the concept of legal reality. Philos. Law General Theory Law **1–2**, 320–327 (2016)
6. Zadeh, L.A.: Fuzzy sets. Inform. Control **8**, 338–353 (1965)
7. Zadeh, L.A., Aliev, R.A. (eds.): Fuzzy Logic Theory and Applications. Part I and Part II. World Scientific, New Jersey (2018)
8. Zadeh, L.A.: The roles of soft computing and fuzzy logic in the conception, design and deployment of information/intelligent systems. Artif. Int. News **2–3**, 7–11 (2001). (in Russian)
9. Zadeh, L.A.: The Concept of a Linguistic Variable and its Application to Making Approximate Reasoning. Mir, Moscow (1976). (in Russian)

10. Flyvbjerg, B.: Making Social Science Matter: Why Social Science Fails and How it Can Succeed Again. Cambridge University Press, Cambridge (2001)
11. Kristeva, J.: Word, Dialogue and Novel. In the Kristeva Reader, edited by Toril Moi. Columbia University, New York (1986)
12. Wroblewski, J.: Problems of Ontological Complexity of Law. Universidad del Pais Vasco, Centro de Análisis, Lógica e Informática Iuridica (1986)
13. Griffiths, J.: What is legal pluralism? J. Legal Pluralism Unofficial Law **24**, 1–56 (1986)
14. Merry, S.E.: Legal pluralism. . Law Soc. Rev. **22**(5), 869–896 (1988)
15. Maksimov, S.I.: Concept of legal reality. In: Chestnov, I.L. (ed.) Postclassical Ontology of Law. SPb, pp. 23–59 (2016). (in Russian)
16. D'Amato, A.: Legal uncertainty. California Law Rev. **71**, 1–55 (1983)
17. Wroblewski, J.: Fuzziness of legal system. In: Kangas, U. (ed.) Essays in Legal Theory in Honor of Kaarle Makkonen, Monographic issue of Oikeustiede, vol. 16, pp. 311–330 (1983)
18. Richard, C., Calfee, J.E.: Deterrence and uncertain legal standards. J. Law Econ. Organ. **2**, 279–303 (1986)
19. Philipps, L.: Vague legal concepts and fuzzy logic. An attempt to determine the required period of waiting after traffic accidents. Fuzzy Logic Neural Net. **2**, 37–51 (1993)
20. Endicott, T.: Vagueness in Law. Oxford University Press, New York (1997)
21. Waldron, J.: Vagueness in law and language. Calif. Law Rev. **82**, 509–533 (1994)
22. Soames, S.: Vagueness in the law. In: Marmor, A. (ed.) The Routledge Companion to Philosophy of Law, pp. 95–108. Routledge, New York (2012)
23. Vlasenko, N.: Uncertainty in law: nature and forms of expression. J. Russ. Law. **2**, 3–43 (2013)
24. Marmor, A.: Varieties of vagueness in the law. Legal Studies Research Paper Series. USC Gould School of Law. 12–8 (2013)
25. Perez, O.: Fuzzy law: a theory of quasi-legal systems. Can. J. Law Jurisprudence **28**(2), 343–370 (2015)
26. Menski, W.: Fuzzy law and the boundaries of secularism. Potchefstroom Electronic Law Journal/Potchefstroomse Elektroniese Regsblad **13**(3) (2010)
27. Reisinger, L.: Legal reasoning by analogy. A model applying fuzzy set theory. In: Proceedings Artificial Intelligent Legal Informatics, Florence. I, pp. 151–163 (1981)
28. Mazzares, T.: Fuzzy logie and judicial decision-making: a new perspective on the alleged norm-irrationalism. Informatica e diritto, XIX annata **2**(2), 13–36 (1993)
29. Legrand, J.: Some guidelines for fuzzy sets application in legal reasoning. Artif. Intel. Law **7**, 235–257 (1999)
30. Davis, K.E.: The Concept of Legal Uncertainty. https://www.law.umich.edu/centersandprograms/lawandeconomics/workshops/Documents/Paper%201.KDavis.The%20Concept%20of%20Legal%20Uncertainty.pdf
31. Damirli, M.A.: Law and History: Epistemological Problems. Publishing house of St, Petersburg University, St. Petersburg (2002).(in Russian)
32. Ragin, Ch.C., Pennings, P.: Fuzzy sets and social research. Sociological Methods Res. **33**(4), 423–430 (2005)
33. Dewey, J.: Experience and Education. Touchstone, New York (1997)
34. Adams, E.S., Spaak, T.: Fuzzifying the natural law-legal positivist debate. Buff. L. Rev. **43**, 85–119 (1995)

Ecological and Economic Security Processes Modeling with Cloud-Based Soft Computing

Yana V. Fedorova$^{(\boxtimes)}$ ⓘ, Ludmila K. Popova ⓘ,
Tatyana N. Sharypova ⓘ, and Ekaterina N. Lozina ⓘ

Rostov State University of Economics, Rostov-on-Don, Russia
fyv21@mail.ru, popova_plk@mail.ru, tnt@mail.ru,
lozinakn@mail.ru

Abstract. In practice the environmental and economic security is a simple analytics in the form of the ratings such as "the best regions to live" or "investment attractiveness", or vice versa, the regions with the least favorable environment. Thus, both the analysis and research are aimed at revealing weak point of the territorial environmental and economic system functioning. While compiling and analyzing the mathematical models for different economic phenomena and processes, it is necessary to recognize clearly the integral structures and systems that compose them as well as to understand the correlation interconnections of the considered processes and their characteristics. The ecological and economic model presented in the form of a function of the time dependence of the state of an object characterized by vector dynamic quantities is considered in the article. The research of the industrial enterprises in relation of their level of the ecological and economic security based on three states: normal, critical and emergency was conducted. Then on the basis of these data the integral indicator of environmental and economic safety of an industrial enterprise using fuzzy sets was identified. The experts' quantitative assessments were used as the intermediate data. Based on the resulting data the information system structure was worked out to support the decisions taken in the sphere of the environmental and economic security with a cloud-based database.

Keywords: Modeling · Fuzzy sets · Soft computing · Cloud technology · Fuzzification · Defuzzification

1 Introduction

While studying the ecological processes and economic phenomena the economic and mathematical models are usually considered in correlation with the target economic systems and recognize them as some integral structures that are mathematical models of environmental and economic systems [1–7]. The environmental and economic model can be presented as the function of the dependence of the state of the object on time:

$$O(t) = f(X, A, G, Y, O, I)$$

where A – initial characteristics of the state of the object; X – object control parameters; G, Y – result characteristics of the state of the object; Q – characteristics of external influences.

The object's characteristics are presented in the form of dynamic vector quantities.

The research and the solution of the set task are related to the experimental process and the identification of the object of the research is an important factor.

In the regions of the Russian Federation, environmental and economic security is specific in the manifestation of the authority of the state authorities in the transition of the economy to the so-called "green economy". This fact shows the dependence of the ecological and economic security on the natural environment, on the quality of the ecological and economic policy as well as a quality of population. In the figure the characteristic of the component "The quality of the population" includes such relevant characteristics of the population (N) as life expectancy, educational level, skill level of workers, demographic level, etc. The component of the group "the Quality of the natural environment" (C) involves monitoring and analysis the degree of the pollution of the air, water and soil environment. According to this classification the indicators-coefficients (kC) are a certain quantitative measurement of the corresponding component (Fig. 1).

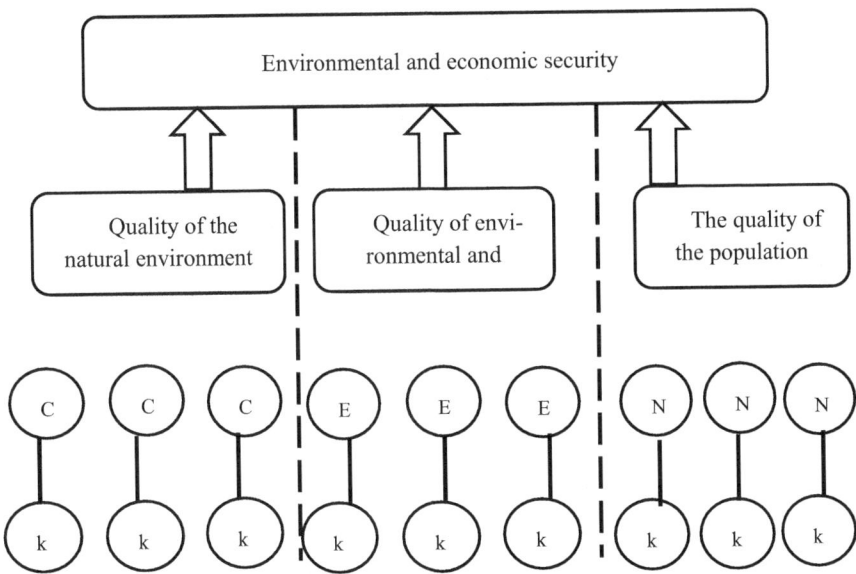

Fig. 1. Classification of factors of the function of dependence of ecological and economic security by subject-object composition.

2 Experimental Part

2.1 The Application of Fuzzy Sets to Determine the Level of Environmental and Economic Security

The testing of the process is conducted on the example of water, soil, atmosphere resources, radiation spaces and industries.

While researching the level of environmental and economic security of the industrial enterprises, three states are taken as a basis: normal, critical and emergency. The normal state is when almost all characteristics correspond to normative or permissible values. The critical state is characterized by various irregularities in the enterprises' activity or clear danger of their occurrence. The state of emergency is characterized by the presence of serious failures that threaten the reliable functioning of the economic systems. The boundary between the normal and critical state (that is, the worst of the acceptable values of the indicator or the best of the acceptable values) is taken as a critical threshold, between critical and emergency - as an emergency threshold. Then the summary indicator of environmental and economic security of an industrial enterprise is identified.

Further on, the stage of transition into fuzzy set, called fuzzification with the subsequent stages of aggregation, activation and accumulation, is carried out. Then before the data output the stage of defuzzification of output variables takes place, whose purpose is to obtain a result in quantitative measurement [1].

In accordance with the considered classification many of the ecological and economic states are differentiated into three fuzzy subsets in this research, each of them possesses its own membership function. The integral indicator of the environmental and economic security of the industrial enterprise is based on this division, and the higher this indicator, the more stable its state.

Then comparing the current values of indicators with their critical values, the assessment of the current state of the ecological and economic security of the analyzed object is carried out. Further on, a generalized assessment of the environmental and economic security level of the enterprise is given.

The adequacy analysis of the obtained fuzzy model means the building a surface of fuzzy inference, and it allows assessing and correcting the effect of changes in the input data on the values of output variables of fuzzy type. The tools, implementing the fuzzy assessment of the indicators of the ecological and economic security, are produced by Fuzzy Logic Toolbox in the MatLab environment.

Development and optimization of the mathematical models of the ecological and economic systems demand to create and adapt the economic and mathematical methods allowing taking into consideration the incompleteness and complexity of the formalization of the source data of the considered environmental and economic systems.

Fuzzy logic is one of the most prospective areas of scientific research in the field of analysis, forecasting and modeling of economic phenomena (Matlab fuzzy logic).

While modeling many tasks it is necessary to take into consideration random factors and perturbations. In this case the apparatus of the theory of probability, a mathematical science that studies patterns in random phenomena, is the most suitable [5].

The research of the ecological and economic security in terms of a systematic approach allows considering the process of its provision as a kind of integrity manifested within the framework of economic systems [6]. Ensuring the environmental and economic security as a system is characterized by a big number of the performed functions, parameters and results of functioning; by complex behavior of the system, reflected in intertwined and overlapping relationships among the variables and time-varying external influences; by permanent spatial and temporal connection manifested while the system elements interact and is fixed in a form of a certain structure; by reflection of the views, goals and values of business entities; by absence of dependence of structure and nature of interconnections among the elements on a level and a type of the economic system development.

The content of the complex assessment of the ecological and economic security level is a development of an integral indicator which allows obtaining the quantitative assessment of the enterprise's security level, which displays the diverse influence of a big number of indicators. Maximum permissible and high values for each indicator were identified at the first stage of ecological and economic security modeling. The formalization of indicators, set at a qualitative level should also be carried out on the basis of membership functions.

The maximum permissible, critical and high values of the ecological and economic security level were identified to determine the membership function form associated with each variable at the next stage ecological and economic security modeling.

Further on, such tuples, that give different from each other values of the premises, are to be considered. As a result the relevant rule bases will be developed together with experts. The surface fuzzy inference is developed to analyze the general adequacy of the fuzzy model, it will allow assessing and correcting the influence of change in the input data on the values of the output fuzzy variables. The gradation of the levels of ecological and economic security of industrial enterprises is displayed in the following form.

Thus, mathematical modeling use on the basis of the fuzzy conclusion, implemented in MatLab 6.8 environment, allowed us to assess the ecological environment, to control the ecological and economic security of the enterprise.

2.2 Decision Support System Structure Development

The further process is reduced to the development of the information system structure to support the decision taking in the sphere of ecological and economic security with using data base with cloud technologies.

The main advantages while using cloud services are the following:

1. Operational access to data from any access point where there is coverage of the Internet access network.
2. Most of services provide a rather good amount of memory to store data.
3. Data security and information protection.
4. Significant physical space and hard drive memory savings to increase the reading information speed from the hard drive.

Disadvantages of cloud service use:

1. There is a possibility of the unauthorized access to the information when it is transmitted.
2. Since the third party is engaged in the information transmitting there is a possibility of data leakage.

The key role of the system is to form the list of recommendations in the decision making process, which have a direct impact on the ecological security development of the region taking into account its economic factors.

The cloud storage of the information is an on-line store where all data are stored in different net hubs, and where the internal structure is invisible for a client and are stored in the so-called cloud that is an abstract global virtual server.

3 Decision Making Support System Structure

To implement the algorithm the object oriented approach was applied. The source code is written in the programming language Visual Basic. The diagram (Fig. 3) shows the most significant links and relations among classes involved in the algorithm.

Rules consist of conditions and conclusions, which are, in their turn, are fuzzy statements. The fuzzy statements include linguistic variable and a term, which is presented by a fuzzy set. A membership function is defined on a fuzzy set, the value of which can be obtained using the method get value. This is the method defined in the Fuzzy SetIface interface. When executing the algorithm, it is necessary to use an activated fuzzy set, which in some way redefines the membership function of a fuzzy set. A unit of fuzzy set is also used in the algorithm.

Mamdani Algorithm is used with fuzzy approach and it implements a rule base as input data. A rule base is a set of rules where each hypothesis is associated with a specific weight coefficient. The rule base can be as follows:

RULE_n: IF «Condition_k» THEN «Conclusion_(q-1)» (Fq-1) AND «Conclusion_q» (Fq).

Symbols:

n—number of rules; m—number of input variables; s—number of output variables; k—number of conditions; q—number of conclusions.

Then the stages of translation into fuzzy sets, called fuzzification with the subsequent stages of aggregation, activation and accumulation take place. After that, before the data outputting, the stage of defuzzification of the output variables takes place, whose purpose is to obtain a quantitative value of the data.

In the algorithm for its initialization, the center of gravity method is used, where the result of defuzzification is calculated by the formula:

$$y_i = \frac{\int\limits_{Min}^{Max} x \cdot \mu_i(x)dx}{\int\limits_{Min}^{Max} \mu_i(x)dz}.$$

where: $\mu_i(x)$—membership function of a fuzzy set E_i; Min и Max—upper and lower bounds of fuzzy variables; y_i—result variable. Membership function is a mapping $\mu_{\tilde{A}}(x) \rightarrow [0, 1], x \in X$.

The procedure of membership function creating is a stage of fuzzification of many prerequisites whose specified values determine of the consequences deduced in the fuzzy inference procedure.

In the process of the computing automation: of the decision making support system in the sphere ecological and economic security the mathematic algorithm includes such calculations as:

The volume of emissions according to the type of the pollutant in the enterprise

$$Q_p^{(rp)} = \sum_j \sum_i K_{ij}^{(rp)}$$

where $K_{ij}^{(rp)}$ – number of emissions i-th product of all types of pollutants for the reporting period.

In the fuzzy logic the conclusion is based on a set of possible factors whose emergence is determined by the membership function. Thus, the generalized rule modus ponens Fuzzy Modus Ponens—FMP) for the fuzzy systems is the following: $\frac{A'',A \supset B}{B''}$.

The sets A' and A are not necessarily coincide. If A' and A are close to each other it is possible to match them and get output B'.

The output B' is determined from the convolution max-min of the fuzzy set A' and the relation R interpreted as a vector-matrix logical multiplication, where the algebraic sum will correspond to taking the maximum in accordance with, and multiplication - taking a minimum. The operations of fuzzy conjunction and fuzzy disjunction of fuzzy sets are defined as obtaining the minimum and maximum degrees of membership of a compound statement.

Defining the elements of sets of factors $\{x_i\}$ as prerequisites in the rules, it would be natural to assume that the target variables are output – "Pollution coefficient", "Economic effect". Let us consider the system of the fuzzy rules as a management system, where y is a controlled variable, and many inputs $\{x_i\}$ are considered as control. Then the task of the logic output is reduced to the determining the "control" values, which can promote decision making. Actually such approach means the conclusion of explanatory reasons that cause certain state of the target variable. And the possibilities of the fuzzy approach to resolving the output on a set of fuzzy rules remain valid. On the other hand, this interpretation results in the necessity to change the position of the variables in the fuzzy rules: conditional part will contain fuzzy statements in relation to a variable y and its logic relation with other variables, and in the right side of the rules – some combination of variables from the set $\{x_i\}$. The system is ranked by the values of the term of the linguistic variable y.

Statistical analysis of the retrospective basis allowed to ground the quantity of the significant factors influencing on the target functions and, thus, to form the reduced base as a working base for further researches.

Let us take plurality $X = \{x_1, x_2, x_3, x_4, x_5, x_6\}$ factors the most relevant reasons influencing on the quality of the ecological and economic system as universum pre-requisites. Many conclusions or manifestations of the quality characteristics $Y = \{y_1, y_2, y_3\}$ are considered the second universum. The following result variables "Pollution coefficient", "Economic effect" were defined for this model.

It is known that fuzzy properties are represented by two notions and their properties: a fuzzy variable and a linguistic variable.

Thus, for factor $x1$ – the increased additional preparation – in short "idp" - the following fuzzy characteristics are defined: linguistic variable β = «idp» has domain of definition $X = [0–2.5]$; term-set of values of a linguistic variable $T = \{$«unsatisfactory», «satisfactory», and «good»$\}$. For each component of the term-set T, representing a fuzzy variable a_i, $(i = 1, 2, 3)$, a fuzzy set A_i should be built. Possible values of a fuzzy variable a_i are the components of this set. The belonging of these values to the set is defined by the semantics of the term a_i is given by the membership function.

In accordance with this algorithm the membership function for the linguistic variable («idp», T, X) is built, where

$T = \{$«unsatisfactory», «satisfactory», «good»$\}$;

$X = \{0, 0.25, 0.5, 0.75, \ldots, 2.5\}$ – base set.

The group expert assessment method was used to develop the universal interface of the decision support systems (DSS) to identify the most important indicators (quantitative and qualitative ones) applicable to the ecological and economic security.

Analysts, programmers, users and network administrators with working experience of three or more years in this sphere were the experts. Expert assessment was carried out with Delphi procedures. The indicators' qualitative work with the interface was ranked in accordance with their importance for this area.

The survey was conducted in three rounds. The summary questionnaire analysis was carried out according to the methods of coordination and disagreement of the expert group's opinions. Rather high degree of consistency was observed since the first round. The degree of consistency the experts' opinions was defined with Spearman's rank correlation coefficient:

$$\rho = 1 - \frac{6 \cdot \sum t_j^2}{n^3 - n}$$

where t_j – the difference between the ranks of the factors; n – number of factors.

Calculated coefficients are put in the result matrix. Rank correlation coefficients matrix ρij shows close are connections between i and j experts. Such matrix is square, dimensioned equal to the number of experts; it is symmetrical to the diagonal consisting of units as the degree of consistency is always maximum.

Matrix ρ is converted to matrix $\rho 0$ according to the following principle:

$$\rho^0 = \begin{cases} 1, \text{если } \rho \geq \varepsilon_\rho, \\ 0, \text{если } \rho < \varepsilon_\rho \end{cases}$$

where ε_ρ - threshold value for the rank correlation coefficients matrix. The threshold value is defined as $\varepsilon_\rho = 0.98$. Using the matrix $\rho 0$, a graph of expert opinions is constructed. The conclusion is based on the graph. The graph of consistency of expert opinions is shown in Fig. 2.

The experts' survey resulted in ranked list of key indicators of the universal interface quality to provide ecological and economic security.

The analysis of the source information and study of normative characteristics and requirements allowed us to distinguish the main functionally meaningful operations, which are important for the work of the ecological and economic security departments and are to be implemented in the automated decision support system.

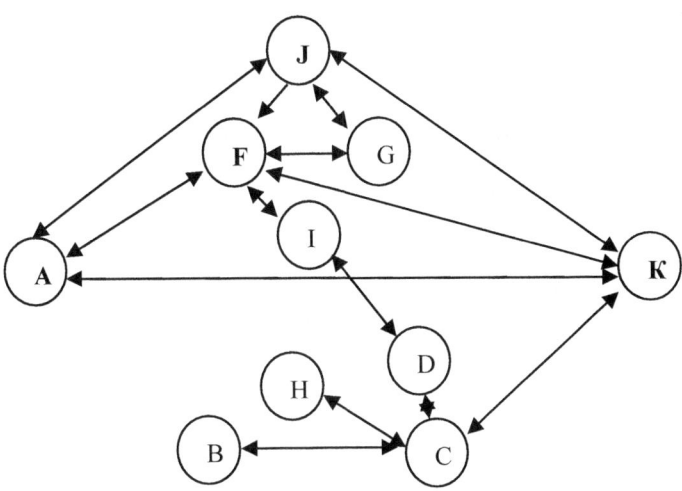

Fig. 2. Graph of expert opinion approvals

Such indicators as an employee's productivity, speed, and system operating costs depend on the interface quality. The suggestions to optimize the user interface are formulated on the basis the analysis of the number of executions of various functional operations. Variants of the decisions execution were tested in the relevant department and modified according to their proposals.

References

1. Environmental benefits from the international environmental and economic policies coordination based on modeling in the digital economy Skiter N.N., Ketko N.V., Gagarin A.G., Kostikova A.V. In the collection: The strategy for the development of agriculture in modern conditions - the continuation of the scientific heritage of Listopad G.E., academician of the Agricultural Academy of Agricultural Sciences (RAAS), Doctor of Technical Sciences, professor at the National Scientific and Practical Conference, pp. 321–328 (2019)

2. Soft computing methods application and cognitive modeling in the tasks of forecasting the ecological security of construction Sanzhapov B.Kh., Sadovnikova NP Ecology of urban areas, vol. 4, pp. 36-40 (2011)
3. Buletova, N.E.: Conceptual foundations of the study of ecological and economic security and their application in the regional economy [Text]. N.E. Buletova. National interests: priorities and security, no. 41 (182), pp. 10–24 (2012)
4. Akaev, A.O.: Strategy of integrated modernization of the Russian economy up to 2025. Econ. Issue **4**, 97–116 (2012)
5. Kuzmin, V.A.: Modeling of the integral indicator of economic security of the region by means of FuzzyLogicToolbox of the MatLab environment. Sci. J. Bull. State Polar Acad. State Polar Acad. Saint Petersburg (2013). A. F. Rogachev, V. A. Kuzmin, E. p. Sila
6. Ivantsova, E.A., Kuzmin, V.A.: Management of environmental and economic security of industrial enterprises. Bull. Volgograd State Univ. Series 3: Econ. Ecol. **5**(28), 136–146 (2014)
7. Alekseychik, T.V., Bogachev, T.V., Karasev, D.N., Sakharova, L.V., Stryukov, M.B.: Fuzzy method of assessing the intensity of agricultural production on a set of criteria of the level of intensification and the level of economic efficiency of intensification. Adv. Intell. Syst. Comput. **896**, 790–798 (2019)

Transformation of Mathematical Methods for Assessing the State of Economic Systems in Digitalization Based on the Theory of Fuzzy Sets (Review)

G. Akperov Imran[1]([⊠]) [iD] and Sakharova Lyudmila[2] [iD]

[1] Private Educational Institution of Higher Education "SOUTHERN UNIVERSITY (IMBL)", Rostov-on-Don, Russia
rector@iubip.ru
[2] Rostov State University of Economics, Rostov-on-Don, Russia
l_sakharova@mail.ru

Abstract. In the article, we analyzed existing methods for assessing the state of complex economic systems using the example of assessing their sustainable development. The analysis includes two parts. The first part is devoted to the classification of existing methods for assessing the stability of the economic system by the number of model parameters, as well as by the type of indicators used. For each class of methods, we carried out an analysis of literary sources, mainly in Russian literature, lists the tasks that can be solved with their help, and highlights the main advantages and disadvantages. The second part of the review is devoted to the analysis of the tendency to apply the theory of fuzzy sets in recent years in solving the problem of assessing the stability of economic systems. We have identified directions for the modification of classical methods for assessing the state of economic systems, which can be solved by using the fuzzy logic apparatus.

Keywords: Sustainability · Economic systems · One-factor methods · Multi-factor methods · Fuzzy-multiple model

1 Introduction

In modern conditions of transition to digital information processing, the inevitable transformation of classical mathematical methods of modeling to new ones takes place. New methods are focused on working with significantly larger amounts of data, as well as structured according to much more complex principles. A far from complete list of requirements for mathematical models, methods and algorithms in modern conditions includes: a systematic approach to modeling, high mobility and transformability, taking into account the probabilistic uncertainty of external conditions, as well as the uncertainty and subjectivity of expert estimates. Of course, the existing apparatus of mathematical methods and models in economics has not lost its relevance. However, considering the new requirements, it should be substantially modified by means of new, rapidly developing branches of mathematics, primarily due to the theory of fuzzy sets.

R. A. Aliev et al. (Eds.): ICAFS 2020, AISC 1306, pp. 215–222, 2021.
https://doi.org/10.1007/978-3-030-64058-3_26

Classical algorithms and decades-long approaches to assessing the state and forecasting of economic systems should be rethought and supplemented with the latest developments in data mining, which can be used to solve problems identified in a new round of the development of economic sciences.

This article is devoted to the analysis of existing methods for assessing the state of complex economic systems using the example of assessing their sustainable development. The analysis consists of two parts. The first part includes the classification of existing methods for assessing the stability of an economic system by the number of model parameters (one-factor and multifactor methods), as well as by the type of indicators used (static and dynamic). For each class of methods, we carried out an analysis of literary sources, mainly in Russian literature. We have listed the tasks that can be solved with their help, and also noted their main advantages and disadvantages. The second part of the review is devoted to the analysis of the tendency to apply the theory of fuzzy sets in recent years in solving the problem of assessing the stability of economic systems.

2 Classical Approaches to Assessing the Level of Sustainable Development of Economic Systems

Modern economic studies pay serious attention to the problem of sustainable economic development. The term "sustainable development" was first coined in the scientific world in 1987, in a report by the Commission on Environment and Development, led by the Prime Minister of Norway, G.Kh.Brutland [1]. The concept of sustainable development came in the 90s replaced the concept of economic growth. It involves taking into account not only the economic aspects of economic development, but also social as well as environmental ones [2, 3]. This approach applies not only to the economy as a whole, but also to its components: regions, industries, individual enterprises [4]. There are various approaches to both determining the economic sustainability of a region [5–14] and assessing the level of sustainable development of economic systems. The lack of a unified assessment methodology makes it necessary to study and systematize approaches to assessing the sustainability of the development of economic systems to form an integrated approach to ensuring their sustainable development.

The primary task is to determine the totality of indicators based on which the stability of the system is determined. General recommendations to determine the criteria for selecting indicators are given in expanded form in the work of H. Bossel [15]. The criteria are developed, justified and translated into a practical format in the work of a number of researchers. Various researchers have proposed different approaches to the selection of sets of indicators, and, based on them, developed their own selection methods [16–25]. Once the set of indicators is determined, it is necessary to develop a methodology for assessing on their basis the stability of the economic system under consideration. A detailed analysis of the methods for constructing assessments of the sustainability of regional development was performed by such authors [26–28].

Displayed in accordance with the study, we identified the following groups of methods for assessing the economic sustainability of the economic system:

1. Depending on the number of parameters in the stability model, we can distinguish: 1) one-factor methods based on the calculation of integral indicators of stability; 2) multivariate methods based on the use of a complex of grouped or ungrouped indicators, without determining the integral index of stability.
2. Depending on the type of indicators used to assess the sustainability of indicators, we can distinguish: 1) methods using static indicators; 2) methods using dynamic indicators.

At the same time, some authors build dynamic models based on a comparison of growth rates of indicators, while others propose the calculation of an integral indicator of enterprise development sustainability based on the calculation of indices of changes in particular indicators. Let us consider these methods in more detail.

One-factor methods based on the calculation of integral indicators of sustainability (sustainable development) of an enterprise using static indicators. The advantages of the approaches presented in the works of Sobchenko N.V. [29], Sarajeva O.V. [30], Khandazhapova L.M., Lubsanova N.B. [31], Karpushkina A.V., Voronina S.V. [32], Gonova O.V. [33], Vinogradova N.A. [34], Chernova T.V. [35], Babkova E.G. [36] are their multidimensional nature, allowing to take into account the components of economic stability, as well as a complex nature, which allows calculating both local indicators of economic stability in the context of structural elements, and an integral indicator of economic stability, which gives the assessment results additional informative significance. The disadvantage of all the above approaches is the use of static indicators that do not allow to assess the dynamics of the development of enterprises, as well as the level of balance of indicators in terms of stability components. One-factor methods based on the calculation of integrated indicators of sustainability (sustainable development) of an enterprise using dynamic indicators. Works by Kolobov A.A., Omelchenko I.N., Orlov A.I. [37], Vasiltsova V.M., Tsvetkov P.S. [38], Vasilyeva N. K. [39], Suslova S.A. [40], Chudilina G.I. [41], Ryabovoy I.V. [42] are devoted to the construction of integral estimates based on the assessment of the dynamics of the development of agricultural production in the region. Author's point of view Tretyakova EA and Osipova M.Yu. [43] lies in the need for the simultaneous use of static and dynamic approaches in the process of assessing regional development. The static approach allows us to assess the level of sustainable development of the regional socio-economic system at any given time. The dynamic approach allows us to assess the balance between the dynamics of the temporal characteristics of indicators both within the social, economic, and environmental components of sustainable development, and between them over a certain time interval. The proposed approach seems to be very valuable and allows modification due to systems of fuzzy-logical conclusions, which allow aggregating normalized indicators of both static and dynamic types.

Multivariate methods based on the use of a complex of grouped or ungrouped indicators using static indicators. Pletnev D.A. [18] proposed a set of static indicators for assessing three forms of corporate sustainability: financial, environmental, and structural. Anischenko L.I. [19] proposed a set of indicators characterizing the six components of the economic sustainability of the enterprise (organizational, production and technical, environmental, financial, social, innovative sustainability). Baranenko S. P. and Shemetov V.V. [20] proposed a set of indicators characterizing the three components of the economic stability of the enterprise (technological, financial, organizational).Menshchikova V.I., Sinopolets N.V. [44] argue that in order to identify bottlenecks in the economic potential of the region on the path to sustainable development, these indicators should be divided into three groups of production, financial and socio-economic.

Multivariate methods based on the use of a complex of grouped or ungrouped indicators using dynamic indicators. Platonov A.M. and Pleshkov S.Yu. [45] for the construction company established nineteen regulatory ratios of dynamics indicators. Kozin M.N. [46] proposed a dynamic model for assessing the production and economic sustainability of an enterprise. Despite the significant advantages of the considered dynamic methods for assessing the level of sustainable economic development of economic entities, they have the following drawback: they do not provide for an assessment of the structural elements of economic sustainability of development.

Thus, summing up, we can note: all of the above methods, despite their obvious practical relevance, have a number of disadvantages, the most significant of which are:1) locality, focus on a specific task, the need to develop a new model when changing the problem statement, including changing a set of indicators; 2) low universality, lack of a single concept for constructing estimates for various tasks of assessing the state of economic systems and mathematical methods for managing them; 3) the lack of a unified method of dimensioning and standardizing the parameters used to form estimates and, as a result, significant limitations on the number of parameters studied (introducing new parameters into consideration makes the model more complicated); 4) the lack of assessments of the structural elements of the system, allowing to assess the state of the subsystems of the system in question, including the underdevelopment of the apparatus for weighting coefficients for various indicators; 5) the impossibility of "cascading" integration (aggregation) of assessments that make it possible to consider economic entities as a "system of systems", that is, to build industry assessments based on assessments of its individual enterprises, regional assessments—based on assessments of individual municipalities, etc.; 6) lack of a clear concept of combining into an integral assessment of static and dynamic indicators; 7) the impossibility of flexible consideration in the models of subjective opinions of experts, the lack of the possibility of varying weighting factors.

One of the ways to overcome these problems is the use of fuzzy logic to assess the sustainable development of economic systems [47, 48].

3 Fuzzy-Multiple Models for Assessing Sustainable Development

An analysis of the literature shows that the possibility of these modifications has already been considered by experts in a number of studies on the stability of economic systems. The theory of fuzzy sets is applied to assessing the level of economic development of the constituent entities of the Russian Federation in the work of Poltavsky S.A. [49]. The author suggests identifying fuzzy clusters and using them in the practice of regional analysis, using the theory of fuzzy sets for this. The advantages of this method are, firstly, the possibility of obtaining a reasonable classification of subjects according to the level of socio-economic development; secondly, the allocation of "erosion bands" will allow us to identify subjects that cannot be assigned to any of the classes; thirdly, on the basis of fuzzy classification, a clear classification can be obtained, then obtaining a fuzzy classification from a clear one is impossible [50].

In the work Gul T.N. [51] an algorithm is proposed for assessing the level of socio-economic development of the Central Federal District. Using the Fuzzy Logic Toolbox (a package of applications included in the MATLAB environment), the author obtained a fuzzy breakdown of regions according to the level of socio-economic development using the k-means method. The advantage of the described method is the low degree of subjectivity, the minimum intervention of experts in the study (only at the initial stage when choosing a set of features). The author also noted the efficiency and accuracy of the approach, the ability to perform operations with large amounts of data. Yachmeneva V.M. [52] offers an assessment of the level of economic sustainability of an enterprise based on fuzzy logic. Karpova N.A. [53] developed a methodology for assessing the stability of consolidated groups of companies, based on a combination of fuzzy-plural and cognitive approaches to modeling. For modeling, the algorithms developed by O.A. Nedosekin to assess the risk of bankruptcy of a company [48].

A few works [54–60] are devoted to the construction of models for assessing the financial condition and stability of economic systems based on fuzzy logic. The tools used in them include various combinations of the fuzzy logic methods described above, implemented based on standard software tools.

Concluding the review, we should also note that currently fuzzy logic has a significant impact on the development of various methods of data mining (Data Mining), widely used in the study of economic systems [61]. Including, appear:

- fuzzy neural networks (fuzzy-neural networks), which draw conclusions based on the apparatus of fuzzy logic, but the parameters of membership functions are configured using training algorithms for neural networks.
- adaptive fuzzy systems (adaptive fuzzy systems), in which it is not necessary to attract experts to formulate rules and membership functions, since in such systems the selection of parameters is carried out in the learning process on experimental data.
- fuzzy queries to databases (fuzzy queries) - a tool that allows you to formulate queries in a natural language.
- fuzzy associative rules allow you to extract patterns from databases that are formulated in the form of linguistic statements.
- fuzzy cognitive maps (fuzzy cognitive maps) used to assess the causal relationships identified between the concepts of a certain area; fuzzy clustering, which allows the same object to belong to several clusters at the same time [62].

4 Conclusion

The analysis shows that methods for assessing the stability of the economic system can be classified in at least two ways: by the number of model parameters (one-factor and multifactor methods), as well as by the type of indicators used (static and dynamic).

The selection of a parameter and its justification is a separate, developed area of research. The corresponding methods for constructing quantitative assessments of the stability of economic systems, with their undoubted practical value and relevance, have several disadvantages. Its are characteristic of classical economic models and insurmountable within the framework of the traditionally used mathematical apparatus. At the same time, they can be overcome using a fuzzy logic apparatus, including fuzzy logic inference systems. Currently, active work is underway to modify traditional mathematical methods based on fuzzy logic, which will allow for the foreseeable future to achieve significant success in developing a modern mathematical apparatus that meets the challenges of the digital economy.

References

1. Brutland, G.K.: Our Common Future, Report of the UN Commission on the Environment and Development. Progress, p. 50 (1987)
2. Gorshenin, E.V., Khomyachenkova, N.A.: Monitoring the sustainable development of an industrial enterprise. Russ. Bus 1(2), 63–67 (2011)
3. Yarullina, G.R.: Methodology for ensuring the sustainable development of an industrial enterprise: Monograph. Publ. house Kazan Univ., Kazan (2010)
4. Gnatyuk, S.N.: Indicators of sustainable development of the region. Polit. Econ. Innov. 5 (2016)
5. Antonova, M.A.: Theoretical and methodological foundations of the study of sustainable development of regions. Soc. Polit. Econ. Law 4 (2013)
6. Alferova, T.V.: Conceptual modeling of the definition of the category "sustainable development". J. Econ. Theory 4, 46–52 (2012)
7. Akhmetishina, A.R.: Regulation of economic and environmental relations in the framework of sustainable development of the economic system: abstract of the dissertation of the doctor of economic sciences, Kazan, 47 p. (2011)
8. Gutman G.V.: Management of the regional economy, 176 p. (2001)
9. Danilov-Danilyan, V.I.: Sustainable development (theoretical and methodological analysis). Econ. Math. Methods 39, 123–135 (2003)
10. Kalinchikov, M.Yu.: Theoretical and methodological foundations of the concept of sustainable development of the region. Region. Econ. Theory Pract. 9(24) (2005)
11. Kuznetsov, O.L.: Sustainable development: the scientific basis of design in the system nature - society - man. SPb, 240 p. (2008)
12. Pikovsky, A.A.: Sustainable development and culture. SPb (2002)
13. Rosenberg, G.S.: Sustainable development: myths and reality. Tolyatti, 191 p. (1998)
14. Sigov I.I.: Regional economy (conceptual apparatus). St. Petersburg, pp. 65–66 (2000)
15. Bossel, H.: Earth at a crossroads: Paths to a sustainable future, 338 p. Camb. Univ. Press, Cambridge (1998)
16. Garipov, R.I., Garipova, E.N.: On the issue of assessing the sustainable development of the regional economic system. Manage. Mod. Syst. 1 (2013)
17. Zotova, A.I., Kirichenko, M.V.: Stability of the financial system of the region: essence, factors, indicators. Theory Pract. Soc. Dev. 5 (2017)
18. Pletnev, D.A.: Criterion and indicators for assessing corporate sustainability in line with a systematic approach. Electron. Sci. Econ. J. 2(2), 21–26

19. Anischenko, L.I.: Key aspects and tools for achieving sustainable development. Management of economic systems. Electron. Sci. J. (2015). (in Russian). https://www.uecs.ru/uecs-73-732015/item/3330-2015-01-28-08-25-40
20. Baranenko, S.P., Shemetov, V.V.: Strategic sustainability of the enterprise, Tsentrpoligraf, 496 p. (2004)
21. Borisova, I.S.: To the question of a model for managing the sustainable development of the regional economy with the predominance of a separate type of economic activity. Bull. Inst. Econ. RAS **3** (2018)
22. Savvateev, E.V.: Regional specificity of managing spatial economic systems in a market economy. Stat. Econ. 3–2 (2012)
23. Shaburova, D.P.: Analysis and evaluation of socio-economic processes in the regions - the basis of the mechanism of sustainable development (for example, Khabarovsk Territory). Power Manage. East of Russia 2(87), 9 (2018)
24. Elyashev, D.V.: Sustainability indicators of the dynamics of agricultural production in the Leningrad region. Izvestiya SPbGAU 2(47) (2017)
25. Lukomets, A.V.: Statistical support of the problem of sustainability of sunflower production in the Krasnodar Territory. Oilseeds **2**, 151–152 (2012)
26. Zhurova, L.I., Toporkov, A.M.: A comparative analysis of approaches to assessing the sustainable development of economic systems. Bull. VUiT **4** (2017)
27. Afonichkin, A.I., Toporkov, A.M.: Theoretical aspects of the formation of parameters of the strategy for sustainable development of economic systems. Bull. VUiT **1**(33) (2015)
28. Tretyakova, E.A., Alferova, T.V., Pukhova, Yu.I.: Analysis of methodological tools for assessing the sustainable development of industrial enterprises. Bull. PSU **4**(27) (2015)
29. Sobchenko, N.V.: A comprehensive methodology for assessing the economic sustainability of an enterprise based on innovative activity. Sci. J. KubSAU 67(03) (2011)
30. Sarajeva, O.V.: Sustainable development of the economy of economic entities of Russia. Probl. Mod. Econ. 3(35) (2010)
31. Khandazhapova, L.M., Lubsanova, N.B.: The scientific basis for the study of the sustainability of the economy of the border region. Region. Econ. Theory Pract. **15**(390) (2015)
32. Karpushkina, A.V., Voronina, S.V.: Sustainable development of the region: theoretical and methodological aspects. UEKS **10**(70) (2014)
33. Gonova, O.V.: Socio-economic development of the region: rating assessment models. Modern high technology. Region. Appl. **3** (2010)
34. Vinogradova, N.A.: Integrated Regional Development Index. Region. Econ. Theory Pract. **2** (425), (2016)
35. Chernova, T.V.: Approaches to the statistical assessment of the socio-economic situation of municipalities. Account. Stat. 4(20) (2010)
36. Babkova, E.G., Panakhov, A.U.: Rating assessment of the balanced development of the regions of the Central Federal District. Econ. Environ. Manage. **2** (2018)
37. Kolobov, A.A., Omelchenko, I.N., Orlov, A.I.: High tech management. Integrated production and corporate structures: organization, economics, management, design, efficiency, sustainability. AST. Exam (2008)
38. Vasiltsova, V.M., Tsvetkov, P.S.: Methodological approaches to assessing the economic sustainability of enterprises. News Southwestern S. Uni. **5**(56), 147–151 (2014)
39. Vasilieva, N. K.: Methods for assessing the sustainability of production in the agricultural sector. Space Econ. **4**(2005)
40. Suslov, S.A., Gromova, I.V.: Methodology of regional assessment of economic sustainability of agricultural production. Bull. NIIEI **5** (2012)

41. Chudilin, G.I. On the state and methodology for assessing the sustainability of agricultural production. Bull. ChSU **1** (2006)
42. Ryabova, I.V.: Assessment of the sustainability of agricultural production in the territorial system of food security. Bull. NIIEI **9**(64) (2016)
43. Tretyakova, E.A., Osipova, M.Yu.: The combination of static and dynamic approaches in assessing the sustainable development of regional socio-economic systems. Bull. PSU. Ser.: Econ. **2**(19) (2016)
44. Menschikova, V.I., Sinopolets, N.V.: The system of indicators for assessing the sustainable development of the regional economy. Socio-Econ. P. Process. 5–6 (2011)
45. Platonov, A.M., Pleshkov, S.Yu.: Ways and methods of ensuring economic sustainability of a construction enterprise based on a dynamic model of economic sustainability. Econ. Region **4**, 240–244 (2008)
46. Kozin, M.N.: A dynamic model for assessing the industrial and economic stability of a defense enterprise. Audit Financ. Anal. **4** (2007)
47. Zade, L.A.: The concept of a linguistic variable and its application to making approximate decisions. Mir, 165 p. (1976)
48. Nedosekin, A.O.: Mathematical foundations of modeling financial activities using fuzzy-plural descriptions. Econ. Sci. (2003). https://www.mirkin.ru/_docs/doctor005.pdf
49. Poltavsky, S.A.: Methodology for assessing the differentiation of the constituent entities of the Russian Federation by the level of economic development (based on the optimization fuzzy c-means Bezdeck-Dann algorithm). Region. Econ. Theory Pract. **17**(74), 25–36 (2008)
50. Radyukova, Y.Yu., Shamaev I.N.: Economic security of the country as a multi-level system of elements and relations. Socio-Econ. P. Process. Tambov **2** (2011)
51. Gul, T.N.: Assessment of the sustainability of the development of the region. Socio-Econ. P. Process. **10** (2011)
52. Yachmeneva, V.M.: Presentation of the economic sustainability of the enterprise. Econ. Manage. **107–112**, 4–5 (2007)
53. Karpova, N.A.: Application of fuzzy logic methods in assessing and forecasting the financial stability of consolidated groups of companies. B. Eurasian Sci. **5**(30) (2015)
54. Soloviev, D.B., Kuzora, S.S., Merkusheva, A.E.: Using fuzzy inference algorithms for a preliminary assessment of participants in a cluster approach. Innovation **5**(235) (2018)
55. Baychenko, A.A., Baychenko, L.A., Areth, V.A.: The use of fuzzy logic in the management of an enterprise in the food industry. Econ. Environ. Manage. **3** (2014)
56. Gupanova, Y.E.: The use of fuzzy logic tools in assessing the quality of customs services. Econ. Anal.: Theory Pract. **1**(448) (2016)
57. Biketov, A.N., Glebova, O.V., Melnikova, O.Yu.: Risk assessment system based on the use of fuzzy logic. Volga Sci. Bull. (40) (2014)
58. Rogachev, A.F., Skiter, N.N., Kuzmin, V.A.: Modeling of environmental and economic security using fuzzy logic tools. UEKS **12**(60) (2013)
59. Biryulin, V.I., Kudelina, D.V.: Fuzzy inference system for evaluating the effectiveness of regional energy. Sci. Reports Belgorod S. Univ. Series: Econ. Inform. **13**(210) (2015)
60. Barykina, Ya.O.: Assessment of financial stability of a commercial enterprise using fuzzy logic methods. Sci. Notes Young Res. **6** (2016)
61. Lebedeva, M.E.: Fuzzy logic in the economy - the formation of a new direction. Ideas ideals **1** (2019)
62. Paklin, N.: Fuzzy logic - mathematical foundation. BaseGroupLabs (2005). (in Russian)

Fuzzy Regulation Model of the Interaction Between the State and Economic Subject Within the Shadow Economy and Tax Field

Akif F. Musayev[1,2(✉)] [iD], Mirali S. Kazimov[3,4] [iD],
Samir S. Rustamov[5,6] [iD], Nazim K. Aliyev[3,7] [iD],
and Shahzada G. Madatova[3,8,9] [iD]

[1] Institute of Economics, ANAS, H. Javid ave., 115, AZ1143 Baku, Azerbaijan
akif.musayev@gmail.com
[2] Azerbaijan University, Nasimi distr., Jeyhun Hajibeyli str., 71,
AZ1007 Baku, Azerbaijan
[3] Training Center of Ministry of Taxes of the Republic of Azerbaijan,
Agha Nematullah str. 44, AZ1033 Baku, Azerbaijan
[4] Baku State University, Acad. Zahid Xalilov str. 23, AZ1148 Baku, Azerbaijan
vergi3m@gmail.com
[5] ADA University, Ahmadbay Agha-Oglu str., 61, AZ1008 Baku, Azerbaijan
srustamov@ada.edu.az
[6] Institute of Control Systems, ANAS, Bakhtiyar Vahabzadeh str., 9,
AZ1141 Baku, Azerbaijan
[7] Institute of Law and Human Rights, ANAS, H. Javid ave., 115,
AZ1073 Baku, Azerbaijan
aliyevnazim2020@gmail.com
[8] The Academy of Public Administration under the President of the Republic
of Azerbaijan, Lermontov str. 74, AZ1001 Baku, Azerbaijan
[9] Azerbaijan State Oil and Industry University, Azadlig ave. 34,
AZ1000 Baku, Azerbaijan
shahzademedetova@gmail.com

Abstract. In this article, economic system is reviewed as an active system; and evaluation and reduction of the process emergence rate of shadow economy via optimization methods is being analyzed within economic relations between the state as a tax authority and economic subject as a taxpayer. Profitability of economic activity as a main influencing factor, expenses for staying in the shadow, contribution of tax and revenue for general utility are evaluated through fuzzy approach. Selecting profitability as a main factor is related with the process of not declaring the real level of profitability to tax authority by economic subject.

Keywords: State and economic subject · Shadow economy · Optimization · Proitability · Tax evasion · Mathematical model · Fuzzy approach

R. A. Aliev et al. (Eds.): ICAFS 2020, AISC 1306, pp. 223–229, 2021.
https://doi.org/10.1007/978-3-030-64058-3_27

1 Introduction

State as an instance of tax authority and economic subject have aims respectively with different directions as tax collection and accomplishment of tax responsibilities. At the same time each of them makes effort to somehow benefit as a result of their own activity. If the utility of the state is expressed by the amount of taxes collected to the state budget, then the utility of the economic subject can be expressed with the left amount of its revenue after accomplishment of tax responsibility.

On the other hand, if we consider that state is socially responsible for the sustainability of the activity of economic subject, then another component of benefit of the state becomes obvious and this is reflected in the left revenue of the economic subject.

At the same time economic subject also owns a utility function of two components. Economic subject should both try to save the left revenue for its social security and accomplish the commitment of corporate responsibility to the state budget.

Thus, both them possess utility function depending on the left revenue and tax amount paid to the state budget. At the same time state should control truthfulness of the tax base of economic subject and the existence of its risk for the emergence of shadow economy. It is clearly seen that within the process of investigation of interaction between state and economic subject several economic and social factors should be taken into consideration. In this case investigation of the emergence of shadow economy and the interaction of its measurement with other economic factors is not an exception, either.

In this article economic system is reviewed as an active system; and evaluation and reduction of the process emergence rate of shadow economy via optimization methods is being analyzed within economic relations between the state as a tax authority and economic subject as a taxpayer.

In this case, profitability of economic activity as a main influencing factor, expenses for staying in the shadow, contribution of tax and revenue for general utility are evaluated through fuzzy approach. Selecting profitability as a main factor is related with the process of not declaring the real level of profitability to tax authority by economic subject.

2 The Reasons of the Emergence of Shadow Economy and Its Assessment Methods

Currently there are numerous approaches to the forms of emergence of shadow economy, the reasons for this and its assessment methods [1–5]. Different aspects of the considered problem are reflected in these approaches and within related models, but their universality level is far away from the perfection. Some reasons for this are substantiated within scientific literature through wide variety of research papers [6–8]. Main consequence reflected here explains that considering tax as only a way of filling the state budget is not satisfactory. Taxes are the tools for positive and negative impact on economic processes formed between the state and economic subject. That is why in order to create a model reflecting activity of both economic subject and supervisory authority economic factors allowing the research on emergence process of shadow economy while allocating resources between them and; the ones paving the way for competitive environment which can help to find factors influencing the emergence of shadow economy should be preferred [8].

3   The Behavioral Model of the State

The utilization function of the state depends on the left revenue after paying all the taxes and overall amount of taxes paid to the state budget; and it is written as following:

$$D = \beta \ln[G(1-u)r_0 + G(r-r_0)] + \alpha \ln Gur_0,$$

Here, r – refers to real profitability; r_0 – refers to declared profitability of the economic subject; Gr_0 – refers to revenue; $G(r-r_0)$ – refers to revenue left in the shadow; $G(1-u)r_0 + G(r-r_0)$ - refers to the revenue of tax payer; Gur_0 – refers to the amount of tax receipts; α – refers to the weight of tax receipts in the utility function; β – refers to the special weight of the revenue of tax payer in the utility function after tax payments.

In this case if we consider P to be the forecast task on tax receipts mathematical model of the state as a tax authority can be written as following:

$$D = \beta \ln[G(1-u)r_0 + G(r-r_0)] + \alpha \ln Gur_0 \rightarrow \max$$
$$Gur_0 + G u(r-r_0)(1-k) \geq P; \tag{1}$$

$$0 \leq r_0 \leq r; \alpha, \beta, u \in (0,1); \alpha + \beta = 1; \alpha, \beta, k \geq 0 \tag{2}$$

Here, k refers to the costs for staying in the shadow (or penalty for tax evasion). Restriction condition (1) explains that economically the amount of tax receipts cannot be less than the amount of forecast task. (2) refers to the necessary restriction conditions.

4   Mathematical Model of the Economic Subject

Defining and fulfilling the tax liability is not less important than the efforts of the economic subject for getting maximum revenue which seems to be the main aim of its activity. In most cases the amount which is separated and paid to the state budget is considered to be a lost and attempts for decreasing this amount are made both via legal and illegal methods by the economic subject. As a result some or whole amount of revenue can be hidden this way.

And this activity reflects in the declaration of economic subject in the way it knows. For this reason, utility gained regarding this activity consists of two parts: first part is the one declared and all related tax paid, the other part occurs to be the hidden one in which tax liability is not accomplished. It becomes obvious that utility function of the economic subject should include both parts. We can write following referring to afore given legend for the previous formula:

$$S = \beta \ln[G(1 - r_0)u + G(r - r_0)(1 - k)] + \alpha \ln Gur_0,$$

Here, $G(1 - r_0)u + G(r - r_0)(1 - k)$ – refers to the revenue of the economic subject;

The amount of tax payment of each economic subject cannot exceed its tax potential and declared profitability can't be larger than real profitability. Tax payer makes effort to maximize its utility within these conditions.

In this case if consider the tax potential of the tax payer as VP, we can write the mathematical model of the economic subject as following:

$$S = \beta \ln[G(1 - u)r_0 + G(r - r_0)(1 - k)] + \alpha \ln Gur_0 \rightarrow max$$
$$Gur_0 + G u(r - r_0)(1 - k) \leq VP; \tag{3}$$

$$0 \leq r_0 \leq r \; ; \; \alpha, \beta, u \in (0, 1); \; \alpha + \beta = 1; \; \alpha, \beta, k \geq 0 \tag{4}$$

Here restriction condition (3) explains that tax payments of the economic subject can't exceed its tax potential. (4) refers to necessary restriction conditions.

We should mention that as economic subject is inclined to stay in the shadow its α, β contribution to general utility changes in dependence of k – which refers to the costs for staying in the shadow.

5   Establishing IF-THEN Fuzzy Model

As mentioned above in the first stage the economic subject declares some part (r_0) of its real profitability (r) and misappropriates the left part $(r - r_0)$. Tax authority is aware about the r as they know the average price level of the field in which economic subject operate. That is why tax authority defines k – fine to eliminate this suspense within the report of the economic subject and informs economic subject about it. Economic subject takes the fine rate into account and in order not to lose the revenue declares a new and higher price, etc.

Let us analyze the evaluation of r_0 as profitability of economic activity playing the role of main factor, k as cost for staying in the shadow, α, β contribution of tax and revenue to general utility with the help of IF-THEN fuzzy approach [9–11]. For this, in accordance with tax potential of economic subject, considering the conditions given in (2) and (4), to ease the formula, let us assume that $\beta = k$ and $\alpha = 1 - \beta$; to find appropriate rate for k we can present fuzzy approach as given below:

– let us exaggerate k and r_0 via fuzzification methods:
– let us define linguistic quantity for the rates of k and α:

"weak", "average", "strong"
We should also express the rates of r_0 with three linguistic quantity:
"low", "average", "high"
In accordance with these linguistic variables IF-THEN conditions can be presented as following:

IF k = "weak" and α = "strong" Then r_0 = "high"
IF k = "average" and α = "average" Then r_0 = "average"
IF k = "strong" and α = "weak" Then r_0 = "low"
IF k = "strong" and α = "strong" Then r_0 = "high"
IF k = "weak" and α = "weak" Then r_0 = "low"

Fragment from calculations of suggested model and fuzzy approach is given in an example below:

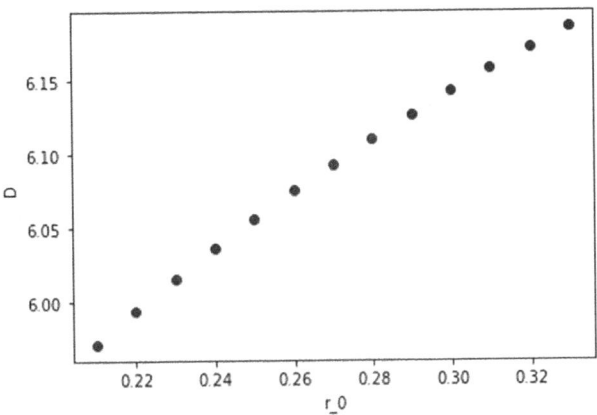

Fig. 1. The rate of utility function of state (D) increases in accordance with the growth of r_0

Initial data: let us assume that, G = 4000\$, P = 150\$, VP = 500\$,

$$u = 0.2 \quad r = 0.33, \quad r_{min} = 0.21, \quad k = 0.2$$

The growth in the utility function of state depending on the increased declared profitability of the economic subject is provided within the graph given below:
While the cost for staying in the shadow is increasing utility function of the state decreases. For k = 0.4 case change in rate variation depending on its declared profitability is shown in the below given graph:

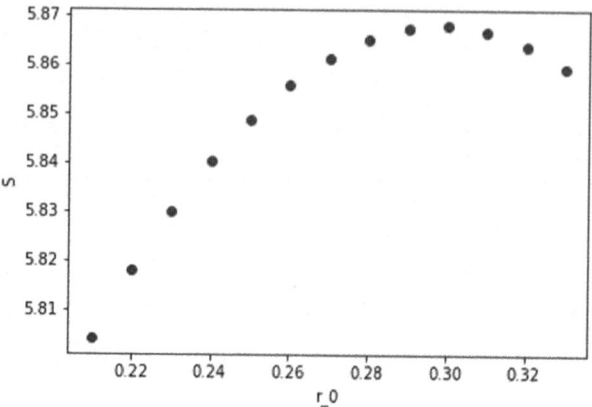

Fig. 2. Variation in the rate of utility function of an economic subject (S) depending on r_0

As seen from the picture, the rate of fine for staying in the shadow reaches to a such level that afterwards staying in the shadow does not seem to be beneficial, it even increases the costs.

6   Conclusion

- Achieved consequences show that suggested instrumentation reflects the emergence of shadow economy real enough;
- While analyzing these graphs it is easily observed that k – increased fine rate respectively increases the inclination for leaving the shadow economy of economic subject;
- As declared profitability rate r_0 becomes similar to real profitability rate r the tax receipts of the state budget increase (Fig. 1).
- In order not to have decreased rate for utility function economic subject is being forced to declare higher rates of r_0 to the state (Fig. 2).
- In some rate of k-fine for staying in the shadow, the rate of r_0 becomes extremely closer to rate of real profitability r.
- While rate for fine k is increasing the shadow revenue noticeably decreases.

References

1. Schneider, F., Buehn, A.: Estimating the Size of the Shadow Economy: Methods, Problems and Open Questions, Discussion Paper No. 9820 (2016). ftp.iza.org/dp9820.pdf
2. Zadeh, L.: Fuzzy sets. Inform. Control **8**, 338–353 (1965). https://www-liphy.ujfgrenoble.fr/pagesperso/bahram/biblio/Zadeh_FuzzySetTheory_1965.pdf
3. Schneider, F.: Estimating the size of the danish shadow economy using the currency demand approach: an attempt. Scand. J. Econ. **88**(4), 643 (1986). https://doi.org/10.2307/3440435

4. Torgler, B., Schneider, F., Schaltegger, C.A.: Local autonomy, tax morale, and the shadow economy. Publ. Choice **144**, 293–321 (2010). https://doi.org/10.1007/s11127-009-9520-1
5. Buehn, A., Schneider, F.: Corruption and the Shadow Economy: A Structural Equation Model Approach. IZA Discussion Paper No. 4182. https://ssrn.com/abstract=1409286
6. Lackó, M.: The hidden economies of visegrad countries in international comparison: a household electricity approach. In: Halpern, L., Wyplosz, Ch. (eds.) Towards a Market Economy, Hungary. Cambridge University Press, Cambridge (Mass.) (1998)
7. Aigner, D.J., Schneider, F., Ghosh, D.: Me and my shadow: estimating the size of the U.S. hidden economy from time series data. Dyn. Econ. Model. 297–334 (1998). https://doi.org/10.1017/cbo9780511664342.015
8. Medina, L., Schneider, F.: Shadow Economies Around the World: What Did We Learn Over the Last 20 Years? IMF Working Paper (2018). ISBN/ISSN: 9781484338636/1018-5941
9. Musayev, A.F., Madatova, S.G., Rustamov, S.S.: Evaluation of the impact of the tax legislation reforms on the tax potential by fuzzy inference method. In: 12th International Conference on Applied Fuzzy System Soft Computing (2016). Procedia Comput. Sci., **102**, 507-514. https://doi.org/10.1016/j.procs.2016.09.435
10. Musayev, A., Madatova, S., Rustamov, S.: Mamdani-type fuzzy inference system for evaluation of tax potential. In: Zadeh, L.A., Yager, R.R., Shahbazova, S.N., Reformat, M.Z., Kreinovich, V. (eds.) Recent Developments and the New Direction in Soft-Computing Foundations and Applications. SFSC, vol. 361, pp. 511–523. Springer, Cham (2018). https://doi.org/10.1007/978-3-319-75408-6_39
11. Rustamov, S., Musayev, A., Madatova, S.: Evaluation of the impact of state's administrative efforts on tax potential using sugeno-type fuzzy inference method. In: Aliev, R.A., Kacprzyk, J., Pedrycz, W., Jamshidi, M., Sadikoglu, F.M. (eds.) ICAFS 2018. AISC, vol. 896, pp. 352–360. Springer, Cham (2019). https://doi.org/10.1007/978-3-030-04164-9_47

Prediction of Daily Solar Irradiation Using CNN and LSTM Networks

Nuray Vakitbilir[1]([⊠]) ⑩, Adnan Hilal[1] ⑩, and Cem Direkoğlu[2] ⑩

[1] Sustainable Environment and Energy Systems, Middle East Technical University, Northern Cyprus Campus, 99738 Mersin 10, Turkey
{nuray.vakitbilir, adnan.hilal}@metu.edu.tr
[2] Electric and Electronics Engineering Program, Middle East Technical University, Northern Cyprus Campus, Kalkanlı, 99738 Mersin 10, Turkey
cemdir@metu.edu.tr

Abstract. Greenhouse gas emissions from conventional energy sources are accelerating the global warming. To alleviate this issue, countries are focusing on renewable energy sources, especially solar enery, to meet the increasing energy demand. Forecasting solar irradiation on different time-horizons is crucial for the integration of the solar energy to the existing or future electricity grids. In this paper, we focus on solar irradiation prediction, and present performances of two different methods for Kalkanlı region of Cyprus. We use a one-dimensional Convolutional Neural Network (1D-CNN) and a Long-Short Term Memory netwok (LSTM) separately for prediction. In particular, 1D-CNN and LSTM networks are employed with two different time-series input datasets to predict one-day ahead global horizontal irradiation (GHI) for Cyprus. Performances of the networks are evaluated and compared.

Keywords: Solar irradiation forecasting · Deep learning · One-dimensional convolutional neural networks · Long short-term memory network

1 Introduction

It is foreseen that global energy demand is likely to increase immensely concerning the growth of the world's population [1]. As a result of global warming caused by greenhouse gas emissions from conventional energy sources, countries are focusing on renewable energy sources to meet increasing energy demand [2]. More countries are expected to integrate renewable energy sources, specifically solar energy, into their energy supply [2–4]. Among the renewable energy sources, solar energy is the main focus of interest as there is tremendous growth in photovoltaic (PV) panel installation in many countries [4, 5]. However, the output of PV panels varies throughout the day, month and year. As its energy output is unstable, it makes the difficult task of balancing demand and supply of electricity in the isolated electrical grids even more challenging [3, 5]. The growth in solar energy production leads to a need for information on solar radiation [4, 6] as solar radiation and the PV output are proportional [4]. Solar radiation information is also useful in agricultural activities such as in crop growth models to estimate the effect of climate change on agriculture [7].

R. A. Aliev et al. (Eds.): ICAFS 2020, AISC 1306, pp. 230–238, 2021.
https://doi.org/10.1007/978-3-030-64058-3_28

Extensive integration of the solar energy to the existing or future electricity grids is one of the reasons for solar radiation prediction as the radiation data is not always readily available due to high cost [3, 8]. For the sake of grid reliability, design and optimisation of solar systems, power system scheduling, congestion management, and trading of the produced electricity in the electricity market, it is a crucial and ongoing task to forecast solar radiation on different time-horizons [3, 9]. Thus, new methods are being developed for this task. These methods include cloud image processing models, empirical models, statistical models, Numerical Weather Prediction (NWP) and machine learning models. Cloud-based models combined with time-series data of solar radiation [6, 10–12] provide better results in forecasting solar radiation [8]. Other approaches mainly use weather-based data [4, 8, 9, 13–15].

Machine learning methods to predict solar radiation by weather data has been used in recent years in many works. Various of these researches have used different parameters such as temperature, wind speed and direction, pressure, daily solar radiation and more. Khosravi et al. [15] compared results of different machine learning networks with inputs as weather data and time-series data in predicting global solar radiation for Abu Musa Island. In weather-based network, they have used pressure, temperature, wind speed, relative humidity and local time as inputs. Salcedo-Sanz et al. [4] analysed the performance of hybrid evolutionary-ELM model to predict daily global solar radiation for Spain. K-nearest neighbour classifier has been investigated by Demirtaş [13] with pressure, temperature, humidity and solar radiation as input for Turkey. Since machine learning algorithms learn from data, it is important to choose and prepare the right data [3]. The local climate, weather and geographic position are the main factors affecting the availability of solar radiation [6, 9]. Therefore, it is crucial to obtain a prediction model designed for a particular part of the world.

In this study, we demonstrate the performance of a Convolutional Neural Network (CNN), and a Long Short-Term Memory network (LSTM), to estimate a day ahead prediction of global solar radiation for Kalkanlı, Cyprus using weather data and time-series data of solar radiation as input. In Sect. 2, forecasting networks used in the study are introduced. The study area and the data are described in Sect. 3, while detailed information on the forecasting models and the forecasting evaluations and the discussion are presented in Sect. 4. Finally, we conclude our study in Sect. 5.

2 Theoretical Overviews

In this section, CNN, LSTM network models and model evaluation methods that are used in this study to forecast global horizontal irraduance (GHI) are explained.

2.1 Convolutional Neural Networks (CNN)

Various architectures of CNN are available, that are made up of four main types of layers; convolutional, pooling, fully connected and logistic regression layers [16]. One of the main advantages of CNN is its powerful ability to features extraction [17].

Convolutional Layer. The convolutional layer generates feature maps from the input data. This layer contains several convolutions or filters. Each kernel is used to generate one feature map. An activation function is applied to introduce nonlinearity to the convolutional layer. ReLu, Sigmoid and tanh are the most used algorithms as the activation functions in CNN. Each kernel represents a different weight matrix. The weight values and a bias term are updated during the training phase. The mathematical formula of the convolutional layer is shown in Eq. (1) [2, 18, 19].

$$y_{i,j,k}^l = F((w_k^l)^T x_{ij}^l + b_k^l) \tag{1}$$

Where the weight and bias of kth convolutional kernel in the lth layer are represented as w_k^l and b_k^l, respectively. x_{ij}^l is the input patch in the lth layer, concentrated at the location (i,j). $F()$ represents the activation function. All regions of the input are shared with the weight w_k^l which reduces the training time and the complexity of the network. In Fig. 1-a, the convolutional layer is presented to show all the mentioned steps to produce feature maps. In our study, we have five convolutional layers and softplus is used as an activation function in all five of the convolutional layers.

Pooling Layer. The main aim of this layer is to decrease the resolution of the feature map. Usually, this layer is used between two convolutional layers. The mathematical presentation of the pooling layer is shown in Eq. (2) [18, 19],

$$P_{i,j,k}^l = Pool\left(y_{m,n,k}^l\right) \tag{2}$$

Where, $(m, n) \in$ Ri, j which represent the region around the location (i, j). Pool () describes the type of pooling operation used. Max pooling and average pooling are the pooling operations that are used commonly. Pooling layer usually increases the network accuracy while reducing the number of parameters in the network [19].

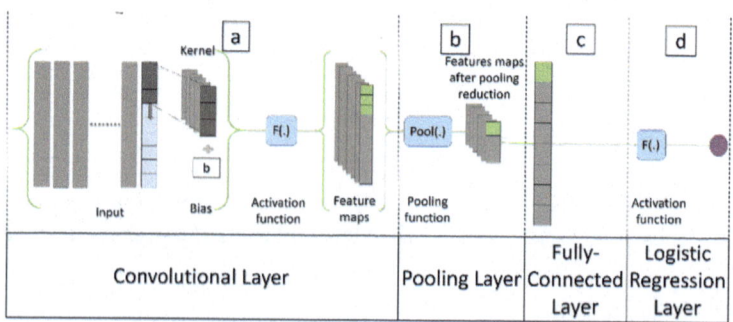

Fig. 1. The architecture of a general Convolutional Neural Network (CNN)

Fully Connected Layer. The main task of this layer is to perform high-level reasoning by transporting the learned feature in the network to one space, as shown in Fig. 1-c. Typically one or more fully connected layers can be used in CNN models [2, 18].

Logistic Regression Layer. The last layer of CNN architecture is the logistic regression layer which is used to give the network output as illustrated in Fig. 1-d. Between the fully connected layer and the logistic regression layer, there exists an activation function, where Rectified Linear Unit (ReLu) is employed in this study.

There are various branches of CNN. One dimensional CNN (1D-CNN) and two-dimensional CNN (2D-CNN) are commonly used in literature. Where the former is widely used to process numerical data such as weather data and energy production, while the latter is frequently used for image and text processing. Both types of CNN models are composed of the same main layers. The main difference occurs in the convolutional layer, where the kernels slide in two dimensions on the input data in the 2D-CNN while in 1D-CNN, the sliding happens only in one dimension [17]. In this study, 1D-CNN is employed, where network details are explained in Sect. 4.

2.2   Long-Short Term Memory (LSTM)

LSTM is one of recurrent neural network (RNN) architectures. RNN models have the capability to derive relations between consecutive events; however, they become insufficient when it comes to relating the long-range events because of gradient vanishing or gradient exploding [18]. Gradient vanishing refers to the fast-exponential decrease of the gradient norm to zero, resulting in a network that cannot learn from long data sequences [20]. LSTM is introduced to overcome this problem with its memory cell, first introduced by Hochreiter and Schmidhuber [21], and extra forget gate included by Gers et al. [22]. Memory blocks in LSTM include input, output and forget gate allow updating and controlling the flow of information in separate blocks [2].

Figure 2 show a sample LSTM block used in this study.

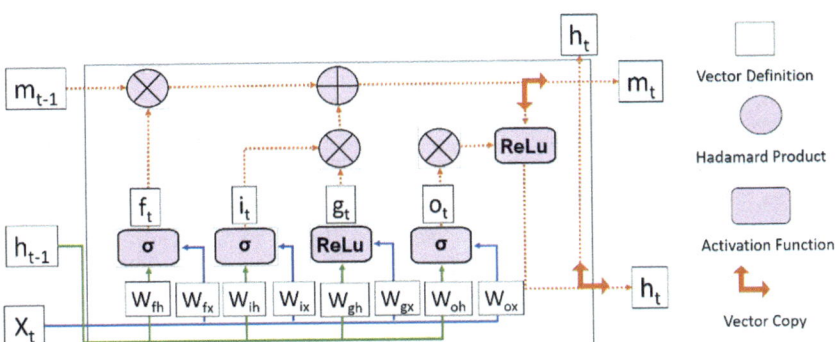

Fig. 2. Sample structure of Long-Short Term Memory (LSTM) block that is used in this study

Forget gate f_t, input gate i_t, intermediate state g_t and output gate o_t formulated using Eq. (3) to Eq. (6);

$$f_t = \sigma\left(W_{fx}X_t + W_{fh}h_{t-1} + b_f\right) \tag{3}$$

$$i_t = \sigma(W_{ix}X_t + W_{ih}h_{t-1} + b_i) \tag{4}$$

$$g_t = ReLu\left(W_{gx}X_t + W_{gh}h_{t-1} + b_g\right) \tag{5}$$

$$o_t = \sigma(W_{ox}X_t + W_{oh}h_{t-1} + b_o) \tag{6}$$

Where σ refers to the nonlinear activation function (sigmoid function), W_x and W_h are the weight matrices, and b is the bias of the relevant gates, X_t refers to input of the current time-step while h_{t-1} is the output of the previous time-step. Forget gate decides on which information to keep from the previous memory cell (m_{t-1}), while input gate determines on information to preserve in the current memory cell (m_t), calculated as,

$$m_t = g_t \odot i_t + m_{t-1} \odot f_t \tag{7}$$

Where $\odot$ refers to Hadamard product. Then, the output gate decides of which memory cell to pass as output (h_t) as formulated in Eq. (8),

$$h_t = ReLu(m_t) \odot o_t \tag{8}$$

The process given from Eq. (3) to (8) continues taking place in the next time steps. Weights and biases are adjusted during the training by minimising the differences between the actual data and the predicted LSTM output. The predicted output of the LSTM ($\bar{y}_t$) is calculated by Eq. (9),

$$\bar{y}_t = W_y h_t \tag{9}$$

2.3   Model Evaluation

The most common score metrics in regression model evaluation are adapted in this study to evaluate the performance of the prediction models. These metrics are Mean Absolute Error (MAE) in Wm^{-2}, Mean Absolute Percentage Error (MAPE) in %, Root Mean Square Error (RMSE) in Wm^{-2} and Coefficient of Determination (R^2), the mathematical formulations are illustrated in Eq. (10) to (13), respectively [2].

$$MAE = \frac{1}{N}\sum_{I=1}^{N}|GHI_r - GHI_p| \tag{10}$$

$$MAPE = \frac{1}{N}\sum_{I=1}^{N}\left|\frac{GHI_r - GHI_p}{GHI_r}\right| \tag{11}$$

$$RMSE = \sqrt{\frac{1}{N}\sum_{I=1}^{N}\left(GHI_r - GHI_p\right)^2} \qquad (12)$$

$$R^2 = \frac{\sum_{i=1}^{N}\left(GHI_p - GHI_m\right)^2}{\sum_{i=1}^{N}\left(GHI_r - GHI_m\right)^2} \qquad (13)$$

Where GHI_r, GHI_p and GHI_m are the ith measured (real), predicted and mean GHI values and N is the number of data points. *RMSE* and *MAE* being close to each other as the forecast model value means has only small deviations from the real data [23].

3   Materials

3.1   Study Area

Cyprus is an island located at 35°N and 33°E in the Mediterranean Sea. Mediterranean climate is dominant over the island, which results in average day temperatures of 30 °C and 13 °C in summer and winter, respectively. Cyprus has excellent potential for receiving global solar irradiation. The yearly average GHI potential is 5.4 kWh/m². In Northern Cyprus, until recently, the only source of electricity has been a conventional power plant. The advances and affordability in the PV panels have resulted in many households to install PV panels over their rooftops. There are also two PV-farms placed in METU–NCC, Kalkanlı and Serhatköy. In this study, we focus Kalkanlı.

3.2   Data

The dataset that is used in this study is obtained from NASA, available at [24]. The dataset includes the date, continuous daily GHI values, maximum and minimum temperature, wind speed and direction, humidity, and pressure data for Kalkanlı, Cyprus between 1st June 1983 and 31st August 2019. The GHI values are summed over the hourly values and divided by 24. The average GHI for the study area is 215 W/m². The dataset is used as two different forecasting input. One of the input set includes all the weather data, including the GHI values and month and day information. The other set is only made of time-series data of GHI. All features, except the target data, are normalised between zero and one using min-max normalisation.

4   Experimental Evaluation

4.1   Model Setup

Table 1 shows the parameters of the forecasting models as well as data types. The dataset is split into 9204 weeks for training and 4000 days for testing, while 10 percent of the training test is used for validation to guide the parameter optimisation.

1D-CNN and LSTM models use two different input dataset to predict GHI, as mentioned in Sect. 3.2, summing up to a total of four different forecasting networks. All the forecasting models predict day-ahead GHI value using previous 7-days time series data. 1D-CNN models with radiation time-series input data and the weather input data have five convolutional layers with a kernel size of 3, each followed by batch normalisation. There are no pooling layers in CNN models as the pooling layer was compromising the accuracy. Both LSTM models, has four cells. Learning rate for all models is kept at 0.01. Network details are shown in Table 1.

Table 1. Training parameters and input data information for the 1D-CNN and LSTM models

Model	Data	Batch size	Epoch	Training units/ filters	Optimiser type	Computation time
CNN	Radiation	40	120	(700, 68, 68, 68,68)	Adam	00:00:08
CNN	Weather	40	120	(700, 68, 68, 68, 68)	SGD	00:00:08
LSTM	Radiation	150	200	(40,30,20,20)	Adam	00:00:01
LSTM	Weather	150	200	(40,30,20,20)	Adam	00:00:02

4.2   Experimental Results and Discussion

For each model, results of evaluation by *MAE, MAPE, RMSE,* and R^2 are calculated and tabulated in Table 2. Results of the model evaluation show that the LSTM and CNN models with weather data have slightly lower *MAE* and *RMSE* compared to the other two models. As mentioned in Sect. 2.3, when *MAE* and *RMSE* are close to each other as a value, the prediction outcome has small deviations from the real data.

Table 2. GHI prediction model error summary

Algorithm	1D - CNN		LSTM	
	Radiation data	Weather data	Radiation data	Weather data
	Test	Test	Test	Test
MAE (W/m^2)	24.79	21.77	25.29	21.41
MAPE (%)	81.00	76.02	19.79	18.03
RMSE (W/m^2)	38.30	33.25	35.76	32.62
R^2	0.81	0.86	0.83	0.86

Figure 3 illustrates the histogram of the model evaluation results using test data from *MAE* generated by each model, where y-axis shows the frequency in percentage, and x-axis represents *MAE* ranges in W/m^2. 70% of our *MAE* results fall under less than 30 W/m^2 for each model, as shown in Fig. 3.

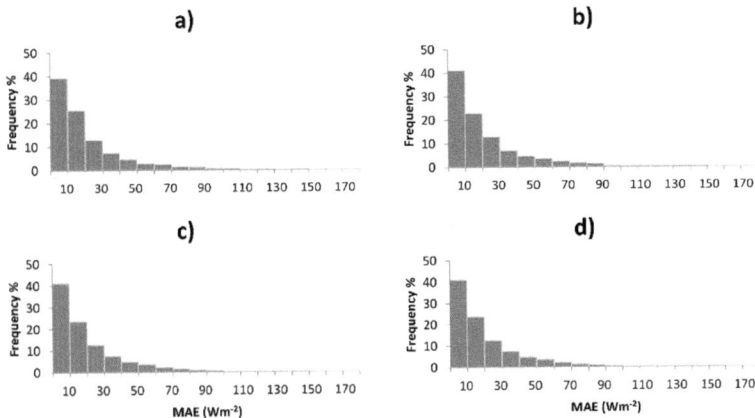

Fig. 3. MAE frequency histograms for the forecasting models. (a) CNN with weather data, (b) CNN with radiation data, (c) LSTM with weather data, and (d) LSTM with radiation data.

In CNN models, the required computational time is slightly longer than that of LSTM. However, when compared to the results of CNN models, it is the trade-off to consider for better performance.

5   Conclusion

In this study, CNN and LSTM networks are employed to predict the day-ahead GHI with two different datasets. All models produce slightly various accuracy, and 70% of results have less than 30 W/m^2 *MAE*. However, the computing time differs on the network type and the input data. Using an LSTM network, such as the one used in this study, with time-series data of only solar radiation could create a convenient solution for applications where the computing time or the weather data availability is critical.

References

1. World Energy Outlook 2019 – Analysis - IEA. https://www.iea.org/reports/world-energy-outlook-2019#. Accessed 07 May 2020
2. Ghimire, S., Deo, R.C., Raj, N., Mi, J.: Deep solar radiation forecasting with convolutional neural network and long short-term memory network algor. Appl. Energy **253**, 113541 (2019)
3. Voyant, C., et al.: Machine learning methods for solar radiation forecasting: a review. Renew. Energy **105**, 569–582 (2017)
4. Salcedo-Sanz, S., Casanova-Mateo, C., Pastor-Sánchez, A., Sánchez-Girón, M.: Daily global solar radiation prediction based on a hybrid coral reefs optimization - extreme learning machine approach. Solar Energy **105**, 91–98 (2014)

5. Elliston, B., MacGill, I.: The potential role of forecasting for integrating solar generation into the Australian National electricity market. In: Australian Solar Energy Society (2010)
6. Lazzaroni, M., Ferrari, S., Piuri, V., Salman, A., Cristaldi, L., Faifer, M.: Models for solar radiation prediction based on different measurement sites. Meas. J. Int. Meas. Confed. **63**, 346–363 (2015)
7. Podestá, G.P., Núñez, L., Villanueva, C.A., Skansi, M.A.: Estimating daily solar radiation in the Argentine Pampas. Agric. Meteorol. **123**, 41–53 (2004)
8. Fan, J., et al.: Comparison of support vector machine and extreme gradient boosting for predicting daily global solar radiation using temperature and precipitation in humid subtropical climates: a case study in China. Energy Convers. Manage. **164**, 102–111 (2018)
9. Ferrari, S., et al.: Illuminance prediction through extreme learning machines. In: 2012 IEEE Workshop on Environmental Energy and Structural Monitoring Systems (EESMS), pp. 97–103 (2012)
10. Kromer, P., Musilek, P., Pelikan, E., Krc, P., Jurus, P., Eben, K.: Support vector regression of multiple predictive models of downward short-wave radiation. In: Proceedings of International of Joint Conference on Neural Networks, pp. 651–657 (2014)
11. Pedro, H.T.C., Coimbra, C.F.M.: Nearest-neighbor methodology for prediction of intra-hour global horizontal and direct normal irradiances. Renew. Energy **80**, 770–782 (2015)
12. Reikard, G.: Predicting solar radiation at high resolutions: a comparison of time series forecasts. Sol. Energy **83**(3), 342–349 (2009)
13. Demirtas, M., Yesilbudak, M., Sagiroglu, S., Colak, I.: Prediction of solar radiation using meteorological data. In: International Conference on Renewable Energy Research and Applications (ICRERA), pp. 1–4 (2012)
14. Marquez, R., Coimbra, C.F.M.: Forecasting of global and direct solar irradiance using stochastic learning methods, ground experiments and the NWS database. Solar Energy **85**(5), 746–756 (2011)
15. Khosravi, A., Koury, R.N.N., Machado, L., Pabon, J.J.G.: Prediction of hourly solar radiation in Abu Musa Island using machine learning algorithms. J. Clean. Prod. **176**, 63–75 (2018)
16. Gu, J., et al.: Recent advances in convolutional neural networks. Pattern Recognit. **77**, 354–377 (2018)
17. Wang, F., et al.: Generative adversarial networks and convolutional neural networks based weather classification model for day ahead short-term photovoltaic power forecasting. Energy Convers. Manage. **181**(2018), 443–462 (2019)
18. Kong, W., Dong, Z.Y., Jia, Y., Hill, D.J., Xu, Y., Zhang, Y.: Short-term residential load forecasting based on LSTM recurrent neural network. IEEE Trans. Smart Grid **10**(1), 841–851 (2019)
19. Burkov, A.: The Hundred-Page Machine Learning Book (2019)
20. Pascanu, R., Mikolov, T., Bengio, Y.: On the difficulty of training recurrent neural networks (2013)
21. Hochreiter, S., Schmidhuber, J.: Long short-term memory. Neural Comput. **9**(8), 1735–1780 (1997)
22. Gers, F.A., Schmidhuber, J., Cummins, F.: Learning to forget: continual prediction with LSTM. IEE Conf. **2**(470), 850–855 (1999)
23. Gensler, A., Henze, J., Sick, B., Raabe, N.: Deep learning for solar power forecasting - an approach using autoencoder and LSTM neural networks. In: 2016 IEEE International Conference on Systems, Man, and Cybernetics (SMC), pp. 2858–2865 (2016)
24. Power Data Access Viewer. https://power.larc.nasa.gov/data-access-viewer/?fbclid=IwAR1yPlfK_3RPZbL3RWwHIrizUeq8SugivFCDN7ASnIeuC8lfO-3TJSlrlRg

Imprecision of Semantic Meaning in a Natural Language

Farida Huseynova[(✉)] [iD]

Azerbaijan State Oil and Industry University, Azadlig Avenue, 20,
AZ1010 Baku, Azerbaijan
farida_hus@hotmail.com

Abstract. Most of linguistic computing models cause loss of information due to the approximation processes and imprecision in the results. The BoW scheme is a simple and popular scheme, but it suffers from numerous drawbacks. Traditionally, the Bag of Words (BoW) representation is used to model the documents in a vector space. It also fails to capture the semantics contained in the documents properly as automatic classification and clustering are carried out in the most common operations. We tried to use Fuzzy logic and its extension computing with words to analyze semantical meanings of a set of terms given in a natural language.

Keywords: Semantic meaning · Natural language · Synonyms · Sensitivity · Inner states · Feeling · Instinct · Mood

1 Introduction

Human language is a system of complex elements for expressing thoughts and ideas in communication process, which is based on human cognition and human intelligence. It is obvious that natural language includes a variety of vague and ambiguous phrases and statements that correspond to imprecision in the underlying cognitive concepts. However, without the richness of meaning inherent in such phrases, human communication would be severely limited, and it is therefore incumbent on us (to attempt) to include such facility within reasoning systems" [1].

Computers today try to mediate human interactions, and perform a wide variety of tasks in understanding complex languages by intelligent analysis; however, they still face difficulties in processing the natural language. There are many semantic theories of languages dealing with the problem, and they are often based on set theories, not on a semantical logic, which is the main bias in translation process. These are the specific drawbacks in their suitability in computer-automated semantic interpretation. Zadeh [2] considered that "Human beings make their most decisions by manipulation of perceptions, and computing with words plays a key role in foundation for a computational theory of perceptions –as it deals with perception-based rational decisions in an environment of imprecision, uncertainty and partial truth".

Understanding of certain semantic aspects of words in computer text processing and meaning representation is very important today. Many new theories appear to resolve this problem. WordNet [3], a lexical database containing synonym sets and

R. A. Aliev et al. (Eds.): ICAFS 2020, AISC 1306, pp. 239–245, 2021.
https://doi.org/10.1007/978-3-030-64058-3_29

other lexical concept, has been used for classification improvement. Word Net associates a unique identifier to each possible sense (meaning) of a word, where the wordsare organised into *synsets*. Each synset is identified by a unique synset number. Each word belongs to one or more synsets with each instance corresponding to different senses of the word and are numbered according to their frequency of occurrence in real world usage. The semantic proximity of a concept in a number of semantic relations are used to connect a synset to other related *sysnets*. The synset is a structure containing sets of words with synonymous meanings, which represents a specific meaning of a word. Since meanings are represented by synsets, semantic relations are pointers between synsets. The synsets together with the semantic relations connecting them can be considered to be a hybrid of a dictionary and thesaurus.

In [4], the synonymy part in WorldNet has been used to expand term lists for each text category, enhancing the accuracy of the text classifier significantly. In [5], text classification based WorldNet's word meanings has been attempted to be learned.

Unfortunately, the task of selecting the correct sense for a word in Word Net dictionaries causes some problems, as there are enormous amount of synonyms with a mass of competing interpretations in the Bag of Words (BoW). The bag of words representation used for these clustering methods is often unsatisfactory as it ignores relationships between important terms that do not co-occur literally. In addition, sophisticated semantics create difficulties in exploiting context and conveying meaning. Because semantically related words co-occur more than 'just by chance' principles, and groups of words hardly exhibit a 'cohesion' between them.

The learning process aims to construct a profile for selecting and recommending relevant semantic items.

2 Methods

The assignment of terms to concepts in WorldNet is ambiguous. Therefore, adding or replacing terms by concepts may have an impact on a loss of information. The proposed approach here uses lexical categories information to reduce the size of vector space and present semantic relationships between words. The aim is to improve a knowledge-based word sense disambiguation technique that makes full use of WordNet as sense inventory. It is necessary to perform a progressive improvement for the clustering accuracy based on word meanings, which helps to overcome the loss of information. Specifically, the content-based filtering approach [6] analyzes a set of words and builds a model or profile of user interests based on the features that describe the target objects.

Our hypothesis is that sensitivity maps can determine which terms are consistently important, hence, likely to be of general use for classification relative to terms that are of low or highly variable sensitivity. Furthermore, adding or replacing some terms by concepts may add noise to the representation and may induce a loss of information. Therefore, we have investigated how the choice of a "most appropriate «concept from the set of alternatives may influence the clustering results. We use a dictionary based method in this research. Dictionary based WSD methods disambiguates a word with reference to a dictionary.

There are so many concise descriptions of commonalities and distinctions in the word clusters. As an example, let's consider the word "happy". A number of ambiguities is faced in understanding of the precise meaning of the word happy and in its realization.

The word *happiness* contains many synonyms that are common across contexts, and with a number of ambiguities in their meanings. Which word would be chosen as an equivalent or synonym? This would be difficult task, as abstract information is often available during machine translation.

First of all, it is necessary to clarify what is happiness itself, and then analyze ambiguities and those uncertainties about the precise meanings of word happy for establishing a meaningful vocabulary by specific judgments (Table 1).

Table 1. Mapping the word happy

Synset	Sysnet
Happy	*gratified satisfied, pleasant, relieved, contented, glowing, thrilled, relieved, overjoyed fulfilled, pleased, mellow, merry, jolly, glad, cheerful, elated, lucky, exhilarated*

It is obvious that everybody hopes to feel happy; therefore, he/she spends much energy in searching of it. We observed people's behavior and provided questionnaires for learning what truly adds happiness to your life? Or what things actually serve a purpose and bring that happiness?

Some people answered that, it is much easier to be happy when a life is full of excitement and our pleasurable experiences are frequent. Hence, different people enjoy life differently and happiness is also understood differently - as having large families, healthy children, a good job achievement, age or material possession, etc. Because when you think about it, people are happiest when in flow, when they're absorbed in something out in the world, when they're with other people, when they're active, engaged, focused on loved ones and things.

In order to argue that happiness is one word, we need to assume the word happiness is one lexical item with many senses (polysemy). Happiness which is based on our pleasures are very ancient, therefore we are always looking for positives. These are our innate pleasures, and one person's happiness is quite different from another person's happiness. Sometimes happiness can be called as "A new science". The degree of happiness depends on our responses to the natural world. Therefore, the occurrences of the synonyms in the different sentences clearly denote different meanings and people interpret it differently. The result is that, due to synonymy, relevant information can be missed if the profile does not contain the exact words in the documents while, due to polysemy, wrong documents could be deemed as relevant.

Most human concepts have specified structures, which are context-dependent. Unfortunately, the identity of the specific meaning that a word assumes in context is only apparently simple. Each word can be interpreted in multiple ways depending on the context in which they occur. For instance, consider the following sentences:

The girl was remembered as a **jolly** little angel, full of life and mischief.
She was all **excited to** go to Japan.

The word "Happy" and its linguistic variables in the above-mentioned sentences were associated with the related terms. Each word can be interpreted in multiple ways depending on the context in which they occur. The occurrences of the synonyms in the two sentences clearly denote different meanings. Due to the approximation processes, the words- jolly and excited may refer to intense happiness. Though "jolly" in the sentence means - full of fun or high-spirited, and "excited" is a state that is higher than the normal condition. It is feeling, showing pleasure or contentment. A number of ambiguities causes these different meanings. One evident problem with the BoW scheme is that, it treats the word "happy" and its linguistic variables in different situations as a single dimension. The main aim here is determining whether happiness and its synonyms have the same meanings. Do all people interpret these words in the same way?

Mendel pointed out that, the human beings accomplish interpretation of true meaning of an information differently because words mean different things to different people and so there is uncertainty associated with words, which means that fuzzy logic (FL) must somehow use this uncertainty when it computes with words [7, 8]. Computing with words needs to consider for the uncertainties associated with the meanings of words, and that these uncertainties require to improve the accuracy and under-standability of the processes. He proposed a specific architecture [9] for making subjective judgments by CWW, which is called a perceptual computer.

Our intention was to determine which word could be the most appropriate, but simply whether word sense disambiguation is needed at all. But the sense in which the word is used in the sentences is considerably different, for example:

The Boy is Overjoyed.

The Boy is Satisfied.

The Boy is Excited.

In which sentence we see more happiness? The words cheerful, satisfied, excited which may seem synonymous, but lead to imprecise results in meanings. In the first sentence overjoyed refers to character of the boy, while in the second and third sentences satisfied and excited refers to an event or activity affecting a sudden feeling.

It is difficult to classify feelings like "happiness" into equally deep and well-defined relationships. The way of understanding the real meaning of a word is an unconscious process that relies on our inner states and, on our knowledge about language itself. Therefore, the way we understand language is heavily based on meaning and context. In other words, we need to show how various meanings of the happiness in synonyms are somehow related to each other. However, according to variations of individual perceptions the same words may indicate quite different meanings. For this purpose, we tried to compare the lexical chain-based document features against the classical bag of words and evaluate their performance on document clustering.

It is required to develop a methodological approach for semantic interpretation in natural language processing (NLP) systems, which may be of great importance not only theoretically, but also have a practical significance. Network of fuzzy constraints, according to this semantics is regarded as the formation from the collection of propositions of natural language [10–14].

3   Observations

Our approach consists of two steps. First, we had to determine what is happiness itself? According to the remarkable ability of manipulation of perceptions- the word "happy" may be a state of mind, an ability of sensory organs. In the first step, we had had to clarify; if everyone wants to be happy, then, does it happen according to our common senses, as instinct, feeling or mood?

The surveys which were provided among 20 people of different ages, helped us to analyze and determine what truly adds value to our lives.

If happiness is a state of mind, an ability of sensory organs, then how it can be interpreted? The fuzziness of perceptions reflects the happiness as an instinct, mood, and feeling. In Fig. 2, the instinct, feeling, mood – the answers are grouped in a framed structure with an inner hierarchy.

Making sense of their similarities and differences of these three factors can really help us manage details of happiness and its linguistic variables (Table 2).

Table 2. Happiness and inner states

Slot	#words English 875	#synsets English 763
Intinct	642	642
Feeling	2,535	2,535
Mood	25,846	21,932
Total	28,898	26,132

4   Evaluation Measure

The word "happy" contains a number of ambiguities that is faced in understanding the precise meaning and its realization. As a second step, our aim was to identify and clarify several of these ambiguities of the word "happy". Besides, the word happiness contains several meaning, at the same time many of its synonyms that may be common or different across contexts. The so- called universal attitude toward "happiness" was the theme of discussion for during the almost long periods. The criteria for happiness don't really apply today. The technique of analysis of the word "happy" and its variables according to our common senses, as instinct, feeling or mood helped us to create imaginary frame structure by the help of its synonyms and find out differences to express our intense emotions.

Handling ambiguities and those uncertainties about the precise meanings of word happiness for establishing a meaningful vocabulary by specific judgments is important. By means of an associated survey, the word happy and its linguistic variables are rank ordered.

Our experiments were carried out to bring clarity to the meaning of real happiness. An assessment of the level of a variable helped the process of establishing a meaningful vocabulary according to specific judgments. Human is able to understand certain terms in a context-free situation, and are able to apply them to other contexts, and we believe that it ought to be possible for computers to do likewise. According to answers of the questionnaire which word shows more happiness and *how much uncertainty should be associated with the words in the list (17 words) of synonyms they answered by- very little, a small amount, a bit, some, a moderate amount, a good amount, a sizeable amount, a large amount, a lot, an extreme amount.*

By the approximation process, the words have been clustered in class- based model according to inner states (instinct, feeling, mood), the degree of happiness and semantic similarity (Table 3).

Table 3. Features used to represent "happiness"

Happy		
INSTINCT	FEELING	MOOD
A sense that comes from the inside (brains and body)	Something, that influences sensory organs	Affect reflects mental state
Happiness is high	**Happiness is Medium**	**Happiness is Low**
Merry, jolly, glad, cheerful, lucky, joy, fortunate	Gratified, contented, glowing, thrilled, overjoyed	Satisfied, pleasant, relieved, fulfilled, pleased, exhilarated

We turn now to the computation of various semantic relations that hold between words. We saw that such relations include synonymy and similarity and tried to find out a relation between words; We could determine that these words are either synonyms or not. For most computational purposes we use instead a looser metric of word similarity or semantic distance. Two words are more similar if they share more features of word similarity semantic distance meaning or are near-synonyms. Two words are less similar, or have greater semantic distance, if they have fewer common meaning elements.

Although we have described them as relations between words, synonymy, similarity, and distance are relations between word senses.

5 Conclusion

The interpretation capabilities of a language-understanding system depending on the semantic theory must be improved and fully accepted satisfactory in all respects. In our paper, the examples are chosen carefully to illustrate and demonstrate the applications of natural language processing environment for every reader. We think, accuracy of content extraction in natural language is necessary for overcoming the intelligence gap in such a simple manner in processing of vagueness in semantics.

References

1. Friedenberg, J., Silverman, G.: Cognitive Science: An Introduction to the Study of Mind. SAGE Publications (2006)
2. Zadeh, L.A.: From computing with numbers to computing with words: from manipulation of measurements to manipulation of perceptions. IEEE T Circ.-I **4**, 105–119 (1999). https://doi.org/10.1109/81.739259
3. Cognitive Science Laboratory at Princeton University. Wordnet 2.0. https://www.cogsci.princeton.edu/wn/ (2003)
4. de Buenaga Rodriguez, M., Gomez-Hidalgo, J., Diaz-Agudo, B.: Using WordNet to complement training information in text categorization. In: 2nd International Conference on Recent Advances in Natural Language Processing, pp. 353–364. John Benjamin Publishing Co., Amsterdam (1997)
5. Kehagias, A., Petridis, V., Kaburlasos, V.G., Fragkou, P.: A comparison of word- and sense-based text categorization using several classification algorithms. J. Intell. Inf. Syst. **21**, 227–247 (2003). https://doi.org/10.1023/A:1025554732352
6. Pazzani, M.J., Billsus, D.: Content-based recommendation systems. In: Brusilovsky P., Kobsa A., Nejdl W. (eds.) The Adaptive Web. LNCS, vol 4321, pp. 325–341. Springer, Berlin, Heidelberg (2007). https://doi.org/10.1007/978-3-540-72079-9_10
7. Mendel, J. M.: Computing with words, when words can mean different things to different people. In: Third International ICSC Symposium on Fuzzy Logic and Applications, pp. 158–164. Rochester Univ., New York (1999)
8. Mendel, J.M.: Uncertain Rule-Based Fuzzy Logic Systems: Introduction and New Directions. Prentice-Hall, Upper Saddle River (2001)
9. Mendel, J.M.: An architecture for making judgments using computing with words. Int. J. Appl. Math. Comput. Sci. **12**, 325–335 (2002)
10. Zadeh, L.A.: Toward a perception-based theory of probabilistic reasoning with imprecise probabilities. J. Statist. Plan. Inference **10**, 233–264 (2002)
11. Aliev, R.A., Fazlollahi, B., Aliyev, R.R.: Soft Computing and its Applications in Business and Economics. Springer, Heidelberg (2004)
12. Aliev, R.A., Pedrycz, W.: Fundamentals of a fuzzy-logic-based generalized theory of stability. IEEE Trans. Syst. Man Cybern. Part B (Cybern.) **39**(4), 971–988 (2009)
13. Aliev, R.A., Huseynov, O.H., Zeinalova, L.M.: The arithmetic of continuous Z-numbers. Inf. Sci. **373**, 441–460 (2016)
14. Aliev, R., Tserkovny, A.: Systemic approach to fuzzy logic formalization for approximate reasoning. Inf. Sci. **181**(6), 1045–1059 (2011)

Stable Iterative Neural Network Training Algorithms Based on the Extreme Method

N. R. Yusupbekov$^{(\boxtimes)}$ ⓘ, H. Z. Igamberdiev ⓘ, O. O. Zaripov ⓘ,
and U. F. Mamirov ⓘ

Tashkent State Technical University, Tashkent, Uzbekistan
dodabek@mail.ru, ihz.tstu@gmail.com, o.zaripov@edu.uz,
uktammamirov@gmail.com

Abstract. The article deals with the construction of neural network training algorithms using the extreme method. For the error propagation method to function, it is necessary to perform initial initialization and train of the neural network neural network when in the operational mode. As a method of training a neural network at the stage of network initialization, an extreme training method is implemented. Pseudo-conversions of matrices from the used recurrent algorithms based on the bordering method are also considered. These relations allow us to perform recurrent pseudo-circulation of matrices in the neural network training algorithm and thus implement the vectors of the input and output weights of the hidden layer neuron using the extreme training method.

Keywords: Input and output weights of a hidden layer neuron · Neural network training · Error back propagation method · Extreme training method · Initial initialization of the neural network

1 Introduction

Neural networks technology is becoming more widespread in the creation of neuro-computers and high-speed information processing systems. High performance due to parallelization of input information in combination with the trainability of neural networks makes this technology very attractive for creating control devices (controllers) in automatic systems. Training on any given operating principle allows you to create automatic control systems that are optimal in terms of performance, energy consumption, and so on. However, it is possible to implement several operating principles and switch from one to the other [1–5].

Attempts to use neural network structures to improve performance when using classical control algorithms seem unconvincing. New structures of intelligent controllers also require new algorithms that take into account both the specifics of the control task and the specifics of the controller [4, 5].

The inclusion of a new type of controller (neural network controller) in the circuit of an automatic control system will naturally require special studies of the dynamics of such systems, the influence of perturbations on the control accuracy, and so on. A special problem is the synthesis of the structure and parameters of the controller itself – it is important to choose the type of neural network reasonably, minimizing the

R. A. Aliev et al. (Eds.): ICAFS 2020, AISC 1306, pp. 246–253, 2021.
https://doi.org/10.1007/978-3-030-64058-3_30

number of layers and the number of neurons in the layer. The simpler the structure of the synthesized neural network controller is, the easier its subsequent hardware implementation will be [1–3].

The learning ability of a neural network is an important prerequisite for the creation of a self-training automatic control systems. The primary task here is to solve the problem of functional identification of control objects and study its results for adjusting the parameters of the neural network controller. The solution of this set of problems will allow creating an intelligent control system for the executive level of management, i.e. essentially an intelligent drive [2–4].

At the same time, the use of several neural network structures for solving direct and inverse kinematic problems can provide the organization of a tactical level of control of a complex multi-link object, for example, a robot. It is obvious that here, too, the tasks of synthesizing the structure and parameters of neural network control structures, as well as the methods of their training, come to the fore.

In general, the hierarchy of intelligent management of a complex dynamic object, in addition to the executive and tactical levels, includes a strategic level that solves a set of tasks that determine the behavior of the object. For this level, given the complexity and volume of the tasks being solved, the use of neurocomputers is very promising [4–7].

Thus, there are prerequisites for creating all levels of the hierarchy of intelligent control on a single, neural network technology, and, consequently, there is a need to develop a new element base based on unified neuroelements and neuroprocessors.

2   Problem Definition

The following control scheme is considered in this paper (Fig. 1).

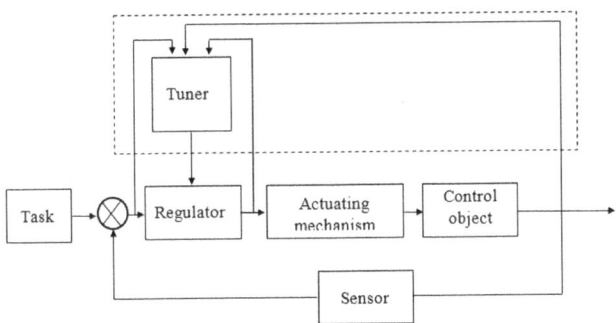

Fig. 1. Structure of the control system with setting of the controller parameters

The task of the control loop (Fig. 1) is to track the task schedule, which is a set of step changes to the task so that the transition process meets the quality criteria for overshoot, oscillation, and static error in steady-state mode.

Neural network training is offered in two stages: the stage of initial initialization of the neural network. Its purpose is to train the neural network to the current parameters of the controller, with which the control object functions at the time of setting the adjuster; the training stage of the neural network when in operational mode [7–10].

For the operation of the method of error back propagation [8], it is necessary to perform initial initialization of the neural network, which is the setting of the initial values of weight coefficients and offsets. After initial initialization, when the current values are applied to the network inputs, the values of the regulator coefficients currently used by the control system should appear at the network output. Failure to meet this condition will lead to a sharp deterioration in the quality of management at the initialization stage of the neural network adjuster.

To solve this problem, it is necessary to initialize the neural network using the extreme learning method [9–11]. The method of extreme learning (ELM, Extreme Learning Machine) allows you to train a neural network without an iterative procedure. The structure of a neural network trained by extreme learning is shown in Fig. 2.

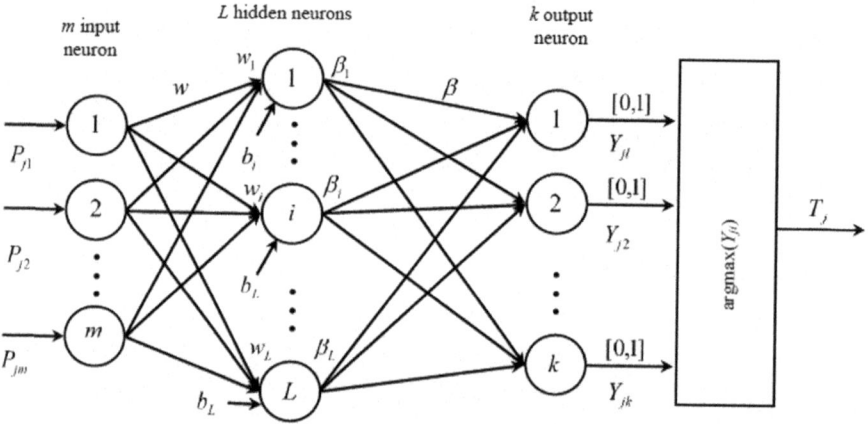

Fig. 2. Structure of a neural network trained by extreme learning

Training is performed by submitting n pre-prepared sets of input values, each of which has a pair of vectors set (p_j, y_i), $j = 1, ..., m$, $i = 1, ..., n$ [10, 11]. The output weight vector of the network y_i is calculated as:

$$y_i = \sum_{i=1}^{l} \beta_i F(w_i p_j^T + b_i), \quad j = 1, ..., n \tag{1}$$

where $p_j \in R^l$, $y_i \in R^n$ – vector of input and output variables; vector of output values; $w_i \in R^l$–vector of input and output weights i-th neuron of the hidden layer, $i = 1, ..., m$.

If the extreme learning method is used, the input weights are w_i and offset b_i for each i-th hidden layer neurons are set using a random variable generator:

$$w_i = random(-1...1), \ b_i = random(0....1).$$

Based on (1), we write the following relations in matrix form:

$$y = H\beta \tag{2}$$

where H – matrix of output values of hidden layer neurons with dimension $n \times m$:

$$H = \begin{bmatrix} F(w_1p_1 + b_1) & \cdots & F(w_mp_1 + b_m) \\ \cdots & \cdots & \cdots \\ F(w_1p_n + b_1) & \cdots & F(w_mp_n + b_m) \end{bmatrix},$$

$$F(w_ip_j + b_i) = \frac{1}{1 + \exp(w_ip_j + b_i)}.$$

It is necessary to calculate the pseudo solution in general:

$$\beta = H^+ y \tag{3}$$

where $H^+ = (H^TH)^{-1}H^T$ – pseudoinverse matrix H. As is known [12], the calculation β equivalent to solving normal systems of equations:

$$H^THB = H^Ty \tag{4}$$

Matrix H, as can be seen from the structure of its construction, it has a large dimension. To solve such systems of equations, one can, in general, use any of the known methods for solving systems of linear equations [12–16]. However, their application in this case is associated with significant difficulties: the inability to solve the problem in one step and reduce the volume of intermediate results, etc. This explains the practical interest in using computational algorithms that are more adapted to solving systems of equations of the type under consideration.

3 Pseudo-inversion of Matrices from the Used Recursive Algorithms

The most time-consuming operation in the multistep estimation algorithm under consideration is the matrix reversal operation H. The accuracy and computational stability of the estimation algorithm significantly depends on the quality of this procedure. In (2–4) the matrix H is an overdetermined matrix of order $n \times m$, perhaps of incomplete rank $r < m$. It is known [17–19] that for each real $(n \times m)$-matrices H there is a single valid pseudoinverse matrix U, satisfying the following properties:

$$UHU = U, \ (HU)^T = HU, \ HUH = H, \ (UH)^T = UH \tag{5}$$

In practice, it is not uncommon for additional information in the form of $(n \times 1)$-matrixes h_{n+1} joins an existing one $(n \times m)$-matrix H_n as its last row:

$$H_{n+1} = \left| \begin{matrix} H_n \\ h_{n+1}^T \end{matrix} \right| \tag{6}$$

It is known [13, 20] that the pseudo-inverse matrix U_{n+1} for the matrix H_{n+1} in form (6), it can be represented as

$$U_{n+1} = \left\| V_n^T \vdots v_{n+1} \right\| \tag{7}$$

where

$$V_n^T = U_n - v_{n+1} h_{n+1}^T U_n.$$

From the first relation (5) using the partition U_{n+1} in form (7), you can write

$$U_{n+1} = U_{n+1} H_{n+1} U_{n+1} = [V_n^T H_n + v_{n+1} h_{n+1}] U_{n+1} \tag{8}$$

From here

$$I = V_n^T H_n + v_{n+1} h_{n+1}^T.$$

Based on (8), we find $U_n H_n = V_n^T H_n + v_{n+1} h_{n+1}^T U_n H_n$, from which we can arrive at the expression

$$V_n^T = U_n - v_{n+1} h_{n+1}^T U_n \tag{9}$$

Substituting (9) in (7), we find.

$$U_{n+1} = U_n - v_{n+1} h_{n+1}^T U_n \vdots v_{n+1} \tag{10}$$

for $n = 1$ we have $U_1 = (h_1^T h_1)^{-1} h_1$. Then v_{n+1} can be defined from the expression [12, 20, 21]:

$$v_{n+1}^T = (h_{n+1}^T U_n U_n^T h_{n+1} + 1)^{-1} h_{n+1}^T U_n U_n^T \tag{11}$$

Entering a designation

$$S_n = U_n U_n^T = (H_n^T H_n)^{-1} \tag{12}$$

and substituting (12) in (11), we find.

$$v_{n+1} = S_n h_{n+1} (h_{n+1}^T S_n h_{n+1} + 1)^{-1}$$

where

$$S_{n+1} = S_n - v_{n+1} h_{n+1}^T S_n,$$

In the case when the line h_{n+1}^T increases the rank of the matrix H_n, column v_{n+1} can be defined based on the expression [13, 18]:

$$v_{n+1} = (I - U_n^T H_n^T) h_{n+1} [h_{n+1}^T (I - U_n^T H_n^T) h_{n+1}]^{-1}.$$

Entering a designation

$$T_n = I - U_n^T H_n^T$$

find

$$v_{n+1} = T_n h_{n+1} (h_{n+1}^T T_n h_{n+1})^{-1}.$$

In this case, the matrix T_n you can define it as follows.

$$T_{n+1} = T_n - v_{n+1} h_{n+1}^T T_n.$$

by $m = 1$ have $T_1 = I - U_1 H_1$.

In expressions (6)–(12) U_{n+1} – pseudo-return matrix for a matrix composed of the first $n + 1$ lines of the original; h_{n+1}^T there is $(n + 1)$-th row of the source matrix; v_{n+1} there is $(n + 1)$-th column of the pseudo-return matrix; T_{n+1}, S_{n+1} – a symmetric matrix when the vector-column v_{n+1} with a vector string h_{n+1}^T; I – identity matrix; T –

transposition sign; $\vdots$ – division sign n matrix column U_{n+1} from $(n + 1)$-th column.

The considered pseudo-circulation algorithm is direct and uses the advantages inherent in the bordering method [12, 21].

To train the neural network in the operational mode, the method of error back propagation is implemented [5]. When choosing the training method, the main requirement was the requirement for the duration of one training cycle: the training cycle should be less than the time of the call to the neural network adjuster [22]. However, due to the high inertia of the control object, the call time of the adjuster has the order of seconds, which is much more than the training duration. In this regard, the choice of the training method was not critical and the decision was made to apply the classical method of back propagation of the error [5].

4   Conclusion

These relations allow us to perform recurrent pseudo-circulation of matrices in the neural network training algorithm and thus implement the vectors of the input and output weights of the hidden layer neuron using the extreme training method.

References

1. Vasiliev, V.I., Ilyasov, B.G.: Intelligent Control Systems. Theory and Practice. Radio Engineering, Moscow (2009)
2. Makarov, I.M., Lokhin, V.M.: Intelligent Automatic Control Systems. Fizmatlit, Moscow (2001)
3. Egupov, N.D.: Methods of Robust. Neuro-Fuzzy and Adaptive Control. Bauman Mosc. State Tech. Uni, Moscow (2002)
4. Omatu, S., Khalid, M., Yusof, R.: Neuro-Control and its Applications. Springer, London (1995)
5. Terekhov, V.A., et al.: Neural network control systems. Book 8. (Neurocomputers and their application). IPRZHR, Moscow (2002)
6. Haykin, S.: Neural Networks and Learning Machines. 3rd edn. Prentice Hall (2009)
7. Yusupbekov, N.R., Gulyamov, Sh.M., Kasimov, S.S., Usmanova, N.B.: Knowledge-Based Planning for Industrial Automation Systems: The Way to Support Decision Making. In: Aliev, R., Kacprzyk, J., Pedrycz, W., Jamshidi, M., Babanli, M., Sadikoglu, F. (eds.) 10th International Conference on Theory and Application of Soft Computing, Computing with Words and Perceptions - ICSCCW-2019. ICSCCW-2019. Adv. Intel. Syst. Comput., **1095**. 873–879, Springer, Cham. (2019). https://doi.org/10.1007/978-3-030-04164-9_115.
8. Patrick, P.: Minimisation methods for training feedforward neural networks. Neural Netw. 7 (1), 1–11 (1994)
9. Huang, G.B., Zhu, Q.Y., Siew, C.K.: Extreme learning machine: theory and applications. Neurocomputing 70(1), 489–501 (2006)
10. Huang, G.B., Wang, D.H., Lan. Y.: Extreme learning machines: a survey. Int. J. Mach. Learn. Cybern. 2(2), 107–122 (2011)
11. Yudin, D.A., Magergut, V.I.: Application of extreme learning machines for classification of image areas. His. Ser. Polit. Sci. Econ. Comput. Sci. 7(150), 95–103 (2013)
12. Gantmacher, F.R., Brenner, J.L.: Applications of the Theory of Matrices. Courier Corporation (2005)
13. James, W.D.: Applied Numerical Linear Algebra. University of California, Berkeley, California (1997)
14. Verzhbitsky, V.M.: Computational Linear Algebra. Higher school of Economics, Moscow (2009)
15. Yusupbekov, N.R., Igamberdiev, H.Z., Mamirov. U.F.: Algorithms of sustainable estimation of unknown input signals in control systems. J. Multiple Value Logic Soft Comput. 33(1–2), 1–10 (2019). https://www.oldcitypublishing.com/pdf/9291.
16. Yusupbekov, N.R., Igamberdiev, H.Z., Sevinov, J.U.: Formalization of identification procedures of control objects as a process in the closed dynamic system and synthesis of adaptive regulators. J. Adv. Res. Dyn. Cont. Sys. **12**(06), 77–88 (2020). https://doi.org/10.5373/JARDCS/V12SP6/SP20201009

17. Mamirov, U.F.: Algorithms of stable control of a matrix object in the conditions of parametric uncertainty. Chem. Technol. Control Manage. **1**, 164–168 (2018). https://uzjournals.edu.uz/ijctcm/vol2018/iss1/27
18. Golub, G.H., Van Loan, C.F.: Matrix Computations. Baltimore and London Johns Hopkins University Press, London (1989)
19. Charles, L.L., Richard, J.H.: Solving Least Squares Problems. Prentice-Hall, Englewood Cliffs (1986)
20. Roger, A.H., Charles, R.J.: Matrix Analysis. Camb. Uni. Press, USA (1985)
21. Matveev, A.A.: An algorithm for the pseudoinversion of matrices. USSR Comput. Math. Mat. Phys. **14**(2), 208–212 (1974). https://doi.org/10.1016/0041-5553(74)90053-6
22. Yeremenko, Yu.I., Glushchenko, A.I.: On the development of a method for selecting the structure of a neural network for solving the problem of adapting the parameters of linear regulators. Managing Large Syst. **62**, 75–123 (2016)

Synthesis of Optimal Technological Parameters of "Iron-Cast-Glass" Grinding Composite Materials Using Fuzzy Logic and Big Data Concepts

T. G. Jabbarov[✉] and N. A. Gurbanov

Department of Mechanical and Materials Science Engineering, Azerbaijan State Oil and Industry University, 20 Azadliq Avenue, AZ1010 Baku, Azerbaijan
tahir-cabbarov@mail.ru, nurlangurbanov@asoiu.edu.az

Abstract. The basis of these technological processes of grinding metallurgy is pressing and baking, their application in the production processes of grinding products highlights the need to optimize the technological parameters of these processes. Until recently, before the production of products from composite materials, the parameters of technological processes were studied on the basis of experimental values of the study of the impact on the complex properties of the products, especially mechanical properties. Often, classical techniques of study are used. Heterogeneous structure, complexity and uncertainty of big experimental data mandates the use modern mathematical models for computational analysis synthesis of materials.In this paper we consider construction of fuzzy IF-THEN rules from huge number of experiemental data to describe relation between material properties and technological parameters under uncertainty.

Keywords: Fuzzy logic · Composite materials · Iron-cast-glass · Pressing pressure · Cooking temperature · Cooking time

1 Introduction

Unlike known iron-based composite materials, "iron-cast-glass" system grinding composite materials have different levels of "heterogeneity". The first level is between the metallic matrix and the glass particles, and the second is inside the matrix, which consists of hard and soft phases.

In the formation of this complex structure plays an important role in the formation of technological parameters - pressing pressure, cooking temperature, cooking time, finding optimal values, in general, the properties of the material.

It is important to find the optimal values of these parameters with computer programs in order to avoid experimental errors in the specification of the technological parameters in the formation of the structure and properties of "iron-cast-glass" grinding composite materials.

To solve such problems, White [1] used a computer and a database in the synthesis of new material. He noted the use of databases by various scientists on the prices of

R. A. Aliev et al. (Eds.): ICAFS 2020, AISC 1306, pp. 254–259, 2021.
https://doi.org/10.1007/978-3-030-64058-3_31

available materials and their technological parameters. This kind of database leads to the spread of materials about new materials [2, 3].

Rakkugila and colleagues [4] used machine learning algorithms in the synthesis of new hybrid-class materials, and they believe that this method can replace experimental methods.

The application of fuzzy logic to model and plan properties helps to develop material production and improve results [5–7].

Due to comlexity and uncertainty of experimental data, the use of fuzzy logic for modeling material properties helps to better describe the development of material production [5–7]. Particularly, it can be used to derive models from big data (huge set of structured and unstructured data from different sources [9]) on material properties and production process parameters.

In this paper we use fuzzy If-Then rules to describe relation between material characteristics and technological parameters under uncertainty. The rules are derived by using fuzzy clustering of experimental data.

The paper is organized as follows. In Sect. 2 it is given preliminary material used in the paper. In Sect. 3 we describe a problem of deriving a model of relation between material properties and technological parameters. The model is described in a form of fuzzy If-Then rules derived from experimental data by using fuzzy clustering. In Sect. 4 we consider application to the case of "iron-cast-glass" grinding composite materials. Section 5 concludes.

2   Preliminaries

Definition 1 A Fuzzy Number. A fuzzy number is a fuzzy set A on $\mathbb{R}$ which possesses the following properties: a) A is a normal fuzzy set; b) A is a convex fuzzy set; c) α-cut of A, A^{α} is a closed interval for every $\alpha \in (0, 1]$; d) the support of A, supp(A) is bounded.

Definition 2 Fuzzy C-Means (FCM) Clustering [4]. FCM clustering problem is a problem of partitioning the dataset $X = \{x_1, \ldots, x_n\}$ into c fuzzy clusters formulated as follows:

$$J_m = \sum_{i=1}^{n} \sum_{j=1}^{c} u_{ij}^m \left\| x_i - v_j \right\|^2 \rightarrow \min$$

subject to

$$u_{ij}^m = \frac{1}{\sum_{j=1}^{c} \left(\frac{\|x_i - v_j\|}{\|x_i - v_k\|} \right)^{\frac{2}{m-1}}},$$

$$v_j = \frac{\sum\limits_{i=1}^{n} u_{ij}^m x_i}{\sum\limits_{i=1}^{n} u_{ij}^m},$$

$$1 \le m < \infty,$$

where u_{ij} is the degree of membership of data point x_i to the j-th fuzzy cluster, $\|\cdot\|$ is a norm, m is a fuzzifier which defines curvature of membership functions of obtained clusters, v_j is a center of the j-th fuzzy cluster.

3   Construction of Fuzzy Model of Relationship Between Material Composition and Properties

Suppose that big data on influence of technological parameters on mechanical properties of materials éxist (Table 1).

Table 1. Structure of big experimental data.

Exp	Technological parameters						Mechanical properties	
#	Parameter 1, y_1	Parameter 2, y_2	...	Parameter n, y_n	Properties1, z_1	...	Properties m, zm,	
1	y_{11}	y_{12}	...	y1n	z_{11}	...	z_{1m}	
...	...	...	...	...	...	...	...	
S	y_{s1}	ys2	...	y_{sn}	z_{s1}	...	z_{sm}	

Real-world experimental data are characterized by uncertainty. Our aim is to construct a formal model of relation between mechanical properties and technological parameters to account for uncertainty. In view of this, we construct fuzzy IF-THEN rules to summarize experimental data in Table 1. For deriving rules, we use fuzzy C-means clustering (Definition 2) of such data in $n \times m$ dimensional space. As a result, data are partitioned into fuzzy clusters C_k, $k = 1,...,K$. By projecting any fuzzy cluster on axes of multidimensional space, we uncover a fuzzy rule:

IF y_1 is A_{k1} and,..., and y_n is A_{kn} THEN z_1 is B_{k1} and,..., and z Medium is B_{km}, $k = 1,..., K$

$A_{k1},...,A_{kn}$ and $B_{k1},...,B_{km}$ are fuzzy terms (projections of C_k $n \times m$ on axes of dimensional space) describe imprecise values of technological parameters and mechanical properties respectively, K is a number of rules. Such a model provides an intuitive description of complex relationship under uncertainty.

The fuzzy IF-THEN rules-based model can be used for analysis of impact of technological parameters to material properties, and for material synthesis (determination of values of technological parameters that induce required property levels).

4   An Application

Let us consider a problem of modeling of "iron-cast-glass" material properties dependence on material technological parameters and test temperature. A fragment of related experimental data is shown in (Table 2):

Table 2. A fragment of the big data on "iron-cast-glass" obsolete powder composite materials.

Technological parameters			Properties			
$y1$ (Pressining pressure, P.MPa)	$y2$ (Sintering temperature, T °C)	$y3$ (Sintering duration, $\tau -$ 1×10^3 s)	$z1$ (σ_u, MPa)	$z2$ (σb, MPa)	$z3$ (hardness, HRB)	$z4$ (resilience, kJ/m^2)
720	1150	7.920	428	405	92	48
736	1155	8.136	422	399	94	46
			.			
			.			
			.			
752	1160	8.352	416	393	96	46
768	1165	8.568	410	387	98	45

The following IF-THEN rules are derived by FCM clustering:

IF y_1 is low and y_2 is low and y_3 is low THEN z_1 is High and z_2 is High and z_3 is High and z_4 is High,

IF y_1 is Medium and y_2 is Medium and y_3 is Medium THEN z_1 is Very High and z_2 is Very High and z_3 is Very High and z_4 is Very High,

IF y_1 is Very High and y_2 is Very High and y_3 is Very High THEN z_1 is Very Low and z_2 is Low and z_3 is Very Low and z_4 is Very Low.

The graphical description of these rules is shown in Fig. 1.

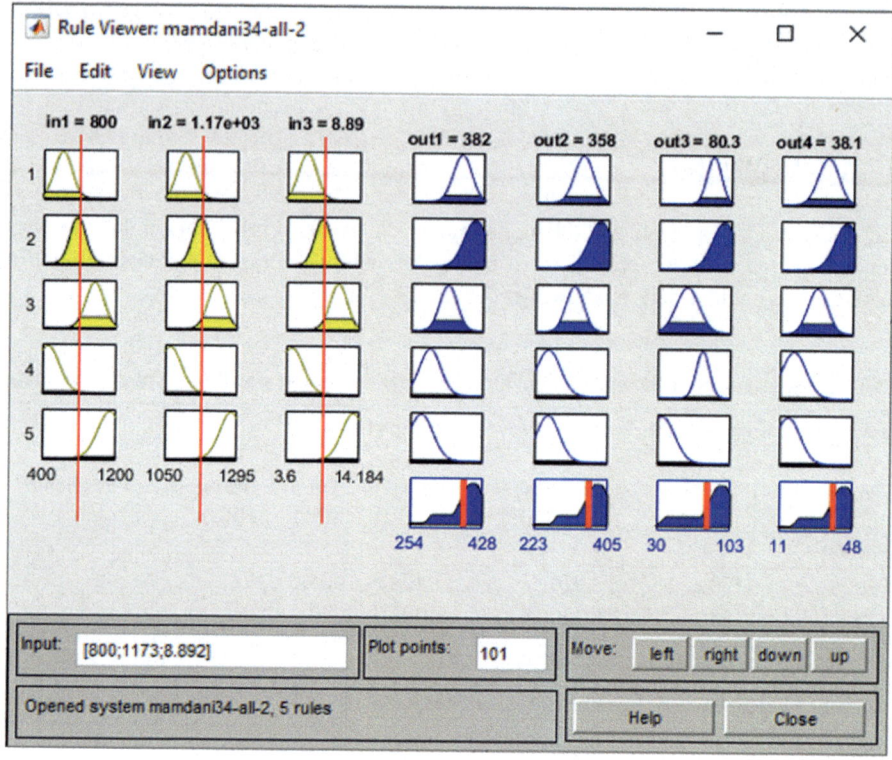

Fig. 1. Graphical description of fuzzy rules

5 Summary

We used FCM clustering of experimental data to derive fuzzy If-Then rules describing relation between material properties and technological parameters. As an example, "iron-cast-glass" material is considered. This fuzzy model may be used for the purposes of material analysis and synthesis.

References

1. Nicola, N.: Can artificial intelligence create the next wonder material. Nature **533**, 22–25 (2016). https://doi.org/10.1038/533022a
2. Babanli, M.B.: Synthesis of new materials by using fuzzy and big data concepts. In: 9th International Conference on Theory and Application of Soft Computing with Words and Perception, ICSCCW 2017. Procedia Computer Science 120, pp. 104–111, Elsevier (2017). https://doi.org/10.1016/j.procs.2017.11.216
3. What is big data analytics? https://www.ibm.com/analytics/hadoop/big-data-analytics

4. Raccuglia, P., Elbert, K.C., Adler, P.D.F., Falk, C., Wenny, M.B., Mollo, A., Zeller, M., Friedler, S.A., Schrier J., Norquist. A.J.: Machine-learning-assisted materials discovery using failed experiments. Nature **533**, 73–76 (2016). https://doi.org/10.1038/nature17439
5. Babanli, M.B., Huseynov, V.M.: Z-number-based material selection problem. In: 12th International Conference on Application of Fuzzy System sand Soft Computing, ICAFS 2016, 29–30 August 2016. Procedia Computer Science 102, pp. 183–189, Elsevier (2016). https://doi.org/10.1016/j.procs.2016.09.387
6. Chen, D., Li, M., Wu, S.: Modeling of microstructure and constitutive relation during super plastic deformation by fuzzy-neural network. J. Mater. Process. Technol. **142**, 197–202 (2003). https://doi.org/10.1016/S0924-0136(03)00598-3
7. Chen, S.-M.: A new method for tool steel materials selection under fuzzy environment. Fuzzy. Sets Syst. **92**, 265–274 (1997). https://doi.org/10.1016/S0165-0114(96)00189-3

Application of Machine Learning Algorithms to Determine the Authorship of Text Fragments

Nadezhda Yarushkina(iD) and Vadim Moshkin$^{(\boxtimes)}$(iD)

Ulyanovsk State Technical University, Ulyanovsk 432027, Russia
{jng, v.moshkin}@ulstu.ru

Abstract. The article describes the main approaches to the automated determination of authorship of texts based on the analysis of copyright styles. The architectures of a convolutional neural network, a multilayer perceptron, and LSTM neural network were proposed to solve this problem. Also, experiments were conducted in which the effectiveness of each of the proposed approaches was evaluated using the example of the task of determining the authorship of English-language poems.

Keywords: Machine learning · Neural network · Binary classification of texts · Authorship definition

1 Introduction

Currently, the task of determining the authorship of a text is solved by analyzing the characteristic features of the language and copyright techniques using syntactic, lexical, phraseological, and stylistic analysis of the text. This is a time-consuming process, including an analysis of the author's texts by an expert to identify the features of his style. Hence the need for automation of this process.

In most cases, the following approaches are most often used to determine authorship of texts: mathematical statistics and probability theory, neural networks, cluster analysis, pattern recognition theories, etc.

The purpose of this work is to assess the applicability of machine learning methods to solve the problem of determining the authorship of the text and to compare models of neural networks in solving this problem.

Formal methods for determining authorship of texts are divided into two large groups: statistical methods and machine learning. The application of machine learning methods to determine the author of the text in the framework of the current study was investigated [1]. Machine learning algorithms include several categories:

- genetic algorithms
- neural networks,
- Bayesian classifier,
- decision trees, etc. [2]

The application of statistical methods for analyzing the authorship of the text was carried out in [3–5]. The authors of the study proposed a method based on taking into account statistics on the use of syntagmatic chains in the text. The accuracy of this method was less than 65%.

The "Linguoanalyzer" system [5] determines the author of the text by applying compression algorithms and Markov chains. This software system could determine the author with an accuracy of 70 to 89 percent depending on the length of the text.

The SMALT system [6, 7] is based on dependency trees and types of relationships and statistical methods for analyzing a literary text. However, the authors were unable to achieve a sufficiently accurate determination of the author of the text using both 16 and 156 attributes.

The basis of the software system "Autologist" is the construction of a binary classifier. In this method, all texts from the training and test sets are expanded into a very large vector indexed by words. In this case, the texts are two sets of points from the training set in multidimensional space - some of them belong to the author, while others do not belong. This software system has achieved an average accuracy of 88% [8, 9].

Despite the results obtained by researchers in the field of detecting authorship of texts of various styles, the use of machine learning methods to solve such problems will allow one to obtain better results, taking into account their effectiveness in working with natural language.

2 Models and Implementation

The dataset for research is a collection of poems in English. This collection is publicly available in the Kaggle system [10].

The data set consists of 15638 poems by 3310 authors. Poems were chosen for the study, as the author's style is more expressed in them. English texts were selected.

The development was carried out in the online service Google Colab in Python [11]. The following libraries were used:

- Keras library [12] for building models of neural networks;
- pandas library for working with selections
- spaCy library for word processing
- matplotlib library for graphing.

We assume that there is a target author of the training and its reference texts (poems) and many other authors with their texts (poems). From here the binary classification problem is formed. For machine learning, it is necessary to divide the data set into training and test sets. In this implementation, the division was made 80% to 20% in the training and test sets, respectively.

It is very important that the relative number of poems of the target author in the sets be approximately the same, the lack of poems of the target author in the training set will lead to poor quality, and in the test set, it will interfere with correctly assessing the quality of the trained model. The specifics of the task and the data set are such that the number of texts of the target author is relatively small. In real problems of determining the authorship of reference texts, the number of texts of the target author may be even less.

Therefore, it is necessary to train the model in conditions of unbalanced classes, for this, it is necessary to adjust the class weights.

It is important to present the poems in the form necessary for learning. Standard approaches (for example, the "word bag" algorithm) will not help to solve a similar problem since they violate the structure of the text and reduce everything to the frequency analysis of words in the text. Obviously, the author's style is not contained in specific words, with the exception of expressions inherent in a particular author. In addition, a trained neural network in concrete words will be useless if you use text with a new context for it.

One of the most important components of the author's style is the syntactic construction of sentences in its texts. It is on this idea that this study is based.

The spaCy library is used for parsing text. SpaCy encodes all strings into hash values to reduce memory usage and increase efficiency.

The schematic parsing of the proposal by the spaCy library is shown in Fig. 1.

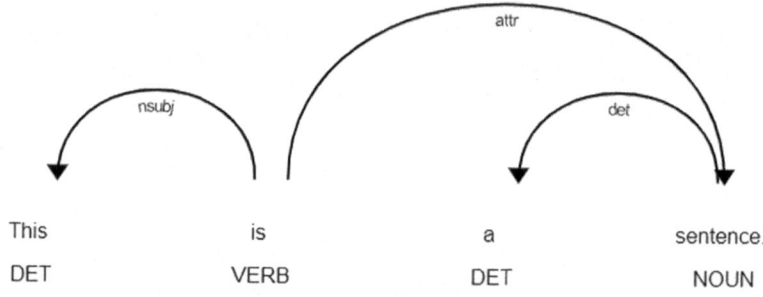

Fig. 1. Parsing schema

The algorithm for vectorizing poems includes the following stages:

- syntactic analysis of the text is performed;
- all strings are encoded into hash values;
- parts of speech in sentences are determined;
- dependencies of words in a sentence are determined;
- a vector of text is built. Vector elements are words and word ratios in the text in numerical form.

Each poem in the dataset is represented as a sequence of hash values of these dependencies (with this approach, the text is represented not by specific words that depend on the context, but as mutual relations of words in the text).

For experiments, the following neural network models were selected:

- multilayer perceptron;
- recurrent neural network LSTM;
- convolutional neural network (CNN).

The multilayer perceptron model consists of the following layers:

1. Layer Embedding;
2. Layer Flatten;
3. A fully connected Dense layer with 256 neurons and a relu activation function.
4. A fully connected Dense layer with 64 neurons and a relu activation function.
5. A fully connected Dense layer with one neuron and sigmoid activation function.

The neural network LSTM model consists of the following layers:

1. Layer Embedding;
2. LSTM layer with 128 neurons;
3. LSTM layer with 64 neurons;
4. A fully connected Dense layer with one neuron and sigmoid activation function.

The convolutional neural network (CNN) model consists of the following layers:

1. Layer Embedding;
2. Layer Conv1D with 256 neurons and relu activation function;
3. Layer Conv1D with 64 neurons and relu activation function;
4. Layer GlobalMaxPooling1D;
5. A fully connected Dense layer with 30 neurons and a relu activation function.
6. A fully connected Dense layer with one neuron and sigmoid activation function.

All models use the «adam» optimizer, the «binary_crossentropy» loss function, and «the accuracy» metric. In all models, the embedding layer is a fully connected layer consisting of 16 ordinary neurons. The last layer of each model has a sigmoid activation function, which is one of the best in solving the binary classification problem.

3 The Results of the Experiments

All models of neural networks were trained on the poems of William Shakespeare, as this author has the largest number of poems in the data set (85 poems).

3.1 Multilayer Perceptron

The multilayer perceptron has been trained for 8 eras. Learning speed 2 s per era. The training schedule is shown in Fig. 2.

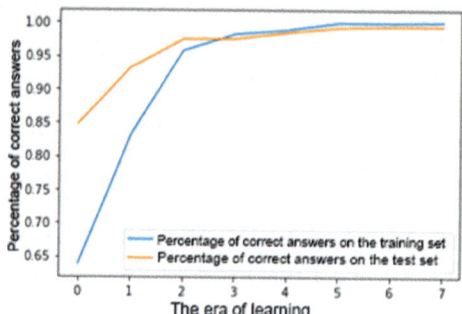

Fig. 2. Perceptron training graph

This graph shows a correlation between accuracy in the training and test sets. This suggests that perhaps the model has extracted the correct features. Also on this graph, the accuracy of the training set is constantly growing, which means the model is successfully trained.

The discrepancy matrix for the predicted values in the test set is presented in Fig. 3 1. The final accuracy was 94.03%, while the multilayer perceptron correctly identified 42.18% of the poems belonging to the author, and correctly identified 94.32% of the poems not belonging to the author.

		Forecast		
		+	-	
Actual	+	7	10	41,18%
		0,22%	0,32%	58,82%
	-	177	2937	94,32%
		5,65%	93,8%	5,68%
		3,8%	99,66%	94,03%
		96,2%	0,34%	5,97%

Fig. 3. The matrix for the predicted values

3.2 LSTM Neural Network

LSTM neural network trained 8 eras. The training time averaged 16 s per era. The training graph is shown in Fig. 4. In this graph, there is a correlation between accuracy in the training and test sets, but it is much weaker, which means that the model found incorrect signs. Accuracy in the training set does not tend to grow continuously, therefore, the model does not learn. Thus, the model is incorrectly constructed, the hyperparameters of the model are incorrect, or the LSTM network is poorly suited for this form of data representation.

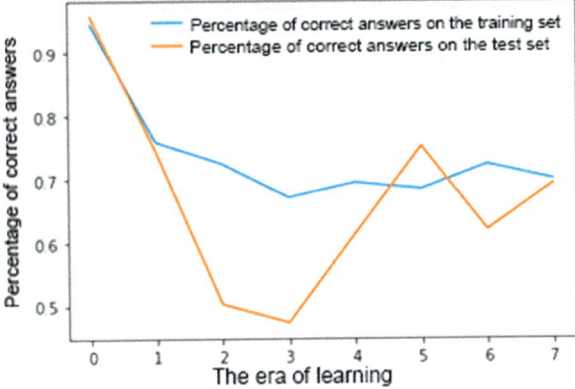

Fig. 4. LSTM network training schedule

The nonconformity matrix is presented in Fig. 5. The total accuracy was 69.08%. The LSTM network correctly identified 76.47% of the poems belonging to the author and correctly identified 69.04% of the poems not belonging to the author.

			Forecast		
			+	-	
Actual	+		**13**	**4**	76,47%
			0,42%	0,13%	23,53%
	-		**964**	**2150**	69,04%
			30,79%	68,67%	30,96%
			1,33%	99,81%	69,08%
			98,67%	0,19%	30,92%

Fig. 5. The matrix for the predicted values

3.3 Convolutional Neural Network

The convolutional neural network (CNN) model has been trained for 16 eras. Learning speed averaged 1 s per era. The training graph is shown in Fig. 4. This graph shows a correlation between the accuracy of the training and test sets. This suggests that perhaps the model has extracted the correct features. Also on this graph, the accuracy of the training set is constantly growing, which means the model is successfully trained.

The convolutional neural network (CNN) training schedule is shown in Fig. 6.

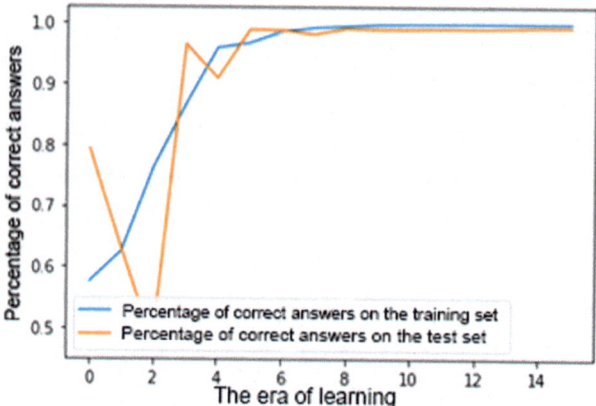

Fig. 6. Schedule for convolutional network training (CNN)

The non-compliance matrix is presented in Fig. 7. The final accuracy was 93.23%. Multilayer CNN correctly identified 94.12% of the author's poems and correctly identified 93.22% of non-author poems.

		Forecast		
		+	-	
Actual	+	**16**	**1**	94,12%
		0,51%	0,03%	5,88%
	-	**211**	**2903**	93,22%
		6,74%	92,72%	6,78%
		7,05%	99,97%	93,23%
		92,95%	0,03%	6,77%

Fig. 7. The matrix for the predicted values

4   Conclusion

As a result of the project, a text presentation method was proposed for computing neural networks with partial preservation of the author's style. In addition, models of neural networks were implemented such as a multilayer perceptron, LSTM, a convolutional neural network. An assessment was made of the quality of work of these models in the task of determining the authorship of poems and a comparison was made based on the results of which a rating of models can be made:

1. convolutional neural network;
2. multilayer perceptron;
3. LSTM neural network.

You can make an assessment and comparison based on the results of quality assessment of trained models. The multilayer perceptron showed good results, but its accuracy in determining the poems belonging to the author is small. The LSTM neural network better defines poems belonging to the author, however, the accuracy of determining non-author poems is too low. In general, the LSTM model of the neural network has been trained much longer and worse. The convolutional neural network (CNN) turned out to be the best, albeit slightly inferior in terms of final accuracy to the multilayer perceptron.

Thus, the task of determining the authorship of texts can be solved by machine learning methods, and the proposed form for presenting texts really preserves the author's style.

Acknowledgement. This study was supported by the Russian Foundation for Basic Research (Grants No. 18–47-732007, 18–47-730035 and 19–07-00999).

References

1. Mukha, A.V., Rozaliev, V.L., Orlova, Yu.A., Zaboleeva-Zotova, A.V.: An automated approach to determining the authorship of a text. Bull. Volgograd State Tech. Univ. **14**, 51–54 (2013)
2. Shrestha, P., Sierra, S., González, F., Montes, M., Rosso, P., Solorio, T.: Convolutional Neural Networks for Authorship Attribution of Short Texts, pp. 669–674 (2017). https://doi.org/10.18653/v1/E17-2106.
3. Ge, Zh., Sun, Yu., Smith, M.: Authorship attribution using a neural network language model. In: Proceedings of the Thirtieth AAAI Conference on Artificial Intelligence, pp. 4212–4213 (2016)
4. Kukushkina, O.V., Polikarpov, A.A., Khmelev, D.V.: Definition of authorship of the text using alphabetic and grammatical information. Probl. Inf. Transfer. **37**(2), 96–109 (2001)
5. Khmelev, D.V.: Recognition of the author of the text using chains of A. A. Markov. Tomsk State Univ. J. Ser. 9: Philol. **2**, 115–126 (2000)
6. Rogov, A.A., Sidorov, Yu.V., Korol, A.V.: Automated system for processing and analysis of literary texts SMALT. Transaction Material II Intern Congress Russian Language Researchers "Russian Language: Historical Fates and the Present", pp. 485–486. Moscow State University (2004)
7. Rogov, A.A., Gurin, G.B., Kotov, A.A., Sidorov, Yu.V., Surovtsova, T.G.: Software package SMALT. In: Digital Libraries: Advanced Methods and Technologies, Electronic Collections: Proceedings of X All-Russia. Scientific Conference "RCDL'2008', pp. 155–160. Dubna (2008)
8. Romanov, A.S.: Technique of identification of the author of the text based on the apparatus of support vectors. Reports Tomsk State Univ. Control Syst. Radioelectron. (19), 1–2 (2009)
9. Romanov, A.S.: Methods and software for identifying the author of an unknown text: Abstract. dis. ... cand. tech. sciences. Tomsk (2010)
10. Kaggle library. https://www.kaggle.com/johnhallman/complete-poetryfoundationorg-dataset . Accessed 6 June 2020
11. Python 3.8.1 documentation. https://docs.python.org/3/ - Download from the screen. Accessed 6 June 2020
12. Keras Documentation. https://keras.io/ - Headline from the screen Accessed 6 June 2020

Predicting Porosity Through Fuzzy Logic Based Methods from South Caspian Basin Data

R. Y. Aliyarov⓾, L. A. Gardashova$^{(\boxtimes)}$⓾, and N. I. Hasanli⓾

Azerbaijan State Oil and Industry University,
Azadliq Avenue 20, Baku, Azerbaijan
r.aliyarov@asoiu.edu.az, latsham@yandex.ru,
narmin.mammadova@gmail.com

Abstract. The problem of porosity prediction has been an object of analysis by researchers in the area of geology. In this paper, it is analyzed Fuzzy logic based methods to forecasting porosity, which of the most important characteristics for modeling reservoir. The used methods are based on the fuzzy recurrent neural network trained with the method of error backpropagation. Fuzzy rule extraction from numerical data is used for creating fuzzy inference-based forecasting system. Several methods are implemented for rule extraction, such that k-means clustering method, fuzzy c-means clustering method, sub clustering, grid-partition method. Core material-CC, F1 + F2, F3, F4 and depth are selected for the initial data for forecasting porosity. Experimental results show efficiency of the proposed method of forecasting compared to the existing classical methods.

Keywords: Forecasting of the porosity · Error backpropagation · Regression · Adaptive neuro-fuzzy inference system · K-means clustering method · C-means clustering method · Sub clustering

1 Introduction

Characterization and analysis of porosity is discussed in scientific works of last years. Porosity is one of the basic characteristics in reservoir and it depends on fluid storage in aquifers, oil and gas fields and geothermal systems. Forecasting porosity is a problem, that uses all accessible data to provide precise models in order to estimate reservoir performance. In literature, there are some methods to determine this property, such as statistical methods and novel techniques like intelligent methods.

Relationship between the input logs and the petrophysical properties is analyzed in [1]. Permeability and porosity prediction from wireline logs using Neuro-fuzzy technique is discussed in [2]. Related works, such as A Study on Prediction of Output in Oilfield Using Multiple Linear regression is explained in literature [3].

Prediction of permeability and porosity from well log data using the nonparametric regression with multivariate analysis and neural network is discussed in [4].

© The Author(s), under exclusive license to Springer Nature Switzerland AG 2021
R. A. Aliev et al. (Eds.): ICAFS 2020, AISC 1306, pp. 268–274, 2021.
https://doi.org/10.1007/978-3-030-64058-3_33

Jafari and etc. [5] works is explained relationships between permeability, porosity and pore throat size in carbonate rocks using regression analysis and neural networks and advantage properties of neural network is also shown.

Porosity and Permeability Estimation using Neural Network Approach from Well Log Data is given in [3, 6].

Prediction of porosity in crystalline rocks using artificial neural networks and an example from the Chinese Continental Scientific Drilling Main Hole is interesting work in this subject area [7].

A review of observations, theory, and experiments about compaction and porosity reduction in carbonates is discussed in [8]:

ANN and ANFIS for Prediction of Porosity and Sand Fraction is described in [9].

In our previous work prediction of multivariable properties of reservoir rocks are discussed in [10–12] by using fuzzy clustering.

However, essential lack of all these works is application of modifications of traditional forecasting methods which cannot take into account uncertainty factor completely. Other essential lack of these works are no combination of quantitative and qualitative methods of forecasting without what are very difficult for achieving creations of such method of forecasting which the exact and transparent result is permission and accessibility for broad audience of users.

The paper contains 5 section. In Sect. 2 is about preliminaries. Statement of the problem is described in Sect. 3. Forecasting the porosity by using neural networks and clustering is the main task of Sect. 4. The obtained results and conclusion of the paper is shown in Sect. 5.

2 Preliminaries

Definition 1 Fuzzy Number [13]. A fuzzy number is a fuzzy set A on R which possesses the following properties: a) A is a normal fuzzy set; b) A is a convex fuzzy set; c) α-cut of A, A^α is a closed interval for every $\alpha \in (0, 1]$; d) the support of A, A^{+0} is bounded.

Definition 2 Mamdani Reasoning [14]. Fuzzy If-Then rules represents as follow::

$$\text{IF } x_1 \text{ is } A_{11} \text{ and } \dots \dots x_n \text{ is } A_{1n} \text{ THEN y is } B_1$$
$$\text{IF } x_1 \text{ is } A_{21} \text{ and } \dots \dots x_n \text{ is } A_{2n} \text{ THEN y is } B_2$$
$$\dots \dots \dots \qquad \dots \dots$$
$$\text{IF } x_1 \text{ is } A_{m1} \text{ and } \dots \dots x_n \text{ is } A_{mn} \text{ THEN y is } B_m$$

where $x_j, j = 1\dots n$-number of input variables, y- is output variable. A_{ij} and B_i are linguistic values in accordance with input and output variables of the rules.

Mamdani reasoning algorithm is given below:

1. For each rule truth level is determining as following:

$$\alpha_i = \min_{j=1}^{n}[\max_{X_j}(A_j'(x_j) \wedge A_{ij}(x_j))$$

where $A_j'(x_j)$-are value of new input variables.

2. Individual outputs is defined as:

$$B_i'(y) = \min(\alpha_i, B_i(y))$$

3. Calculation the aggregative output:

Definition 3 Fuzzy C-means Algorithm [15]. This algorithm works by assigning membership to each data point corresponding to each cluster center on the basis of distance between the cluster center and the data point. More the data is near to the cluster center more is its membership towards the particular cluster center. Clearly, summation of membership of each data point should be equal to one. After each iteration membership and cluster centers are updated.

Main objective of fuzzy c-means algorithm is to minimize:

$$J(U, V) = \sum_{i=1}^{n} \sum_{j=1}^{c} (\mu_{ij})^m \|x_i - v_j\|^2$$

where, '$\|xi - vj\|$' is the Euclidean distance between ith data and jth cluster center.

3   Statement of the Problem

In this paper, a new method is made of a combination of a quantitative method - calculations based on crisp data set by application of the recurrent fuzzy neural networks trained with the help of algorithm error backpropagation.

The numerous factors influencing porosity have not been taken into account in the model of forecasting multivariable properties of reservoir rocks by using fuzzy clustering.

From this point of view work is provided to solve the following problems:

- Development of effective architecture of fuzzy neural networks, a method of its training, for the data processing;
- A choice of parameters corresponding algorithm (for instance, membership function, method training);
- Comparison of the obtained results of forecasting with results of traditional methods.

For the decision of these problems the effective architecture of fuzzy recurrent neural networks with five inputs, and one output has been developed. The network is trained on the basis of error backpropagation method. Such neural network is characterized by smaller computing complexity and good ability to training.

4 Forecasting of the Porosity by Using Fuzzy Neural Network and Clustering

For forecasting the porosity by using neural networks with 5 inputs and one output have been used. In calculations, used 211 data and 2/3 of it was used for training, and 1/3 data for test of FRNN. The fragment of training data is given in Table 1.

Table 1. Fragment of training data

Field_name	Cc	F1 + F2Cor	F3Cor	F4Cor	Depth	Por.
Darvin	0,028	0,668	0,166	0,138	1253	0,314
Neft-dashlari	0,038	0,699	0,118	0,144	1820	0,286
Darvin	0,046	0,641	0,164	0,149	1210	0,298
Neft-dashlari	0,046	0,571	0,172	0,211	328	0,300
Gum Deniz	0,045	0,765	0,125	0,065	2850	0,181
Guneshli	0,047	0,543	0,309	0,101	3644	0,269
Darvin	0,048	0,565	0,242	0,146	1952	0,259
Neft-dashlari	0,097	0,414	0,272	0,218	435	0,320
...	...	...	...	...	...	...
Guneshli	0,052	0,640	0,238	0,070	3211	0,214
Gum Deniz	0,065	0,636	0,210	0,089	2656	0,209
Guneshli	0,044	0,484	0,331	0,141	3323	0,272
Gum Deniz	0,030	0,653	0,241	0,077	2636	0,200
Darvin	0,038	0,532	0,239	0,191	762	0,252
Guneshli	0,055	0,541	0,314	0,091	3402	0,246
Sang_Duv_ Xara_Zire	0,084	0,475	0,311	0,129	3424	0,267

Training error over selected method is described in Table 2:

Table 2. Training error over selected methods

Data numbers	Generate.fis	Optimization method	Train epocs	Cluster size	Membership function	Error
148	Sub clustering	Hybrid	5000	10	gaussmf	0,0313608
148	Sub clustering	Back propagation	5000	10	gaussmf	7.21652
148	Sub clustering	Hybrid	5000	10	trimf	0,0313608
148	Sub clustering	Back propagation	5000	10	trimf	7.21652

Fragment of the obtained clusters [12] is given Fig. 1.

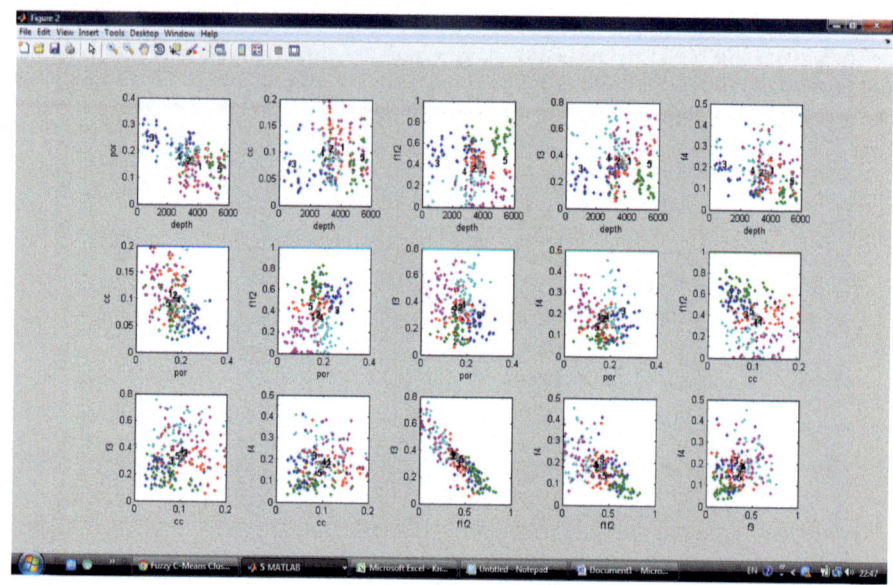

Fig. 1. Fragment of the obtained clusters

Fragment of testing data for fuzzy c-means and fuzzy inference based method is described in Table 3:

Table 3. Fragment of testing data for fuzzy c-means and y inference based method

Field_name	Cc	F1 + F2Cor	F3Cor	F4Cor	Depth	Porosity	Forecast porosity
Sang_Duv_Xara_Zire	0,127	0,27936	0,378882	0,214758	4477	0,208	0,198
Djanub	0,182	0,516976	0,168508	0,132516	3212	0,142	0,177
Sang_Duv_Xara_Zire	0,074	0,353732	0,340768	0,2315	3975,5	0,178	0,177
8Mart	0,067	0,334947	0,516882	0,081171	5642,5	0,181	0,164
Bulla	0,065	0,348755	0,40766	0,178585	5353	0,173	0,167
Sang_Duv_Xara_Zire	0,146	0,317688	0,332206	0,204106	4573	0,177	0,17
...	...	...	...	...	...	...	...
Djanub	0,145	0,03078	0,671175	0,153045	3252	0,203	0,176
Sang_Duv_Xara_Zire	0,094	0,192072	0,57078	0,143148	2540,5	0,171	0,178
Sang_Duv_Xara_Zire	0,144	0,01284	0,609472	0,233688	2561	0,198	0,172
Sang_Duv_Xara_Zire	0,144	0,009416	0,609472	0,237112	2523,5	0,198	0,172
Guneshli	0,042	0,25866	0,535522	0,163818	2902,5	0,152	0,184
Djanub	0,078	0,00922	0,704408	0,208372	3195,5	0,186	0,172

RMSE = 0,044879

5 Conclusion

In this paper, it is analyzed Fuzzy logic based methods to forecasting porosity, which of the most important characteristics for modeling reservoir. Several methods is used for rule extraction, such us k-means clustering method, c-means clustering method, sub clustering, grid-partition method.

The initial data for forecasting porosity is selected core material-CC, F1 + F2, F3, F4 and depth. Experimental results show efficiency of the proposed method of forecasting as compared to the existing classical methods. All calculations are made in MATLAB environment, With C++ and MS Excel.

Comparison results is given below. Root Mean Square Error (RMSE) of the forecast various methods has made: a discriminant analysis: linear discriminant analysis -0.07332, regression-0.082604, fuzzy neural network -0.068331%, fuzzy c-means and Mamdani fuzzy inference system based method -0.044879.

Why is Mamdani method and only Mamdani method applied in the proposed work?

Mamdani inference method is applied in this paper for the reasons such that output membership function is present, expressive power and interpretable rule consequent, distribution of output, it has more accuracy in security evaluation block cipher algorithm, possess less flexibility in the system design, it is using in MISO (Multiple Input and Single Output) and MIMO (Multiple Input and Multiple Output) systems and Mamdani inference system is well suited to human input.

The received results give the basis to approve, that for today forecasts made by using of Soft Computing Technology are more reliable and exact.

Acknowledgement. This study was supported by the "Science Development Foundation at the President of the Azerbaijan Republic (Grant No. EİF / MQM / Elm-Tehsil-1–2016-1 (26)).

References

1. Chaki, S., Verma, A.K., Routray, A., Jenamani, M., Mohanty, W.K., Chaudhuri, P.K., Das, S.K.: Prediction of Porosity and Sand Fraction from Well Log Data using ANN and ANFIS a comparative study (2013). https://www.spgindia.org/10_biennial_form/P419.pdf
2. El-Shahat, W., Afify, W., Hassan, A.: Permeability and porosity prediction from wireline logs using Neuro-fuzzy technique. Ozean J. Appl. Sci. **3**(1), 157–175 (2010)
3. Mustafar, I.B., Razali, R.: A study on prediction of output in oilfield using multiple linear regression. Int. J. Appl. Sci. Technol. **1**(4), 107–113 (2011)
4. Hassi, R., Baouche, R., Baddari, K.: Prediction of permeability and porosity from well log data using the nonparametric regression with multivariate analysis and neural network. Egypt. J. Petrol. **26**, 763–778 (2017)
5. Jafari, A., Kazemzadeh, E., Rezaee, M.R.: Relationships between permeability, porosity and pore throat size in carbonate rocks using regression analysis and neural networks. J. Geophys. Eng. **3**, 370–376 (2006)
6. Verma, A.K., Cheadle, B.A.: Porosity and Permeability Estimation using Neural Network Approach from Well Log Data. https://www.researchgate.net/publication/236150010

7. Konaté, A.A., Pan, H., Khan, N., Ziggah, Y.Y.: Prediction of porosity in crystalline rocks using artificial neural networks: an example from the Chinese continental scientific drilling main hole. Studia Geophys. Geodaetica. **59**, 113–136 (2015)

8. Croize, D., Renard, F., Gratier, J.P.: Compaction and porosity reduction in carbonates: a review of observations, theory, and experiments. Adv. Geophys. **54**, 181–238 (2013). https://doi.org/10.1016/B978-0-12-380940-7.00003-2

9. Sadeghi, R.A., Kadkhodaie, A., Rafiei, B., et al.: A committee machine approach for predicting permeability from log data: a case study from a heterogeneous carbonate reservoir, Balal Oil Field. Persian Gulf. J. Geol. Petrol. **1**(2), 1–0 (2011)

10. Aliyarov, R.Y., Ramazanov, R.A.: Prediction of multivariable properties of reservoir rocks by using fuzzy clustering. Proc. Comput. Sci. **102**, 434–440 (2016)

11. Aliyarov, R.Y., Hasanov, A.B., Samadzada, G.A.: Application of fuzzy linear regression model to forecasting of reservoir rocks properties. In: Advances in Intelligent Systems and Computing Series, pp. 835–841. Springer (2019)

12. Gardashova, L.A., Ramazanov, Dj.I., Qurbanova, T.H: Fuzzy neural-network and fuzzy c-means method based forecasting of efficiency of the oil wells from energy consumption point of view. In: Seventh International conference on Soft Computing/ Computing with Words and Perceptions in System Analysis, Decision and Control, Turkey, pp. 329–335 (2013)

13. Aliev, R.A.: Fundamentals of the Fuzzy Logic- Based Generalized Theory of Decisions. Springer, Heidelberg (2013)

14. https://home.dei.polimi.it/matteucc/Clustering/tutorial_html/cmeans.html

15. https://nl.mathworks.com/help/fuzzy/what-is-mamdani-type-fuzzy-inference.html

Qualimetry of Socio-Economic Systems Based on Fuzzy-Multiple Aggregation of Factor Hierarchies

Ludmila Gebrovskaya[1] , Galina V. Lukyanova[1] ,
Victoriya S. Kokhanova[2]([⊠]) , Elena Filimonova[1] ,
and A. Zhabrova Tamara[1]

[1] Rostov State University of Economics, Rostov-on-Don, Russia
zamnach@rsue.ru, lukyanova.g@yandex.ru,
elena-gamilton@mail.ru, tamarazhabrowa@yandex.ru
[2] Private Educational Institution of Higher Education «Southern University
(IMBL)», Rostov-on-Don, Russia
kohanovavs@yandex.ru

Abstract. The aim A methodology has been developed for a comprehensive assessment and ranking of the level of social efficiency of agriculture in the Rostov region based on time series of statistical data. The substantiation of the choice as a mathematical apparatus for the formation of the assessment of the universal tool of fuzzy logic is carried out: systems of fuzzy-logical conclusions - standard five-level [0, 1] -classifiers. The possibility of a fuzzy classification of properties, as well as qualimetry based on aggregation of hierarchies of factors, made it possible to assess the level of social efficiency of agriculture in the region based on a set of heterogeneous indicators, and also to rank the regions of the region by their level of success.

Keywords: Social efficiency of agriculture · Set of indicators · Fuzzy-logical conclusion

1 Introduction

Research methods of economic systems can be divided into several types. Econometric modeling is considered classic, in which statistical data are collected and analyzed using economic statistics methods. Another type of toolkit is the model of general economic equilibrium. They are based on the concept of an equilibrium state in the economy, achieved through equal supply and demand in all markets. A special niche is occupied by simulation models, indispensable in the description of complex systems with elements of random behavior developing over time. Finally, we mention the balance models, the main feature of which is the correspondence of the "receipt" of a resource and its "distribution". Each class of methods includes many varieties of modeling tools. All methods have their strengths and weaknesses and a certain range of tasks that they solve.

Currently, fuzzy logic tools are widely used in economic research. The mathematical theory of fuzzy sets, proposed in 1965 by Lotfi A. Zade (L.A. Zadeh born February 4, 1921, Novkhany, Azerbaijan SSR), professor of technical sciences at the

© The Author(s), under exclusive license to Springer Nature Switzerland AG 2021
R. A. Aliev et al. (Eds.): ICAFS 2020, AISC 1306, pp. 275–282, 2021.
https://doi.org/10.1007/978-3-030-64058-3_34

University of California at Berkeley, allows us to describe fuzzy concepts and knowledge, operate with this knowledge and draw fuzzy conclusions [1, 2]. More recently, the application of the methods of fuzzy logic in solving economic problems was considered by economists as a kind of newfangled excess. But today we can talk about a turning point in the assessment of this scientific situation [3–5].

2 The Ideas of Fuzzy Logic in Demand in Socio-Economic Research

2.1   Idea 1. Fuzzy Classification

The following ideas of fuzzy logic are most in demand. All levels of economic parameters can be measured not only quantitatively, but also qualitatively. To do this, it is necessary to define the linguistic variable "Parameter X Level", the carrier of which is the domain of X parameter definition, and the term-set of values are fuzzy subsets of "Very Low Level, Low Level, Average Level, High Level, Very High Level" of Parameter X. For pentascals, it is necessary to construct a system of membership functions of the carrier X to the corresponding fuzzy subsets. The simplest way to set is a system of trapezoidal fuzzy numbers [6]. This idea corresponds to the use of a system of fuzzy-logical conclusions - standard fuzzy multi-level [0, 1] - classifiers.

The pentascal is optimal in most cases [7, 8], but in some cases it is advisable to use the simplest case of a binary scale such as High, Low or Bad, Good. Depending on the statement of the problem, the number of terms can vary: for example, in accordance with the established classification of assessing the state of systems, three, four or even ten terms can be used [9–12]. In the case of an arbitrary number of terms, it is possible to use not trapezoidal, but sigmoid membership functions possessing smoothness sufficient for mathematical rigor and defined by the formulas:

$$\mu_k = \exp\left(-\left(x - \frac{k-1}{4}\right)^{2n}/\sigma^{2n}\right), \quad k = 1, 2, 3, 4, 5 \tag{1}$$

Here, $n \in N$ is a natural number characterizing the steepness of the graph of the membership function. The nodes of the classifier, respectively, are located at the points:

$$\overline{g_k} = \frac{k-1}{4}, \quad k = 1, 2, 3, 4, 5 \tag{2}$$

The parameter σ is selected based on the completeness of the constructed system of functions. Indeed, we require that at the intersection points of two adjacent graphs the following relationship holds:

$$\mu_k\left(\frac{k-1}{4} + \frac{1}{8}\right) = \mu_{k+1}\left(\frac{k-1}{4} + \frac{1}{8}\right) = \frac{1}{2}$$

Then from formulas (1–2) we get:

$$\sigma^{2n} = \frac{1}{8^{2n} \ln 2}.$$

Therefore, the final form of membership functions has the form:

$$\mu_k = \exp\left(-\left(x - \frac{k-1}{4}\right)^{2n} 8^{2n} \ln 2\right), \quad k = 1, 2, 3, 4, 5 \tag{3}$$

The advantage of functions (3) selected in this way is that the value of each of them is negligible (in the context of the problem under study) at the top of the neighboring function. Indeed, for already $n = 2$

$$\mu_k(g_k) = \mu_k\left(\frac{k}{4}\right) = \left(\frac{1}{2}\right)^{16} \approx 1,53 \cdot 10^{-5}$$

It was established, that on the interval between two vertices, up to an infinitely small value, the following equality holds:

$$\mu_k(x) + \mu_{k+1}(x) = 1, \quad \frac{k-1}{4} \le x \le \frac{k}{4},$$

and in the vicinity of the intersection point, the graphs are approximated by tangents to them. Thus, the system of functions (3) satisfies the completeness condition and can be used to construct a five-level classifier. Obviously, the formulas can be generalized to the K-level classifier by the formulas:

$$\mu_k = \exp\left(-\left(x - \frac{k-1}{4}\right)^{2n} 2^{2n}(K-1)^{2n} \ln 2\right), \quad \overline{g}_k = \frac{k-1}{K-1}, \quad k = 1, 2, ..., K.$$

2.2 Qualimetry Based on Aggregation of Factor Hierarchies

Let some property of an economic, social or any other system be represented as a tree-like hierarchy of factors, moreover:

- within the framework of the hierarchy, systems are defined for the relationship of preference of one sub-property to another for one level of the hierarchy;
- the sub-properties that make up the lower levels of the hierarchy can be measured both quantitatively and qualitatively (including verbally).

In this case, it is possible to carry out a comprehensive assessment of the system properties if: 1) to make all measurements on a qualitative basis, making a fuzzy classification of quantitative factors according to the scheme described above; 2) to simulate Fishburn weight systems for modeling preference systems; 3) to aggregate normalized values of indicators in the framework of two-dimensional convolution, where one of the systems of weights is the weight of the factors, and the other with the system of weights is the nodal points of the standard fuzzy multilevel [0, 1] - classifiers (Fig. 1). When a numerical value of a comprehensive assessment of a system property is obtained, linguistic recognition of the qualitative level of a given complex property can be made based on the corresponding fuzzy classifier.

Also, on the basis of the obtained assessment of the system's properties, it is possible to perform a paired analysis of the quality of the object and some of its additional properties (for example, price) in order to distinguish such objects for which, for example, a maximum of quality is achieved at a fixed price, or, conversely, a minimum price at fixed level of quality. Selected objects form the Edgeworth-Pareto set.

2.3   Values like Fuzzy Numbers, Fuzzy Sequences, and Fuzzy Functions

The input parameters of the model, as well as the studied values can be fuzzified, that is, presented in the form of fuzzy triangular numbers. If research planning is carried out in discrete time (for example, financial flows), then the combination of flows and their results forms a sequence of fuzzy numbers. If planning is carried out in continuous time, then we should talk about fuzzy functions.

The universality of these three ideas of fuzzy logic is proved by their wide application in solving such problems as strategic planning, a comprehensive analysis of the

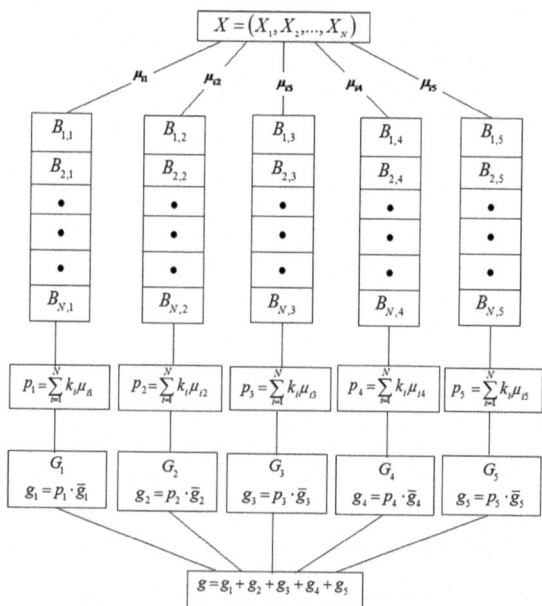

Fig. 1. Aggregation of normalized values of indicators in the framework of two-dimensional convolution based on standard fuzzy multilevel [0, 1] – classifiers

state of the corporation, analysis of the creditworthiness of a bank borrower, risk assessment of an investment project, optimizing a stock portfolio, assessing the investment attractiveness of securities, forecasting stock indices, choosing a manager companies, real estate appraisal, transport logistics, choice of corporate information system, analysis of news background, post oenie sociograms and analysis, collective behavior.

3   Evaluation and Ranking of the Level of Social Efficiency of Agriculture Based on a System of Fuzzy-Logical Conclusions - Standard Five-Level [0, 1] –classifiers

The Let us demonstrate how, based on the first two ideas of fuzzy logic, an assessment and ranking of the level of social efficiency of agriculture can be carried out on the basis of time series of statistical data. As a mathematical apparatus for the formation of estimates, we use a system of fuzzy-logical conclusions - standard five-level [0, 1] - classifiers.

The indicators of the social efficiency of agriculture are traditionally evaluated by such indicators as the share of the rural population in the total population of the region, the rate of natural and migration growth of the population, the level of registered unemployment, the level of economic activity of the population, the disposable resources of the rural population on average per household member per month, specific the weight of the population with cash incomes below the subsistence level, the number of state and municipal educational institutions in rural capacity of outpatient clinics in rural areas, based on 10,000 rural population, etc. The totality of these indicators is used for a comprehensive analysis of the state of the social subsystem of agriculture; however, in its pure form, it does not allow an integral assessment of the municipality, to rank the districts according to the level of effectiveness of the social sphere.

We offer a methodology for assessing the social efficiency of agriculture in areas of a given region. The methodology for a comprehensive assessment of the social efficiency of an agricultural region is based on previously developed methods for assessing the efficiency of agricultural production according to the set of criteria of two groups: the level of intensification of production and the level of economic efficiency of intensification of production in agriculture [7–9]. In addition, the methods used to assess areas for compliance with the principles of environmental management [10–12].

The following indicators (the weight of all indicators 1/8) were taken as a set of indicators for the formation of an assessment of the social efficiency of agriculture (by region): 1) average monthly wage of employees of organizations, ruble (from 2010 to 2019); 2) the amount of social payments to the population and taxable cash incomes to the population, thousand rubles; 3) total fertility rate, per mille (2010–2018); 4) general mortality rate, per mille (2010–2018); 5) the number of doctors of all specialties (2008–2013); 6) the number of nursing staff (2008–2013); 7) the number of medical beds (2008–2013); 8) the commissioning of individual residential buildings in the territory of the municipality, sq.m. of total area.

The calculation of the aggregated values of all the studied indicators for N years is carried out on the basis of a scheme that integrates the time series of data for each of the indicators and takes into account the significance of different time periods due to weight factors; while the numbering of time periods is carried out in reverse order. The methodology was tested in the districts of the Rostov region, while the time series of indicators were taken from the Rosstat base [8]. Linguistic variables with a carrier in the form of a numerical segment [0, 1] and a term-set G = {G1, G2, G3, G4, G5} correspond to the values of the indicators, as well as the final grade, where G1 is the "steady tendency to decrease growth"; G2 – "growth tendency"; G3 – "stagnation tendency"; G4 – "upward trend"; G5 – "steady growth trend".

Table 1 shows the results of calculating the aggregated values of the indicators (the numbering corresponds to the above), as well as comprehensive estimates of social efficiency calculated in their areas in the districts. The last column of the table indicates the term to which a comprehensive assessment of the relevant region belongs.

Table 1. Aggregated values of parameters and integrated estimates of social efficiency calculated in their areas in the regions

No	1	2	3	4	5	6	7	8	Total	Term
1	1	0.909	0.336	0.373	0.509	0.327	0.336	0.855	0.5502	G - 3
2	0.945	0.909	0.345	0.573	0.6	0.6	0.555	0.364	0.5821	G - 3
3	0.691	0.909	0.318	0.409	0.755	0.455	0.518	1	0.6122	G - 4
4	0.909	0.909	0.227	0.6	0.327	0.327	0.364	0.527	0.507	G - 3
5	0.455	0.909	0.609	0.664	0.764	0.6	0.427	0.818	0.6634	G - 4
6	0.964	0.909	0.373	0.664	0.418	0.291	0.436	0.727	0.5904	G - 3
7	0.8	0.909	0.227	0.427	0.545	0.527	0.418	0.6	0.554	G - 3
8	1	0.909	0.282	0.4	0.636	0.418	0.4	0.655	0.554	G - 3
9	0.764	0.909	0.309	0.536	0.6	0.418	0.509	0.655	0.5816	G - 3
10	0.636	0.909	0.427	0.545	0.436	0.491	0.491	0.8	0.5972	G - 3
...	...	...	...	...	...	...	...	...	...	...
30	1	0.909	0.318	0.373	0.545	0.473	0.645	0.927	0.624	G - 4
31	0.455	0.909	0.518	0.564	0.7	0.7	0.427	0.545	0.5957	G - 3
32	0.455	0.909	0.518	0.564	0.627	0.327	0.427	0.655	0.5656	G - 3
33	0.818	0.909	0.391	0.482	0.527	0.455	0.336	0.691	0.574	G - 3
34	1	0.718	0.3	0.482	0.436	0.382	0.455	0.582	0.536	G - 3
35	0.636	0.718	0.309	0.445	0.445	0.364	0.391	0.564	0.4862	G - 3
36	0.527	0.718	0.627	0.482	0.373	0.373	0.445	0.618	0.5214	G - 3
37	0.6	0.718	0.318	0.482	0.518	0.491	0.418	0.255	0.4794	G - 3
38	0.709	0.718	0.391	0.518	0.436	0.4	0.373	0.564	0.5034	G - 3
39	1	0.718	0.373	0.664	0.473	0.4	0.473	0.745	0.5913	G - 3
40	0.655	0.718	0.264	0.645	0.464	0.309	0.436	0.618	0.5377	G - 3
41	0.855	0.718	0.391	0.482	0.364	0.382	0.382	0.527	0.5029	G - 3
42	0.618	0.718	0.282	0.264	0.4	0.418	0.373	0.8	0.4889	G - 3
43	0.455	0.718	0.536	0.464	0.409	0	0.455	0.855	0.516	G - 3

where 1) Azov; 2) Aksay; 3) Bagaevsky; 4) Belokalitvinsky; 5) Bokovsky; 6) Upper donsko; 7) Veselovsky; 8) Volgodonsk; 9) Dubovsky; 10) Egorlyksky; 11) Zavetinsky; 12) Zernogradsky; 13) Zimovnikovsky; 14) Kagalnitsky; 15) Kamensky; 16) Kashar; 17) Konstantinovsky; 18) Krasnosulinsky; 19) Kuibyshevsky; 20) Martynovsky; 21) Matveevo-Kurgan; 22) Millerovsky; 23) Milyutinka; 24) Morozovsky; 25) Myas-nikovsky; 26) Neklinovsky; 27) Oblivsky; 28) October; 29) Oryol; 30) Sandstone; 31) Proletarian; 32) Remontnensky; 33) Rodionovo; 34) Nesvetaysky; 35) Salsky; 36) Semikarakorsky; 37) Soviet; 38) Tarasovsky; 39) Tatsinsky; 40) Ust-Donetsk; 41) Tselinsky; 42) Tsimlyansky; 43) Sholokhovskiy. The final assessment in the Rostov region, where the weight of the population is the share of the municipality, is 0.532, which corresponds to the term G3 - the average level of social efficiency.

4 Conclusion

The analysis showed the universality of the mathematical apparatus of fuzzy logic, namely, a system of fuzzy-logical conclusions - standard five-level [0, 1] -classifiers.

The proposed method for constructing a comprehensive assessment of the social efficiency of agriculture allows us to assess the state of the regions of the region based on a combination of series of intensification indicators for an arbitrary number of time periods.The technique is easy to use, easily formalized in the form of software systems. It can be easily modified by introducing new indicators into consideration and changing their significance in the hierarchy by recalculating weighting factors.

References

1. Zadeh, L.A.: The concept of a linguistic variable and its application to making approximate decisions. Inform. Sci. **8**(3), 199–249 (1975)
2. Zadeh, L.A.: Fuzzy sets. Inform. Control **8**, 338–353 (1965)
3. Nedosekin, A.: Fuzzy financial management. AFA Library, Moscow (2003)
4. Nedosekin, A.: Application of the fuzzy sets to the problems of financial management. Audit and financial analysis (2000). https://www.cfin.ru/press/afa/2000-2/08.shtml
5. Nedosekin, A., Kozlovsky, A., Abdulaeva, Z.: Analysis of branch economic stability by fuzzy-logical methods. Econ. Manage.: Probl. Solutions **5**, 10–16 (2018)
6. Konysheva, L.K., Nazarov, D.M.: Fundamentals of the theory of fuzzy sets: a training manual, St. Petersburg: Peter, 192 p. (2011)
7. Alekseychik, T.V., Bogachev, T.V., Karasev, D.N., Sakharova, L.V., Stryukov, M.B.: Fuzzy method of assessing the intensity of agricultural production on a set of criteria of the level of intensification and the level of economic efficiency of intensification. Adv. Intell. Syst. Comput. **896**, 790–798 (2019)
8. Vovchenko, N.G., Stryukov, M.B., Sakharova, L.V., Domokur, O.V.: Fuzzy-logic analysis of the state of the atmosphere in large cities of the industrial region on the example of Rostov region. Adv. Intell. Syst. Comput. **896**, 709–715 (2019)
9. Albekov, A.U., Arapova, E.A., Karasev, D.N., Stryukov, M. B. Sakharova, L.V.: A program for assessing the intensity of agricultural production through a fuzzy 5-point classifier. Federal Service for Intellectual Property (2018)

10. Arapova, E.A., Lukyanova, G.V., Sakharova, L.V., Akperov, G.I.: Fuzzy-logic analysis of the level of comfort and environmental well-being of the urban environment on the example of large cities of Rostov region. In: ICAFS-2018, Advances in Intelligent Systems and Computing, Scopus, vol. 896, pp. 643–650 (2018).
11. Sakharova, L.V., Alekseychik, T.V., Bogachev, T.V., Arapova, E.A.: Assessment of the state of the atmosphere in the region using fuzzy modeling. Bull. Rostov State Univ. Econ. (RINH). Sci. Practical J. **3**(63), (2018).
12. Albekov, A.U., Arapova, E.A., Karasev, D.N., Stryukov, M. B. Sakharova, L.V.: A program for assessing the level of pollution in the region based on fuzzy-multiple analysis of statistical data. Certificate of registration of a computer program (2018)

Fuzzy Optimal Control of Coke Production

K. R. Aliyeva$^{(\boxtimes)}$ ⓘ

Department of Instrument-Making Engineering, Azerbaijan State Oil
and Industry University, 20 Azadlig Avenue, AZ1010 Baku, Azerbaijan
kamalann64@gmail.com

Abstract. In this paper coke unit in oil-refinery plants considered as complex dynamic system with huge degree of uncertainty. For optimization of coke unit regime we use fuzzy multistage decision making approach. Solution of the problem is based on Bellman-Zadeh fuzzy decision-making method and fuzzy dynamic programming. Application of suggested dynamic model and control algorithm for optimization coke unit shows their effectiveness and validity.

Keywords: Fuzzy optimal control · Fuzzy matrix · Dynamic modeling · Dynamic programming · Decision making

1   Introduction

To optimization and control devoted different studies [7, 8]. In particular in existing literatures there are some works on development some measurement devises, controllers for different regime parameters etc. Unfortunately, applications of these research results in practice are very scare. In this study we investigate the multistage decision-making problem for terminal control of unit in oil refinery plants.

In order to handle an optimal control problem with fuzzy process, in the paper we will introduce and deal with a fuzzy optimal control problem by using dynamic programming in coke production. In Sect. 3, we will introduce a fuzzy optimal control problem of coke production, and present the principle of optimality for fuzzy optimal processing. In Sect. 4, we will obtain a fundamental result called the equation of optimality in fuzzy optimal control in coke production. In the last section, we give conclusion of the problem.

2   Preliminaries

Definition 1. **A fuzzy set** $\tilde{A}$ in a universe of discourse X is characterized by a membership function $\mu \tilde{A}\,(x)$ which associates with each element x in X a real number in the interval [0, 1]. The function value $\mu\tilde{A}(x)$ is termed the grade of membership of x in $\tilde{A}$ [1].

A fuzzy number [2] is a fuzzy set A on $\mathcal{R}$ which possesses the following properties: a) A is a normal fuzzy set; b) A is a convex fuzzy set; c) α-cut of A, A^{α} is a closed interval for every $\alpha \in (0, 1]$; d) the support of A, supp(A) is bounded. There is an equivalent parametric definition of fuzzy number A fuzzy number A is a pair $(\underline{\mu_A}, \overline{\mu_A})$ of functions $\underline{\mu}_A(\alpha)$, $\overline{\mu}_A(\alpha)$; $0 \le \alpha \le 1$ with following requirements:

© The Author(s), under exclusive license to Springer Nature Switzerland AG 2021
R. A. Aliev et al. (Eds.): ICAFS 2020, AISC 1306, pp. 283–288, 2021.
https://doi.org/10.1007/978-3-030-64058-3_35

1. $\underline{\mu}_A(\alpha)$ is a bounded monotonic increasing left continuous function;
2. $\bar{\mu}_A(\alpha)$ is a bounded monotonic decreasing left continuous function;
3. $\underline{\mu}_A(\alpha) \leq \bar{\mu}_A(\alpha),\ 0 \leq \alpha \leq 1$.

A Fuzzy Mapping. A fuzzy mapping $f : X \times Y$ is a fuzzy set on $X \times Y$ with membership function $\mu_f(x, y)$.

Definition 2. A fuzzy function [4, 5] $f(x)$ is the fuzzy mapping of the fuzzy set, with membership function $\mu_{f(x)}(y) = \mu_f(x, y)$.
  The fuzzy set $f(A)$ on Y is a fuzzy mapping of a fuzzy set defined in the form:

$$\mu_{f(A)}(y) = \max_{x \in X}\{\min(\mu_A(X), \mu_f(x, y))\}, \quad \forall y \in Y$$

Definition 3. Fuzzy solution [3, 6]. Let the fuzzy goal G and the fuzzy restriction U be in the space of alternative X of tasks. Then the fuzzy set D formed by the intersection of G and U is called the solution. In the symbolic form $D = G \cap U,\ \mu_D = \mu_G \wedge \mu_U$.

3 Statement of the Problem

The problem of fuzzy optimal control in coke production is formulated as follows. To determine such values of the recirculation coefficient (U_1) for recycled raw materials (U_2), the temperature of light gas oil (U_3) and the temperature at the top of the reaction chamber (U_4), taking into account the technological limitations and coking within the given (X_1) density and (X_2) raw material consumption conditions to ensure that the amount of Y maximal coke during the transition from one state to another for a stable operating time of the reaction chamber.
  Mathematically, this problem can be formulated as follows: at given values of excitation effects X_1 and X_2, it is necessary to find such a fuzzy control sequence

$$\{U_0 = (U_{10}, \ldots, U_{40}), U_1 = (U_{11}, \ldots, U_{41}), \ldots, (U_{N1}, \ldots, U_{4N})\} \tag{1}$$

within the limits

$Y_{k+1} = 0.00005Y$
$+ (0.5/13.327 + 0.8/16.593 + 1.0/24.76 + 0.8/26.4177 + 0.5/19.915)Z_{1k}$
$+ (0.5/0.0035 + 0.8/0.015 + 1.0/-0.0236 + 0.8/-0.023 + 0.5/-0.0065)Z_{2k}$
$\quad + (0.5/-0.8245 + 0.8/-0.0465 + 1.0/-0.824 + 0.8/-0.820 + 0.5/-0.616)U_{1k}$
$+ (0.5/0.0052 + 0.8/0.004 + 1.0/0.0038 + 0.8/0.0041 + 0.5/0.0077)U_{2k}$
$+ (0.5/0.005 + 0.8/0.0051 + 1.0/0.005 + 0.8/0.0051 + 0.5/0.007)U_{3k}$
$+ (0.5/0.0041 + 0.8/0.006 + 1.0/0.0053 + 0.8/0.0045 + 0.5/0.0092)U_{4k}$
$Y_0 = 0$

$$\tag{2}$$

maximize the function

$$J = Y_N \tag{3}$$

U_{1k} - approximately should be in the interval $(1.2; 1.8)$;
U_{2k} - approximately should be in the interval $(495; 520)$;
U_{3k} - approximately should be in the interval $(515; 520)$;
$U_{4k} = 455$
$$\tag{4}$$

Or in the general

$$J = Y_N \rightarrow \max,$$

$$Y_{k+1} = f(X_{1k}, X_{2k}, U_{1k}, \cdots, U_{4k}, Y)$$

$Y_0 = 0$
U_{1k} - approximately should be in the interval $(1.2; 1.8)$;
U_{2k} - approximately should be in the interval $(495; 520)$;
U_{3k} - approximately should be in the interval $(515; 520)$;
$U_{4k} = 455$
$$\tag{5}$$

   Problem (2)–(4) is a fuzzy terminal control problem with a fixed time. The problem of optimal control of the coking process - (2)–(4) is solved in the following sequence. First, we define the membership functions of fuzzles and sets that express fuzzy goals. The solution of the problem of optimal control of the coking process - (2)–(4) is carried out in the following sequence. First, we define the affiliation functions of the fuzzy sets G^N and U_K, which express the fuzzy goal (2) and the constraints (4). Set the membership function as follows:

$$\mu_{G^N}(Y_N) = \exp(-\tau_0|Y_N - b_0|), \ \tau_0 > 0$$

Anologically, construct the membership function of the fuzzy constraint, $\mu_u(U_{ik})$:
$$\mu_u(U_{ik}) = \{1 + d_t(U_{ik} - b_i)^{r_i}\}^{-1}, \ r_i \succ 0$$
$$d_i(i = 1, 4) = \text{const}, \ r_j(j = 1, 4) = \text{const}$$

$$\mu_{G^{N-V}}(Y_{N-V}) = \max U_{N-V} \min\{\mu_{U_{N-V}}(U_{N-V}), \mu_{G^{N-V+1}}(Y_{N-V+1})\},$$
$$Y_{N-V+1} = f(Y_{N-V}, U_{N-V}), V = \overline{1, N}$$

4 Solution of the Problem

In this case, the solution of the problem (2)–(4) is defined as follows:

$$D = U_1 \cap U_2 \cap \cdots U_{N-1} \cap G^N \tag{6}$$

Where, G^N, $U_1, \ldots U_{N-1}$- is the reflection of Y_N in the area of situations.

In terms of the membership function, the solution of (6) can be expressed as follows:

$$\mu_D(U_D, \cdots, U_{N-1}) = \min\{(\mu_u(U_0), \cdots, \mu_{N-1}(U_{N-1}), \mu_{G^N}(Y_N)\} \tag{7}$$

In (7), the position of Y_N can be expressed as $Y_0, U_0, \cdots, U_{N-1}$ using Eq. (2).

Then the above problem can be expressed as follows: In (7) we find the sequence $\mu_{u_0}, \cdots, \mu_{u_{N-1}}$ that maximizes μ_D.

It is convenient to express the solution of Eq. (7) in the form $U_K^* = \prod(Y_K)$, $K = 0, \overline{N-1}$.

This solves the following problem.

$$\mu_D(U_0^*, \cdots, U_{N-1}^*) = \max u_0, \cdots, \max u_{N-1} \min\{(\mu_{u_0}(U_0), \cdots, \mu_{U_{N-2}}(U_{N-2}), \mu_{U_{N-1}}(U_{N-1}), \mu_{G^N}(f(Y_{N-1}, U_{N-1}))\} \tag{8}$$

We use a fuzzy dynamic programming method to solve Eq. (8). In this case, the system of recurrent equations takes the following form:

$$\mu_{G^{N-V}}(Y_{N-V}) = \max U_{N-V} \min\{\mu_{U_{N-V}}(U_{N-V}), \mu_{G^{N-V+1}}(Y_{N-V+1})\},$$
$$Y_{N-V+1} = f(Y_{N-V}, U_{N-V}), V = \overline{1, N} \tag{9}$$

Thus, maximization solution of $U_0^*, \cdots, U_{N-1}^*$ in the (10), is determined using the possible maximization values of U_{N-1}^*, in condition when this function defined as, U_{N-1}^*, Y_{N-V}. Equation (9) is solved by a fuzzy dynamic programming method.

The calculation of the optimal values of $U_i(i = \overline{1, 4})$ control parameters in the coking process is determined at the beginning of the reactor operation. The results of solving the problem (2)–(4) using (9) reflect the current and recommended optimal values of the parameters of the coking process (Table 1).

Table 1. Current and recommended optimal values of the parameters of the coking process

Stage N	Fuel density	Fuel consump	Recircul coefficient	Recircul temper	Gas oil temper	Upper chamber temper	Coke Production
Current regime							
23	0.987	165	1.14	505.0	489.0	445.0	
Optimal solution							
13	0.995	160	1.35	496.0	515.0	448.0	Approx.25
15	0.995	160	1.35	496.0	515.0	448.0	Approx.25
17	0.995	160	1.35	496.0	515.0	448.0	Approx.66
19	0.995	160	1.35	496.0	515.0	448.0	Approx.66
21	0.995	160	1.35	496.0	515.0	448.0	Approx.106
23	0.987	165	1.35	496.0	515.0	448.0	Approx.146
1	0.987	165	1.35	501.0	485.0	441.0	Approx.186
3	0.987	165	1.35	501.0	485.0	441.0	Approx.226
5	0.987	165	1.35	501.0	485.0	441.0	Approx.266
7	0.987	165	1.35	501.0	485.0	441.0	Approx.306
9	0.987	165	1.35	501.0	485.0	441.0	Approx.346
11	0.987	165	1.35	501.0	485.0	441.0	Approx.386
13	0.987	165	1.35	501.0	485.0	441.0	Approx.426
15	0.987	165	1.35	501.0	485.0	441.0	Approx.466

5   Conclusion

In this paper we have formulated the problem of optimal control of the coking process. At given density and the consumption of raw materials, it was determined such recycling factors as temperature of secondary raw materials, light gas oil temperature and top temperature of the reaction chamber that satisfies the technological constraint and ensures the transition of the coking process from one state to another for a fixed operating time reaction chamber, the maximum amount of coke accumulated in the reaction chamber. Numerical values of all optimal control parameters values of coke unit regime for every stage of coke processing are defined.

References

1. Zadeh, L.A.: Fuzzy sets. Inf. Control **8**, 38–353 (1965)
2. Bellman, R.E., Zadeh, L.A.: Decision making in a fuzzy environment. Manage. Sci. **17**(4), B 141–B 164 (1970)
3. Aliev, R.A., Aliyev, R.R.: Soft Computing and its Applications.World Scientific Publishing Company (2001)
4. Zadeh, L.A.: Similarity relations and fuzzy orderings. Inform. Sci. **3**(2), 177–200 (1971)
5. Zadeh, L.A.: Fuzzy sets as a basis for a theory of possibility. Fuzzy Sets Syst. **1**, 3–28 (1978)

6. Aliev, R.A., Mamedova, G.A., Aliyev, R.R.: Fuzzy Sets Theory and its Application, Tabriz University Press (1993)
7. Mamedova, G.A., Aliyeva, K.R.: Fuzzy optimal control of coke production. J. "Oil gas" Moscow **6,**80–84(1987)
8. Mamedova, G.A., Aliyeva, K.R. Fuzzy optimal coke production control. J. "Oil and gas" Moscow,**1,** 71–75(1988)

A Review on Recent Deep Learning-Based Computer-Aided Systems for Breast Cancer Diagnosis

Ali Işın[1(✉)] , Şerife Kaba[2] , and Ahmet İlhan[3]

[1] Department of Biomedical Engineering, Cyprus International University, Nicosia, Mersin 10, TRNC, Turkey
aisin@ciu.edu.tr
[2] Department of Biomedical Engineering, Near East University, Near East Boulevard, P.O. Box: 99138, Nicosia, Mersin 10, TRNC, Turkey
serife.kaba@neu.edu.tr
[3] Department of Computer Engineering, Near East University, Near East Boulevard, P.O. Box: 99138, Nicosia, Mersin 10, TRNC, Turkey
ahmet.ilhan@neu.edu.tr

Abstract. As one of the main unsolved problems of the current healthcare, cancer and one of its most frequent forms for women, breast cancer, is a popular research area in medicine. Early diagnosis is critical for improving the survival chances of breast cancer patients. Manual/visual diagnosis performed by a clinician is prone to many difficulties like high error rates and subjectivity. In addition, the time cost of the diagnosis procedure is very high and negatively effects the clinical routine. Automatic Computer-Aided Diagnosis (CAD) can be a solution to these problems. Recent research implementing deep learning-based techniques for developing an effective computer-aided breast cancer detection system proved to be very successful. In this paper, those recent deep learning-based automatic breast cancer detection systems and their performances are reviewed in detail for guiding the further research efforts in the field.

Keywords: Breast cancer · Deep learning · Computer aided diagnosis

1 Introduction

Cancer is one of the unsolved problems of the current generation of medicine. There are many different types of cancer, but especially for women around the world breast cancer is one of the most abundant and lethal forms [1, 2]. Although treatment methods for cancer improved over the recent years, early detection and diagnosis of breast cancer are still very important for improving patient treatment possibilities and for increasing the patient survival rates.

In current medical routine mammography scanning is the most used medical imaging modality for breast cancer diagnosis. In addition, histopathology images can also be used to perform diagnosis or to support the mammography-based diagnosis. Mammography-based diagnosis is carried out by the radiologists. Radiologist goes over the Mammography scans of the patient and tries to detect the cancerous tissue manually/visually. Since Radiologists manual (very detailed) involvement is required, the whole diagnosis procedure is very time consuming, heavily based on radiologist expertise and knowledge, and prone to errors with high inter-observer (between radiologists) variability. In order to provide fast and objective breast cancer detection and diagnosis, development of computer-aided breast cancer detection algorithms to assist the radiologist throughout the diagnosis procedure became popular research is in the field of computerized medicine [3].

Manual detection of breast cancer over mammography scans can be very problematic because the cancerous tissue can cover very small volume in the breast tissue thus in the corresponding scan. In addition, contrast levels representing the cancerous tissue on the scan can be very insignificant with respect to the surrounding tissue contrast levels. All these factors make visual detection of the cancerous tissue by the radiologist a very difficult process, which can result in many false positives and false negatives (which is very dangerous considering the case of cancer). Because of these disadvantages of manual procedures, the assistance provided by a robust and accurate automatic cancer detection system can assist radiologists heavily in the very critical case of a breast cancer diagnosis.

In previous years most of the common image processing methods, like edge detection, filtering and so on, along with traditional classification and traditional machine learning approaches have been implemented to solve the automatic cancer detection problems [3]. Although some of these methods have yielded effective results, with the emergence of deep learning-based methods they have been quickly outperformed and deep learning-based CAD methods have become the current state-of-the-art not only in cancer diagnosis but almost in all of the current medical diagnosis research [3–5].

In recent years, the leading deep learning techniques, outperformed most of the previous traditional automatic diagnosis methods and became the current state-of-the-art technique [3]. Deep learning techniques have very deep (many) processing layers when compared to traditional neural nets. These deep layers allow the network to learn very representative and complex features which showing better classification performances of cancer detection systems leading to better diagnosis results. Developing and implementing an effective method for the given goal is not an easy task for the researches. It requires high expertise and it is a very time-consuming process. Considering that the deep learning techniques carry out the given task automatically by itself, the burden of handcrafted processes is discarded, highly improving the automatic diagnosis performance.

In accordance with the aforementioned information, in this paper, recent advances in those state-of-the-art deep learning-based breast cancer diagnosis methods and their performances are reviewed to generate a knowledge base for the future research and improvement in this field.

2   Deep Learning-Based Breast Cancer Diagnosis Systems

Al-antari et al. [6] presented a CAD system for breast cancer diagnosis by a Deep Belief Network (DBN) which automatically determines the region of the breast masses and identifies them as benign, malignant or normal. In this study, a conventional digital mammography database is used to evaluate the DBN-based CAD framework for the diagnosis of breast cancer. 2 ROI extraction techniques are used. These are all mass regions of interest: ROIs and multiple mass ROIs. In the prior technique, 4 ROIs are randomly subtracted and the pixel size of a detected mass is equal to 32 * 32. In the following technique, all the determined breast mass is used. Both techniques have a total of 347 statistical characteristics to train and analyze the suggested CAD system. For classification purposes, quadratic order discriminant analysis, neural network classifiers and linear discriminant analysis are used as traditional techniques. In the final stage, the results obtained are compared using DBN. In this work, the Digital Database for Screening Mammography is used to perform the suggested system. This dataset is obtained from the University of South Florida and is accessible online for research objectives. The results show that the proposed DBN performs better than traditional classifiers. For the 2 ROI techniques, the overall accuracy of a DBN is calculated as 92.86% and 90.84%, respectively.

Al-antari et al. [7] presented a fully integrated CAD system for screening digital X-ray mammograms including classification, segmentation and detection of breast masses through deep learning approaches. In this study, the regional deep learning approach You-Only-Look-Once (YOLO) is used to determine breast mass from all mammograms. In the most difficult cases, YOLO approach can be applied to detect masses from all mammograms and perform better than other methods. A new deep network model, the Full-resolution Convolutional Network (FrCN) is suggested and used to segment breast masses. In this study, because of the proposed deep model segmentation capability, it can perform a better classification result with the CAD system by applying the FrCN method. In the final stage, a deep CNN is applied to identify the breast mass and categorize it as cancerous (malignant) or non-cancerous (benign). For the evaluation of the integrated CAD system, the publicly available INbreast database is used for classification, detection and segmentation accuracy. 4-fold cross-validation analyses are performed at each stage with verification, test and training datasets that are achieved by layered classification, to determine that each mammogram is tested equivalently and bias errors are prevented. The outcomes of the 4-fold cross-validation tests of the suggested CAD framework indicate that 97.62% Matthews Correlation Coefficient (MCC), 99.24% F1 score and 98.96% mass determination accuracy are obtained on the INbreast database. In addition, the breast mass segmentation outcomes with FrCN generated an overall accuracy of 92.97%, Dice (F1-score) of 92.69%, MCC of 85.93% and Jaccard similarity index based on 86.37%. The segmented and detected breast masses are analysed by CNN. 95.64% overall accuracy, 96.84 F1 score, 94.78% area under the receiver operating characteristic curve (AUC) and 89.91% MCC are obtained.

Saikia et al. [8] applied and analysed CNN approaches via classification models and a comparative study for breast Fine Needle Aspiration Cytology (FNAC) cell diagnosis. FNAC is mainly used to diagnose breast cancer, by conventional clinical practice in subjective visual evaluation of breast cytopathology cell sample images by microscope to assess the status of different cytological (pertaining to cell biology) properties. It provides rapid and minimally invasive tissue diagnosis without preserving the histological structure. In addition, the emergence of digital imaging and computation in the diagnosis may develop diagnostic accuracy and allow pathologists to decrease workload. Consequently, there are many difficulties to maintain the reproducibility and consistency of the findings. In this paper, a variety of CNN comparisons based on fine-tuned transfer learning strategy are used to identify and classify cell sample images. The proposed strategy is tested via GoogLeNet-V3, VGG19, VGG16 and ResNet-50, known as CNN architecture. FNAC cell images used in this study are collected from the laboratory of the health center with the help of medical doctors working at Ayursundra Healthcare Pvt. Ltd, Guwahati in India. This article provides a comparative evaluation of models that bring a new dimension to the FNAC application, which showed that GoogLeNet-V3 (fine-tuned) reached 96.25% accuracy.

Singla et al. [9] presented a fine-tuned pre-trained CNN Inception-v3 model via reverse active learning is applied to classify cancerous and healthy breast tissues. In this study, Optical Coherence Tomography (OCT) images are used. OCT is a three-dimensional, non-invasive, non-contact, high-resolution imaging technique approaching histopathology. Recently, various pilot analyzes have been performed to detect breast cancer using OCT, that involves imaging of human tissues, preclinical studies for the animal, and intraoperative clinical applications. OCT is also used in breast cancer margin evaluation. To assist OCT systems, automatic measurement of breast tissues is created with a machine learning model. The proposed method in this study is 91.7%, 90% and 90.2%, respectively, via testing datasets obtained for specificity, accuracy and sensitivity.

Feng et al. [10] presented a CAD system to classify breast cell nuclei as benign and malignant in histopathology images. This system can be summarized in five stages. The first stage is the sampling which divides the unlabeled images taken from both subsets of the dataset (benign and malignant) to 15 * 15 sized patches. The second stage is the feature extraction which extracts the high-level features of sampled patches using deep NN based Stacking Denoising Autoencoder (SDAE). The third stage is the classification of the sampled images as with and without cell nuclei using Softmax classifier which is the last layer of the SDAE. In the fourth stage, the network is re-trained with the labeled data which are based on ground truth patches to update the weights using Softmax classifier. In the last stage, the whole network is fine-tuned by the Back-propagation algorithm using the updated weights. The proposed system is tested on the Breast Cancer Cell (BCC) dataset which is obtained from the University of California. Four experiments are carried out to test the system. In the first experiment, the feature extraction performance of the system is measured. The result is 98.28% for the benign and 90.54% for the malignant cases. In the second experiment, the performance of the system is compared with 7 other methods. The result is that the system is clearly better than the other methods with 98.27% for benign and 90.54% for malignant cases. As the third experiment, the second experiment is repeated using insufficient data to test the performance of the system. 10% of the data used in the second experiment as input. The result is 97.98% for benign and 88.37% for malignant cases. Forth experiment is conducted by adding noise to the input data to test the robustness of the system. The result is that the proposed is better than all other classification methods using additional noise ratio between 0.1–0.5 and random pixel corruption ratios between 0.1–0.5.

Wang et al. [11] presented a CAD system to detect and classify breast masses in mammography images. This system can be summarized in three stages. The first stage is the preprocessing which enhances the images using adaptive mean filter and contrast enhancement algorithms. By using these algorithms, noises are reduced and the contrast between the surrounding tissues and the masses in question are increased. The second stage describes the proposed mass detection algorithm. This algorithm is divided into four parts. In the first part, the first and last non-zero pixels of each row and column are found using sequential scanning to determine the ROI. In the second part, the ROI is segmented into fixed rectangular (48 * 48) sub-regions using a sliding window. In the third part, deep features of the sub-regions are extracted using CNN. In the last part, Unsupervised Extreme Learning Machine (US-ELM) algorithm is used to cluster deep features as suspicious and non-suspicious mass areas. The last stage is divided into two parts. In the first part, in addition to deep features, morphological, texture and density features of breast masses are extracted to achieve a better classification rate. The second part is the classification of the breast masses as benign and malignant using ELM classifier. To evaluate the proposed system performance, accuracy, sensitivity, specificity, benign accuracy, malignant accuracy and AUC are calculated as 86.50%, 85.10%, 88.02%, 88.50%, 84.50% and 92.3% respectively.

Li et al. [12] presented a CAD system to classify breast histology images as normal, benign and malignant (in situ carcinoma or invasive carcinoma) using CNN. This system can be summarized in three stages. In the first stage, different sized patches named as smaller (128 * 128) and larger (512 * 512) including cell-level and tissue-level features are extracted from the images respectively. In the second stage, discriminative patches are extracted using patches' screening method which is based on the k-means clustering algorithm. The reason of this is that some of the sampled cell-level patches do not include sufficient information to match the image tag. In the last stage, a classifier is created to train each of the wholes of the image for classification. The proposed system is tested on Bioimaging Challenge 2015 Breast Histology dataset. To evaluate the system performance; precision, recall and F-score are calculated for each case. The performances of the system for normal cases are 87.5%, 78% and 82.5% respectively. The performances of the system for benign cases are 75%, 100% and 85.7% respectively. The performances of the system for in situ cases are 100%, 89% and 94.2% respectively. The performances of the system for invasive cases are 100%, 89% and 94.2% respectively. The overall accuracy of the system is 88.89%.

Ragab et al. [13] presented a CAD system to detect and classify breast tumors as benign and malignant in mammography images. This system can be summarized in three stages. The first stage is the image processing which includes an image enhancement technique and two different tumor segmentation approaches. Contrast-limited Adaptive Histogram Equalization (CLAHE) is used to enhance the images by improving the contrast of the image. In the first segmentation approach, the tumor region is cropped manually from the enhanced image using circular contours. In the second segmentation approach, thresholding and region-based segmentation techniques are applied. By using these techniques, the grayscale image is converted to the binary image and all objects excluding the largest one (tumor region) is removed by counting the number of pixels of each object. In the second stage, the object removed image is multiplied with the original image to convert the tumor region to its original pixel values. The proposed system is tested on the Digital Database for Screening Mammography (DDSM) and the Curated Breast Imaging Subset (CBIS) of DDSM. Above mentioned techniques are applied for only DDSM. The reason of this is that the images in CBIS-DDSM are already segmented. The last stage is divided into two parts. The first part is the feature extraction which extracts the features of the tumors using deep CNN. The second part is the classification of the tumors. In this part, the Support Vector Machine (SVM) is joined with the fully connected layer of the deep CNN to achieve a better classification rate. To evaluate the proposed system performance, AUC and accuracy are calculated. The AUC for both segmentation approaches applied to DDSM is 88%. The AUC for the CBIS-DDSM is 94%. The accuracy of the system for both segmentation approaches applied to DDSM is 79.1% and 80.9% respectively. The accuracy of the system for CBIS-DDSM is 87.2%.

Duraisamy and Emperumal [14] presented a CAD system to classify breast tumors in digital mammograms. This system can be summarized in three stages. The first stage is the ROI (tumor region) segmentation using the Chan-Vese level set based segmentation method. The second stage is divided into two parts. The first part is the preprocessing which includes cropping, patch extraction, normalization and data augmentation. The ROI is cropped from the image using the bounding box algorithm. The cropped ROI is divided into patches with varying sizes as 9 * 9, 13 * 13, 17 * 17 and 22 * 22. The patches are normalized by subtracting the mean value of the pixels from each pixel of the image. For data augmentation, additional images are created from the original input data by rotating the images 90°, 180° and 270° orientations and flipping them. This is needed because CNN requires a large dataset for training. In the second part, the features of the patches are extracted using deep CNN. The last stage is the classification of the tumors using Fully Complex-valued Relaxation Network (FCRN) classifier. To evaluate the proposed system performance, accuracy, sensitivity, specificity and AUC are calculated as 99%, 98.75%, 100% and 98.15% respectively.

Ribli et al. [15] presented a faster Region-CNN based CAD system to detect and classify lesions using mammogram images. The data used in this study consist of two public databases. These databases are Digital Database for Screening Mammography (DDSM) and INbreast database at Semmelweis University in Budapest. The database is used for the training include benign or malignant lesions. In this paper, the heart is chosen as the model, and the object detection frame is a version of the Faster R-CNN37. Faster Region-CNN relies on a CNN via additional constituents to detect, localize and classify objects in an image. In this study, the model is created using VGG16, which is named a 16-layer deep CNN. The last layer can detect malignant or benign lesions in images. For each identified lesion, the output of the model is a bounding box and a score reflecting confidence in the lesion class. The maximum score of all malignant lesions detected in the image is calculated to define a single-score image. Average scores of specific images are taken for various images of the same breast. The proposed approach determined the classification performance in the public INbreast database with an AUC equal to 0.95. In the final, validation dataset and ranked 2nd in the Digital Mammography DREAM Challenge with an AUC equal to 0.85.

3 Conclusion

Breast cancer still poses a great danger for women health all around the world. Early diagnosis is life-saving and integration of CAD methods into the current clinical routine is an important step towards reducing the time and overall costs of diagnosis, providing objective diagnostic results along with improving overall diagnosis performances of the radiologists. In this paper, as the current state-of-the-art for CAD, deep learning-based methods for breast cancer detection are collected under a review article in detail guiding the interested researchers in their future research. As recently investigated literature proves, deep learning will lead towards the development of a commercially successful CAD system and remain very effective and popular for solving the problems faced in the field of CAD.

References

1. American Cancer Society: Breast Cancer Facts & Figures. American Cancer Society Inc, Atlanta (2015)
2. Eurostat. Health Statistics: Atlas on Mortality in the European Union; Office for Official Publications of the European Union: Luxembourg (2009)
3. Işın, A., Direkoğlu, C., Şah, M.: Review of MRI-based brain tumor image segmentation using deep learning methods. Procedia Comput. Sci. **102**, 317–324 (2016). https://doi.org/10.1016/j.procs.2016.09.407
4. Işın, A., Ozdalili, S.: Cardiac arrhythmia detection using deep learning. Procedia Comput. Sci. **120**, 268–275 (2017). https://doi.org/10.1016/j.procs.2017.11.238
5. Işın, A., Sharif, T.: Deep learning for lung lesion detection. In: Aliev, R.A., Kacprzyk, J., Pedrycz, W., Jamshidi, M., Sadikoglu, F.M. (eds.) ICAFS 2018. AISC, vol. 896, pp. 799–806. Springer, Cham (2019). https://doi.org/10.1007/978-3-030-04164-9_105
6. Al-antari, M.A., Al-masni, M.A., Park, S.-U., Park, J., Metwally, M.K., Kadah, Y.M., Han, S.-M., Kim, T.-S.: An automatic computer-aided diagnosis system for breast cancer in digital mammograms via deep belief network. J. Med. Biol. Eng. **38**(3), 443–456 (2017). https://doi.org/10.1007/s40846-017-0321-6
7. Al-antari, M.A., Al-masni, M.A., Choi, M.T., Han, S.M., Kim, T.S.: A fully integrated computer-aided diagnosis system for digital X-ray mammograms via deep learning detection, segmentation, and classification. Int. J. Med. Inform. **117**, 44–54 (2018). https://doi.org/10.1016/j.ijmedinf.2018.06.003
8. Saikia, A.R., Bora, K., Mahanta, L.B., Das, A.K.: Comparative assessment of CNN architectures for classification of breast FNAC images. Tissue Cell **57**, 8–14 (2019). https://doi.org/10.1016/j.tice.2019.02.001
9. Singla, N., Dubey, K., Srivastava, V.: Automated assessment of breast cancer margin in optical coherence tomography images via pretrained convolutional neural network. J. Biophotonics **12**(3) (2019). https://doi.org/10.1002/jbio.201800255
10. Feng, Y., Zhang, L., Yi, Z.: Breast cancer cell nuclei classification in histopathology images using deep neural networks. Int. J. Comput. Assist. Radiol. Surg. **13**(2), 179–191 (2017). https://doi.org/10.1007/s11548-017-1663-9
11. Wang, Z., Li, M., Wang, H., Jiang, H., Yao, Y., Zhang, H., Xin, J.: Breast cancer detection using extreme learning machine based on feature fusion with CNN deep features. IEEE Access **7**(105), 146–105158 (2019). https://doi.org/10.1109/ACCESS.2019.2892795

12. Li, Y., Wu, J., Wu, Q.: Classification of breast cancer histology images using multi-size and discriminative patches based on deep learning. IEEE Access **7**, 21400–21408 (2019). https://doi.org/10.1109/ACCESS.2019.2898044
13. Ragab, D.A., Sharkas, M., Marshall, S., Ren, J.: Breast cancer detection using deep convolutional neural networks and support vector machines. PeerJ 7 (2019). https://doi.org/10.7717/peerj.6201
14. Duraisamy, S., Emperumal, S.: Computer-aided mammogram diagnosis system using deep learning convolutional fully complex-valued relaxation neural network classifier. IET Comput. Vis. **11**(8), 656–662 (2017). https://doi.org/10.1049/iet-cvi.2016.0425
15. Ribli, D., Horváth, A., Unger, Z., Pollner, P., Csabai, I.: Detecting and classifying lesions in mammog rams with deep learning. Sci. Rep. **8**(1), 1–7 (2018). https://doi.org/10.1038/s41598-018-22437-z

Hybrid Modeling Method for the Study of Socio-Economic Systems

Imran G. Akperov[(⊠)] [ID]

PEI HE SU (IMBL), Rostov-on-Don, Russia
rector@iubip.ru

Abstract. The use of models with a hybrid architecture for data mining in complex socio-economic systems is considered. Approaches to identifying the properties of geo-information space: adaptability, self-learning, self-adjustment and sustainability of development are proposed.

Keywords: Complex systems · Geo-information space · Interoperability · Soft computing · Hybrid models

1 Introduction

The main advantages of the geo-information spatial approach in the study of distributed economic, social, environmental, and other processes include the possibility of using the advantages of measurement connectivity, a systematic approach, and the effect of self-organization.

Currently, there is a serious instrumental base of simulation of the spatial organization of human activity, the main regularities of individual information sectors of the space [1, 2], transformation of its individual components in a competitive digital economy.

P. A. Minakir noted that: "Spatial economy is a form of existence of the economy as a set of interacting economic agents, distributed in a certain way in geographical space (GIP). At the same time, an economic agent means an individual who participates in at least one of the processes of production, exchange, and consumption" [3].

In the course of system analysis, "methods of lexical and semantic modeling of cognitive knowledge structures were identified that allow us to take into account the features of the geo-information environment" [4].

Methods and approaches of *fuzzy lexical and semantic modeling*, as shown in the work [3], can be successfully applied in the design of the architecture of the information space, the formation of the structure of technical and software tools for the intellectual analysis of geo data and the integration of thesaurus-type linguistic support in a single GIP.

This approach is designed to study methods and approaches of computational linguistics to create a new generation of information technology (IT) GIS-oriented. However, there is a principle of incompatibility L. Zade: "As the complexity of a system increases, our ability to formulate precise, meaningful statements about its behavior decreases to a certain threshold, beyond which accuracy and meaning become mutually exclusive" [5].

At the same time the assessment of the adequacy of the model to the actual object of management in socio-economic systems is related to the set of accepted restrictions on the studied (in fact, fuzzy) dynamic system.

System studies of the socio-economic component (SEC) of the UGIS suggest that "as a result, both the conditions for its formation will be optimized, and also the effectiveness of further functioning and development" [6]. Special attention in SEC research "should be paid to the study of the causal relationships of the behavior of the socio-economic system and the identification of its structure and properties that will ensure the effective implementation of the goals of the activities" [7].

2 Chaos in the Socio-Economic System

Real socio-economic systems have an almost complete set of "NON-factors: inaccuracy, vagueness, incompleteness, undefinability, etc." [8]. However, not all of them can be taken into account using traditional probability theory and mathematical statistics [9.10]. For complex modern systems such as System of Systems (SOS), which, of course, include SEC, which have "strange" attractors in their phase portraits (so-called «butterfly effect») [11], there is a chaos caused by "deterministic randomness", and, consequently, by the nonlinearities of the general model and a certain set of initial conditions [11, 12]. Even the slightest disturbance of such a complex system as the SEC, for example, the weather, not to mention the economic crisis or coronovirus, can lead to a chain of events leading to complete unpredictability. There is no alternative to using soft models in this situation.

In the course of modeling, individual properties can be aggregated, for example, by manageability, to ensure clarity and speed in obtaining the result. Among the properties of complex nonlinear SEC, the properties of dynamism, flexibility and adaptability are of particular importance, which, in turn, determine stability and, ultimately, efficiency [11].

The research of UGIS should be based on the principle of consistency in order to consider each layer as an aggregate of components of semantically related types that have heterogeneous properties but co-exist in a certain cognitive space.

All types of spaces of UGIS have "a number of common properties: the length in different directions, the mutual location of space objects, nodes (centers), networks, etc. The most important advantage of the spatial approach is the possibility of a multi-dimensional representation of a spatially localized system" [6].

3 Lyapunov's Time

Lyapunov time is the time during which the system comes to a state of chaos. In other words, this is the time during which you can predict the behavior of the system (the "non-chaos" time). It is possible to calculate if to use the Lyapunov's exponent for the dynamic SEC, that is, the speed with which the two points in phase space converge or move away from each other:

$$\Lambda = \lim_{n \to \infty} \frac{1}{n} \sum_{k=0}^{n} \log_2 \left| \frac{df(x_k)}{dx_k} \right|.$$

The meaning of this indicator: when the distance $df(x_k)$ changes at the k-step in comparison with the corresponding parameter dx_k in a larger direction, the value of the Lyapunov's exponent $\Lambda > 0$, it means that there is chaos or instability of the SEC system. When $df(x_k)$ is at the k-step, in comparison with the corresponding parameter dx_k in the smaller direction, the logarithm of a number less than one is negative, $\Lambda \leq 0$ - SEC - is stable.

Modeling of Socio-economic Systems
The solution of this type of multi-criteria modeling problems in the UGIS involves the use of mathematical systems that describe the main processes of functioning of this SoS.

As a hybrid, they usually use [4, 13]:

– neuro computing + fuzzy logic (NF);
– fuzzy logic + chaos theory (FCh);
– neural networks + chaos theory (NCh), etc.

4 Modeling

Let's consider an example of chaos formation in an absolutely deterministic Lorentz model in the Cauchy form [2], given by a system of three nonlinear first-order differential equations.

«The next slide shows a phase portrait of the system's behavior, where the presence of a strange attractor is obvious even visually. Such qualitative research is convenient for rapid analysis of the state of society for the operational forecast of chaos.

For a more accurate forecast, it is desirable to obtain information about the stability margin of the current trend.

The algorithm for quantifying the state of the system model assumes the following sequence of actions» [15]:

1) forming the Cauchy model, for example:

$$\begin{cases} \dot{x} = -k(x - y) \\ \dot{y} = -xz - y + px \, ; \\ \dot{z} = -qz + xy \end{cases}$$

2) obtaining a Jacobi matrix composed of partial derivatives of the right-hand sides of the corresponding differential equations

$$
\begin{bmatrix}
-k & k & 0 \\
-z+p & -1 & -x \\
y & x & -q
\end{bmatrix}
=
\begin{bmatrix}
-10 & 10 & 0 \\
-z+28 & -1 & -x \\
y & x & -11
\end{bmatrix};
$$

3) substitution of initial conditions

$$
\begin{bmatrix}
-10 & 10 & 0 \\
-z_0+28 & -1 & -x_0 \\
y_0 & x_0 & -11
\end{bmatrix};
$$

4) finding the eigenvalues of the resulting numerical matrix - for $n = 3$ there will be 3 of them: $\mu 1$, $\mu 2$, $\mu 3$;

5) finding first-order Lyapunov exponents as the real part of the eigenvalue $\lambda_j^1 = \mathrm{Re}\ \mu_j^1$;

6) finding a one-dimensional Lyapunov exponent $\lambda_1 = \max_j \lambda_j^1$;

7) finding $\lambda_1^2 = \lambda_1^1 + \lambda_2^1$; $\lambda_2^2 = \lambda_1^1 + \lambda_3^1$; $\lambda_3^2 = \lambda_3^1 + \lambda_2^1$;

8) finding $\lambda_2 = \max_j \lambda_j^2$;

9) finding $\lambda_1^3 = \lambda_1^1 + \lambda_2^1 + \lambda_3^1$

In this case, the information space of the socio-economic system is formed, within which the zones of its predictive behavior are found, using Lyapunov indicators [4].

5   A Model for Evaluating UGIS as a Complex Information SoS

The complexity of modeling SEC as a subsystem of the UGIS is due to [3.14]:

- the complexity of the research object, the non-linearity and undefinability of processes and initial conditions, the presence of threshold effects, bifurcations and time lags (differential models of SEC with a delay);
- the effect of interaction of SEC model variables, which mostly implement NON-factors;
- the complexity of measuring fuzzy variables of the model;
- fuzzy and unstable relationships in the model;
- significant influence of the human factor on all socio-economic processes.

As a result of "hybridization of methods of intellectual data processing, combining several artificial intelligence technologies, the term soft computing appeared" [8], which was introduced by L. Zadeh in 1994. "Soft computing is a set of computational methodologies that provide a framework for understanding, designing, and developing intelligent systems. Soft Computing combines areas such as probabilistic reasoning and evolutionary algorithms, artificial neural networks (NN), and fuzzy logic (FL). These areas complement each other and are used in various combinations to create hybrid intelligent systems" [9].

Hybrid models that combine the main advantages of soft computing can be used to reliably assess the stability of geo-information SEC. "Among them, we will highlight such methods that can implement the properties of adaptability and the ability to learn, self-tune. These are primarily neural networks and fuzzy logic. Both technologies are modeling tools and work after the learning or knowledge extraction stage. Neural networks are used in cases when dependent and independent variables are connected by complex nonlinear relations" [13], therefore, they have the ability to generate chaos.

The general "structure of a system using fuzzy logic and neural networks contains the main blocks, the synergetic effect of their joint interaction determines the intelligence of the system: the knowledge base, the decision block, the blocks of fuzzification $((\mu_A^i(x_j)))$, aggregation and defuzzification $(y_k\,(x))$" [6] (Fig. 1).

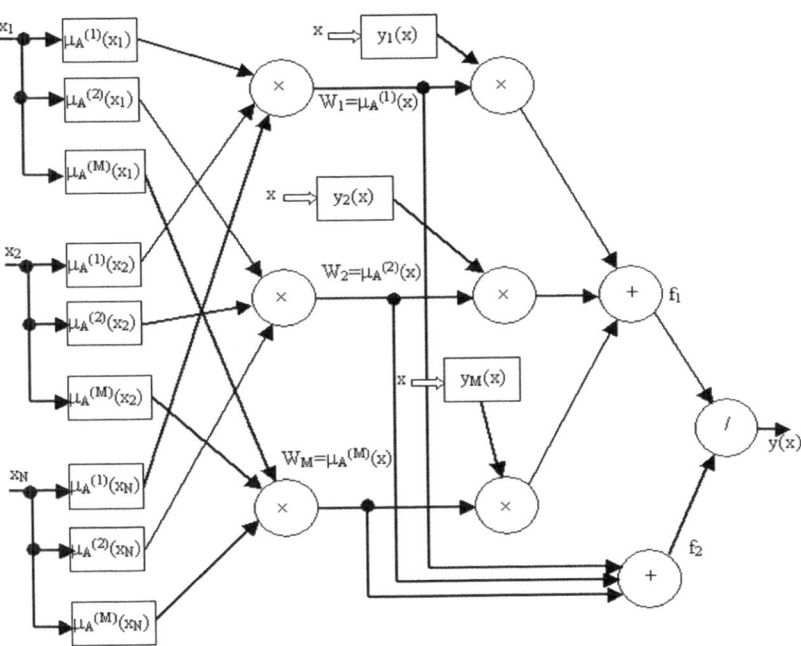

Fig. 1. Classic neuro-fuzzy (NF) process research model [2, 3]

6 Reservoir Calculations

Classical neural networks consist of an input vector, several layers of neurons connected to each other in a certain way, and an output vector. Each neuron is a function of a linear combination of inputs (Fig. 2a).

The network learning process - is a reduction in the network output error relative to the expected output of the training sample [26]. The problem of reducing the learning error or optimization is solved by adjusting the coefficients of a linear combination on each of the neurons, using a training sample, using one of the modifications of the gradient descent method. The more layers, the longer the setup takes.

Reservoir calculations are based primarily on the use of output, final layers in a multi-layer neural network. In other words, you don't need to configure the internal layers. There was a so-called "dynamic reservoir" of nonlinear neurons connected to each other, in general, randomly. The tank has an entrance and exit. The output is a simple layer of linear neurons.

And the reservoir, in fact, is an extensive set of different nonlinear functions, from which you can "collect" any function that is needed at the moment.

This approach has many interesting properties. For example, "you can attach different output layers to the same reservoir and thus solve different tasks. In other words, the reservoir itself is homogeneous, not configured for a specific task, and can be used for anything. It is assumed" [15] that our brain works in a similar way.

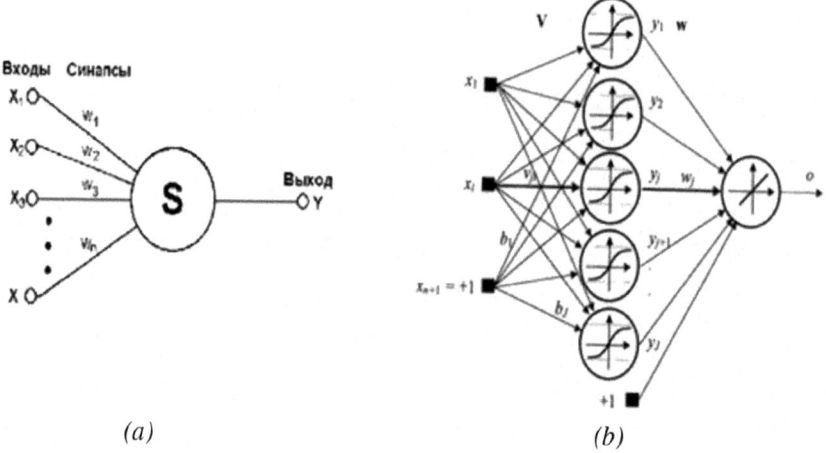

(a)   (b)

Fig. 2. Neural network as a vector operator [3]

7 Conclusion

A single geographic information space can be considered as a set of interconnected and interacting layers in the course of its functioning. Since UGIS is a complete system, its model in the form of a knowledge base must also have this property.

Research has shown that the use of hybrid models with fuzzy, chaotic, and neural models can provide these conditions.

References

1. Bagiev, G.L., Pinchuk, A.V., Serova, E.G., Shulga, A.O.: On the formation of the concept of spatial interaction. Probl. Modern Econ. **4**(44), 219–225 (2012)
2. Akperov, I., Khramov, V., Lukasevich, V., Mityasova, O.: Fuzzy methods and algorithms in data mining and formation of digital plan-schemes in earth remote sensing. In: 9th International Conference on Soft Computing, Computing with Words and Perceptions, pp. 120–125. Hungary (2017)
3. Minakir, P.A.: Economic analysis and measurement in space. Spatial Econ. **1**, 12–39 (2014)
4. Khramov, V.V.: Decision-making problems in the study of multi-criteria objects in a fuzzy information environment. In: The Proceedings: Problems of Ensuring the Effectiveness and Stability of Complex Technical Systems. Materials of the XXI Interdepartmental Scientific and Technical Conference, pp. 120–124. Russia (2002)
5. Zadeh, L.A.: Fuzzy logic, neural networks, and soft computing. Commun. ACM **37**(3), 77–84 (1994)
6. Suslov, V.I.: Measuring the Effects of Interregional Interactions: Models, Methods, Results. Nauka, Novosibirsk (2014)
7. Aliev, R.A., et al.: IUS of Gas Field Facilities: Current State and Development Prospects. Nedra, Moscow (2014)
8. Narinyani, A.S.: Underdetermination in systems of representation and processing of knowledge. Izv. USSR Academy of Sciences. Techn. Cybern. **5**, 3–28 (1986)
9. Khramov, V.V.: Intellectual Information Systems. Data Mining. Rostov-on-Don, FSBEI of HE RSTU (2012)
10. Khramov, V.V.: Method of aggregation of several sources of fuzzy information. Izvestiya TSURE **3**(21), 52–53 (2001)
11. Serova, E.G.: Formation of a stable adaptive architecture of the digital information system in the conditions of spatial economy. Collection of scientific articles for the scientific and practical round table "Synergy and logistics in the innovative development of the Russian economy", vol. 179. Scientific works of the Free economic society (2013)
12. Khramov, V.V.: Theory of information processes and systems: educational and methodological guide. Rostov-on-Don (2011)
13. Khramov, V.V.: Computer Modeling. Manual for course and diploma design. M, MO (1992)
14. Akperov, I.G., Kramarov, S.O., Povkh, V.I., Khramov, V.V., Radchevsky, A.N.: Patent RU No. 2612326, Method for forming a digital plan-scheme of agricultural objects and a system for its implementation. http://www.spsl.nsc.ru/fulltext/EXPOS/%D0%A7%D0%97_7/Ukaz atel%20patentov.pd
15. Khramov, V.V.: Methodology of representation of territories in target sounding of the Earth from space. Collection of scientific works "Public-private partnership and public procurement in the system of implementation of import substitution policy in Russia" (based on the materials of the V International socio-economic Forum Intellectual resources-regional development). Part 2, pp. 142–148. Rostov-on-Don, SU (IMBL) (2016)

On the Analytical Investigation of Hybrid Transmission NOMA (H-NOMA)

Hüseyin Haci[1(✉)] and Joydev Ghosh[2]

[1] Department of Electrical and Electronic Engineering, Near East University, Mersin 10, Turkey
huseyin.haci@neu.edu.tr
[2] School of Computer Science and Robotics, National Research Tomsk Polytechnic University, Tomsk, Russia
joydev.ghosh.ece@gmail.com

Abstract. Non-orthogonal multiple access (NOMA) is shown to be the optimal channel access method and a strong candidate to be employed at the fifth generation (5G) and beyond networks. This paper studies direct transmission (DT) and cooperative transmission (CT) modes of operations in NOMA communications and proposes an investigation on evolving a cooperative transmission NOMA (C-NOMA) into a Hybrid transmission NOMA (H-NOMA) that can be used for design and deployment of relay based wireless networks, such as networks for Internet of Things (IoT) applications.

Keywords: Rician fading channel · NOMA · Direct transmission mode · Cooperative transmission mode · Shannon capacity · Cumulative Distribution Function (CDF)

1 Introduction

Smart world applications such as Internet of Things (IoT) and Cloud based applications require high-capacity and reliable data transmissions between end-points and gateways [1]. Channel access method is a crucial technology that determines the performance of communication systems. Non-orthogonal multiple access (NOMA) is shown to be the capacity achieving channel access method [2]. Thus, NOMA is envisioned as a strong candidate for the channel access method of the fifth generation (5G) and beyond networks [3]. Cooperative communications are the concept of user-equipment (UE) not acting only as the receiver of the data transmissions but also as a helper (i.e. relay for multi-hop communications). Cooperative communications can significantly increase the communications performance through the diversity gain [4]. Further, NOMA can be employed together with cooperative communications without need for any hardware change. Employment of NOMA can increase the diversity gain further, since each UE can relay multiple signals simultaneously by superimposing them on a single symbol. This makes cooperative NOMA communications a popular research topic. In [4] researchers proposed a novel transmission scheme for cooperative NOMA communications where a priori information available to the user with better channel condition is exploited to assist in other NOMA users' symbol detection. It has been shown that the

© The Author(s), under exclusive license to Springer Nature Switzerland AG 2021
R. A. Aliev et al. (Eds.): ICAFS 2020, AISC 1306, pp. 304–311, 2021.
https://doi.org/10.1007/978-3-030-64058-3_38

outage probability and diversity order achieved are significantly improved compared to conventional cooperative communications. In [3] authors studied optimal power allocation for cooperative NOMA for cellular multiple-input multiple-output (MIMO) downlink transmissions. User pairing shown to decrease the complexity. The performance investigation showed that by coordinating inter-beam and inter-frequency power allocation, the cell-edge users' performance can be significantly improved with a minor trade-off to cell-average users' performance. In [2] researchers proposed a cooperative relay system using NOMA and a sub-optimal power allocation scheme that can approach to the optimal scheme's performance when SNR is high. A significant increase in spectral efficiency is demonstrated compared to the cooperative relay systems that use orthogonal multiple access.

Despite interesting research, all the aforementioned researchers studied Rayleigh fading channels. Rayleigh channels are a special category of Rician fading channels and can only be practical when there is no line-of-sight signal. However, in today's networks there are many Femtocell and other types of Smallcell base-stations deployed where there is line-of-sight signal with high probability. Thus, providing mathematical analysis that can be realized for channels with line-of-sight signals is important. Rician fading is a more generalized channel model compared to Rayleigh fading and can be used for channel models where there is a line-of-sight signal. Accordingly, the contributions of this paper are listed as follows:

- This paper studies and derives mathematical analysis for H-NOMA communications for Rician fading channels.
- Also, proposes a performance investigation model that is developed by using the conditional modes of operations for NOMA communications. Cumulative distribution function (CDF) analyses are shown to match very well with the computer simulations.

The rest of the paper is organized as follows. In Sect. 2, the method is described. The results are provided in Sect. 3 and relevant discussion is done in Sect. 4. Section 5 concludes the paper.

2 System Model

In Fig. 1, we assume a two-user NOMA scenario with a source (S) and two users for downlink transmission, where S_1 is used to denote near user and S_2 is used to denote far user. Therefore, S_1 and S_2 are considered as strong and the weak users, respectively. The performance is evaluated over Rician fading channels with the channel co-efficient sh_1,h_2, h_3 for the links S to S_1, S to S_2 and S_1 to S_2, respectively, where $h_i \sim \mathcal{CN}(0, \Omega_i)$.

The transmission principle of the proposed Hybrid transmission NOMA (H-NOMA) scheme follows two conditional modes of transmissions, such as direct transmission (DT) mode and cooperative transmission (CT) mode. At first, S broadcasts a reference signal to S_1 and S_2 to estimate the channel state information (CSI) of the received signals. By following the received signal, the comparison of estimated CSI (denoted by δ_{CSI}) between S_2 and S with set threshold[1] has been computed and

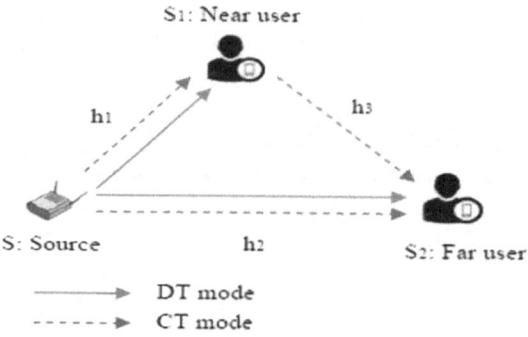

Fig. 1. The system model of H-NOMA transmission

Then S_2 decides whether it can successfully decode its required symbol via direct transmission (DT) mode. If channel estimation is higher than γ_{th}, then S_2 sends 1-bit binary acknowledgments[2] to S and S_1 to adapt DT mode of operation. And if channel estimation is lower than γ_{th} which implies it is unsuitable to decode its required symbol successfully, then S_2 sends 1-bit binary acknowledgments to S and S_1 to adapt CT mode of operation.

2.1 Channel Distribution

In this paper, all estimated channel coefficients are modelled under Rician fading condition as it can mitigate self-interference (SI) link due to the short distance between antennas. Therefore, the probability density function (PDF) in the context of Rician fading is formulated as,

$$f_{pdf}((\rho|v), \varphi) = \begin{cases} \frac{2\rho}{\varphi}\exp\left(\frac{-\rho^2-v^2}{\varphi}\right)I_0\left(\frac{2\rho v}{\varphi}\right) for\ \rho \geq 0, \varphi \geq 0 \\ \frac{1}{\sqrt{\pi\varphi}}\exp\left(-\frac{(\rho-v)^2}{\varphi}\right) for\ K \gg 1 \end{cases} \tag{1}$$

where, $I_0(.)$ is a zero-order modified Bessel function. The shape parameter, $K = \frac{v^2}{\varphi}$ defined as the ratio of the power contributions by line-of-sight path to the remaining multipaths.

The Rician distribution can be written as, $R \sim Rice(|v|, \varphi)$ if $R = \sqrt{X^2 + Y^2}$ where $X \sim N(v\cos\theta, \varphi)$ and $Y \sim N(v\sin\theta, \varphi)$ are statistically independent normal random variables and $\theta = \{\frac{\pi}{2}, \frac{\pi}{3}, \frac{\pi}{4}\}$.

2.2 DT Mode of Operation

Initially S broadcasts $(\sqrt{a_1 P_t}x_1 + \sqrt{a_2 P_t}x_2)$ to S_1 and S_2, where x_1 and x_2 are the unit power signals for S_1 and S_2 respectively, P_t denotes total transmit power, a_1 and a_2 are the corresponding power allocation coefficients of S_1 and S_2 respectively. We also assume that $a_1 \geq a_2$ with $a_1 + a_2 = 1$ to keep improved user fairness alongside quality of service (QoS). Hence, the received signals at S_1 and S_2 during DT mode of operation can be respectively given by,

$$y_1 = h_1\left(\sqrt{a_1 P_t}x_1 + \sqrt{a_2 P_t}x_2\right) + N_{01}, \tag{2}$$

$$y_2 = h_2\left(\sqrt{a_1 P_t}x_1 + \sqrt{a_2 P_t}x_2\right) + N_{02}, \tag{3}$$

Where N_{01} and N_{02} denote the additive white Gaussian noise (AWGN) with zero mean and variance σ^2 at S_1 and S_2 respectively. The S_2 only decodes x_1 from (3) by treating x_2 as an unwanted signal, while S_1 acquires x_2 from (1) applying successive interference cancellation (SIC) technique. Therefore, the SINR for x_1 and SNR for x_2 at S_1 can be respectively expressed as,

$$\gamma_1 = \frac{|h_1|^2 a_1 P_t}{|h_1|^2 a_2 P_t + \sigma^2}, \tag{4}$$

$$\gamma_2 = \frac{|h_1|^2 a_2 P_t}{\sigma^2}. \tag{5}$$

The SINR for x_1 at S_2 can be expressed as,

$$\gamma_3 = \frac{|h_2|^2 a_1 P_t}{|h_2|^2 a_2 P_t + \sigma^2}. \tag{6}$$

2.3   CT Mode of Operation

In this mode of operation transmission is generally perform in two phases. From Fig. 1, the overlapping arrows between DT mode and CT mode are assumed to be 1st phase, else 2nd phase. In the 1st phase, the SINR for x_1, SNR for x_2 at S_1 and SINR for x_1 at S_2 are all that remain same as (3), (4) and (5). In the 2nd phase, S_1 forwards the re-encoded x_2 to S_2. The received signal at S_2 in the 2nd phase is given by,

$$y_3 = h_3\sqrt{P_t}x_2 + N_{03}. \tag{7}$$

Hence, the SNR for x_2 at S_2 can be expressed as,

$$\gamma_4 = \frac{|h_3|^2 P_t}{\sigma^2}. \tag{8}$$

2.4   Shannon Capacity Analysis

Based on DT Mode of Operation, we can determine the capacities C_1^{DT} and C_2^{DT} of x_1 and x_2 respectively as,

$$C_1^{DT} = \frac{B}{2} min\{\log_2(1+\gamma_1), \log_2(1+\gamma_3)\} \tag{9}$$

$$C_2^{DT} = \frac{B}{2} \log_2(1+\gamma_2) \tag{10}$$

Likewise, based on CT Mode of Operation we can determine the capacities C_1^{CT} and C_2^{CT} of x_1 and x_2 respectively as,

$$C_1^{CT} = \frac{B}{2} min\{\log_2(1+\gamma_1), \log_2(1+\gamma_3)\} \tag{11}$$

$$C_2^{CT} = \frac{B}{2} min\{\log_2(1+\gamma_2), \log_2(1+\gamma_4)\} \tag{12}$$

By following the H-NOMA scheme, the total capacity of the systems can be expressed by,

$$C_{total} = \frac{B}{2} \left[\beta_{ACK}\{C_1^{DT} + C_2^{DT}\} + \beta_{ACK}\{C_1^{CT} + C_2^{CT}\} \right] \tag{13}$$

3   CDF Analysis

With the consideration of $h_i \sim \mathcal{CN}(0, \sigma_i)$ the probability of successfully decoding x_1 at S_1 can be expressed as,

$$P(\gamma_1 > \gamma_{th}) = P\left(|h_1|^2 > \frac{\gamma_{th}}{a_1 P_t - a_2 P_t \gamma_{th}} \right) = \begin{cases} e^{-\frac{\gamma_{th}}{(a_1 - a_2 \gamma_{th})P_t \sigma_1^2}}, \gamma_{th} < \frac{a_1}{a_2} \\ 0, \gamma_{th} < \frac{a_1}{a_2} \end{cases} \tag{14}$$

Now, the CDF of γ_1 is given by,

$$f_{\gamma_1} = \frac{a_1}{(a_1 - a_2 \gamma_{th})^2 P_t \sigma_1^2} e^{-\frac{\gamma_{th}}{(a_1 - a_2 \gamma_{th})P_t \sigma_1^2}} \tag{15}$$

4   Results and Discussion

Figure 2 (a) shows the probability density function (PDF) of the Rician channel versus K for the exact expression and the approximation given by (1). K is the ratio of the power contributions by line-of-sight signal to the remaining multipath signals. It can be seen from the figure that both curves match very well for the whole considered region of $0 \leq K \leq 18$. This proves that the approximation can be used without significant divergence from the results that would be obtained using exact expression.

Figure 2(b) shows the spectral efficiency, i.e. capacity in bits per seconds per Hertz, versus power allocation coefficient of s_2, changing from 0 to 0.5, for the total system, s_1 and s_2. The total system capacity is seen to be stable with respect to the change in power allocation coefficient of s_2. This is because, as the power allocation coefficient increases the capacity of s_2 increases with a logarithmic scale and the capacity of s_1 decreases in a similar scale. Power allocation coefficient can be used to adjust the capacity of s_1 and s_2 without degrading the total system capacity.

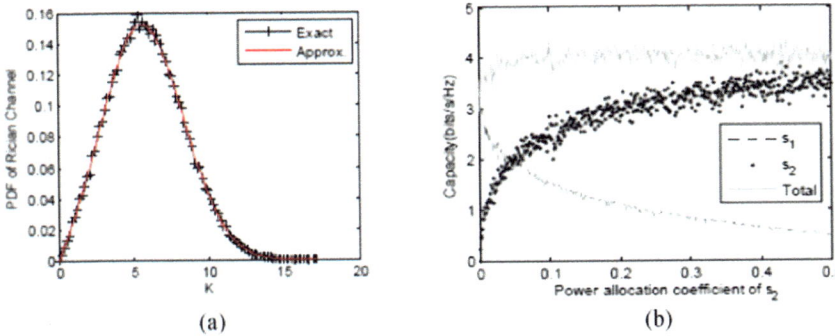

(a)   (b)

Fig. 2. PDF of Rician channel versus K, $0 \leq K \leq 18$, for the exact and approximation expressions (Fig. 2a), The capacity (bits/s/Hz) versus power allocation coefficient of s_2 for the total system (Fig. 2b).

Figure 3(a) displays the capacity of s_1 in bits/s/Hz versus power allocation coefficient of s_2 changing from 0 to 0.5 and received SNR of $s_2 = \{10, 100, 400\}$ dB. It can be seen that the capacity curve of s_1 is not much affected with respect to the change in SNR of s_2 from 10 to 100 and 400 dB. This is because, by employing H-NOMA the diversity is increased at the receiver and the signal of s_2 can be successfully detected with high probability even with SNR = 10 dB. And the interference from signal of s_2 to the signal of s_1 can be suppressed through successive interference cancellation (SIC) [1].

The capacity of s_2(bits/s/Hz) versus its power allocation coefficient ranging from 0 to 0.5 and SNR of $s_2 = \{10, 100, 400\}$ dB is shown in Fig. 3(b). As the SNR increases the capacity of s_2 increases with a higher degree of change from 10 to 100 dB compared to 100 to 400 dB. This shows that the capacity of s_2 may saturate after a level of increase in its SNR, since the residual interference from s_1 may influence this performance.

Figure 4(a) shows the total capacity of the system (bits/s/Hz) versus power allocation coefficient ranging from 0 to 0.5 and received SNR = $\{10, 100, 400\}$ dB of s_2. In compliance with Fig. 3, the total capacity of the system is shown to be independent of the power allocation coefficient since the capacity loss of s_1 is compensated with the capacity gain of s_2 as this coefficient increases. With the increase of the received SNR of s_2, the total capacity of the system increases due to s_2 achieving higher capacity and not interfering with the signal of s_1 through performing SIC.

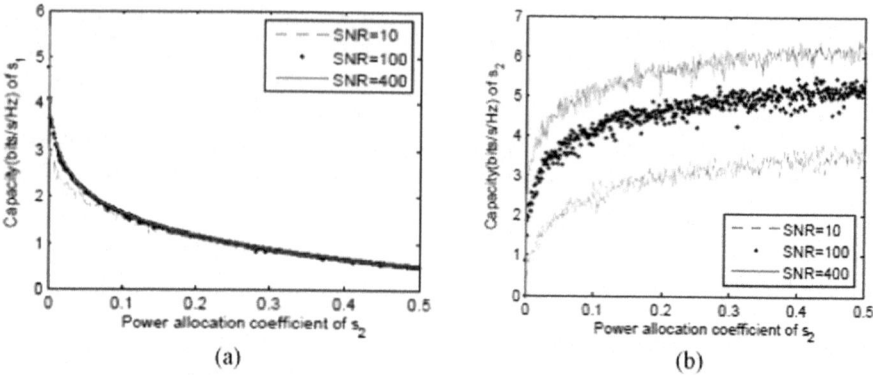

(a)   (b)

Fig. 3. The capacity of s_1 (bits/s/Hz) versus power allocation coefficient of s_2 changing from 0 to 0.5 for SNR = {10, 100, 400} dB (Fig. 3a), the capacity (bits/s/Hz) versus power allocation coefficient ranging from 0 to 0.5 and received SNR = {10, 100, 400} dB of s_2 (Fig. 3b).

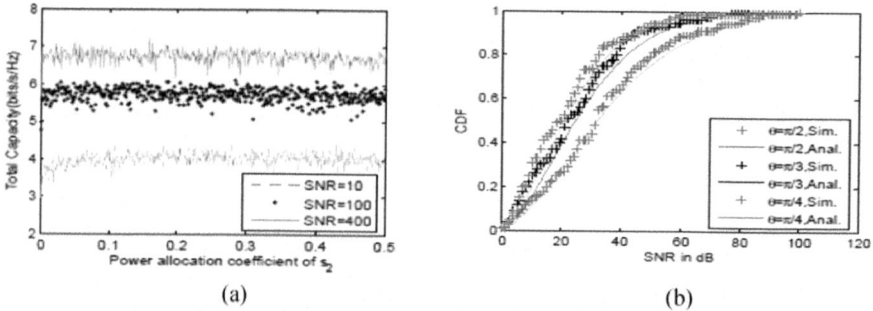

(a)   (b)

Fig. 4. Total capacity of the system (bits/s/Hz) versus power allocation coefficient ranging from 0 to 0.5 and received SNR = {10, 100, 400} dB of s_2 (Fig. 4a), computer simulation and analytical results for CDF of SNR for $\theta = \{\pi/2, \pi/3, \pi/4\}$ (Fig. 4b).

Figure 4(b) shows the computer simulation and analytical results for CDF of SNR for $\theta = \{\pi/2, \pi/3, \pi/4\}$. It can be seen from the figure that both curves match very well for all the considered cases of θ. Accordingly, mathematical analyses developed for the Rician fading channel model are valid for a wide range of channel scenarios. Further, it is clear that θ has a significant effect on the SNR performance.

Figure 5 shows the computer simulation and analytical results for the CDF of SNR for $K = \{3, 4, 5\}$. It is clear from the figure that for all the considered cases of K the computer simulations and analytical results match well. Thus, the mathematical analyses developed are valid for a wide range of shape parameters (i.e. K) for Rician fading channel model. The change of K has less effect on the CDF of SNR compared to the effect of θ shown by Fig. 4(b).

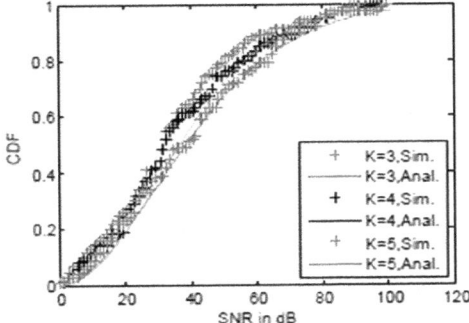

Fig. 5. Computer simulation and analytical results for CDF of SNR for $K = \{3, 4, 5\}$

5 Conclusion

This paper studies cooperative NOMA communications and proposes novel mathematical analysis for performance investigation over Rician fading channels. A probabilistic model is proposed that derives the CDF of SNR for a user-pairing scheme for cooperative NOMA communications. It is shown that the computer simulations and analytical results match very well. CDF of SNR analysis can be used as a tool for design and deployment of 5G and beyond networks where cooperative communications are useful, such as IoT applications.

References

1. Haci, H., Zhu, H., Wang, J.: Performance of non-orthogonal multiple access with a novel asynchronous interference cancellation technique. IEEE Trans. Comm. **65**, 1319–1335 (2017)
2. Kim, J., Lee, I.: Capacity analysis of cooperative relaying systems using non-orthogonal multiple access. IEEE Commun. Lett. **19**, 1949–1952 (2015)
3. Hayashi, Y., Kishiyama, Y., Higuchi, K.: Investigations on power allocation among beams in non-orthogonal access with random beamforming and intra-beam SIC for cellular MIMO downlink. In: 78th Vehicular Technology Conference (VTC Fall), pp. 1–5. IEEE, Las Vegas (2013)
4. Ding, Z., Peng, M., Poor, H.V.: Cooperative non-orthogonal multiple access in 5G systems. IEEE Commun. Lett. **19**, 1462–1465 (2015)

Weight Determining of Attributes in Multicriterial Decision Making for Software Selection

Nihad Mehdiyev[(✉)] [ID]

Department of Instrument-Making Engineering, Azerbaijan State Oil
and Industry University, 20 Azadlig Ave., 1010 Baku, Azerbaijan
nihadmehdi@gmail.com

Abstract. Selection the true software packages is very important for growing
any company and have a significant influence on company competitiveness.
There is a different affecting factor in the selection of software for company and
to solve this problem was used multi-criteria decision-making problem. An
important issue of a decision-making problem is determining the weights of
attributes that are given by experts. Since different experts have different
importance in decision-making, it is important to find a set of appropriate
weights. In this paper, a pair of most compromising and least compromising
solutions is derived from individual judgments of decision-makers and then,
these solutions are applied as the bases to determine the amount of individual
alignment with the experts opinion by using a closeness coefficient approach.
Determining the weights of decision-makers, decision-making problem of
software selection is then solved.

Keywords: Multiple criteria decision-making · Software selection · Weights ·
Multi-attribute · Fuzzy eigenvectors · Fuzzy number

1 Introduction

Software selection may include a lot of needs and products and it is needed to focus on
the basic factors affecting the selection process [1, 2]. Simulation is very important tool
to assume new systems and evaluation, observation system's actions. At the present
time, the market suggests a multiple simulation software packages [3]. So, an organized
and standardized type of simulation software selection is needed. Many companies
choose to set of tools without establishing formal estimation factors. It is very
important to determine ways to consistently and objectively estimate a tool's utilization
[4]. Tewoldeberhan et al. [5] described an exhaustive discrete-event simulation software selection technique for big foreign firms. Dorado et al. [6] used the AHP technique for choosing software in education. The results of these works represented that
some generic selection methods and AHP method is largely used as a decision tool for
software selection. King and Newman [7] evaluated software for business simulation
based on factors that are researching methods for estimating potential software and
increasing the experience. Hlupic [8] represented a software tool that select simulation

© The Author(s), under exclusive license to Springer Nature Switzerland AG 2021
R. A. Aliev et al. (Eds.): ICAFS 2020, AISC 1306, pp. 312–319, 2021.
https://doi.org/10.1007/978-3-030-64058-3_39

software given for the basic aspects. Hlupic et al. [9] developed a detailed framework for simulation software estimation where size is above 300 factors. Fuzzy-based methods make possible decision makers to use linguistic language for estimating criteria easily and intuitively. It improves software selection process by placing the uncertainty and ambiguity occurred during human decision making [10]. Using fuzzy methodology in software selection process can also decries uncertainties specific in experts' judgments. In this paper, a methodology based on AHP and TOPSIS is employed as MCDM methods. The fuzzy analytical hierarchy process is used to define weights of the basic factors, and fuzzy technique for order preference by similarity to ideal solution is used for ranking of alternatives. AHP helps decision makers to state a selection problem into a hierarchy. It is a very important method for handling categorical and numerical multi-criteria problems. In addition, AHP procedures are appropriate to individual and group decision making. TOPSIS is a largely accepted multi-criteria method because its sound logic, simultaneous consideration of the ideal and the non-ideal solutions. This paper is stated as follows: Sect. 2 represents proposed techniques, and conclusion of these methodology are discussed. In Sect. 3, the results of proposed techniques are tested through the sensitivity analysis. Results of this work are described in Sect. 4.

2 Preliminaries

Definition 1. $M \in F(R)$ is a fuzzy number, if the following two conditions are satisfied:

(1) $x_0 \in R$ where $\mu_M(x_0) = 1$
(2) For any $0 \leq \alpha \leq 1$, $A_\alpha = \left[x, \mu_{A_\alpha}(x) \geq \alpha\right]$ is a closed interval where F(R) represents a family of all fuzzy sets and R is the set of real numbers.

Definition 2. A continuous number [11, 12] is an ordered pair $C = (A, B)$ where A is a continuous fuzzy number playing a role of a fuzzy constraint on values that a random variable X may take:

$$X \text{ is } A$$

And B is a continuous fuzzy number with a membership function $\mu_B : [0, 1] \rightarrow [0, 1]$ playing a role of a fuzzy constraint on the probability measure of A :

$$P(A) = \int_R \mu_A(x)p(x)dx \text{ is } B$$

Definition 3. A PCM (C_{ij}) is a square matrix:

$$\left(C_{ij} = \left(A_{ij}, B_{ij}\right)\right) = \begin{pmatrix} (C_{11} = (A_{11}, B_{11})) & \cdots & (C_{1n} = (A_{1n}, B_{1n})) \\ \cdot & \cdots & \cdot \\ (C_{ij} = (A_{n1}, B_{n1})) & \cdots & (C_{nn} = (A_{nn}, B_{nn})) \end{pmatrix}$$

$C_{ij} = \left(A_{ij}, B_{ij}\right)$, $i, j = 1, \cdots, n$ describes partially reliable information on degree of preference for $i - th$ alternative (criterion) against $j - th$ one.

Definition 4. $C = (A, B)$ is characterized by fuzzy number A, fuzzy number B and underlying set of probability distributions G, we propose to define distance between continues numbers $D(C_1, C_2)$. Distance between A_1 and A_2 is computed as

$$D(A_1, A_2) = \sup_{\alpha \in (0,1)} D\left(A_1^{\alpha}, A_2^{\alpha}\right),$$

$$D\left(A_1^{\alpha}, A_2^{\alpha}\right) = \left| \frac{A_{11}^{\alpha} + A_{12}^{\alpha}}{2} - \frac{A_{21}^{\alpha} + A_{22}^{\alpha}}{2} \right|.$$

A^{α} denotes α-cut of A. Distance between B_1 and B_2 is computed analogously.

We also find distance between the sets G_1 and G_2 of probability distributions p_1 and p_2 underlying C_1 and C_2.. The distance between p_1 and p_2 can be expressed as

$$D(G_1, G_2) = \inf_{p_1 \in G_1, p_1 \in G_1} \left\{ 1 - \int_R ((p_1 p_2)^{\frac{1}{2}} dx)^{\frac{1}{2}} \right\}$$

Given $D(A_1, A_2)$, $D(B_1, B_2)$ and $D(G_1, G_2)$ the distance for continues numbers is defined as

$$D(Z_1, Z_2) = \beta D(A_1, A_2) + (1 - \beta) D_{total}(B_1, B_2)$$

where $D_{total}(B_1, B_2)$ is a distance for reliability restriction computed as

$$D_{total}(B_1, B_2) = w D(B_1, B_2) + (1 - w) D(G_1, G_2)$$

$\beta, w \in [0, 1]$ are DM's assigned importance degrees.

Below we propose a definition of inconsistency index matrix adopted from the index introduced for real-valued matrix in [13].

Definition 5. An inconsistency index K for PCM (C_{ij}) is defined as follows:

$$K\left((C_{ij})\right) = \max_{\substack{\\ i<j<k}} \min \left\{ D\left(C(1), \left(\frac{c_{ik}}{c_{ij}c_{jk}}\right)\right) D\left(C(1), \frac{c_{ij}c_{jk}}{c_{ik}}\right) \right\}$$

where the components of $C(1) = (A, B)$ are fuzzy singletons $A = 1$ and $B = 1$.

3 Statement and Solution of the Problem

Three alternatives of software - A_1, A_2, A_3 and four basic criteria - C1 –vendor, C2 – functionality, C3 –reliability, C4–portability have been used. Decision matrix D, which consists of alternatives $S_i, i = \overline{1,3}$ and criteria $C_j, j = \overline{1,4}$ is given in Table 1. Elements of matrix D are TFN describing linguistic terms rating of the alternative S_i according C_j.

Table 1. The comparison matrix for criteria's

Criteria's	C_1	C_2	C_3	C_4
C_1	(1, 1, 1)	(1.5, 2, 2.5)	(2, 2.5, 3)	(1, 1.5, 2)
C_2	(1/2.5, 1/2, 1/1.5)	(1, 1, 1)	(1, 1.5, 2)	(1.5, 2, 2.5)
C_3	(1/3, 1/2.5, 1/2)	(1/2, 1/1.5, 1)	(1, 1, 1)	(1/2.5, 1/2, 1/1.5)
C_4	(1/2, 1/1.5, 1)	(1/2.5, 1/2, 1/1.5)	(1.5, 2, 2.5)	(1, 1, 1)

Let us consider a problem of generation of consistent PCM $\left(C'_{ij} \right)$ most similar to a given inconsistent PCM $\left(C_{ij} \right)$. The elements of inconsistent matrix $\left(C_{ij} \right)$ will be considered as a perturbation of the elements of matrix $\left(C'_{ij} \right)$ for which reciprocity and consistency are verified. We have to change elements of $\left(C_{ij} \right)$ in order to arrive at $\left(C'_{ij} \right)$.. The problem is formulated as follows:

$$J = \sum_{i=1}^{n}\sum_{j=1}^{n} D(C_{ij}, C'_{ij}) \to \max$$

Multiplicative reciprocity:

$$C'_{ij} C'_{ji} = C \quad (1)$$

multiplicative transitivity:

$$C'_{ij} C'_{jk} = C'_{ik}$$

non-negativity:

$$C'_{ij} \geq C(0), \ i,j = 1, \cdots, n$$

For finding eigenvalues and eigenvectors, determining consistency of the matrix there are some steps:

1. *Finding crisp maximum Eigenvalue and Eigenvectors.*
   Find maximum Eigenvalue by solving equation: $\det(A - \lambda I) = 0$. Here matrix A contains only medium values of fuzzy elements. Minimizing DE objective function $CF(\lambda) = D^2(\lambda) - \varepsilon\lambda$, where λ is the parameter to optimize, $D(\lambda) = \det(A - \lambda I)$ is a function dependent of λ, ε is a constant (may need to be adjusted for concrete problem, in our case was set to 1). (DE search space dimension is 1).
   Construct new DE objective function to find Eigenvectors:
   $CF(X) = \text{VectorDistance}(AX, \lambda X)$. Parameters to search are (N-1) elements of vector X (except the 1st one, which is set to 1). (DE search space dimension is N-1)
2. *Finding fuzzy Eigenvalues.*
   Use fuzzy A matrix. Construct new DE objective function as follows:

$$CF(\lambda_L, \lambda_R) = \text{FuzzyDistance}(AX, (\lambda_L, \lambda_M\lambda_R)X),$$

where "FuzzyDistance" is sum of fuzzy distances between corresponding elements of vectors. Vector X consists of fuzzy singletons, produced from crisp Eigenvectors of step 1. Only left and right components of fuzzy Eigenvalues are subject to optimize (Medium component is taken from step 1). Use fuzzy A matrix and fuzzy Eigenvalues. Construct DE objective function as follows:

$$CF(X) = \text{FuzzyDistance}(AX, \lambda X).$$

Parameters to search are N elements of fuzzy vector X. Only left and right components are searched, middle component is taken from step 1. (DE search space dimension is 2*N). The Found Crisp and Fuzzy Eigenvalues and Eigenvectors:

MAX LAMBDA = 4.12683169439773
Eigen Vector = [1 0.690669443092956 0.364211788538241 0.556625212009703]
Fuzzy LAMBDA = [3.32105295372562, 4.12683169439773, 5.03471342425143]
Fuzzy Eigen Vectors = [0.00000 1.00000 1.00809], [0.00000 0.69067 0.71495], [0.00000 0.36421 0.36421] [0.00000 0.55663 0.59366]

Parameters of DE for crisp solution:

F = 0.9, Cr = 1.0
VectorDimension = 16
PopulationSize = 500
MaxGenerations = 10000
Parameters of DE for fuzzy solution:
F = 0.9, Cr = 1.0
VectorDimension = 48
PopulationSize = 1500
MaxGenerations = 10000

For finding weights of criteria's we normalize eigenvectors by using formula

$$Z = \frac{X - minX}{maxX - minX}$$

Normalized Fuzzy Eigen Vectors $= \begin{bmatrix} [0.00000\ 1.00000\ 1.00809],\ [0.00000\ 0.51\ 1], \\ [0.0\ 0.0\ 0.0],\ [0.00000\ 0.29\ 0.35]. \end{bmatrix}$

Table 2. Fuzzy decision matrix

	C_1- vendor	C_2-functionality	C_3- reliability	C_4-portability
W	0.38	0.25	0.22	0.15
a_1	0.4, 0.5, 0.6, 0.7	0.6, 0.7, 0.8, 0.9	0.8, 0.9, 1, 1	0.6, 0.7, 0.8, 0.9
a_2	0.4, 0.5, 0.6, 0.7	0.8, 0.9, 1, 1	0.4, 0.5, 0.6, 0.7	0.6, 0.7, 0.8, 0.9
a_3	0.6, 0.7, 0.8, 0.9	0.4, 0.5, 0.6, 0.7	0.8, 0.9, 1, 1	0.8, 0.9, 1, 1

After normalizing decision matrix from Table 2, we get fuzzy normalized decision matrix which is shown in Table 3:

Table 3. Fuzzy normalized decision matrix

	C_1- vendor	C_2-functionality	C_3- reliability	C_4-portability
a_1	0.15, 0.19, 0.23, 0.27	0.15, 0.18, 0.2, 0.23	0.18, 0.2, 0.22, 0.22	0.09, 0.1, 0.12, 0.14
a_2	0.15, 0.19, 0.23, 0.27	0.2, 0.23, 0.25, 0.25	0.09, 0.11, 0.13, 0.15	0.09, 0.1, 0.12, 0.14
a_3	0.23, 0.27, 0.31, 0.34	0.1, 0.13, 0.15, 0.18	0.18, 0.2, 0.22, 0.22	0.12, 0.14, 0.15, 0.15

In next step, we define $a*$ and a^- solutions. The fuzzy positive ideal and negative ideal solutions are shown in Table 4.

Table 4. The fuzzy positive ideal and negative ideal solutions.

	C_1- vendor	C_2-functionality	C_3- reliability	C_4-portability
$a*$	0.23, 0.27, 0.31, 0.34	0.2, 0.23, 0.25, 0.25	0.18, 0.2, 0.22, 0.22	0.12, 0.14, 0.15, 0.15
a^-	0.15, 0.19, 0.23, 0.27	0.1, 0.13, 0.15, 0.18	0.09, 0.11, 0.13, 0.15	0.09, 0.1, 0.12, 0.14

Using the Euclidean distance, we can determine the fuzzy positive ideal solution and the fuzzy negative ideal solutions. The results are presented in Table 5.

Table 5. Proximity distance.

	C_1			C_2			C_3			C_4		
	a_1	a_2	a_3	a_1	a_2	a_3	a_1	a_2	a_3	a_1	a_2	a_3
$a*$	0.16	0.16	0	0.09	0	0.19	0	0.17	0	0.06	0.06	0
a^-	0	0	0.14	0.1	0.19	0	0.17	0	0.17	0	0	0.06

Proximity to the fuzzy positive ideal solution is calculated for each alternative:

$$C_1^* = S_1 - /(S_1^* + S_1^-) =$$

$$= [0/(0.16+0) + 0.1/(0.1+0.09) + 0.17/(0.17+0) + 0/(0+0/06)]/4 = 0.38,$$

$$C_2^* = [0/(0+0.16) + 0.19/0.19 + 0/0.17 + 0/0.06]/4 = 0.25,$$

$$C_2^* = [0.14/0.14 + 0/0.19 + 0.17/0.17 + 0.06/0.06]/4 = 0.75.$$

Finally, we compare the alternatives:

$$a_3 > a_1 > a_2$$

Thus, we determined that a_3 is the best and a_2 is the worst alternative.

4   Conclusion

A methodology based on hybrid AHP and TOPSIS is employed as multi-criteria decision-making methods. Analytical hierarchy process is used to define weights of the basic factors, and fuzzy technique for order preference by similarity to ideal solution is used for ranking of alternatives. AHP helps decision makers to state a selection problem into a hierarchy. It is a very important method for handling categorical and numerical multi-criteria problems.

References

1. Bresnahan, J.: Mission impossible, CIO Magazine, October, p. 15 (1996)
2. Franch, X.: On the lightweight use of goal-oriented models for software package selection. In: Proceedings of the 17th Conference on Advanced Information Systems Engineering CAISE 2005. Lecture Notes in Computer Science, vol. 3520, pp. 551–566. Springer, Heidelberg (2005)

3. Gupta, A., Verma, R., Singh, K.: Smart sim selector: a software for simulation software selection. Int. J. Eng. (IJE) **3**(3), 175–185 (2009)
4. Chikofsky, E.J., Martin, D.E., Chang, H.: Assessing the state of tools assessment. IEEE Softw. **9**(3), 18–21 (1992)
5. Tewoldeberhan, T.W., Verbraeck, A., Hlupic, V.: Implementing a discrete-event simulation software selection methodology for supporting decision making at Accenture. J. Oper. Res. Soc. **61**(10), 1446–1458 (2010)
6. Dorado, R., Gómez-Moreno, A., Torres-Jiménez, E., López-Alba, E.: An AHP application to select software for engineering education. Comput. Appl. Eng. Educ. (2011). https://doi.org/10.1002/cae.20546
7. King, M., Newman, R.: Evaluating business simulation software: approach, tools and pedagogy. Horizon **17**(4), 368–377 (2009)
8. Hlupic, V.: Simulation software selection using SimSelect. Simulation **69**(4), 231–239 (1997)
9. Hlupic, V., Irani, Z., Paul, R.: Evaluation framework for simulation software. Int. J. Adv. Manufact. Tech. **5**(5), 366–382 (1999)
10. Jadhav, A.S., Sonar, R.M.: Framework for evaluation and selection of the software packages: a hybrid knowledge-based system approach. J. Syst. Softw. **84**(8), 1394–1407 (2011)
11. Aliev, R.A., Huseynov, O.H., Aliyev, R.R., Alizadeh, A.V.: The Arithmetic of Z-Numbers: Theory and Applications, 316 p. World Scientific, Singapore (2015)
12. Aliev, R.A., Huseynov, O.H.: Decision Theory with Imperfect Information, p. 444 World Scientific, Singapore (2014)
13. Zbigniew, D., Koczkodaj, W.: Generalization of a new definition of consistency for pairwise comparisons. Inform. Process. Lett. **52**(5), 273–276 (1994)

Educational-Methodical Developments Projects Management Based on Fuzzy Logic

Sergey Glushenko$^{(\boxtimes)}$ ⓘ, Sergey Shcherbakov ⓘ, Vera Grechkina ⓘ, and Nadegda Misichenko ⓘ

Rostov State University of Economics, Rostov-on-Don, Russia
gs-gears@yandex.ru, sergwood@mail.ru, wwwera@list.ru,
misssnadia@rambler.ru

Abstract. The article demonstrates the importance of using risk analysis in the implementation of an educational-methodological project (EMP) and confirms the feasibility of using fuzzy logic for risk assessment. The use of fuzzy models allows to consider both quantitative and qualitative characteristics, as well as to represent fuzzy descriptions using fuzzy sets and linguistic variables.

The fuzzy production model described (FPM) contains ten input linguistic variables, that characterize the risk factors, five output linguistic variables, that characterize the risks of EMP, including an integral risk indicator, that characterizes the quality level of the educational program in the as a whole and five rule bases. EMP allows to remove the restrictions on the number of input variables taken into consideration and to integrate both qualitative and quantitative approaches to assess the quality of methodological documents.

The process of implementing the fuzzy modeling of the rule base of the specialized Fuzzy Logic Toolbox package of the MATLAB software is described. The mechanism for obtaining estimates based on the Mamdani algorithm allows getting a numerical value of the quality level, its linguistic description, as well as the degree of confidence of the expert in the resulting assessment.

Keywords: Educational project · Risk · Fuzzy set · Term-set · Fuzzy production model · Linguistic variable · Rule base · Membership function

1 Introduction

In the past few years, innovations in the process of organizing the educational process have been introduced into higher education institutions, such as: the competency-based approach, federal state standards (FSS) of the third generation (as well as FSS 3) and a modular-rating system. These changes were primarily aimed at ensuring high quality education [7].

The Federal Law "On Education in the Russian Federation" provides the following wording for the quality of education: "The quality of education is a comprehensive characteristic of educational activities and student training, expressing the degree of their compliance with federal state educational standards, educational standards, federal state requirements and (or) the needs of an individual or legal entity in whose interests

© The Author(s), under exclusive license to Springer Nature Switzerland AG 2021
R. A. Aliev et al. (Eds.): ICAFS 2020, AISC 1306, pp. 320–327, 2021.
https://doi.org/10.1007/978-3-030-64058-3_40

educational activity is carried out, including the degree of achievement of the planned results of the educational program" [3].

On the other hand, the above innovations have led to a significant increase in the volume of educational and methodological activities. The development processes of educational and methodological support affect performers in all departments of the university and are characterize by significant labor costs. This primarily refers to the faculty, which provide the required quality of training. In many cases, the level of load of teachers exceeds the permissible limit, which contributes to the development of negative situations, namely: lack of time for preparation for classes and scientific research, a sharp deterioration in the quality of educational and methodological support and the educational program (EP) as a whole [8].

Thus, it is obvious that the development process of educational and methodological support and the draft educational program of the university is accompanied by risk, namely the risk of consumer quality. This risk is manifested in the development of a program that does not meet the requirements of stakeholders.

In the process of analyzing the draft educational program and assessing consumer-quality risk, the well-known principle of the linguistic approach can be applied, in which the assessment is carried out in terms of "low risk", "permissible risk" and "high risk". However, it is difficult for a decision maker (program manager or project author) to give them an accurate (objective) quantitative assessment and describe them using a mathematical language, which affects the effectiveness of decisions. Increased efficiency can be achieved by applying methods and models that take into account the existing uncertainties in the project. Often, accounting and analysis of project uncertainties and risks is carried out by analytical and expert methods. However, analytical methods require a large amount of statistical data and are usually focused on quantitative indicators, and expert methods are difficult to apply in the operational assessment of uncertainties and risks, since they require highly qualified specialists and a lot of time.

The use of methods and models based on fuzzy knowledge is devoid of the above disadvantages. They allow EMP, when receiving risk assessments, to use both quantitative characteristics that are objectively characterized by uncertainty, and qualitative, subjective expert assessments expressed by fuzzy concepts, as well as formalize fuzzy descriptions using fuzzy numbers, sets, linguistic variables and fuzzy evidence [1, 9].

2 Fuzzy Production Risk Assessment Model

To analyze the consumer quality of educational-methodological development (EMD) projects, fuzzy models are presented in the form of fuzzy production networks (FPN), elements and sets of elements of which realize various components of fuzzy models and stages of fuzzy inference [2].

The basis of the fuzzy production model are fuzzy products:

$$\textbf{IF } x \text{ is } A, \textbf{ THEN } y \text{ is } B, \tag{1}$$

where x is the input variable, $x \in X$; X is the domain of definition of the antecedent of the fuzzy rule; A is a fuzzy set defined on X; $\mu_A(x) \in [0, 1]$ is the membership function of the fuzzy set A; y is the output variable, $y \in Y$; Y is the consequent domain of the fuzzy rule; B is a fuzzy set defined on Y; $\mu_B(y) \in [0, 1]$ is the membership function of the fuzzy set B.

An analysis of EP projects involving experts leading specialists in this subject area revealed the factors that may be sources of EMP risk (see Table 1). When setting linguistic variables characterizing risk factors, the following term sets can be used that determine the levels of factors [4]:

- T2 = {Low (L); High (H)};
- T3 = {Low (L); Medium (M); High (H)};
- T4 = {Very Low (VrL); Low (L); Medium (M); High (H)};
- T5 = {Very Low (VrL); Low (L); Medium (M); High (H); Very High (VrH)}.

Table 1. Risk factors for educational-methodological projects

Designation	Name of linguistic variable
x_1	Project goal
x_2	Project boundaries
x_3	Correctness of document forms
x_4	Completeness of the documentation
x_5	Project deadlines
x_6	Correctness of formal parameters (hours, competencies, etc.)
x_7	Staffing
x_8	The correctness of interdisciplinary logic
x_9	The correctness of the national development strategy
x_{10}	Original content

In the process of analyzing risk factors, indicators were identified that can characterize the risks of EP (see Table 2). When setting linguistic variables characterizing risk indicators, the following term sets are used that define risk indicators:

- T1 = {Evidence at low risk (ELR); Evidence at medium risk (EMR); Evidence at high risk (EHR)};
- T2 = {Evidence at very low risk (EVrLR); Evidence at low risk (ELR); Evidence at
- medium risk (EMR); Evidence at high risk (EHR); Evidence at very high risk (EVrHR)}.

Table 2. Risk indicators of educational-methodological projects

Designation	Name of linguistic variable
y_1	Risk of the correctness of the project goal
y_2	Risk of violation of requirements
y_3	Risk of poor-quality educational material
y_4	Risk of low attractiveness of the program
y_5	The integrated risk indicator of the educational program

The relationship between factors (antecedent) and risk indicators (consequent) is a binary fuzzy relation on the Cartesian product of the corresponding fuzzy sets. A fuzzy causal relationship between the antecedent and consequent is set in the form of fuzzy products [5]. The following Table 3 gives a production rules.

Table 3. Fuzzy production rules of the model EMP (fragment)

Rule designation	Antecedent	Consequent
Base Rule R1		
R1.1	$x1 = L \land (x2 = H \lor x2 = M)$	$y1 = EVrHR$
R1.2	$x1 = L \land x2 = L$	$y1 = EHR$
R1.3	$x1 = M \land x2 = L$	$y1 = EMR$
R1.4	$x1 = H \land (x2 = H \lor x2 = M)$	$y1 = ELR$
R1.5	$x1 = H \land x2 = L$	$y1 = EVrLR$

A fuzzy production model allows you to remove restrictions on the number of input variables used and to include both qualitative and quantitative risk assessment methods.

3 Method Implementation

As an example of the implementation of the proposed methodology, we consider the task of assessing the achievement of the goals of the ER project. The linguistic variable "Project Goal" is used as the input linguistic variable (x_1), which has the following term set: T1 = {"not enough", "limited", "completely"}.

The term "insufficient" (L) - corresponds to a situation where the goal of the ER project does not adequately match the goals of the educational organization. The term "limited" (M) - corresponds to a situation where the goal of the ER project corresponds to the goals of the organization with certain restrictions. The term "fully" (H) - corresponds to a situation where the goal of the ER project is fully consistent with the goals of the organization.

Another linguistic variable (x_2) is "Project Boundaries" with a term-set: T2 = {"minimum", "normal", "redundant"}.

The term "minimum" (L) corresponds to the situation when the boundaries of the project of the EP determine the minimum set of topics for the direction. The term "*normal*" (M) corresponds to the situation when the boundaries of the project of the EP correspond to the required set of topics for the direction. The term "redundant" H corresponds to a situation, where the boundaries of the project of the EP have an excess or inaccurately defined set of topics for the direction.

The output variable (y1) is the linguistic variable *"Risk of compliance with the project goal"*, which has the following term set: T3 = {Low risk evidence (L), Average risk evidence (M), High risk evidence (H)}.

The developed structure of a fuzzy production network for analysis and risk assessment of consumer quality of the ER project is shown in Fig. 1.

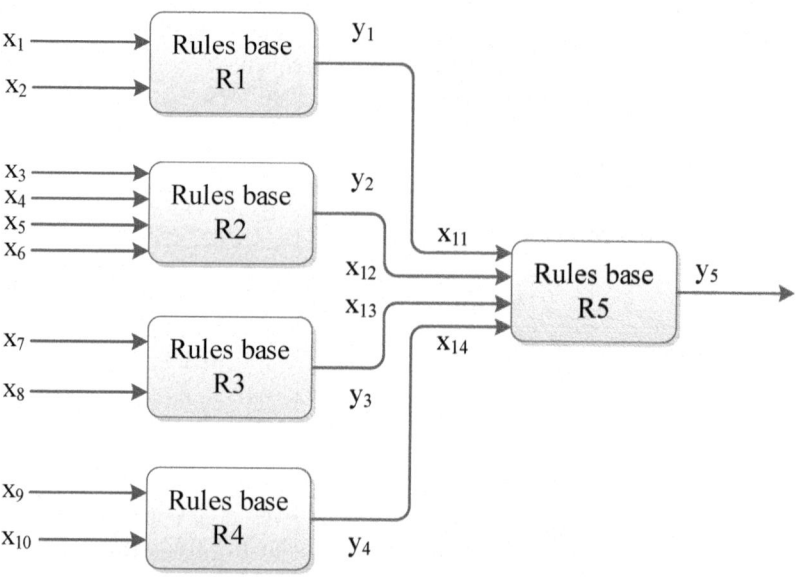

Fig. 1. The structure of the fuzzy production network EMP

The process of fuzzy modeling of the rule base is implemented with the specialized Fuzzy Logic Toolbox software package MATLAB [6]. Fuzzy inference is implemented based on the Mamdani algorithm.

Step 1. Fuzzification - the introduction of fuzziness. At this step, it is necessary to define membership functions for the term sets of input and output linguistic variables:

- input1 variable in the model is consistent with the linguistic variable "Project goal" – x_1;
- input2 variable in the model is consistent with the linguistic variable "Project boundaries" – x_2;
- output1 variable in the model is consistent with the linguistic variable "Compliance with the project goal" - y_1.

Triangular membership functions were defined as the main functions (Fig. 2). In the general case, the triangular membership function has the following form:

$$\mu_{\Delta}(x, a, b, c) = \begin{cases} 0, & x \leq a \\ \frac{x-a}{b-a}, & a \leq x \leq b \\ \frac{c-x}{c-b}, & b \leq x \leq c \\ 0, & c \leq x \end{cases} \quad (2)$$

where a, b, c are numerical parameters characterizing the base of the triangle (a, c) and its vertex (b), and the condition a ≤ b ≤ c must be fulfilled.

The membership functions of fuzzy term sets of the input linguistic variable "Project Goal" will have the following form:

$$\mu_{\Delta}^{L}(x; 0; 0; 0,5), \ \mu_{\Delta}^{M}(x; 0; 0,5; 1,0), \ \mu_{\Delta}^{H}(x; 0,5; 1,0; 1,0). \quad (3)$$

Figure 2 shows the graphs of the membership functions of the term sets of the linguistic variable input1 - "Project Goal".

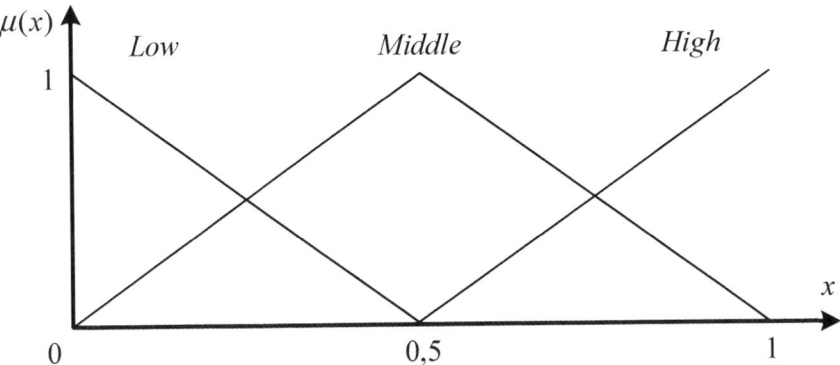

Fig. 2. Membership functions for the input variable input1

The membership functions of fuzzy term sets of the input linguistic variable "Project Boundaries" will have the following form:

$$\mu_{\Delta}^{L}(x; 0; 0; 0,4), \ \mu_{\Delta}^{M}(x; 0; 1,0,5; 0,9), \ \mu_{\Delta}^{H}(x; 0,6; 1,0; 1,0). \quad (4)$$

The membership functions of fuzzy term sets of the output linguistic variable "Correspondence to the project goal" will have the following form:

$$\mu_{\Delta}^{L}(x; 0; 0; 0,5), \ \mu_{\Delta}^{M}(x; 0; 0,5; 1,0), \ \mu_{\Delta}^{H}(x; 0,5; 1,0; 1,0). \quad (5)$$

Step 2. Making fuzzy rules. In the Mamdani algorithm, the rule base should be defined as the structure of the corresponding model rule base. For Base rule R1, this is a structure with two inputs and one output (see Fig. 1).

Step 3. Accumulation of the conclusion according to all the rules was carried out using the tach – disjunction operation. During defuzzification, the center of gravity method was used for a discrete set of values of membership functions:

$$y' = \frac{\sum_{r=1}^{Y_{max}} y_r \mu_{B'}(y_r)}{\sum_{r=1}^{Y_{max}} \mu_{B'}(y_r)} \tag{6}$$

where Y_{max} is the number of y_r elements in the domain Y discretized for calculating the "center of gravity".

Using the Model. With the classifier value input1 = 0.1, the value of the linguistic variable x_1 - "Project Goal" corresponds to the term L - "does not adequately match the goals of the educational organization" with a level of confidence $\mu_{x_1}^L = 0{,}8 \ \mu$. With the classifier value input2 = 0.6, the value of the linguistic variable x_2 - "Project boundaries" corresponds to the term M - "corresponds to the required set of topics for direction" with a level of confidence $\mu_{x_2}^M = 1{,}0 \ \mu$.

According to the given initial conditions, rules 2 and 5 are activated. The resulting value of the classifier of the output variable output1 corresponds to a value of 0.74, which determines the value of the linguistic risk variable of the project y1 – "Compliance with the project goal" equal to High – "High evidence of risk" with a level of confidence $\mu_{y_1}^H = 0{,}92 \ \mu$.

The implementation of fuzzy inference rules for Base rule R1 is shown in Fig. 3.

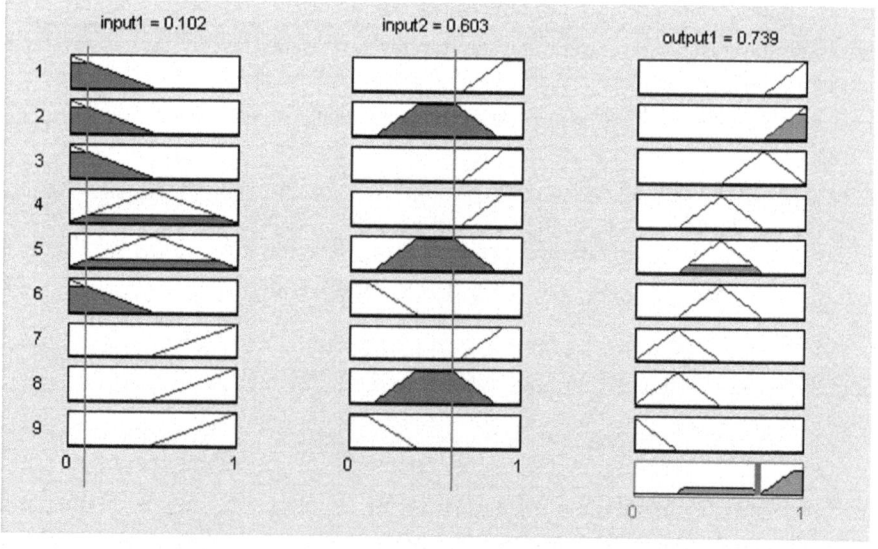

Fig. 3. Implementation of fuzzy inference rules EMP

The Fuzzy Logic Toolbox interface offers the possibility to control the "quality" of the output mechanism. Therefore, smooth and monotonous graphs of the dependencies of the given "inference surface" testify to the good "quality" of the inference mechanism and the sufficiency and consistency of the inference rules used.

4 Conclusion

In the fuzzy model developed for assessing the risks to the quality of consumers of educational-methodological development projects, ten input linguistic variables, that characterize the risk factors were identified, five output linguistic variables that characterize the risks of EMP. The model contains five bases of rules and allows the linguistic analysis of the risks to the quality of consumers of projects that represent potential threats to the process of developing an educational program. The FPN allows to remove the restrictions on the number of input variables taken into consideration and to include both qualitative and quantitative methods for assessing the quality of the guidance documents, as well as to set the priorities of the risks (from very high to insignificant) required by the program manager for decision making.

References

1. Borisov, V.V., Kruglov, A.S., Fedulov, A.S.: Fuzzy Models and Networks. Hotline-Telecom, Moscow (2012). (in Russian)
2. Dolzhenko, A.I.: A model for analysis of consumer quality risks of economic information systems (in Russian). Herald North-Caucasus Federal Univ. 1(18), 129–134 (2009)
3. Federal Law of December 29, 12, No. 273-FZ "On Education in the Russian Federation". http://www.consultant.ru/document/cons_doc_LAW_140174
4. Glushenko, S.A.: An adaptive neuro-fuzzy inference system for assessment of risks to an organization's information security. Bus. Inform. 1(39), 68–77 (2017). https://doi.org/10.17323/1998-0663.2017.1.68.77
5. Glushenko, S.A., Doljenko, A.I.: Fuzzy self-tuning model for analysis of project risks. KnE Eng. 271–279 (2018). https://doi.org/10.18502/keg.v3i4.2250
6. Leonenkov, A.V.: Fuzzy Modelling in MATLAB and FuzzyTECH. BVH-Petersburg, St. Petersburg (2005). (in Russian)
7. Shcherbakov, S., Saveleva, N., Veretennikova, E., Aruchidi, N.: Educational methodical activity in the university: simulation and economics. In: International Multi-Conference on Industrial Engineering and Modern technologies "FarEastCon2019", pp. 396–405. Atlantis Press (2019)
8. Veretennikova, E.G., Scherbakov, S.M., Miroshnichenko, I.I., Savelyeva, N.G.: The management processes of educational and methodological support in higher education (in Russian). Black Sea-Caspian Cooperation Forum: Security. Sustainability. Development: Materials of the International Discussion Platform, pp. 696–701. Rostov-on-Don: IPK RSEU (RINH) (2017)
9. Zade, L.A.: A concept of linguistic variable and its application for fuzzy decision-making. Mir, Moscow (1976). (in Russian)

ANN Based Solution to Maintain Connection Continuity Problems of Mobile Users in the Buildings Elevators

Fahreddin Sadikoglu[ID] and Jamal Fathi[✉][ID]

Near East University, North Cyprus, Mersin 10, Turkey
fahreddin.sadikoglu@neu.edu.tr,
jamalfathi2004@gmail.com

Abstract. Mobile coverage services inside elevators became an important issue to be considered between owners of buildings and their tenants. Moreover, the overall necessity of the mobile's signal coverage inside the elevators affects the reliability of wireless applications. Unluckily, several solutions were discussed to decrease the propagation lose inside the elevators: Antennas affixed within the elevator's pivot failed to provide stable coverage within the buildings, as much as the antenna available in the lobby is considered as the cause of the handoff for the persons who are using their mobiles inside the elevators, which leads to call drops and disconnections while download process. This paper is arranged and incremented an Artificial Neural Network to predict the changes in the IP addresses as an important point to maintain a stable IP address in the multi-floors building with elevator. Finally, we proved the accuracy of our model as a perfect solution for wireless propagation lose inside the elevators of buildings.

Keywords: Coverage area · Elevators · Access point wireless local area networks · Neural network

1 Introduction

This paper is involved in Wireless Local Area Network (WLAN) localization with the combination of Artificial Neural Network (ANN) to predict the changes in IP addresses of the access point located in buildings including elevators. The existence of ANN is to prevent changes in the IP addresses of the access point, so that to be assured of continuity of connection. In the elevators-based floor plan, the set of anticipated occupant's paths is intensified into sides. These links are unaddressed components that are related by loops in accordance to their real-world connectivity. It is assumed that the mobile-user can be anywhere.

A traditional scheme as in Abuhasnah [1] to be a combination of WLAN re-locating base stations or incrementing additional splitters to the base stations. However, most mobile device users that one may want to use it as indoor or outdoor are accoutered with low-quality Inertial Navigation System (INS), which makes reliable Pedestrian Dead Reckoning (PDR) difficult or impossible. While in Wireless Communications (WC), the process of communication is described that the required

© The Author(s), under exclusive license to Springer Nature Switzerland AG 2021
R. A. Aliev et al. (Eds.): ICAFS 2020, AISC 1306, pp. 328–336, 2021.
https://doi.org/10.1007/978-3-030-64058-3_41

information is transmitted from one end to another, propagating as Electromagnetic (EM) waves. The receiving subscriber's connections with base station terminal is generally minor, regularly short as a couple of meters.

To estimate the received signal in log-domain as used in Abuhasnah [1]:

$$Y = 24.5 + 33.8\log(d) + 4.0K_{floor} - 16.6S_{win} - 9.8G_{G/1} - 0.25A_{elv}$$

Where, d is the distance between transmitter and receiver, K_{floor} is the number of floors, and S_Q is the number of building sides seen by the transmitter are the received power at the receiver and the gain of the antenna in the transmitter's direction, d represents the distance between transmitter and receiver; K_{floors} is the number of floors separating transmitter and receiver; the variable S_{win}, also called sight, represents the amount of signal leaving and returning to the building, complemented with some considerations on the ability of the signal to propagate on the floor where the transmitter was located; $G_{G/1}$ represents the tendency exhibited by the signals to be higher on the first two floors of a building; and A_{elv} is the floor area. In addition, some other variables had either a positive or negative influence, or even no effect at all on the signal strength.

2 Related Research

AbuHasna, [1], investigated the process of propagation loss inside and outside the buildings using frequencies ranged between 900–1800 MHz as much as, applied linear prediction filter to the received signal. In Fathi et al. [2], prepared a wide study on the propagation loss inside elevators of the NEU buildings. Moreover, compared all existing systems with the proposed system, and proved that the proposed system could be valid for all conditions, including elevators inside the buildings. While, Fathi, [3], provided a recommendation to rearrange the positions of the base stations in order to obtain a higher level and more stable mobile signals. As much as studying the standard systems and their propagation path loss inside the buildings. Moreover, Obeidat et al. [4], improved a path loss prediction model and compared it with the other indoor prediction models using simulation data and real-time measurements. Where, Bakinde et al. [5], improved a new path loss model that is dependent upon Adaptive Neuro-Fuzzy Interference System (ANFIS) using more than one transmitter so that to be valid for VHF bands. In Zreikat et al. [6], analyzed path loss as free space, Cost-231 Hata, Empirical Hata, Walfisch-Ikegami, Stanford University Interim (SUI) and Ericson (9999) models. Moreover, it proved that SUI and Empirical Hata models are the most accurate and best choice prediction models. As much as, Bose et al. [7], improved a new mathematical model in order to minimize the handover time delay. Otherwise, Elechi et al. [8], surveyed the quality of mobile subscriber's signal, so that the collected measurements in Nigeria GSM operators as MTN, GLO, Arial and Etisalat using RF tracker. Miura et al. [9], surveyed to improve and targeted the minimum propagation loss is a process, in order to improve and maximize communication progress through WiFi and internal loops. While, Jadhav et al. [10], investigated Hata, SUI, ECC 33 and Cost231 W-I models and designated the use of Cost-231 Hata's model using

2375 MHz for the selected suburban in India. Where, Beauregard et al. [11], trained a neural network (NN) to obtain an accumulated approximately 2% to the combination of PDR and GPS investigated system in order to get a step length prediction for relative positioning. Moreover, Zyad et al. [12], in order to characterize indoor propagation including Access Point, presented a powerful tentative path loss so that the covered models using wide RSS measurements obtained by LOS and dimensional LOS. As much as the developed models under two conditions as quasi and RSS. In Zyad et al. [13], investigated propagation loss in both cases, inside and outside buildings, as much as identified the points where there is a loss in the transmitted signal. Improved the efficiency of the proposed algorithm. While, Fathi et al. [14], used Friis and Okumura models in order to set out a predicted system of UHF. The results of the used system is compared with different systems and obtained a new system suitable for the city. As much as, Majed et al. [15], developed an algorithm in order to prove the accuracy of the predicted system for wireless coverage. Finally, Perez et al. [16], examined a system used in the indoor communication with 1.8 GHz, the exponent used in the model as a random variable is treated.

3   Proposed Model

In the configuration of a representative point-to-point (PTP) wireless ground-work. The proposed model is shown in Fig. 1.

The important point to be considered is that the placement of the transmitter and receiver. The signal transmitted by the transmitter sits on for the building with a low-loss RF through a cable to the WLAN. The characteristics of the communication process in the LOS varies between some meters to kilometers, so that it depends on many factors. These factors are including the gain of the transmitter's antenna, the performance of the bridge, path loss propagation, and the location of the amplifier of the RF.

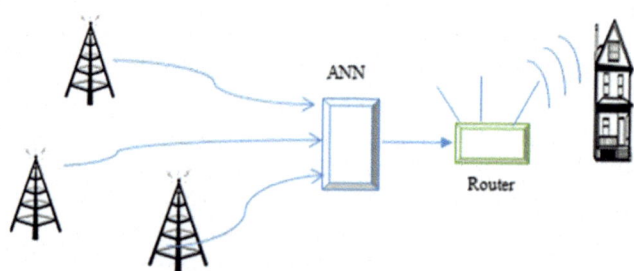

Fig. 1. Proposed mode

3.1   ANN Prediction Process

Regarding the use of ANN in this work, assuming the mobile users are moving and the important point is that the access points are changing the IP addresses automatically, which yields to disconnection process while it's searching for a suitable one to connect and re-start transmission process.

3.1.1   Used Algorithm

1. Assigned target is arranged to be the required IP address
2. Inputs are the received signals from the nearest Mobile base stations
3. As usual, the target is assigned as d, y represents the perceptron, and n represents data. Used formula as

$$E_j(n) = d_j(n) - y_i(n). \tag{1}$$

4. Adjustment process for the weights by

$$\varepsilon(n) = 0.5 \sum e_j^2(n). \tag{2}$$

5. Process of adjustment to detect the changes in weights as

$$\Delta w_{ij}(n) = -\eta \left(\frac{\partial \varepsilon(n)}{\partial v_j(n)} \right) y_i(n), \tag{3}$$

where η is the learning rate, y_i is the previous neuron.

6. Adjustment process is done by

$$-\frac{\partial \varepsilon(n)}{\partial v_j(n)} = \varphi(v_j(n)) \sum \left(\frac{\partial \varepsilon(n)}{\partial v_k(n)} \right) w_{kj}(n) \tag{4}$$

In the above algorithm, we are managing the stability of the IP addresses transmitted through the access point in each floor of the building that includes elevators without any disconnections. The ANN training results are tabulated in Table 1.

Table 1. Training results

Items	Results
Input	9 stations
Output	1 IP address
Iterations	19
Time	00.00.01 ms
Performance	1.25
Gradient	2.28
Mu	0.00100
Accuracy	90.7447

According to the obtained results shown in Table 1, the best validation performance is found to be 0.083486 at epoch 3 as shown in Fig. 2, while Fig. 3 shows the histogram versus instances, so that Error equals to Target minus Outputs, so that accuracy of the system reached 90.7447.

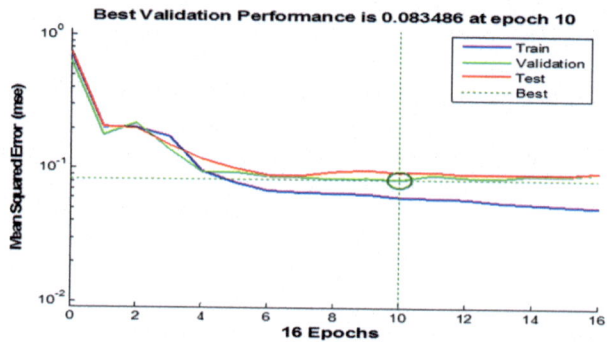

Fig. 2. Mean square error

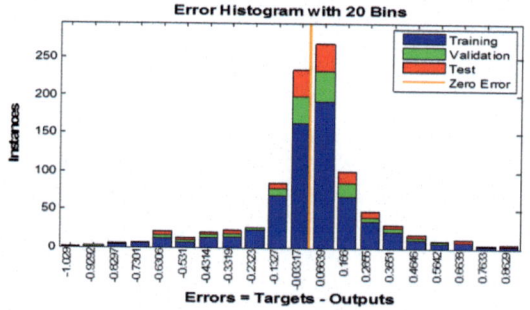

Fig. 3. Error histogram

Using the obtained results by the ANN to estimate the path loss using the proposed formula, this results in an accurate estimation map of the signal strength within multiwall buildings. This is valid for 2D and 3D LOS system, in order to relocate the position of the access point, which is used as an amplifier for the signal inside multiwall buildings between both transmitter antenna and receiver. The proposed formula is a combination of all previous standard models just with an additional term which is the elevators [1–3, 7]. This model is valid for any multiwall building by applying the blueprint image using image processing technologies in order to detect the walls inside the building. Using Hough-Transform (HT) to detect the multi-wall in the building, as shown in Fig. 4.

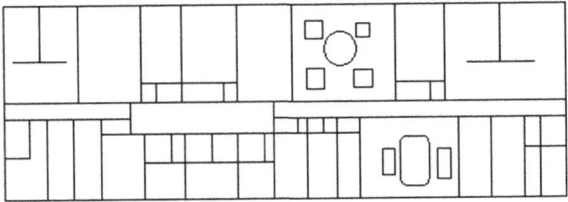

Fig. 4. First floor in the faculty of health sciences

Regarding the use of Eq. (1) and the obtained results in [2, 3, 7], so that the building includes an elevator, several floors (Fig. 4). The wave propagation case in the mentioned building is studied, and all possible cases are taken into consideration. As much as, the case of adding extra splitters to the base stations, which was another solution to solve the coverage areas for the mobile users to do their requested works in the download process. In this paper, the process of estimating the accurate location of the access point to be installed and pointed so obtaining maximum coverage without any loss in the signal. The application of the proposed model in order to determine the accurate location of the access point inside the building that includes the elevator. Within the available base stations inside the campus, the most suitable one is selected according to its average signal strength, average load and average handover time.

The collected data from the nearest base station and used to detect the accurate location is tabulated in Table 2.

Table 2. Collected data from the nearest base station

Base station ID	Hand-over time	Strength dB	Load	Ratio
2	196.933	−50.787	43.625	0.228
3	196.982	−50.787	43.625	0.228
4	198.013	−50.787	43.625	0.228
5	197.968	−50.787	43.625	0.228
6	197.676	−50.787	43.625	0.228
7	197.480	−50.787	43.625	0.228
8	156.144	−50.866	40.520	0.203
9	156.451	−50.866	40.520	0.203
10	156.611	−50.866	40.520	0.203
11	156.026	−50.866	40.520	0.203
12	156.047	−50.866	40.520	0.203
13	156.509	−50.866	40.520	0.203
14	156.616	−50.866	40.520	0.203
15	156.194	−50.866	40.520	0.203

Regarding the selected base station and its data, the average signal strength versus load is plotted, as shown in Fig. 5.

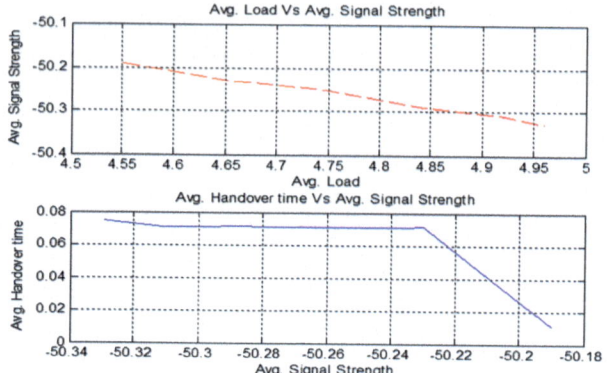

Fig. 5. Average signal strength versus load

The data mentioned above is used to locate the accurate position of the access point as the transmitter point (Tx), to keep the strength of the signal inside the elevator. Presenting the walls by straight-lines, then selecting any floor in the building that includes the elevator, and selecting two points as two points on the borders of the chosen floor. Moreover, selecting the position of the access point and applying Eq. (1). The obtained result is shown in Fig. 6, which is the accurate location of the Tx inside the building.

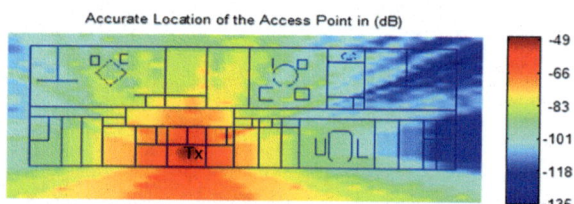

Fig. 6. Accurate location of the Tx inside the building

4 Conclusion

In this paper, to eliminate the path loss propagation in buildings with elevators, and avoiding disconnection issues which leads to the loss in the transmitted data packets is studied. Moreover, the use of the ANN in this system has the significant advantage in the stability of the IP addresses so that the continuity of the transmission process is achieved. Regarding the proposed model and the multi-wall building with elevator, the accurate position of the access point so that the stable signal with minimum handover issues and cancellation in the path loss propagation is achieved too. The obtained results allowed to maintain the connection continuity inside buildings, including the artificial neural network to provide continuous communication of the mobile user with the signal to noise ratio 0.203 in the presence of the handover.

References

1. Jamal, F.A.B.: Estimating coverage of radio transmission into and within buildings for line of sight visibility between two points in terrain by linear prediction filter. In: Third Mosharaka International Conference on Communications, Signals and Coding. MIC-CSC, pp. 1–5 (2009)
2. Jamal, F.A.B., Firudin, Kh.M.: Direction prediction assisted handover using the multilayer perception neural network to reduce the handover time delays in LTE networks. Procedia Comput. Sci. **120**, 719–727 (2017)
3. Jamal, F.: Improvement in strength of radio wave propagation outside the coverage area of the mobile towers for cellular mobile WiFi. In: Aliev, R., Kacprzyk, J., Pedrycz, W., Jamshidi, M., Sadikoglu, F. (eds.) 13th International Conference on Theory and Application of Fuzzy Systems and Soft Computing ICAFS. Advances in Intelligent Systems and Computing, vol 896, pp 464–471. Springer, Cham (2019)
4. Obeidat, H.A., Asif, R., Ali, N.T., Dama, Y.A., Obeidat, O.A., Jones, S.M.R., et al.: An indoor path loss prediction model using wall correction factors for wireless local area network and 5G indoor networks. Radio Sci. **53**, 544–564 (2018)
5. Bakinde, N.T.S., Faruk, N., Popoola, S.I., Salman, M.A., Oloyede, A.A., Olawoyin, L.A., Calafate, C.T.: Path loss predictions for multi-transmitter radio propagation in VHF bands using Adaptive Neuro-Fuzzy Inference System. Eng. Sci. Technol. Int. J. **21**, 679–691 (2018)
6. Zreikat, A., Djordjevic, M.: Performance analysis of path loss prediction models in wireless mobile networks in different propagation environments. In: Proceedings of the 3rd World Congress on Electrical Engineering and Computer Systems and Science (EECSS 2017), Rome, Italy, 5–6 June (2017)
7. Bose, A., Chuan, H.F.: A practical path loss model for indoor WiFi positioning enhancement, 1-4244-0983-7/07. IEEE (2007)
8. Elechi, P., Otasowie, P.O.: Path loss prediction model for GSM fixed wireless access. Eur. J. Eng. Res. Sci. EJERS **1**(1) (2016)
9. Miura, Y., Yasuhiro, O.D.A., Tokio, T.: Outdoor-To-Indoor propagation modeling with the identification of path passing through wall openings, Wireless Laboratories, NTT DoCoMo, Inc. 3–5 Hikari-no-oka, Yokosuka-shi, Kanagawa, 239-8536, Japan, 0-7803-7589-0/02. IEEE (2002)
10. Jadhav, A.N., Kale, S.S.: Suburban area path loss propagation prediction and optimization using Hata model at 2375 MHz. Int. J. Adv. Res. Comput. Commun. Eng. **3**(1) (2014)
11. Beauregard, A., Haas, H.: Pedestrian dead reckoning: a basis for personal positioning. In: Proceedings of the 3rd Workshop on Positioning, Navigation and Communication (WPNC 2006) (2006)
12. Zyad, N., Julius, D., Jamal, F.: A new algorithm to enhance radio wave propagation strength in dead spots for cellular mobile WiFi downloads. IEEE (2014). 978-1-4577-1343-9/12
13. Zyad, N., Juliusm, D., Jamal, F.: A new selected points to enhance radio wave propagation strength outside the coverage area of the mobile towers in the dead spots of cellular mobile WiFi downloads. In: Long Island Section Systems, Applications and Technology Conference. IEEE (2015)

14. Fathi, J., Sadikoglu, F.: Radio wave propagation model for enhancing wireless coverage in elevator of buildings. In: Aliev, R., Kacprzyk, J., Pedrycz, W., Jamshidi, M., Babanli, M., Sadikoglu, F. (eds.) 10th International Conference on Theory and Application of Soft Computing, Computing with Words and Perceptions – ICSCCW 2019. Advances in Intelligent Systems and Computing, vol 1095. Springer, Cham (2020)

15. Majed, M.B., Abd Rahman, Th., Abdul Aziz, O., Hindia, M.N., Hanafi, E.: Channel characterization and path loss modeling in indoor environment at 4.5, 28, and 38 GHz for 5G cellular networks. Int. J. Antennas Propag. **2018**, Article ID 9142367 (2018). https://doi.org/10.1155/2018/9142367

16. Perez, C.V., Garcia, J.L.G., Higuera, J.M.: A simple and efficient model for indoor path-loss prediction. Meas. Sci. Technol. **8**, 1166–1173. UK. PII: S0957-0233(97) 81245-3 (1997)

Current State and Diagnostics of the Consumer Goods Market Using Neural Networks

Mikayilova Rena[1,2](✉) [ORCID]

[1] Digital Economy and ICT, Azerbaijan State Economic University,
AZ1147 Baku, Azerbaijan
`rana.mikayilova@unec.edu.az`
[2] Azerbaijan State Oil and Industry University,
Azadliq Avenue 20, Baku, Azerbaijan

Abstract. This work is devoted to the study of the possibilities of using neural networks to solve the problem of predicting the volume of retail turnover of the consumer goods market per capita in order to determine the further level of development of the population's security. The difficult point in forecasting the retail turnover of the consumer goods market is the choice of forecasting methods that differ from other commodity markets. The solutions to the problem posed using methods for forecasting time series using the Neural Network Toolbox package of the MATLAB system are considered.

Keywords: Consumer goods market · Forecasting methods · Neural networks · Neural networks with a time lag · The volume of retail turnover

1 Introduction

The final stage of the formation of the commodity market (after the industrial and agricultural commodity markets) is the consumer commodity market, which is subdivided into the retail consumer goods market, the public catering market and the paid services market [1, 2].

A feature of the consumer goods market is that it is formed by the ratio of supply and demand on a permanent trading floor. An important place in the diagnostics of the consumer goods market is occupied by the analysis of methods for forecasting retail turnover. There are various approaches in this direction. So, for example, IS Novikova [3] believes that the formation of a new shopping facility should thoroughly study the main indicators that affect the efficiency of its work: turnover, income, costs, profitability, etc. As noted above, our opinion is significant the indicator on which the efficiency of a commercial firm depends is the turnover. Therefore, it becomes necessary to study the main methods of forecasting retail turnover.

The retail turnover of a commercial enterprise consists of two important categories - the flow of consumers (number of checks) and the average check, which fluctuate under the pressure of various factors. They can be both internal and external.

At the same time, not all factors of a commercial enterprise can be influenced equally. The commercial enterprise practically does not influence the environmental factors, they should be taken into account when predicting economic results.

R. A. Aliev et al. (Eds.): ICAFS 2020, AISC 1306, pp. 337–345, 2021.
https://doi.org/10.1007/978-3-030-64058-3_42

When opening a new business enterprise, you should give a correct forecast assessment of the company. Hence, for an effective selection of techniques and max rational calculation of the forecast, the entire set of methods and approaches of innovation should be studied.

It is known that forecasting is a probabilistic scientifically grounded judgment about the trends of the phenomenon under study in the future.

When forecasting, one should consider as many factors as possible.

This especially applies to complex factors affecting trade: location and location of a commercial enterprise, target audience, volume and capacity of the market, competition, regional characteristics, etc.

In practice, such forecasting methods and techniques are mainly used as the method of expert assessment; balance method; normative method; extrapolation method; modeling method; method of economic forecasting [3].

The use of the balance method makes it possible to implement the principle of proportionality and balance, which consists in the consistency of the needs of a commercial firm in various types of raw materials, material, financial and labor resources with the possibilities of producing and selling goods. Since the total needs are higher than the available resources, in the process of forming the balances, the needs are distributed according to the degree of importance: into priority and less significant.

The balance method is used in forecasting estimates both at the macro level and at the micro level of firms, however, it is difficult to apply it when making a forecast of the formation of a new commercial enterprise. In this case, the method will not include many characteristics of the market - demand, competition, market trends.

The normative method is widely used in the consumer market. Which includes predictive values based on taking into account physiological norms of consumption, scientifically grounded rational norms, determining the values of their achievement.

Norms and standards are developed in advance based on the relevant standards. The norm is the maximum permissible fluctuation rate of a particular phenomenon. When forecasting, the norms are laid down in all significant indicators.

The advantage of the method is the minimum time spent on finding a solution to the problem. At the same time, the complexity of the regulatory method lies in the definition and development of these norms and their objective validity.

In the consumer market, a modeling method that includes constructing a model based on an initial study of a commercial enterprise, highlighting its main features, content, and development trends. The simulation identifies the important factors that need to be taken into account and indicates how these factors might affect the business. Forecasting using models consists of development, practical analysis, comparison of the results of the initial forecast calculations with the actual data of the state of the process in a commercial enterprise.

The method of economic forecasting in the consumer market is based on the development of economic and mathematical models. They are based on applied mathematics, especially the sections of linear programming, mathematical statistics, game theory and others. Economic and mathematical forecasting contributes to the determination of many factors that influence the model, to find connections between indicators, to choose the best options and solutions for the development of the model. With the help of diagnostics, it is possible to identify the essence of this process, as well as to determine the patterns of its change in the future, to comprehensively assess the ways to achieve the set goals.

Extrapolation is used when it is required to obtain an approximation of a function outside a known interval.

The difficult point in forecasting the retail turnover of the consumer product market is the choice of forecasting methods that differ from other product markets [1, 3].

Studies have shown that not all criteria and indicators of forecasting methods lend themselves to mathematical formulation, and in most cases they require their inclusion in forecasting, assessment and decision-making models as the most desirable and important factors.

On the other hand, having a large amount of information on hand, one should analyze their retrospectiveness and in the future predict their dynamics of change based on learning algorithms based on neural networks [4].

In this regard, the paper proposes to use artificial intelligence methods, including, for our case, neural networks (ANN) [5, 6] to predict retail turnover in the consumer market. Neural networks have come into practice wherever it is necessary to solve problems of forecasting, classification or automation. There are several reasons for this success. 1. Neural networks are a powerful modeling technique that allows you to reproduce extremely complex dependencies. Neural networks are nonlinear in nature. In addition, neural networks cope with the "curse of dimension", which does not allow modeling linear dependencies in the case of a large number of variables. 2. Neural networks learn by example. A neural network user picks up representative data and then runs a learning algorithm that automatically perceives the data structure. Of course, the user is required to have some kind of heuristic knowledge about how to select and prepare data, choose the desired network architecture and interpret the results, but the level of knowledge required for the successful application of neural networks is much more modest than, for example, when using traditional methods of statistics.

2   Formulation of the Problem

For a long time, scientists have been discussing a number of indicators of the consumer goods market. The subject of the dispute is such indicators as the capacity and volume of the consumer product market, which in one case imply an indicator of the market capacity by a broader category, the volume of the product market. Note that for characterizing the consumer market there is a set of indicators, the use of which depends on the level of competition in the market, the characteristics of goods in the consumer market, market saturation with one or another product, the number and structure of buyers.

Forecasting retail turnover in the consumer goods market, we can determine the trend in the development of retail turnover for food products and non-food products.

Consider the problem of approximating and predicting a number of dynamics, i.e. building a function for a finite set of points, based on indicators of retail turnover (Table 1) for 2003-20017 (data are given by year, N = 15 - sample size). Figure 1 shows a graph of the dynamics of the volume of retail trade turnover per capita.

When starting the development of a neural network solution, as a rule, one faces the problem of choosing the optimal architecture of a neural network. Since the areas of application of the most well-known paradigms overlap, different types of neural networks can be used to solve a specific problem, and the results may be the same. Whether this or that network is better and more practical depends in most cases on the conditions of the problem.

Let us solve the posed problem using time series forecasting methods using the Neural Network Toolbox software package of the MATLAB system, in particular, using nntarintool [5].

3 Application of Neural Networks to Solve the Problem of Forecasting the Volume of Retail Trade

Let us consider the possibilities of using neural networks to solve the problem of forecasting the volume of retail turnover per capita.

The essence of the time series forecasting method is as follows: there is a time series Y, represented by its values, lags (reflecting the lag or advance in time of one phenomenon in comparison with others) in the previous moments of time Yt − k, Yt − k + 1, Yt, where t is the current moment in time, k is the depth of the historical sampling. It is necessary to find the magnitude of the change in its value, which will occur at the next moment in time, based on the analysis of past values, as well as the history of changes in other factors affecting the dynamics of the predicted value. In this case, it can be important both the lowest possible value of the forecast error and the

Table 1. The volume of retail trade turnover per capita.

Year	Retail turnover per capita (x1)	Including	
		Prod. products	Unproductive products
2003	369,8	255,9	113,9
2004	444,2	303,6	140,6
2005	542,7	366,1	176,6
2006	666,0	442,1	223,9
2007	862,9	567,3	295,6
2008	1 214,1	789,0	425,1
2009	1 302,0	847,4	454,6
2010	1 484,4	960,1	524,3
2011	1 754,1	1 070,6	683,5
2012	1 913,6	1 090,7	822,9
2013	2 114,3	1 129,9	984,4
2014	2 336,8	1 170,8	1 166,0
2015	2 699,2	1 350,7	1 348,5
2016	3 132,5	1 564,4	1 568,1
2017	3 623,2	1 837,2	1 786,0

correct determination of the nature of the further change in the value of the predicted value (increase or decrease).

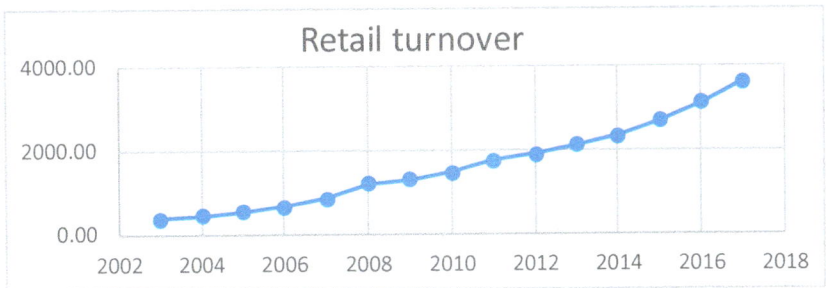

Fig. 1. Graph of the dynamics of the volume of retail trade turnover per capita.

When solving the forecasting problem, it is required to find the value of Y at a future point in time based on the known previous values

$$\overline{Y}, \overline{V}, \overline{C} : Yt + 1 = g(\overline{Y}, \overline{V}, \overline{C})$$

$$Y_{t+1} = g(Y_{t-k} \ldots Y_t, V1_{t-k} \ldots V1_t, \ldots, Vn_{t-k} \ldots Vn_t, C1_{t-k}, \ldots, C1_t, \ldots, Cm_{t-k} \ldots Cm_t) \tag{1}$$

Expression (1) is the main relationship that determines the forecast.

The forecasting problem is considered by us as an input-output time series problem, for the solution of which we use a neural network with a time delay, the general architecture of which is shown in Fig. 2.

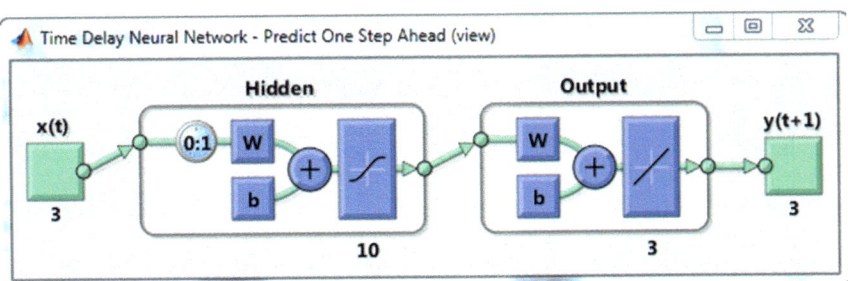

Fig. 2. General architecture of time delay neural network.

As you can see from the figure, the neural network has the following parameters: IW: {2 × 1 cell} containing 1 input weight matrix; LW: {2 × 2 cell} containing 1 layer weight matrix; b: {2 × 1 cell} containing 2 bias vectors.

The general forecasting algorithm using a neural network consists mainly of the following steps: - obtaining a time series with an interval of the selected time iteration; - row smoothing; - formation of a table of "windows" with the immersion depth of time intervals; - adding additional data to the table (data for previous years); - determination of training and test samples; - selection of neural network parameters; - training a neural network; - testing the performance of the neural network in real conditions.

The results of the neural network for forecasting are shown in Figs. 3, 4 and 5.

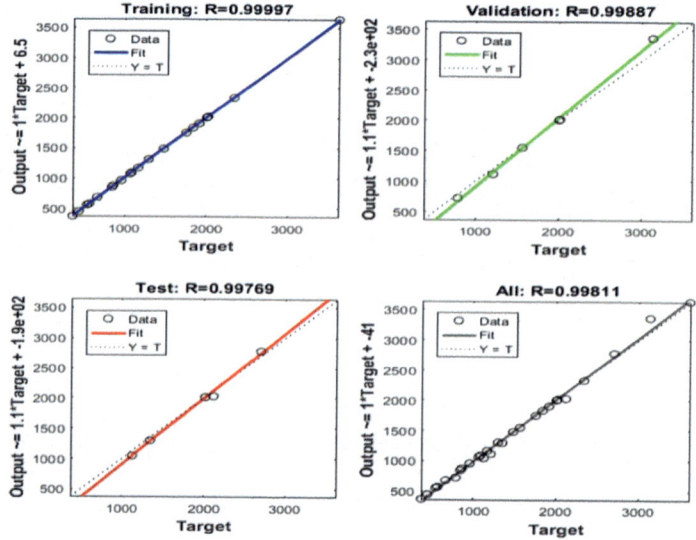

Fig. 3. Neural network training regression

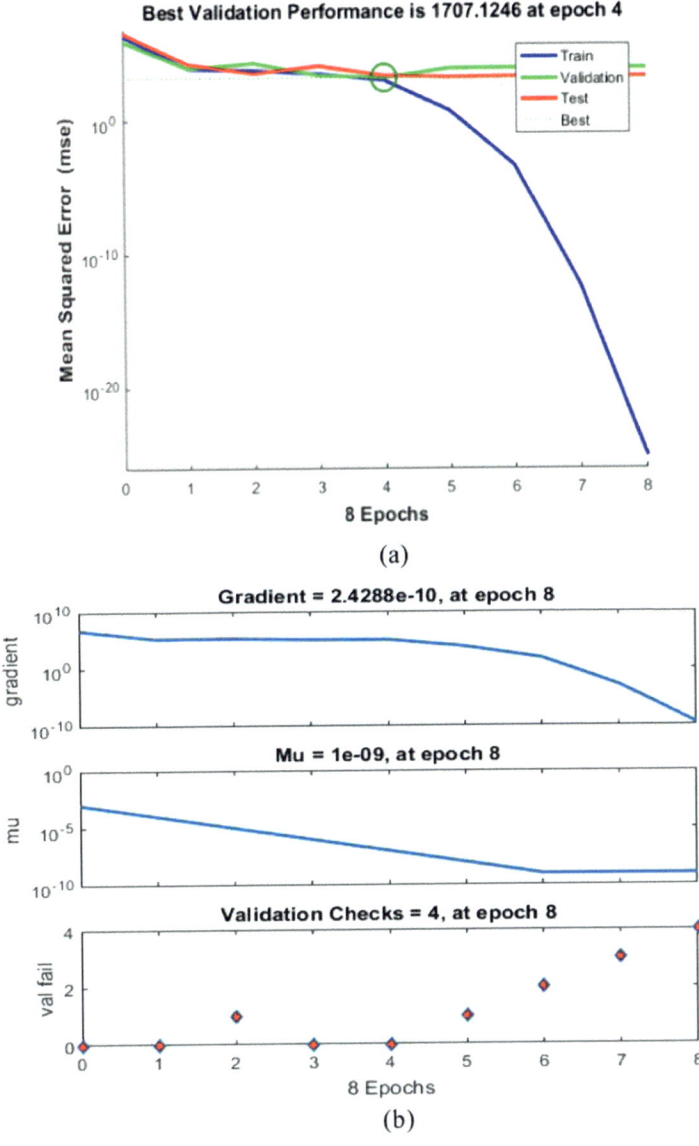

Fig. 4. a) Validation performance b) Neural network training state

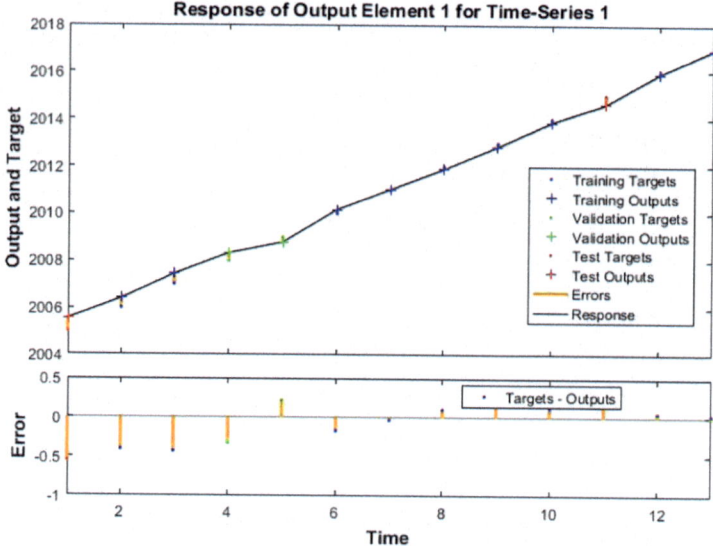

Fig. 5. Response of output element

As can be seen from Fig. 5, after training, the results of the predicted and real values of the experimental data almost coincide, which is confirmed by the fact that the coefficient of determination for all samples of the neural network R2 is greater than 0.997 (Fig. 3). This proves the feasibility of using time series forecasting methods to solve the problem of forecasting the volume of retail turnover for the consumer goods market.

4   Conclusion

Analysis of the forecasting results showed that:

- according to the rate of development, the total volume of the republic's trade will increase according to the trend shown in Fig. 1;
- a positive trend is observed in the indicators of retail turnover of food products and non-food products. The volume of retail trade turnover per capita will significantly increase, which is an indicator of a high level of income of the population;
- a difficult point in the study of the consumer product market is the choice of forecasting methods that differ from other product markets;
- we propose to use the method of forecasting the consumer product market based on the use of artificial intelligence methods, in particular, for example, the use of a neural network with a time delay as a special case for forecasting a time series.

References

1. Samigulin, E.V.: Economic law and parameters of the retail product market. AUCA Acad. Rev. 117–125 (2008)
2. Petrova, A.V.: Analysis of the product market and assortment. Yekaterinburg, UrGUPS (2016)
3. Novikova, I.S.: Analysis of methods for forecasting the turnover of a trading facility. www.c: Premier/Downloads/EPM
4. Aliev, R.A., Bijan, F., Aliev, R.R.: Soft Computing and its Applications in Business and Economics, Springer, Heidelberg (2004)
5. Beale, M.H., Hagan, M.T., Demuth, H.B.: Neural Network Toolbox. User's Guide. Math Works, Inc., Natick (2014)
6. Kruglov, V.: Mathematical extension packages MATLAB. Special reference, SPb, Piter (2001)

The National Technological Initiative as a Digital "Lego Bricks" for Industries: Some Prospects of the Modernization New Wave in the Russia's Economy

Vovchenko Natalia, Andreeva Olga, Dmitrieva Valeria$^{(\boxtimes)}$, and Kaptsova Valeriya

Rostov State University of Economics, 69, B. Sadovaya str.,
344002 Rostov-on-Don, Russian Federation
nat.vovchenko@gmail.com, olvandr@ya.ru,
riyachan2807@gmail.com, kaptsovavalera@mail.ru

Abstract. Last 30 years, Russian economists have been talking about the modernization or transformation of the economy of Russia. And during this period of time, all attempts to modernize it was interrupted with the economic and political crises. So, now, the NTI program has become the response of this challenge. NTI is a Russian mega-project aimed at creating favourable conditions for the implementation of innovative projects by providing financial and non-financial support at various stages of the technology companies' development. Like a Lego set for kids, which can be used to assemble anything, NTI provides regions with the opportunity to choose the direction of the economic modernization. Regions can "construct" any product and any market they need, can use any support tools and measures. This article suggests to popularize the NTI program as an effective set of tools for regional transformations. We proposed a fuzzy-multiple algorithm for assessing the effectiveness of the project. It is based on an algorithm using indicators reflecting the achievement of goals and objectives (significant benchmark results) with the actually achieved level of budget expenditures aimed at achieving them.

Keywords: National technological initiative · Innovation economy · Digital transformation · NTI markets · Modernization

1 Introduction

For about 30 years, Russian economists have been talking about the need to go from a "raw material" model of the Russian economy to an innovative one, however, the income of the oil and gas sector still compensate the most part of the federal budget. For the same 30 years, attempts have been made to modernize the economy [1–10].

The modernization first wave in the Russian economy, which took place in 1999–2008, was actually interrupted. At the same time, due to the excessive depreciation of fixed assets it was impossible to carry out the technological modernization properly. Moreover, according to some experts, the share of worn-out equipment was up to 64% [10].

R. A. Aliev et al. (Eds.): ICAFS 2020, AISC 1306, pp. 346–353, 2021.
https://doi.org/10.1007/978-3-030-64058-3_43

According to G. Khanin and D. Fomin during the post-Soviet privatization, the evaluation of fixed assets was carried out taking into account not objective factors, but the interests of definite political groups [10–13]. Thus, the authors note that in 1992, during the inventory of fixed assets, it was admitted that the results of that evaluation would not affect the privatization. At the same time, the gap between the balance sheet and recovery values of fixed assets was the same throughout the post-Soviet years and reached 8 times in 2015 [10]. The underestimation of the fixed assets cost, coupled with an inefficient privatization mechanism, as well as traditional systemic problems of the Russian economy, became one of the most serious obstacles to the first stage of modernization.

After a long recession caused by the consequences of the global financial crisis of 2008–2009, there was renewed an interest in the problem of modernizing the Russian economy. This was the beginning of the modernization second wave, but it had not the same high growth rates of economic indicators as the first one, and the upward trend was interrupted by political events in 2014 [9]. High dependence on import typical for the recent history of Russia, as well as a number of other problems in 2014, in the context of economic sanctions against Russia, actually contributed to the unfolding of a systemic crisis. Among the negative consequences of the imposition of sanctions are:

- pressure on the domestic market, which led to reducing in production;
- reducing the labour market and the growth of the army of unemployed;
- reducing the investment in fixed assets due to reduced foreign investment inflows and lower effective demand from households;
- the growth of an inefficient public sector and the weakening of market and legal institutions;
- falling confidence in the government and low business activity, etc.

From 2015, in the context of economic restrictions, the need for import substitution, as well as other traditional problems, the modernization third wave of the Russian economy began. It was based not only on the need for technical re-equipment of production and, first of all, digital transformation, but also on the idea of stimulating scientific and inventive activity as an economic growth driver.

Structural imbalances in the Russian economy, the unequal socio-economic development of Russian regions, the weak diversification of regional industry complexes, as well as traditionalism, inertia and weak receptivity to innovation are problems that cannot be solved by transferring capital from one industry to another, from one region to another. It was necessary to rethink the industry policy at the federal level. The response to this challenge was the formation of a long-term program to support innovation, called the "National Technological Initiative" (hereinafter – NTI).

The NTI program is a mega-project aimed at creating favourable conditions for the implementation of innovative projects by providing financial and non-financial support at various stages of the technology companies' development [20]. At the same time, digital solutions and the commercialization of breakthrough technologies during the implementation of the NTI program are in priority.

One of the most important problems of the project is to evaluate its effectiveness. The paper presents the corresponding fuzzy-multiple algorithm, similar to that used to assess the implementation of departmental target programs, using a set of performance indicators - indicators for assessing the achievement of goals and objectives (significant control results) with the budget expenditures actually achieved to achieve them.

2 Literature Review

Technological innovations play a key role in fast-growing economies [14–25]: they provoke and direct structural changes, expanding sectors associated with manufacturing high-tech products, producing knowledge and services, they accelerate the business cycle, contribute to industrial opportunities, create the new stable competitive environment, as well as include some regions and countries in global value chains [7]. In this approach, the development of knowledge and competencies of the individual and the entire organization should be considered as the basis for future changes.

Beginning with significant works by F. Machlup "The Production and Distribution of Knowledge in the United States" and K.J. Arrow "The Economic Implications of Learning by Doing", published in 1962, the field of education and research becomes especially interesting for economists, and changes in knowledge are included in neo-classical models of the economic growth: if the amount of spent labour or capital can change in positive and negative ways, the amount of knowledge only increases, and it is important that all the changes occur in a geometric progression. In his later works KJ. Arrow consistently developed the idea of including behavioural parameters in the economic growth models: information is transmitted not only through prices, but also through the entire system of the production and the popularization of knowledge [4], and technological progress is a direct consequence of learning by doing [3]. This was followed by numerous works of outstanding economists such as P. Romer [21, 22], N. Mankiw, D. Weil [15], R. Lucas [14], P. Aghion and P. Howitt [2], G. Grossman and E. Helpman [12], R. Barro and X. Sala-I-Martin [5] et al., presented the most detailed explanation of technological progress, including the production of innovations and the increase in human capital.

The problem of developing NTI programs is related not only to the distribution of knowledge and technology as the embodiment of this knowledge. Modernizing or even complete transforming the Russian economy as a whole and integrating it into global fast-growing markets [11] is the most difficult task facing researchers and practical economists.

The modernization of the Russian economy is rather difficult in the context of a complex geopolitical situation, sanctions and a slowdown in the economic growth, demographic shifts, and a number of other reasons. According to M. Dabrowski [6], Russia will not be able to return to the economic growth rates typical for the economy of the ninetieth.

It is noted that economic transformations within the modernization third wave should lead to an increase in the investment attractiveness of the Russian economy, the infrastructure development, supporting innovation and more efficient markets, strengthening international relations and the increased participation of Russia in international organizations. In fact, the NTI program has the same purposes. In the scientific literature, the issues of the NTI formation are presented rather narrow, researchers take into account only general issues of the NTI program formation and infrastructure functioning [1, 23, 24].

3   The Regional Participation in NTI Versus NTI as the Regional Development Factor

With new standards for working with data, information security technologies, the IT infrastructure development, robotics and human resources and supporting the ambitious goals of the national projects "Science", "Education" and "Digital economy", the national technology initiative gradually and purposefully forms the new economy based on network structures. Developing inter-organizational network structures are characterized by the following:

- all the participating organizations are independent;
- they are able to quickly adapt to changes in market conditions;
- independently approve the most optimal form of the network structure itself;
- they focus the activities of each participating organization on its best competencies and capabilities within specific projects;
- they reduce transaction and control costs;
- they form effective systems of knowledge and technology transfer within the structure.

With their legal independence, participating organizations are involved in solving certain problems and tasks, "linking" the territory of the region (or even several regions), thereby having a significant impact on the life standard and competitiveness of the region as a whole. A significant advantage of forming inter-organizational networks is to create a powerful and comprehensive socio-economic effect triggering a regional multiplier, as well as to create and keep a competitive environment [7, 8].

The NTI program is initially aimed at finding new markets in which Russia already has or may gain a competitive advantage in the future through the introduction of innovative technologies and safety standards, new forms of interaction with consumers, and the formation of new competencies and new professions on the basis of digital technologies. New markets include [16, 18]: Aeronet; Autonet; Marinet; Technet; Energynet; Neuronet; Healthnet; Safenet; Foodnet; Medianet; Fashionnet; Edunet; Gamenet; Econet; Homenet.

These markets cannot exist within a separate traditional industry. Any of the above markets is formed at the intersection of industries. And this feature is the most difficult obstacle to their development due to the inertia of thinking and the low propensity to change management systems. As a result, literally "permeated" with end-to-end technologies, NTI markets are actually configured over traditional industries, but in the future they should completely replace them. An enterprise in a traditional industry focuses on the production of goods and services, while in NTI markets, achieving network and synergy effects is more important than generating profits from the direct production. The benefits of joint activity, the distributed production and the project management, strengthening inter-regional relationships, the continuous development of employee competencies, the knowledge and technology transfer are just some of the advantages of new markets. In traditional industries, often regional producers are not competitive even in the interregional economic space. However, the NTI program, on the one hand, allows regions to become involved in the innovation production process, and on the other – in global value chains.

Like a Lego set for kids, which can be used to assemble anything, NTI provides regions with the opportunity to choose the direction of the new transformation. The set of traditional industries is important, but it doesn't matter: new industries and markets not related to the basic ones can be developed. Foreign experience in implementing innovation policy shows that a combination of traditional and exclusively innovative industries can be a rather effective solution to the problem of territories' development.

Sweden is an illustrative example here, due to its small size and difficult climate, which is forced to concentrate most of its economic activity in its Southern part. This country preserves the unique traditional small-scale production, but like other member states of the European Union, at the same time it focuses on the technological innovation.

Within the framework of the NTI program, regions get access to a set of cases of the best world and domestic practices in implementing high-tech projects, development tools (such as government support, expertise, PR and GR, consierge-service, acceleration and so on [25]) and public resources (including financial resources) [20]. Thus, the region can choose which new markets to join, and organizations can form an optimal network structure that will allow them to concentrate necessary resources, including financial ones [26].

The greatest obstacle to implement NTI programs is the inertia of the Russian economy. Russian regions have joined the NTI agenda, choosing Technet, Neuronet, Energynet and Foodnet as their main promising markets, i.e., actually traditional industries. The number of innovative projects passed the NTI examination and accepted for the implementation is also small and it is only 92. The NTI "Foresight 2.0" initiative [16], launched in 2020, as a form of combining the efforts of science, business and the state to solve the Russian economic development problems, should help regions rethink the advantages of including NTI markets in their work. Even by focusing on their own strong core industries, "permeated" with end-to-end technologies, regions can revive their own economy and start creating a competitive product.

To evaluate the effectiveness of the implementation of the action plan (roadmap) of the NTI as a whole, an algorithm similar to that used to evaluate the implementation of departmental target programs can be used, using a set of performance indicators - indicators for assessing the achievement of goals and objectives (significant control results) with the budget expenditures actually achieved aimed at their achievement for the reporting period (year) [19]. Each specific goal and objective of the plan's activities is evaluated by indicator I_i, calculated in the following order:

1) if the positive result is the excess of the actual indicator over the planned one, the performance indicator is equal to:

$$I_i = i_f / i_p$$

where I_f - the actual value of the indicator; i_p - planned value of the indicator;

2) if a positive result is a decrease in the actual indicator compared to the planned one, the calculation has the form:

$$I_i = i_p / i_f$$

The calculation of the assessment of the implementation of the tasks provided by the plan is as follows.

Each of the indicators is associated with a linguistic fuzzy variable Y = "implementation efficiency (by indicator)". It corresponds to a term-set of three terms $Y = \{Y_1, Y_2, Y_3\}$, where Y_1 = "ineffective implementation", Y_2 = "insufficiently effective implementation", Y_3 = "effective implementation".

As the membership functions of the terms, we take fuzzy trapezoid numbers: $a_1 = \langle 0.0, 0.45, 0.0, 0.55 \rangle$; $a_2 = \langle 0.55, 0.85, 0.45, 0.95 \rangle$; $a_3 = \langle 0.95, 1.0, 0.85, 1.0 \rangle$. Membership functions are defined in accordance with the generally accepted classification: 90 points and above - effective implementation; from 50 to 90 points - insufficiently effective implementation; less than 50 points - ineffective implementation. After that, experts set weight coefficients ri for each of the indicators. The total effectiveness is estimated based on the double convolution formula [17]:

$$I = \sum_{k=1}^{3} y_k \sum_{i=1}^{n} r_i \cdot \mu_k(I_i)$$

Here $\mu_k(I_i)$ are the values of the membership function of the three terms for each of the indicators. Then, for the resulting linguistic variable, it is recognized.

4   Conclusion

In conclusion, it must be admitted that the NTI program is no longer a point-based development strategy, and it is much less a strategy for overcoming regional inequality or sectoral imbalances. This is a strategy for the comprehensive innovative development of the Russian economy, to which each region can contribute.

Advantages of NTI programs for regions are:

- their proactive nature. NTI programs are directive in a way, but, nevertheless, organizations, their network associations and regions in general have a choice: to choose the NTI market, their own development strategy, NTI tools and measures to support NTI;
- the complexity of decisions: NTI programs are developed jointly by representatives of science and education, business and government agencies;
- their cross-disciplinary and cross-sectoral nature: it is extremely difficult to connect the interests of different industries representatives outside of the proactive NTI projects;
- the ability to constant update of educational programs to meet the requirements for new professions in the changing world;

- the ability to reorganize and even resuscitate industries that the region cannot restore on its own;
- the accelerated digitalization;
- network and synergistic effects.

The initiative nature of the NTI program is its main advantage and disadvantage at the same time, as this initiative nature makes the NTI formation "optional" in the region.

In general, it can be noted that, in fact, the NTI program is a new program for modernizing the economy, taking into account the formation of new employee competencies and new requirements and standards for work and safety.

References

1. Afonina, I.A., Sibirskaya, E.V., Oveshnikova, L.V.: Study of the infrastructure of the National Technology Initiative. Microeconomics **5**, 5–11 (2017)
2. Aghion, P., Howitt, P.: A model of growth through creative destruction. Econometrica **60**, 323–351 (1992). https://doi.org/10.3386/w3223
3. Arrow, K.J.: Information and economic behaviour. Voprosy Ekonomiki **5**, 30–42 (1995)
4. Arrow, K.J.: The economic implications of learning by doing. Rev. Econ. Stud. **29**(3), 155–173 (1962). https://doi.org/10.2307/2295952
5. Barro, R., Sala-i-Martin, X.: Technological Diffusion, Convergence and Growth. NBER Working Paper, 5151 (1995). https://doi.org/10.1023/a:1009746629269
6. Dabrowski, M.: Factors determining Russia's long-term growth rate. Russ. J. Econ. **5**, 328–353 (2019). https://doi.org/10.32609/j.ruje.5.49417
7. Dmitrieva, V.D.: Distribution of innovations and knowledge in the system of economic clusters. In: Interdisciplinarity in Modern Socio-humanitarian Knowledge-2019 (Knowledge as a Goal, Means and Catalyst for Social Development in the Digital World): Materials of the 4th Internatioal Science Conference, Publishing House of the Southern Federal University, Rostov-on-Don (2019)
8. Dmitrieva, V.D.: Problems of managing innovative territorial clusters in Russia. Effective management. In: The Collection of Materials of the 4th Scientific and Practical Conference Dedicated to the Memory of the Honored Professor of Moscow University M.I. Panova, Moscow, Russia, 27 October 2017. Polygraph Service, Moscow (2018)
9. Federal State Statistic Service. https://www.gks.ru. Accessed 02 May 2020
10. Fomin, D.A., Khanin, G.I.: The dynamics of capital assets in the economy of the Russian Federation over the post-Soviet period (1992–2015). Stud. Russ. Econ. Dev. **28**(4), 373–383 (2017). https://doi.org/10.1134/S1075700717040062
11. Gissin, V.I., Mekhantseva, K.F., Putilina, T.I., Surzhikov, M.A.: Green Economy: emerging national models, estimations and trends in the EU and the CIS. ERSJ **21**(1), 156–166 (2018). https://doi.org/10.35808/ersj/1168
12. Grossman, G., Helpman, E.: Innovation and Growth in the Global Economy MIT Press, Cambridge (1991). https://doi.org/10.1111/j.1467-9485.1993.tb00652.x
13. Khanin, G.I., Fomin, D.A.: Institutions and statistics (for example, statistics of fixed capital). Terra Economicus **4**, 33–45 (2017). https://doi.org/10.23683/2073-6606-2017-15-4-33-45
14. Lucas, R.: Making a miracle. Econometrica **2**, 251–271 (1993). https://doi.org/10.2307/2951551

15. Mankiw, N., Romer, P., Weil, D.A.: Contribution to the empirics of economic growth. QJE **2**, 407–437 (1992)
16. National Technological Initiative. Foresight NTI 2.0. https://nti2035.ru/nti_new. Accessed 02 May 2020
17. Nedosekin, A.O.: Methodological foundations of modeling financial activities using fuzzy-multiple descriptions, SPbSUEF, Saint Petersburg (2003)
18. NTI. Interactive Map. https://map.ntinews.ru. Accessed 02 May 2020
19. Rastova, Yu.I.: Assessment of the effectiveness of the organization of risk management processes. Manag. Sci. Mod. World **1**, 454–459 (2018)
20. Regional Standard of NTI. https://www.rvc.ru/eco/regions/regstandart. Accessed 30 Sept 2019
21. Romer, P.: Endogenous technical change. J. Polit. Econ. **5**, 71–102 (1990). https://doi.org/10.3386/w3210
22. Romer, P.: Idea gaps and object gaps in economic development. J. Monet. Econ. **3**, 543–573 (1993). https://doi.org/10.1016/0304-3932(93)90029-F
23. Siberian, E.V., Oveshnikova, L.V.: NTI as a strategic direction of the technological development of Russia. Stat. Econ. **1**, 34–41 (2018)
24. Siberian, E.V.: Transformation of the economy in the context of the formation of the National Technological Initiative. Bull. VolSU **3**(3), 21–30 (2017)
25. Support Tools for NTI Companies. http://projects.nti2035.ru/documents/docs/Instrumenty_podderzhki_kompaniy_NTI_2019.pdf. Accessed 20 Sept 2019
26. Vovchenko, N.G., Andreeva, O.V.: The financial circuit of a national technology initiative. Finance **12**, 8–13 (2019)

Analysis of Relationship Between Learning Outcomes and Student's Exam Results Using Association Rule Mining and Fuzzy Inference Rules

M. A. Salahli[1]([⊠])(iD), T. Gasimzadeh[2], F. Alasgarova[3],
and A. Guliyev[3]

[1] Department of Computer and Instructional Technologies Education,
Çanakkale Onsekiz Mart University, Çanakkale, Turkey
msalahli@comu.edu.tr
[2] Department of Instrument Making Engineering, Azerbaijan State Oil
and Industry University, 20 Azadlig Avenue, AZ 1010 Baku, Azerbaijan
tgasimzade@silkwaywest.com
[3] Department of Information Technology and Natural Sciences,
Azerbaijan Tourism and Management University, Koroglu Rehimov Street,
822/23, AZ 1072 Baku, Azerbaijan
flora.aleskerova@gmail.com, akber_guliyev@yahoo.com

Abstract. The academic success of students largely depends on the correct selection and evaluation of learning outcomes. The most effective tool for measuring student knowledge is exams. In this context, the scores of the students in the exams are the important indicators of their learning achievements. In this study the relationship between learning outcomes and exam results of students is analyzed. The research was carried out based on mathematics exam scores of students studying in the 6th grade of secondary school and the data related to the learning outcomes of this course. To reveal this relationship, an approach based on fuzzy logic and data mining methods is considered. The weights of the learning outcomes in the questions were analyzed, then the relationships between the outcomes and scores were determined using the Apriori algorithm. Then fuzzy inference rules were created based on the most frequent items determined as the results of the Apriori algorithm.

Keywords: Mathematics learning outcomes · Apriori algorithm · Fuzzy inference rules

1 Introduction

Developing students' skills in problem solving, decision making and reasoning is an important educational challenge for the 21st century educational system. However, the acquisition of these skills largely depends on the learning outcomes determined for the courses taken by the students. The University of Toronto Innovation Center has defined learning outcomes as follows: "Learning outcomes define the knowledge or skills students need to acquire at the end of a particular assignment, class, lesson or program, and that knowledge and these skills will be useful for them" [1]. The evaluation of students' achievement of the learning outcomes specified in the curriculum is an important part of the teaching-learning process. Learning outcomes are the knowledge and skills to be gained on a concrete topic. The most effective tool for evaluating these knowledge and skills and thus determining the level of achievement of learning outcomes are exams and tests. In this context exam and test questions should be able to adequately express the knowledge and skills targeted by learning outcomes. Learning outcomes are considered an important tool for student-centered teaching and learning [2].

The main purpose of this study is to reveal learning outcomes which have more impact on students' failure in mathematics lessons. The results of the research can help math teachers pay more attention to related topics and make improvements in textbooks and curricula. Association rules method was used to determine the learning outcomes that adversely affect student achievements. In our study, the impact values of learning outcomes and student grades were expressed as item sets.

Since exams are the most effective way to evaluate students' learning outcomes, intensive researches are conducted to investigate the relationship between student achievements and exam grades. The results of these studies have been used in order to make improvements in the curriculum and content on the related topics and to use more effective methods and tools to achieve learning objectives. The authors of the study [3] showed the main reasons for performance measuring and assessing student learning outcomes like program evaluation, student success evaluation, measure effectiveness, and others. The relationship between test frequency and student success in eighth mathematics was investigated in [4]. The study [5] contains a modern understanding of the assessment of learning outcomes in higher education, an overview of recent achievements and prospects for further development. The importance of learning outcomes is indicated in [5]. The relationship between students' achievement in primary schools and the results of central exams has been investigated in [6].

Educational data mining (EDM) methods are widely used to analyze statistical data and to create prediction models for evaluating student achievements. Application of Educational Data Mining to Improve Learner Performance in primary Schools is the subject of the study [7]. One of the widely used EDM methods is the association rules method. Association Rules Mining is used for identifying all associations and correlations between attribute values.

Many software tools have been developed to apply Data Mining methods. In terms of its simplicity and accessibility, the WEKA data mining application tool is popular among researchers. WEKA is a collection of machine learning algorithms for solving real-world data mining problems. It contains implementations of algorithms for classification, clustering, and association rule mining, along with graphical user interfaces and visualization utilities for data exploration and algorithm evaluation [8]. In [9], the data set of student academic records was processed using Naive Bayes, BayesNet, JRip and J48 classification algorithms using the Wiki tool. Students were classified according to their cities, schools, grades and Math results. The classifiers were compared and the best performing classification algorithm was determined. In the study [10] students' academic performance were evaluated by classification methods using the WEKA application.

Fuzzy logic is widely used in education, especially in evaluating students' academic performance [11–16]. In [11] a fuzzy logic system to help the teacher deciding the final mark of critical students is presented. In the study [14] authors use fuzzy logic to solve the problem of Evaluating Student learning achievement. Fuzzy Logic approach was used in [15] for evaluation of prospective math teachers' ability to enter graduate education.. In the study [16] a performance evaluation method based on fuzzy logic systems is proposed. Fuzzy logic and association rules methods were used together in many studies [17–20]. In the study [17] authors propose fuzzy association rule mining algorithm for fast and efficient performance on very large data sets. Fuzzy association rules use fuzzy logic to convert numerical attributes to fuzzy attributes. In the study [20] a multilevel association rule mining using fuzzy concepts is proposed. Studies in the related field show that although association rules and fuzzy logic methods are used for various problems of education. However, there is no study in the literature where fuzzy logic and data mining methods are used together to evaluate learning outcomes.

2   Research Problems

The purpose of this study is to increase the level of students' attainment of targeted learning outcomes by researching the effect of learning outcomes on exam results. To achieve this purpose, the following sub-problems should be solved. The influence of each learning outcomes on exam questions should be determined. The effect of each learning outcome on the student's exam scores should be determined. The solution of these sub-problems will help identify the problems that students and lecturers should focus on during the course.

3   The Method

The schematic expression of the method is shown in Fig. 1. The method we propose to determine the effect of learning outcomes on exam questions can be expressed as follows:

1. Learning outcomes of the course are determined. Let our set of learning outcomes be L = {l_1, l_2,..., l_n}. Set of exam questions are defined: $Q = \{q_1, q_2, ..., q_m\}$.
2. For each member of the set L and for each member of the set Q, the impact of each member of set L on each member of set Q - w_{l_i, q_j}, is evaluated. Where $l_i \in L, q_j \in Q$. For this, opinions of experts are asked.
3. The learning outcomes that most affect student academic performance and the relationship between these outcomes are determined by the association rules, which is a popular data mining method.
4. The association rules are used produce the fuzzy inference rules that express fuzzy relationships between learning outcomes and students' exam grades.

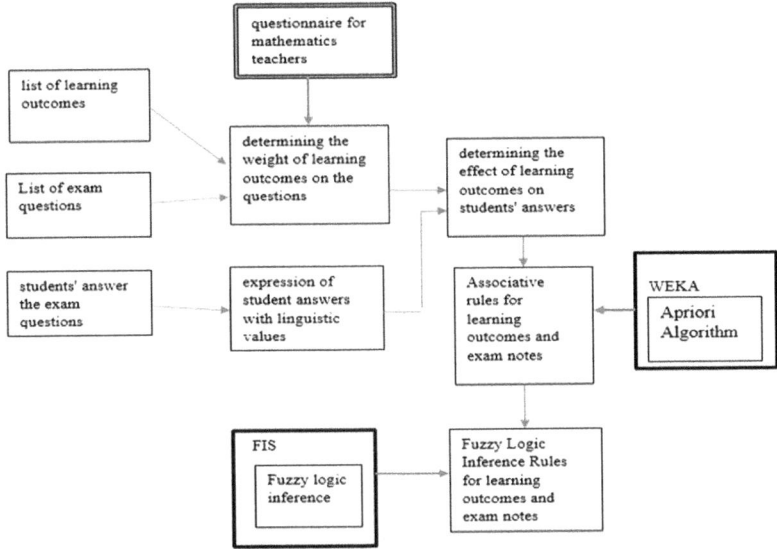

Fig. 1. A schematic expression of the method used to solve the research problem

Learning outcomes and exam question results are needed to solve the problem. The list of learning outcomes for the 6th grade mathematics course was formed according to the program determined by the Ministry of National Education of Turkey (Table 1). 26 exam results were examined in our study. The answers given by students to the exam questions have been rated with 10 points. In order to classify the achievements levels of students, these numerical scores have been converted into linguistic values as shown in Table 2.

Table 1. The learning outcomes

Name	Description of Learning Outcomes
EXP	Writes a repetitive product of a natural number as an exponential expression and calculates its value
OPR	It takes four arithmetic operations with natural numbers, considering the priority of the transaction
BRD	It performs operations to apply the common multiplier bracketing and dispersion feature in natural numbers
PRO	Solves and installs problems that require four operations with natural numbers
MRM	Determines multipliers and multiples of natural numbers
DİV	Explains and uses the rules of divisibility without division into 2, 3, 4, 6, 7, 8, 9 and 10
PRN	Determines prime numbers by their properties
PRM	Determines the prime multiples of natural numbers
CDM	Determines the common divisors and common multiples of two natural numbers and solves related problems
CMP	Compares and sorts whole numbers

In order to determine the weight of learning achievements in exam questions, the mathematics teachers of the school were asked to evaluate the 3 most important learning outcomes required for the solution of each exam question as low, medium and high according to their weight in the question (Table 3).

These data were used to determine the relationship between the student answers and the learning outcomes. The relationship between the learning outcomes for each question and the grades the student received from these questions was analyzed.

Table 2. Numerical grades and their linguistic values

Numerical grade	Linguistic value of the grade
0–4	Low
5–7	Medium
8–10	Good

Table 3. Learning Outcomes and related Queries

Queries	Learning Outcomes ant its weightings on the queries									
	EXP	OPR	BRD	PRO	MRM	DIV	PRN	PRM	CDM	CMP
Q1			Medium					High	Low	
Q2	High								Medium	Low
Q3	High	Medium			Low					
Q4						High	Low	Medium		
Q5		Low	Medium	High						
Q6		Low	High	Medium						
Q7						High			Medium	Low
Q8		High		Medium		Low				
Q9						High		Medium		
Q10		High	Medium						Low	
Q11	High			Low					Medium	
Q12		Medium	Low	High						
Q13	High				Medium		Low			
Q14			Medium		High		Low			Low
Q15		Low		High						Medium
Q16						High			Medium	Low

To reveal these relations association rules method - Apriory algorithm was used. Association rules method determine data items that are frequently used together. If we express the learning outcomes and student grades as data items, then we would learn which of these outcomes (or its combinations) most impact the student grades.

The Apriori algorithm revealed the learning gains most effective in student failures in exams. Learning outcomes need to be examined in relation to each other, not in separation. A student who fails a learning outcome is likely to fail another related outcome. The Apriori algorithm revealed the learning gains most effective in student failures in exams. Learning outcomes need to be examined in relation to each other, not in separation. A student who fails a learning outcome is likely to fail another related acquisition.

In this context, it is important for us to investigate the association rules produced by the Apriori algorithm, which expresses the relationships between the learning outcomes. Some association rules generated by Apriori algorithm is given below:

1. IF ERM = "low" THEN BRD = "low".
2. IF CDM = "low" THEN EXP = "low".
3. IF OPR = "low" AND CDM = "low" THEN EXP = "low".
4. IF EXP = "low" AND BRD = "low" AND PRO = "low" THEN CPR = "low".

The linguistic low, medium and high values used in learning rules for learning outcomes are interval values. By expressing these values as fuzzy values, we can express the relationships between learning outcomes and student grades more effectively. For this reason, the association rules generated with the Apriori algorithm were converted into fuzzy logic inference rules in the next stage of our research.

FisPro (Fuzzy Inference System Professional) application [21] was used to create fuzzy inference rules.

Some of fuzzy inference rules, generated from FisPro are:

1. Rule: IF OPR = "low" THEN BRD = "low"
2. Rule: IF OPR = "med" AND PRM = "low" THEN BRD = "low"
3. Rule: IF OPR = "med" AND PRM = "low" THEN BRD = "med"

These rules express the relationships between the outcomes OPR and PRM and the outcome BRD. For example, rule 1 indicates that if the student's knowledge of OPR learning outcome is "low", the student's knowledge of BRD outcome will also be "low". The screenshot showing the fuzzy inference rules, obtained from implementation of the FisPro application is given in Fig. 2. In Fig. 3 the graphic expression of fuzzy inferences for the rules 1–6 is given. Mamdani fuzzy inference method is used in our application.

Rule	Active	IF opr	AND prm	THEN brd
1	☑	low		low
2	☑	mid	low	low
3	☑	mid	mid	mid
4	☑	mid	low	low
5	☑	good	mid	mid
6	☑	good	good	good

Fig. 2. Fuzzy inference rules, generated from FisPro application (The screenshot)

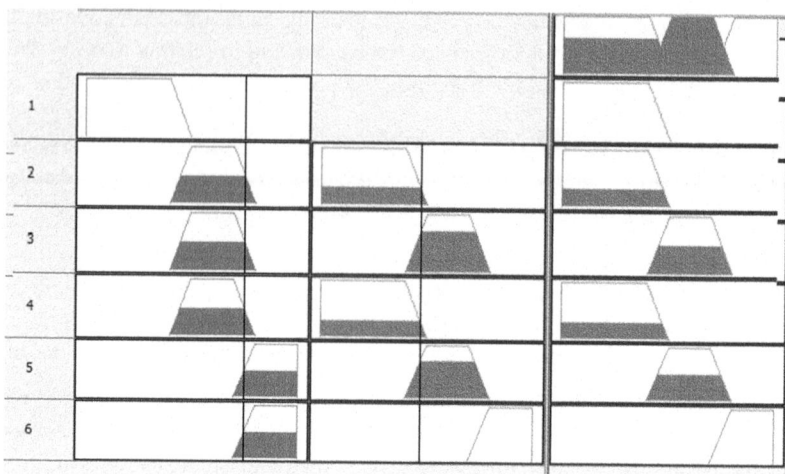

Fig. 3. Graphic expression of fuzzy İnferences for the rules 1–6 showing in Fig. 2

4 Conclusion

In this study, the effects of secondary school 6th grade mathematics achievements on students' exam results were analyzed. As a result of analysis, learning outcomes that have more impact on students' achievements have been determined. The research was carried out with the method in which association rules and logical inference methods are used together. Association rules were created with the Apriori algorithm in the WEKA data mining tool. FisPro software was used for the production of fuzzy inference rules.

References

1. Developing Learning Outcomes. https://teaching.utoronto.ca/teaching-support/course-design/developing-learning-outcomes/what-are-learning-outcomes/
2. Divjak, B., Ostroški, M.: Learning Outcomes in mathematics: case study of their implementation and evaluation by using elearning. In: 2nd Internatioanl Science Colloquium Mathematics Children (Learning outcomes), Element, Zagreb, pp. 65–76 (2009)
3. Introduction to Student Learning Outcomes Assessment. https://oira.unc.edu/files/2017/07/Introduction-to-SLO-Assessment.pdf
4. Guven, U.: The Relationship between Testing Frequency and Student Achievement in Eighth-Grade Mathematics: An International Comparative Study Based on TIMSS 2011 (Doctoral dissertation, Duquesne University) (2017). https://dsc.duq.edu/etd/131
5. Coates, H.: Assessment of learning outcomes. In: Curaj, A., Matei, L., Pricopie, R., Salmi, J., Scott, P. (eds.) The European Higher Education Area. Springer, Cham (2015)
6. Ayral, M., Ozdemir, N., Findik, L.Y., Ozarslan, H., Unlu, A.: The relationship between the students' achievement of Turkish language class and the central exam score. Procedia - Soc. Beh. Sci. **143**, 721–725 (2014)

7. Ramaphosa, K.I.M., Zuva, T., Kwuimi, R.: Educational data mining to improve learner performance in Gauteng primary schools. In: 2018 International Conference on Advances in Big Data, Computing and Data Communication Systems (icABCD), Durban, pp. 1–6 (2018)
8. Bouckaert, R.R., Frank, E., Hall, M.A., Holmes, G., Pfahringer, B., Rautemann, P., Witten, I.: WEKA–experiences with a java open-source project. J. Mach. Learn. Res. **11**, 2533–2541 (2010)
9. Yeasmin, S.: Analysis of StudentPerformance using Data Mining. B.Sc. in Computer Science and Engineering thesis. https://library.mist.ac.bd:8080/bitstream/handle/123456789/147/mistugthesis.pdf?sequence=1
10. Chawla, T.: Analysis of student's academic performance using classification algorithm in weka. Int. J. Adv. Res. Comp. Sci. **8**, 1245–1248 (2017). https://doi.org/10.26483/ijarcs.v8i7.4576
11. Annabestani, M., Rowhanimanesh, A., Mizani, A., Rezaei, A.: Descriptive evaluation of students using fuzzy approximate reasoning (2019)
12. Semerci, Ç.: The influence of fuzzy logic theory on students' achievement. Turkish Online J. Educ. Tech. **3**, 56–61 (2004)
13. Montero, J.A., Alsina, R.M., Morán, J.A., Cid, M.: Fuzzy logic system for students' evaluation. In: Cabestany, J., Prieto, A., Sandoval, F. (eds.) Computational Intelligence and Bioinspired Systems. IWANN 2005. Lecture Notes in Computer Science, vol. 3512. Springer, Heidelberg (2005)
14. Ingoley, S.N., Bakal, J.W.: Evaluating students' performance using four-node fuzzy controller. In: 2013 Nirma University International Conference on Engineering (NUiCONE), pp. 1–6 (2013)
15. Bahadır, E.B.: Evaluation of prospective math teachers' ability to enter graduate education with fuzzy logic along with various components. Eur. J. Educ. Stud. **5**(8), 61–87 (2018)
16. Gokmen, G., Akinci, T.C., Tektas, M., Onat, N., Kocyigit, G., Tektas, N.: Evaluation of student performance in laboratory applications using fuzzy logic. Procedia – Soc. Behav. Sci. **2**(2), 902–909 (2010)
17. Mangalampalli, A., Pudi, V.: Fuzzy association rule mining algorithm for fast and efficient performance on very large datasets. In: International Conference on Fuzzy Systems, Jeju Island, pp. 1163–1168 (2009). https://doi.org/10.1109/FUZZY.2009.5277060
18. Lekha, A., Srikrishna, C.V., Vinod, V.: Fuzzy Association rule mining. J. Comput. Sci. **11**(1), 71–74 (2015)
19. Hong, T.P., Lee, Y.C.: An overview of mining fuzzy association rules. In: Bustince, H., Herrera, F., Montero, J. (eds.) Fuzzy Sets and Their Extensions: Representation, Aggregation and Models. Stud. Fuzz. Soft Comp., vol. 220. Springer, Heidelberg (2008)
20. Rani, U., Prakash, R.V., Govardhan, D.A.: Mining multi level association rules using fuzzy logic. Int. J. Emerg. Techn. Adv. Eng. **3**(8), 2250–2259 (2013)
21. FisPro: An open source portable software for fuzzy inference systems FisPro. https://www.fispro.org/download/documentation/fispro36inline.pdf

Effects of Euro on Balance of Trade in the Euro Area

Hüseyin Özdeşer[(✉)] [iD]

Near East University, Nicosia, North Cyprus, Turkey
huseyin.ozdeser@neu.edu.tr

Abstract. The introduction of the euro aimed to improve the efficiency of the economic integration in the EU. Since 2002, the euro has assumed a strong position in the international money markets after the dollar. After the introduction of the euro, three main money anchors have been established in the international money markets, namely the dollar, euro and yen. Nevertheless, there are still uncertainties regarding the future value of the euro compared to the dollar and yen, the low economic performance of some EMU member countries and the incomplete integration process of the EU. In previous studies, it was shown that the introduction of the euro affected real interest, per capita income, and inflation rate in a positive way in European Monetary Union (EMU) member countries. Inevitably, these positive effects will also have impacts on the trade in EMU member countries. In the present study, the impacts of the euro on trade in the euro area will be investigated. In the study, the impacts of the euro on the trade balance of the three largest economies will be analysed, which are Germany, France, and Italy.

Keywords: Euro · Euro area · Import · Export · Trade balance · Effect

1 Introduction

In 1999, the current account in the euro area was negative, while in 2001, it became positive and continued to rise. Even though there were falling rates, it remained in positive values [1].

The current account imbalances in a country mainly emerge as a result of the difference in the international competitiveness of the country. Current Account (CA) imbalances may materialise if the national income in a country is greater than the sum of total consumption demand, investment demand and government expenditures, which means that the country will have a current account surplus. On the other hand, if total demand is greater than national output, then the country will experience a current account deficit; this means that the savings rate in the country is less than investment. If private and public savings in a country are greater than private investment and government expenditures, this means that there is excess saving and a current account surplus is recorded. In the opposite case, a current account deficit is observed [1].

The most important motivation in regard to the introduction of the euro as a single currency was based on the expectation that the euro would enrich the macroeconomic stability, achieve developments in investment and trade, would lead to increased

R. A. Aliev et al. (Eds.): ICAFS 2020, AISC 1306, pp. 362–369, 2021.
https://doi.org/10.1007/978-3-030-64058-3_45

efficiency in EU market integration to be more efficient and economic and political relations between the member states would be strengthened [2].

As the EMU member countries are using the euro as a single currency, this means that any uncertainty in exchange rates has been eliminated; hence, the EMU will be a more competitive region, which is an important element for increasing its trading capacity with the rest of the world. For the trade of a country to be stable, it is necessary to have macroeconomic stability and this was one of the greatest aims of the intro-duction of the euro. Currently, a total of 19 countries are using the euro, and this figure will continue to rise. Consequently, the trade potential of the euro area will tend to rise through the addition of new members.

The imports and exports in all the countries have important impacts on growth and development. All the countries around the world should be a part of the globalised world, as no country can remain self-sufficient. The global natural resources are not equally distributed between countries, which means that those countries who are unable to produce certain goods are becoming dependent on others. Within the last forty years, national economies have become more integrated. Inevitably, this has affected people's lifestyles in addition to policymaking. Certainly, this presents many opportunities and numerous professionals are working to understand these opportunities [3] "The Common Commercial Policy of the EU is based on Article 113 of the Treaty of Rome and entails a common Community tariff regime and common trade agreements with third countries" [4]. The EU is an important example of the economic integration process as it aims to remove all the economic and political restrictions between the member countries. The EU member countries apply common trade agreements with third parties. The trade potential of the EU is as follows: exports to the USA 20%, Norway 2.7%, Russia 4.6%, China 10.5%, Japan 3.2%, Turkey 4.5%, India 2.2%, South Korea 2.7%, and Switzerland 8%. Furthermore, in terms of imports, the potential is: USA 13.8%, Norway 4.2%, Russia 7.8%, China 10.5%, Japan 3.7%, Turkey 3.8%, India 2.4%, South Korea 2.7%, Switzerland 5.9%, Belgium, Luxembourg, the Netherlands, France, Germany, and Italy established the European Community with the Treaty of Rome in 1957. The aim of the treaty was to form a common market and to develop the less developed parts within the community. The main goal of the treaty was to establish a unified Europe and create a stronger rival against the USA [5].

For the time being, the borders of the EU are not certain and it is not known where the borders of the EU will end; the enlargement process is acting in a dynamic way. Inevi-tably, this is creating a risk for the future of the EU and also uncertainty about the future character of the euro in the international money markets as the single currency of the EU.

2 Methodology

Inflation rate, interest rate, and per-capita income are the important factors that affect the imports and exports of a country. The study will analyse how the exports and imports in the three largest economies of the euro area are affected by the introduction of the euro, namely Germany, France and Italy (In 2016 GDP-current US dollars for Germany 3.478 trillion, for France 2.465 trillion and for Italy 1.859 trillion). In an economy, when the level of per capita income rises, then this will encourage consumers to purchase more imported products, because people always aim to buy something that does not exist in their own economy. On the other hand, a fall in the inflation rate in the economy will cause the domestic consumers to buy more domestic goods due to falling inflation and the country will also be more competitive in the international markers where the exports of the country will tend to rise. Additionally, a fall in the real interest rates will improve the investment capacity and subsequently the production capacity of the country, where the production of the country can be expanded based on exporting. To examine the precise effects of the introduction of the euro, the comparison will be made between two periods: the first period is the 16 years before the euro was adopted as the single currency (1986–2001) and the second time period is the 16 years after (2002–2017) the euro was adopted as legal tender in the economy. The aim of the study is to analyse the clear effects of the euro on the balance of trade in the euro area. The euro has created positive impacts on the inflation rate, interest rate and per capita income in the three largest economies in the euro area (where per capita income increased, while real interest and inflation rate have fallen) [6]. Furthermore, this study will analyse whether positive changes in the inflation rate, interest rate and per capita income as a result of the introduction of the euro have affected the current account in the euro area positively or not.

3 Export and Import Analysis Between Two Periods

Before Germany became a member of the EMU the average level of exports was 23.90% of the GDP (382.478%/16 = 23.90%). After Germany became an EMU member, the average level of exports was recorded as 41.574% of the GDP

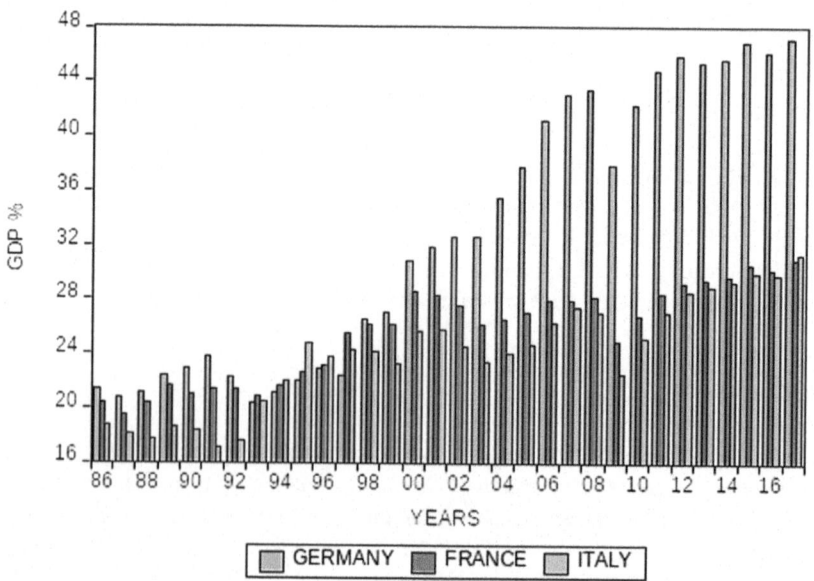

Fig. 1. Export of Germany, France and Italy Before and After EMU Membership (1986–2017).

(665.198/16 = 41.574%). The average (percentage) change between the two periods is 73.94% (41.574% − 23.90%/23.90%*100 = 73.94%). Hence, on average, comparing the level of exports in Germany in the 16 years before the country became a member of the EMU and the 16 years after, it can be seen that it increased by 73.94% (Fig. 1, 2 and 3).

Before France joined to EMU the average level of exports was 23.022% of the GDP (368.366%/16 = 23.022%). After the country gained EMU membership, the average level of exports was recorded as 28.185% of the GDP (450.962/16 = 28.185%). The average(percentage) change between the two periods is 22.42% (28.185% − 23.022%/23.022*100 = 22.42%). Therefore, on average, comparing the level of exports in France in the 16 years before the country became a member of the EMU and the 16 years after, it can be seen that it increased by 22.42%.

The average level of exports was 21.239% of the GDP (339%/16 = 21.239%) for Italy before joining to EMU. After the membership of Italy to the EMU the average level of exports is recorded as, 26.850% of the GDP (429%/16 = 26.850%).The average (percentage) change between two periods is 26.41% (26.850% − 21.239%/21.239*100 = 26.41%). So as an average, the level of exports compare 16 years before Italy became a member of the EMU and 16 years after Italy became a member of the EMU increased as 26.41%.

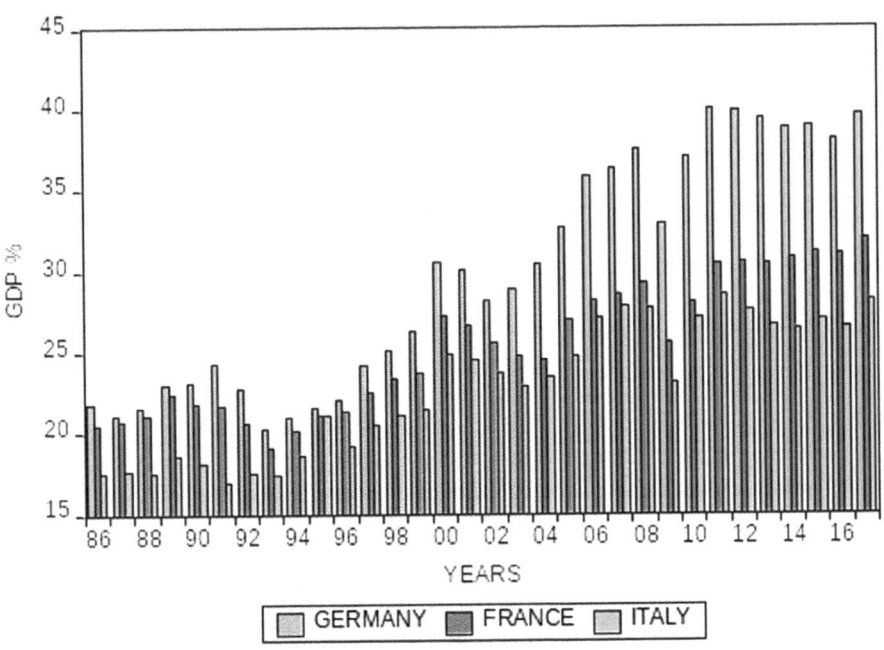

Fig. 2. Imports of Germany, France and Italy Before and After EMU Membership (1986–2017)

Before EMU membership the average level of imports was 23.667% of the GDP (378.683%/16 = 23.667%) for Germany. After Germany became a member, the average level of imports was recorded as 35.907% of the GDP (574.523%/ 16 = 35.907%). The average(percentage) change between the two periods is 51.71 (35.907% − 23.667%/23.667%*100 = 51.71%). So as an average, the level of imports compares 16 years before Germany became a member of the EMU and 16 years after Germany became a member of the EMU increased as 51.71%. Therefore, on average, comparing the level of imports in Germany in the 16 years before the country became a member of the EMU and the 16 years after, it can be seen that it increased by 51.71%.

For France before a member of the EMU the average level of imports was 23.984% of the GDP (383.748%/16 = 23.984%). After France achieved membership, the average level of imports was recorded as 28.659% of the GDP (458.545%/ 16 = 35.907%). The average (percentage) change between the two periods is 19.49% (28.659% − 23.984%/23.984%*100 = 19.49%). Therefore, on average, comparing the level of imports in France in the 16 years before the country became a member of the EMU and the 16 years after, it can be seen that it increased by 19.49%.

In Italy before the EMU membership the average level of imports was 19.531% of the GDP (312.504%/16 = 19.531%). After Italy became a member of the EMU, the average level of imports was recorded as 26.172% of the GDP (418.767%/ 16 = 26.172%). The average (percentage) change between the two periods is 14% (26.172% − 19.531%/19.531%*100 = 34%). Therefore, on average, comparing the level of exports in Italy in the 16 years before the country became a member of the EMU and the 16 years after, it can be seen that it increased by 14%.

4 Trade Potential of the Euro Area

The most important factors that play an important role in the trade potential of the euro area are the value of the euro against the other foreign currencies and also the low income and high-income players in the global markets.

Between 1999 and 2006, China rapidly increased its share of world trade and this caused a fall to occur in the export market of the euro area [7].

Between the member states of the EMU, there have been important differences since the establishment of the euro area. The differences are due to potential output per capita, output growth rates and trade balances. These imbalances can mostly be seen in low income countries. The highest imbalances occur in potential output per capita and trade balances. When a member state has an imbalance in one, then they have an imbalance in the other two as well. However, when a member state lowers the imbalance in one, then the fall in the imbalance does not act in a parallel manner in the others. In Greece, Ireland, Portugal, and Spain the imbalances have been particularly high. Furthermore, the imbalances tend to be residual and the imbalances a larger in those EMU member countries with lower levels of potential output per capita and smaller population sizes [8].

The imbalances that occur between the member countries of the EMU in potential output per capita, output growth rates and trade balances create uncertainty about the efficiency of the applied monetary policies in the euro area, which in turn effects the macroeconomic stability; where macroeconomic stability is important for the international trade stability of a country.

Due to the sizes of banks in the euro area, they play an important role in the international lending market, which increases the efficiency of the transmission of the global shocks to the euro area. Also, it is important to note that the euro area is not an economic area that only receives global shocks, as it can also be the producer and distributor of the global crises [9].

It is also important that Brexit should be mentioned in this context, as Brexit certainly has a negative impact on the trade potential of the EU and also the euro area. Brexit not only created an uncertain atmosphere in the euro area; but, also for the whole EU. The separation of the United Kingdom will cause a loss for both sides. The UK is one of the top three trading partners for the euro area. The euro area has a surplus with the UK and also the UK has a surplus in the trade of financial services. Within the last two decades, the total volume of exports of goods and services between the euro area and the UK is equal to 6 percent of the GDP in the euro area. For Ireland, the Netherlands, Belgium and Luxembourg in particular, trade with the UK is a significantly important factor. At the same time, the UK is a supplier of financial services to the euro area [10].

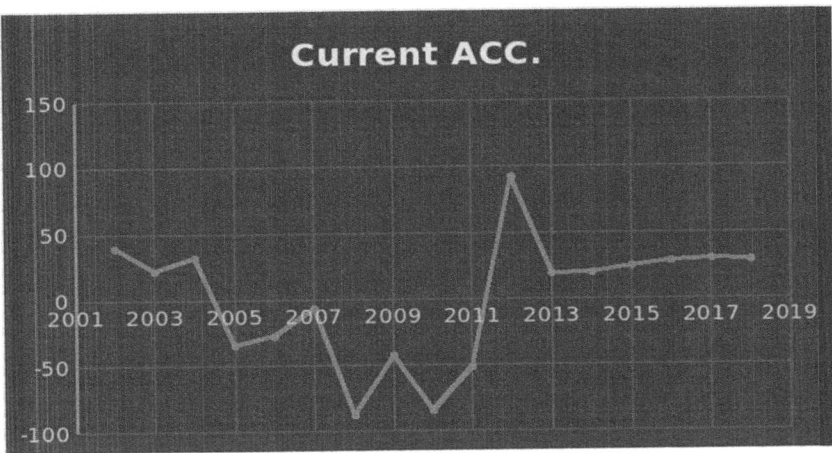

Fig. 3. Current Account Analysis of the Euro Area Between 2002–2018 (The values are given as billions of Euro)

For each year between 2002–2018, the given data for CA was taken as an average for 12 months except for 2004 and 2018 due to lack of data availability, where for these years the average was taken from January to August. The starting year is 2002, which represents the introduction year of the euro. The CA surplus value for the years between 2002–2018 is 184 billion euro and the total value of CA deficit is 36.2 billion euro, while the difference between CA surplus and deficit is 147.8 billion euro. The value of the CA surplus has been much higher than the CA deficit in the euro area since the introduction of the euro. This result openly shows that the positive impacts created by the euro on per capita income, GDP deflator and real interest rate [6] positive affected CA hence balance of payments in the euro area.

5 Conclusion

When the two periods representing the 16 years before and after EMU membership are examined for Germany, France and Italy, it can clearly be seen that there were increases in the average level of exports and imports as % of their GDP for Germany, France and Italy. The total change in the exports of the three economies are as follows:

Germany: 73.94%
France: 22.42%
Italy: 26.41%,

Thus the total change for the three countries combined is;
73.94% + 22.42% + 26.41% = 122.77%. Germany, as the largest economy in the EMU compared with two other countries, has the highest change in the level of exports as a % of its GDP. The second highest change took place in Italy and the third in France. The percentage increases in France and Italy are relatively similar.

In the case of imports, when the two periods are compared, there were also increases in the percentage of imports for the three countries. The percentage changes of the three economies are as follows;

Germany: 51.71%
France: 19.49%
Italy: 14.0%;

In total the total change in the three countries combined is;
51.71% + 19.49% + 14.00% = 85.2%.

The highest change in imports occurred in Germany as in the case of exports. The second highest increase in imports took place in France and the third in Italy.

According to the given values, the total change in balance of trade for the three largest economies of the EMU is:

Rise in Exports	Rise in Imports
For Three Countries	For Three Countries
(Comparing two periods)	(Comparing two periods)

122.77% − 85.2% = 37.57%, The difference is calculated to be 37.57%. This means that, as a result of the introduction of the euro, the level of exports increased by 37.57% more than imports. Therefore, the introduction of the euro in these three largest economies caused the balance of trade to be affected in a positive way. Thus, the use of the euro as a single currency, as mentioned in the study of Özdeşer, lowered the inflation and real interest rates in these three countries; resultantly, there was a positive impact on the balance of trade, as should be the case according to economic theory. It was found in the study that the CA surplus was 147.8 billion euro higher than the CA deficit in the euro area between the years 2002–2018.

References

1. Zestos, G.K.: European Monetary Integration the Euro, Thomson South-Western, USA (2006)
2. Bukovsak, M., Andrijana, C., Pavic, N.: Adoption of the Euro in Croatia: Possible Effects on International Trade and investment. Creation National Bank, Zagreb **1** (2017)
3. Daniels, J., Vanhouse, D.: International Monetary and Financial Economics. Pearson Education International, USA (2005)
4. Nello, S.: The European Union Economics, Policies, and History, 3rd edn., vol. 512, p. 393. McGraw-Hill Higher Education, U.K. (2005)
5. Yarbrough, B.V., Yarbrough, R.M.: The World Economy-Trade and Finance. Harcourt College Publishers, USA (2006)
6. Ozdeser, H.: Analysing the economic impacts of the euro for the three largest economies in the EMU(European Monetary Union) and the Place of the Euro in Global Economics. In: 13th International Conference on Theory and Applications of Fuzzy Systems and Soft Computing-ICAFS-2018, pp. 2194–5357, 361–370. Springer, Cham (2019)
7. Baumann, U., Mauro, F.: Globalization and Euro Area Trade-Interactions and Challenges, Occasional Paper Series, **14**(55), European Central Bank, Eurosystem (2007)
8. Mink, M., Jacobs, J., Haan, J.: Euro Area Imbalances. DNB Working Paper, no. 540, pp. 23–25 (2016)
9. Habib, M., Venditti, F.: The global financial cycle: implications for the global economy and the euro area. ECB Econ. Bull. **6** (2018)
10. IMF Country Reports No.18/24. Euro Area Policies. International Monetary Fund, Washington, D.C. (2018)

Analysis of Economic Condition of the Enterprises of IT-Branch of the Rostov Region Based on Open Internet Sources and Fuzzy Logic

Nikolay Kuznetsov[1] , Irina Kuznetsova[2],
Victoriya Kokhanova[2(✉)] , and Sergey Rogozhin[1]

[1] Rostov State University of Economics, B. Sadovaya str., 69,
344002 Rostov-on-Don, Russia
kuznecov@rsue.ru, svrogozhin@gmail.com
[2] Private Educational Institution of Higher Education
"SOUTHERN UNIVERSITY (IMBL)", Rostov-on-Don, Russia
ikuzn@iubip.ru, kohanovavs@yandex.ru

Abstract. We have developed a methodology for assessing the financial and economic condition of a given industry in a region based on an analysis of data on enterprises in the industry taken from open Internet sources. As an experimental material for our study, we used the data of coefficient analysis taken from the TestFirm website. As a basis for our fuzzy-multiple methodology, we took the spectrum-point methodology Audit-IT, which assesses the financial condition of an organization based on indicators of the financial position and efficiency of the organization. At the same time, the Audit-IT methodology uses the summation of points awarded for the values of nine standard coefficients. The Audit-IT methodology is easy to use, but not suitable for assessing the financial and economic condition of an entire industry. Therefore, we modified it and presented our algorithm for assessing the financial and economic condition of the industry based on the aggregation of relevant data on enterprises in the industry under study. For data aggregation, we used a system of fuzzy inference, multilevel [0, 1] - classifiers. The technique has been tested at IT enterprises in the Rostov region (OKVED 62). The analysis carried out on the basis of the methodology allowed us to identify the main trends in the financial and economic development of the IT industry in the Rostov region.

Keywords: Analysis of financial statements · Comprehensive assessment of the financial and economic condition of the industry

1 Introduction

The relevance of the research topic is determined by the fact that the integrated assessment of enterprises in a given industry is of great importance for government bodies that shape investment and tax policies in the region. For the IT industry, it is especially important to know which of the enterprises (small, medium, large) are most dynamically and rapidly developing in the region, what is their average financial and economic condition (in particular, is there an increased loan load). This will help to formulate the optimal policy of the authorities in relation to the respective enterprises, to provide the basis for the most rapid and intensive development of the IT industry in the region. Currently, there are various methods for the formation of integrated assessments of the financial condition of enterprises, however, all in pure form are not applicable to achieve these goals. So, for example, the method of integrated assessments allows you to take into account all the relationships between the financial performance of the enterprise, to predict the state of the enterprise in the future [1]. However, these techniques have a number of significant drawbacks, which are, first of all, in the subjective approach to determining expert assessments, as well as the neglect of the specific features of the enterprise and industries [2].

As a tool for financial analysis, financial ratios are used that allow for a comprehensive study of the state of the enterprise by evaluating their ratios. Work is underway to develop criteria that allow tribute to a qualitative assessment of the financial and economic condition of the enterprise; for this purpose, as a rule, integral point estimates are used [3]. The model of Nedosekin A.O. was widely known; it designed to assess the risk of bankruptcy of an enterprise based on standard five-point [0, 1] - classifiers [4]. The model was generalized to a set of indicators that go beyond the scope of financial reporting and are taken into account in Argentina's quantitative scale [5]; including, in addition to the traditional block of the finance level, the author considered the enterprise management block, which includes the level of top management, financial management, marketing and advertising divisions, the development of the distribution network and branches, etc.

However, as already mentioned, all of the above methods cannot be used directly to build a comprehensive assessment of a particular industry. Therefore, we proposed a proprietary methodology based on aggregating data for individual enterprises from open Internet sources ("For an honest business"), as well as studying them using classical methods of financial analysis, such as Audit-IT, "Your Financial Analyst" and TestFirm (Audit company "Avdeev and K") [6, 7], with subsequent generalization.

The novelty of the approach used is as follows. On the basis of well-known methods designed to assess the financial condition of individual organizations, we assess the financial condition of an entire industry in the region. For each group, based on fuzzy inference systems, we aggregate estimates for individual enterprises. As a result, we obtain integrated quantitative assessments of financial stability, solvency and performance, as well as a generalizing (integral) assessment of the financial condition of the industry in the region. The constructed assessments allow drawing conclusions about the general trends in the financial and economic condition of each of the five groups, as well as assessing the corresponding groups of enterprises as potential credit partners in the region under consideration.

2   Materials and Methods

Based on the parsing of the site "For Honest Business", we compiled a database of IT enterprises in the Rostov region (OKVED-62). The analysis showed that enterprises can be distributed, in accordance with the number of employees working for them, into

5 groups: micro-enterprises (1–5 people), mini-enterprises (6–10 people), small enterprises (11–15 people), medium-sized enterprises (16–50 people), large enterprises (51 and more people). At the same time, we established (based on the data from the TestFirm website) that the information necessary for analysis is available, by groups: micro-enterprises - 175 enterprises, mini-enterprises - 58 enterprises; small enterprises - 17 enterprises; medium-sized enterprises - 53 enterprises; large enterprises - 9 enterprises. As a criterion, we chose the opportunity to calculate the coefficients of indicators of three groups: financial stability (coefficients of autonomy, provision with own circulating assets, coverage of investments), liquidity (coefficients of current, quick and absolute liquidity) and operational efficiency (profitability of sales, rate of net profit and return on assets), as well as the final score of the enterprise. We took the coefficient values for each enterprise on the TestFirm website. We designed the database in the form of an excel table for further analysis.

Scheme of Audit-IT research of the financial condition of an enterprise based on coefficient analysis

We took the Audit-IT methodology as the basis for our methodology. In accordance with it, the study is carried out according to three groups of indicators: financial stability, solvency, and efficiency. The Audit-IT methodology evaluates the financial stability of an enterprise based on three coefficients: the autonomy coefficient, the equity ratio, and the investment coverage ratio. Assessment of the company's solvency is based on three indicators: current liquidity ratio; quick liquidity ratio; absolute liquidity ratio. Evaluation of the effectiveness of the enterprise is carried out on the basis of three coefficients: profitability of sales; net profit rates; return on assets (ROA). The criteria for assessing the listed indicators adopted for the IT industry are given in Table 1.

Table 1. Evaluation criteria for coefficient ratios adopted for the IT industry

№	Coefficient	Evaluation criteria
1.	Autonomy ratio	∞ < Critical $\leq$ 0 < Unsatisfactory < 0.45 $\leq$ Well < 0.55 $\leq$ Excellent < 0.7 $\leq$ Well < ∞
2.	The ratio of own working capital	∞ < Critical < -0.2 $\leq$ Unsatisfactory < 0.1 $\leq$ Well < 0.15 $\leq$ Excellent < ∞
3.	Investment coverage ratio	∞ < Critical < 0.5 $\leq$ Unsatisfactory < 0.7 $\leq$ Well < 0.8 $\leq$ Excellent < ∞
4.	Current (total) liquidity ratio	∞ < Critical < 1 $\leq$ Unsatisfactory < 2 $\leq$ Well < 2.1 $\leq$ Excellent < ∞
5.	Quick ratio (intermediate) liquidity	∞ < Critical < 0.5 $\leq$ Unsatisfactory < 1 $\leq$ Well < 1.1 $\leq$ Excellent < ∞
6.	Absolute liquidity ratio	∞ < Critical < 0.05 $\leq$ Unsatisfactory < 0.2 $\leq$ Well < 0.25 $\leq$ Excellent < ∞
7.	Return on sales	∞ < Critical < 0 $\leq$ Unsatisfactory < 0.08 $\leq$ Well < 0.2 $\leq$ Excellent < ∞
8.	Net profit margin	∞ < Critical < 0.02 $\leq$ Unsatisfactory < 0.01 $\leq$ Well < 0.04 $\leq$ Excellent < ∞
9.	Return on assets (ROA)	∞ < Critical < 0 $\leq$ Unsatisfactory < 0.12 $\leq$ well < 0.24 $\leq$ Excellent < ∞

The generalizing (integral) assessment of the financial condition of the organization is formed on the basis of assessments of the financial position (the first six indicators, in order) and the assessment of the effectiveness of the organization (the remaining three indicators). In this case, the gradation of estimates is used, given in Table 2. In accordance with the TestFirm comparative analysis methodology, all financial ratios calculated for the enterprise are compared with the median values of indicators of all organizations in the Russian Federation and organizations within the industry. The comparison is carried out for nine financial ratios indicated in Table 1. Depending on the hit of each value in the corresponding interval, the indicator is assigned a point from -2 to $+2$, while: -2 - "bad", -1 - "unsatisfactory", $+1$ - "good"; $+2$ – "excellent"; 0 - the value deviates from the median by no more than 5%, "normal". To form a conclusion based on the results of the analysis, the scores are summarized with an equal weight of each indicator, as a result, an estimate is also obtained from -2 to $+2$. On the basis of the points of financial position and performance, a generalized assessment is calculated - the point of financial condition, as the sum of the point of financial position multiplied by 0.6, and the score of financial results multiplied by 0.4. Thus, the gradation of the generalized assessment of the financial condition is also described in Table 2.

Table 2. Gradation of grades (TestFirm "Your Financial Analyst" methodology)

Score		Symbol (rating)	Qualitative characteristic of financial condition
From	To (on)		
2,0	1,0	+2	Excellent
1,0	0,1	+1	Well
0,1	-0,1	0	Normal
−0,1	−1,0	−1	Unsatisfactory
−1,0	−2,0	−2	Bad

Thus, the Audit-IT methodology uses the conventional spectrum-scoring model. The described methodology is applicable to assessing the financial condition of one enterprise but is not applicable to their totality. To obtain an aggregated estimate for each of the described indicators, we applied data aggregation based on fuzzy inference systems, standard multi-level [0, 1] classifiers [8, 9].

Fuzzy-multiple modification of the Audit-IT method for analyzing the financial condition of the industry in the region

For each subgroup, we built four aggregated assessments: financial stability (a set of indicators: the coefficients of autonomy, provision with own circulating assets, coverage of investments), liquidity (a set of indicators: ratios of current, quick and absolute liquidity) and operational efficiency (set of indicators: profitability of sales, net profit margin and return on assets), as well as the final score of the subgroup's financial condition.

For this purpose, in accordance with the theory of standard multilevel [0, 1] - classifiers, we have assigned a linguistic variable to each of the coefficients. The universal set of each variable is a unit segment, and the term set, in accordance with

Table 1, consists of four terms characterizing the state of the enterprise according to the investigated indicator: "critical", "unsatisfactory", "good", "excellent". We compared linguistic variables constructed according to the same scheme to aggregated estimates of financial stability, liquidity and performance. For the aggregated assessment of the final score, we compared the linguistic variable with a term-set of five terms, in accordance with Table 2: "bad", "unsatisfactory", "normal", "good", "excellent".

The process of constructing an aggregated estimate included the following stages.

1. Construction of a fuzzy set "Indicator value in a group", based on the relative frequencies of terms in each subgroup.
2. Aggregation of the constructed fuzzy sets using fuzzy multilevel [0, 1] -classifiers. The result is the numerical value of the linguistic variable, enclosed in a unit segment, and a verbal description in the form of an indication of a term.
3. Multiplying the constructed value by four for assessing financial stability, liquidity, and performance, and by ten for the final score. Multiplication is performed for clarity of results.
4. Presentation of the estimates obtained in the form of tables, analysis of the situation.

3 Results

The research results are presented in Tables 3, 4, 5 and 6. The analysis shows that, on average, the state of IT enterprises can be estimated as average, which meets the estimates of "satisfactory", "normal" and "positive".

Table 3. Comparative analysis of assessing the overall financial condition of enterprises by groups

Group	Group average score	Qualitative assessment of condition, term
Microenterprises	6,46	Normal
Mini enterprises	6,99	Positive
Small enterprises	6,54	Positive
Medium-sized enterprises	6,10	Normal
Large enterprises	4,63	Satisfactory

Microenterprises and medium-sized enterprises meet the "normal" assessment, and large enterprises - the "satisfactory" assessment, which corresponds to the normal (satisfactory) financial condition of enterprises, in which most of the indicators fit into normative values. Therefore, in accordance with generally accepted terminology, microenterprises and medium-sized IT enterprises - branches of the Rostov Region, as a whole, are potential partners in the relationship with which a prudent approach to risk management is necessary. These companies may qualify for credit resources, however, the decision in each case must be taken individually (neutral creditworthiness).

Microenterprises and small enterprises IT - branches of the Rostov Region, in general, meet the rating "positively". This testifies to their good (positive) financial condition of enterprises, as well as their ability to meet their obligations in the short term. Therefore, the respective enterprises can be classified as credit partners for which the probability of obtaining credit resources is quite high (good creditworthiness).

Table 4. Comparative analysis of indicators of financial stability by groups

Grades				
Group	Estimation of autonomy and term coefficient	Evaluation of the security ratio with own circulating assets and terms	Evaluation of the coefficient of investment and term coverage	The average assessment of financial stability by group and term
Microenterprises	2,86	2,75	2,92	2,84
	Well	Well	Well	Well
Mini enterprises	2,48	2,32	2,52	2,44
	Unsatisfactory	Unsatisfactory	Well	Unsatisfactory
Small enterprises	2,52	2,38	2,87	2,59
	Well	Unsatisfactory	Well	Well
Medium-sized enterprises	2,33	2,31	2,41	2,35
	Unsatisfactory	Unsatisfactory	Unsatisfactory	Unsatisfactory
Large enterprises	2,21	3,23	1,88	2,44
	Unsatisfactory	Unsatisfactory	Unsatisfactory	Unsatisfactory

Table 5. Comparative analysis of solvency indicators by groups

Grades				
Group	Assessment of current liquidity ratio and term	Evaluation of quick liquidity ratio and term	Assessment of absolute liquidity ratio and term	Average solvency rating per group and term
Microenterprises	2,18	2,05	2,76	2,33
	Unsatisfactory	Unsatisfactory	Well	Unsatisfactory
Mini enterprises	2,41	2,74	2,44	2,53
	Unsatisfactory	Well	Unsatisfactory	Well
Small enterprises	2,31	2,31	2,25	2,29
	Unsatisfactory	Unsatisfactory	Unsatisfactory	Unsatisfactory
Medium-sized enterprises	2,39	2,28	2,16	2,28
	Unsatisfactory	Unsatisfactory	Unsatisfactory	Unsatisfactory
Large enterprises	2,68	3,12	3,23	3,01
	Well	Well	Well	Well

Table 6. Comparative analysis of performance indicators for groups

Grades				
Group	Evaluation of the profitability of sales and term	Valuation of net profit and term	Return on Assets and Term	Average performance rating for group and term
Microenterprises	2,38	2,67	2,63	2,56
	Unsatisfactory	Well	Well	Well
Mini enterprises	2,59	2,55	2,63	2,59
	Well	Well	Well	Well
Small enterprises	2,19	2,26	2,13	2,19
	Unsatisfactory	Unsatisfactory	Unsatisfactory	Unsatisfactory
Medium-sized enterprises	2,26	2,32	2,22	2,27
	Unsatisfactory	Unsatisfactory	Unsatisfactory	Unsatisfactory
Large enterprises	3,67	3,34	3,12	3,38
	Well	Well	Well	Well

At the same time, the analysis of financial stability showed for enterprises of microenterprises, medium-sized enterprises and large enterprises of IT - the industry of the Rostov region a high degree of dependence on borrowed capital (borrowing). The analysis of solvency indicators allowed us to reveal, on average, the insufficient solvency of enterprises of micro-enterprises, small enterprises and medium enterprises. The analysis of performance indicators showed insufficient business activity, on average, of small enterprises and medium-sized enterprises of IT - the industry of the Rostov region.

4 Conclusion

A method for assessing the financial and economic condition of enterprises in a given industry in the region is developed on the basis of open Internet sources. As material for the study, we used the data of coefficient analysis taken from the TestFirm website. The spectrum-point methodology Audit-IT was taken as the basis for the fuzzy-multiple assessment methodology, which assesses the financial condition of the organization based on indicators of the financial position and efficiency of the organization; at the same time, the Audit-IT methodology uses the summation of points awarded for the values of nine standard coefficients. The Audit-IT methodology is easy to use, but not suitable for assessing the financial and economic condition of an entire industry. Therefore, we modified it and presented our own algorithm for assessing the financial and economic state of the industry based on the aggregation of data on enterprises in the industry. For data aggregation, a system of fuzzy multilevel [0,1] classifiers are used. The technique has been tested at IT enterprises in the Rostov region

(OKVED 62). The analysis carried out based on the methodology allowed us to identify the main trends in the financial and economic development of the IT industry in the Rostov region.

References

1. Kuvshinov, M.S.: Innovative tools for forecasting the assessment of the financial condition of the enterprise. B. South Ural State U. Ser. Econ. Manag. **30**, 56 (2012)
2. Smelova, T.A., Merzlikina, G.S.: Assessment of economic viability in crisis management of the enterprise. Volgstu, Volgograd, Russia (2003)
3. Hoarse, F.P., Hoarse, A.F.: Comparative analysis of methods for assessing the financial condition of an organization. Polythemat. Netw. Electron. Sci. J. Kuban State Agrarian Univ. **81**, 22 (2012)
4. Nedosekin, A.O.: Fuzzy financial management. AFA Library, Moscow, Russia (2003)
5. Nedosekin, A.O.: Business risk assessment based on fuzzy data: monograph, St. Petersburg, Russia (2005)
6. Audit-IT: Financial analysis. Audit firm "Avdeev and K": audit and accounting services, 1999–2019 (2019). https://www.audit-it.ru
7. TestFirm Audit firm "Avdeev and K": audit and accounting services, 1999–2019. https://www.testfirm.ru/
8. Alekseychik, T.V., Bogachev, T.V., Karasev, D.N., Sakharova, L.V., Stryukov, M.B.: Fuzzy method of assessing the intensity of agricultural production on a set of criteria of the level of intensification and the level of economic efficiency of intensification. In: Advances in Intelligent Systems and Computing, vol. 896, pp. 790–798 (2019)
9. Vovchenko, N.G., Stryukov, M.B., Sakharova, L.V., Domokur, O.V.: Fuzzy-logic analysis of the state of the atmosphere in large cities of the industrial region on the example of Rostov region. In: Advances in Intelligent Systems and Computing, vol. 896, pp. 709–715 (2019)

Digital Marketing Technologies Selection Under Z-Environment

Gunay E. Imanova[1]([⊠]) and Gunel Imanova[2]

[1] Azerbaijan State Oil and Industry University, 20 Azadlyg Avenue,
Baku, Azerbaijan
gunayimanovaa@gmail.com
[2] Department of Business Administration, Near East University,
North Cyprus via Mersin 10, Nicosia, Turkey
gunelimanova97@gmail.com

Abstract. As a result of advanced technological changes, the customer-company relationship becomes more digital. Recent digital marketing technologies have profoundly changed the: products produced; prices offered to customers to boost the profit gained; places of brick-and-mortar to virtual ones; interaction and communication way to enhance the customer satisfaction, and personalized promotion to influence customer buying decision; and targeting customer segments. As a result, companies can influence the consumer buying behavior at the right time, with the right product/service, in an enhanced manner. In this paper, we investigate the selection of optimal digital marketing technology among Artificial Intelligence (AI), Internet of Things (IoT), and Virtual or Augmented Reality (VR/AR). Our research is based on consistency driven approach for determination of the weights of criteria that are product, price, place, promotion and targeting the customer, and distance between ideal positive and negative solution under Z-number. A numerical example is provided in order to check the validity of the considered approach.

Keywords: Digital marketing · Artificial Intelligence (AI) · Internet of Things (IoT) · Virtual Reality (VR) · Augmented Reality (AR) · Z-valued matrix · Consistency of pairwise comparison matrix

1 Introduction

For the dynamic marketing industry, especially during pandemic, digital marketing is inevitable for the enhanced brand equity and company revenue. As the access to internet is rising it constitutes opportunities for the companies to use digital marketing technologies in order to effectively and efficiently serve its customers. Digital marketing provides competitive advantage through cost efficiency, better two-way communication with customer, etc. Furthermore, more accessible information about product/service, better targeting ability through customized promotions to influence the customers in a shorter time frame are just to mention some of them [1]. Most probably one of the great changes in consumer behavior confronts us when we consider increased usage rate of online social media and web-based platforms that can shape the buying decision of the consumers [2]. As a result of such a change, customer relations, buying habits, interaction between customers oblige the marketers for a change by directing them to implement novel digital marketing technologies [3]. Following the technological changes in the marketing environment may enable the firms to gain competitive advantage over rivalry through better understanding of customers, developing more effective and efficient targeting strategies, deep engagements with customers, and as a result, building long term relationships accordingly [4]. As the technology becomes more advanced, different powerful digital marketing technologies are aroused to be leveraged. The main issue is to evaluate those digital marketing technologies in order to select the best alternative to be used. Unfortunately, there are so restricted researches about the mentioned evaluation. There are a few researches that is based on evaluation of different digital marketing technologies and tools through multi-criteria decision-making (MCDM) analysis using Analytical Hierarchy Process (AHP) and CORPAS integration [5], fuzzy-AHP [6] and fuzzy-TOPSIS integration [7] methods for determining the weights of criteria, and for selecting the best alternative among the considered technologies, besides, showing the priority ranking for them. Unfortunately, with AHP the uncertainty of the variables of interest is not taken into account and it is a very important point that the reliability related to values of considered variables are ignored.

The remaining of this study is structured as follows. Section 2 represents the preliminary information for further explanation, and it is followed by the statement of MCDM problem and the suggested solution method based on Z-numbers, in Sect. 3. For the next part, Sect. 4 involves an illustrative problem with considered criteria and alternatives in order to implement the suggested method to a real-world digital technology selection problem. At the end, Sect. 5 represents the conclusion points for the study.

2 Preliminaries

*Definition 1. **Fuzzy Numbers** [8]:* In a real-world decision-making environment experts or DMs are usually uncertain with possibility of the given values, and the possible degrees of this impreciseness or uncertainty is also unknown to them. In a fuzzy set there is an assigned degree for each possible value of x, $\mu(x) \in [0, 1]$.

$$x_0 \in R, \text{ where } \mu_M(x_0) = 1$$

For any $0 \leq \alpha \leq 1$, $A_\alpha = \left[x, \mu_{A_\alpha}(x) \geq \alpha\right]$ is a closed interval, where F(R).

*Definition 2. **Random Variables** [8]:* Random variables can be classified into two different types which are called continuous random variables and discrete random variables. Continuous one which can be represented by X, may take infinite numerical outcomes of possible values x. In contrast, for a discrete random variable X can only take countable distinct values.

Definition 3. Triangular Fuzzy Numbers [9]: We can define a triangular fuzzy number which is denoted by $\tilde{A}$ as a triplet, shown as (a_1, a_2, a_3), where, the membership is identified using the following formula:

$$\mu_A(x) = \begin{cases} 0 & x \in (-\infty, a_1) \\ \frac{x-a_1}{a_2-a_1} & x \in [a_1, a_2] \\ \frac{c-x}{a_3-a_2} & x \in [a_2, a_3] \\ 0 & x \in (a_3, +\infty) \end{cases}$$

Definition 4. Z-Numbers [10]: In order to make a robust decision data or information acquired must be reliable. Z-number principle is based on this reality. Z-number is described as a concept consisting of two parts A and B, or Z = (A, B). In this ordered pair of representation, A indicates the fuzzy constraint or restriction on the real values that a random variable X can take; where B, the second part indicates the fuzzy value for the probabilistic measure of certainty or reliability on the first part A. Here, both A and B are presented in natural language, NL and they are fuzzy numbers.

Definition 5. Z-Valued Matrix [11]: It is a matrix for elements of Z-numbers, which identifies partial reliability of the information based on values of the random variables $X_{ij}, i, j = 1, \ldots, n$:

$$(Z_{ij} = (A_{ij}, B_{ij})) = \begin{pmatrix} Z_{11} = (A_{11}, B_{11})) & \ldots & Z_{1n} = (A_{1n}, B_{1n})) \\ & \ldots & \\ Z_{n1} = (A_{n1}, B_{n1})) & \ldots & Z_{nn} = (A_{nn}, B_{nn})) \end{pmatrix}$$

3 Statement and Solution of the Problem

In our research, every alternative, $A = \{a_1, a_2, a_3\}$, is characterized by 5 criteria $C = \{c_1, c_2, c_3, c_4, c_5.\}$. The purpose of this study is to evaluate the digital marketing technologies to achieve a sound decision for a company, through selecting the better digital marketing technology among the alternatives of Artificial Intelligence (AI), Internet of Things (IoT), and Augmented/Virtual Reality (AR/VR), that feeds all the 5 criteria given below:

- C1- Product/service: Proposed mix of goods and services.
- C2- Price: Money that must be paid by the customers to buy the product/service, i.e. cost of purchase for a customer.
- C3- Place/distribution: Practices that are carried out to make the product available to the customer, in other words, where to sell the product, how to distribute them are the main decisions to be considered.
- C4- Promotion: Communication practices to be pursued in order to influence and encourage the customers to buy the proposed product or service.
- C5- Targeting *customer segments*: Segments or group of customers to be targeted in order to serve them effectively and efficiently [4].

Some literature-based explanations for the decision alternatives of this study are given below:

Artificial Intelligence (AI): AI is a program, machine or computer that can easily mimic or imitate the real man decision-making process and achieve effective results within short time frame [12–14], which is the reason for their "smart" tags [15]. According to Forbes's research [16], AI may provide a competitive advantage through dynamic pricing concept. The two best examples are the taxi, ride service Uber and the giant online retailer Amazon. Here, Uber increases supply and price during peak demand periods for taxi, and provide the best price for the customers. Furthermore, Amazon gathers the customers data through purchase history of individual customer to target them effectively and efficiently with the products and promotions generated by the related seller. As a result, AI helps to understand which customers to offer the product/service, at the optimal prices and the right time [16, 17]. Besides, chatbots are the AI-based machines that use natural language-based user interfaces for communication with customers in online platforms or social media channels such as Facebook. So, chatbots have the potential to build enhanced customer relationships through interactive communication and timely reply capability that provides better customer satisfaction [18].

Internet of Things (IoT): IoT involves interconnected programs or systems that provide a new product or enhance the customer engagement and loyalty through personalized communication [19]. IoT is used in smart products such as smart homes and cars, home security, smart TVs, smartwatches, healthcare devices such as MRI, etc. [20]. Companies use IoT to fix any break down or improve the device by sending signals to supplier if product need an instant modification. For example, Tesla car updates itself as the new software update must be maintained [20]. Huge information about customers makes targeting effective and timely. Promotion is personalized according to collected information from the internet, website or apps, associated with the individual preferences. IoT connected programs or machines may use some basic pricing strategies through smart data pricing model [21] however, pricing strategy is not well developed as in AI.

Augmented/Virtual Reality (AR/VR): AR is a computer-aided technology that permits digital or virtual content such as audio, video files, 2D or 3D objects to overlay with the real-world environment information or the customer's perception of the real-world in real-time. As a result, virtual or artificial world-scenes are achieved [22, 23]. We can implement them in different areas such as e-commerce, games, healthcare, marketing, etc. [22]. AR is used in digital marketing to promote a product. For instance, IKEA, a giant furniture company allows customers to visualize the furniture for sale by showing how the selected furniture will look like at home [23]. VR is a related concept to AR and it uses computer-aided technology to enable a simulation like customer experience in order to make the customer feel like he/she visits or virtually engage with the product/service or brand itself [15]. Some common examples are virtual test drives of car brands, 3D product show, virtually visiting different places (for tourism companies), and so forth. Companies that are using AR and VR try to increase the customer satisfaction through better communication (promotion) and positive word-of-mouth accordingly [15]. Recently, some brands such as Gucci provides try-on service for its sneakers through AR to reach customers who can't visit the brick-and-mortar/stores (place) due to COVID-19 pandemic. Also, during such a crisis optimal price offering can be taken as more important criterion for the digital technology seeking firms.

Solution method is based on consistency driven approach for determining t weights of criteria that are product, price, place, promotion and targeting the customer, and identification of distance between ideal positive and negative solution under Z number.

At the first step we investigate consistency of DM's preferences by using consistency driven approach to Z-valued preferences. Step 1 assigns weights of all the criteria. Here, linguistic terms enable the DM to characterize a certainty degree B, on the first part of evaluation, A, that are described by triangular fuzzy numbers.

In Step 2 we formulate ideal positive and negative Z solutions through the weights of criteria obtained from the previous step. Furthermore, we proceed to ranking of alternatives A_1, A_2, A_3.

4   A Practical Example

Pairwise comparison for each criterion is performed. Pairwise comparison matrix is deployed in Table 1. Besides, optimal Z-matrix is given in Table 2.

Table 1. Pairwise comparison matrix

	C_1	C_2	C_3
C_1	(1, 1, 1), (0.8, 0.9, 1.0)	(1/4, 1/3, 1/2), (0.6, 0.7, 0.8)	(1/3, 1/2, 1/1), (0.4, 0.5, 0.6)
C_2	(2, 3, 4), (0.6, 0.7, 0.8)	(1, 1, 1), (0.8, 0.9, 1.0)	(1, 2, 3), (0.4, 0.5, 0.6)
C_3	(1, 2, 3), (0.4, 0.5, 0.6)	(1/3, 1/2, 1/1), (0.4, 0.5, 0.6)	(1, 1, 1), (0.8, 0.9, 1.0)
C_4	(1, 2, 3), (0.6, 0.7, 0.8)	(1/3, 1/2, 1/1), (0.4, 0.5, 0.6)	(1/3, 1/2, 1/1), (0.2, 0.3, 0.4)
C_5	(1, 2, 3), (0.2, 0.3, 0.4)	(1/4, 1/3, 1/2), (0.4, 0.5, 0.6)	(1/4, 1/3, 1/2), (0.4, 0.5, 0.6)
	C_4	C_5	
C_1	(1/3, 1/2, 1/1), (0.6, 0.7, 0.8)	(1/3, 1/2, 1/1), (0.2, 0.3, 0.4)	
C_2	(1, 2, 3), (0.4, 0.5, 0.6)	(2, 3, 4), (0.4, 0.5, 0.6)	
C_3	(1, 2, 3), (0.2, 0.3, 0.4)	(2, 3, 4), (0.4, 0.5, 0.6)	
C_4	(1, 1, 1), (0.8, 0.9, 1.0)	(1, 2, 3), (0.4, 0.5, 0.6)	
C_5	(1/3, 1/2, 1/1), (0.4, 0.5, 0.6)	(1, 1, 1), (0.8, 0.9, 1.0)	

Table 2. Optimal Z-matrix

	C_1	C_2	C_3
C_1	(1.000, 1.000, 1.000), (0.998, 0.998, 1,000)	(0.581, 0.581, 0.581), (0.492, 0.997, 0.997)	(0.578, 0.578, 0.578), (0.005, 0.005, 0.498)
C_2	(1.711, 1.711, 1.729), (0.974, 0.974, 0.974)	(0.999, 1.001, 1.001), (0.999, 0.999, 1.000)	(1.010, 1.010, 1.010), (0.498, 0.989, 0.989)
C_3	(1.701, 1.707, 1.716), (0.979, 0.90, 0.980)	(0.996, 0.996, 0.996), (0.977, 0.979, 0.979)	(1.000, 1.001, 1.001), (0.999, 0.999, 1.000)
C_4	(1.363, 1.363, 1.363), (0.977, 0.983, 0.983)	(0.801, 0.801, 0.801), (0.493, 0.996, 1.000)	(0.812, 0.812, 0.812), (0.005, 0.006, 0.497)
C_5	(0.981, 0.981, 0.988), (0.971, 0.973, 0.973)	(0.589, 0.589, 0.589), (0.993, 0.998, 0.998)	(0.579, 0.579, 0.579), (0.500, 0.997, 0.997)
	C_4	C_5	
C_1	(0.713, 0.713, 0.713), (0.985, 0.990, 0.990)	(1.016, 1.016, 1.016), (0.490, 0.991, 0.991)	
C_2	(1.246, 1.246, 1.255), (0.976, 0.979, 0.979)	(1.760, 1.760, 1.760), (0.976, 0.978, 0.978)	
C_3	(1.236, 1.240, 1.248), (0.980, 0.982, 0.982)	(1.742, 1.742, 1.743), (0.978, 0.980, 0.980)	
C_4	(0.999, 1.001, 1.001), (0.998, 0.998, 1.000)	(1.403, 1.403, 1.403), (0.489, 0.992, 0.992)	
C_5	(0.713, 0.713, 0.717), (0.980, 0.983, 0.983)	(1.000, 1.001, 1.001), (1.000, 1.000, 1.000)	

Let's consider a problem for generation of consistent PCM $(C'_{i}j)$ most similar to a given inconsistent PCM (C_{ij}). The elements of inconsistent C-matrix (C_{ij}) will be considered as a perturbation of the elements of matrix $(C'_{i}j)$ for which reciprocity and consistency are verified. Elements of (C_{ij}) must be changed to reach $(C'_{i}j)$. It is formulated as:

$$J = \sum_{i=1}^{n} \sum_{j=1}^{n} D(C_{ij}, C'_{ij}) \to max$$

Multiplicative reciprocity: $C'_{ij} C'_{ji} = C(1)$

Multiplicative transitivity: $C'_{ij} C'_{jk} = C'_{ik}$ Non negativity: $C'_{ij} \geq C(0)$, $i, j = 1, \cdots, n$

Optimal fuzzy matrix (CF = 61.6715523538376, EF = 0.663943188691947):
Inconsistency of original matrix = 0.312666822312678
Inconsistency of optimal matrix = 0.00275001346704826

Finding Crisp Maximum Eigenvalue and Eigenvectors. Maximum Eigenvalue is found by solving the equation: $\det(A - \lambda I) = 0$. Matrix A involves only medium values of fuzzy elements. DE objective function is minimized, $CF(\lambda) = D^2(\lambda) - \varepsilon\lambda$, where λ is the parameter to optimize, $D(\lambda) = \det(A - \lambda I)$ is a function dependent of λ, ε is a constant (may need to be adjusted for concrete problem, in our case it was set to 1). A new DE objective function is constructed in order to find Eigenvectors: $CF(X) =$ VectorDistance $(AX, \lambda X)$. Here, parameters to search are $(N - 1)$ elements of vector X, (except the first one, which is set to 1) and DE search space dimension is $N - 1$.

Finding Fuzzy Eigenvalues. Fuzzy A matrix is used. A new DE objective function is constructed as follows: $CF(\lambda_L, \lambda_R) =$ FuzzyDistance$(AX, (\lambda_L, \lambda_M \lambda_R)X)$, where "FuzzyDistance" is sum of fuzzy distances between corresponding elements of vectors. Vector X involves fuzzy singletons, produced from crisp Eigenvectors of step 1. Only left and right components of fuzzy Eigenvalues are the subject for optimization (Medium component is taken from step 1) DE search space dimension is 2.

Finding Fuzzy Eigenvectors. Fuzzy A matrix and fuzzy eigenvalues are used to construct DE objective function as follows: $CF(X) =$ FuzzyDistance$(AX, \lambda X)$. Parameters to search are N elements of fuzzy vector X. Only left and right components are searched, and we take middle component from the first step. DE search space dimension is 2 * N.

- MAX LAMBDA = 5.16005013572706
- Eigen Vector = 1   3.79822723718122   2.71100546608386   1.85201348451787   1.22492577367661
- Correctness Check: A * X = 5.16005 19.59904 13.98892 9.55648 6.32068
- Lambda * X = 5.16005 19.59904 13.98892 9.55648 6.32068
- Fuzzy LAMBDA = [3.37291502986634, 5.16005013572706, 7.36441755699047]
- Fuzzy Eigen Vectors = [0.00000 1.00000 1.00000] [0.00000 3.79823 3.79823]
- [0.00000 2.71101 2.88024] [0.00000 1.85201 2.12760] [0.00000 1.22493 1.28750]
- Correctness Check: A * X = [0.00000 5.16005 9.19445] [0.00000 19.59904 27.97173] [0.00000 13.98892 21.21126] [0.00000 9.55648 15.66856] [0.00000 6.32068 9.75433]
- Lambda * X = [0.00000 5.16005 7.36442] [0.00000 19.59904 27.97173] [0.00000 13.98892 21.21126] [0.00000 9.55648 15.66856] [0.00000 6.32068 9.48166]

To find the criteria weights, we normalize eigenvectors using the following formula:

$$Z = \frac{x - \min x}{\max x - \min x}$$

$NFE = $ [0.000000.000000.00000], [0.000001.000001.00000], [0.000000.611460.67193], [0.000000.304480.40296], [0.000000.0803820.10274]

where NFE represents Normalized Fuzzy Eigenvectors.

Accordingly, fuzzy decision and fuzzy normalized decision matrix are constructed as in Table 3 and 4, respectively. Furthermore, $a*$ and a^- solutions are defined using TOPSIS method in Table 5. It shows positive ideal and negative ideal solutions.

Table 3. Fuzzy decision matrix.

	C_1-product	C_2-price	C_3-place	C_4-promotion	C_5-targeting
w	0.00	0.48	0.31	0.17	0.04
a_1 - AI	0.6, 0.7, 0.8	0.8, 0.9, 1	0.2, 0.3, 0.4	0.6, 0.7, 0.8	0.8, 0.9, 1
a_2 - IoT	0.8, 0.9, 1	0.4, 0.5, 0.6	0.4, 0.5, 0.6	0.4, 0.5, 0.6	0.6, 0.7, 0.8
a_3-VR/AR	0.4, 0.5, 0.6	0.0, 0.1, 0.2	0.6, 0.7, 0.8	0.8, 0.9, 1	0.2, 0.3, 0.4

Table 4. Fuzzy normalized decision matrix.

	C_1-product	C_2-price	C_3-place	C_4-promotion	C_5-targeting
a_1 - AI	0.0, 0.0, 0.0	0.384, 0.432, 0.48	0.062, 0.093, 0.124	0.102, 0.119, 0.136	0.032, 0.036, 0.04
a_2 - IoT	0.0, 0.0, 0.0	0.192, 0.24, 0.288	0.124, 0.155, 0.186	0.068, 0.085, 0.102	0.024, 0.028, 0.032
a_3- VR/AR	0.0, 0.0, 0.0	0.0, 0.48, 0.096	0.186, 0.217, 0.248	0.136, 0.153, 0.17	0.08, 0.012, 0.016

Table 5. Fuzzy positive ideal and negative ideal solutions.

	C_1-product	C_2-price	C_3-place	C_4-promotion	C_5-targeting
$a*$	0.0, 0.0, 0.0	0.384, 0.432, 0.48	0.186, 0.217, 0.248	0.136, 0.153, 0.17	0.032, 0.036, 0.04
a^-	0.0, 0.0, 0.0	0.0, 0.48, 0.096	0.062, 0.093, 0.124	0.068, 0.085, 0.102	0.08, 0.012, 0.016

Closeness of each alternative to the fuzzy positive ideal solution is calculated as:

$$C_1^* = S_1^- / (S_1^* + S_1^-)$$

Table 6. Closeness coefficient of each alternative and ranking.

Alternatives	S_i^*	S_i^-	CC_i	Ranking
a_1 - AI	0.158	0.383	0.71	1
a_2 - IoT	0.33	0.306	0.48	2
a_3-VR/AR	0.349	0.196	0.35	3

As a consequence, AI is the optimal technology to be used according to achieved results. The last regarded choice may be AR/VR (Table 6), but especially during pandemic outbreak they may be useful to reach the customer. Results are shown as; $a_1 > a_2 > a_3$.

5 Conclusion

As the digital marketing constantly embraces us, it is a vital decision to select an important digital marketing technology. In this paper, AI is selected as an optimal choice, that precedes IoT and AR/VR, respectively. Future researches may involve different criteria and alternatives for distinct purposes of different companies.

References

1. Durmaz, Y., Efendioglu, I.H.: Travel from traditional marketing to digital marketing. Global J. Manage. Bus. Res. **16**(2), 35–40 (2016)
2. Crittenden, V., Crittenden, W.: Digital and social media marketing in business education: implications for the marketing curriculum. J. Mark. Edu. **37**, 71–75 (2015)
3. Tiago, M.T.P.M.B., Veríssimo, J.M.C.: Digital marketing and social media: why bother? Bus. Horiz. **57**(6), 703–708 (2014). https://doi.org/10.1016/j.bushor.2014.07.002
4. Kotler, P., Armstrong, G., Opresnik, O.M.: Harlow: Principles of Marketing. Pearson, England (2018)
5. Mukul, E., Büyüközkan, G., Güler, M.: Evaluation of digital marketing technologies with MCDM methods. In: 6th International Conference on New ideas in Management, Economics and Accounting, France, Paris (2019)
6. Leung, K.H., Mo, D.Y.: A fuzzy-AHP approach for strategic evaluation and selection of digital marketing tools. In: IEEE International Conference on Industrial Engineering and Engineering Management, (IEEM), Macao, Macao, pp. 1422–1426 (2019)
7. Şengül, Ü., Eren, M.: Selection of digital marketing tools using fuzzy AHP-fuzzy TOPSIS. In: Fuzzy Optimization and Multi-criteria Decision Making in Digital Marketing, pp. 97–126. IGI Global (2016)
8. Aliev, R.A., Alizadeh, A., Aliyev, R.R., Huseynov, O.H.: Arithmetic of Z-numbers, the: theory and applications. World Scientific (2015)
9. Aliev, R.A., Aliev, R.R.: Soft Computing and Its Application. World Scientific (2001)
10. Zadeh, L.A.: A note on Z-numbers. Inf. Sci. **181**(14), 2923–2932 (2011)
11. Aliyeva, K.: Eigensolution of 2 by 2 Z-matrix. In: 10th International Conference on Theory and Application of Soft Computing, Computing with Words and Perceptions - ICSCCW-2019. Advanced Intelligent System Computing, vol. 1095. Springer, Cham (2020)
12. Mishkoff, H.: Understanding Artificial Intelligence. Instrument Learning Centre. Dallas, Texas (1985)
13. Imanov, E., Daniel, E.: Knowledge base intelligent system of optimal locations for safe water wells. In: International Conference on Theory and Application of Soft Computing, Computing with Words and Perceptions, pp. 519–526. Springer, Cham (2019)
14. Imanov, E., Altıparmak, H., Imanova, G.E.: Rule based intelligent diabetes diagnosis system. In International Conference on Theory and Applications of Fuzzy Systems and Soft Computing, pp. 137–145, Springer, Cham (2018). https://doi.org/10.1007/978-3-030-04164-9
15. Crittenden, W.F., Biel, I.K., Lovely III, W.A.: Embracing digitalization: Student learning and new technologies. J. Mark. Educ. **41**(1), 5–14 (2019)
16. Forbes. https://www.forbes.com/sites/forbesfinancecoucil/2018/11/02/power-pricing-in-the-age-of-ai-and-analytics/#8a516d0784a3

17. Martínez-López, F.J., Casillas, J.: Artificial intelligence-based systems applied in industrial marketing: an historical overview, current and future insights. Ind. Market. Manag. **42**(4), 489–495 (2013)
18. Brandtzaeg, P. B., Følstad, A.: Why people use chatbots. In: International Conference on Internet Science, pp. 377–392. Springer, Cham (2017). https://doi.org/10.1145/3236669
19. Nguyen, B., Simkin, L.: The Internet of Things (IoT) and marketing: The state of play, future trends, and the implications for marketing. J. Market. Manag. **33**, 1–6 (2017). https://doi.org/10.1080/0267257X.2016.1257542
20. Spilotro, C.E.: Connecting the Dots: How IoT is Going to Revolutionize the Digital Marketing Landscape for Millennials. Undergraduate Honors theses (2016)
21. Niyato, D., Hoang, D.T., Luong, N.C., Wang, P., Kim, D.I., Han, Z.: Smart data pricing models for the internet of things: a bundling strategy approach. IEEE Network **30**(2), 18–25 (2016)
22. Yuen, S.C.Y., Yaoyuneyong, G., Johnson, E.: Augmented reality: an overview and five directions for AR in education. J. Educ. Tech. Dev. Exchange (JETDE) **4**(1), 11 (2011). https://doi.org/10.18785/jetde.0401.10
23. Scholz, J., Smith, A.N.: Augmented reality: designing immersive experiences that maximize consumer engagement. Bus. Horiz. **59**(2), 149–161 (2016)

Prediction of Osteometabolic Disorders Due to Diabetes Using Decision Support Systems

Sain S. Safarova[1]([✉]) [iD] and Saadat S. Safarova[2] [iD]

[1] Department Internal Medicine II, Azerbaijan Medical University,
Baku AZ1000, Azerbaijan
safarovasain@gmail.com
[2] Department Obstetrics and Gynecology I, Azerbaijan Medical University,
Baku AZ1000, Azerbaijan
dr.safarovas@gmail.com

Abstract. To investigate whether a deep learning model can predict the bone mineral density (BMD) of lumbar vertebrae and levels of bone remodeling markers from unenhanced biochemical laboratory analysis. For the initial preparation of the model, we used data from laboratory and instrumental studies of 317 patients with diabetes mellitus. The development of the proposed methodology is due to the need to train artificial neural networks for intelligent decision support systems in order to process results of some laboratory data to predicting the values of indicators characterizing the qualitative and quantitative state of the bone. The results showed that the values obtained using the neural network diagnostic model reproduce the clinical research picture with a high degree of adequacy, which allows building a diagnostic algorithm for stratification impaired bone metabolism in diabetes. Therefore, ANNs could be a useful way to better understand the relationships between the numerous different variables that play a role in bone disorders in diabetes.

Keywords: Artificial neural network · Diabetes · Reparative osteogenesis

1 Introduction

Diabetes mellitus (DM) commonly known as a chronic metabolic disease with a wide range of systemic complications, which also includes osteometabolic disorders. Diabetes-induced disorders play a key role in the development of significant loss of bone mass and to changes in its microarchitectonics that leading to an increased risk of fragility fractures associated with low bone mineral density (BMD) and strength [3–5].

Clinicians are currently available a wide range of information, ranging from clinical symptoms of the disease to various types of biochemical data, the results of instrumental methods and hardware research. Each type of data provides information that should be evaluated and assigned to a specific pathology during diagnostic analysis.

In order to increase the adequacy and effectiveness of the diagnostic process and to avoid mistakes in everyday medical practice, decision-making systems in diagnostics based on artificial intelligence methods (including artificial neural networks) are becoming more widely used. Such systems make it possible to increase the

R. A. Aliev et al. (Eds.): ICAFS 2020, AISC 1306, pp. 388–394, 2021.
https://doi.org/10.1007/978-3-030-64058-3_48

effectiveness of clinical analysis by processing complex and interrelated arrays of medical data and integrate them into the results of diagnostics performed by a clinician [2].

Artificial Neural Networks (ANN) is a modern area of the artificial intelligence methodology, which is increasingly used in decision support systems (DSS) in various fields. The effectiveness of the use of this mathematical apparatus in medical practice is due to its wide capabilities for modeling complex multi-parameter systems, processing large and interconnected arrays of medical data due to the formalization of the experience and intuition of a doctor/researcher [6], which avoids costly diagnosis and the selection of an inappropriate treatment methodology. Currently, applications of ANN in medicine mainly solve the problems of image recognition (ultrasound, X-ray, ECG) [2, 8].

The doctor in his practice constantly solves the problem of diagnosis, which is the problem of identifying diseases and choosing treatment tactics, based on complex, multifactorial observations and applying accumulated knowledge (personal and collective), intuition. All this determines the feasibility and relevance of the development of DSS, imitating medical reasoning, based on the formalization of the experience and intuition of a specialist in difficult to diagnose clinical cases. In this case, the input data are a set of symptoms/signs, as well as laboratory and instrumental analysis data, and the output is calculated data that allows you to make a final diagnosis. Analysis of studies [10] shows that at the moment, the implementation of such DSS is effective based on the methodology of artificial neural networks.

Since cause-effect relationships are important from the point of view of modeling an algorithm for identifying risk groups associated with reparative osteogenesis disorders [1], we investigate whether a mathematical method will based on the use of DSS based on an artificial neural network [8, 9], capable of calculating the dependence of bone mineral density (BMD) and bone remodeling markers on a some of biochemical laboratory parameters to predict the risk of osteoporotic bone changes in patients diabetes based on the model.

2 Patients and Methods

The survey was conducted from November 2015 to July 2017. All patients gave their written consent to participate in this study. The research methodology was approved by the ethics committee of the medical university, on the basis of which it was conducted.

2.1 Patients

A cross-sectional study evaluating the data of 98 patients with type 1 diabetes (women - 57, men - 41) and 137 patients with type 2 diabetes (women - 85, men - 52) who have not previously been diagnosed with bone metabolism disorders and osteoporosis was evaluated. The age of the patients varied from 40 to 69 years.

Exclusion criteria: persons previously treated for osteoporosis or having a history of fracture, as well as patients with diseases of the endocrine system, liver and kidneys of a non-diabetic nature, with a history of stage 4–5 diabetic nephropathy.

2.2   Methods

The state of bone formation was judged by the activity of total alkaline phosphatase (ALP) and the content of the aminoterminal propeptide collagen type I (PINP) in blood serum. The level of bone resorption was judged by the content of the C-terminal telopeptide (b-CTx). All patients underwent dual-energy X-ray absorptiometry (DXA) (densitometer HOLOGIC, model Discovery QDR 4500A. USA) of the lumbar spine (L1–L4).

The connection between the results of laboratory research and the parameters of bone metabolism was revealed by developing a method for the analysis of risk factors for osteoporosis. The analysis included the study of a number of hormones, indicators of the functional state of the kidneys, ionic blood balance, markers of bone remodeling, as well as the results of dual x-ray absorptiometry of the patient. The study of the above patient data gave the researchers a list of 30 variables, including the BMD value for each of the patients. All of the variables considered, according to previous medical studies, have an impact on the development of osteoporosis. These variables also included factors such as age, gender, height, body mass index (BMI), menopause duration in women, glycemic index, type and duration of diabetes mellitus, etc.

2.3   Statistics

We develop ANN model based DSS, to offer some insight regarding the complex biological connections between variables on study, which displays a set of input data in a set of output data, a multilayer perceptron (MLP). For the initial preparation of the model, we used data from laboratory and instrumental studies of 317 patients. The construction of the neural network was carried out using MATLAB 8.6 (R2015b) [7]. The Bayesian regularization algorithm was used to train the neural network. The implementation of the approach was verified by conducting a comparative study, which included testing a dynamically constructed network and presenting a comparative analysis of the classification results.

The methodology used to construct a self-learning predictive system using neural networks for intelligent DSS consisted of the following stages: task setting, input data preparation, creation and training of a neural network including: selection of the type of neural network, formation of a training data supply circuit that determines the number of input signals and their corresponding input synapses, as well as the answer, enable or disable the rationing of input data.

Rationing of input signals is one of the types of pre-processing and is extremely important in the methodology for creating neural network systems. During normalization, not the value of the parameter (for this example), but its equivalent, obtained by recalculation according to a certain scheme, is fed to the input synapse. The technological aspect was as follows. The applied technology for training neural networks provides a universal structure and training algorithms for clinical data of any nature. However, as a result of our numerous experiments, it was found that the most versatile and fast-learning architecture of a fully connected sigmoid (having the characteristic function of neurons) neural network works optimally when the input parameters are in the range from -1 to 1. Each input signal is recalculated before being fed to the synapse the formula:

$$NX_i(t) = \frac{X_i(t) - X_i^{\min}}{X_i^{\max} - X_i^{\min}}$$

Where $X_i(t)$ - is the initial signal, $NX_i(t)$ - is the normalized signal obtained, $X_i^{\min}$ and $X_i^{\max}$ - respectively, the minimum and maximum values of the interval of input parameters in the field supplied to synapse i.

In our case, neural network training was an automatic process, which only after its completion required the participation of a specialist to evaluate the results. Undoubtedly, an adjustment was required - the creation of additional networks with other parameters, etc., in order to be able to evaluate the system at any stage of training by testing a control sample. Network training continued as long as it was able to produce the best possible results on independent data. The model of the developed neural network is schematized in Fig. 1.

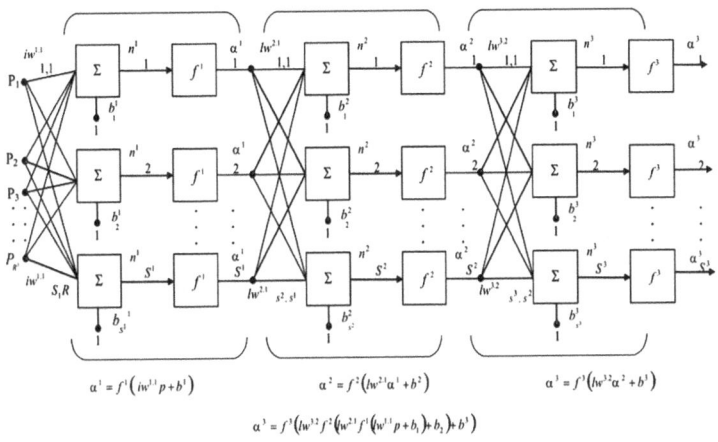

Fig. 1. Multilayer neural network

Given the high speed of learning the neural network, in the framework of the proposed methodology, a strategy was developed that allows to circumvent the above compromise, due to the longer time required for training. In order to verify the quality of training of the neural network, it was tested. Testing of the sample was carried out with the well-known answers of the examples. Thus, it was checked whether the network correctly determines the answers of the examples and how confidently it does it. The network-defined response of the example was compared with a previously known one. First, testing was conducted on the sample on which the network was trained. When testing the same training sample, the answers of all examples were determined correctly. Further, a similar sample was tested with pre-known answers, but examples that did not participate in the training of the network. After the implementation of the stages discussed above, the neural network is ready for the last stage - forecasting indicators.

3   Results

Artificial neural network algorithm determined the relationship between input variables and BMD, bone remodeling markers. The practical effect of the constructed model for predicting BMD and values of bone remodeling markers in diabetes based on the analysis of a number of laboratory parameters has been proved. The model was used to determine which patients should undergo densitometry and analysis of bone remodeling markers, to check the qualitative and quantitative characteristics of the bone and, thus, prevent some risks associated with osteoporosis.

The topology of the model consisted of an input layer, a hidden layer, and an output layer. A model with final ANN parameters was trained using data from 80% of patients from a randomly selected database. Data from the remaining 20% of patients were used to verify the results. The average value of the absolute measurement error in these patients was 2.09%. As a result, some adjustments were made to the model settings to increase its adequacy. Further training is achieved during its practical operation. The learning process continued until errors were reduced for all examples and stopped at the moment when the error in the control sample began to increase.

4   Discussion

Artificial neural networks demonstrate the ability to model complex relationships between variables to identify groups at risk of developing osteoporosis or fractures from the general category of people with diabetes. The study described in this article demonstrates the usefulness of machine learning methods, such as artificial neural networks, to study the relationship between BMD and bone remodeling markers with a number of factors related to diabetes (glycemic control, ion homeostasis and kidney function). An artificial neural network was used to build a mathematical model that determined the relationship between input variables and BMD, as well as bone remodeling markers. As a result, some adjustments were made to the model settings to increase its adequacy. Further training is achieved during its practical operation. The learning process continued until errors were reduced for all examples and stopped at the moment when the error in the control sample began to increase. For ease of use, a visual interface was created as shown in the Fig. 2.

The constructed neural network model is capable of predicting BMD and values of bone remodeling markers in patients with diabetes mellitus in accordance with the results of their laboratory analyzes. This model can be used to determine which patients should undergo densitometry and analysis of bone remodeling markers to check bone quality and prevent some of the risks associated with osteoporosis.

Comparative analysis of this approach showed that the values obtained using the neural network diagnostic model reproduce the clinical research picture with a high

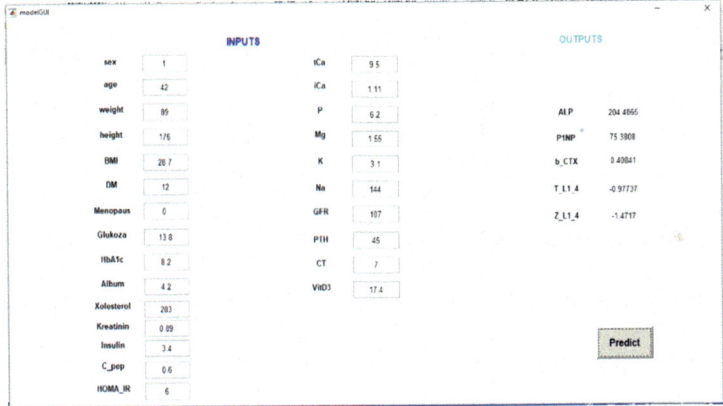

Fig. 2. An example of a neural network calculation

degree of adequacy, which allows building a diagnostic algorithm for stratification impaired bone metabolism in diabetes.

Case study research was developed methodology of intelligent decision support system using artificial neural networks, which allows predicting the values of indicators characterizing the qualitative and quantitative state of the bone based on the measurement results of several laboratory variables, for early diagnosis of the risks of structural and functional bone changes in patients with diabetes.

Funding Statement. This research did not receive any specific grant from any funding agency in the public, commercial or non-profit sector.

Data Availability. The clinical material data used to support the findings of this study is available from the corresponding author upon request.

Conflicts of Interest. The author declares that there is no conflict of interest regarding the publication of this paper.

References

1. Abdel-Mageed, S.M., Bayoumi, A.M., Mohamed, E.I.: Artificial neural networks analysis for estimating bone mineral density in an Egyptian population: towards standardization of DXA measurements. Am. J. Neural Netw. App. **1**, 52–56 (2015). https://doi.org/10.11648/j.ajnna.20150103.11
2. Cruz, A.S., Lins, H.C., Medeiros, R.V.A., et al.: Artificial intelligence on the identification of risk groups for osteoporosis, a general review. BioMed. Eng. OnLine **17**, 12 (2018). https://doi.org/10.1186/s12938-018-0436-1
3. Jiao, H., Xiao, E., Graves, D.T.: Diabetes and its effect on bone and fracture healing. Curr. Osteop. Rep. **13**(5), 327–335 (2015)

4. Jørgensen, H.S., Winther, S., Bøttcher, M., Hauge, E., et al.: Bone turnover markers are associated with bone density, but not with fracture in end stage kidney disease: a cross-sectional study. BMC Nephrol. **18**, 284 (2017)

5. Oei, L., Rivadeneira, F., Zillikens, M.C., Oei, E.H.G.: Diabetes, Diabetic Complications, and Fracture Risk. Curr. Osteop. Rep. **13**(2), 106–115 (2015)

6. Liu, Q., Cui, X., Chou, Y.C., et al.: Ensemble artificial neural networks applied to predict the key risk factors of hip bone fracture for elders. Biomed. Sig. Process Control **21**, 146–156 (2015). https://doi.org/10.1016/j.bspc.2015.06.002

7. Math Works: MATLAB (2017). www.mathworks.com

8. Pouliakis, A., Karakitsou, E., Margari, N., et al.: Artificial neural networks as decision support tools in cytopathology: past, present, and future. Biomed. Eng. Comput. Biol. **7**, 1 (2016). https://doi.org/10.4137/BECB.S31601

9. Shioji, M., Yamamoto, T., Ibata, T., et al.: Artificial neural networks to predict future bone mineral density and bone loss rate in Japanese postmenopausal women. BMC Res. Notes **10**, 590 (2017). https://doi.org/10.1186/s13104-017-2910-4

10. Yu, X., Ye, C., Xiang, L.: Application of artificial neural network in the diagnostic system of osteoporosis. Neurocomputing **214**, 376–381 (2016). https://doi.org/10.1016/j.neucom.2016.06.023

Method for Selecting Protective Technological Information Measures Based on the Apparatus of Fuzzy Sets

O. V. Serpeninov[1]([⊠])[iD], N. E. Sheydakov[1][iD],
G. A. Lopatkin[1][iD], and A. S. Voskovskaya[2][iD]

[1] Rostov State University of Economics, B.Sadovaya Street,
69, 344002 Rostov-on-Don, Russia
`serpeninov53@mail.ru`, `sheidakov@mail.ru`,
`lopatkin@mail.ru`
[2] Financial University under the Government of the Russian Federation,
Moscow, Russia
`ASVoskovskaya@fa.ru`

Abstract. In modern conditions, increasing the competitive ability of high-tech products is ensured by innovative approaches in creating products with higher quality compared to the analogue due to the emergence of new qualities and properties. Because of increased competition, there is a need to intensify efforts to counteract the conduct of industrial intelligence for developed products. The article considers a method of protecting important technological information, based on the formation of a distorted image of developed products among competitors through the use of protection measures in the form of hiding a certain amount of protected information and bringing misinformation to competitors. The developed method supplements well-known approaches to ensuring the protection of technologically important information for the developer by methodical concealment and technical misinformation of competitors. It is based on the well-known mathematical apparatus of probability theory and mathematical statistics, the theory of fuzzy sets. The given illustrative example confirms the efficiency of the proposed method. Using the proposed method can increase the effectiveness of long-term investments in the development of complex high-tech products by eliminating the emergence of similar products on the market, and determine the least costly measures to shield the protected information. The method is quite easily implemented in practice with the participation of qualified experts; its use is carried out in combination with other methods of protecting technologically important information.

Keywords: Competitive advantage · Industrial intelligence · Misinformation · Cover object · Damage · Fuzzy sets · Protection measures

1 Introduction

In modern conditions, there is a significant increase in the share of high-tech products, the creation and production of which require significant investment. Paying back new products and making profits are provided by competitive advantages in comparison

R. A. Aliev et al. (Eds.): ICAFS 2020, AISC 1306, pp. 395–402, 2021.
https://doi.org/10.1007/978-3-030-64058-3_49

with the products offered on the market or developed by competitors. However, at the same time, a conflict situation develops on the market because of opposition to the development and promotion of the products by competing structures. The counter-measures taken are usually proportional to the potential danger of the developed products to competitors. It is possible to assess the degree of danger of the products being developed only on the basis of obtaining information about their quality and properties extracted in various ways, including methods of industrial espionage. The intensity of the counteraction depends on the availability of competitors with infor-mation about the planned properties and characteristics of the developed products, the innovative technologies used and technical developments.

Due to the impossibility of directly affecting competitors in order to force them to abandon the opposition, developers are forced to use various methods to protect their interests. The developer is obliged to protect the important technological information about the developed products by methodical concealment and misinformation of competitors, which are well known and quite fully described in many papers [1–4]. The aim of the investigation is to develop a method for choosing effective measures to protect technological information from competitors, based on its degree of importance and the possibility of obtaining protected information about the developed products.

2 Development and Efficiency Confirmation of the Proposed Method

To ensure a stable position in the market, the organization carries out the development of high-tech products with improved properties and higher quality indicators, which surpasses analogues in both domestic and foreign markets. The return on investment and profit are expected after the products are sold on the market at a point in time T. It is planned to ensure the competitive ability of the developed products due to its higher quality:

$$K[Z^m(t)] > K[Z^a(t)], \tag{1}$$

where $K[Z^m(t)]$ – the quality of the developed products at time T of its appearance on the market;

$K[Z^a(t)]$ – the quality of the available analogue at the same point in time;

$Z(t)$ – a set of qualitative characteristics of products.

The fulfillment of this condition is achieved by improving a certain set of product characteristics $z_i^m(t) \geq z_i^a(t)$ $(i = \overline{1,I}$, where I – quantity of product qualitative char-acteristics), and also the emergence of new qualitative characteristics.

In order to minimize losses from actions of the competitors by obtaining infor-mation about the developed products, it is necessary to convince competitors of the inappropriateness of these actions by methodical concealment and misinformation, i.e. they will consider the developed products to be of insufficiently high quality and, therefore, these products will not be able to make significant competition for similar products.

This is achieved by bringing to the competitors the underestimated characteristics of the object, the synthesis of which is aimed at misinforming competitors and hiding the real characteristics of the developed products. For this purpose, a new appearance of the developed products is formed, which differs from the real one – the cover object. In this case, the fulfillment of the requirement should be met to have the quality of the formed sample $K[Z^o(t)]$ to satisfy the ratio:

$$K[Z^m(t)] > K[Z^o(t)] > K[Z^a(t)]. \tag{2}$$

Therefore, the task of protecting economically important industrial information is reduced to the synthesis of a set of characteristics of the cover object in the form of development of similar products by time T, the receipt of information about which by competitors, will minimize the damage from their implementation of countermeasures $Q\{K[Z^m(t)]\}$.

It is associated with a decrease in the quality of developed products relative to competitors' products, depending on their implementation of countermeasures:

$$\Delta K[Z^m(t)] = K[Z^m(t)]|_{n=0} - K[Z^m(t)]|_{n=n_j}, \tag{3}$$

where n_j is understood as the used option of counteraction from competitors.

Since the intensity of the counteraction directly depends on the information $U[Z^m(t)]$ received by competitors about the set of product characteristics, the ratio $\Delta K[Z^m(t)] = F\{U[Z^m(t)]\}$ is valid.

Therefore, it is necessary to determine the optimal list of characteristics of the cover object $Z^o(t)$ and for each of the selected characteristics to determine its numerical value, providing the solution to the problem of information protection and minimizing damage from competitors.

Determining the optimal list of the cover object characteristics depends on the correct determination of the importance of information about each of the characteristics and ensuring the given reliability of bringing false information about the developed products through various channels. Bringing the information prepared by the developer to competitors can use various sources: media, advertising products and specially organized channels for information leakage. In this case, it is necessary to take into account the degree of trust in information sources. The solution to most of these problems is based on informal-heuristic methods of the apparatus of fuzzy sets [5].

To assess the effectiveness of the synthesized cover object, it is necessary to determine the set of possible countermeasures by competitors $N = \{n_j\}$, $(j = \overline{1,J}$, where J is the number of countermeasures) and the potential danger of each of them to reduce competitive ability in violation of the developed products privacy in the form of destructive actions.

Analyzing the set of possible countermeasures by competitors $N = \{n_j\}$ allows us to obtain estimated costs for the implementation of each of them $C(n_j)$, as well as to determine the time for its implementation.

Let all possible threats be identified $Y = \{y_k\}$, ($k = \overline{1, K}$, where K is the number of threats to the breach of confidentiality of information about the developed products) that appear and are realized during the period T with a certain probability $P_k(t), t \in [1, T]$. As a result of the threat implementation, a set of illegal actions that violate the confidentiality of information is performed. The damage from the implementation of a particular l destructive action can be calculated: it is G_l and, besides, the unacceptable damage G_o is known. The value $V_l = G_l/G_o$ will characterize the share of damage from the implementation of the l destructive action, which is relatively unacceptable. The probability of the realization of the l destructive action is determined through the probability of the realization of those threats that are associated with this destructive action, and is determined from the ratio:

$$P_l(t) = 1 - \prod_{y \in Y_l} [1 - P_y(t)], \tag{4}$$

where Y_l – many threats, the implementation of which implements the l destructive action.

The value $W_l = V_l \cdot P_l(t)$ is called the hazard coefficient of l destructive action, and then the damage $W_{\sum}(t)$ from the implementation of the set of destructive actions (the danger coefficient of the set of threats) will be

$$W_{\sum}(t) = \sum_{l-1}^{L} W_l \cdot P_l(t). \tag{5}$$

Let's call the degree of security as the value supplementing the hazard coefficient to 1, which is a measure of the security of the protected object from the totality of threats:

$$\gamma(t) = 1 - W_{\sum}(t). \tag{6}$$

The determination of the hazard coefficients of destructive actions is based on the ratios linking these coefficients with indicators of the importance (value) of information. Such indicators in the form of coefficients of importance can be determined on the basis of a heuristic analysis of information about the developed products and their categorization by importance (value). In this case, information about the developed products is presented in the form of a set of informative features; the category of importance (value) of the information contained is set by the user for each of them. For example, four categories of information importance can be introduced: 1 – especially important, 2 – very important, 3 – important, 4 – not very important.

When determining the hazard coefficients of destructive actions in the form of fuzzy values, it is necessary to conduct an expert analysis of the danger of destructive actions. Since there are too many factors influencing the estimated values, different experts assess the value of the estimated parameter differently and often find it difficult to set a specific number. The apparatus of fuzzy sets is used to correctly parry such situations, while the hazard coefficients are set in the form of fuzzy numbers that can take their values from a certain given interval (set) with different values of membership functions [6–8]. The most commonly used ones are triangular and trapezoidal fuzzy numbers, while triangular numbers are often preferred in solving economic problems as predicted values of the measured parameter.

The hazard coefficient of each threat in the form of a triangular representation [7] is determined by a triple of numbers (a, b, c): a – a number less than which its value cannot be; b – a number about which it can be said with maximum certainty that a quantity takes precisely this value; c – a number greater than which the coefficient cannot be. The values of membership functions $\mu_A(x)$ for these points are set so that for the midpoint this value is the largest, and the other two are zero. Further, when processing the results of a survey of experts, the number may lose its triangular appearance.

For the case of specifying hazard risk factors with fuzzy numbers, formula (5) is written in the form:

$$W_{\sum}(t) = \sum\nolimits_{l-1}^{L} W_l \circ P_l(t). \tag{7}$$

where the operation "$\circ$" of multiplying fuzzy numbers is carried out according to the rule [5, 9]

$$\mu_D(x) = \sup_{a \cdot b = x} \min \mu_A(a) \cdot \mu_B(b), \tag{8}$$

the operation of adding fuzzy numbers is as follows:

$$\mu_D(x) = \sup_{a + b = x} \min \mu_A(a) \cdot \mu_B(b). \tag{9}$$

Determining the degree of protection (6) in the absence of $\gamma_o(t)$ and the application of the technical information protection measures $\gamma_{PD}(t)$, we can evaluate the effectiveness of the technical information protection by the formula:

$$E(t) = \{[\gamma_{PD}(t) - \gamma_o(t)]/\gamma_{PD}(t)\} \cdot 100\%. \tag{10}$$

Knowing the fuzzy values and membership functions of the hazard coefficients of destructive actions, it is possible to calculate the fuzzy values of the degree of protection of an object of informatization from a complex of threats taking into account the probabilities of implementing threats according to the rules of the theory of fuzzy sets.

To confirm the efficiency of the proposed method it is to the point now to consider the illustrative example. Let threats be possible in the form of unauthorized access to the network server and interception of information through the technical leakage channel during its processing in the information system. The number of destructive actions, the implementation of which is possible in relation to information resources about the developed products, is 3.

The hazard coefficients of these destructive actions (Fig. 1), presented in the form of fuzzy numbers, for the considered period of time are respectively equal:

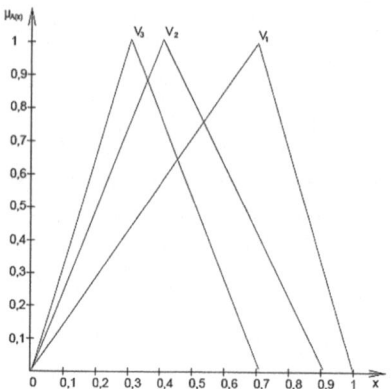

Fig. 1. The hazard factors

$$V_1(t) = 0/0 + 1/0,7 + 0/1; V_2(t) = 0/0 + 1/0,4 + 0/0,9;$$
$$V_3(t) = 0/0 + 1/0,3 + 0/0,7.$$

In this case, the threat associated with unauthorized access to the network server causes only the first destructive action, the threat of interception of information through the technical leakage channel is the second destructive action, and the third destructive action can be caused by both the first and second threats.

Calculations showed that the probabilities of each threat without protection measures being 0.2 and 0.1 respectively. The probabilities of the implementation of threats that cause the corresponding destructive consequences will be determined according to formula (4):

$$P_1(t) = 1 - (1 - 0,2) = 0,2; P_2(t) = 1 - (1 - 0,1) = 0,1;$$
$$P_3(t) = 1 - (1 - 0,2) \cdot (1 - 0,1) = 0,28.$$

The total hazard coefficient $W_{\sum}(t)$ in the absence of information security measures, taking into account formula (8) and formula (9), will be determined as follows:

$$W_{\sum}(t) = (0/0 + 1/0,14 + 0/0,2) + (0/0 + 1/0,04 + 0/0,09) + (0/0 + 1/0,084 + 0/0,196) = (0/0 + 1/0,264 + 0/0,486)$$

Then, in the case of a clear dependence of the degree of protection on the hazard coefficient of the totality of threats, the value of the degree of protection is determined by formula (6):

$$\gamma(t) = (0/0,514 + 1/0,736 + 0/1).$$

To protect against unauthorized access to the server, identification and authentication of devices, including stationary, mobile and portable, were introduced, as a result of which the probability of an access threat was reduced to 0.05.

Organizational measures taken in the form of control and management of physical access to hardware, information security tools and means of ensuring their functioning have reduced the likelihood of information interception through the technical leakage channel to 0.01.

It is necessary to evaluate the effectiveness of the protective measures taken when there is a clear relationship between the security indicator and a fuzzy set of hazard risk factors.

The values of the total hazard coefficient and the security index when applying protective measures are calculated similarly, and for the values of the probabilities of threats

$$P_1^{M3}(t) = 0,05; P_2^{M3}(t) = 0,01; P_1^{M3}(t) = 1 - (1 - 0,05) \cdot (1 - 0,01) = 0,06,$$

it is defined as:

$$\gamma(t) = (0/0,899 + 1/0,943 + 0/1).$$

A defuzzified security index at a zero cut-off level without protective measures $\gamma^0(t) = 0,75$, and under the conditions of application of measures $\gamma^{M3}(t) = 0,947$. Then, according to formula (10), the efficiency indicator is equal to:

$$E(t) = (0,947 - 0,750/0,947 \cdot 100\% = 20,8\%$$

Thus, as a result of the application of the proposed method of choosing measures to protect technological information based on the apparatus of fuzzy sets, one of the possible options for constructing a system for protecting protected information about the developed products is obtained, which confirms the efficiency of the proposed method. It should be noted that there can be a lot of such options, since the values of the cover object characteristics brought to competitors are within a continuous interval, and the number of possible communication channels is limited. This allows you to get many sets of values of the characteristics of the cover object, and this makes it possible to assess the technological feasibility of their implementation and economic viability.

The described approach allows us to assess the security of important technological information about innovative developed products in the face of a fuzzy idea of the danger degree of a complex of anthropogenic and man-made threats. To take into account the dynamics of threat realization, it is necessary to develop mathematical models for calculating the probability of threat realization.

3 Conclusion

The developed method complements the basic approaches for choosing measures to protect information about developed products and the innovative approaches used by hiding a certain amount of protected information about the developed products and bringing misinformation to competitors. The method is implemented with the assistance of qualified experts and is based on the mathematical apparatus of the theory of fuzzy sets, probability theory and mathematical statistics. The illustrative example of information protection for increasing the competitiveness of the developed products which has been presented in the article, on the one hand, proves the effectiveness of the described method and, on the other hand, shows the direction of the development of this issue in further research.

Further development of the proposed method of choosing measures to protect technological information on the development of high-tech innovative products can go through the creation of standard applied methods for assessing potential damage from competitors, assessing the level from awareness of the current state of development, as well as choosing means of hiding protected information and sources of bringing the required misinformation to competitors.

References

1. Korchenko, A.G.: Building Information Security Systems on Fuzzy Sets. MK-Press, Kiev (2006). (in Russian)
2. Lapsar, A.P., Lapsar, S.A.: Ensuring the safety of innovative developments in a competitive confrontation. Finan. credit **23**(1), 49–62 (2017). (in Russian)
3. Serpeninov, O.V., Fedorova, Y.V., Lozina, E.N.: Application of the apparatus of fuzzy sets for ensuring the information infrastructures security. In: Akperov, I.G. (ed.) Soft Models of Management in Terms of Digital Transformation: Monograph/Under The General. PEI HE SU (IUBiP), Rostov-on-Don (2019)
4. Volkov, E.I.: Cybersecurity of the digital business of a small enterprise. In: Intellectual Resources for Regional Development, vol. 5, no. 1, pp. 35–40. Southern University (IUBiP), Rostov-on-Don (2019). (in Russian)
5. Zade, L.: The Concept of a Linguistic Variable and Its Application to Making Approximate Decisions. Mir, Moscow (1976). (in Russian)
6. Efimova, E.V., Serpeninov, O.V., Cherkezov, S.E., Sheydakov, N.E.: Analysis of factors of economic efficiency of virtual business relations in the digital economy. In: Intellectual Resources for Regional Development, vol. 4, no. 1, pp. 184–191. Southern University (IUBiP), Rostov-on-Don (2018). (in Russian)
7. Tishchenko, E., Serpeninov, O., Zhilina, E.: Comparative analysis of the computerized testing systems in education. In: Innovative Approaches to the Application of Digital Technologies in Education and research. Stavropol, 20–23 May 2019. https://easychair.org/cfp/SLET-2019. Accessed 03 May 2020
8. Zhilina, E.V., Popova, L.K., Rutta, N.A., Sheydakov, N.E.: Fuzzy model of functioning of educational-laboratory and production capacities of the educational cluster in the information security field. In: Aliev, R.A., Kacprzyk, J., Pedrycz, W., Jamshidi, M., Babanli, M.B., Sadikoglu, F.M. (eds.) ICSCCW 2019, vol. 1095, pp. 704–711. Springer, Cham (2020). https://doi.org/10.1007/978-3-030-35249-3_91
9. Novak, V., Perfilieva, I., Mochkrozh, I.: Mathematical Principles of Fuzzy Logic. Fizmatlit, Moscow (2006). (in Russian)

Analysis of the Gas Pipelines Operation Based on Neural Networks

G. G. Ismayilov[(⊠)] ⓘ, E. Kh. Iskandarov ⓘ, F. B. Ismayilova ⓘ,
and S. G. Hacizade ⓘ

Azerbaijan Oil and Industry University, Baku, Azerbaijan
asi_zum@mail.ru, e.iskenderov62@mail.ru,
fidan.ismayilova.2014@mail.ru,
syleymanhajizade@gmail.com

Abstract. The article deals with the diagnostics of complications that can occur during transportation of oil through submarine pipelines via the application of artificial neural network being one of the modern elements of information technology. In this regard, artificial neural network was constructed for the diagnostics of small (underground) oil leakages being difficultly detected in complicated submarine oil pipeline system. A number of "neurons" were included in the developed neural networks core that can be freely functioning and interacting, enabling to choose the best model according to information abundance.

Keywords: Pipeline · Oil leakages · Artificial neural network · Algorithm · Input signals · Weight factor · Output parameters · Mean square error

1 Introduction

At that time, the observation indicators of that process were used to adapt and select the best model characterizing the real process.

The main criterion for selecting the best model is quadratic mean error being minimal.

As the developed artificial neural networks are self-informed, while including new observation indicators of the pipeline modes, these networks reselect the models and reflect all the changes occurring in the system, as well as, various oil leakages by determining the dependence of the weight coefficients on the number of observations for selected models.

The paper is organized as follow. Section 2 is devoted to the preliminaries which related to the considered topic. In Sect. 3 we consider numerical example and results of computer simulation related to the considered problem. Section 4 presents our conclusion.

2 Adaptive Neuro-Fuzzy Inference System

Adaptive Neuro-Fuzzy Inference System (ANFIS) [1] being adaptive fuzzy logic system is isomorph to fuzzy knowledge base. In neuro-fuzzy network smooth membership functions are used and it allows to use quick training algorithms when making neuron networks.

ANFIS serves to minimize the difference and experimental data. Let's look through architecture of fuzzy system and layers. ANFİS is carried out in the form of 5 layers network of Mamdani and Sugeno type fuzzy logics. At this time 1-st layer carries out terms of variables 2-th layer antisedents of fuzzy rules, 3-rd layer normalizing of performing levels of this rule, 4-th layer completing of this rule, 5-th layer combining of the obtained results according to various rules. In the Fig. 1, 2 input and 1 output, 4 rules based network is described.

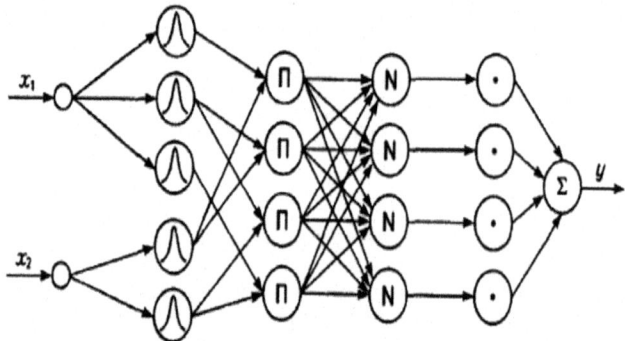

Fig. 1. Adaptive neuro-fuzzy inference system

3   Solution and Discussion of the Problem

In the solution of the problem, the following functional dependencies of pressure losses were used in case of oil leakage and non-leakage as a major transmission function for the purpose of constructing artificial neural network on the base of accepted principles for a simple pipeline operating in stationary mode [2–8].

For non-leakage case

$$\Delta_{p_0} = AQ_0^2 \tag{1}$$

However, for leakage case

$$\Delta_{p_1} = A_1 Q_0^2 + B_1 (Q_0 - q)^2 \tag{2}$$

Here $A = K \cdot L$; $A1 = K \cdot X$; $B1 = K(L - X)$. K- relatively are pressure losses in case of oil leakage and non-leakage in the pipeline; Q_0 and q relatively are oil consumption flowing in the pipeline and the volume of the oil leaking from the pipeline at a single time; λ is hydraulic resistance coefficient; p_0 is the density of oil; L and D relatively are the length and diameter of the pipeline; X is the distance of oil leakage site from pipeline beginning.

In order to adapt the artificial neural networks to the processes occurring in the pipelines, as mentioned above, these pipelines modes should have the abundance of observation indicators (the abundance of initial data), (for example, the abundance of the values of the pressure at the beginning and end of pipeline and oil consumption measured within a few months, etc.). Of course, the more the number of the measured values, the high the accuracy of network informing. A part of that abundance (usually most part) is used to inform the network, and the other part is used to test it.

The results of informing and testing of the artificial neural network for the simple oil pipeline are given below. For this purpose, the results of the mathematical experiment given in the literature for the simple oil pipeline were used.

According to the results of that experiment, the dynamics of observation indicators (total 186 observations) of a stationer mode for a simple pipeline is shown in Figs. 2 and 3.

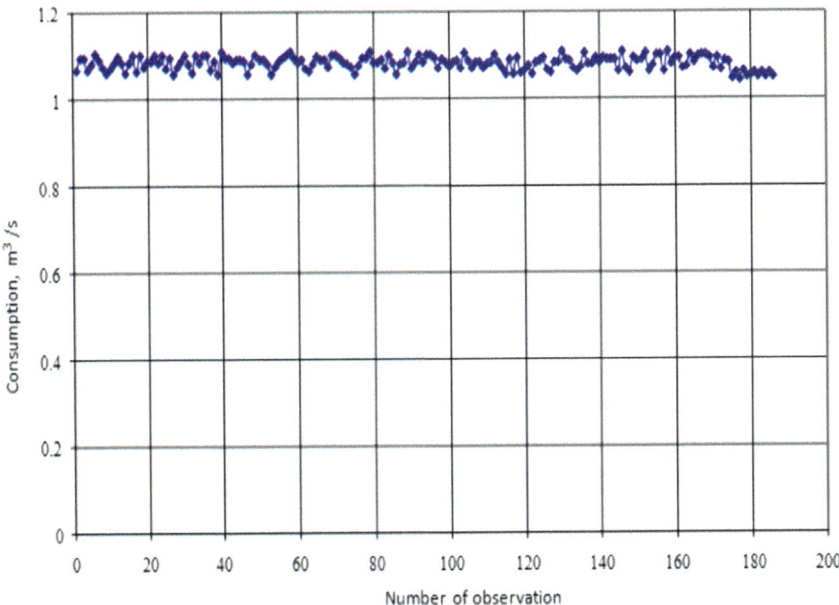

Fig. 2. The dynamics of oil consumption signal in the entrance of neural network.

Of them 150 indicators were used to inform the proposed artificial neural network, and 24 were used to test them. 12 observation indicators, shown at the end of the graphics in those figures, have been added by the author sand characterize the oil leakage from oil pipelines.

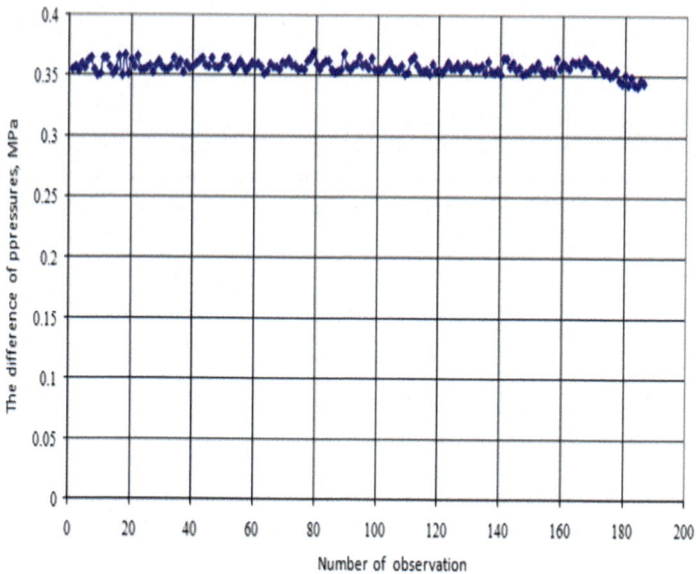

Fig. 3. The dynamics of pressure difference signal in the entrance of neural network.

At this time, the oil consumption being included in the simple oil pipeline, has been stable, however, due to the oil leakage, the average consumption of the pipeline has decreased by 2.9% and the pressure difference by 3.08%.

Taking into account that the maximum difference from average values in case of oil consumption and pressure loss in the pipeline relatively is 4.98% and 4.57%, it becomes clear that in case of leakage, varying the difference of consumption and pressure at the mentioned ranges practically can be undetected or detected after a considerable period of time. This can cause considerable oil loss in the pipeline.

The weight coefficient values of the model (1) has been obtained as $A = 0.614$ for informing stage and $A = 0.5862$ for the test stage.

As a result, the difference between weight coefficients determined for the informing and testing stages is $\varepsilon(A) = 4.53\%$ so that, this is one of error ranges in practice.

Thus, it is considered that the artificial neural network has already been developed to characterize the oil flow process in the provided simple pipeline and can be used to fully characterize the operation mode of that pipeline. In order to use the developed neural network for other simple oil pipelines, it is necessary to reconfigure the network according to observation indicators of that pipeline, that is, to adapt it to the processes occurring in the pipeline [9–12].

Now let's consider the case of oil leakage from the pipeline. In order to clarify the problem, first of all, it is necessary to include the data characterizing oil leakage into the neural network. It should be taken into account that, if oil leakage starts in the pipeline, it is not always possible to be detected according to observation indicators. Therefore, it is necessary to determine whether network status is changed by adding observation indicators individually. The correlation coefficient sharply varies between weight coefficients characterizing the state of neural network and initial indicators included in the network and the indicators obtained at the output as the oil leak occurs in the pipeline.

Clearly, it can be displayed in the processing example of the observation indicators described in the simple pipeline. The changes of oil consumption and pressure difference signals included into the entrance of the artificial neural network depending on the analogue number of the observations have been illustrated in Figs. 4 and 5. The ratio of the values obtained at the output of fourfold neural network's of those signals analogues to the average values relatively was $6.5 \cdot 10^8$ and $1.7 \cdot 10^9$.

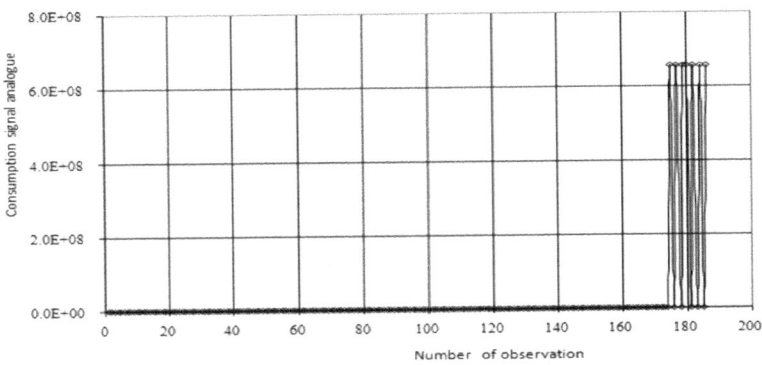

Fig. 4. Dynamics of consumption signal analogue at fourfold neural network output

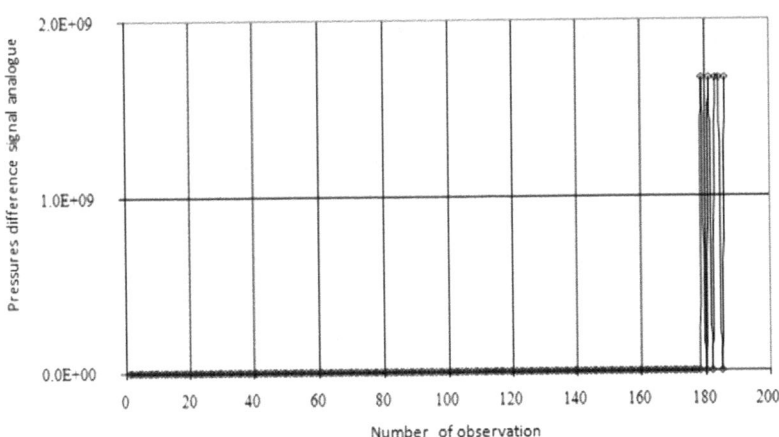

Fig. 5. Dynamics of pressures difference signal analogue at fourfold neural network output

4 Conclusion

Thus, if there are any changes in the regime parameters of the oil pipeline, it can be detected with high accuracy, efficiency and reliability with the help of a developed artificial neural network. Since the application of artificial neuron technologies does not require any special research, additional labor and material resources, it can be considered very efficient and expedient to detect oil leaks, including underground leaks, from pipelines currently operating onshore and offshore.

References

1. Aliev, R.A., Fazlollahi, B., Aliev, R.R.: Soft Computing and Its Applications in Business and Economics. Springer, Heidelberg (2004)
2. Rozenblatt, F.: Principles of Neuro-dynamics (Perceptron and Theory of Brain Mechanisms), Mir (1965)
3. Hopfield, J.J.: Neural networks and physical systems with emergent collective computational abilities. Proc. Natl. Acad. Sci. **79**, 2554–2558 (1982)
4. Hinton, G.E.: Connectionist Learning Prosedures. Artif. Intell. **40**, 185–234 (1989)
5. Widrow, V.: Adaptive sampled-data systems, a statistical theory of adaptation. IRE WESCON Convention Record, part. 4. Institute of Radio Engineers, New York (1959)
6. Widrow, V., Hoff, M.: Adaptive switching circuits. IRE WESCON Convention Record. Institute of Radio Engineers, New York (1960)
7. Barsky, A.B.: Neural essence and artificial intelligence. Inf. Technol. **1** (2003)
8. Ibishov, B.H., Ismailov, B.G.: Diagnostics of oil leakage from pipelines to the environment. In: Materials of "Azerbaijan - After Independence" International Conference, Baku, pp. 134–135 (2003)
9. Ibishov, B.H., Ismailov, B.G.: Diagnostics of oil leakage from submarine pipelines on the base of artificial neural networks. In: Reports of "Khazargasyatag-2004" Scientific–Practical Conference, Baku, pp. 334–335 (2004)
10. Ogibalov, P.M., Mirzajanzade, A.Kh.: Mechanics of Physical Processes. Moscow University (1976)
11. Korotaev, Y.P., Margulova, R.D., Nedra, M. (eds.): Extraction, Preparation and Transportation of Natural Gas and Condensate. Reference manual in 2 volumes (1984)
12. Iskenderov, E., Ismayilov, B.G.: About evaluation of the changes regime parameters in the gas pipeline according to neural technology. Theor. Appl. Mech. Interuniv. Sci. Tech. J. **1–2**, 134–140 (2017)

Using Deep Learning as Prediction Model in Poll-Driven Social Science Applications

Adnan Khashman[1] ⓘ, Gunay Sadikoglu[2(✉)] ⓘ,
and Zeliha Khashman[3] ⓘ

[1] Final International University, Girne, Mersin 10, Turkey
adnan.khashman@ecraa.com
[2] European Centre for Research and Academic Affairs (ECRAA), Lefkosa,
Mersin 10, Turkey
gunay.sadikoglu@neu.edu.tr
[3] Near East University, Lefkosa, Mersin 10, Turkey
zeliha.khashman@ecraa.com

Abstract. A novel application of artificial intelligence in social sciences is discussed in this paper. The work describes an intelligent deep learning model that is designed and trained to anticipate people's expectations regarding reaching a political settlement for the ongoing Cyprus conflict. Our model utilizes deep learning neural networks architecture to associate people's information and views on important issues related to the ongoing peace process mediations, together with their expectation for signing a peace deal by the end of year 2020. We collected the database through completed surveys and polls by a group of university students studying international relations in Northern Cyprus. The obtained database undergoes digital coding prior to using it to train and test the proposed prediction model. A deep learning fully connected dense neural network model (DNN) with five hidden layers is designed, implemented, and its performance is then compared to that of the classic back propagation neural network. The obtained results suggest that the proposed novel method based on deep learning architecture, can be successfully applied to predict similar real-life applications where people's reactions can be recorded, digitized and used to predict their perceptions.

Keywords: Deep learning · Neural networks · Social sciences · Computational politics · Cyprus problem · Prediction models

1 Introduction

Merging different disciplines and research fields has recently gained rapid advancement in many emerging artificial intelligence applications; often leading to the emergence of unique intelligent applications with varying success results. The complexity of the required tasks; together with inevitable technical limitations in computing power, are usually the cause of low success rates for such applications. However, this does not slow down the progress in developing such artificially intelligent tools; and with the continuous rise in computational power, strange and novel applications continue to emerge with the universal objective of imitating the way our minds perceive information and, accordingly, makes decisions. One such emerging intelligent application field in social sciences is computational politics. This is a rapidly developing application area; which has been successfully addressed in few recent works [1–7].

Deep learning is a branch of machine learning where algorithms are inspired by the structure and function of the brain. These biologically inspired neural networks have been efficiently implemented in the past years to solve different tasks in many application areas including engineering, medicine, economy, and energy [8–17]. The basic difference between deep learning and traditional neural learning is that deep learning uses more than one hidden layer within its neural network structure. Such a learning paradigm represents a more biologically-inspired architecture; where deep networks get trained in a layer-wise fashion and rely on more distributed and hierarchical learning of features as it is found in the human visual cortex, thus allowing the representation of highly nonlinear functions, discovery of more interesting features in training data, and better modeling of complex tasks [18].

There are different deep learning algorithms; such as convolutional neural networks (CNNs) and stacked denoising autoencoder (SDAEs) which have shown their success in image classification tasks as shown in our previous works [19, 20]. In this work, a fully connected deep learning dense neural network (DNN) will be considered and investigated. The DNN has several hidden layers that are fully interconnected with the input and output layers. As a result of this configuration, these deep networks allow for more synaptic weights; which consequently increases and improves their learning memory during arbitration. Unfortunately, such a configuration could have the disadvantages of overfitting, underfitting, higher computational expense, and longer training time. The later problem is more evident when parallel programming and GPUs are not available, and learning is performed using a single CPU. To encounter these problems, optimization of the DNN architecture is necessary. This can be reasonably achieved by maintaining lower numbers of hidden neurons in each layer and keeping the number of hidden layers to a minimum.

In this work, we seek an optimum design and configuration for a deep learning neural network and aim to:

- investigate and evaluate the performance of a deep learning neural network (DNN), and compare its performance to that of the classic back propagation neural networks (BPNN).
- present a novel intelligent application in computational politics utilizing our developed deep learning model.
- demonstrate the success of the prediction system by using it in a real-life application; which is to model people's perception and expectations regarding the Cyprus problem and its potential solution.

The differences between the DNN and BPNN models are in: the number of hidden layers, the number of neurons in each hidden layer, and the values of learning parameters. The performance of both prediction models will be evaluated and then compared. The implementation of these models uses a dataset which we collected via a detailed survey on the Cyprus conflict and on the basic issues which are important for Cypriots. The survey records the personal information and opinions of 80 persons on these issues, as well as their prediction of a settlement to the Cyprus problem by end of 2020.

In this paper, Sect. 2 describes briefly the application case and explains the attributes of the dataset and people's opinions, and their coding into digital data. Section 3 presents the design and architecture of the DNN and BPNN neural models; and describes the adopted learning schemes. Section 4 discusses the experimental results and draws a performance evaluation and comparison. Finally, Sect. 4 concludes this work.

2 Application Case Description and Data Coding

Here we briefly introduce the application case; i.e. the Cyprus conflict, and explains the dataset attributes and the coding method of these attributes into binary data.

2.1 About the Cyprus Case

The Mediterranean island of Cyprus became independent from Great Britain in 1960. The republic's constitution emphasized power sharing between Turkish and Greek Cypriots to soften the historical unrest between the two ethnicities.

Unfortunately, violence broke out among the two communities between 1963–1974, bringing conflict for the newly declared united Cyprus republic. As a result of ethnic clashes and violence the Turkish Cypriot small communities were driven to enclaves where they continued to live under harsh conditions until the Turkey intervened in 1974 and annexed the northern part of the island, thus creating a safer territorial zone for Turkish Cypriots. As a result, the island was partitioned into northern Turkish and southern Greek regions [21]. In addition to the UN buffer zone that separates the Turks and Greeks today, two political entities are in operation; Greek Cypriot state and Turkish Cypriot state. The Greek state is recognized worldwide as the Republic of Cyprus, despite the lack of domestic legitimacy as it acts as though it represents the whole island at international levels. Whereas, the Turkish state; which is known as the Turkish Republic of Northern Cyprus, is only recognized by Turkey; and it does not accept the legitimacy of the Greeks on the entire island. During the long years of conflict, both sides insist on their stand in the conflict and thus could not reach a an agreement for unifying the island. The critical issues that have been on the negotiation table include 'governance and power sharing', 'EU matters', 'property', 'refugees', 'security and guarantees', 'economy', and territorial adjustment'. Unfortunately, these important issues became the major obstacles in securing a peace deal [3, 4, 22].

2.2 Data Collection and Digital Coding

The dataset comprises collected information data from volunteering participants personal details as well as their views on 14 important issues regarding the Cyprus peace mediation process; that do have an effect on the possibility of securing a solution to the conflict by the end of year 2020 [3, 4]. The attributes which contain participant's personal information in addition to their responses to the survey questions; together with the allocated number of binary digits for each attribute are listed in Table 1. Responses to questions were classified into five classes (i.e. five digits); namely, *strongly disagree, disagree, no idea, agree, strongly agree.* The information and the opinions of the participants were transformed into digital code by assigning a true value or binary '1' to the selected responses of the participants, while assigning all other values into false or binary '0'.

Table 1. Binarization of the participants' information and questions.

	Input attributes	Binary digits
Participant details:		
1.	Gender	2
2.	Age	4
3.	Birth Place	7
4.	Family Residing Place	7
5.	Nationality	7
6.	Religion	6
7.	Current Education	4
8.	Education Finance	5
9.	Academic Field	2
Conflict-related questions:		
10.	Status quo acceptable	5
11.	Conflict is about military invasion	5
12.	Communal rights are more important than individual rights	5
13.	Territorial integrity of both sides required	5
14.	Cypriot culture should be preserved	5
15.	Threat that emanates from the other community should end	5
16.	Economic benefits will be achieved	5
17.	Refugee issue will be solved	5
18.	Guarantors remain the same as in "Treaty of Guarantee"	5
19.	Freedom to reside and work on both sides	5
20.	Foundation for solution is bi-zonal bi-communal federation	5
21	Both states must be internationally recognized	5
22.	Third parties have interests in the conflict	5
23.	UN has been unsuccessful	5
24.	Annan peace plan was acceptable	5
	Output attributes	**Binary digits**
25.	A peace deal could be secured by the year 2020	5

3   The Prediction Models

Modeling deep networks faces difficulty in training due to performance degradation with the increase in hidden layers [23]. Poor performance has been associated with neuron saturation and over-fitting due to the increase in neuron numbers as more hidden layers are formed [24]. However, successful modeling using deep architecture can still be achieved if the design is optimized.

In this paper our deep learning model is based on the fully connected deep learning dense neural network architecture, where error back propagation is fully applied to all deep layers. The sigmoid transfer function is adopted to activate all processing neurons within the deep model; and additional bias neurons with true values of '1' are utilized to trigger activation of these processing neurons during training. The deep layers weights (i.e. learning memories) are updated the delta rule. In addition to input and output layers, our proposed deep learning prediction model (DNN) will have five hidden layers. Figure 1 display the design of the proposed DNN that has seven layers in total. The input layer has 119 input neurons receiving the binarized input data that represents a person's personal information and thoughts on the Cyprus conflict. The number of neurons in input layer is determined by the total number of digits allocated for each attribute as shown in Table 1. Therefore:

119 = 2 * Gender + 4 * Age + 7 * BirthPlace + ... + 5 * AnnanPeacePlanWas Acceptable

Using the same method, the number output layer nodes is five neurons; which is also equal to the total number of potential answers. The number of neurons in each deep hidden layer is initialized with one neuron and then increased until optimum performance is obtained.

For performance evaluation and comparison purposes we also train and implement the popular and successful back propagation neural model (BPNN) for the same prediction task and using similar design for the input and output layers.

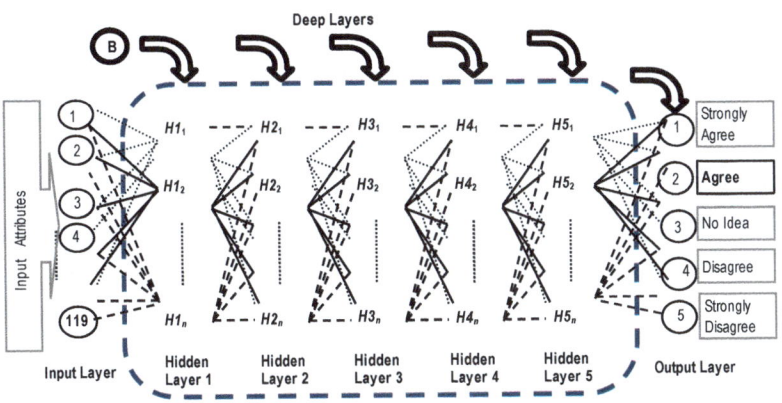

Fig. 1. Topology of DNN deep prediction model with 5 deep hidden layers and bias B-neurons.

4   Experimental Results and Discussions

The presented DNN and BPNN models were simulated using C-language and the Dev-C++ 5.11 compiler. The implementation was made using a PC with Intel Core i3-7100 CPU 3.90 GHz, 8 GB RAM, and 'Windows 10 Pro' 64-bit operating system. The program goes through several loops of a range of possible values for the critical parameters in training the neural models and searches for the optimum values. Training the neural prediction models was done using (62.5%:37.5%) training-to-testing dataset ratio, i.e. 50 dataset instances were used for training; and the remaining 30 instances were used for testing. Table 2 shows a detailed listing of the experimental results.

The highest obtained overall correct prediction rate (CPR) was 80%, which is considered sufficient because such a prediction task uses a small number of instances. This successful outcome was obtained when using the deep learning neural network model (DNN). When comparing this result to the conventional backpropagation neural model (BPNN), it can also be seen that deep learning architecture has an advantage over the classic model. Naturally, the classic BPNN has less computational costs due to having a single hidden layer, however, the improvement in correct prediction results justifies the additional computational time.

Table 2. Best obtained values and prediction rates.

Neural model	Classic BPNN	Deep DNN
Hidden layers	1	5
Hidden neurons in each layer	H_{L1}: 3	H_{L1}: 2, H_{L2}: 4, H_{L3}: 3, H_{L4}: 5, H_{L5}: 4
Input neurons	119	119
Output neurons	5	5
Learning rate	0.004	0.005
Momentum rate	0.80	0.50
Obtained error	0.00796	0.00966
Iterations	1	1950
Training time (s)[1]	0.129153	283.897049
Testing time for 30 instances (s)[1]	0.000889	0.967035
CPR[2] – Training	(50/50) 100%	(50/50) 100%
CPR[2] – Testing	(12/30) 40%	(14/30) 46.67%
CPR[2] – Overall	(62/80) 77.5%	(64/80) 80%

[1] With Dev-C++ 5.11 compiler. PC specification: Intel Core i3-7100 CPU 3.90 GHz, 8 GB RAM, and 'Windows 10 Pro' 64-bit operating system. [2] CPR: Correct Prediction Rate.

5   Conclusion

In this article we presented a novel application of both deep learning and classic neural networks in the field of social sciences. The proposed neural prediction models comprise a deep learning dense neural network with five hidden layers, as well as a classic backpropagation neural network for comparison purposes. The developed neural models were trained to predict people's expectations regarding reaching a political settlement for the ongoing Cyprus conflict; by associating people's views and thoughts on important issues that are related to the ongoing peace process mediations, together with their expectations for having a peace deal by the end of year 2020.

The paper also described how to transform information obtained via polls reflecting people's thoughts into digital binary data that can be used in arbitrating neural networks. The obtained experimental results show that the deep learning model (DNN) has advantage in correct prediction rates, and disadvantage of higher computational costs when compared to the classic back propagation neural (BPNN) model. The increase in time costs is directly related to the increased number of the deep hidden layers within the deep network architecture. The DNN training and testing was fast and it achieved an 80% overall correct prediction rate. These results suggest that the proposed deep learning neural model, can be successfully applied in similar real-life applications, where people's thoughts can be binarized and utilized to predict their expectations. Future work will explore the use of the developed DNN models in other emerging applications in social sciences.

References

1. Jill, G.S.: Election result forecasting using two layer perceptron network. J. Theor. Appl. Inform. Tech. **4**(11), 1019–1024 (2008)
2. Borisyuk, R., Borisyuk, G., Rallings, C., Thrasher, M.: Forecasting the 2005 general election: a neural network approach. Br. J. Politics Int. Relat. **7**(2), 199–209 (2005)
3. Khashman, Z., Khashman, A.: Modeling people's anticipation for Cyprus peace mediation outcome using a neural model. Procedia Comput. Sci. **120**, 734–741 (2017)
4. Khashman, Z., Khashman, A.: Intelligent Modelling in Computational Politics: Case of Cyprus Mediation Process. In: Proceedings of the MACOS'2016 Conference, Brasov, Romania (2016)
5. Khashman, Z., Khashman, A.: Anticipation of political party voting using artificial intelligence. Procedia Comput. Sci. **102**, 611–616 (2016)
6. Yu, B., Kaufmann, S., Diermeier, D.: Classifying party affiliation from political speech. J. Inform. Tech. Politics **5**(1), 33–48 (2008)
7. Monterola, C., Lim, M., Garcia, J., Saloma, C.: Feasibility of a neural network as classifier of undecided respondents in a public opinion survey. Int. J. Public Opinion Res. **14**(2), 222–229 (2002)
8. Khashman, A., Sekeroglu, B.: Multi-banknote Identification Using a Single Neural Network. In: Blanc-Talon, J., Philips, W., Popescu, D., Scheunders, P. (eds) Adv. Concepts Intel. Vision Syst. (ACIVS2005), LNCS, **3708**, 123–129. Springer, Heidelberg (2005)
9. Khashman, A., Dimililer, K.: Medical radiographs compression using neural networks and haar wavelet. In: EUROCON 2009, EUROCON'09. IEEE, 1448–1453. (2009)
10. Khashman, A.: Investigation of different neural models for blood cell type identification. Neural Comput. Appl. **21**(6), 1177–1183 (2012)
11. Khashman, A.: Intelligent local face recognition. Recent Advances in Face Recognition, IntechOpen (2008)
12. Khashman, A.: Intelligent face recognition: local versus global pattern averaging. In: AI 2006: Advanced Artificial Intelligence, pp. 956–96 (2006)
13. Khashman, A.: An emotional system with application to blood cell type identification. Trans. Inst. Measur. Control SAGE **34**(2–3), 125–147 (2012)
14. Khashman, A., Abbas, H.H.: Acute lymphoblastic leukemia identification using blood smear images and a neural classifierIn: In: International Work-Conference on Artificial and Natural Neural Networks, pp. 80–87 (2013)

15. Olaniyi, E.O., Oyedotun, O.K., Ogunlade, C.A., Khashman, A.: In-line grading system for mango fruits using GLCM feature extraction and soft-computing techniques. Int. J. Appl. Pattern Recog. **6**(1), 58–75 (2019)
16. Khashman, A., Sadikoglu, G.: Data coding and neural network arbitration for feasibility prediction of car marketing. International Conference on Theory and Applications of Fuzzy Systems and Soft Computing, pp. 249–255 (2018)
17. Khashman, A., Carstea, C.G.: Oil price prediction using a supervised neural network. Int. J. Oil, Gas Coal Tech. **20**(3), 360–371 (2019)
18. Kruger, N., Janssen, P., Kalkan, S., Lappe, M., Leonardis, A., Piater, J., Rodriguez-Sanchez, A.J., Wiskott, L.: Deep hierarchies in the primate visual cortex: what can we learn or computer vision? IEEE Trans. Pattern. Anal. Mach. Intel. **35**(8), 1847–1871 (2013)
19. Oyedotun, O.K., Olaniyi, E.O., Khashman, A.: Deep learning in character recognition considering pattern invariance constraints. Int. J. Intel. Syst. App. **7**(7), 1–10 (2015)
20. Oyedotun, O.K., Khashman, A.: Iris nevus diagnosis: convolutional neural network and deep belief network. Turk. J. Elec. Eng. Comput. Sci. **25**(2), 1106–1115 (2017)
21. Khashman, Z.: Wither Ripeness Theory in Cyprus Conflict. ASN Convention, New York, 23–25 April 2015
22. Khashman, Z.S.: Creating conditions for peacemaking: the Cyprus case. Percept.: J. Int. Affairs **8**(3), 1–18 (2003)
23. Pierre, B.: Autoencoders, unsupervised learning, and deep architectures. In: Workshop on Unsupervised and Transfer Learning, vol. 27, pp. 37–50 (2012)
24. Glorot, X., Bengio, Y.: Understanding the difficulty of training deep feedforward neural networks. In: Proceedings of 13th International Conference on Artificial Intelligence and Statistics, pp. 249–256 (2010)

Development of an Expert Information System for Sports Selection and Orientation Using Fuzzy Logic Methods

Vitchenko Olga[(✉)] [iD], Shcherbakov Sergey [iD], Bykov Nikolay [iD], and Manuilenko Eleonora [iD]

Don State Technical University,
Gagarin Square, 1, 344000 Rostov-on-Don, Russia
`owinf@mail.ru`

Abstract. The article describes the development of an information and expert system of sports selection and orientation of athletes with the author's methodology for ranking their results. Before describing the methodology, the identified problems in the development of this information and expert system are indicated and possible ways of their solution are given when designing the system architecture and the data analysis unit. The following are the theses of fuzzy logic used in the development of the methodology, a set of indicators of athletes is determined, consisting of six groups: the level of physical development; health status; functionality; mental features; level of development of psychomotor qualities, sports result. For each group of indicators, a quantitative and qualitative characteristic is given. After that, the process of ranking the indicators of athletes on the basis of fuzzy-logical conclusions systems is described in detail. As a result of the application of this technique, it is supposed to obtain a rating table, which contains the unified data for individual indicators. Athletes rating can be obtained both current and forecast. For each athlete, it is also expected to obtain a generalized assessment for all groups of indicators, which can be observed in dynamics. The approach to the development of an information and expert system of sports selection and orientation, outlined in the article, can serve as the basis for additional research of determining the effectiveness of sports training methods and their implementation with modern digital technologies.

Keywords: Sports selection · Information and expert system · Fuzzy logic · Ranking technique

1 Introduction

A distinguishing feature of modern sports training systems is the combination of training and competitive activity of athletes to the limit of human capabilities, which, firstly, requires careful professional selection of those wishing to engage in a particular sport, and secondly, obtaining accurate and reliable information about the state of all the systems of athlete's body and body reactions to training and competitive loads in the process of sports training.

According to experts, up to 30% of athletes at the stages of initial sports training, specialization, and improvement of sportsmanship do not engage in "their own" sport, which is one of the reasons for the occurrence of professional diseases and injuries of an athlete, stagnation of sports results and, as a result, retirement of talented children and teenagers from sports. Considering that the cost of training an athlete per year reaches hundreds of thousands of rubles or more, the economic damage (direct and indirect losses to society as a whole) is huge.

At present, identifying the athlete's prospectivity rating for professional occupations in a particular sport is an urgent scientific problem [2]. The number of significant indicators reflecting the state of the athlete's body, which must be taken into account, both individually and in interconnections, can be several tenths. Naturally, the analysis and interpretation of such an array of information on only one athlete requires a wide range of competencies from the doctor, trainer and other specialists. Therefore, there is a need to develop intelligent systems that allow you to quickly and accurately (regardless of the number of indicators analyzed) determine the condition of the athlete, identify his strong and limiting links of preparedness, as well as the degree and rating of prospects at the moment, to offer the best options for the correction of the training and competitive activity of athletes [1]. The further use of these systems as the means of computer and simulation modeling, will optimize the system of sports training [7].

2 Expert Information System of Sports Selection and Orientation Architecture

The information and expert system of sports selection and orientation "INEX SOO" (EIS), developed at our university, is aimed at solving the mentioned above problems. The software implementation of the EIS meets the following criteria [3, 4]:

- flexibility and extensibility (the ability to adapt to different stages of selection for different sports);
- orientation to various input methods;
- friendly user interface and clear presentation of the results.

Based on the above criteria, the user and computing architecture of the system was selected (see Fig. 1). Information expert system elements, relations and working principle illustrated with UML.

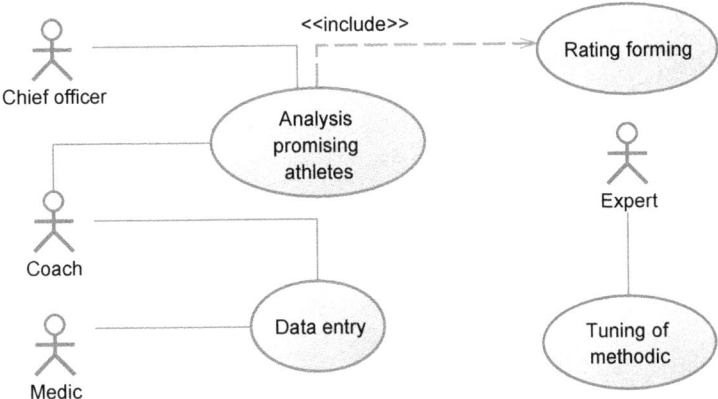

Fig. 1. USECASE diagram of the expert information system

The main blocks of the system are shown in the Fig. 2.

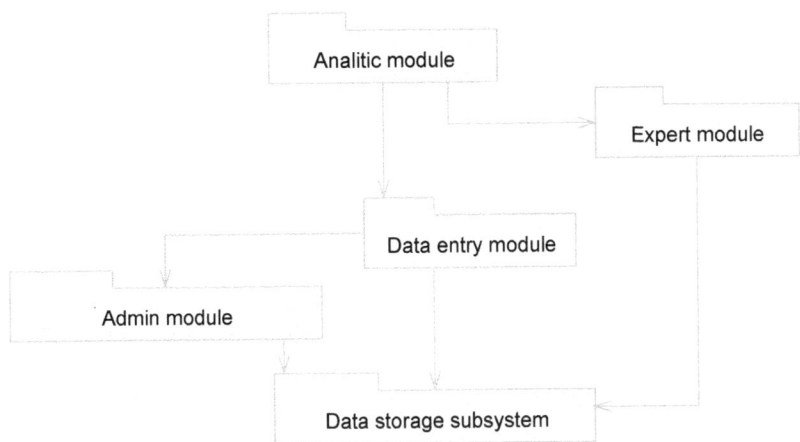

Fig. 2. Block package diagram of the expert information system

In addition to system and supporting components, the system includes the following blocks:

– a block for inputting the initial data of athletes, the results of their expectations, as well as their sports results - creates an array of data for the further development and tuning of machine learning models;
– the analytical unit allows to process and display ranked information about the training of athletes, their current sports results, the dynamics of indicators, including the health reserve at various stages of preparation in order to support decision-making for sports selection and team building;

– the expert unit is designed to configure the ranking parameters of athletes through the determination of key indicators, norms, weighting factors, and allows to set alternative ranking methods.

In order to provide flexibility and the possibility of the expert information system functionality expanding, the following class structure was chosen (see Fig. 3).

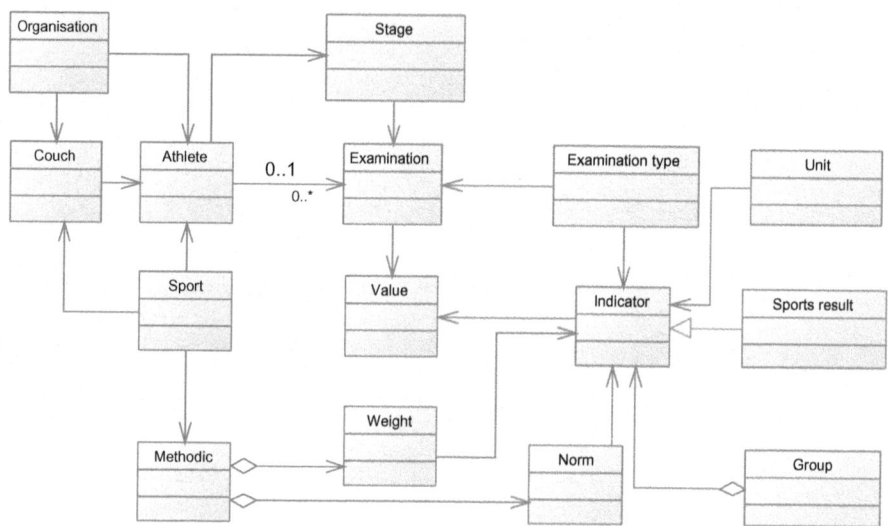

Fig. 3. Class diagram of the expert information system.

As can be seen from Fig. 3, the class "Indicator" has a key role in the EIS, which is responsible for a specific indicator - medical, sports or informational. Different types of surveys include different sets of indicators. In this case, for the examination of the athlete, a certain value of the indicator is fixed for a certain date. A rating of athletes is being built as a result of processing the data set, taking into account their weights and standard values [6].

3 Indicators and a Ranking Technique of Athletes Based on Fuzzy-Logical Conclusions Systems

It is known that a sports result depends not only on the level of talent and preparedness of an athlete, but also on many other factors (random and non-random, objective and subjective, external and internal), therefore, to judge the degree of its prospects on the basis of one or even several sports results (especially in childhood and adolescence) is incorrect.

Therefore, the ranking of athletes in terms of their prospects for further professional sports should be based on a comprehensive assessment: the sports result, the functional state of all body systems, the dynamics of these indicators.

According to our methodology, when ranking athletes, 6 groups of indicators characterizing various aspects of the athlete are taken into account:

1. The level of physical development $\{X1\}$;
2. Health status $\{X2\}$;
3. Functionality $\{X3\}$;
4. Mental features $\{X4\}$;
5. The level of development of psychomotor qualities $\{X5\}$;
6. Sports result $\{X6\}$.

When a child is enrolled in a sports school, as well as at the stage of initial sports training, the 6th group of indicators is not taken into account. In essence, the 6th group of indicators "Sports Result" is a function of all five groups of indicators, which are interconnected to one degree or another, since they reflect certain aspects of the athlete's vital activity, which can be reflected by the formula

$$\{X6\} = f\{X1, X2, X3, X4, X5\},$$

But to achieve the maximum sports result, two more conditions are necessary:

– competent sports preparation for competitions (methodology, support, social conditions, etc.);
– objective refereeing of competitions.

Therefore, it is advisable to consider the sports result when determining the athlete's prospectivity rating as one of the indicators, the significance of which increases with the athlete's mastery.

Each group of indicators can include from one (Xij) to several indicators affecting the quality of an athlete's professional activity. The significance of each indicator characterizing a certain function or property of an organism, necessary to achieve high sports results, differs significantly for a certain sport and specialization, as well as for different stages and periods of sports training.

Therefore, at each stage of sports training, and sports selection, as well as at any given moment of ranking, the degree of significance of each indicator is set by experts (specialists in this sport), in the form of weighting factors - Aij.

Before starting the ranking of athletes, all indicators, except personal data, are normalized to the initially set maximum value of each indicator.

It is impossible to determine a reliable forecast of an athlete's prospects based on the results of one survey or the results of one competition, therefore, we believe that in order to determine the athlete's development trend, at each stage of sports training, it is necessary to determine the value and character of the dynamics of indicators based on the results of four surveys and the four best sports results during one and a half - two years.

It is proposed to determine the athlete's prospects in the EIS being developed on the basis of fuzzy-logical conclusions [8] as follows.

1. Convert athlete data based on the results of the current survey:

- identification (receipt from the customer) of the values of indicators of body functions (Kij);
- rationing the obtained Kij values in order to obtain dimensionless values - dividing the obtained indicator value by the reference value of this Kijo indicator, if an increase in the indicator leads to improvement in sports results; if on the contrary, then the reference value of the indicator is divided by the value of the current indicator of the athlete - we get NKij;
- rationing of the obtained NKij according to the significance of the indicator for achieving high sports results depending on the sport and specialization, as well as the preparation stage, that is, it is determined by multiplying NKij by the weight coefficient Aij.

2. Determination of the athlete's current rating by the set of indicators received from the customer based on one or more convolution methods {NKij} (for example, by summing the received NKij).
3. Determining the trend of the athlete rating based on the analysis of the results of two, three, etc. surveys (calculation of the absolute and relative changes in the values of each of the indicators NKij and their combination {NKij}) to optimize the management of the training process.
4. Presentation and transformation of the athletic performance of athletes into a standardized indicator NKij.

It should be noted that the values of indicators of body functions can be both quantitative and qualitative, representing in last case a linguistic variable. For example, the indicator "potential of the nervous system" has four meanings: "strong", "medium", "medium-weak", "weak". In this case, we use the fuzzy logic tool to process the data, which allows us to aggregate large amounts of information and build on their basis complex assessments of systems'state - fuzzy multi-level [0,1] - classifiers [5].

The linguistic variable P_i = "level of the i-th indicator" $(i = 1, 2, ...)$ is introduced. The set of values of the variable P_i is the term set of n terms $P = \{P_1, P_2, P_3, P_4, ..., P_n\}$: for example, P_1 is "strong"; P_2 - "medium; P_3 - "medium-weak"; P_4 - "weak".

Each researched indicator will be associated with the value of membership functions that relate it to the corresponding term of the linguistic variable. The construction of the membership function is the main problem that is proposed to be solved using hybrid-fuzzy neural networks. It is proposed to configure parameters of accessory functions on the basis of hybrid-fuzzy neural networks, using the Golub-Pereira method.

To assess the dynamics of athletes' indicators, fuzzy multilevel [0,1] classifiers of the second type, or dynamic classifiers can be used. The dynamics of athletes' indicators is subjected to analysis in the directions established by experts: g1; g2; g3; etc.

Each of the estimates is a linguistic variable, with a universal set [0,1] and a term set of n terms p = {P_1, P_2, P_3, P_4, ..., P_n}.

Calculation of normalized values of the studied indicators for the considered period N of certain time intervals (for example, weeks, months, at each stage of the survey) is carried out on the basis of a scheme that takes into account the significance of different time periods due to weight factors:

$$x_i = 0,5\left(1 + \sum_{i-1}^{N-1} k_i I_i\right), \ k_i = \frac{2(N-i)}{(N-1)N},$$

where k_i – weighting coefficients determined by the Fishburn rule; numbering of time periods is carried out in reverse order. I_i – integer-valued functions defined in such a way that the value "1" corresponds to an increase in the i-th indicator (worsening of the situation); the value "-1" - a decrease in the i-th indicator; value "0" - stabilization, no change. Herewith, the term-sets of indicators have the same form as above, and the membership functions can be standard uniform, for example, trapezoidal. Thus, the proposed methodology for monitoring the performance of allows to determine their dynamics and development trends.

4 Conclusion

The described architecture and ranking technique for athletes based on fuzzy-logical inference systems are copyrighted, determine the flexibility of the IES, the ability to implement alternative rating methods, which is relevant for different sports.

The developed analytical tools of the information system make it possible to use it for decision-making in the selection process, the picking of national teams, the formation of crews, etc.

In addition to calculating the overall rating, ratings can also be formed for individual subsets of indicators, for example, "without a sports result", "a combination of physical and mental qualities", etc., depending on the needs of decision makers.

For a particular athlete, a profile can be built showing its development by the main groups of indicators (see Fig. 4).

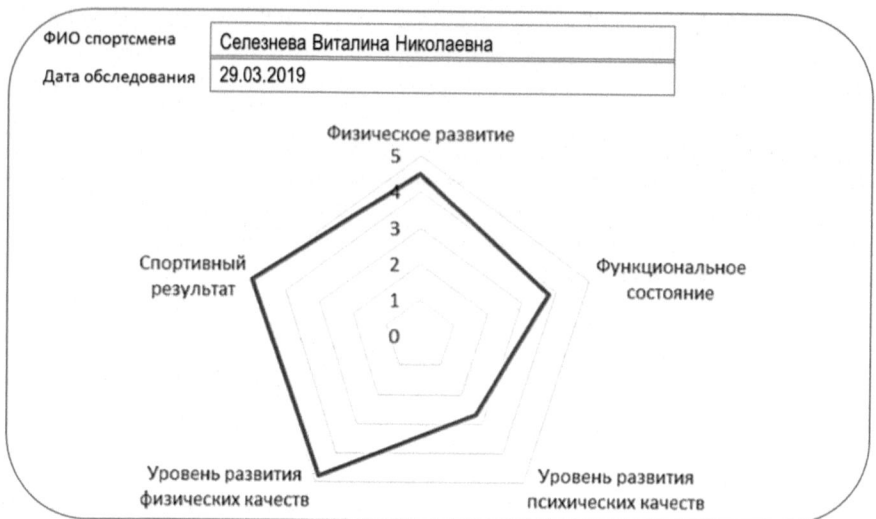

Fig. 4. Athlete profile

The rating dynamics can also be obtained, allowing to evaluate the effectiveness of training and predict the possible growth of the athlete's results.

Prospects for the development of IES are as follows:

- adaptation of the system of various sports, including complex coordination, high-speed power and game;
- forecasting the effectiveness of athletes based on the collected data array using machine learning methods;
- issuing recommendations on the training of athletes based on production models;
- creation of an automated sports training management system for athletes, teams, organizations, development of interactive panels (dashboards) for heads of sports institutions.

References

1. Bykov, N.N.: Smart sport, or sports intellectualization. In: Innovative Transformations in the Field of Physical Culture, Sports and Tourism Collection of Materials of the XXI All-Russian Scientific-Practical Conference, pp. 27–30 (2018)
2. Bykov, N.N, Shlyapnikov, A.V.: Organizational fundamentals of the modern system of sports selection and orientation. In: Physical Education and Sport: Actual Issues of Theory and Practice Collection of articles of the All-Russian Scientific and Practical Conference, pp. 188–192 (2019)
3. Egorov, A.A., Loginov, S.I., Shentsev, I.V., Ermakov, V.A.: Information system for data mining in sport. In: Prospective Studies in Physical Culture, Sports and Tourism Materials of an International Scientific and Practical Conference, pp. 28–32 (2014)

4. Evtin, A.B., Kaloshina, A.V., Shibaeva, M.Yu., Kuryatnikov, D.S.: Information systems in Physical culture and sports. In: Socio-Economic Problems of the Development of Municipalities Materials and Reports of the XXII International Scientific and Practical Conference, pp. 99–102 (2017)
5. Ledenyova, T.M., Moiseev, S.A.: Formalization of the properties of interpreted linguistic scales and terms of fuzzy models. Appl. Inform. **4**(40), 126–132 (2012)
6. Vitchenko, O.V., Stryukov, M.B., Dashko, YuV: Big data analysis as a method of analytics in business and education. Intell. Resour. Reg. Dev. **2**, 19–24 (2019)
7. Yarmolinsky, V.I., Gubkin, S.V.: Digital technologies in medicine and physical culture. Health improving physical culture of youth: actual problems and prospects. In: Materials of the III International Scientific and Practical Conference, pp. 119–125 (2018)
8. Zadeh, L.A.: Fundamentals of a new approach to the analysis of complex systems and decision-making processes. Mathematics today. Znanie. Moscow (1974)

Fuzzy-Logic Modelling of the Relationship Between Work Motivation and Conflict Solving Skills of Academicians Regarding Life Balance Model of Positive Psychotherapy

Memmedova Konul[✉] and Guldal Kan Şebnem[iD]

Department of Psychological Counseling and Guidance, Near East University,
98010 Lefkosa, Mersin 10, TRNC, Turkey
{konul.memmedova, sebnem.guldal}@neu.edu.tr

Abstract. Analysis of the relationship between conflict resolution skills of the balanced model of Positive Psychotherapy (PTT) and work motivation is a focus of researcher recent years. However, in all publications it is assumed that the results of measurement of the balance model dimensions and work motivation are precise and for modelling the input-output relationship is used a statistical approach. In this study is considered an application of fuzzy-logic approach to establish the relationship between the balance model variables and motivation of academicians. The application of the fuzzy logic approach enables to handle of uncertainty and imprecision of the measurement. For measurement of the balance model variables is used a 4 four-dimensional ("body", "achievement", "relationship" and "spirituality") balanced life scales. The work motivation of academicians is measured by using a multidimensional work motivation scale (MWMS).

Data collected from survey are used in fuzzy-logic model performing the human like decision making based on computations with words.

Keywords: Positive psychotherapy · Balance model · Job motivation · Fuzzy logic

1 Introduction. Background of Problem

Following to PPT [1–3], the life balance of any individual is defined by 4-dimensional (spiritual, emotional, menta and physical dimensions) structure shown in Fig. 1.

The healthy and productivity of any individual is related with balancing level among above mentioned 4-dimensions [1–3]. The violation of the balance between different components negatively effect on life of the individuals and their conflict solving skills and motivation and satisfaction. Balanced life scale and the Balanced life basic skills scales in the Context of PPT Balance Model were adopted to Turkish culture by Apay and Kara [4].

R. A. Aliev et al. (Eds.): ICAFS 2020, AISC 1306, pp. 426–432, 2021.
https://doi.org/10.1007/978-3-030-64058-3_53

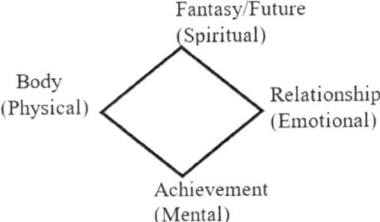

Fig. 1. Life balance model

2 Analysis of the Existing Publications

The academicians in a university are working in intense tempo including teaching, research, and performing the service activities for society. The motivation of academicians is influenced by external and internal factors and become controversial [5–7]. This situation effects on life balance of academicians. Therefore, it forces both internal and external motivation and pushes them. Positively affected force allows to achieve the academicians work satisfaction in university environment. Increasing satisfaction enables an academician to do what he/she does lovingly, to interact positively with management and colleagues, and for devotion and high performance. The statistical modeling the association between work motivation and conflict solving skills of academicians regarding life balance model of positive psychotherapy is given in [5]. In [8] the author provides basic concepts of positive psychology, character empowerment and preventive effects in the context of psychological counselling and guidance. In [9] is presented the interaction between the educational managers and the teachers by evaluating their positive psychological states. The WIESBADEN PTT and family therapy inventories have been adapted to Turkish culture in [10]. In [11] developed the Adolescent Life Goal Profile scale for measurements of life goals among university students in the context of PTT.

The Internationally recognized Work Motivation Scale that has been applied in 7-languages [12] and has been adapted to Turkish culture in [13]. This scale is used in this paper for data collection purpose.

As shown from the overview mentioned above, there are enough publications on psychological aspects of PPT balance model and its relations with conflict resolution skill and work motivation of individuals. In all of the publications, for modelling is used the statistical approach. The statistical approach is based on the assumption that the collected data are accurate, and the information obtained by the measurement is accurately quantified. In other words, the statistical approach does not handle imprecision and uncertainty of input data.

The publications related to applications of fuzzy-logic in psychology and educational research are given below.

In [14], the author gives the fuzzy logic approach for student admissions to graduate programs.

Application of advanced fuzzy-logic approach (Z number) in educational and psychological research is presented in [15]. Doğanalp in [16] uses a fuzzy multi-criteria decision-making approach for evaluating and rating faculty members based on fuzzy TOP SIS management algorithm. Fuzzy-logic modelling the impact of Pilates exercises on academic achievement through psychological variables is given in [17]. Konul gives [18] effect of technology using on technology on anxiety and aggression levels of students. The modelling the association between personality and colour preferences by using fuzzy-logic is devoted [19]. In [20] authors describe the measurement of job satisfaction using fuzzy-logic. The paper [21] is devoted to the model of the travellers' shopping motivation and their buying behaviour using fuzzy-logic.

As shown from the above mentioned, this paper is the 1st devoted to finding the relationship between the life balance model of PPT and motivation using the fuzzy-logic approach.

3 Set of Problem

The association of the life balance model of PPT with the fuzzy-logic model is a statement of the problem. This association allow establishing the relationship between four dimensions (Fantasy-Relationship-Achievement-Body) of balance model and work motivation of academicians. The developed conceptual model is given in Fig. 2.

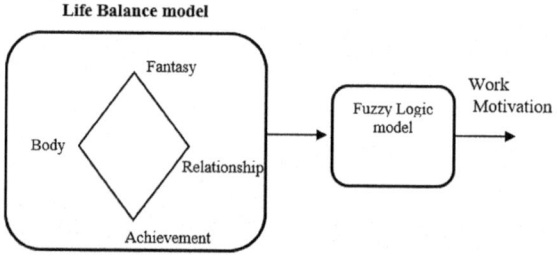

Fig. 2. Conceptual presentation of the problem

3.1 Data Collection

The data collection tools consist of life balance model and multiscale work motivation scales (questionnaires). In the study is used 4-dimensional balanced life scale designed by authors. This scale involves 23 items having Gronbach alpha 0.83, KMO value 0.806 and load factor between 0.66–0.78 depending on dimensions.

As work motivation scale is used mentioned above Çivilidağ & Şekercioğlu's multidimensional work motivation scale adopted in Turkey, involving 19 items and covering 6-dimensions of work motivation (Gronbach alpha 0.92).

The questionnaires were distributed among 250 participants, and them 206 (70% male, 30% female) returned the completed questionnaires. Each item of the Life balance model questionnaire was evaluated using five-point Likert Scales: 1-Never (N), Rarely (R), Sometimes (S), Often (O), Always (A). The multiscale work motivation questionnaire evaluated by 7-points Likert scales:

1 – Very low (VL), 2 – L (L), 3 – Medium low (ML), 4 – Medium (M), 5 – Medium high (MH), 6 – High (H), 7 – Very High (VH).

4 Design Fuzzy-Logic Model

Fuzzy-logic model is represented in the Fig. 3. The variables Fantasy, Relationship, Achievement and Body are transformed into fuzzy sets.

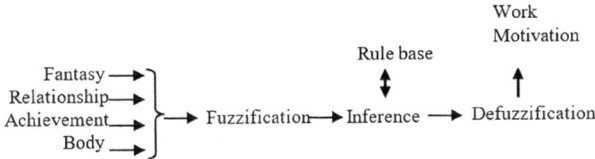

Fig. 3. Conceptual presentation of the fuzzy-logic modelling

In the Fig. 3 and Fig. 4 are presented results of fuzzification of inputs using *FuzzyTech* software. After fuzzifications of the input variables we have fuzzy sets of linguistic variables "fantasy", "relationship", "achievement" and "body" defined by 5-membership functions ranging from "Never" .... "Always".

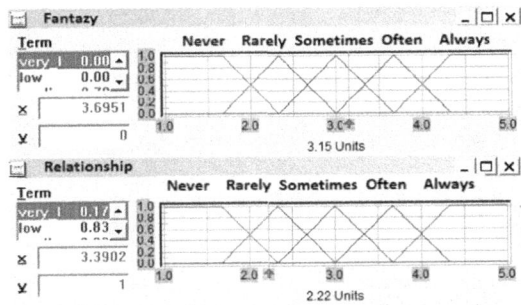

Fig. 4. Fuzzifications of the "Fantasy" and "Relationship"

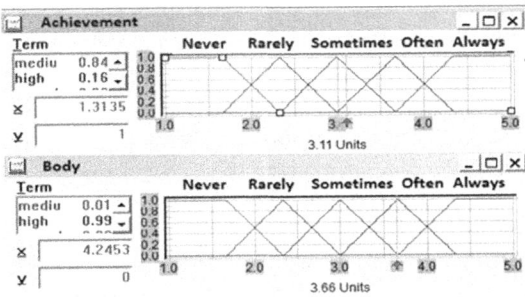

Fig. 5. Fuzzifications of the "Achievement" and "Body"

Regarding the current values of fuzzy input variables, the inference engine selects a convenient rule from Rule base. The output part of the selected rule is fuzzy motivation which is converted into crisp value by defuzzification.

Each rule consists of 2 parts (Input and Output) related to IF-Then operators. The total number of rules is defined as (5)4 × 7 = 4375. But regarding the firing degree of rules were chosen 40 rules. Fragment of the list of rules is given in Table 1.

Examples of linguistic interpretation of rules 2 and 5 is given below:

Rule 2. If Fantasy is rarely and Relationship is rarely, and Achievement is often, and Body is sometimes Then motivation is medium-low.

Rule 5. IF fantasy is sometimes and Relationship is sometimes, and Achievement is often, and Body is often Then motivation is high.

Table 1. Fragment of IF-THEN rules

Rules	IF				THEN
№	Fantasy	Relationship	Achievement	Body	Motivation
1	Sometimes	Rarely	Rarely	Sometimes	High
2	Rarely	Rarely	Often	Sometimes	Medium low
3	Often	Often	Often	Sometimes	Medium
4	Always	Sometimes	Sometimes	Often	High
5	Sometimes	Sometimes	Often	Often	High
6	Sometimes	Sometimes	Sometimes	Rarely	Medium
7	Sometimes	Rarely	Rarely ·	Sometimes	Medium
9	Often	Rarely	Always	Sometimes	Very high
10	Sometimes	Sometimes	Often	Often	Medium
-	.	.	.	.	-

The variable editors are shown in Figs. 4, 5, 6 correspond to rule 6.

IF Fantasy is *sometimes* and Relationship is *rarely,* and Achievement is *sometimes,* and Body *often* rare Then motivation is *medium.*

The red arrows under the horizontal axis correspond to the example of numerical values of the 4-dimensions balance model variables. For fantasy this value is 3.16, for relationship 2.22, for achievement 3.11, for body 3.66 and for motivation 4 units. Taken into account these values, for Rule 6 we can write:

IF Fantasy   is *3.16* and   Relationship   is *2.22* and   Achievement   is *3.11* and Body *is* 3,66, Then motivation is *4.0.*

The "Term" boxes in Fig. 4, 5, 6 indicate membership degrees. Red arrow in Fig. 6 shows the defuzzied value of motivation.

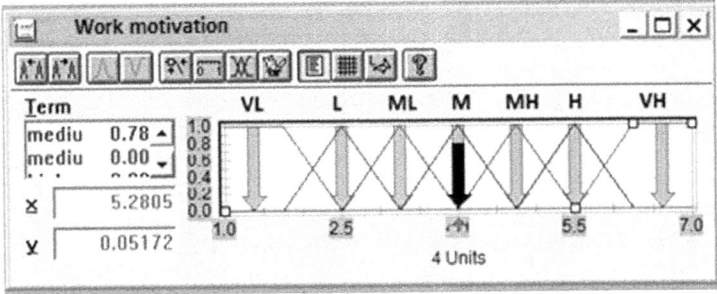

Fig. 6. Defuzzification of work motivation

motivation = {very low = 0.0, low = 0.0. medium low = 0.0, medium = 0.78, medium high = 0. 0.0, high = 0.0, very high = 0.0}

5 Conclusion

The relationship between the work motivation and the PPT balance model based on fuzzy-logic approach was found. If the academician sometimes dreams, often successful, the relationship is rare, and his satisfaction with the body is sometimes, then his/her motivation is also high. As the work motivation is high, the success rate of individuals increases.

Application of fuzzy-logic allows handling imprecision and uncertainty inherent in input data.

One of the main advantages of fuzzy-logic modelling is its ability to map balance model of each individuals with his/her work motivations, that is very difficult or impossible with the statistical approach.

Simulation of the model using the variety of rules listed demonstrates the adequacy of the proposed approach.

References

1. Peseschkian, N.: Positive Psychotherapy. Theory and Practice of a New Method. Springer, Heidelberg (1987)
2. Peseschkian, N.: Positive Psychotherapy of Everyday Life: Training in Partnership and Self Help With 250 Case Histories, 1st Edn. (2002)
3. Tayyab, R., Martin, S.: Positive Psychotherapy: book (Series in Positive Psychology), Oxford (2019)
4. Aypay, A., Kara, A.: Development of the balanced life scale and the balanced life basic skills scale in the context of positive psychotherapy balance model. Sakarya Univ. J. Educ. **8**(3), 63–79 (2018). https://doi.org/10.19126/suje.408531

5. Kan Gürdal, Ş.: Investigation of the relationship between business motivation of conflict-solving skills according to the balance model of academicians in positive psychotherapy (in Turkish). Ph.D. Thesis, Near East University (2020)
6. Gümüş, S., Sezgin, B.: The Effect of Motivation on Organizational Commitment and Performance (in Turkish). Hiperlink publications, Istanbul (2012)
7. Eryilmaz, A.: Examining the relationships between adult subjective well-being and primary and secondary abilities in the context of positive psychotherapy. J. Clin. Psychiatry **14**, 17–28 (2011)
8. Eryilmaz, A.: Use of positive psychology in the context of developmental and preventive services in the field of psychological counseling and guidance. J. Happiness Well-Being **1** (1), 1–22 (2013)
9. Öznacar, Behcet, Güldal Kan, Şebnem, Şensoy, Şeniz: An evaluation of positive psychological state of educational administrators in a tolerance context. Qual. Quant. **52** (2), 1093–1104 (2017). https://doi.org/10.1007/s11135-017-0559-7
10. Sari, T., Eryılmaz, A., Varlıklı-Öztürk, G.: Adaption of WIESBADEN inventory of positive psychotherapy to Turkish culture. In: World Congress on Positive Psychology, İstanbul, Turkey (2010)
11. Eryilmaz, A.: Investigating of psychometric properties the scale of setting life goals with respect to positive psychotherapy on university students (in Turkish). Klinik Psikiyatri **15**, 166–174 (2012)
12. Gagné, M., et al.: The multidimensional motivation scale: validation evidence in 7th languages and nine countries. Eur. J. Work Organ. Psy. (2014). http://dx.doi.org/10.1080/1359432X.2013.877892
13. Çivilidağ, A., Şekercioğlu, G.: Studying of adaptation to Turkish culture the multidimensional motivation scale. Mediterr. J. Hum. **7**(1), 143–156 (2017). https://doi.org/10.13114/MJH.2017.326
14. Özmen, C.: Fuzzy logic approach for student admissions to graduate programs. Black Sea J. Soc. Sci. **11**(20), 111–137 (2019)
15. Sadikoglu, F., Huseynov, O., Memmedova, K.: Z-regression analysis in psychological and educational researches. Procedia Comput. Sci. **102**, 385–389 (2016)
16. Doğanalp, B.: Fuzzy multi-criteria decision making and faculty valuation study. Kastamonu Univ. J. Faculty Econ. Admin. Sci. **12**(2), 498–517 (2016)
17. Memmedova, Konul: Quantitative analysis of effect of Pilates exercises on psychological variables and academic achievement using fuzzy logic. Qual. Quant. **52**(1), 195–204 (2017). https://doi.org/10.1007/s11135-017-0601-9
18. Memmedova, K.: Fuzzy logic modelling of the impact of using technology on anxiety and aggression levels of students. Procedia Comput. Sci. **120**, 495–501 (2017)
19. Memmedova, K.: Modelling the association between personality and color preferences by using fuzzy logic. In: Book Series: Advances in Intelligent Systems and Computing (2020)
20. Abiyev, RH., Sadikoğlu G., Abiyeva, E.: Fuzzy Evaluation of Job Satisfaction of Hotel Employees. In: World Congress in Computer Science, Computer Engineering, and Applied Computing (2015)
21. Sadikoglu, G.: Modeling of the travelers' shopping motivation and their buying behavior using fuzzy logic. Procedia Comput. Sci. **120**, 805–811 (2017)

One Approach to Volatile Time Series Forecasting

Ramin Rzayev[1]([⊠]) [iD], Tahir Mehdiyev[1] [iD], and Parvin Alizada[2] [iD]

[1] Institute of Control Systems of ANAS,
B.Vahabzadeh str. 9, 1141 Baku, Azerbaijan
raminrza@yahoo.com, tahir.mehdiyev@gmail.com
[2] Baku State University, Z.Xalilov str. 23, 1148 Baku, Azerbaijan
palizade@inbox.ru

Abstract. An algorithm to form a universe covering weakly structured historical data of the time series and trapezoidal membership functions used fuzzification method are proposed. According to the obtained results, the proposed approach provides more appropriate (adequate) 1st-order fuzzy model for arbitrary volatile time series. Nevertheless, this model does not pretend to have absolute predicting accuracy, which can be achieved and/or achieved by other more complex higher-order models. The purpose of the paper is to show that the use of simple fuzzy prediction models of the 1st-order reserves the opportunity for further improvement of the predicting technology of weakly structured time series. Prediction results of the arbitrary time series demonstrate that the combination of universe establishment algorithms and historical data fuzzification using trapezoidal membership functions, the construction of 1st-order internal cause-effect relations and the method of defuzzification of outputs of the used fuzzy model can still be superior in quality predicting and prediction reliability not only similar 1st-order models, but other models of higher order as well.

Keywords: Volatile time series · Weakly structured data · Fuzzy set · Membership function · Fuzzy model · Time series forecasting

1 Introduction

As is known, in the framework of constructing a fuzzy inference system the fuzzification procedure (or the introduction of fuzziness) is understood as the process of finding the values of membership functions of fuzzy sets that describe the terms of input linguistic variables. In other words, the correspondence between a specific value (term) of an individual input linguistic variable of the fuzzy inference system and the value of the corresponding membership function is established by fuzzification. At the same time, two groups of methods for identification of membership functions are distinguished: *direct* and *indirect*. In particular, these methods are studied in [1–4], where it is shown how membership functions can be formed by one or a group of experts. Direct methods are characterized by the fact that the construction of membership functions is directly carried out by experts, which possess knowledge in the subject area. Examples of direct fuzzification methods are presentation of membership

R. A. Aliev et al. (Eds.): ICAFS 2020, AISC 1306, pp. 433–441, 2021.
https://doi.org/10.1007/978-3-030-64058-3_54

functions in tabular, graphical or analytical forms. Indirect methods involve the choice of values of the membership function, which is performed in accordance with the conditions previously formulated by experts. Among these methods are methods based on pairwise comparisons of relevant statistical data [5], based on expert ranking estimates, etc. As it is not difficult to notice, all methods in one way or another bear on heuristic knowledge and, therefore, differ in portions of subjectivity, which inherent for any expert judgment. Thus, the importance and actuality of the study of identification methods for membership functions in the process of fuzzification of input data within the framework of fuzzy modeling and prediction of weakly structured time series become apparent.

2 Problem Definition

The object of study is a volatile time series $\{x_t\}_{k=1}^n$, where x_t is weakly structured historical data (HD), which should be represented as a appropriate fuzzy set A_t, characterized by a tuple [5]: $\{x_t/\mu_{A_t}(x_t)\}$, $\mu_{A_t}(x_t) \rightarrow [0, 1]$, $t = 1 \div n$. It is necessary to identify the membership functions $\mu_{A_t}(x_t)$ that allows the most adequately to describe the weakly structured HD by corresponding fuzzy sets. The final goal of this study is to contribute to a more adequate fuzzy modeling of weakly structured time series for its forecasting.

3 Fuzzy Modeling and Forecasting of Volatile Time Series

Existing approaches to fuzzy modeling of time series require the following steps: 1) forming a coverage of all HD in the form of universe; 2) fuzzification of HD; 3) establishment of internal relations in the form of fuzzy relations and their division into groups; 4) finding the outputs of the applied fuzzy model and their defuzzification.

So, to introduce the fuzziness, first of all, it is necessary to determine the universe. In the case of a time series, the coverage of the data range is suitable basis. Thus, to construct such covering let us use the following step-by-step procedure proposed in [6].

Step 1. To simplify the calculations, sorting the HD x_t ($t = 1 \div n$) into the increasing sequence $\{x_{p(i)}\}$, where p is a permutation that sorts the values of HD in ascending order.

Step 2. Calculation of the ensemble mean for all pairwise distances $d_i = |x_{p(i)} - x_{p(i+1)}|$ between any two consecutive values of $x_{p(i)}$ and $x_{p(i+1)}$ according to the formula:

$$AD(d_1, d_2, \ldots, d_n) = \frac{1}{n-1} \sum_{i=1}^{n-1} |x_{p(i)} - x_{p(i+1)}| \tag{1}$$

and standard deviation according to the formula

$$\sigma^2_{AD} = \frac{1}{n-1} \sum_{i=1}^{n-1} (d_i - AD)^2. \qquad (2)$$

Step 3. Identification and elimination of anomalies (sharply distinguished values) to be ejected. Here, the values of both the average distance AD and the standard deviation σ_{AD} are used. In this case, the values of pairwise distances must be eliminate, if they do not satisfy the condition: $AD - \sigma_{AD} \leq d_i \leq AD + \sigma_{AD}$.
Step 4. Recalculating the average distance between any two consecutive values from the totality of values remaining after sorting taking into account emissions.
Step 5. Establishing the universe U. After finally finding the ensemble mean AD for all pairwise distances between successive values, the universe is established in the form of a segment $U = [D_{min} - AD, D_{max} + AD] = [D_1, D_2]$, where D_{min} and D_{max} are respectively the minimum and maximum values on all data set.

There are different ways to identify membership functions that restore fuzzy subsets of the given universe. In particular, one of such methods is the construction of symmetric trapezoidal membership functions as follows (see Fig. 1):

$$\mu_{A_k}(x) = \begin{cases} 0, & x < a_{k1} \\ (x - a_{k1})/(a_{k2} - a_{k1}), & a_{k1} \leq x \leq a_{k2}, \\ 1, & a_{k2} \leq x \leq a_{k3}, \\ (a_{k4} - x)/(a_{k4} - a_{k3}), & a_{k3} \leq x \leq a_{k4}, \\ 0, & x > a_{k4}, \end{cases} \qquad (3)$$

with parameters satisfying the conditions: $a_{k2} - a_{k1} = a_{k3} - a_{k2} = a_{k4} - a_{k3}$, where $k = 1 \div n$, n is the total number of fuzzy sets A_k describing the HD of the time series. According to [6], n is calculated by the formula: $n = (D_2 - D_1 - AD)/(2 \cdot AD)$.

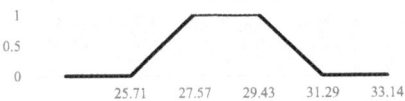

Fig. 1. A symmetric trapezoidal membership function.

As an example, a volatile time series is chosen (see Table 1 and Fig. 2). This time series was considered in [7] in the context of solving this problem.

Table 1.

Year	HD	Year	HD	Year	HD	Year	HD	Year	HD	Year	HD
1984	9	1989	60	1994	63	1999	34	2004	21	2009	51
1985	31	1990	49	1995	14	2000	37	2005	44	2010	46
1986	23	1991	31	1996	55	2001	56	2006	31	2011	63
1987	24	1992	27	1997	11	2002	57	2007	12	2012	32
1988	36	1993	37	1998	17	2003	62	2008	18	2013	44

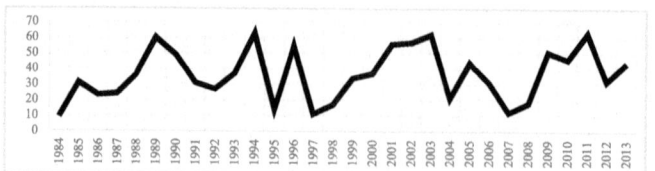

Fig. 2. Volatile time series.

According to (1) and (2), for a given time series we accordingly have: $AD = 1.8621$ and $\sigma_{AD} = 1.5023$. After eliminating x_i, which are not satisfied the condition: $0.36 \approx 1.8621 - 1.5023 \leq x_i \leq 1.8621 + 1.5023 \approx 3.36$, the final average value is obtained as $AD = 1.8571$. In this case, the segment $U = [D_2 - D_1] = [9 - 1.8571, 63 + 1.8571] \approx [7.14, 64.86]$ is universe, and the total number of its fuzzy subsets for describing the HD is calculated as follows: $n = (64.8571 - 7.1429 - 1.8571)/(2 \cdot 1.8571)$.

Thus, to describe the HD of the considered time series in the form of fuzzy subsets of the universe $U = [7.14, 64.86]$ the fifteen appropriate membership functions are identified (Fig. 2), the parameters of which are summarized in Table 2.

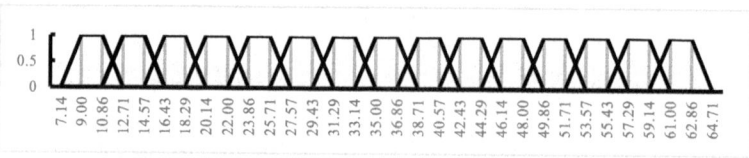

Fig. 3. Symmetric trapezoidal membership functions.

Table 2. Fuzzy sets for describing the HD of time series.

Fuzzy set	Parameters of the trapezoidal membership function				Fuzzy set	Parameters of the trapezoidal membership function			
	a_{k1}	a_{k2}	a_{k3}	a_{k4}		a_{k1}	a_{k2}	a_{k3}	a_{k4}
A_1	7.14	9.00	10.86	12.71	A_9	36.86	38.71	40.57	42.43
A_2	10.86	12.71	14.57	16.43	A_{10}	40.57	42.43	44.29	46.14
A_3	14.57	16.43	18.29	20.14	A_{11}	44.29	46.14	48.00	49.86
A_4	18.29	20.14	22.00	23.86	A_{12}	48.00	49.86	51.71	53.57
A_5	22.00	23.86	25.71	27.57	A_{13}	51.71	53.57	55.43	57.29
A_6	25.71	27.57	29.43	31.29	A_{14}	55.43	57.29	59.14	61.00
A_7	29.43	31.29	33.14	35.00	A_{15}	59.14	61.00	62.86	64.71
A_8	33.14	35.00	36.86	38.71					

Now, the HD can be fuzzified by the above identified symmetric trapezoidal membership functions according to the principle: the HD x_t is described by the fuzzy set to which it belongs with the greatest degree. In those cases, when the HD belongs to the interval $[a_{k2}, a_{k3}]$, it is relatively easy to find its fuzzy analogue. In other cases, additional calculations are necessary. In particular, according to (3) for value 31 we have: $\mu_{A6} = 0.1538$ and $\mu_{A7} = 0.8462$ (see also Fig. 3). Therefore, the fuzzy set A_7 is chosen as the fuzzy analogue. As result of these calculations, obtained fuzzy analogies of HD x_t are presented in Table 3.

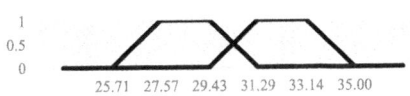

Fig. 4. Fragment of adjacent membership functions.

Table 3. The fuzzy analogy of time series.

Year	HD	Fuzzy set	Year	HD	Fuzzy set	Year	HD	Fuzzy set
1984	9	A_1	1994	63	A_{15}	2004	21	A_4
1985	31	A_7	1995	14	A_2	2005	44	A_{10}
1986	23	A_5	1996	55	A_{13}	2006	31	A_7
1987	24	A_5	1997	11	A_1	2007	12	A_2
1988	36	A_8	1998	17	A_3	2008	18	A_3
1989	60	A_{14}	1999	34	A_7	2009	51	A_{12}
1990	49	A_{12}	2000	37	A_8	2010	46	A_{11}
1991	31	A_7	2001	56	A_{13}	2011	63	A_{15}
1992	27	A_6	2002	57	A_{14}	2012	32	A_7
1993	37	A_8	2003	62	A_{15}	2013	44	A_{10}

Within the fuzzy time series, the internal relations are showed corresponding cause-effect relations by fuzzy implicative rules of the form "If < ... >, then < ...>". For example, the 1st internal fuzzy relations are grouped according to the principle: if the fuzzy set A_8 is connected, for example, sequentially with A_{13}, A_{14} and A_{15}, then the corresponding group G8 is localized as $A_8 \Rightarrow A_{13}, A_{14}, A_{15}$. According to this principle the groups of the 1st and 2nd orders are presented, respectively, in Table 4 and Table 5.

Table 4. Fuzzy relations of the 1st order.

Group	Relation	Group	Relation	Group	Relation
G1	$A_1 \Rightarrow A_3, A_7$	G6	$A_6 \Rightarrow A_8$	G11	$A_{11} \Rightarrow A_{15}$
G2	$A_2 \Rightarrow A_3, A_{13}$	G7	$A_7 \Rightarrow A_2, A_5, A_6, A_8, A_{10}$	G12	$A_{12} \Rightarrow A_7, A_{11}$
G3	$A_3 \Rightarrow A_7, A_{12}$	G8	$A_8 \Rightarrow A_{13}, A_{14}, A_{15}$	G13	$A_{13} \Rightarrow A_1, A_{14}$
G4	$A_4 \Rightarrow A_{10}$	G9	$A_9 \Rightarrow \varnothing$	G14	$A_{14} \Rightarrow A_{12}, A_{15}$
G5	$A_5 \Rightarrow A_5, A_8$	G10	$A_{10} \Rightarrow A_7$	G15	$A_{15} \Rightarrow A_7$

Table 5. Fuzzy relations of the 2nd order.

Group	Relation	Group	Relation	Group	Relation	Group	Relation
G1	$A_1, A_3 \Rightarrow A_7$	G8	$A_5, A_5 \Rightarrow A_8$	G15	$A_8, A_{13} \Rightarrow A_{14}$	G22	$A_{13}, A_1 \Rightarrow A_3$
G2	$A_1, A_7 \Rightarrow A_5$	G9	$A_5, A_8 \Rightarrow A_{14}$	G16	$A_8, A_{14} \Rightarrow A_{12}$	G23	$A_{13}, A_{14} \Rightarrow A_{15}$
G3	$A_2, A_3 \Rightarrow A_{12}$	G10	$A_6, A_8 \Rightarrow A_{15}$	G17	$A_8, A_{15} \Rightarrow A_2$	G24	$A_{14}, A_{12} \Rightarrow A_7$
G4	$A_2, A_{13} \Rightarrow A_1$	G11	$A_7, A_2 \Rightarrow A_3$	G18	$A_{10}, A_7 \Rightarrow A_2$	G25	$A_{14}, A_{15} \Rightarrow A_4$
G5	$A_3, A_7 \Rightarrow A_8$	G12	$A_7, A_5 \Rightarrow A_5$	G19	$A_{11}, A_{15} \Rightarrow A_7$	G26	$A_{15}, A_2 \Rightarrow A_{13}$
G6	$A_3, A_{12} \Rightarrow A_{11}$	G13	$A_7, A_6 \Rightarrow A_8$	G20	$A_{12}, A_7 \Rightarrow A_6$	G27	$A_{15}, A_4 \Rightarrow A_{10}$
G7	$A_4, A_{10} \Rightarrow A_7$	G14	$A_7, A_8 \Rightarrow A_{13}$	G21	$A_{12}, A_{11} \Rightarrow A_{15}$	G28	$A_{15}, A_7 \Rightarrow A_{10}$

In particular, assuming that x_i is the value of the HD for the current i-th year, and x_{i+1} is the value of the HD for the next $(i + 1)$-th year, then the fuzzy relation of the 1st order $A_4 \Rightarrow A_{10}$ can be interpreted as the fuzzy implication: "If x_i is A_4, then x_{i+1} is A_{10}". Or, for example, the 1st order fuzzy relation $A_8 \Rightarrow A_{13}, A_{14}, A_{15}$ can be interpreted as the fuzzy implicative rule: "If x_i is A_8, then x_{i+1} is A_{13} or x_{i+1} is A_{14} x_{i+1} is A_{15}". Accordingly, the fuzzy relation of the 2nd order, for example, $A_{15}, A_2 \Rightarrow A_{13}$ can be interpreted as the fuzzy implicative rule: "If x_i is A_{15} and x_i is A_2, then x_{i+1} is A_{13}".

Various methods are used to determine the fuzzy predicts and their defuzzification. In particular, according to [8, 9] the essence one of approaches is as follows. If HD x_i is described by the fuzzy set A_j, which has only one internal relationship, for example, $A_j \Rightarrow A_k$, then the predict for the next $(i+1)$-th year will be the fuzzy set A_k. If there is the group of relations, for example, $A_j \Rightarrow A_{k1}, A_{k2}, \ldots, A_{kp}$, then the sum of fuzzy sets $A_{k1} \cup A_{k2} \cup \ldots \cup A_{kp}$ will be the fuzzy predict for the $(i+1)$-th year.

For defuzzification of outputs, the following two principles are applied [8].

Principle 1. In the case of the fuzzy relation $A_i \Rightarrow A_j$, where A_i is the fuzzy analogue of the ID for the i-th year, a crisp (defuzzified) predict for the next $(i+1)$-th year is determined as the abscissa of the center of the upper base of corresponding trapezoid. For example, applying the formula $F(A) = (1/\alpha_{max}) \int_0^{\alpha_{max}} M(A_\alpha) d\alpha$, where $A_\alpha = \{u|\ \mu_A(u) \geq \alpha,\ u \in U\}$ are the α-level sets $(\alpha \in [0, 1])$; $M(A_\alpha)$ are cardinal numbers of corresponding α-level sets, which are calculated by formula: $M(A_\alpha) = \sum_{k=1}^n u_k/n$, $u_k \in A_\alpha$, for fuzzy predict $A_{10} = \{0/40.57; 1/42.43; 1/44.29; 0/46.14\}$ (see Table 2) we have: $F(A_{10}) = \int_0^1 M(A_{10\alpha}) d\alpha \approx M(A_{10\alpha}) \Delta\alpha = 43.36 \cdot 1 = 43.36$, where for $0 < \alpha < 1$, $\Delta\alpha = 1$, $A_{10\alpha} = \{42.43; 44.29\}$, $M(A_{10\alpha}) = (42.43 + 44.29)/2 \approx 43.36$.

Principle 2. In the case of the fuzzy relation $A_i \Rightarrow A_j, A_t, A_p$, where A_i is the fuzzy analogue of the ID for the i-th year, a crisp (defuzzified) predict for the next $(i+1)$-th year is determined as the arithmetical mean value of abscissas of the center of the upper base of corresponding trapezoids of fuzzy sets A_j, A_t and A_p. In particular, the predicts for 1989, 1994 and 2001 caused by the relations $A_8 \Rightarrow A_{13}, A_{14}, A_{15}$ are calculated as follows: $[(53.57 + 55.43)/2 + (57.29 + 59.14)/2 + (61.00 + 62.86)/2]/3$.

The all outputs of the 1st order time series model are summarized in Table 6 as predicts, and the geometric interpretation of this model is presented in Fig. 4 against the background of the original time series. Due to too many criteria for evaluating the HD (total 15 fuzzy sets), there is no need to apply relationships of the 2nd and higher orders, because, for example, the fuzzy model of the 2nd order coincides with the fuzzy time series established in Table 3. Nevertheless, we considered it necessary to show its geometric interpretation in Fig. 4 and Fig. 6.

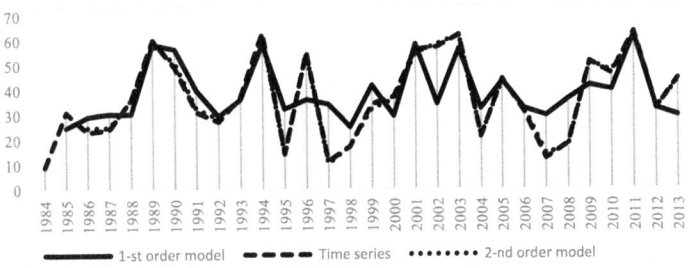

Fig. 5. Time series models of the 1st and 2nd orders.

Table 6. The time series model of the 1st order.

Year	HD	Fuzzy output	Predict	Year	HD	Fuzzy output	Predict
1984	9		–	1999	34	$A_7. A_{12}$	41.50
1985	31	$A_3. A_7$	24.79	2000	37	$A_2. A_5. A_6. A_8. A_{10}$	29.24
1986	23	$A_2. A_5. A_6. A_8. A_{10}$	29.24	2001	56	$A_{13}. A_{14}. A_{15}$	58.21
1987	24	$A_5. A_8$	30.36	2002	57	$A_1. A_{14}$	34.07
1988	36	$A_5. A_8$	30.36	2003	62	$A_{12}. A_{15}$	56.36
1989	60	$A_{13}. A_{14}. A_{15}$	58.21	2004	21	A_7	32.21
1990	49	$A_{12}. A_{15}$	56.36	2005	44	A_{10}	43.36
1991	31	$A_7. A_{11}$	39.64	2006	31	A_7	32.21
1992	27	$A_2. A_5. A_6. A_8. A_{10}$	29.24	2007	12	$A_2. A_5. A_6. A_8. A_{10}$	29.24
1993	37	A_8	35.93	2008	18	$A_3. A_{13}$	35.93
1994	63	$A_{13}. A_{14}. A_{15}$	58.21	2009	51	$A_7. A_{12}$	41.50
1995	14	A_7	32.21	2010	46	$A_7. A_{11}$	39.64
1996	55	$A_3. A_{13}$	35.93	2011	63	A_{15}	61.93
1997	11	$A_1. A_{14}$	34.07	2012	32	A_7	32.21
1998	17	$A_3. A_7$	24.79	2013	44	$A_2. A_5. A_6. A_8. A_{10}$	29.24

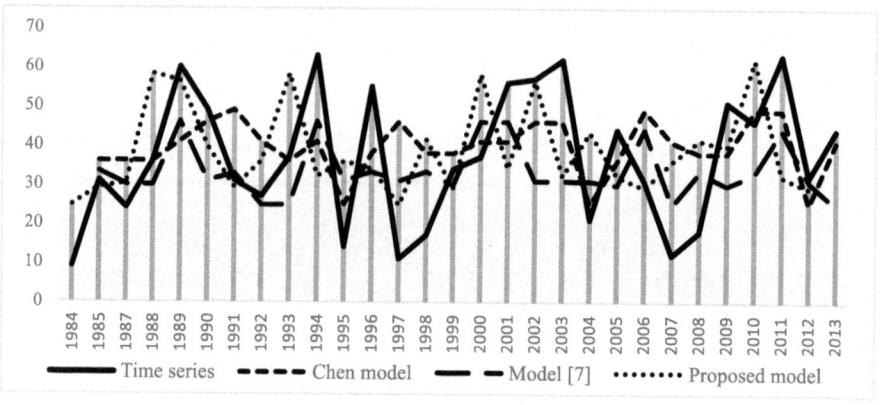

Fig. 6. Time series and its models of the 1st order.

Table 7. Comparative analysis of the 1st order models.

Year	HD	The Chen model	The model established in [7]	Proposed model	Year	HD	The Chen model	The model established in [7]	Proposed model
1984	9	–	–	–	2000	37	41	46.0706	29.24
1985	31	36	33.2953	24.79	2001	56	41	46.0706	58.21
1986	23	41	24.7750	29.24	2002	57	46	30.9644	34.07
1987	24	36	29.9126	30.36	2003	62	46	30.9644	56.36
1988	36	36	29.9126	30.36	2004	21	25	30.9644	32.21
1989	60	41	46.0706	58.21	2005	44	36	29.9126	43.36
1990	49	46	30.9644	56.36	2006	31	49	44.0833	32.21
1991	31	49	32.8618	39.64	2007	12	41	24.7750	29.24
1992	27	41	24.7750	29.24	2008	18	38	33.2953	35.93
1993	37	36	24.7750	35.93	2009	51	38	29.9126	41.50
1994	63	41	46.0706	58.21	2010	46	49	32.8618	39.64
1995	14	25	30.9644	32.21	2011	63	49	44.0833	61.93
1996	55	38	33.2953	35.93	2012	32	25	30.9644	32.21
1997	11	46	30.9644	34.07	2013	44	41	24.7750	29.24
1998	17	38	33.2953	24.79	MAPE (%)		51.94	42.72	35.90
1999	34	38	29.9126	41.50	MSE		229.48	221.71	116.34

4   Conclusion

In [7], on the basis of given time series for the limited number of evaluation criteria, the 1st order models were constructed using Chen's model [8, 9], as well as using our own approach. To estimate the degree of adequacy of the time series models proposed in [7] and in this paper, the statistical criteria MAPE and MSE are used. The results of forecasts and estimates of their reliability for all models (see Fig. 5) are presented in Table 7. The corresponding values of MAPE and MSE are also indicated there. According to the criteria of MAPE and MSE, the approach proposed in the paper significantly improves the forecasting quality of the selected volatile time series.

References

1. Saaty, T.L.: Decision making with the analytic hierarchy process. Int. J. Serv. Sci. 1(1), 83–98. https://doi.org/10.1504/ijssci.2008.017590
2. Pospelov, D.A.: Fuzzy Sets in Control and Artificial Intelligence Models. Nauka, Moscow (1986). (in Russian)
3. Borisov, A.N., Alekseyev, A.V. Merkuryeva, G.V., Slyadz N.N.: Fuzzy Information Processing in Decision Making Systems. Radio i svyaz [Radio and communication], Moscow (1989). (in Russian)
4. Korchenko, A.G., Ryndyuk V.A.: The study of methods of forming membership functions based on quantitative pairwise comparisons. Sci. Techn. J. "Zakhist Informatsii" 3, 10–17 (2003). (in Russian). http://er.nau.edu.ua/handle/NAU/36381
5. Rzayev, R.R.: Neuro-Fuzzy Modeling of Economic Behavior. Lambert Academic Publishing, Saarbrücken, Germany (2012). (in Russian)
6. Ortiz-Arroyo, D., Poulsen, J.R.: A Weighted fuzzy time series forecasting model. Indian J. Sci. Technol. 11(27), 1–11 (2018). https://doi.org/10.17485/ijst/2018/v11i27/130708
7. Rzayev, R.R., Jamalov, Z.R., Mehdiyev, T.Z., Hasanov, V.I.: Time series modeling based on fuzzy analysis of position-binary components of historical data. Nechetkie Sistemy i Myagkie Vychisleniya [Fuzzy Systems and Soft Computing] 10(1), 35–73 (2015). (in Russian)
8. Chen, S.M.: Forecasting enrollments based on fuzzy time series. Fuzzy Sets Syst. 81, 311–319 (1996). https://doi.org/10.1016/0165-0114(95)00220-0
9. Chen, S.M.: Forecasting enrollments based on high-order fuzzy time series. Cybern. Syst: Int. J. 33, 1–16 (2002). https://doi.org/10.1080/019697202753306479

Assessment of the Accessibility of Digital Educational Platforms Based on Fuzzy - Logical Analysis Methods

Evgeniy N. Tishchenko and Elizabeth A. Arapova[✉]

Rostov State University of Economics, B. Sadovaya str., 69,
344002 Rostov-on-Don, Russia
celt@inbox.ru, dist_edu@ntti.ru

Abstract. The paper we have considered the concept of accessibility of a software system as one of the characteristics of its quality. On the basis of the international standard of the Web-accessibility WCAG we have identified the most important for the evaluation criteria. We have developed a methodology for a comprehensive assessment of the availability of educational resources based on fuzzy-multiple aggregation of accessibility estimates for certain types of violations.

Keywords: Software quality · Accessibility · WCAG · Inclusive education · Distance learning

1 Introduction

The availability of a software system is defined in [1–3] as one of the basic characteristics of quality. Accessibility characterizes the ability of the system to support the work of people with special needs, taking into account all factors that may limit their capabilities [4]. The low availability of digital educational platforms creates barriers to their use in inclusive educational practice [5].

Massive Open Online Courses [6] (MOOC) are becoming increasingly popular at the present stage of development of Internet and SMART- education.

Assessment of their accessibility can be constructed based on international standards of accessibility Web-content WCAG (Web Content Accessbility Guidelines, versions 2.0–2.1) [7, 8]. WCAG contains more than two hundred requirements for the development of a web resource, taking into account all the possible problems that a person may encounter when using it. On the basis of these requirements, we have identified most important criteria (Table 1) (taking into account the possible use by the student of assistive technologies) [9].

Thus, for a comprehensive assessment, we identified 5 groups of indicators that determine the availability of educational content for four categories of people with health problems. These indicators have different significance, which depends on the severity of the violation. Accordingly, they have a different effect on the possibility of information perception of educational material.

R. A. Aliev et al. (Eds.): ICAFS 2020, AISC 1306, pp. 442–450, 2021.
https://doi.org/10.1007/978-3-030-64058-3_55

Table 1. Classification of criteria for the availability of educational content by type of violation.

	Criteria	Impairment			
		Of view	Hearing	Musculos keletal system	Cognit ive
K_1	**Presentation of non-textual (graphic, audio, video) of content**				
K_{11}	Text version of the graphic elements	■			■
K_{12}	Text version of the pre-recorded audio and video content	■			■
$K_{1.3}$	Subtitles for media (audio, video) of content		■		■
K_2	**Presentation of the structure and sequence of study information**				
K_{21}	Text alternative representations of important semantic units	■			■
K_{22}	Text alternative touch release information	■			■
K_{23}	Text alternative color highlighting information	■			■
K_{24}	Providing additional meta information about the items on the page and the links between them	■			■
K_3	**Methods for control and navigation**				
K_{31}	Management of the entire content functionality via the keyboard			■	
K_{32}	The ability to disable or override the actions of keys	■		■	
K_{33}	Manage functionality by moving the device			■	■
K_4	**Semantic Content Model**				
K_{41}	Links for quick access to the main content	■		■	
K_{42}	Possible to determine the sequence of reading material	■			■
K_{43}	Page titles, chapters describe their theme or purpose	■			■
K_{44}	The link value follows from the link text itself	■			■
K_{45}	The procedure for transferring the focus corresponds to the semantic content model	■		■	
K_{46}	Providing text labels and, if necessary, prompts for input elements	■			
K_5	**The visual presentation of content**				
K_{51}	The contrast (minimum): Contrast ratio> = 4.5: 1	■			
K_{52}	Changing text size (except for titles and text images) in the range up to 200%	■			■
K_{53}	Mechanism pause or stop the audio-video recording, as well as the volume control (the chance to reduce it to zero).		■		
K_{54}	Lack of elements, flashing more than 3 times per second	■			■
K_{55}	Ability to shut down and setting time constraints in studying the test material or control	■		■	■

In this paper, we offered complex estimation technique availability web-oriented educational system-based methods fuzzy-logic analysis [10].

A comprehensive assessment of accessibility can be formed on the basis of fuzzy-multiple aggregation of relevant accessibility estimates for four basic categories of disorders: hearing, vision, musculoskeletal system, mental sphere.

2   General Description of the Method

General principles of fuzzy - logical analysis are described in the [11]. The proposed method includes the following steps:

> *Step 1.* The formation of a comprehensive indicator of accessibility for each type of violation (hearing, vision, musculoskeletal system, mental sphere) gi, i = 1.4.
> *Step 2.* Aggregation of the grades obtained in the final assessment of the availability of educational resources.

The calculation of the value of each indicator is carried out by the following algorithm:

1. We introduce the linguistic variables gi. The universal set for each linguistic variable is the interval [0,1], and the set of values of all five variables g_1, g_2, g_3, g_4, g - term set G = {G_1, G_2, G_3, G_4, G_5}.
   For each of the terms we define a trapezoidal membership function, in accordance with the theory of standard five-level fuzzy [0,1]- classifiers.
2. We highlight the criteria (sub-criteria) of accessibility for each type of disorders K_{ij} (Table 1). The importance of each criterion is ranked using weighting coefficients k_{ij}.
3. We perform the calculation of values xi (i = 1 ... N) of each indicator. The value of xi (0 $\ll$ x_i $\ll$ 1) represents the proportion of educational content elements for which the criterion is satisfied.
4. We carry out the transition from the numerical values of the parameters to a numeric value estimates. The rule of transition from the values of indicators to the weights of terms of linguistic variables g_i:

$$ g_i = \sum_{k=1}^{5} p_k\, g_k, \quad p_k = \sum_{k=1}^{N} k_i \cdot \mu_{ik}(x_k) $$

where k_i - the weights of indicators calculated in Step 2;
g_k – mid gaps, are carriers of the terms;
$\mu_{ik}(x_k)$- the values of the membership functions calculated for numeric metrics.

5. We perform Linguistic recognition obtained numerical rating in accordance with the definition of the term-sets G = {G_1, G_2, G_3, G_4, G_5}.

3 Practical Part

Using the proposed methodology, we performed a comprehensive assessment and ranking by the level of accessibility of six national mass open educational (MOOC) platforms [12, 13]: "Open Education" (openedu.ru), online university "Intuit" (intuit. ru), "Stepik" (stepik. org), "Lectorium" (lektorium.tv), "Universarium" (universarium. org), "Teachpro" (teachpro.ru).

To audit the accessibility indicators, we used the following methods for testing the studied resources [14]:

1. Expert audit of accessibility.

It is carried out by a specialist who is browsing a resource and looking for possible accessibility problems (including using assistive tools).

2. Automated testing accessibility using specialized software.

As such means we used WAVE program (Web Accessibility Evaluation Tool) and Total Validator.

For each category of violations, we identified important criteria. The weighting factors k_{ij} reflect the importance of each K_{ij} taking into account the severity of the violation. Example calculation coefficients kij for persons with hearing lesion shown in Table 2.

Table 2. Calculation of weighting coefficients for availability indicators for the hearing impaired (in accordance with the International Classification of hearing loss [15]).

Criteria (sub-criteria)		I degree of hearing loss (mild)	II degree of hearing loss (mean)	III degree of hearing loss (heavy)	Deafness	Weight, k_{ij}
K_{12}	Text version of the pre-recorded audio and video content	0	1	1	1	0,250
K_{13}	Subtitles for media content	1	1	1	1	0,333
K_{53}	pause mechanism to the audio-video recording	0	1	1	1	0,250
K_{55}	The possibility of switching off and setting time limits	0	0	1	1	0,167

Next, we introduced the linguistic variables g_i ($i = 1 \ldots 4$) - an assessment of the accessibility of the educational resource for people with certain impairments (hearing, vision, motor, cognitive). The universal set for each linguistic variable is the interval [0,1], and the set of values is the term set $G = \{G_1, G_2, G_3, G_4, G_5\}$. The x_{ij} parameters, on the basis of which the g_i score is generated, represent the proportion of educational content elements for which the K_{ij} criterion is implemented. The results of the availability audit and fuzzy multiple aggregation of ratings are presented in Tables 3 4, 5, 6.

Table 3. Evaluation of the availability of educational platforms for persons with hearing damage

Education platform	K_{12}	K_{13}	K_{53}	K_{55}	g_1	Term	Availability level, resource usage limits
Open education	1	1	1	1	**0,885**	G_5	High accessibility, for the deaf
Intuit	0	0	1	0	**0,315**	G_2	Low, no more than I degree of hearing loss
Stepik	0	0	1	1	**0.441**	G_3	Average, no more than II degree of hearing loss
Lectorium	0	0	1	0	**0,315**	G_2	Low, no more than I degree of hearing loss
Universarium	0	0	1	0	**0,315**	G_2	Low, no more than I degree of hearing loss
TeachPro	0	1	1	0	**0.441**	G_2	Low, no more than I degree of hearing loss

Table 4. The results of the audit of the accessibility of educational platforms for people with visual impairment (in accordance with the International Classification of visual impairment [16]).

Criteria	k_{ij}	Open education	Online University	Stepik	Lectorium	Universarium	TeachPro
K_{11}	0,058	1	0.8	0.7	0.5	0.5	0.5
K_{12}	0,058	1	0	0	0	0	0
K_{13}	0,058	1	0	1	0	0	0
K_{21}	0,058	0	0	0	0	0	0
K_{22}	0,058	0	0	0	0	0	0
K_{23}	0,058	0	0	0	0	0	0
K_{24}	0,058	0.5	0.1	0.8	0.2	0	0.5
K_{31}	0,058	0.5	0.5	0.5	0	0	0.5
K_{32}	0,058	0	0	0	0	0	0
K_{41}	0,058	1	1	1	1	1	1
K_{42}	0,058	0	0	0	0	0	0

(*continued*)

Table 4. (*continued*)

Criteria	k_{ij}	Open education	Online University	Stepik	Lectorium	Universarium	TeachPro
K_{43}	0,058	1	1	1	1	1	1
K_{44}	0,058	1	0.5	1	0.5	0.5	1
K_{45}	0,058	1	0.5	0.5	0.5	0.5	0.5
K_{46}	0,058	0.3	0.5	0.2	0.1	0.1	0.2
K_{51}	0,019	0	0	0	0	0	0
K_{52}	0,019	0	0	0	0	0	0
K_{53}	0,019	1	1	1	1	1	1
K_{54}	0,019	1	0	1	1	1	1
K_{55}	0,058	1	0	1	0	0	0
g2 =		**0.561**	**0.379**	**0.354**	**0.312**	**0,308**	**0.379**
Term		**G3**	**G2**	**G2**	**G2**	**G2**	**G2**
Availability level, resource usage limits		G3-average, visual acuity of at least 0.3			G2-low, only for a small visual impairment (visual acuity of at least 0.7)		

Table 5. Results of evaluation of the accessibility of educational platforms for persons with lesions musculoskeletal.

Criteria	k_{ij}	Open education	Intuit	Stepik	Lectorium	Universarium	TeachPro
K_{31}	0,167	0.5	0.5	0.5	0	0	0.5
K_{32}	0,167	0	0	0	0	0	0
K_{33}	0,167	0	0	0	0	0	0
K_{41}	0,167	1	1	1	1	1	1
K_{45}	0,167	1	0.5	0.5	0.5	0.5	0.5
K_{55}	0,167	1	0	1	0	0	0
g3 =		**0,625**	**0.377**	**0.503**	**0.314**	**0.314**	**0.377**
Term		G4	G2	G3	G2	G2	G2
Availability		Tall enough	Low	Middle	Low	Low	Low

Comprehensive assessment of educational content availability indicator is generated based on the aggregation obtained above g_1–g_4 ratings. We introduced a linguistic variable: g = "Comprehensive assessment of the availability of educational resources."

The weighting coefficients ki of each indicator gi are calculated according to the formula:

$$k_i = \frac{n_i}{\sum_{m=1}^{4} n_m}$$

where k_i - weighting factor index g_i; n_i – the number of sub-criteria for the indicator g_i

Table 6. The results of evaluation of availability of educational platforms for persons with cognitive impairment.

Criterion	k_{ij}	Open education	Intuit	Stepik	Lectorium	Universarium	TeachPro
K_{11}	0,100	1	0.8	0.7	0.5	0.5	0.5
K_{12}	0,100	1	0	0	0	0	0
K_{13}	0,100	1	0	1	0	0	0
K_{43}	0,100	1	1	1	1	1	1
K_{44}	0,100	0	0	0	0	0	0
K_{51}	0,100	0	0	0	0	0	0
K_{52}	0,100	0	0	0	0	0	0
K_{53}	0,100	1	1	1	1	1	1
K_{54}	0,100	1	0	1	1	1	1
K_{55}	0,100	1	0	1	0	0	0
$g_4 =$		**0.657**	**0.343**	**0.562**	**0,390**	**0,390**	**0,390**
Term		G4	G2	G3	G2	G2	G2
Availability		tall enough	low	middle	low	low	low

The results of the calculation and ranking of educational systems are presented in Tables 7, 8.

Table 7. Overall evaluation availability educational platforms.

	g1	k1	g2	k2	g3	k3	g4	g4	g
Open education	0,885	0.1	0.561	0.5	0,625	0.15	0.657	0.25	0,622
Intuit	0,315	0.1	0.379	0.5	0.377	0.15	0.343	0.25	0.337
Stepik	0.441	0.1	0.354	0.5	0.503	0.15	0.562	0.25	0.408
Lectorium	0,315	0.1	0.312	0.5	0.314	0.15	0,390	0.25	0.320
Universarium	0,315	0.1	0,308	0.5	0.314	0.15	0,390	0.25	0.320
TeachPro	0.441	0.1	0.379	0.5	0.377	0.15	0,390	0.25	0,375

Thus, according to the results of research rather high figure demonstrates only available platform "Open Education» (openedu.ru), which (with some restrictions) can be used in training people with disabilities. Other systems have serious accessibility problems that create barriers to their use in inclusive education.

Table 8. Ranking of educational platforms for comprehensive accessibility assessment for persons with disabilities.

Education platform	Value
G_1 – "No accessibility at all"	
G_2 – "low level of accessibility for persons with disabilities"	
Online University "Intuit"	g = 0,337
Lectorium	g = 0,320
Universarium	g = 0,320
Teachpro	g = 0,375
G_3 – "average level of accessibility for persons with disabilities"	
Stepik	g = 0,408
G_4 – "a fairly high level of accessibility for persons with disabilities"	
Open education	g = 0,622
G_5 – "a high level of accessibility for people with disabilities"	

4 Conclusion

Evaluation of the quality of information systems in education of people with disabilities involves accessibility analysis. We have developed a methodology that forms a comprehensive assessment of the availability of a software system based on estimates of indicators for certain types of violations. The methodology allows you to evaluate the value of the indicator, identify existing problems of accessibility, and also identify the boundaries of the use of educational resources for people with disabilities of varying severity.

References

1. ISO/IEC TR 9126-2: Software engineering - Product quality - Part 2: External metrics (2003). https://www.iso.org/standard/22750.html
2. ISO/IEC TR 9126-2: Software engineering - Product quality - Part 2: External metrics (2003). https://www.iso.org/standard/22891.html
3. ISO/IEC TR 9126-4: Software engineering - Product quality - Part 4: Quality in use metrics (2004). https://www.iso.org/ru/standard/39752.html
4. The Russian Federation National Standard GOST R 52872-2012 Internet resources. Accessibility Requirements for the visually impaired. http://docs.cntd.ru/document/1200103663. Accessed 19 Mar 2020)
5. The Russian Federation National Standard GOST R 52872-2012 Internet resources. Accessibility Requirements for the visually impaired. http://docs.cntd.ru/document/1200103663. Accessed 19.03.2020

6. The old man Access to education for people with disabilities//Theory and practice of social development, February 2011
7. Kushnaryov, S.E.: MOOC as a form of distance education//Scientific notes of the Crimean Federal University named after VI Vernadsky. Philol. Sci. 2(2), (2016)
8. Guidelines for Web Content Accessibility (WCAG) 2.0. https://www.w3.org/Translations/WCAG20-ru/
9. Web Content Accessibility Guidelines(WCAG)2.1. https://www.w3.org/TR/WCAG/
10. Klochkova, E.N., Darda, E.S.: Market assistive technology and devices: a statistical assessment of the state. Russia: Trends Prospects 12(1) (2017)
11. Tishchenko, E.N., Zilina, E.V., Sharypova, T.N., Palyutina, G.N.: Fuzzy models for information security training programs for the development of the results. Intell. Resour. Reg. Dev. 4(1) (2018)
12. Arapova, E.A., Lukyanova, G.V., Sakharova, L.V., Akperov, G.I.: Fuzzy-Logic analysis of the level of comfort and environmental well-being of the urban environment on the example of large cities of Rostov region. Adv. Intel. Syst. Comput. 896, 643–650 (2018)
13. The Russian educational online projects. https://hr-media.ru/19-krupnejshih-rossijskih-onlajn-obrazovatelnyh-proektov/. Accessed 24 Mar 2020
14. Husyainov, T.M.: The main characteristics of massive open online course (MOOC) as an educational technology, Science. Thought: Mag. 2 (2015)
15. Study on the availability of Runet Internet resources for people with disabilities (HIA). https://perspektiva-inva.ru/userfiles/download/Accessibility_of_Runet_2013.Pdf. Accessed 15 Jan 2020)
16. Golovchits, L.A., Vlados, M.: Preschool sign language education (2010)

Soft Computing for Prediction of Secondary Structure of the Protein

Elbrus Imanov[1](✉) ⓘ and Ritta Shaheen[2] ⓘ

[1] Department of Computer Engineering, Near East University,
via Mersin 10, North Cyprus, Turkey
elbrus.imanov@neu.edu.tr
[2] Department of Biomedical Engineering, Near East University,
via Mersin 10, North Cyprus, Turkey
ritta.shaheen@gmail.com

Abstract. The significance of determining the structure of proteins comes up from the importance of their role in the body, making proteins the primary target of drugs and key to developing new drugs. There are several experimental methods to determine PS 'Protein structure', but a reliable and direct prediction method to determine PS is not yet available. Hence, it was necessary to use intelligent systems that predicted PS based on the sequence of amino acids. Predicting the SSP "secondary structure of protein" is a very important step. It has increasingly become the basis for a number of ways to predict protein structure and thus know its function.

The aim of this research is to design and implement a high-performance method for predicting SSP from the AAS "amino acid sequence" and calculating prediction accuracy based on recorded results. To achieve this goal, the work was done in two phases: first, to construct an SSP system, based solely on the AAS for the protein using ANFIS "Adaptive Neuro-Fuzzy Inference Systems" This system achieves accuracy of up to 65% and is a good accuracy compared to many SSP systems.

The second stage was the construction of two SSP systems to study the effect of coding the data used on the system's input on its accuracy. Both systems had same designs and structures using artificial neural network. The input in the second is coded using Profiles, which showed a significant improvement in system accuracy exceeded 10%, bringing the accuracy of total prediction up to 70.5%.

Keywords: Artificial intelligence · Neuro-fuzzy · Fuzzy inference system · Neural network · Secondary structure protein · Amino acids

1 Introduction

The first Artificial intelligence is a science that studies meanings, concepts and methods that make the machines intelligent. Zadeh conceived that the traditional AI would be more effective in following its objectives, if it is not restricted solely to process symbolic information [1]. That is why Artificial intelligence is concerned with studying the "intelligent" behavior of a human being in order to model or automate this behavior. Neural network is an evolutionary computation fuzzy system that offers matchless benefits in managing imprecision and uncertainty within information, and delivering unique solutions to the complicated problems [2]. Human brain's information processing ability is mimicked by the ANN [3]. Artificial intelligence is related to several sciences such as cognitive science, logic, neuroscience, informatics, and other sciences [4]. In the recent years the use of neural networks, fuzzy and genetic algorithms increased notably to solve many problems in field of biological sciences. The neural network consists of elements of a process called artificial neurons, which correspond to natural neurons, and are connected to a network [5]. The system becomes more stable beside its improved output expression capacity, after conducting fuzzy processing function information to the neural network [6]. In Fuzzy logic everything has a degree of belonging or membership, each logical system can be modeled, and knowledge is translated as a set of variables. The conclusion is presented as a logical processing of an expanded set of flexible conditions [7]. Since that, these techniques are capable to deal with linear and continuous data. This type of complex data is in its common in biological data sets, and therefore these techniques can help produce new solutions to the most difficult problems in bioinformatics. Bioinformatics work on the knowledge that belong to the broad areas of computer science and biology in order to use this large amount of data in practical applications of biological and medical studies. Soft Computing, which is a special field in computer science, is able to analyze a complex medical data [8]. The bioinformatics researches include many different areas; the most important one is the one concern with one of the biggest problems in Biology which is the prediction of protein structure from the sequence of the amino acids.

This research aims to build a system using a Soft Computing technique to forecast the protein's secondary shape from the amino acid sequence, where the system will be trained to find the secondary structure of the protein. Proteins consisted primarily from amino acids that are linked with each other within the multi-peptide series which folds within the spatial structure forming protein [9]. Proteins inside the human body are made of amino acids; only 20 amino acids contribute to the formation of these proteins despite their different functions and forms. Proteins are very important molecules within the cells of the organism, where proteins enter most of the cell's functions. The proteins differ in their spatial structure and thus in their function, which is determined by this spatial structure [10]. Diverse tasks carried out by the protein; however, the proteins can be divided into several major categories such as "Enzymes, Structural proteins, Antibodies, contractile proteins, bonding proteins etc." [11]. Proteins are organized so that hydrophobic side chains toward the center and hydrophilic-side chains on the surface. Hydrogen bonds form regular secondary structures called alpha-helix and beta-sheets [12]. Diseases caused by protein folding defect the idea of life can be presented as the process of coordinating the activity of proteins, and the disease may be caused by a defect in one protein. Improper folding may result in impaired protein function [13]. It is believed that protein folding dysfunction is the leading cause of Alzheimer's illness, Parkinson's illness, and spongiform encephalopathy [14]. Smooth the smallest coins in the volume or possessions of the amino acid series border or even the deletion or addition can alter or stop the protein function [15]. Machine learning techniques usually used to solve this problem. Machine learning is continually improved and advanced for the implementation of artificial intelligence model [16]. Lot of techniques and tools was produced to predict protein structure. One of the most important findings that we have achieved through the work carried out in this research, which was shown to us by the recorded results, is the following points: Effectiveness of "adaptive fuzzy neural systems ANFIS" to solve important difficulty in bioinformatics which is predicting the secondary shape of protein. The prediction accuracy of the proposed system based on this technique, which is a good performance result compared to the other existing secondary structure prediction systems. The importance of using additional biological information when training a secondary protein prediction system, such as protein profile.

2   The Theoretical Method

The Theoretical method we used in process, in addition the used material such as training and testing data set, beside the theoretical and practical basis which we railed on in this research, like bioinformatics, artificial neural network and fuzzy system. We can divide the work into two phases. Phase one design and build predication program to predict the secondary structure of protein using one of the artificial intelligent methods, the system was built and trained to predict the related SS to each AA in the protein order, that's meaning finding if the AA is a part of Alfa helices, beta sheet or any other type of the secondary structure, then calculate the accuracy of the designed system basing on the registered result. Fuzzy Neural Network ANFIS is used to build this system. Phase two construct SSP systems using one of the artificial intelligence methods which is the artificial neural network. We will compare the two systems as an attempt to distinguish the effect of the used input code on the accuracy. In the second system we used additional biological information to encode AA in a protein chain. Flowchart for the developed system of Phase I and Phase II are shown in Fig. 1.

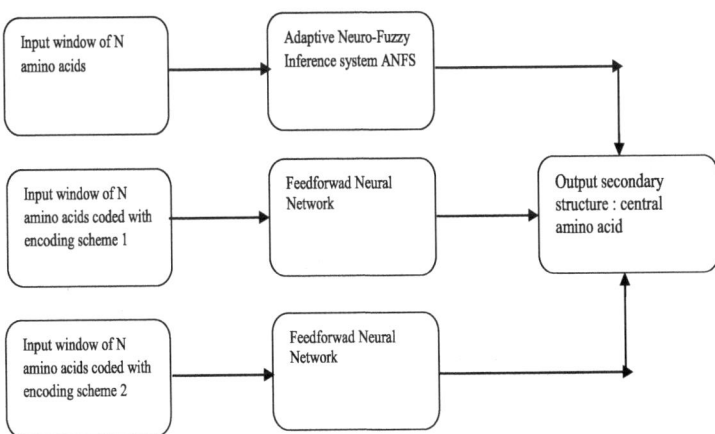

Fig. 1. Flowchart for developed system of Phase I and Phase II.

Neural networks solve problems with self-learning and self-regulation, and their intelligence derives from the behavior of simple mathematical units in multiple neurons [5]. A special biological cell which is called neuron processes the information to other neurons through the aid of some chemical end electrical changes [17]. The neural network represents a set of nodes and connections. The node refers to the neuron while the arrows point to the links through which the signals pass between the neurons. Processing of neural system is shown Fig. 2.

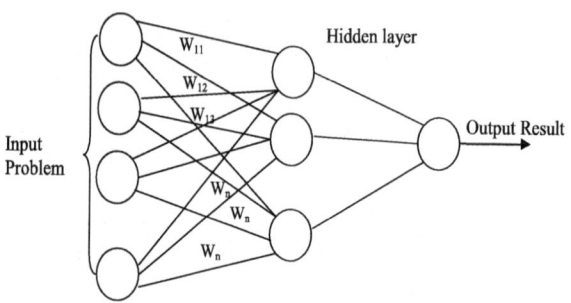

Fig. 2. Processing of neural system.

Fuzzy inference system was applied only to modeling systems that contained a predefined basic rule structure by interpreting the variables' properties of the model by the users. After collecting the data space, the numbers of "fuzzy rules" and "membership" are determined. Linear squares are used to determine the output of the membership functions, which produces an effective fuzzy inference system. ANFIS learns and determines the membership functions using the estimation of lower squares and back propagation. Hence, the fuzzy inference system through ANFIS minimizes the training errors [18]. Training data set contains all PDB "proteins in Protein Data Bank". It was collected by Pollastri. [19]. Testing Data Set is a group of protein chains collected by Rost and Sander and possesses the following specifications: Contains 126 protein series comprising 24395 amino acid [20]. The length of all the chains in which there are no more than 80 amino acids. The Similarities between the strings of this group does not exceed 25%. The "Profile" is a series PSSM "Position-Specific Scoring Matrix", which is obtained after performing a multiple alignment of a protein chain or more with a database of protein chains [21]. The training and testing data were coded using the tendency values of the 20 amino acids that Chu and Fasman calculated. The matrix consists of 43128 amino acids, encoded according to the values of the Chu and Fasman calculated for alpha-helices. Each AA is expressed by a window with a length of 15 amino acids. The amino acid takes position for eight of the windows, surrounded by seven amino acids from the right and seven on the left, corresponding to the amino acid sequence of the protein studied. To solve the secondary protein prediction problem, we used MATLAB R2018a Neural Network toolbox to build a feed-forward neural network, include single input, second and output layer. The network's input depends on the sliding window technique which applied along the protein chain. Seventy-five sequences were chosen randomly from TRAIN set were used to train and test the two systems, which consist of 18730 amino acids. The training and testing data were encoded using a 20-bit binary matrix, where each amino acid corresponds to 20 cells of the values [0, 1]. The coding of the secondary protein structure subtypes was done using a binary system.

3   Experimental Results

For the Phase I system specification in this system, we chose to use a 15-amino acid input window, so each position of the window coded by Chu and Fasman tendency value, and the input layer consists of 15 input units. As mentioned before, Table 1 lists the 1unit of five ANFIS specification.

Table 1. 1Unit of the five ANFIS specification.

Sub. clustering				Mem. functions	Optim. method	Error tolerance	Epochs
Rang of influence	Squash factor	Accept ratio	Reject ratio				
1.3	1.45	0.5	0.15	3	Hybrid	0	100
1.3	1.45	0.5	0.15	4	Hybrid	0	100
1.3	1.35	0.5	0.15	4	Hybrid	0	100

Now, we introduce the results of testing suggest system that we have constructed. These results reflect the accuracy of the Q3 prediction system and are calculated according to the following equation:

$$Q_3 = \frac{P_\alpha + P_\beta + P_\tau}{N} * 100\%,$$

where P_α, P_β, P_τ, are the numbers of amino acids of alpha-type, beta-Sheet and loops zones respectively and correctly predicted, and N denotes the complete symbol of amino acids in the test group protein chains. Here are the results of the first system test divided down by the test set used: Testing using the R126 test group is considered fairly standard, and has been used in the evaluation of several SSP systems in order to compare the performance of these systems. Table 2 shows the performance of the unit of the first system after testing with R126.

Table 2. Performance of system units with R126.

Unit	Q3	Qa	Qb	Qc
Unit	59.7%	58.11%	66.5%	58.5%

Table 3 shows the performance of the first system after being tested using a part of train.

Table 3. Performance of first system unit with part of TRAIN.

Unit	Q3	Qa	Qb	Qc
Unit	62.23%	67.1%	59.31%	54.64%

For Phase I in this research, an SSP system was developed that relies only on the sequence of the AAS of the protein without using any additional information. The carefully selected TRAIN dataset, using the tendency value of Chu and Fasman values, was used to train five units, each unit containing three "ANFIS". Each one predicts only one of the secondary protein subtypes, after which the highest value is found between the previous three ANFIS values and the corresponding classification is the unit output. Comparing the accuracy recorded in the proposed system test using the R126 test group, and using part of TRAIN groups for testing. The accuracy of the prediction in the second case is increased. This increase is due to the similarity between the TRAIN series, which is no more than 50%, while the R126 series is no more similar to 25%. The effect of the similarity between the training group chains and the test group on the accuracy of the structure in predicting the secondary shape of the protein is evident. Table 4 shows the comparison of performance.

Table 4. Comparison of performance of the first phase system with R126 and TRAIN.

Group test	Majority vote	Unanimous vote
R26	60.32%	64.5%
TRAIN	62.4%	65%

For Phase II first system specification in this system, we used a window of 15 amino acid input; each window position coded using a 20-bit binary matrix, from which the input layer consisted of 20 * 15 input units.

Second System specification as in the previous system, a window of input of 15 AAS was used. The input is a matrix of "Profile", so that each amino acid in the input window encoded by corresponding 20 values. The input layer consists of 20 * 15 input units. Phase II at this stage, the effect of the input coding of the SSP system was studied on its accuracy. Two systems were constructed for both the "artificial neural network" structure and the training and test data set. Table 5 shows a comparison between the results of both systems for the accuracy of the total prediction and the accuracy of the tolerance of each subtype of protein.

Table 5. Comparison of results of the first and second phase II systems.

Training data	Q3	Qa	Qb	Qc
Protein chain	60.1%	70.2	49.2%	59.5%
"Profiles"	70.5%	74.9%	62.9%	71.3%

The first point is the simplicity of the system we designed compared to the research mentioned as we used the simplest structure of the "Feedforward" networks, without adding any complexity to this structure. The second point is the relatively small number of protein chains that have been used, which have reached 75 training and testing series together that may result in a lack of knowledge.

4   Conclusion

In the early protein-building field, the researchers realized that the three-dimensional shape of the protein is set on next to the main succession of amino acids. This observation made it possible to predict protein structure from the sequence of amino acids with high accuracy. The SSP plays a large role in dig up as much information as possible from the amino acid sequence of the protein. Hence, the demand for finding useful and reliable methods for predicting secondary structure was principal investigation in bioinformatics. The aim of this paper was to develop and implement a highly effective method for predicting protein SS from the AAS. The work accomplished can be divided into two main parts. First, the development of a prediction system based solely on the sequence of amino acids for the protein chain without the use of any additional information. The basis of this system is ANFIS, where the system was trained using a carefully selected protein sequence. Our developed system achieves accuracy of up to 65%, which is a good accuracy compared to the accuracy of the other researches. The experimental results of our research on protein sequences from well-known databases, demonstrated the effectiveness of using ANFIS to solve major difficulty in bioinformatics which is the prediction of protein secondary structure. Second a comparison between the two prediction systems of protein secondary structure, which have the same structure and are based on of the feedforward neural networks, and trained and tested with the same data set, but with different coding. The result showed a significant improvement in accuracy by +10% when using additional information, and thus we reached a prediction accuracy of 70.5%. Taking into consideration that increasing a few percentage points in the accuracy of protein secondary prediction is a very important achievement.

References

1. Aliev, R.A., Fazlollahi, B., Aliev, R.R.: Soft Computing and its Application in Business and Economics, vol. 157, p. 110. Springer, Heidelberg (2004). https://doi.org/10.1007/978-3-540-44429-9

2. Zadeh, L.A.: Fuzzy set. Inform. Control **8**(3), 338–353 (1965). https://doi.org/10.2307/2272014

3. Maind, S.B., Wankar, P.: Research paper on basic of artificial neural network. Int. J. Recent Innov. Trends Comput. Commun. **2**(1), 96–100 (2014). https://doi.org/10.17762/ijritcc.v2i1.2920

4. Bibel, W.: On a scientific discipline (once) named AI. In: IJCAI, pp. 5143–5149 (2018)

5. Maass, W.: Networks of spiking neurons: the third generation of neural network models. Neural Netw. **10**(9), 1659–1671 (1997). https://doi.org/10.1016/S0893-6080(97)00011-7

6. Ni, X.: Research of data mining based on neural networks. World Acad. Sci. Eng. Technol. **39**, 381–384 (2008)

7. Zadeh, L.A.: Fuzzy logic. Computer **21**(4), 83–93 (1988). https://doi.org/10.1109/2.53

8. Yardimci, A.: Application of soft computing to medical problems. In: International CONFERENCE 2009 on Intelligent System Design and Applications, pp. 614–619 (2009). https://doi.org/10.1109/ISDA.2009.168

9. Hunter, L.: Molecular biology for computer scientists, pp. 18–24 (2014)

10. Clark, D.P., Pazdernik, N.J., McGehee, M.: Molecular Biology. Elsevier Science, Amsterdam (2018)

11. Lodish, H., Berk, A.: Molecular Cell Biology. WH Freeman and Company, New York (2004)

12. Voet, D., Voet, J., Pratt, C.: Principles of Biochemistry. 3rd International student edition edn. Wiley, New York (2009)

13. Miller, N.: The misfolding diseases unfold. Beremans, UK (2011)

14. Chaudhuri, T.K., Paul, S.: Protein-misfolding diseases and chaperone-based therapeutic approaches. FEBS J. **273**(7), 1331–1349 (2006). https://doi.org/10.1111/j.1742-4658.2006.05181.x

15. Khan, S., Vihinen, M.: Spectrum of disease-causing mutations in protein secondary structures. BMC Struct. Biol. **7**(1) (2007). https://doi.org/10.1186/1472-6807-7-56

16. Imanov, E., Anwar, A.: Artificial neural network for the left ventricle detection. In: 10th International Conference 2019 on Theory and Application of Soft Computing, Computing with Words and Perceptions (2019)

17. Sharma, V., Rai, S., Dev, A.: A comprehensive study of artificial neural networks. Int. J. Adv. Res. Comput. Sci. Softw. Eng. **2**(10), 278–284 (2012)

18. Wei, M., Bai, B., Sung, A.H., Liu, Q., Wang, J., Cather, M.E.: Predicting injection profiles using ANFIS. Inform. Sci. **177**(20), 4445–4461 (2007). https://doi.org/10.1016/j.ins.2007.03.021

19. Pollastri, G., Przybylski, D., Rost, B., Baldi, P.: Improving the prediction of protein secondary structure in three and eight classes using recurrent neural networks and profiles. Proteins: Struct. Funct. Bioinf. **47**(2), 228–235 (2002)

20. Rost, B., Sander, C.: Prediction of protein secondary structure at better than 70% accuracy. J. Mol. Biol. **232**(2), 548–599 (1993)

21. Gribskov, M., Mclachlant, A.D., Eisenberg, D.: Profile analysis: detection of distantly related proteins. Proc. Natl. Acad. Sci. **84**(13), 4355–4358 (1987). https://doi.org/10.1073/pnas.84.13.4355

Application of Theoretical Aspects of Sociology and Psychology of Management in Economic Systems on the Basis of Fuzzy Logic

Mammedzade Elnur$^{(\boxtimes)}$ 🆔, Harun Öztas 🆔, and Mirzayeva Kemale 🆔

Odlar Yurdu University, Baku AZ1072, Azerbaijan
elnur.memmedzade.97@bk.ru, harun_oz_tas@hotmail.com,
kama_ahmadova@live.ru

Abstract. When studying the problems of using social psychological methods of management in production and commercial enterprises, first of all, it is necessary to determine the scientific orientation of the sociological and psychological foundations of management. Management plays an important role in the modern stage of research of scientific directions, and in this field the study of the inner world of man, his worldview on the basis of the environment on the basis of fuzzy logic plays an important role. In the scientific sphere, this direction is called the sociology and psychology of management and defined as a field of social knowledge that studies the systems and processes of management in a society of social relations. The peculiarity of the sociology of management is that it belongs to the active sociology. The sociology of management allows for real change at different levels of the employee hierarchy. In short, the sociology of management covers the process of formation, activity and development of certain spheres of life, social change, social relations, patterns of social activity and the mechanism of behavior in the process of management. The sociology and psychology of management examines certain areas of the management system and includes a set of work on the selection, placement and formation of management staff. In this study, a fuzzy logic model was used to assess the employee's perception of the environment, and the results were applied to the input and output parameters in the Fuzzy Toolbox section of the Matlab program.

Keywords: Fuzzy logic · Economy · Sociology · Psychology · Management · Matlab

1 Introduction

It is known that any scientific direction is based on objective laws and specific laws, as well as reflected in separate principles.

The process of sociology management is no exception, and the basic laws of this field include the following [185]:

- The law of unity of social management conditioned by cultural, political and economic factors of social development;

- The law of proportionality, which provides for a rational ratio within the social system, governed and managed subsystems, as well as between them;
- The law of optimal ratio of centralization and decentralization functions of social management. The level of centralization of governance changes in the process of social development, and this acts as a law of social governance;
- The law on the participation of various segments of the population in social governance, increasing its effectiveness [1–3].
- As for the principles of social governance, they place certain demands on the system, structure, process and mechanism of social governance. In this regard, the principles of social management must meet the following requirements:
- It must be based on the laws of development of society, the social and economic laws of its management;
- It should be in line with the goals of social management, reflect its main features, connections and features;
- Exclude time and regional aspects of social governance processes;
- have a legal structure, or rather, be reflected in various normative documents.

The main principles of social management are:

- The principle of unity of state, economic and socio-psychological management;
- The principle of socio-psychological unity of the field and territory;
- Scientific principle of social management;
- The principle of optimal selection, training, placement and use of staff;
- The principle of economy and efficiency of social management;
- The principle of structure;
- the necessary principle of diversity and flexibility;
- The principle of direct and feedback.

This is not a complete list of social governance, as there are other principles related to social governance.

It should be noted that the object of this study is the socio-psychological methods of management in production and commercial enterprises, or rather in economic systems, based on fuzzy logic, so we are primarily interested in issues related to economic sociology.

The object of economic sociology is the study of socio-economic activity of individuals [4–7].

Socio-psychological economics and sociology are closely intertwined, so no economic activity can be carried out without taking into account the functional features of social structures, and in turn, no social activity can be carried out without calculating its economic effects.

Moreover, as a social phenomenon, the object of economic sociology is the economic system, its sources, evolution, and place in society.

From this point of view, the following can be included in the field of research of economic sociology: social regularities of economic development of the country; social efficiency of decisions made and economic efficiency of management decisions; issues of social relations in the internal spheres of production and commercial enterprises.

Economic sociology is aimed at diagnosing the interaction of economic and social spheres.

The conceptual nature of economic sociology can be defined on the basis of fuzzy logic by defining the boundaries that separate economic sociology from other spheres, such as economic theory and general sociology.

2 Fuzzy Logic

Fuzzy Logic, the concept of fuzzy logic, is a theory first introduced by Lotfi Zadeh. In 1965, the foundations of the idea of creating a functioning of rules and transferring it to the machine were laid by Zadeh in 1965 by making use of human life experiences and all kinds of knowledge. Fuzzy logic can be defined as a decision mechanism design in its simplest and simplest form. However, it should be taken as an example that looks more like a human than a machine.

Founder Lütfi Ali Askerzade from Baku, known as Zadeh, made the discovery during his studies in Berkeley University Electrical and Electronics Engineering Research Laboratory. It is a starting point that he published his studies in search of a solution to a technical problem in 1965 in the "Information and Control" magazine under the title "Fuzzy Sets". This new step, which will deeply shake perspectives, will soon gain success in highly progressive application areas in technology. This first step, which causes great changes, becomes a door to innovation in explaining many things that previously seemed inexplicable [8, 10].

When the working principles of classical logic are expressed mathematically, we come across a table consisting of "1" and "0" values. Here, an entity can belong to a particular set. Alternatively, it cannot belong to this cluster. While "true" is expressed as "1", "false" is evaluated as "0". On the other hand, it seems impossible that the natural, verbal and human things we mentioned at the beginning always fit into these categories. That is why classical sets, in other words, approaches of the understanding of definite sets are insufficient. At this point, the alternative, fuzzy logic, comes into play. Between the two main values "1" and "0", opportunities are also given to other possibilities. For hazy logic, 1 and 0 are seen as boundary regions, not absolute values.

3 Assessing an Employee's Worldview Based on Fuzzy Logic

The forthcoming XXI century has radically changed the worldview of man in the management system. The human imagination in modern times is characterized as follows Table 1: skilled and corporate person is preferred; the organization is seen as a living organism made up of staff united by common values; Enterprises must be characterized by constant, consumer-oriented change.

The most modern concept of human development should focus on the development of a creative personality, where staff training costs are seen not as labor costs, but as long-term investments that ensure the high quality of human resources necessary for the organization to prosper.

Table 1. Man's worldview as an object of management in production and commercial enterprises

The staff of the enterprise	Characteristics
Economic staff	The main incentive for any staff is high profits
Consumer staff	The main motive of work is the desire for status and power
Hierarchical staff	It is the freedom of individual choice for staff, social self-determination and advancement in the enterprise hierarchy
Professional staff	The main incentives are to participate in the work of the enterprise
Corporate staff	The staff must be included in the organizational mechanism of the enterprise

In short, it is advisable to test the employee's high level of professionalism, creativity, personal responsibility to people and society, self-management, honesty based on fuzzy logic.

Depending on which personnel management concept the management of the enterprise prefers, a personnel management system is established. The impact of the choice of personnel management concept on personnel policy is presented in Table 2 [5, 6].

Table 2. Personnel policy and personnel management concept

Concept	Staff costs	Staff-capital
Policy on organizational standards and personal factors	The organizational structure, job responsibilities and system of requirements for staff are well established	Creativity and initiative are preferred, the organization is ready for a flexible approach to staff management
Reward policy	An accurate system of remuneration that does not depend on the achievements of the employee	Differentiated payment of labor, everyone can earn as much as they can
Employment policy	Such hiring of cheap and even unskilled workers	Hiring expensive but highly skilled workers
Staff career policy	Search for managers outside the enterprise	Search for a leader in the enterprise
Stabilization policy within the organizational labor market	Search for people who will be fully satisfied with the existing working conditions in the organization	Minimization of staff turnover, development of motivation program
Achievement evaluation policy	Formal assessment of staff	Individual approach to staff evaluation

Modern experience shows that economic reform puts man first among other types of resource factors, so individual management is of great importance in anti-crisis management. At the heart of individual management are new approaches to the organization of management labor in the enterprise, its payment, stimulation of professional development, creativity, innovation, entrepreneurship, as well as the assessment of the employee's personal contribution to the final results of the enterprise [7, 8].

Thus, the criteria shown in Table 3 were applied on the basis of fuzzy logic to assess the employee's sociological and psychological worldview.

Table 3. Evaluation criteria based on fuzzy logic of the employee

Creativity	Professionalism	Responsibility
Poor	Poor	Poor
Average	Average	Average
Good	Good	Good
Very good	Very good	Very good
Excellent	Excellent	Excellent

Based on the above approaches to the evaluation category, a specific report was prepared based on the fuzzy logic of the individual elements.

4   Practical Application of Fuzzy Technologies

The first step in applying fuzzy logic to a system is to determine the inputs and outputs of the system. Given the employee's worldview, the most important expectation is to take into account the level of sociological and psychological optimality. To meet these expectations, the optimality parameters that make up the results of the fuzzy logic model are important. The inputs and outputs of the fuzzy logic model are penalized as shown (see Fig. 1). Membership function numbers, names, lower and upper limits of all parameters are determined by the effect of input and output parameters on the problem to be modeled. (see Fig. 1) shows the membership functions of the input parameters and (see Fig. 2) shows the lower and upper limits of the membership functions of the output parameters [8–10].

After determining the lower and upper limits of the membership functions and the parameters required to build the model, 74 rules were established to establish the necessary relationships between the parameters affecting the system (see Fig. 3).

The value of the efficiency factor is classified into one of five fuzzy sets: poor, average, good, very good and excellent.

This value is between [0–100]. When the process is complete, a list is created and can be used as a template.

Each variable, triangular or trapezoidal has five fuzzy values with interrelationships:

- For input and output variables - (Fig. 1, 2): poor, average, good, very good and excellent.

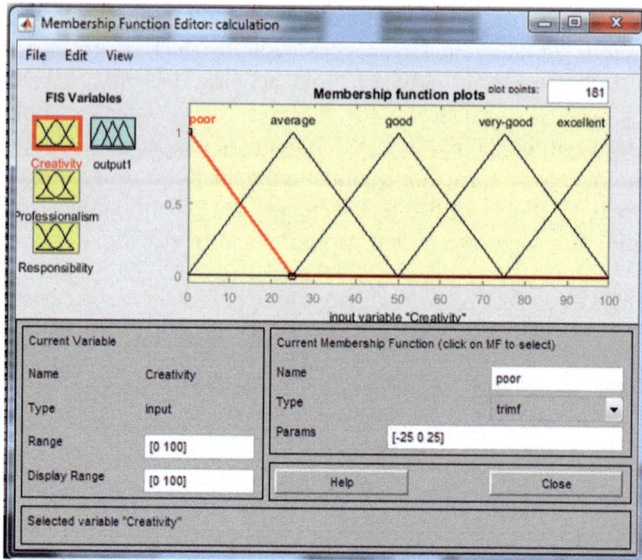

Fig. 1. Input variables

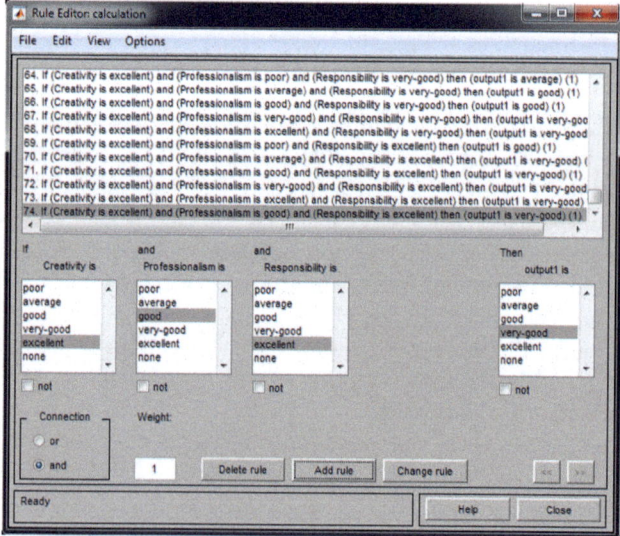

Fig. 2. Rules window

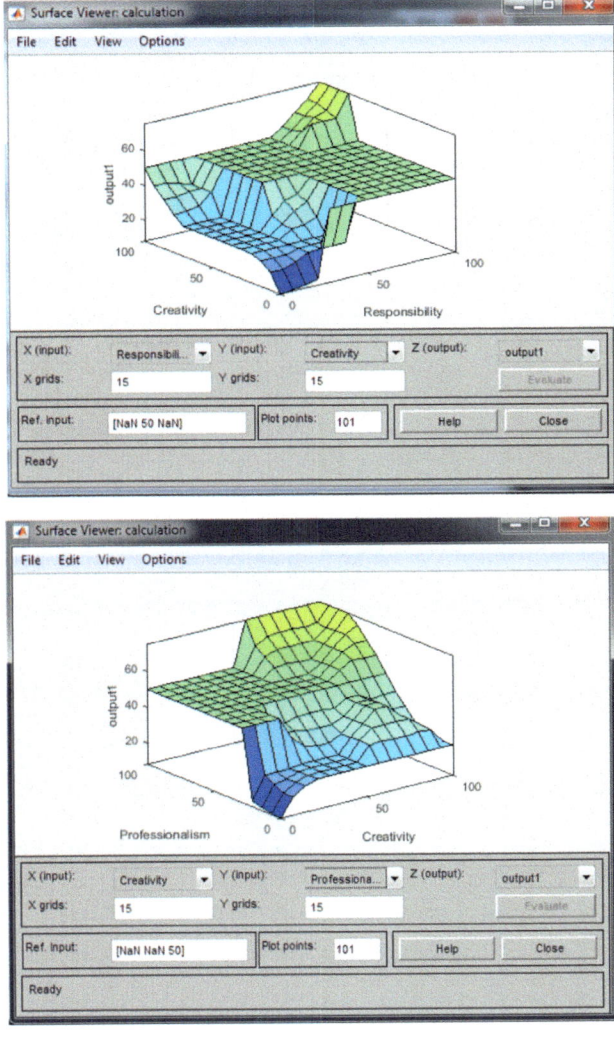

Fig. 3. Output variables.

5   Conclusion

The evaluation of alternatives in Table 1 of the MATLAB\Fuzzy Change Systems note described above allowed the following results to be obtained: Alternatively, the Optimality value was marked as weak, average, good, very good and excellent instead of weak and perfect.

The results are more rhetorical, showing the optimality of the employee's world-view, based on fuzzy complementary theory. On the other hand, given other simple but vague and difficult formation criteria, it is useful to use fuzzy set theory when determining a simple objective method based on perfect personnel selection.

References

1. Gubarets, M.A., Mazilkina, E.M.: Promotion and positioning in marketing (in Russian). Moscow (2018)
2. Grant, R.: Modern strategic analysis (in Russian). Per. from English SPb.: Peter (2018)
3. Burov, M.P.: State regulation of the national economy: modern paradigms and mechanisms of regional development (in Russian). Dashkov IK, Moscow (2018)
4. Samigulin, E.V.: Economic law and parameters of the retail product market. AUCA Acad. Rev. 117–125 (2008). (in Russian)
5. Petrova, A.V.: Analysis of the product market and assortment (in Russian). Yekaterinburg, UrGUPS (2016)
6. Novikova, I.S.: Analysis of methods for forecasting the turnover of a retail facility (in Russian). https://docplayer.ru/35883828-Analiz-metodov-prognozirovaniya-tovarooborota-torgovogo-obekta.html
7. Aliev, R.A., Bijan, F., Aliev, R.R.: Soft Computing and Its Applications in Business and Economics. Springer, Berlin Heidelberg (2004)
8. Beale, M.H., Hagan, M.T., Demuth, H.B.: Neural Network Toolbox. User's Guide. Math Works, Inc., Natick (2014)
9. Kruglov, V.: Mathematical extension packages MATLAB. Special reference, SPb, Piter (2001)
10. Zadeh, L.A.: Fuzzy logic, neural networks, and soft computing. Commun. ACM. **37**, 77–84 (1994)

Deep Learning-Based New Alloy Synthesis

Mustafa B. Babanli[✉] [iD]

Azerbaijan State Oil and Industry University, AZ1010 Baku, Azerbaijan
mustafababanli@yahoo.com

Abstract. Data-driven analytics offers enormous opportunities to accelerate the various stages of the materials synthesis ranging from discovery to manufacturing and deployment. Considering reliable performance of deep learning approaches due to recent advancements in various applications domain, this study is an attempt to explore the applicability of these machine learning techniques for alloy design and synthesis problem. Since the success of such data intensive approaches depends heavily on the data availability, a density tree-based semi-artificial data generation was implemented in this study to obtain the required data basis. A comparative analysis with an alternative machine learning algorithm revealed the superiority of the proposed deep neural networks for the examined alloy synthesis problem.

Keywords: Alloy synthesis · TiNi alloy · Machine learning · Deep learning · Semi-artificial data

1 Introduction

Conventional materials development is a complex and time-consuming process. Many attempts have been made with the purpose of accelerating the required timeline by integrating experimental methods, computational tools and data driven systems approaches [1]. The data-driven analytics is expected to advance materials research once the issues related to data decentralization, restricted access to required data and missing data standards are alleviated [2]. A large number of existing studies have already examined the applicability of the machine learning approaches for material synthesis problems and delivered promising results [3–8].

Conducting machine learning studies within the materials informatics projects, particularly for alloy synthesis problem, requires a thorough understanding of the underlying data structures, carrying out a robust modelling, ensuring a reliable deployment and finally conducting evaluation of applied systems.

This study proposes a deep learning supported alloy synthesis. To surmount the problem regarding the obtained small datasets due to limited number of costly experimentations, we apply a semi-artificial data generation method. After verifying the reliability of the obtained semi-synthetic datasets, the deep learning approach is applied to generate the relevant predictions.

The remainder of the study is structured as follows. Section 2 discusses the applied deep neural networks. Section 3 examines the datasets by introducing the relevant exploratory data analysis visualizations. Section 4 presents the obtained results by the

© The Author(s), under exclusive license to Springer Nature Switzerland AG 2021
R. A. Aliev et al. (Eds.): ICAFS 2020, AISC 1306, pp. 467–471, 2021.
https://doi.org/10.1007/978-3-030-64058-3_58

applied deep learning approach and conducts a comparative analysis with an alternative technique. Section 5 concludes the study.

2   Data-Driven Analytics with Deep Learning

Deep learning techniques comprise multiple layers of information processing stages to learn representations of the data for pattern classification tasks [9]. Although these methods have mainly found their application in machine translation, sentiment analysis, image, video, audio and speech recognition tasks, the recent applications for the manufacturing intelligence have also provided promising results [10]. In this article, the deep neural networks were applied to address the underlying regression problems in the alloy synthesis problem. We deploy the deep maxout feedforward neural networks by using supervised training scheme [11]. The advantage of the maxout neural networks has been documented for various applications [12]. This type of network uses a new type of activation function, maxout unit:

$$f(\alpha 1, \alpha 2) = max(\alpha 1, \alpha 2) \qquad f(\cdot) \in R$$

where α is the weighted combination $\alpha = \sum_i wixi + b$, xi and wi denote the input values and their weights. In addition to learning the relationships between the hidden layers, the maxout technique learns the activation functions of each hidden unit. Maxout networks provide especially higher predictive performance when combined with the dropout technique. Dropout is a recent and promising technique to train the deep neural networks particularly aimed to avoid the overfitting [13]. Furthermore, to avoid the overfitting problem, we adopt the early stopping and L2 weight decay regularization technique which is also known as Tikhonov regularization [14]. In order to find out the suitable amount of the epochs, we force the model building to stop if the Root Mean Squared Error (RMSE) improves by less than 0.01 between scoring epochs. Stochastic gradient descent was used to minimize the chosen loss function and the backpropagation was used to compute the gradients. Furthermore, the adaptive learning rate algorithm was implemented.

3   Experiments and Data

This study attempts to predict three output variables y_1, y_2, y_3 of TiNi alloy (y_1 – conventional ultimate strength, MPa; y_2 – conventional yield strength, MPa; y_3 – unit elongation, %) by using six inputs, namely, *Ni, Ti* (alloy composition, %), *a, b, c* (lengths of the unit cell edges, nm) and β (the angle between sides with length a and b).

Dataset collected from the alloy synthesis process comprises 17 data instances that leads to complications for training, verifying and validating models. In view of this, we generate semi-artificial data by using the data generator approach [15] which constitutes on the density trees [16]. Furthermore, considering its suitability for the regression problem as in the underlying case we adopt this approach to generate 10,000 new instances. The fragment of these data is shown in Table 1.

Table 1. Semi-artificial data for prediction

#	Ti	Ni	a	b	c	β	y_1	y_2	y_3
1	0.494	0.506	0.290	0.412	0.465	97.31	983	247	49.40
2	0.508	0.492	0.290	0.411	0.465	97.34	982	218	49.91
3	0.500	0.500	0.289	0.412	0.465	97.84	978	335	50.20
4	0.500	0.500	0.290	0.412	0.464	96.80	983	235	49.30
5	0.500	0.500	0.287	0.411	0.464	96.80	982	235	50.20
6	0.503	0.497	0.291	0.412	0.461	96.80	987	235	49.40
7	0.503	0.497	0.289	0.412	0.462	97.08	981	227	49.40
8	0.503	0.497	0.291	0.412	0.460	96.98	984	221	49.40
				⋮					
9994	0.503	0.500	0.287	0.411	0.465	97.11	981	221	50.20
9995	0.497	0.507	0.288	0.412	0.466	97.60	987	296	48.36
9996	0.503	0.500	0.289	0.412	0.460	97.40	981	218	49.40
9997	0.500	0.500	0.287	0.411	0.464	97.90	981	235	50.20
9998	0.507	0.500	0.289	0.412	0.466	97.60	990	296	48.00
9999	0.508	0.495	0.287	0.412	0.465	96.80	981	235	49.40
10000	0.495	0.500	0.289	0.415	0.464	97.10	987	339	47.90

A quick check of the statistical properties suggests that semi-artificially generated dataset is similar to the original dataset. A brief statistical summary of each predictor is presented in the Table 2.

Table 2. Statistical summary of the input variables

Measure	Ti	Ni	a	b	c	Beta
Min.	0.493	0.4850	0.2870	0.4108	0.46	96.4
1st Qu.	0.4987	0.4980	0.2880	0.4117	0.4622	96.8
Median	0.5000	0.5000	0.2889	0.4120	0.4635	97
Mean	0.5005	0.4997	0.2889	0.4124	0.4633	97.09
3rd Qu.	0.5030	0.5012	0.2896	0.4124	0.464	97.35
Max.	0.5150	0.5080	0.2909	0.4150	0.466	97.9

4   Results

In this section, we introduce the evaluation results obtained from the Deep Maxout Networks with Dropout for predicting outputs y_1, y_2, y_3. A comparative analysis with an alternative machine learning technique referred to as Random Forest is conducted. As an evaluation measure, RMSE is used. To assess the performance of the applied deep learning technique we followed the hold-out method. 90% of the data was used to train the model and the remaining 10% was used to test the model to examine the performance in the unseen data. The obtained results are shown in Table 3.

Table 3. RMSE-based evaluation results for outputs y_1, y_2, y_3

Model	Output 1	Output 2	Output 3
Random forest	2.618872	35.8184	0.56960
Deep maxout networks with dropout	2.429634	32.8347	0.5099

As one can see, the applied deep learning approach outperforms the random forest approach.

5   Conclusion

This study examines the applicability of the deep neural networks for alloy synthesis problem after enriching the real experiment data by generating new data instances after learning the interdependencies among features and output variables in the original data. The obtained results revealed the strong predictive performance of the proposed approach by outperforming an alternative technique. For the future work, we aim to investigate the applicability of various other deep learning architectures such as long-short term memory networks or convolutional neural networks. Furthermore, to examine the reliability, consistency, and robustness of the delivered prediction outcomes for alloy design and synthesis, we pursue the objective to conduct extensive evaluation studies not only in lab environment but also through field studies for the chosen use-cases.

References

1. White, A.: The materials genome initiative: one year on. MRS Bull. **37**(8), 715–716 (2012)
2. Hill, J., Mulholland, G., Persson, K., Seshadri, R., Wolverton, C., Meredig, B.: Materials science with large-scale data and informatics: unlocking new opportunities. MRS Bull. **41** (5), 399–409 (2016)
3. Babanli, M.B.: Fuzzy logic and fuzzy expert system-based material synthesis methods. In: Volonesku, C. (ed.) Fuzzy Logic. IntechOpen (2020) https://doi.org/10.5772/intechopen. 84493
4. Babanli, M.B., Huseynov, V.M.: Z-number-based alloy selection problem. Proc. Comput. Sci. **102**, 183–189 (2016)
5. Castelli, I.E., Jacobsen, K.W.: Designing rules and probabilistic weighting for fast materials discovery in the Perovskite structure. Model Simul. Mater. Sc. **22**(5), 055007 (2014). https://doi.org/10.1088/0965-0393/22/5/055007
6. Pilania, G., Wang, C., Jiang, X., Rajasekaran, S., Ramprasad, R.: Accelerating materials property predictions using machine learning. Sci. Rep-UK **3**(1), 1–6 (2013)
7. Takahashi, K., Tanaka, Y.: Materials informatics: a journey towards material design and synthesis. Dalton T. **45**(26), 10497–10499 (2016)

8. Babanli, M.B.: Fuzzy Logic-Based Material Selection and Synthesis, p. 276. World Scientific, Singapore (2019)
9. LeCun, Y., Bengio, Y., Hinton, G.: Deep learning. Nature **521**(7553), 436 (2015)
10. Wang, J., Ma, Y., Zhang, L., Gao, R.X., Wu, D.: Deep learning for smart manufacturing: Methods and applications. J. Manuf. Syst. **48**, 144–156 (2018)
11. Candel, A., Parmar, V., LeDell, E., Arora, A.: Deep learning with H2O. H2O Inc. (2016)
12. Goodfellow, I.J., Warde-Farley, D., Mirza, M., Courville, A., Bengio, Y.: Maxout networks (2013). http://proceedings.mlr.press/v28/goodfellow13.pdf. Accessed 1 Aug 2018
13. Srivastava, N., Hinton, G., Krizhevsky, A., Sutskever, I., Salakhutdinov, R.: Dropout: a simple way to prevent neural networks from overfitting. J. Mach. Learn. Res. **15**(1), 1929–1958 (2014)
14. Tikhonov, A.N.: On the stability of inverse problems. Dokl. Akad. Nauk SSSR **39**, 195–198 (1943)
15. Robnik-Sikonja, M., Robnik-Sikonja, M.M.: Package 'semiArtificial' (2019)
16. Ram, P., Gray, A.G.: Density estimation trees. In: Proceedings of the 17th ACM SIGKDD International Conference on Knowledge Discovery and Data Mining, pp. 627–635 (2011)

Decoding Secret Message with Frequency Analysis

Yucel Inan$^{(\boxtimes)}$ (iD)

Department of Computer Engineering, Near East University,
Nicosia, Mersin 10, TRNC, Turkey
yucel.inan@neu.edu.tr

Abstract. In this study, Cryptography systems are emphasized in order for the messages to be encrypted, transmitted and decoded according to the specific system. Caesar Cryptography is one of the easy to solve ciphers in Cryptography. For the frequency analysis of an encryption method, the basic operating principle of the encryption method is based on the frequency of use of letters in the alphabet to which the plain text to be encrypted belongs. In addition, the study aims to reveal the weaknesses and strengths of the software by examining the reliability of the letters against frequency analysis for the tests and the results are evaluated on the graphs and tables.

Keywords: Caesar cipher · Cryptanalysis · Ciphered text · Frequency analysis attacks · Letter weakness and strength of the ciphered text · Python

1 Introduction

Security is a necessary element for human beings to survive [1]. Privacy is also a prerequisite for ensuring security. That is why security and privacy are two concepts that complement each other. The existence of a society is only proportional to the security it can provide. As security decreases, the threat and danger increase. The essence of security is based on "knowledge". In its broadest definition, it is called Cryptology, which is the combination of all the open and confidential messages made for the purpose of securing the security, the secure transmission of the messages and the deciphering of the transmitted messages, the logos which means secret in Greek, meaning Cryptos and science [2]. Information is becoming a more valuable issue today. The general problem of communication and informatics is information security. Virtual shopping, individual banking transactions and e-mail traffic over the Internet force the Internet to become a safer environment. This is to focus on how to achieve a high level of information and communication security. This is the Cryptography science that will provide. The aim here is to understand how to use letter frequency analysis to break certain ciphers. Monogram Frequency counts, Caesar ciphers type ciphers are more effective. The same plain letters are encoded in the same cipher letter. Although the letters have changed, the base letter frequencies do not change. If the plain letter is five frequencies, its cipher letter becomes 5 frequencies. This article aims to test Cryptographic reliability of the Caesar encryption algorithm. The Caesar cipher is easily solved by detecting the frequency of occurrence of each character and then

R. A. Aliev et al. (Eds.): ICAFS 2020, AISC 1306, pp. 472–479, 2021.
https://doi.org/10.1007/978-3-030-64058-3_59

comparing its own frequency to the frequency of the letter in the original message language, observing what it represents. The reliability of the comparisons between ciphers letter and plain letters were analyzed using the Python 2.7.15 language.

Cryptography is all of the techniques used to transform readable information into a form that cannot be read by undesired parties [3]. In Fig. 1, the algorithm "This is a secret message" of the article is applied with the Cryptographic algorithm which is the subject of the article. The purpose of Cryptography is to ensure and protect the confidentiality of important information [4].

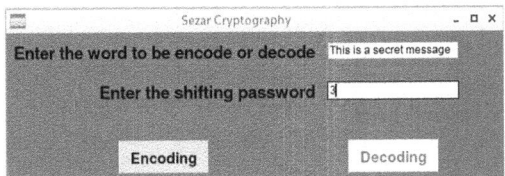

Fig. 1. "Wklv lv d vhfuhw phvvdjh" cipher text is generating when the plain alphabet is shifted to 3-character left.

2 Related Works

Cryptanalysis has been the subject of many investigations. One of the studies published was that [5]. Emphasis is placed on the use of Genetic Algorithms in the Cryptanalysis of classical ciphers. Another publication in, Genetic Algorithm approach was applied for statistical analysis of English language documents in 2015 In [6] author. This publication has implemented encryption and decryption Caesar cipher using the Neural Network in 2014 [7]. This work Using Artificial Neural Network, it provides security for strength encryption against attacks [8].

3 Cryptanalysis

Cryptanalysis is the method of finding the correct text using some techniques from the encrypted, meaningless text [9]. Various methods have been tried to reach the hidden information. Some of these methods are attack techniques that are used against the Cryptographic methods, These are Known Plain text Attacks, Chosen Plain text Attacks, Chosen Cipher text Attacks, Brute Force Attacks, Letter Frequency Attacks, Man in the middle Attacks, Differential Attacks [10].

3.1 Cryptanalaysis of a Caesar Cipher Algorithm

Cryptanalysis, the science of deciphering the encrypted message, has emerged in an effort to break the simple ciphers. Simple cipher systems to solve all the cipher alphabets in an easier way to solve mathematics, statistics and linguistics should have sufficient knowledge of the fields. It is important to carry out important studies in these areas required for Cryptanalysis [11]. The fact that these disciplines were developed led to the discovery of Cryptanalysis by the famous Arab scientist and the first works in this field were written. The first of these is *"Manuscript on Deciphering Cryptographic Messages"* written by Al-Kindi, who lived in the 9th century and is known as the philosopher of Arabs. The first time on Cryptanalysis was the concept of frequency analysis. Al-Kindi technique to solve an encrypted message written in the same language long enough to find a text and calculate the frequency of each letter is necessary to use. The most commonly used letter in the text corresponds to the most commonly used letter in the encrypted message. The same is done for other letters in the order. After this is finished, the letters in the message will appear. Al-Kindi called this Cryptanalysis method frequency analysis [12].

3.2   Calculation of Letter Count

The algorithm was tested using the English text. The text used for the letter frequency consists of all the works of the cipher text. By using python language, this plain text is converted to cipher text by Caesar's cipher method. The spaces in cipher text are calculated without changing the punctuation marks. The count letter of Frequency and Analysis for Caesar Cipher Method has the following functions. Cipher Text Field, Letter Frequency and Letter Frequency Analysis, Delete button, Exit button. Figure 2 is shown Frequency letter count of cipher text.

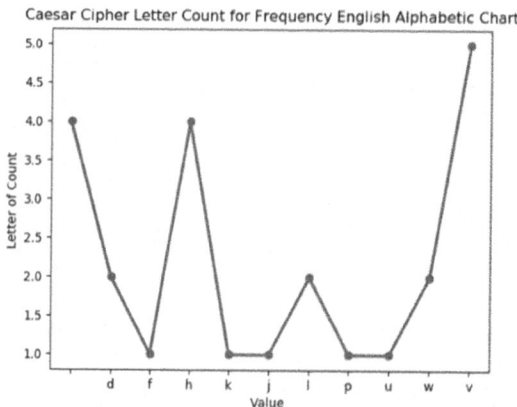

Fig. 2. The relative number of letters in the y axis, the letters used in the x axis are shown.

Cipher text "Wklv lv d vhfuhw phvvdjh" is written in the top box. Clicking on the Letter Frequency button will generate a graph showing the letter frequency above. At this stage, the length of the cipher, the frequency of character formation, working time, interface are presented. This letter frequencies will help you to decode the cipher text. If the language of the message is known, the most common letter in that language and the cipher in the hand is replaced with the most common letter in the assumption that the cipher can be solved.

3.3   Letter Frequency Analysis

One of the Cryptanalysis methods is to detect the character frequency in the cipher text. Therefore, in order to test the cipher text, the results were analyzed. Frequency analysis is the process of counting the number of different encrypted text characters to decrypt the information. The English alphabet also has 26 letters. However, each one does not appear to be equal in the English text. E, T, A, O, I, N are the six most commonly used letters in English. V, K, J, X, Q, Z are rarely used in English text. As the different letters of the alphabet appear often than others, we can perform some frequency analysis of an encrypted message to help us encrypt it (Fig. 3). The English letters are in the frequency values expressed in Table 1, the following frequency values are taken from Table 1, the website and also the frequency result of the cipher text are shown on Table 2.

Table 1. Relative frequencies of English alphabet letters values.

Letter	a	b	c	d	e	f	g	h	i	j	k	l	m	n	o	p	q	r	s	t	u	v	w	x	y	z
Freq %	8.1	1.4	2.7	4.2	12.7	2.2	2.0	6.0	6.9	0.1	3.8	4.0	2.4	6.7	7.5	1.9	0.1	5.9	6.3	9.2	2.7	1.0	5.3	0.2	4.0	0.1

Table 2. Letter frequencies values used in ciphered text.

Letter	a	b	c	d	e	f	g	h	i	j	k	l	m	n	o	p	q	r	s	t	u	v	w	x	y	z
Freq	0	0	0	2	0	1	0	4	0	1	1	2	0	0	0	1	0	0	0	0	1	5	2	0	0	0
Freq %	0.0	0.0	0.0	8.0	0.0	4.0	0.0	16.0	0.0	4.0	4.0	8.0	0.0	0.0	0.0	4.0	0.0	0.0	0.0	0.0	4.0	20.0	4.0	0.0	0.0	0.0

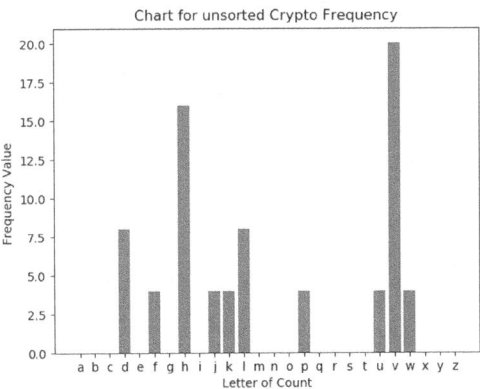

Fig. 3. Ciphered text frequency analysis is shown in y axes, relative frequency percentage.

This designed ciphered text is calculated from frequency analysis on the graphical user interface. The graph is compiled by taking the English written text of the cipher.

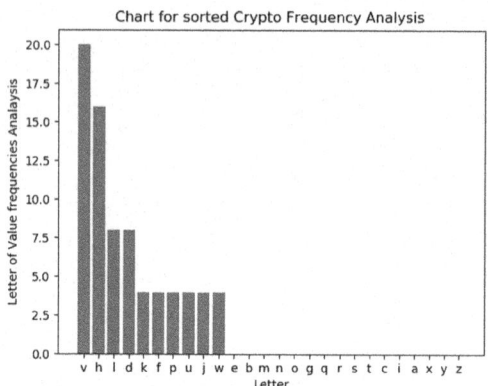

Fig. 4. Ciphered text frequency analysis is shown in y axes, relative frequency percentage.

When "Frequency Analysis" button is clicked, there will be a graph showing the Fig. 4 frequency analysis of the above Figure. This illustrates the graphical formation, the 24 characters of the cipher text, the length of the word, the frequency of the use of letters, the run time is presented in this graphical interface. At this stage, the formula = (f/n) * 100 formula is used. f: Frequency of the letter, n: The total number of letters in the cipher text is n = 24. For example, the letter v is displayed 5 times, the frequency in the formula is the only visible number of the letter v, i.e. (5/24) * 100 = 20. The letter 'E' is the most common letter used in the English Language. The most common letter used in the ciphered text can also be matched with 'V' or not matched. Some predictions can do made, and the text can be decoded by checking whether the encrypted text is meaningful or not. We compared the most commonly used letters in Table 1 of the original plain text with the current original in order to determine whether the letters v, h, l, d would create the most value according to how often the letters can be used, as shown in Table 3, 4, 5 and 6, according to Fig. 4. As a result, one or two trials can be obtained. The strengths and weaknesses of the algorithm are tested, and the results are shown on graphic.

Table 3. Letter frequency analysis matching estimation.

Matching1	e ⟶v	t ⟶h	a⟶ d	o⟶ l	i ⟶f	n⟶j
Matching2	e⟶h	t ⟶v	a⟶ d	o⟶ f	i ⟶l	n⟶j

Table 4. Matching1 ciphered text is "Wklv lv d vhfuhw phvvdjh"

W	k	l	v		l	v		d		v	h	f	u	h	w		p	h	v	v	d	j	h
			e			e		a		e									e	e			
		i			i			a			e			e				e			a		e

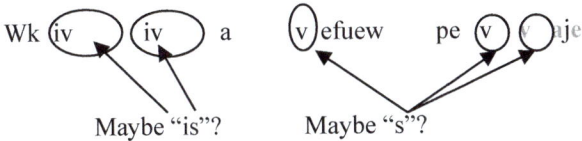

Wk iv iv a   vefuew   pe v vaje

Maybe "is"?   Maybe "s"?

Table 5. Matching2

W	k	l	v		l	v		d		v	h	f	u	h	w		p	h	v	v	d	j	h
		i	s		i	s		a		s	e			e				e	s	s	a		e

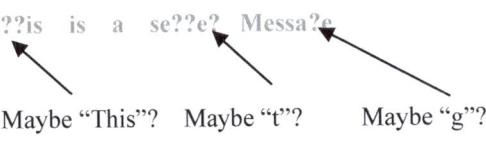

W k is is a se fu et p essag e

Maybe "m"?

Table 6. Matching3

W	k	l	v		l	v		d		v	h	f	u	h	w		p	h	v	v	d	j	h
		i	s		i	s		a		s	e			e			m	e	s	s	a		e

??is is a se??e? Messa?e

Maybe "This"?   Maybe "t"?   Maybe "g"?

This is a se??et Message

Maybe "secret"?

This is a secret Message

3.4   Program Code

In the below Code, this code purpose of the given is to apply Letter Frequency Analysis. Python code, Analysing the Count Letter and Frequency of Encrypted Message.

```
Program Function letter_Frequency(event=None):
    k=""
    for i in message:
        if i.isupper():
            i=i.lower()
        if i.isdigit():
            k=k+str(i)
            continue
        if i<0:
            k=k+""
            continue
        k=k+i
    d={}
    for i in k:
        if i not in d:
            d[i]=1
        else:
            d[i]+=1
    freq=[]
    for i in range(26):
        freq.append(0)
    for i in range(len(k)):
        if k[i] >='a' and k[i]<='z':
            sayi=ord(k[i])-ord('a')
            freq[sayi]=freq[sayi]+1
          alphabet=['a','b','c','d','e','f','g', 'h','i','j','k
                ,'n','o','p','q','r','s','t','u','v','w','x','y','z']
    n=len(message)+0.0
    freq_list=[]
    for x in freq:
        freq_list.append((x/n)*100)
    for i in range(len(freq_list)):
        for j in range(len(freq_list)):
            if freq_list[i]>freq_list[j]:
                temp=freq_list[i]
                freq_list[i]=freq_list[j]
                freq_list[j]=temp
                tempc=alphabet[i]
                alphabet[i]=alphabet[j]
                alphabet[j]=tempc
    list1=[]
    list2=[]
    for i in range(len(alphabet)):
        list1.append(alphabet[i])
        list2.append(freq_list[i])
```

4 Conclusions

In this study, frequency analysis in the Caesar cipher method was examined. A letter frequency attack is applied to evaluate the encryption strength of the method. Longer cipher text give a better approximation to the letter frequency of the original language used, and as a result, each character of the cipher text is determined by comparing the frequencies of the letters in the cipher text and in the original language. For this reason, the encryption method can only guarantee the security of the data for a limited period of time.

References

1. Burman, S.: Cryptography and security - future challenges and issues. In: 15th International Conference on Advanced Computing and Communications (ADCOM 2007), Guwahati, Assam, pp. 547–551 (2007)
2. Gunnels, P.E.: The mathematics of cryptology. http://people.math.umass.edu/~gunnells/talks/crypt.pdf. Accessed 26 July 2019
3. Inan Y.: Comparing image distortion of LSB. In: Aliev, R., Kacprzyk, J., Pedrycz, W., Jamshidi, M., Sadikoglu, F. (eds) 13th International Conference on Theory and Application of Fuzzy Systems and Soft Computing - ICAFS-2018. Advance Intelligent System Computing, vol. 896. Springer, Cham (2018)
4. Purnama, B., Ah, H.R.: A new modified caesar cipher cryptography method with legible ciphertext from a message to be encrypted. Proc. Comput. Sci. **59**, 195–204 (2015)
5. Al-Janabi, S.T., Al-Khateeb, B., Abd, A.J.: Intelligent techniques in Cryptanalaysis: review and Future Directions. UHD J. Sci. Tech. **1**(1), 1–10 (2017)
6. Bhasin, H., Khan, A.B.: Cryptanalysis using soft computing techniques. J. Comput. Sci. Appl. **3**(2), 52–55 (2015)
7. Singh, J., Yadav, S.S.: Implementation of caesar cipher and chaotic neural network by using matlab simulator. Int. J. Recent Dev. Eng. Tech. **2**(6), 16–20 (2014)
8. Volna, E., Kotyrba, M., Kocian, V., Janosek, M.: Cryptography based on neural network. In: Proceedings of 26th European Conference on Modelling and Simulation, Koblenz, Germany, pp. 386–391 (2012)
9. Sun, Y., Chen, L., Xu, R., Kong, R.: An image encryption algorithm utilizing julia sets and hilbert curves. PLoS One **9**(1), 1–9 (2014)
10. Patni, P.: A poly alphabetic approach to ceasar cipher algorithm. Int. J. Comput. Sci. Inform. Tech. **4**(6), 954–959 (2013)
11. Mayur, T., Saraswat, L.: A review on common encryption techniques to brute force shielded technique: honey encryption. Int. J. Sci. Research Dev. **3**(12), 203–204 (2016)
12. Jothy, K.A., Sivakumar, K., Delsey, M.J.: Efficient cloud computing with secure data storage using AES and PGP algorithm. Int. Comput. Sci. Inform. Tech. **8**(6), 582–585 (2017)

Modeling the Phase Diagram of the Tl$_9$SmTe$_6$-Tl$_4$PbTe$_3$-Tl$_9$BiTe$_6$ System

S. Z. Imamaliyeva[1] , G. I. Alekberzade[2], A. N. Mamedov[1,3(✉)] ,
D. B. Tagiev[1], and M. B. Babanly[1]

[1] Nagiyev Institute of Catalysis and Inorganic Chemistry of ANAS,
Baku, Azerbaijan
samira9597a@gmail.com, asif.mammadov.47@mail.ru,
dtagiyev@rambler.ru, babanlymb@gmail.com
[2] National Aerospace Agency of the Republic of Azerbaijan, Baku, Azerbaijan
alakbarzadegi@gmail.com
[3] Azerbaijan Technical University, Baku, Azerbaijan

Abstract. Using the Multipurpose Genetic Algorithm (MGA), analytical models of phase diagrams of the Tl$_9$SmTe$_6$–Tl$_8$Pb$_2$Te$_6$, Tl$_9$SmTe$_6$–Tl$_9$BiTe$_6$ and Tl$_8$Pb$_2$Te$_6$–Tl$_9$BiTe$_6$ systems were obtained as temperature dependence on the composition. The boundaries of the uncertainty band for the liquidus surface of crystallization of solid solutions and the TlSmTe$_2$ compound, as well as the solidus of solid solutions, are determined. According to the model of regular solutions of non-molecular compounds, the thermodynamic functions of mixing solid solutions depending on the composition and temperature are determined. It was found that solid solutions of based on the compounds Tl$_9$SmTe$_6$, Tl$_8$Pb$_2$Te$_6$ and Tl$_9$BiTe$_6$ have thermodynamic stability in the entire concentration range. The analytical multi-3D model of the phase diagram of the Tl$_9$SmTe$_6$-Tl$_8$Pb$_2$Te$_6$-Tl$_9$BiTe$_6$ system was determined and visualized for both the quasi-ternary and nonquasi-ternary high-temperature part of the TlSmTe$_2$ crystallization.

Keywords: Phase diagram · Tl$_9$SmTe$_6$-Tl$_4$PbTe$_3$-Tl$_9$BiTe$_6$ system ·
Multipurpose genetic algorithm · 3D modeling

1 Introduction

Binary and complex chalcogenides of heavy p-elements have several functional properties, such as thermoelectric, photoelectric, optical [1–3]. Moreover, the discovery of topological insulator properties in some layered chalcogenide phases makes them prospective materials for spintronics and quantum calculations [4–9]. The introduction into the crystal lattice of the atoms of magnetic elements of the above-indicated phases leads to an improvement of their functional properties and appearance magnetic properties in them [10]. Currently, the search for new functional materials is carried out in the direction of complicating the composition of known compounds by obtaining solid solutions, doped phases, composites, etc. [4, 8, 9]. Thallium subtelluride Tl$_5$Te$_3$, which crystallizes in a tetragonal structure (Sp.Gr. I4/mcm), is a suitable matrix compound for the fabrication of new complex chalcogenide materials [11]. There are many ternary analogs of this compound of Tl$_4$A^{IV}Te$_3$ and Tl$_9$B^{V}Te$_6$ types (A^{IV}-Sn, Pb, Cu, Mo; B^V-Sb, Bi) [12–14]. These compounds and complex phases based on them also have a number of unique functional properties, which makes them promise for use in various fields of modern technology [15–20]. In [21, 22], new structural analogs of the Tl$_5$Te$_3$ compound of the Tl$_9$LnTe$_6$ type (Ln-Ce, Nd, Sm, Gd, Tb) were obtained, their melting nature and melting points, as well as crystallographic parameters, were determined. Further studies showed that these compounds have thermoelectric and magnetic properties [23, 24]. The development of new multi-component phases of variable composition is requires the study of phase equilibria in systems composed of structural analogs of binary and ternary compounds, since the formation of solid solutions in them is expected [12, 25]. In order to obtain new phases of variable composition with a structure of the Tl$_5$Te$_3$ type, we studied phase equilibria in a number of systems consisting of Tl$_5$Te$_3$ and its ternary analogs [26–29], in which continuous substitutional solid solutions were revealed. The constructed phase diagrams serve as the basis for choosing the composition of melts for growing single crystals of solid solutions with a given composition by directional crystallization.

The purpose of this work is the 3D modeling of the phase diagram of the Tl$_9$SmTe$_6$-Tl$_4$PbTe$_3$-Tl$_9$BiTe$_6$ system, including the boundary systems Tl$_9$SmTe$_6$-Tl$_4$PbTe$_3$, Tl$_4$PbTe$_3$-Tl$_9$BiTe$_6$ and Tl$_9$SmTe$_6$-Tl$_9$BiTe$_6$, using the uncertainty principle for heterogeneous equilibria.

2 Objects and Equations for Modeling

To modeling the Tl$_9$SmTe$_6$(1)-2Tl$_4$PbTe$_3$(2)-Tl$_9$BiTe$_6$(3) ternary system, is necessary analytical approximation of the liquidus and solidus curves of the boundary systems Tl$_9$SmTe$_6$(1)-2Tl$_4$PbTe$_3$(2), Tl$_9$SmTe$_6$(1)-Tl$_9$BiTe$_6$(3) и 2Tl$_4$PbTe$_3$(2)-Tl$_9$BiTe$_6$(3). The lead-containing ternary compound was taken as a2Tl$_4$PbTe$_3$ dimer in order to equalize the number of atoms in molecules. The experimental data of [26, 29] we used for modeling these systems According to the results of this work, these cross sections are characterized by the formation of continuous solid solutions (δ phase) with the Tl$_5$Te$_3$ structure. The Tl$_9$SmTe$_6$(1)-2Tl$_4$PbTe$_3$(2), Tl$_9$SmTe$_6$ (1)-Tl$_9$BiTe$_6$ (3) systems are non-quasi-binary in a certain part due to the incongruent nature of the melting of the Tl$_9$SmTe$_6$ compound. As a result, in a wide range of concentrations (up to 60 mol% Tl$_9$SmTe$_6$), the TlSmTe$_2$ compound crystallizes from the melt, which leads to the formation of two- (L+TlSmTe$_2$) and three-phase (L+TlSmTe$_2$+δ) fields. Below the solidus, all boundary systems are quasi-binary.

For the analytical description of the liquidus and solidus of the Tl$_9$SmTe$_6$(1)-2Tl$_4$PbTe$_3$(2), Tl$_9$SmTe$_6$(1)-Tl$_9$BiTe$_6$(3) and 2Tl$_4$PbTe$_3$(2)-Tl$_9$BiTe$_6$(3)) systems, the uncertainty principle for heterogeneous equilibria were used in [30–32]. Due to the monotonic dependence of the temperature of liquidus and solidus on the composition for the Tl$_9$SmTe$_6$(1)-2Tl$_4$PbTe$_3$(2) and Tl$_9$SmTe$_6$(1)-Tl$_9$BiTe$_6$(3) systems, the following equations were used:

$$T(liquidus) = a + bx + (c \pm \Delta)x(1 - x) \tag{1}$$

$$T(solidus) = a + bx + (d \pm \Delta)x(1 - x) \tag{2}$$

Due to the more complex dependence of the temperatures of liquidus and solidus on the composition for the $2Tl_4PbTe_3$ (2)-Tl_9BiTe_6 (3) system, the following equations were used:

$$T(liquidus) = a + bx + (c_0 + c_1x + c_2x^2)x(1 - x) \tag{3}$$

$$T(solidus) = a + bx + (d_0 + d_1x + d_2x^2 + d_3x^3)x(1 - x) \tag{4}$$

Coefficients a and b are determined based on the melting temperatures of Tl_9SmTe_6, Tl_4PbTe_3 and Tl_9BiTe_6. The coefficients c and d have intervals that are associated with the experimental error. For the liquidus and solidus surfaces of the ternary system $Tl_9SmTe_6(1)$-$2Tl_4PbTe_3(2)$-Tl_9BiTe_6 (3), the following equations were used:

$$T(liquidus) = yT_{liq}(1 - 2) + (1 - y)T_{liq}(1 - 3) + y(1 - y)(1 - x)T_{liq}(2 - 3) \tag{5}$$

$$T(solidus) = yT_{sol}(1 - 2) + (1 - y)T_{sol}(1 - 3) + by(1 - y)(1 - x)^2 \tag{6}$$

Here 1-Tl_9SmTe_6; 2-($2Tl_4PbTe_3$); 3-Tl_9BiTe_6. x is the mole fraction of the component 1-Tl_9SmTe_6, $y = x_2/(1 - x)$; $(1 - y) = x_3/(1 - x)$; x_2, x_3-mole fraction of compounds 2 and 3.

To calculate the free Gibbs energy of formation of solid solutions of boundary systems, the regular solutions model was used, which has been successfully tested in [32]:

$$\Delta G_T^0 = (a + bT)x^m(1 - x)^n + RT[pxln(x) + q(1 - x)\ln(1 - x)] \tag{7}$$

In Eq. (7), the first term represents the enthalpy of mixing of solid solutions in an asymmetric version of the regular solutions model. For solid solutions with unlimited solubility, the mixing parameter is $a < 0$; $b > 0$. The second term is the configurational entropy of formation of solid solutions from ternary compounds [33]; p and q- the number of different atoms in compounds. $R = 8.314$ J mol^{-1} K^{-1}.

3 Solution of Phase Equilibrium Equations by Using the MGA

For determining the coordinates of the boundary of heterogeneous equilibrium, a liquid alloy-solid solution in the systems $Tl_9SmTe_6(1)$-$2Tl_4PbTe_3(2)$, $Tl_9SmTe_6(1)$-$Tl_9BiTe_6(3)$ and $2Tl_4PbTe_3(2)$-$Tl_9BiTe_6(3)$ the method of the multipurpose genetic algorithm (MGA) [34] was used: the search interval for the coordinates of the phase equilibrium was chosen based on the melting temperatures of ternary pure compounds.

MGA changes the temperature ($T_i^m < T < T_j^m$) and the mole fraction of the base component x, taking into account the fact that they describe the uncertainties of the experimental DTA data. A fuzzy logic scheme considers and scales all objective values from 0 to 1. The genetic algorithm is run until all members of the population have reached 1 or reached a state in which the discrepancy does not exceed the error in determining temperatures (T = 1 ÷ 5 K) and mol fraction (x = 0.01 ÷ 0.02). The experimental values of DTA are in the middle of the uncertainty regions of the liquidus and solidus curves of quasi-binary systems (Figs. 1, 2 and 3).

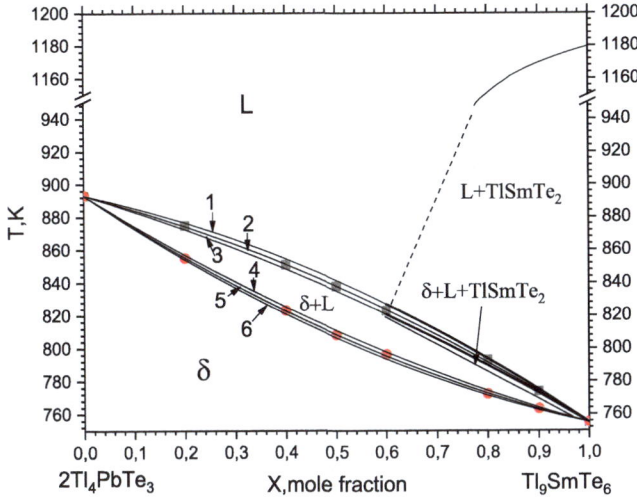

Fig. 1. Phase diagram of the Tl₉SmTe₆-2Tl₄PbTe₃ system. δ-solid solutions.

The obtained analytical dependences for the liquidus and solidus of the quasi-binary system (Eqs. 1–4) are shown in the captions of Figs. 1, 2 and 3. The equations are presented in the version for the computer. Based on the uncertainty principle, liquid uses and solid uses are represented by bands for which the coefficients of Eqs. (1) and (2) for the systems Tl₉SmTe₆-2Tl₄PbTe₃ (Fig. 1), Tl₉SmTe₆ – Tl₉BiTe₆ (Fig. 2) and Tl₈Pb₂Te₆ –Tl₉BiTe₆ have the following intervals, respectively: c = 44 ÷ 70 and d = −52 ÷ −70; c = 39 ÷ 52 and d = −43 ÷ −57; c = 39 ÷ 52 and d = −43 ÷ −57. The coefficients c_i and d_i in the equations of liquidus (Eq. 3) and solidus (Eq. 4) of the Tl₈Pb₂Te₆–Tl₉BiTe₆ system (Fig. 3 and Eq. 3) also have intervals.

In Fig. 1 symbols are experimental data [26]. The curves are described by equations:

1) T(liq)=893-138*x+70*x*(1-x); 2) T(liq)= 893-138*x+56*x*(1-x);
3) T(liq)=893-138*x+44*x*(1-x); 4) T(sol)= 893-138*x-52*x*(1-x);
5) T(sol)= 893-138*x-62*x*(1-x); 6) T(sol)= 893-138*x-70*x*(1-x)

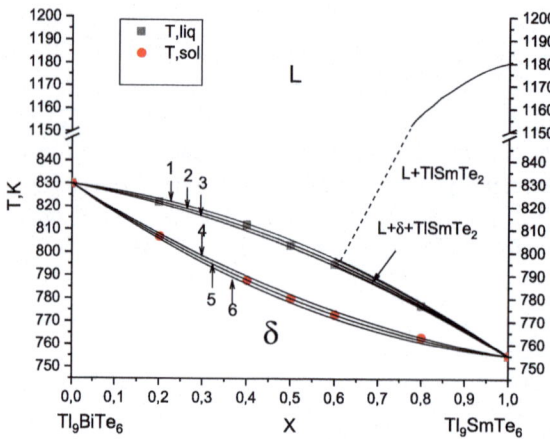

Fig. 2. Phase diagram of the Tl$_9$SmTe$_6$-Tl$_9$BiTe$_6$ system. δ-solid solutions.

In Fig. 2 symbols are experimental data [26]. The curves are described by equations:

1) T(liq)=830-75*x+52*x*(1-x);   2) T(liq)=830-75*x+45*x*(1-x);
3) T(liq)=830-75*x+39*x*(1-x);   4) T(sol)=830-75*x-43*x*(1-x);
5) T(sol)=830-75*x-50*x*(1-x);   6) T(sol)=830-75*x-57*x*(1-x)

In Fig. 3 symbols are experimental data [29]. The curves are described by equations:

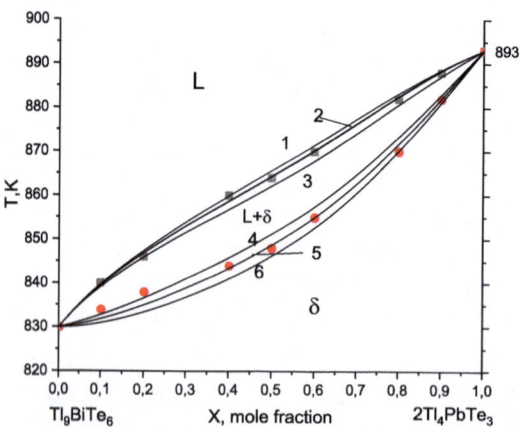

Fig. 3. Phase diagram of the Tl$_8$Pb$_2$Te$_6$–Tl$_9$BiTe$_6$ system. δ-solid solutions.

1) T(liq) = 830 + 63 * x + x * (1 − x) * (46.7 − 100.7 * x + 80.4 * x^2);
2) T(liq) = 830 + 63 * x + x * (1 − x) * (46.7120.7 * x + 98.4 * x^2)
3) T(liq) = 830 + 63 * x + x * (1 − x) * (42.7 − 130 * x + 98.4 * x^2)
4) T(sol) = 830 + 63 * x + x * (1 − x) * (−45 + 68 * x − 184 * x^2 + 125 * x^3)
5) T(sol) = 830 + 63 * x + x * (1 − x)*(−55 + 68 * x − 184 * x^2 + 125*x^3)
6) T(sol) = 830 + 63 * x + x * (1 − x) * (−65 + 68 * x − 184 * x^2 + 125 * x^3)

Due to the presence of areas of uncertainty for liquidus and solidus, to solve the Eq. (7) we used an asymmetric version of the regular solutions model of non-molecular compounds [33]. A multipurpose genetic algorithm was used to find the parameters of this equation [34]. The following conditions were used to carry out the iteration process:

$$x = 0 \div 1; \ a < 0; \ b > 0; \ m > n > 1; \ 750K < T < 900$$

As a result of the calculations an analytical expression for Eq. (7) were determined:

$$\Delta G_T^0 (\text{kJ/mol}) = (-1640 + 2.48T)x^{1.3}(1-x)^{1.1}$$
$$+ \ 8.314T[2x\ln(x) + (1-x)\ln(1-x)] \tag{8}$$

Here x- mole fraction Tl$_9$SmTe$_6$ in solid solution (Tl$_9$SmTe$_6$)$_x$(2Tl$_4$PbTe$_3$)$_{1-x}$.
To determine the thermodynamic stability of solid solutions (Figs. 1, 2 and 3), the Lupis internal stability function [35] was used:

$$F = x(1-x)\frac{d^2(\Delta G/RT)}{dx^2} \tag{9}$$

Substituting dependence (8) into formula (9), we find that in the entire concentration range F > 0.

4   Results and Discussion

For 3D modeling of the liquidus and solidus surfaces of the Tl$_4$PbTe$_3$-Tl$_9$SmTe$_6$-Tl$_9$BiTe$_6$ system, the analytical method which described in [36, 37] was used. Equations (5) and (6) are obtained as (equations are presented in computer version):

$$T(\text{liquidus} Tl SmTe_2) = (-1248 + 4673 * x - 2300 * x^2) * y$$
$$+ (-1248 + 4673 * x - 2300 * x^2) * (1 - y) \tag{10}$$

$$T(\text{liquidus for solid solutions}) = (893 - 138 * x + 56 * x * (1 - x)) * y$$
$$+ (830 - 75 * x + 45 * x * (1 - x)) * (1 - y) + 40 * y * (1 - y) * (1 - x) \tag{11}$$

$$T(\text{solidus}) = (893 - 138 * x - 62 * x * (1 - x)) * y$$
$$+ (830 - 75 * x - 50 * x * (1 - x)) * (1 - y) + 44 * y * (1 - y) * (1 - x)^2$$

$$(12)$$

In Eqs. (10–12): $y = x_2/(1 - x)$. x – mol. fractions of Tl_9SmTe_6; x_2 and x_3, mol. fractions of Tl_8PbTe_3 and Tl_9BiTe_6. In Eq. (10): $x = 0.65 \div 1$; $y = 0 \div 1$. In Eqs. (11), (12): $x = 0 \div 1$; $y = 0 \div 1$.

It follows from Eqs. (10)–(12) that the Tl_9SmTe_6-$2Tl_4PbTe_3$-Tl_9BiTe_6 systems at $100 \div 60$ mol% Tl_9SmTe_6 are non-quasibinary due to the incongruent character of the melting of the Tl_9SmTe_6 compound. Therefore, in the 100–60 mol% Tl_9SmTe_6 concentration range, starting from 1190 K, the $TlSmTe_2$ compound crystallizes from the melt, which leads to the formation of two-L + $TlSmTe_2$ and three-phase L + $TlSmTe_2$ + δ fields. Below solidus, unlimited solid solutions of Tl_9SmTe_6, $2Tl_4PbTe_3$ and Tl_9BiTe_6 compounds are formed. It was found that in the entire concentration range in the temperature range T = 300–900 K, the second derivative of the integral free mixing energy is greater than zero (Eq. 8). Therefore, the values of the stability function (Eq. 9) are also greater than zero (F > 0) in the entire range of concentrations, which indicates the thermodynamic stability of solid solutions.

5   Conclusions

By applications of MGA approach, the boundaries of the heterogeneous equilibrium of liquid - solid in the complex system Tl_9SmTe_6 – $Tl_8Pb_2Te_6$ – Tl_9BiTe_6 in a wide range of temperatures $300 \div 1200$ K have been determined. As the initial data, the melting points and thermodynamic parameters of the compounds Tl_9SmTe_6, $Tl_8Pb_2Te_6$ and Tl_9BiTe_6, coordinates of phase diagrams of boundary quasi-binary systems as well as the minimum number of experimental DTA measurements were used. The boundaries of the uncertainty for the liquidus and solidus determined and modeled based on the determination error of the experimental data DTA and thermodynamic functions of formation of compounds Tl_9SmTe_6, $Tl_8Pb_2Te_6$, Tl_9BiTe_6 and intermediate alloys. The MGA approach, using the thermodynamic condition of internal stability, made it possible to determine the parameters of the analytical dependences of temperature on the composition, thus 3D modeling the crystallization surfaces of solid solutions and the Tl_9SmTe_6 compound.

The information obtained is intended to determine the optimal conditions for the synthesis of solid solutions in the Tl_9SmTe_6-$Tl_8Pb_2Te_6$-Tl_9BiTe_6 system, which are promising materials for producing topological insulators.

Acknowledgments. This work was carried out as part of the scientific program of the international laboratory "Advanced Materials for Spintronics and Quantum Computing", created on the basis of the Institute of Catalysis and Inorganic Chemistry of ANAS (Azerbaijan) and the International Physics Center Donostia (Spain) and partially funded by a grant EİF/MQM/Elm-Tehsil-1-2016-1 (26)-71/01/4-M-33.

References

1. Alonso-Vante, N.: Chalcogenide Materials for Energy Conversion: Pathways to Oxygen and Hydrogen Reactions. Springer, Heidelberg (2018)
2. Liu, X., Lee, S., Furdyna, J., Luo, T., Zhang, Y.-Z.: Chalcogenide: From 3D to 2D and Beyond. Elsevier, Amsterdam (2019). 398 p
3. Chalcogenides: Advances in Research and Applications, 111 p. Woodrow Phillips, Nova (2018)
4. Shvets, I.A., Klimovskikh, I.I., Aliev, Z.S., Babanly, M.B., Chulkov, E.V., et al.: Impact of stoichiometry and disorder on the electronic structure of the $PbBi_2Te_{4-x}Se_x$ topological insulator. Phys. Rev. B **96**, 235124 (2017). https://doi.org/10.1103/PhysRevB.96.235124
5. Hogan, C., Holtgrewe, K.F., Aliev, Z.S., Babanly, M.B., Chulkov, E.V., et al.: Temperature driven phase transition at the antimonene, Bi_2Se_3 van der Waals heterostructure. ACS Nano. **13**(9), 10481–10489 (2019). https://doi.org/10.1021/acsnano.9b04377
6. Banik, A., Roychowdhury, S., Biswas, K.: The journey of tin chalcogenides towards high-performance thermoelectrics and topological materials. Chem. Commun. **54**, 6573–6590 (2018)
7. Pacile, D., Eremeev, V.S., Caputo, M., Pisarra, M., Babanly, M.B., et al: Deep insight into the electronic structure of ternary topological insulators: a comparative study of $PbBi_4Te_7$ and $PbBi_6Te_{10}$. Phys. Status Solidi (RRL) - Rapid Res. Lett. (2018). https://doi.org/10.1002/pssr.201800341
8. Trang, C.X., et al.: Metal-insulator transition and tunable Dirac-cone surface state in the topological insulator $TlBi_{1-x}Sb_xTe_2$ studied by angle-resolved photoemission. Phys. Rev. B. **93**, 165123 (2016)
9. Babanly, M.B., Chulkov, E.V., Aliev, Z.S., Shevel'kov, A.V., Amiraslanov, I.R.: Phase diagrams in thermaterials science of topological insulators based on metal chalcogenides. Russ. J. Inorg. Chem. **62**(13), 1703–1729 (2017). https://doi.org/10.1134/s0036023617130034
10. Otrokov, M.M., Klimovskikh, I.I., Bentmann, H., Amiraslanov, I.R., Babanly, M.B., Mamedov, N.T., Chulkov, E.V., et al.: Prediction and observation of the first antiferromagnetic topological insulator. Nature **576**, 416–422 (2019). https://doi.org/10.1038/s41586-019-1840-9
11. Bhan, S., Shubert, K.: Kristallstruktur von Tl_5Te_3 and Tl_2Te_3. J. Less. Common. Metals **20**(3), 229–235 (1970)
12. Imamaliyeva, S.Z., Babanly, D.M., Tagiev, D.B., Babanly, M.B.: Physicochemical aspects of development of multicomponent chalcogenide phases having the Tl_5Te_3 structure: a review. Russ. J. Inorg. Chem. **63**, 1704–1730 (2018). https://doi.org/10.1134/S0036023618130041
13. Babanly, M.B., Azizulla, A., Kuliev, A.A.: System Tl_2Te-Bi_2Te_3-Te. Russ. J. Inorg. Chem. **30**(9), 2356–2359 (1985)
14. Bradtmöller, S., Böttcher, P.: Crystal structure of molybdenum tetra thallium tritelluride, $MoTl_4Te_3$. Z. Kristallogr. **209**(1), 75 (1994)
15. Heinke, F., Eisenburger, L., Schlegel, R., et al.: The influence of nanoscale heterostructures on the thermoelectric properties of bi-substituted Tl_5Te_3. Anorg. Allg. Chem. **643**, 447–454 (2017). https://doi.org/10.1002/zaac.201600449
16. Arpino, K.E., Wasser, B.D., McQueen, T.M.: Superconducting Dome and Crossover to an Insulating State in $[Tl_4]Tl_{1-x}Sn_xTe_3$. APL Mat. **3**(4), 041507 (2015)
17. Piasecki, M., Brik, M.G., Barchiy, I.R., et al.: Band structure, electronic and optical features of Tl_4SnX_3 (X = S, Te) ternary compounds for optoelectronic applications. J. Alloys Compd. **710**, 600–607 (2017). https://doi.org/10.1016/j.jallcom.2017.03.280

18. Shah, W.H., Khan, A., Waqas, M., Syed, W.A.: Effects of Pb doping on the seebeck co-efficient and electrical properties of $Tl_{8.67}Pb_xSb_{1.33-x}Te_6$ chalcogenide system. Chalcogenide Lett. **14**(2), 61–68 (2017)

19. Wolfing, B., Kloc, C., Teubner, J., Bucher, E.: High performance thermoelectric Tl_9BiTe_6 with an extremely low thermal conductivity. Phys. Rev. Lett. **36**(19), 4350–4353 (2001)

20. Guo, Q., Chan, M., Kuropatwa, B.A., et al.: Enhanced thermoelectric properties of variants of Tl_9SbTe_6 and Tl_9BiTe_6. Chem. Mater. **25**(20), 4097–4104 (2013)

21. Imamalieva, S.Z., Sadygov, F.M., Babanly, M.B.: New thallium –neodymium tellurides. Inorganic Mater. **44**(9), 935–938 (2008). https://doi.org/10.1134/s0020168508090070

22. Babanly, M.B., Imamalieva, S.Z., Babanly, D.M.: Tl_9LnTe_6 (Ln-Ce, Sm, Gd) compounds – the new structural analogies of Tl_5Te_3. Azerbaijan Chem. J. **2**, 121–125 (2009)

23. Bangarigadu-Sanasy, S., Sankar, C.R., Schlender, P., Kleinke, H.: Thermoelectric properties of Tl10-xLnxTe6, with Ln = Ce, Pr, Nd, Sm, Gd, Gd, Dy, Ho and Er, and 0.25 < x<1.32. J. Alloys Compd. **549**, 126–134 (2013). https://doi.org/10.1016/j.jallcom.2012.09.023

24. Bangarigadu-Sanasy, S., Sankar, C.R., Dube, P.A., Greedan, J.E.: Magnetic properties of Tl_9LnTe_6, Ln = Ce, Pr, Tb and Sm. J. Alloys. Compd. **589**, 389–392 (2014). https://doi.org/10.1016/j.jallcom.2013.11.229

25. Babanly, M.B., Mashadiyeva, L.F., Babanly, D.M., Imamaliyeva, S.Z., Taghiyev, D.B., Yusibov, Y.A.: Some aspects of complex investigation of the phase equilibria and thermodynamic properties of the ternary chalcogenid systems by the EMF method. Russ. J. Iniorg. Chem. **13**, 1649–1671 (2019). https://doi.org/10.1134/s0036023619130035

26. Imamaliyeva, S.Z., Alakbarzade, G.I., Mahmudova, M.A., Amiraslanov, I.R., Babanly, M.B.: Phase equilibria in the section of the Tl_4PbTe_3-Tl_9SmTe_6-Tl_9BiTe_6 Tl-Pb-Bi-Sm-Te system. Acta Chem. Slovenica. **65**, 365–371 (2018)

27. Imamaliyeva, S.Z., Alakbarzade, G.I., Mahmudova, M.A., Amiraslanov, I.R., Babanly, M.B.: Experimental study of the Tl_4PbTe_3-Tl_9TbTe_6-Tl_9BiTe_6 section of the Tl-Pb-Bi-Tb-Te system. Mater. Res. **21**(4), e20180189 (2018). https://doi.org/10.1590/1980-5373-mr-2018-0189

28. Imamaliyeva, S.Z., Alakbarzade, G.I., Salimov, Z.E., Izzatli, S.B., Jafarov, Ya.I., Babanly, M.B.: The Tl_4PbTe_3-Tl_9GdTe_6-Tl_9BiTe_6 isopleth section of the Tl-Pb-Bi-Gd-Te system. Chem. Probl. **4**, 495–504 (2018). https://doi.org/10.32737/2221-8688-2018-4-496-504

29. Babanly, M.B., Dashdiyeva, G.B., Huseynov, F.N.: Phase equilibriums n the Tl_4PbTe_3-Tl_9BiTe_6 system. Chem. Probl. **1**, 69–72 (2008)

30. Duong, T.C., Hackenberg, R.E., Landa, A., Honarmandi, A., Talapatra, A.: Revisiting thermodynamics and kinetic diffusivities of uranium–niobium with Bayesian uncertainty analysis. CALPHAD **55**, 219–230 (2016)

31. Mammadov, A.N., Aliev, Z.S., Babanly, M.B.: Study of the uncertainty heterogeneous phase equilibria areas in the binary YbTe-SnTe alloy system. In: Aliev, R.A., Kacprzyk, J., Pedrycz, W., Jamshidi, M., Sadikoglu, F.M. (eds.) 13th Inter. Conf. on Theory and Appl. of Fuzzy Systems and Soft Comp. ICAFS-2018. AISC, vol. 896, pp. 815–822. Springer, Cham (2019). https://doi.org/10.1007/978-3-030-04164-9_107

32. Mammadov, A.N., Alverdiev, I.Dz., Aliev, Z.S. Tagiev, D.B. Babanly, M.B.: Thermodynamic calculation and modeling of the phase diagram of the Cu_2SnS_3-Cu_2SnSe_3 system. In: Aliev, R.A., et al. (eds.) 10th International Conference on Theory and Application of Soft Computing, Computing with Words and Perceptions - ICSCCW, ICSCCW 2019. AISC, Prague, Czech Republic, 27–28 August 2019, vol. 1095, pp. 1–8. Springer (2020). https://doi.org/10.1007/978-3-030-35249-3_118

33. Mamedov, A.N.: Thermodynamics of systems with non-molecular compounds, 124 p. LAP Germany (2015). (in Russian)

34. Preuss, M., Wessing, S., Rudolph, G., Sadowski. G.: Solving phase equilibrium problems by means of avoidance-based multiobjectivization. In: springer Handbook of Computational Intelligence. Part E.58, Evol. Comput., pp. 1159-1169 (2015). https://doi.org/10.1007/978-3-662-43505-2_58

35. Mamedov, A.N., Mekhdiev, I.H., Agaeva, S.A., Gulieva, S.A.: Calculation of the adsorption of binary alloy components using the stability function. Russ. J. Physic. Chem. A. **70**(8), 1455–1457 (1996)

36. Yusibov, Yu.A, Alverdiev, I.Dzh., Ibragimova, F.S., Mamedov, A.N., Tagiev, D.B., Babanly, M.B.: Study and 3D Modeling of the Phase Diagram of the Ag–Ge–Se System. Russ. J. Inorg. Chem. **62**(9), 1223–1233 (2017)

37. Yusibov, Yu.A., Alverdiev, I.Dzh., Mashadiyeva, L.F., Mamedov, A.N., Tagiev, D.B., Babanly, M.B.: Study and 3D modeling of the phase diagram of the Ag–Sn–Se system. Russ. J. Inorg. Chem. **63**(12), 1622–1635 (2018). https://doi.org/10.1134/s0036023618120227

Technique of Automated Classification of "Work Programs of Disciplines" Based on a Neural Network and Self-organizing Maps

A. V. Kurbesov$^{(\boxtimes)}$ ⓘ, I. I. Miroshnichenko ⓘ, N. A. Aruchidi ⓘ, and E. A. Kuharenko ⓘ

Rostov State Economic University (RINH), Rostov-on-the-Don, Russia
akurbesov@yandex.ru, iimo2@ya.ru,
bnatalya2000@mail.ru, Lisa97.mgd@yandex.ru

Abstract. Increases the complexity of educational documentation leads to the need for adequate assessment of the complexity of a particular discipline and the complexity of the teacher labor. Required to consider and create tools to quantify their work. Analysis tool is selected and neural network Kohonen self-organizing maps. The study was developed by the original soft-ware product is invariant to the number of estimated parameters, providing high-quality original data clustering and visualization of the results. The approach has been tested in the working programs of disciplines implemented for areas of "Information systems and technology" and "Applied Computer Science". Data for the analysis were taken over the past three years.

Keywords: Education · Neural network · Cluster analysis · Kohonen self-organizing map · Work programs

1 Introduction

In modern conditions, the activities of university teachers in the preparation of educational and methodological documentation is becoming increasingly important. The volume, time, and laboriousness of creating documentation is constantly increasing, more resources are required to process it and maintain it up to date. In fact, the burden on teachers is constantly increasing during the performance of these works. [3, 4]. The risk of reducing the quality of educational and methodological support is increasing. A number of publications investigate the problem of the quality of educational and methodological support. Thus, the authors of [5–7] note the relevance of this problem, suggest mechanisms for evaluating individual sections of educational and methodological documents, and develop numerical quality assessments in the form of a certain result indicator based on a set of weights. Separately, we note that in [5] approaches to automation of the assessment of the quality of educational and methodical documentation are outlined.

The authors [1, 2] propose various algorithms for assessing the quality of educational and methodological documents in the process of monitoring the educational and methodological support of a university.

R. A. Aliev et al. (Eds.): ICAFS 2020, AISC 1306, pp. 490–497, 2021.
https://doi.org/10.1007/978-3-030-64058-3_61

However, today there is no clear and simple technique that allows classifying work programs of a discipline according to the level of complexity of creation and (or) labor input and, in accordance with this, to evaluate the total labor input of a teacher.

2   Problem Statement

There is an array of work programs for disciplines of a particular specialty (faculty) and, as a result, a complete set of characteristics of each of these programs. The main characteristics used were hours of classroom and self-study, hours for control, data on competencies, information about the literature of the discipline (requiring additional study), etc. It is required to build a neural network that provides automatic classification of these work programs, as well as a graphical display of the results for their better visualization and subsequent interpretation. As the main tool for solving this problem, preference is given to algorithms based on a self-learning neural network and Kohonen self-organizing maps.

3   Kohonen Self-organizing Maps Algorithm

The Kohonen network is used for automatic classification without a teacher. A distinctive feature of this network is the ability to learn the network "teacher free". The task of training is to determine the winner neuron for each input layer and to adjust the weights so that the centers of the vectors match the initial data as much as possible. We describe the classification principle underlying the Kohonen neural network. The essence of the method is to convert the original set into a group of classes, using a transform function of the form:

$$\Psi : D \rightarrow \{1, 2, \ldots, K\}$$

At the input, we have many disordered vectors of dimension N:

$$X_m = \left(X_1^m, X_2^m, X_N^m\right), \ m = 1, 2, \ldots, M.$$

When dividing this set into K - classes, sets of weighting coefficients are given by the formula:

$$W_k = \left(w_1^k, w_2^k, w_N^k\right), \ k = 1, 2, \ldots, K.$$

Let's build a learning algorithm:

1 We carry out the normalization:

$$\text{Max}_n = \max\left(X_n^m\right), \ \text{Min}_n = \min\left(X_n^m\right), \ n = 1, 2, \ldots, N;$$

$$a_n = \frac{1}{Max_n - Min_n}, \ b_n = \frac{-Min_n}{Max_n - Min_n}, \ n = 1, 2, \ldots, N;$$

$$x_N^m = a_n x_n^m + b_n, \ n = 1, 2, \ldots, N, \ m = 1, 2, \ldots, M.$$

2 Initialize the weights $\{w_n^m\}$ with random numbers distributed uniformly between [0.98, 0.39].
3 Set the learning coefficient $\lambda = 0.34$.
4 While $\lambda > 0$, perform steps 5–6.
5 Repeat l times (in our case, l = 12): For each x^m we look for the nearest vector w^k and, for the found vector w^k, correct the components:

$$w_N^k = w_N^k + \lambda\left(x_n^m - w_N^k\right), \ n = 1, 2, \ldots, N.$$

6 Reduce the Learning Coefficient $\lambda = \lambda - \Delta\lambda$, Where $\Delta\lambda = 0.04$

The algorithm admits variability of a number of parameters.
For self-organizing cards, a modification of the above algorithm is used, in the framework of which each neuron has its own geometric position relative to other neurons.

1. We initialize the vectors Wm by random numbers having the same order as the original data.
2. Set the time parameter t equal to "1".
3. Randomly select an arbitrary element from X.
4. We find a neuron whose weights are closest to X in the metric of N-dimensional space. Let this neuron be number m ^ *.
5. We carry out the adjustment of the weights of all neurons according to the following formula

$$\omega_n^m = \omega_n^m + \mu(t)h(t, q(m, m^*))\left(x_n - \omega_n^m\right),$$

where

$$\mu(t) = \mu_0 e^{-at},$$
$$h(t, q) = e^{\frac{q^2}{2q(t)}},$$
$$q(t) = q_0 e^{-bt},$$

Assign t = t + 1.
6. If the maximum t is exceeded, then we exit the algorithm.
7. Otherwise, go to step 3.

After the construction, we evaluate all the values of the input data and mark on the two-dimensional plane (map) neurons whose weights are closest to the estimated input data.

Upon completion of training, to evaluate a new vector of source data, it is necessary to find the one closest to the current neuron and select it on the map. From this location and the location of the measured values, this vector can be estimated.

The configuration of Kohonen's self-organizing maps is based on a random distribution of the initial values of the weights, so each time we will get homogeneous classifications, but the cluster numbering will be random.

4 Conducting Research

Initial data for research should be summarized in one table, stored in csv format with comma delimited format. Fragments of the source data are shown in Table 1. A data file prepared in this way is used for import into the developed software subsystem. After importing the csv file, this information is analyzed in the developed software product.

To initialize the map (set the initial values for network nodes), we used values from eigenvectors uniformly distributed in the range of input data.

Table 1.

	Classroom (contact), h	Individual task, h	Control, h	Competencies, n	Literatura, e	Interactive, p
Life safety	36	72	0	3	5	22
Accounting and analysis	54	18	0	3	5	22
Introduction to the specialty	54	18	0	3	3	10
Geographic Information Systems	54	18	36	2	5	17
Foreign language for working communication	54	18	0	3	8	56
Operating Systems	72	144	0	5	8	24
Advanced Computing Technology	36	36	0	2	5	18
...	...	...	...	...	...	...
Presentation of knowledge in information systems	108	72	36	2	2	14
Software engineering	36	108	0	6	3	14

The total number of assessed disciplines is 118, data for the period of study from 2016–2019 were used.

Consider the results of the analysis of the work programs of the direction "information systems and technologies" over the past two years using Kohonen maps.

The map shown in Fig. 1 shows the input data of work programs, which are included in the IST profile, where each neuron is marked with its own color. Otherwise, each individual cluster is assigned its own color.

Fig. 1. Visualization of IST work programs for 2018–19

The map gives a visual representation of the results that can be presented in tabular form (Fig. 2).

31	mathematical and simulation modeling	6	
32	methods and tools for designing IS and T	6	
33	object-oriented programming	1	
34	organization and structuring of multi-level systems	0	
35	advanced computing technologies	0	
36	database design	0	
37	development and maintenance of software systems	0	
38	mobile app development	0	
39	database systems	0	
40	electronic document management systems	0	
41	creating a web representation	0	
42	standardization and prospects of IT	0	
43	systems theory and system analysis	0	
44	data processing technologies	0	
45	data management	1	
46	corporate systems management	0	
47	information systems project management	0	
48	e-government	2	

Fig. 2. A fragment of the result of the work programs of the IST area for 2018–19

As a result, we see that the first grade includes programs with the highest indicators: literature, hours devoted to the discipline, interactive and competencies, and in the second year the number of programs is growing. From this we can conclude that each year the teacher's complexity increases in the number of indicators in this direction, since this direction is connected and devoted to various technologies and the necessary knowledge grows every year. That is, the teacher complements the curriculum, which increases his time spent on the development of this documentation and conducting the discipline as a whole.

Next, we will analyze the work programs of the "applied informatics" direction over the past three years, where Kohonen maps were also used. The result is shown in Fig. 3 and 4.

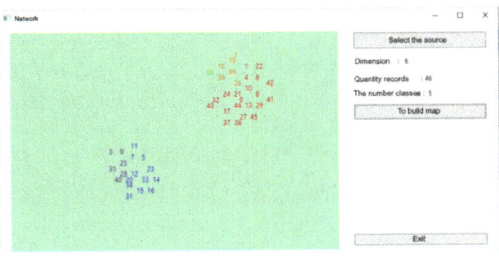

Fig. 3. Visualization of the work programs of the "applied informatics" direction for 2016–17

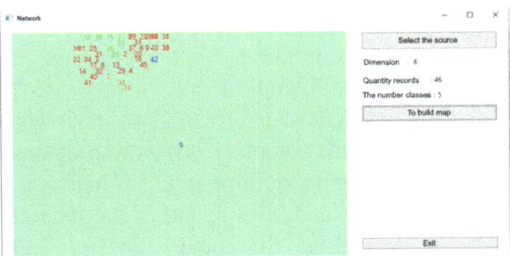

Fig. 4. Visualization of the work programs of the "applied informatics" direction for 2017–18

We present the results in tabular form (Fig. 5, 6)

number	title	className	
0	Life safety	0	
1	Accounting and analysis	0	
2	Introduction to the specialty	1	
3	Geoinformation system	4	
4	Foreign language of professional communication	0	
5	Foreign language	2	
6	Internet marketing	0	
7	Computer science and programming	3	
8	Information and analytical systems	0	
9	Information network	4	
10	Information systems in accounting and audit	0	
11	Information systems in education	3	
12	Information systems and technologies	3	
13	History	0	
14	Computer geometry and graphics	2	
15	Marketing and information business	2	
16	Mathematical and simulation modeling	2	
17	Organization management	0	
18	Methods and systems to support managerial decision-making	1	

Fig. 5. A fragment of the result of the work programs of the "applied informatics" direction for 2016–17

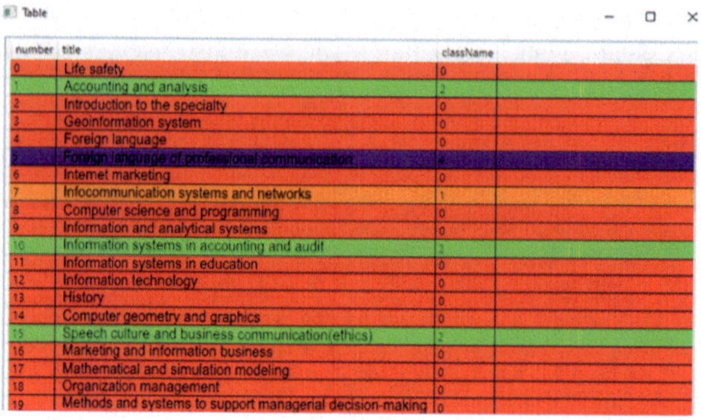

Fig. 6. A fragment of the result of the work programs of the "applied informatics" direction for 2017–18

The results presented demonstrate the changes from 2016 to 2019. Every year, the number of indicators is increasing, especially in specialized programs. We note that in class 1, the main work programs are profile programs, i.e. work programs are divided into working groups based on the affiliation of specialists to a different profile.

Since there is an intensive growth of information volumes when teaching IT disciplines, there is a rapid increase in the complexity of work programs in the time axis of coordinates, this is a direct reflection in work programs. There is a significant increase in the complexity of the disciplines and, as a result, of the time and laboriousness of maintaining documentation, as well as the compilation and understanding of new metamodels of the educational process.

5   Conclusions

The result of the analysis allows us to draw the following conclusions:

1. The effectiveness of the use of neural networks and Kohonen self-organizing maps for the analysis and clustering of documentation of disciplines work programs has been proved;
2. The developed software is invariant to the number of indicators of each individual discipline, provided that it has the same dimension for each evaluated work program;
3. The clustering of the programs of the Faculty of Computer Technology and Information Security has been carried out, which allows to identify the busiest teachers and the complexity and laboriousness of the courses they create and teach, which can be taken into account when rating the faculty.
4. The possibility of assessing the complication (simplification) of the courses of the same type of names when moving along the time axis is shown.
5. With minor adaptation, this technique should become in-variant for the faculty and the profile of specialist training.

A promising development of this study should be recognized as the automation of a computational experiment in the framework of using the developed software package. The analysis carried out in the future will improve the efficiency of compiling "work programs of disciplines" to assess their complexity and reduce labor costs in compiling them.

References

1 Zinchenko, V.O.: Monitoring the quality of the educational process at the university. Bull. Kostroma S Univ. Pedagogy. Psychol. Sociokinet. **22**(4), 188–192 (2016)
2. Ibragimov, G.I., Kamasheva, Yu.L.: Evaluation of the quality of educational and methodological support of educational programs at a higher vocational school, 152 p. Cognition, Kazan (2010)
3. Ishimova, I.N.: Options for optimizing the methodological support of the educational process at the university. Vestnik SUSU Ser. Educ. Pedagog. sci. **6**(3), 41–47 (2014)
4. Klimenko, A.A., Samarskaya, M.V.: Simulation modeling of labor costs for the formation of educational and methodical documentation at the university. In: New Directions of Scientific Thought: Materials of the International Scientific and Practical Conference, 171–174 (2016).
5. Logachev, M.S.: Integral indicator for an automated system for monitoring the quality of the educational process. Vestnik MGUP named after Ivan Fedorov **2**, 30–33 (2016)
6. https://en.wikipedia.org/wiki/Kohonen_Neural_Network
7. Kohonen, T.: Self-organizing Maps. Series "Adaptive and Intellectual Systems". Knowledge Lab, Binom (2017)

Z-Decision Making in Human Resources Department

Konul Jabbarova[1]([⊠]) 🆔 and Akif V. Alizadeh[2] 🆔

[1] Department of Computer Engineering, Azerbaijan State Oil and Industry
University, 20 Azadlig Ave., Baku AZ1010, Azerbaijan
konul.jabbarova@mail.ru
[2] Department of Control and Systems Engineering, Azerbaijan State Oil and
Industry University, 20 Azadlig Ave., Baku AZ1010, Azerbaijan
akifoder@yahoo.com, a.alizadeh@asoiu.edu.az

Abstract. Personnel assessment and selection is a really essential activity for any company. Some of such problems are characterized by both fuzziness and partial reliability of information. In this paper we apply the VIKOR to decision making in personnel selection under Z-number-valued information.

Keywords: VIKOR index · Z-number · Regret measure · Utility measure

1 Introduction

Multi-criteria decision-making (MCDM) is one of the most important activities in personnel selection problem [1]. Staff recruitment and selection of qualified personnel are necessary for the advancement of any organization. The choice of the right individual to the right post is the key point for decision making in staff recruitment. For staff selection process some of the methods of MCDM can be utilized. In [1] a PROMETHEE method based model was developed. In [2] they propose a methodology of selection of human resources. For the avoidance of highly emotional states, which usually happens for the human resources selection process, a fuzzy selection of the most suitable candidates was proposed. Entropy–KEMIRA approach was applied to solving human resources selection problem in [3]. The approach is applied for the three groups of criteria. Weights are calculated by solving optimization problem [3].

Paper [4] addresses fuzzy logic application in MCDM problems in human resources management. In [5] they apply the fuzzy Analytic Hierarchy Process (AHP) method selection of staff. In [6] they apply the Fuzzy AHP-TOPSIS method to the human resource selection problem. The results demonstrated some variation in various regions, indicating that local issues might also impact process of selection.

Real-world problems usually are characterized by bimodal information and complexity [7–15]. But this is not taken into account in existing works on the MCDM problems. Prof. Zadeh introduced the Z-number concept. A Z-number consists of a pair of fuzzy numbers $Z = (A, B)$, where A is a soft constraint on a value of a variable of interest, and B is a soft constraint on a value of a probability measure of A, playing a

© The Author(s), under exclusive license to Springer Nature Switzerland AG 2021
R. A. Aliev et al. (Eds.): ICAFS 2020, AISC 1306, pp. 498–507, 2021.
https://doi.org/10.1007/978-3-030-64058-3_62

role of reliability of A. In this article we consider and application of the VIKOR model to personal selection problem with Z-number-valued information.

The paper is organized as follows. In Sect. 2 we provide a prerequisite material used in the sequel. In Sect. 3, we apply VIKOR method to decision problem under fuzziness and partial reliability of information. Section 4 concludes.

2 Preliminaries

Definition 1. A Discrete Z-number [7–10]. A discrete Z-number is an ordered pair $Z = (A, B)$ where A is a discrete fuzzy number playing a role of a fuzzy constraint on values of a random variable X: X is A. Is a discrete fuzzy number with a membership function $\mu_B : \{b_1, \ldots, b_n\} \to [0, 1], \{b_1, \ldots, b_n\} \subset [0, 1]$, playing a role of a fuzzy constraint on the probability measure of A: $P(A) = \sum_{i=1}^{n} \mu_A(x_i) p(x_i)$ is B.

Definition 2. Operations over Discrete Z-numbers [7–10]: Let X_1 and X_2 be discrete Z-numbers describing information about values of X_1 and X_2. Consider computation of $Z_{12} = Z_1 * Z_2$, $* \in \{+, -, \cdot, /\}$. The first stage is computation of $A_{12} = A_1 * A_2$.

The second stage involves construction of B_{12}. We realize that in Z-numbers Z_1 and Z_2, the 'true' probability distributions p_1 and p_2 are not exactly known. In contrast, fuzzy restrictions represented in terms of the membership functions are available

$$\mu_{p_1}(p_1) = \mu_{B_1}\left(\sum_{k=1}^{n_1} \mu_{A_1}(x_{1k})p_1(x_{1k})\right), \ \mu_{p_2}(p_2) = \mu_{B_2}\left(\sum_{k=1}^{n_2} \mu_{A_2}(x_{2k})p_2(x_{2k})\right).$$

Probability distributions $p_{jl}(x_{jk}), k = 1, .., n$ induce probabilistic uncertainty over $X_{12} = X_1 + X_2$. Given any possible pair p_1, p_2, the convolution $p_{12} = p_1 \circ p_2$ is computed as

$$p_{12}(x) = \sum_{x_1 + x_2 = x} p_1(x_1)p_2(x_2), \ \forall x \in X_{12}; x_1 \in X_1, x_2 \in X_2$$

Given p_{12s}, the value of probability measure of A_{12} is computed:

$$P(A_{12}) = \sum_{k=1}^{n} \mu_{A_{12}}(x_{12k})p_{12}(x_{12k}).$$

However, p_1 and p_2 are described by fuzzy restrictions which induce fuzzy set of convolutions:

$$\mu_{p_{12}}(p_{12}) = \max_{\{p_1, p_2: p_{12} = p_1 \circ p_2\}} \min\{\mu_{p_1}(p_1), \mu_{p_2}(p_2)\}$$

Fuzziness of information on p_{12} induces fuzziness of $P(A_{12})$ as a discrete fuzzy number B_{12}. The membership function $\mu_{B_{12}}$ is defined as

$$\mu_{B_{12}}(b_{12}) = \max \mu_{p_{12}}(p_{12})$$

subject to

$$b_{12} = \sum_{i=1}^{n} \mu_{A_{12}}(x_i) p_{12}(x_i)$$

As a result, $Z_{12} = Z_1 * Z_2$ is obtained as $Z_{12} = (A_{12}, B_{12})$.

A scalar multiplication $Z = \lambda Z_1$, $\lambda \in R$ is a determined as $Z = (\lambda A_1, B_1)$.

Definition 3. Fuzzy Pareto optimality (FPO) principle based comparison of Z-numbers [11]. Fuzzy Pareto optimality (FPO) principle allows to determine degrees of Pareto Optimality of multiattribute alternatives. We apply this principle to compare Z-numbers as multiattribute alternatives – one attribute measures value of a variable, the other one measures the associated reliability. According to this approach, by directly comparing Z-numbers $Z_1 = (A_1, B_1)$ and $Z_2 = (A_2, B_2)$ one arrives at total degrees of optimality of Z-numbers: $do(Z_1)$ and $do(Z_2)$. These degrees are determined on the basis of a number of components (the minimum is 0, the maximum is 2) with respect to which one Z-numbers dominates another one. Z_1 is considered higher than Z_2 if $do(Z_1) > do(Z_2)$.

Let us consider a MCDM problem under Z-valued information.

Definition 4 [12]. A distance between Z-numbers. The distance between Z-numbers $Z_1 = (A_1, B_1)$ and $Z_2 = (A_2, B_2)$ is defined as

$$D(Z_1, Z_2) = \frac{1}{n+1} \sum_{k=1}^{n} \left\{ \left| a_{1\alpha_k}^L - a_{2\alpha_k}^L \right| + \left| a_{1\alpha_k}^R - a_{2\alpha_k}^R \right| \right\} + \frac{1}{m+1} \sum_{k=1}^{m} \left\{ \left| b_{1\alpha_k}^L - b_{2\alpha_k}^L \right| + \left| b_{1\alpha_k}^R - b_{2\alpha_k}^R \right| \right\}$$

where $a_\alpha^L = \min A^\alpha$, $a_\alpha^R = \max A^\alpha$, $b_\alpha^L = \min B^\alpha$, $b_\alpha^R = \max B^\alpha$.

3 Statement of the Problem and a Solution Method

Let us examine MCDM problem with Z-number-valued information. Suppose that a local chemical firm tries to recruit an online manager [13]. The firm's human resources department offers several corresponding selection tests, as the advantage criteria to be estimated. In this problem subjective (interviews) and objective (knowledge, skill tests) criteria are used. Tests of knowledge involve language, professional and safety rule test. Tests of skill consist of professional and computer skills. Interviews involve panel and 1-on-1 interviews [13]. There are 5 skilled candidates in the list. Subjective and objective criteria (only quantitative data here) are given in Table 1. Objective and subjective criteria are described by Z-numbers $f_{ij} = (A_{ij}, B_{ij})$ (Table 1 and Table 2). All importance weights of each objective and subjective criteria are given in Table 3.

Table 1. Z-number-valued decision matrix

	Objective attributes			Subjective attributes			
	Knowledge tests			Skill tests			
	Language test C_1	Professional test C_2	Safety rule test C_3	Professional skills C_4	Computer skills C_5	Panel interview C_6	1-on-1 interview C_7
f_1	(0.4, 0.44, 0.48) (0.8, 0.9, 1)	(0.37, 0.41, 0.45) (0.7, 0.8, 0.9)	(0.48, 0.53, 0.58) (0.7, 0.8, 0.9)	(0.39, 0.43, 0.47) (0.7, 0.8, 0.9)	(0.4, 0.44, 0.48) (0.8, 0.9, 1)	(0.42, 0.47, 0.52) (0.6, 0.7, 0.8)	(0.39, 0.43, 0.47) (0.6, 0.7, 0.8)
f_2	(0.42, 0.47, 0.52) (0.8, 0.9, 1)	(0.34, 0.38, 0.42) (0.7, 0.8, 0.9)	(0.41, 0.46, 0.51) (0.7, 0.8, 0.9)	(0.41, 0.45, 0.5) (0.7, 0.8, 0.9)	(0.39, 0.43, 0.47) (0.8, 0.9, 1)	(0.34, 0.38, 0.42) (0.6, 0.7, 0.8)	(0.39, 0.43, 0.47) (0.6, 0.7, 0.8)
f_3	(0.39, 0.43, 0.47) (0.8, 0.9, 1)	(0.48, 0.53, 0.58) (0.7, 0.8, 0.9)	(0.4, 0.44, 0.48) (0.7, 0.8, 0.9)	(0.41, 0.45, 0.5) (0.7, 0.8, 0.9)	(0.44, 0.49, 0.54) (0.8, 0.9, 1)	(0.48, 0.53, 0.58) (0.6, 0.7, 0.8)	(0.44, 0.49, 0.54) (0.6, 0.7, 0.8)
f_4	(0.39, 0.43, 0.47) (0.8, 0.9, 1)	(0.45, 0.5, 0.55) (0.7, 0.8, 0.9)	(0.38, 0.42, 0.46) (0.7, 0.8, 0.9)	(0.43, 0.48, 0.53) (0.7, 0.8, 0.9)	(0.33, 0.37, 0.41) (0.8, 0.9, 1)	(0.34, 0.38, 0.42) (0.6, 0.7, 0.8)	(0.37, 0.4, 0.45) (0.6, 0.7, 0.8)
f_5	(0.42, 0.47, 0.52) (0.8, 0.9, 1)	(0.36, 0.4, 0.44) (0.7, 0.8, 0.9)	(0.33, 0.37, 0.41) (0.7, 0.8, 0.9)	(0.38, 0.42, 0.46) (0.7, 0.8, 0.9)	(0.44, 0.49, 0.54) (0.8, 0.9, 1)	(0.4, 0.44, 0.48) (0.6, 0.7, 0.8)	(0.41, 0.46, 0.51) (0.6, 0.7, 0.8)

Table 2. Importance weights of each criteria

No	Criteria	Weights
C_1	Language test	0.066
C_2	Professional test	0.196
C_3	Safety rule test	0.066
C_4	Professional skills	0.130
C_5	Computer skills	0.130
C_6	Panel interview	0.216
C_7	1-on-1 interview	0.196

Let us solve this problem by using the algorithm of the VIKOR method under Z-number-valued information.

At the *first stage*, we define ideal point and negative ideal point for every criteria. As all the criteria values are Z-numbers, the ideal point and the negative ideal point will also be defined as Z-numbers: $f_j^+ = (0.98 \quad 0.99 \quad 1)(0.8 \quad 0.9 \quad 1)$, $f_j^- = (0 \quad 0.01 \quad 0.02)(0.8 \quad 0.9 \quad 1)$.

At the *second stage* we need to calculate the values of regret measure $R_i = (A_{R_i}, B_{R_i})$ for each alternative:

$$R_i = \max \left[w_j \frac{\left(f_j^+ - f_{ij}\right)}{\left(f_j^+ - f_j^-\right)} \right], \quad i = 1, \dots, n \tag{1}$$

The operations of subtraction, multiplication and division of Z-numbers in (1) and all the next formulas are performed by using Definition 2.

In order to compute $R_i = (A_{R_i}, B_{R_i})$ for all $f_i, i = 1, \dots, n$, we obtained the following results:

$$\left(f_1^+ - f_{11}\right) = (0.98\ 0.99\ 1)(0.8\ 0.9\ 1) - (0.4\ 0.44\ 0.48)(0.8\ 0.9\ 1)$$
$$= (0.5\ 0.55\ 0.6)(0.68\ 0.82\ 0.97);$$

$$\left(f_1^+ - f_1^-\right) = (0.98\ 0.99\ 1)(0.8\ 0.9\ 1) - (0\ 0.01\ 0.02)(0.8\ 0.9\ 1)$$
$$= (0.96\ 0.98\ 1)(0.71\ 0.84\ 0.99);$$

$$\frac{\left(f_1^+ - f_{11}\right)}{\left(f_1^+ - f_1^-\right)} = \frac{(0.5\ 0.55\ 0.6)(0.68\ 0.82\ 0.97)}{(0.96\ 0.98\ 1)(0.71\ 0.84\ 0.99)} = (0.5\ 0.56\ 0.63)(0.52\ 0.7\ 0.95);$$

$$w_1 \frac{\left(f_1^+ - f_{11}\right)}{\left(f_1^+ - f_1^-\right)} = 0.066 \cdot (0.5\ 0.56\ 0.63)(0.52\ 0.7\ 0.95)$$
$$= (0.033\ 0.037\ 0.042)(0.52\ 0.7\ 0.95);$$

$$\left(f_2^+ - f_{21}\right) = (0.98\ 0.99\ 1)(0.8\ 0.9\ 1) - (0.37\ 0.41\ 0.45)(0.7\ 0.8\ 0.9)$$
$$= (0.53\ 0.58\ 0.63)(0.62\ 0.75\ 0.9);$$

$$\left(f_2^+ - f_{21}^-\right) = (0.98\ 0.99\ 1)(0.8\ 0.9\ 1) - (0\ 0.01\ 0.02)(0.8\ 0.9\ 1)$$
$$= (0.96\ 0.98\ 1)(0.71\ 0.84\ 0.99);$$

$$\frac{\left(f_2^+ - f_{21}\right)}{\left(f_2^+ - f_2^-\right)} = \frac{(0.53\ 0.58\ 0.63)(0.62\ 0.75\ 0.9)}{(0.96\ 0.98\ 1)(0.71\ 0.84\ 0.99)} = (0.53\ 0.59\ 0.66)(0.48\ 0.65\ 0.88);$$

$$w_2 \frac{\left(f_2^+ - f_{21}\right)}{\left(f_2^+ - f_2^-\right)} = 0.196 \cdot (0.53\ 0.59\ 0.66)(0.48\ 0.65\ 0.88)$$
$$= (0.104\ 0.116\ 0.129)(0.48\ 0.65\ 0.88);$$

$$\left(f_3^+ - f_{31}\right) = (0.98\ 0.99\ 1)(0.8\ 0.9\ 1) - (0.48\ 0.53\ 0.58)(0.7\ 0.8\ 0.9)$$
$$= (0.4\ 0.46\ 0.52)(0.6\ 0.74\ 0.89);$$

$$\left(f_3^+ - f_3^-\right) = (0.98\ 0.99\ 1)(0.8\ 0.9\ 1) - (0\ 0.01\ 0.02)(0.8\ 0.9\ 1)$$
$$= (0.96\ 0.98\ 1)(0.71\ 0.84\ 0.99);$$

$$\frac{\left(f_3^+ - f_{31}\right)}{\left(f_3^+ - f_3^-\right)} = \frac{(0.4\ 0.46\ 0.52)(0.6\ 0.74\ 0.89)}{(0.96\ 0.98\ 1)(0.71\ 0.84\ 0.99)} = (0.4\ 0.47\ 0.54)(0.47\ 0.64\ 0.87);$$

$$w_3 \frac{\left(f_3^+ - f_{31}\right)}{\left(f_3^+ - f_3^-\right)} = 0.066 \cdot (0.4\ 0.47\ 0.54)(0.47\ 0.64\ 0.87)$$
$$= (0.027\ 0.031\ 0.036)(0.47\ 0.64\ 0.87);$$

$$\left(f_4^+ - f_{41}\right) = (0.98\ 0.99\ 1)(0.8\ 0.9\ 1) - (0.39\ 0.43\ 0.47)(0.7\ 0.8\ 0.9)$$
$$= (0.51\ 0.56\ 0.61)(0.6\ 0.74\ 0.89);$$

$$\left(f_4^+ - f_4^-\right) = (0.98\ 0.99\ 1)(0.8\ 0.9\ 1) - (0\ 0.01\ 0.02)(0.8\ 0.9\ 1)$$
$$= (0.96\ 0.98\ 1)(0.71\ 0.84\ 0.99);$$

$$\frac{\left(f_4^+ - f_{41}\right)}{\left(f_4^+ - f_4^-\right)} = \frac{(0.51\ 0.56\ 0.61)(0.6\ 0.74\ 0.89)}{(0.96\ 0.98\ 1)(0.71\ 0.84\ 0.99)} = (0.51\ 0.57\ 0.64)(0.47\ 0.64\ 0.87);$$

$$w_4 \frac{\left(f_4^+ - f_{41}\right)}{\left(f_4^+ - f_4^-\right)} = 0.130 \cdot (0.51\ 0.57\ 0.64)(0.47\ 0.64\ 0.87)$$
$$= (0.066\ 0.074\ 0.083)(0.47\ 0.64\ 0.87);$$

$$\left(f_5^+ - f_{51}\right) = (0.98\ 0.99\ 1)(0.8\ 0.9\ 1) - (0.4\ 0.44\ 0.48)(0.8\ 0.9\ 1)$$
$$= (0.5\ 0.55\ 0.6)(0.68\ 0.82\ 0.96);$$

$$\left(f_5^+ - f_5^-\right) = (0.98\ 0.99\ 1)(0.8\ 0.9\ 1) - (0\ 0.01\ 0.02)(0.8\ 0.9\ 1)$$
$$= (0.96\ 0.98\ 1)(0.71\ 0.84\ 0.99);$$

$$\frac{\left(f_5^+ - f_{51}\right)}{\left(f_5^+ - f_5^-\right)} = \frac{(0.4\ 0.44\ 0.48)(0.8\ 0.9\ 1)}{(0.96\ 0.98\ 1)(0.71\ 0.84\ 0.99)} = (0.5\ 0.56\ 0.63)(0.52\ 0.7\ 0.94);$$

$$w_5 \frac{\left(f_5^+ - f_{51}\right)}{\left(f_5^+ - f_5^-\right)} = 0.130 \cdot (0.5\ 0.56\ 0.63)(0.52\ 0.7\ 0.94)$$
$$= (0.065\ 0.073\ 0.08)(0.52\ 0.7\ 0.94);$$

$$\left(f_6^+ - f_{61}\right) = (0.98\ 0.99\ 1)(0.8\ 0.9\ 1) - (0.42\ 0.47\ 0.52)(0.6\ 0.7\ 0.8)$$
$$= (0.46\ 0.52\ 0.58)(0.52\ 0.65\ 0.79);$$

$$\left(f_6^+ - f_6^-\right) = (0.98\ 0.99\ 1)(0.8\ 0.9\ 1) - (0\ 0.01\ 0.02)(0.8\ 0.9\ 1)$$
$$= (0.96\ 0.98\ 1)(0.71\ 0.84\ 0.99);$$

$$\frac{\left(f_6^+ - f_{61}\right)}{\left(f_6^+ - f_6^-\right)} = \frac{(0.46\ 0.52\ 0.58)(0.52\ 0.65\ 0.79)}{(0.96\ 0.98\ 1)(0.71\ 0.84\ 0.99)} = (0.46\ 0.53\ 0.6)(0.41\ 0.57\ 0.78);$$

$$w_6 \frac{\left(f_6^+ - f_{61}\right)}{\left(f_6^+ - f_6^-\right)} = 0.216 \cdot (0.46\ 0.53\ 0.6)(0.41\ 0.57\ 0.78)$$
$$= (0.099\ 0.114\ 0.130)(0.41\ 0.57\ 0.78);$$

$$\left(f_7^+ - f_{71}\right) = (0.98\ 0.99\ 1)(0.8\ 0.9\ 1) - (0.39\ 0.43\ 0.47)(0.6\ 0.7\ 0.8)$$
$$= (0.51\ 0.56\ 0.61)(0.53\ 0.65\ 0.79);$$

$$\left(f_7^+ - f_7^-\right) = (0.98\ 0.99\ 1)(0.8\ 0.9\ 1) - (0\ 0.01\ 0.02)(0.8\ 0.9\ 1)$$
$$= (0.96\ 0.98\ 1)(0.71\ 0.84\ 0.99);$$

$$\frac{\left(f_7^+ - f_{71}\right)}{\left(f_7^+ - f_7^-\right)} = \frac{(0.51\ 0.56\ 0.61)(0.53\ 0.65\ 0.79)}{(0.96\ 0.98\ 1)(0.71\ 0.84\ 0.99)} = (0.51\ 0.57\ 0.64)(0.42\ 0.57\ 0.78);$$

$$w_7 \frac{\left(f_7^+ - f_{71}\right)}{\left(f_7^+ - f_7^-\right)} = 0.196 \cdot (0.51\ 0.57\ 0.64)(0.42\ 0.57\ 0.78)$$
$$= (0.1\ 0.112\ 0.125)(0.42\ 0.57\ 0.78);$$

Thus, we computed value of R_1 as follows:

$R_1 = \max(((0.033\ 0.037\ 0.042)(0.52\ 0.7\ 0.95)), (0.104\ 0.116\ 0.129)(0.48\ 0.65\ 0.88)),$
$\quad ((0.027\ 0.031\ 0.036)(0.47\ 0.64\ 0.87)), ((0.066\ 0.074\ 0.083)(0.47\ 0.64\ 0.87)),$
$\quad ((0.065\ 0.073\ 0.08)(0.52\ 0.7\ 0.94)), ((0.099\ 0.114\ 0.130)(0.41\ 0.57\ 0.78)),$
$\quad (0.1\ 0.112\ 0.125)(0.42\ 0.57\ 0.78)) = (0.104\ 0.116\ 0.129)(0.48\ 0.65\ 0.88).$

Analogously, we computed the regret measures for the other alternatives:

$$R_2 = (0.110\ \ 0.122\ \ 0.135)(0.47\ 0.65\ 0.88);$$
$$R_3 = (0.08\ \ 0.09\ \ 0.106)(0.47\ 0.64\ 0.89);$$
$$R_4 = (0.121\ \ 0.134\ \ 0.149)(0.42\ 0.57\ 0.78);$$
$$R_5 = (0.106\ \ 0.118\ \ 0.131)(0.47\ 0.64\ 0.87);$$

At the *third stage*, we calculate the values of utility measures $S_i = (A_{S_i}, B_{S_i})$:

$$S_i = \sum_{j=1}^{n} \left[w_j \frac{\left(f_{ij}^+ - f_{ij}^+\right)}{\left(f_{ij}^+ - f_{ij}^-\right)} \right] \quad (2)$$

The obtained results:

$S_1 = ((((0.033\ 0.037\ 0.042)(0.52\ 0.7\ 0.95)) + (0.104\ 0.116\ 0.129)(0.48\ 0.65\ 0.88))$
$\quad + ((0.027\ 0.031\ 0.036)(0.47\ 0.64\ 0.87)) + ((0.066\ 0.074\ 0.083)(0.47\ 0.64\ 0.87))$
$\quad + ((0.065\ 0.073\ 0.08)(0.52\ 0.7\ 0.94)) + ((0.099\ 0.114\ 0.130)(0.41\ 0.57\ 0.78))$
$\quad + (0.1\ 0.112\ 0.125)(0.42\ 0.57\ 0.78)) = (0.5\ 0.55\ 0.625)(0.41\ 0.57\ 0.78);$

Analogously, we computed the utility measures for the other alternatives:

$$S_2 = (0.52\ 0.57\ 0.654)(0.42\ 0.57\ 0.78);$$
$$S_3 = (0.43\ 0.5\ 0.573)(0.42\ 0.58\ 0.78).$$
$$S_4 = (0.5\ 0.57\ 0.632)(0.42\ 0.57\ 0.78).$$
$$S_5 = (0.49\ 0.58\ 0.633)(0.41\ 0.57\ 0.77).$$

At the *fourth stage*, we need to compute VIKOR index $Q_i = (A_{Q_i}, B_{Q_i})$ for all alternatives:

$$Q_i = \left[v \frac{(S_i - S^-)}{(S^+ - S^-)} + (1 - v) \frac{(R_i - R^-)}{(R^+ - R^-)} \right] \quad (3)$$

S^-, S^+, R^-, R^+ are to be obtained as the highest and lowest values by comparing the Z-numbers on the basis of the FPO principle (Definition 3):

$$S^- = \min_i S_i; S^+ = \max_i S_i; R^- = \min_i R_i; R^+ = \max_i R_i$$

The computed S^-, S^+, R^-, R^+ are as follows:

$$R^+ = (0.110\quad 0.122\quad 0.135)(0.47\quad 0.65\quad 0.88),$$
$$R^- = (0.08\quad 0.09\quad 0.106)(0.47\quad 0.64\quad 0.89)$$

and

$$S^+ = (0.52\quad 0.57\quad 0.654)(0.42\quad 0.57\quad 0.78),$$
$$S^- = (0.43\quad 0.5\quad 0.573)(0.42\quad 0.58\quad 0.78).$$

The values of VIKOR index are obtained as

$$Q_1 = (-4.8 \quad 0.35 \quad 10.3)(0.08 \quad 0.2 \quad 0.45);$$
$$Q_2 = (-5.5 \quad 0.5 \quad 11.8)(0.17 \quad 0.24 \quad 0.46);$$
$$Q_3 = (-7.35 \quad 0 \quad 7.35)(0.16 \quad 0.28 \quad 0.52).$$
$$Q_4 = (-4.95 \quad 0.5 \quad 10.6)(0.08 \quad 0.19 \quad 0.44).$$
$$Q_5 = (-4.95 \quad 0.57 \quad 10.6)(0.07 \quad 0.19 \quad 0.44).$$

At the *fifth stage*, we compare the values of R, S, Q measures by using Definition 3 and The results of ordering of the alternatives with respect to R, S, Q are as follows (Table 3).

Table 3. The results of ranking with respect to R, S, Q measures

R	S	Q
f_3	f_3	f_3
f_1	f_1	f_5
f_5	f_4	f_1
f_4	f_5	f_4
f_2	f_2	f_2

At the *sixth stage*, according to the VIKOR algorithm, we need to analyze the results of this ordering by verification of the conditions C1 and C2 [14]. For brevity, we don't describe them here (they are described in details in [14]). We have found out that these conditions are satisfied. Now, we need to evaluate the difference between values of Q_1 and Q_2. For computational efficiency we will do this by calculating distance $D(Q_1, Q_2)$ (Definition 4). The obtained result is $D(Q^1, Q^2) = 5.96 > \frac{1}{4} = 0.25$. This implies that the alternative f_1 is a compromise solution.

4 Conclusion

In this paper we apply VIKOR method to human resources management problem. All values of alternatives are described by Z-numbers. Computation of utility, regret measures and VIKOR index are based on computations with Z-numbers. The obtained results show effectively of the approach.

References

1. Afshari, A.R., Anisseh, M., Shahraki, M.R., Hooshyar, S.: PROMETHEE use in Personnel selection. In: Proceedings of the International Conference on ICT Management For Global Competitiveness and Economic Growth in Emerging Economies-ICTM (2016)

2. Timar, D.R., Balas, V.E.: Decision making in Human resources selection methodology. In: 2nd International Workshop on Soft Computing Applications, pp. 123–127. IEEE Xplore (2007). https://doi.org/10.1109/sofa.2007.4318316

3. Krylovas, A., Dadelo, S., Kosareva, N., Zavadskas, E.K.: Entropy–KEMIRA approach for MCDM problem solution in human resources selection task. Int. J. Inf. Tech. Decis. **16**(05), 1183–1209 (2017). https://doi.org/10.1142/S0219622017500274

4. D'Urso, M.G., Masi, D.: Multi-criteria decision-making methods and their applications for human resources. Int. Arch. Photogram. Remote Sens. Spat. Inf. Sci. **VI**, 31–37 (2015). https://doi.org/10.5194/isprsarchives-XL-6-W1-31-2015

5. Afshari, A.R., Nikolić, M., Akbari, Z.: Personnel selection using group fuzzy AHP and SAW methods. J. Eng. Manag. Competit. **7**(1), 3–10 (2017). https://doi.org/10.5937/jemc1701003A

6. Kusumawardani, R.P., Agintiara, M.: Application of fuzzy AHP-TOPSIS method for decision making in human resource manager selection process. Proc. Comput. Sci. **72**, 638–646 (2015). https://doi.org/10.1016/j.procs.2015.12.173

7. Aliev, R.A., Alizadeh, A.V., Huseynov, O.H.: The arithmetic of discrete Z-numbers. Inf. Sci. **290**, 134–155 (2015). https://doi.org/10.1016/j.ins.2014.08.024

8. Aliev, R.A., Alizadeh, A.V., Huseynov, O.H., Jabbarova, K.I.: Z-number-based linear programming. Int. J. Intell. Syst. **30**(5), 563–589 (2015). https://doi.org/10.1002/int.21709

9. Aliev, R.A., Huseynov, O.H.: Decision Theory with Imperfect Information. World Scientific, Singapore (2014)

10. Aliev, R.A., Huseynov, O.H., Aliyev, R.R., Alizadeh, A.V.: The Arithmetic of Z-Numbers. Theory and Applications. World Scientific, Singapore (2015)

11. Aliev, R.A., Huseynov, O.H., Serdaroglu, R.: Ranking of Z-numbers and its application in decision making. Int. J. Inf. Technol. Decis. Mak. **15**(6), 1503–1519 (2016). https://doi.org/10.1142/S0219622016500310

12. Aliev, R.A., Pedrycz, W., Huseynov, O.H., Eyupoglu, S.Z.: Approximate reasoning on a basis of Z-number-valued If–Then rules. IEEE T. Fuzzy Syst. **25**(6), 1589–1600 (2017). https://doi.org/10.1109/TFUZZ.2016.2612303

13. Shih, H.-S., Shyur, H.-J., Lee, S.E.: An extension of TOPSIS for group decision making. Math. Comput. Model. **4**, 801–813 (2007)

14. Chatterjeea, P., Chakraborty, S.: Comparative analysis of VIKOR method and its variants. Decis. Sci. Lett. **5**(4), 469–486 (2016)

15. Aliev, R.A., Huseynov, O.H., Zeinalova, L.M.: The arithmetic of continuous Z-numbers. Inform. Sci. **373**, 441–460 (2016)

Fuzzy Logic-Based Identification Chemical Reaction Rate

Mahsati Jafarli$^{(\boxtimes)}$ (iD)

Azerbaijan State Oil and Industry University,
Azadlig 20, Nasimi, Baku, Azerbaijan
mehseti.babanli@gmail.com

Abstract. In this paper we are focusing on an analysis of a reaction rate based on fuzzy logic. The deficiency of enough knowledge or information about the behavior of chemical processes makes difficulty to control it and needs to be studied. This shortcoming is due to the inability to accurately determine the parameters that characterize the process, which creates great difficulties in the management of processes. Since there is not any solution to this problem among classical methods, such processes are often controlled by the human operator, who makes decisions on the basis of expressing the processes by words, namely, linguistic measurement, but it is not within the researcher's ability to control processes characterized by large amounts of data. This paper examines the fuzzy model to determine the reaction rate. For this purpose, the alkylation reaction, the equations of benzene, alkylbenzene and polymer compositions are analyzed. By obtained results from equation is created a fuzzy model and was studied the rate of the reaction using possibility measure-based method.

Keywords: A reaction rate · Fuzzy logic · Fuzzy inference · Alkyl benzene · Benzene and polymer compositions · Differential equation · Coefficient rate

1 Introduction

There are studies in the scientific literature on the conduct of chemical reactions at different reaction rates [1, 2]. The speed of the reaction depends on several factors (e.g., temperature, particle size, etc.). In such cases, the rate of a chemical reaction can be analyzed using fuzzy logic and approximate reasoning. It is known that chemical reactions have low (minimum) and maximum rates, and these rates can be normal, dangerous (explosion) and so on. Reaction rate analysis is a process characterized by uncertainty.

On the other hand, given that differential equations are an effective tool for analyzing chemical processes, it becomes important to analyze chemical reaction behavior using differential equations. In the scientific literature there are works on the application of first-order differential equations in chemistry. Studying the dynamics and its stability of the first order isothermal reactions described by the fuzzy differential equation is one of the problems to be solved. Thus, there is very little research on modeling chemical reactions with uncertain parameters and variables. It is also urgent to study the dynamics and stability of the first-order isothermal reactions described by fuzzy differential equations by computer simulation and to prove the validity of this approach.

© The Author(s), under exclusive license to Springer Nature Switzerland AG 2021
R. A. Aliev et al. (Eds.): ICAFS 2020, AISC 1306, pp. 508–514, 2021.
https://doi.org/10.1007/978-3-030-64058-3_63

In the study of chemical materials, for example, large amounts of data are processed in chemical reactions in catalytic materials. In this case, the need for the acquisition of useful knowledge from the data and the complexity of the issue make it necessary to use a new method. On the other hand, researches on modeling chemical reactions with uncertain parameters and variables are rare in the scientific literature. The study of the rate constant and stability of isothermal chemical reactions remains one of the main problems of kinetic modeling.

The methods based on fuzzy logic are used to solve such problems [3–7]. These methods allow verbal description. As we know, the information described in words is easier to understand, but its processing is not possible using classical mathematical methods. The scientific literature has shown that different approaches are used to manage processes characterized by uncertainty, depending on the type of uncertainty, such as stochastic approach, fuzzy approach. If uncertainty in the process is characterized by a lack of information, researchers prefer methods based on probability theory, and if uncertainty is associated with inaccuracy of information, they use methods based on fuzzy logic theory.

Despite mathematical analysis of chemical reaction systems have been conducted for a long period, this is still a topical issue, as evidenced by papers published in the last five years. Analysis of isothermal chemical reaction systems under certain and uncertain conditions reveals the problems here. However, the data characterized by uncertainty in chemical reactions require a new approach to this analysis.

One of the most successful technologies for the development of control systems for complex chemical processes has been proven in the scientific literature to be fuzzy logic technology.

Thus, fuzzy logic is a technology that can be used to solve real-life problems and can be used in chemical engineering to evaluate the processes of decomposition, combustion, separation, input and output rates in chemical reactions, and risk management in chemical reactions. Here, the factors that affect the rate of the reaction - temperature, particle size, the values of the factors that affect the stability and instability of the reaction, and so on, may be characterized by uncertainty.

There are considerable conceptual amount of works in this field, in the scientific literature, which are characterized by problems. As can be seen, the use of fuzzy logic in chemical processes is a topical issue and is one of the focus of analysts.

This study examines the alkylation reaction between propylene and benzene by using a fuzzy model.

The paper is structured as follows. In Sect. 2, we present some prerequisite material on fuzzy number, fuzzy inference etc. In Sect. 3 statement of the problem is given. In Sect. 4, we describe solution of the given problem. Section 5 offers some conclusions.

2 Preliminaries

Definition 1. Trapezoidal fuzzy number [1]. A fuzzy number $\tilde{A} = (a, b, c, d)$ is said to be a trapezoidal fuzzy number if its membership function is given by

$$\mu_{\tilde{A}}(x) = \begin{cases} 0, & x < a \\ \frac{x-a}{b-a}, & a < x \leq b \\ 1, & b < x < c \\ \frac{d-x}{d-c}, & c \leq x < d \\ 0, & x > d \end{cases}$$

where $a \leq b \leq c \leq d$.

Definition 2. Possibility measure based fuzzy inference method [4]. This method consist of the following steps:

1. Calculation the truth degree of the given rule as:

$$d_{jk} = Poss(\delta_k/a_{jk}) \cdot cf_k$$

2. Evaluated w_i objects has intend linguistic value defined as (δ_i, cf_i). where δ_i is linguistic value, $cf_k \in]0, 100]$ is confidence degree of the value δ_i. δ_k- linguistic value of the rule object, a_{jk}- current linguistic value (j is index of the rule, k is index of relation) value (for example, A_ir)
3. Calculation $D_j = (\min_j d_{jk}) * CF_j/100$, where CF is the confidence degree of the rule.
4. Designer of the rule determine the firing level (π) and $D_j \geq \pi$ is checked. If the antecedent holds true, then the consequent of rule is calculated.
5. Each evaluated w_i objects has S_i value: $w_i, (\delta_i^1, cf_i^1), \ldots, \ldots, (\delta_i^{S_i}, cf_i^{S_i})$ S_i is the number of the rules in fuzzy inference process.
6. The mean value is defined as follows:

$$\bar{v}_i = \frac{\sum_{n=1}^{S_i} \delta_i^n \cdot cf_i^n}{\sum_{n=1}^{S_i} cf_i^n}$$

3   Statement of the Problem

The reaction process is based on the alkylation reaction between propylene and benzene. Given the ratio between propylene alkylation rate constants and benzene, the product composition equations can be written as [6, 7]:

$$C_0 = 100e^{-k_1 t} \tag{1}$$

$$C_1 = 100(6.67e^{0.85k_1 t} - 6.667e^{-k_1 t}) \tag{2}$$

$$C_2 = 100(2.075e^{-0.28k_1t} - 9.94e^{-0.85k_1t} + 7.87e^{-k_1t}) \tag{3}$$

where C_0, C_1, C_2-are the yields of benzene, alkylbenzene and polymers, respectively.

The dependence of the molar ratio of propylene-benzene on the constant of the reaction rate is described by the equation:

$$m = 4 - 1.125e^{-0.02k_1t} - 1.913e^{-0.285k_1t} - 3.473e^{-0.85k_1t} + 2.110e^{-k_1t} \tag{4}$$

The rate constant k can be determined by various methods [6]. A simple and obvious way to determine it is given in work [6]: knowing the conversion time of half of the benzene involved in the reaction $\tau_{1/2}$, we can calculate the rate constant k by the equation [6]:

$$k = \frac{\ln 2}{\tau_{1/2}} \tag{5}$$

In the case $G'_{cp} = 0.25 mole/hour$, $m' = 1/0.8$ and $G'_{ctl} = 10\%$ to benzene found $k = 0.0055$ in relations to (1–3) the percentage composition of the products at the outlet of the reactor is determined by the Eqs. (6–8).

$$C'_0(t) = 100e^{-0.0055t} \tag{6}$$

$$C'_1(t) = 100(6.67e^{-0.004675t} - 6.667e^{-0.0055t}) \tag{7}$$

$$C'_2(t) = 100(2.071e^{-0.00154t} - 9.941e^{-0.004675t} + 7.87e^{-0.0055t}) \tag{8}$$

In this study the basic problem is to create fuzzy model and analyse the reaction rate depend on the yields of benzene, alkylbenzene and polymers.

4   Solution of the Problem

From the Eqs. (1–3) of the percentage components of benzene, alkylbenzene and polymer are used to create a fuzzy model of the alkylation of benzene with propylene.

For design of IF-Then fuzzy model fuzzy C-means approach [9] is used.

The initial data fragment is given in Table 1. By using data which described in Table 1 is defined centers of the clusters. Dataset contains 33 records: inputs- benzene (C0), alkylbenzene (C1) and polymers (C2), output-cofficient rate (K).

Table 1. Fragment of the initial data (solutions of differential equations)

c_0	c_1	c_2	K
99,999785715	0,302643535	0,099999812	0,0045
99,999778095	0,302737528	0,099999805	0,00466
99,999755238	0,303019505	0,099999785	0,00514
99,999747619	0,303113497	0,099999778	0,0053
99,999709524	0,303583459	0,099999745	0,0061
...	...	...	...
99,999625715	0,304617374	0,099999671	0,00786
99,999602858	0,304899351	0,099999651	0,00834
99,999580001	0,305181328	0,099999631	0,00882
99,999557144	0,305463305	0,099999611	0,0093

The centers of the clusters obtained from the data by the Fuzzy C-means method [9], and the fragment of the membership values are shown in Tables 2 and 3.

Table 2. The centers of the clusters

Clusters	c_0	c_1	c_2	K
Cluster 1	99.9996	0.3045	0.1000	0.0076
Cluster 2	99.9997	0.3041	0.1000	0.0069
Cluster 3	99.9997	0.3037	0.1000	0.0063
Cluster 4	99.9997	0.3041	0.1000	0.0070
Cluster 5	99.9996	0.3044	0.1000	0.0074

Constructed fuzzy model is the following form:

If amount of the benzene is norm and amount of alkyl benzene is very high and amount of the polymer is middle. Then coefficient of rate is high;

If amount of the benzene is high and amount of alkyl benzene is large and amount of the polymer is middle. Then coefficient of rate is low;

If amount of the benzene is high and amount of alkylbenzene is norm and amount of the polymer is middle. Then coefficient of rate is very low;

If amount of the benzene is high and amount of alkylbenzene is large and amount of the polymer is middle. Then coefficient of rate is low;

If amount of the benzene is norm and amount of alkylbenzene is high and amount of the polymer is middle. Then coefficient of rate is middle.

Table 3. Fragment values of the membership function

Cluster 1	Cluster 2	Cluster 3	Cluster 4	Cluster 5
0.1194	0.2002	0.3529	0.1881	0.1393
0.1147	0.1984	0.3665	0.1856	0.1349
0.1093	0.1960	0.3826	0.1825	0.1298
0.1031	0.1927	0.4020	0.1784	0.1238
0.0961	0.1883	0.4258	0.1731	0.1167
....	....	.....	.. ..	.... .
0.0813	0.1775	0.0188	0.3296	0.3929
0.0160	0.0057	0.0010	0.0082	0.9691
0.5529	0.0172	0.0044	0.0225	0.4031
0.9560	0.0045	0.0014	0.0055	0.0325

For realization of the created model is used ESPLAN shell (Fig. 1). Fuzzy model is described in expert system shell as PB-react.skb:

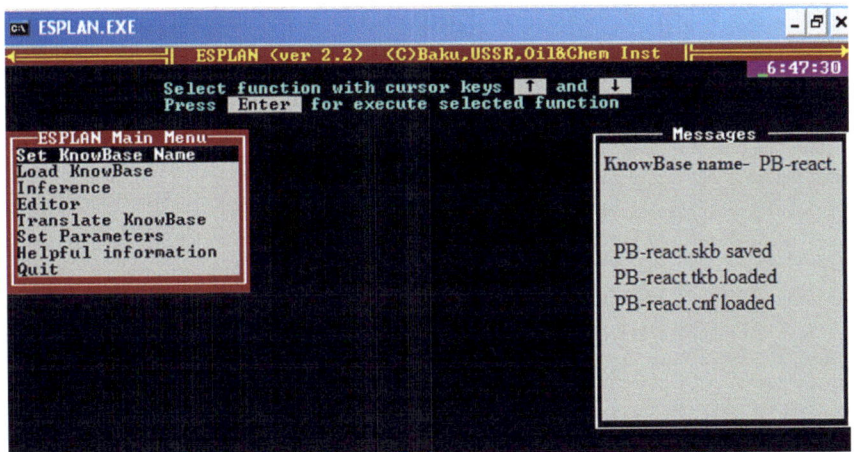

Fig. 1. Initial menu of the ESPLAN shell.

In this model, the ESPLAN knowledge base shell was used to investigate the relationship between the reactants and the rate (velocity).

In the expert system shell ESPLAN other linguistic terms can be used such *as a few, average, more than A, less than A, approximately A, from A to B, strict more than A, strict Less than A* and etc. For every linguistic value, ESPLAN automatically calculates fuzzy number by using the universe. The trapezoidal fuzzy numbers (definition 1) describing the used linguistic terms are given below:

Very-low or less than A: $(0, I, A - Z, Z)$;
Large or approximately A : (Z, A, A, Z);
High or more than a $= (Z, A + Z, S, 0)$;
Middle or neutral $= (Z, I, + 2 * Z, I + 3* Z, Z)$;

where I and S- respectively minimum and maximum value of universe, $Z = (S - I)/5$. For instance, object = "coefficient rate", I = minimum = 0.0045, S = maximum = 0.0095, linguistic term = "high": High = $(Z, A + Z, S, 0)$.

Our aim is to define the coefficient rate by fuzzy linguistic terms.

Test1. If amount of the benzene is high and amount of alkyl benzene is high and amount of the polymer is middle Then

Define the coefficient of rate.

Result: Coefficient rate is approximately 0.008.

5   Conclusion

A fuzzy logical inference method and Fuzzy C-means knowledge method were used to implement the model to study the reaction rate, and computer simulation was performed in MS Excel spreadsheet and MatLab environment, and the accuracy and efficiency of the results were determined. The study of the quality of fuzzy models proves the correctness of the established models and the applied approach. The proposed approach is superior than existing works in that it allows for the consideration of fuzzy uncertainty. This model can be applied to the study of the dynamic characteristics of industrial facilities.

References

1. Aliev, R.A.: Industrial invariant systems of the automatic control (Russian). Moscow (1971)
2. Emine, C., Aylin Bayrak, M.: A new method for solution of fuzzy reaction equation. Commun. Math. Comput. Chem. **73**, 649–661 (2015)
3. Aliev, R., Mamedova, G.: Aliev, R: Fuzzy Set Theory and its Application. Tabriz, Iran (1993)
4. Rafik, A.: Uncertain Computation-Based Decision Theory. World Scientific Publishing (2018)
5. Abdullayev, T.S., Gardashova, L.A., Aliev, B.F., Aliev, A.G., Ismailov, B.I.: Fuzzy expert system ESPLAN and its application in Business, Medicine and technics. In: Seventh International Conference on Application of Fuzzy Systems and Soft Computing, ICAFS-2006, pp. 205–215, Germany (2006)
6. Babanli, M.M.: Evaluation of desirable isothermal reactions rate in uncertain environment. Adv. Intell. Syst. Comput. Ser. **1095**, 366–372 (2019)
7. Dalin, N.A., Markosov, P.I., Shenderova, R.I., Prokofieva, T.V.: Benzene alkylation by olefins. Goskhimizdat, Moscow (1957). (in Russian)
8. Gardashova, L.A.: An algorithm based on a possibility measure and used to evaluate the job satisfaction index. Eastern-Eur. J. Enterprise Technol. 2(4(80)), 11–18 (2016). https://doi.org/10.15587/1729-4061.2016.66450
9. https://sites.google.com/site/dataclusteringalgorithms/fuzzy-c-means-clustering-algorithm

Fuzzy Based Wireless Channel Path Loss Prediction Model

Bülent Bilgehan$^{(\boxtimes)}$ [ID]

Near East University, Nicossia, Cyprus
bulent.bilgehan@neu.edu.edu

Abstract. The accuracy of the path loss in a communication channel is an important issue. There are different path loss models applied in the planning of wireless communication systems. However, the standard models rely on the local atmospheric characteristics of the environments. This research article introduces a combination of artificial intelligence and a fuzzy model named Neuro-fuzzy (NF) to predict the path loss in a wireless channel. The introduced model uses five layers of backpropagation gradient descent algorithm. The process uses the least square error method to estimate the optimal parameter values of the introduced model. The signal strength from the radio transmitter (Cyprus FM), operating at a frequency of 103.4 MHz was measured across three major cities in the country. The results of the newly introduced model compared with the classical models. The performance analysis of the NF path loss prediction model produced the best values in all respects. The newly introduced model produced the minimum Root Mean Square Error (RMSE), mean error (ME) and standard deviation error (SDE) compared with the classical models. The results verify to use the introduced model for better representation of the wireless channel characteristics.

Keywords: Wireless channel · Path loss · Channel characteristics · Artificial model · Fuzzy model

1 Introduction

An accurate representation of the path loss is very significant in the design procedure of the communication systems [1]. The true signal strength at the receiver can only be determined by the accurate path loss representation. Some receivers such as the Global System for Mobile Communications (GSM) mobile phone operate based on the threshold signal strength value [2]. Therefore, it is important to determine accurately the signal strength at a distance. The classical path loss models are simple in the calculation but do not include atmospheric and geometric factors [3]. This produces less sensitive path loss values for different locations [4]. Most of these models are tested in paper [5].

The geometric model in [5] managed to produce a more accurate path loss representation for the recorded data. However, the deterministic models are data-driven, and the recorded data include all the local atmospheric and geometric parameters. Therefore, they produce a better representation of path loss. Yet, they have not produced an exact representation of the recorded data. Further investigation is required to obtain optimum, a more accurate model with less complex calculation. The suitable approach is to use artificial intelligence (AI) to improve the prediction.

Several AI methods can be used in applications. The paper [6] applied an artificial neural network (ANN) for Microcell path loss prediction. The research paper [7] applied ANN for path loss prediction in the urban macrocellular environment. The path loss prediction becomes more difficult to model at the wireless sensors [8]. The ANN application resulted in an accurate wave propagation model in [9]. The accurate path loss model is closely linked with the localization of the transmitting source [10]. The fuzzy-based ANN is applied as in [11].

The performance of the newly introduced model compared with the widely used classical models such as Hata, Egli, ECC-33 and COST 231 models. The measure of the performance was based on the Root Mean Square Error (RMSE), Mean Error (ME), and Standard Deviation (SD) concerning the recorded data.

2   Method

There are two main parts under this section. The first section explains the recorded data and the second section describes the artificial intelligence-fuzzy combination model.

2.1   Signal Strength Measurements

The data were recorded in three different cities in Cyprus. The signal under test had a frequency of 2100 MHz. The data was recorded with the help of a software tool named as Test Mobile System (TEM) 16.3.6. The software displays the signal strength as well as the distance from the base station. The recorded measurement routes are namely Girne, Magusa and Lefkosa. The routes are selected to include different geometric conditions such as mountains and high building blocks. The average height of the buildings was approximately 6m. The total distance under the test was 80 km and recorded 100,000 data. The raw data passed through the filtering process to remove the noise. The testing equipment displayed in Fig. 1.

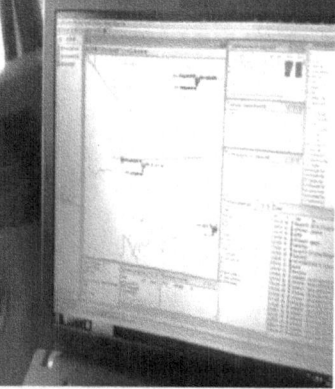

Fig. 1. Signal Measuring instruments.

The signal was measured in Magusa, Girne, Lefkosa, and Güzelyurt in Cyprus. The routes are selected to include the different geographical locations in the country.

The total coverage distance was 80 km and recorded 100,000 data. The raw data passed through a filter to remove the noise.

2.2 Artificial Intelligence-Fuzzy Combined Method

The artificial intelligence and fuzzy-based model produce an efficient mapping of inputs to outputs. The model uses five layers as shown in Fig. 2. The model includes fixed nodes are represented by circular and adaptive nodes are represented by rectangular shapes.

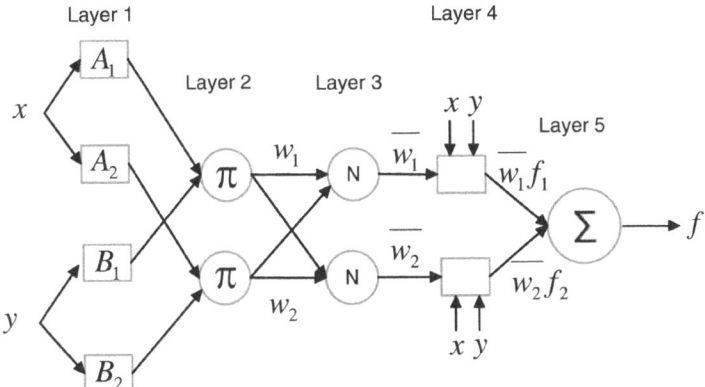

Fig. 2. Artificial intelligence and fuzzy structure.

The introduced structure uses a Sugeno model operated with two rules.

Rule 1: If x is A_1 and y is B_1 then

$$f_1 = p_1x + q_1y + r_1$$

Rule 2: If x is A_2 and y is B_2 then

$$f_2 = p_2x + q_2y + r_2$$

Layer 1: The node is an adaptable type and can be defined as;

$$L_j^1 = \mu A_i(x) \qquad i = 1, 2$$

The variable x denotes the input to the ith node, A_i is the adaptable node having a membership function of $\mu A_i(x)$. This membership function can be taken as;

$$\mu A_i(x) = \cfrac{1}{1 + \left[\left(\frac{x-f_i}{d_i}\right)^2\right]^{e_i}}$$

Equation above is the standard bell membership function (MF) which gives the best result compare with the other membership functions. The variables $\{d_i, e_i, f_i\}$ are the previous values.

Layer 2: This layer holds fixed nodes only. The function of the fixed node is to solve the instantaneous weights of w_1 and w_2. The inputs are multiplied at layer 2 as:

$$L_i^2 = w_i = \mu A_i(x) X \mu B_i(y), \qquad i = 1, 2$$

Layer 3: The nodes in this layer are constants. The outputs of the nodes are determined as:

$$L_i^3 = w_l = \frac{w_i}{\sum w_i}, \qquad i = 1, 2$$

Layer 4: This layer includes the adaptable nodes. The outputs of the nodes are determined as:

$$L_i^4 = w_i f_i = w_i(p_i x + q_i y + r_i), \qquad i = 1, 2$$

The variable values of p, q and r are determined by the least-squares method.
Layer 5: This layer sums all incoming data and the output is:

$$L_i^5 = \sum_{i=1}^{2} w_i f_i = \frac{\sum w_i f_i}{\sum w_i}$$

The recorded data applied as input to the design network. The initial process is to train the network. The training process uses a hybrid method so that it combines the backpropagation gradient descent algorithm and the least square estimation method. The least-square algorithm determines the output parameters. The output parameter values are transferred to the backpropagation algorithm to monitor the output error level until reduced to the pre-set value. This determines the end of the training process.

3 Results

The recorded data processed through the network given in Fig. 2. The output of the network gives the new path loss representation for different locations. The path loss representation of the classical models and the NF plotted in Fig. 3, 4.

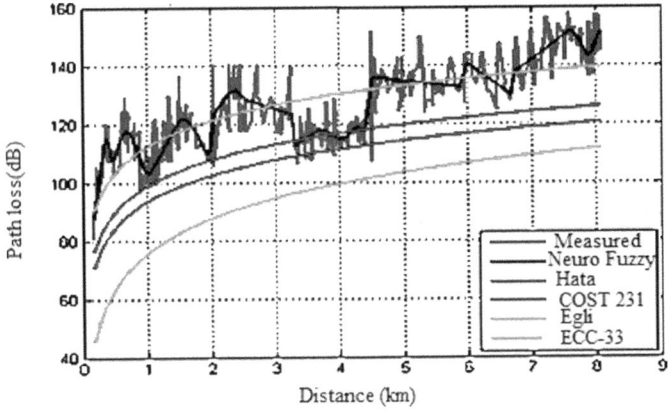

Fig. 3. Path loss in Magusa location.

The plots in Fig. 3 include the classical path loss models, recorded data and the NF in Magusa location. The closest path loss of the real data is NF. The ECC-33 model uses almost the average values and the others have lower performance than the NF model. Fig. 4 represents the data recorded at Lefkosa and the process was repeated as in Fig. 3.

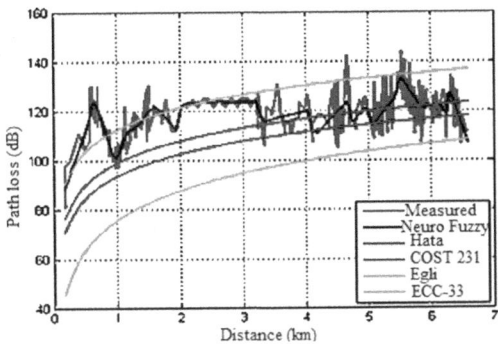

Fig. 4. Path loss in Lefkosa location.

The comparison of the plots in Fig. 4 shows the superiority of the introduced NF model. The ECC-33 model has a close representation at a distance of 2–3 km but not at other ranges. The distance from the transmitter extended to 13.5 km to evaluate the performance of the NF model. The results are plotted in Fig. 5.

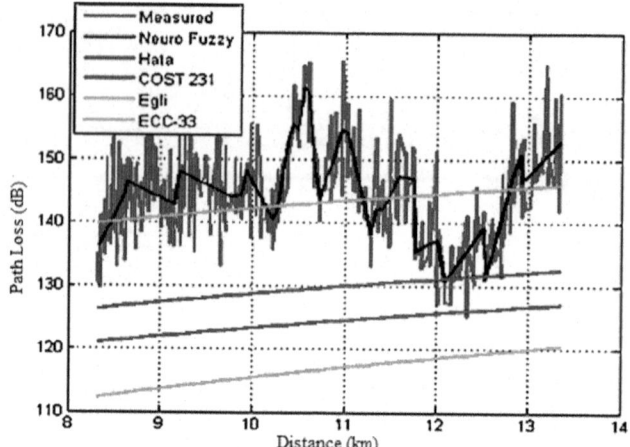

Fig. 5. Path loss in Girne location.

The plots in Fig. 5 shows the NF model to be the best amongst the others. It almost follows the recorded data. The second nearest predicting model is ECC-33.

Table 1. The Performance metrics for the measured locations.

Model		Locations			Average
		Girne	Lefkosa	Magusa	
NF	RMSE(dB)	5.0278	5.3432	5.6733	5.3481
	ME(dB)	−4.66E−06	−6.41E−06	−6.55E−06	−5.87E−06
	SDE(dB)	7.8164	14.6675	6.8821	9.7886
COST 231	RMSE(dB)	17.5156	21.8612	12.5812	17.3193
	ME(dB)	−14.9875	−19.8751	−8.8943	−14.5856
	SDE(dB)	5.9879	12.7649	8.7691	9.1739
HATA	RMSE(dB)	11.9786	16.8239	9.8743	12.8922
	ME(dB)	−11.2315	−15.4361	−5.2381	−10.6352
	SDE(dB)	6.6785	13.2893	9.1094	9.6924
EGLI	RMSE(dB)	22.2134	35.0758	25.3412	27.5434
	ME(dB)	−20.1672	−34.4376	−21.8734	−25.4927
	SDE(dB)	8.9856	17.7623	12.8129	13.1869
ECC-33	RMSE(dB)	8.9123	8.9945	12.4387	10.1151
	ME(dB)	3.8531	−1.7621	9.0721	3.7210
	SDE(dB)	7.0421	14.0351	10.2132	10.4301

Table 1 uses standard evaluation tools (RMSE, ME, SDE) for the error analysis. The evaluation considers all three locations. The acceptable range of RMSE value is between 0–7 dB [12]. The RMSE, ME and SDE are the lowest for the NF model. Such statistical analysis is sufficient to say NF is the best fitting model. A further comparison amongst the classical models the ECC-33 has the lowest values.

The other important aspect is to identify the minimum epoch while considering RMSE. Each route tested and the best selection plotted in Fig. 6.

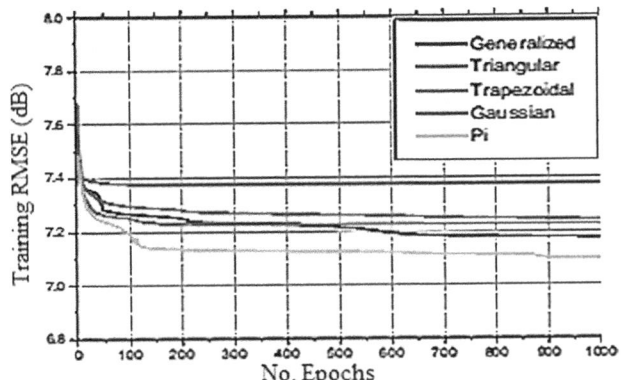

Fig. 6. Membership Function and the training RMSE for location Magusa

The best candidate to produce stability and convergence at a minimum number of iterations was identified to be Magusa location.

4   Conclusion

This work is based on the path loss prediction model using artificial intelligence. The NF path loss model tested, compared with classical models and the errors examined based on the standard statistical analysis methods. The NF model gives the lowest RMSE value of 5.3dB and ME of $-5.87E-06$. Amongst the classical models, ECC-33 gives the best prediction values. Furthermore, the SDE values of NF model are high as expected because it almost follows the measured values. Based on the achievements it can be stated that the NF model is the fittest for path loss prediction in Cyprus.

References

1. Al-Hourani, A., Kandeepan, S., Jamalipour, A.: Modeling air-to-ground path loss for low altitude platforms in urban environments. In: 2014 IEEE Global Communications Conference, pp. 2898–2904 (2014)
2. Vinggaard, N., Arora, A.S.: U.S. Patent No. 5,815,507. U.S. Patent and Trademark Office, Washington, DC (1998)

3. Abhayawardhana, V.S., Wassell, I.J., Crosby, D., Sellars, M.P., Brown, M.G.: Comparison of empirical propagation path loss models for fixed wireless access systems. In: 2005 IEEE 61st Vehicular Technology Conference, vol. 1, pp. 73–77 (2005)
4. Durgin, G., Rappaport, T.S., Xu, H.: Measurements and models for radio path loss and penetration loss in and around homes and trees at 5.85 GHz. IEEE Trans. Commun. **46**(11), 1484–1496 (1998)
5. Bilgehan, B., Ojo, S.: Multiplicative based path loss model. Int. J. Commun. Syst. **31**(17), e3794 (2018)
6. Ostlin, E., Zepernick, H.J., Suzuki, H.: Macrocell path-loss prediction using artificial neural networks. IEEE Trans. Veh. Tech. **59**(6), 2735–2747 (2010)
7. Mom, J.M., Mgbe, C.O., Igwue, G.A.: Application of artificial neural network for path loss prediction in urban macrocellular environment. Am. J. Eng. Res. **3**(2), 270–275 (2014)
8. Li, Z., Hong, T., Ning, W., Hong, Y., Wen, T., Li, J.: Path-loss prediction for radio frequency signal of wireless sensor network in field based on artificial neural network. Trans. Chin. Soc. Agr. Eng. **26**(12), 178–181 (2010)
9. Benmus, T.A., Abboud, R., Shatter, M.K.: Neural network approach to model the propagation path loss for great Tripoli area at 900, 1800, and 2100 MHz bands. In: 2015 16th International Conference on Sciences and Techniques of Automatic Control and Computer Engineering (STA), pp. 793–798 (2015)
10. Bilgehan, B., Abdulbari, A.: Fast detection and DOA estimation of the unknown wideband signal sources. Int. J. Commun. Syst. **32**(11) (2019)
11. Gharghan, S.K., Nordin, R., Jawad, A.M., Jawad, H.M., Ismail, M.: Adaptive neural fuzzy inference system for accurate localization of wireless sensor network in outdoor and indoor cycling applications. IEEE Access **6**, 38475–38489 (2018)
12. Hata, M.: Field strength and its variability in VHF and UHF land-mobile radio service. IEEE Trans. Veh. Tech. **29**(3) (1980)

The Assessing the Effectiveness of Reforestation Based on the Theory of Fuzzy Sets

Mustafa Ansary , Vovchenko Natalia , B. Stryukov Michael, and N. Kuz'minov Alexander$^{(\boxtimes)}$

Rostov State University of Economics, Rostov-on-Don, Russia
mustafa_007ru@yahoo.com, nat.vovchenko@gmail.com,
mstryukov@mail.ru, akuzminov@sfedu.ru

Abstract. The aim of the study is to develop a methodology for assessing the effectiveness of reforestation based in the Rostov region a set of indicators in three groups: the dynamics of indicators of the reforestation, the dynamics of the implementation of measures for thinning and information about patrolling the territory of the forest fund. The objective of the study is to develop an algorithm that allows the system on the basis of fuzzy logic conclusions form a comprehensive evaluation of the effectiveness of reforestation in the region on a range of diverse indicators ranked. Based method that allows forming a comprehensive evaluation of the effectiveness of reforestation of the time series of three groups: the dynamics of indicators of the reforestation, the dynamics of the implementation of measures for thinning and information about patrolling the territory of the forest fund. The contribution of each of the considered parameters is evaluated using the weighting factor reflecting its importance.

Keywords: Standard multilevel fuzzy [0, 1] - classifiers · Evaluation technique regeneration efficiency

1 Introduction

Currently, researchers are actively working to create methods for assessing the effectiveness of reforestation. So, in articles [1, 2] A method for evaluating the mechanism of the integrated use of forest resources in the region. Integral assessment of the relative sustainability of forest management includes the ranks of environmental, social and economic indicators,

$$W_0w = (nA_i - B_i - C - D) \times \frac{1}{2}, \tag{1}$$

where n - number of indicators under the root; Ai - the sum of the squares of the relative performance of system stability; Bi - sum of the relative performance in terms of (reserves); C - employment in the forestry sector in the region to the total number of able-bodied population; D - the share of the forest sector in the region's gross domestic product of the region.

R. A. Aliev et al. (Eds.): ICAFS 2020, AISC 1306, pp. 523–530, 2021.
https://doi.org/10.1007/978-3-030-64058-3_65

The article [3] estimates effects of fire, insects and windfall on the quality of stem wood in forests example Krasnoyarsk region, and also the development of correction factors when calculating the fee for commercial timber perform in plantations damaged fire, insects and windfall. The paper [4] studied the system of coefficients, to assess the effectiveness of regeneration over the years. It was found that for plantations at the age of 1–3 years, the effectiveness of reforestation evaluation criteria are: 1) compliance with work actually performed the designed technology of the plantation; 2) preservation of forest crops area; 3) survival forest crops. For forest cultures in the age of 5 is introduced index "Putting young trees in the wooded forest lands." For plants over the age of five years, introduced a system of indicators: 1) the proportion of forest plantations as part of wooded forest land; 2) the speaker is not covered with forest vegetation of forest land in need of reforestation; 3) coefficient of reforestation; 4) the coefficient of efficiency of regeneration; 5) Input ratio of young trees to the category of commercially valuable plants; 6) dynamic species composition plantations (factor change sawmills). Integral indicator assessment reforestation in this work, as well as in the previous one, is missing. Articles [5–7] are using the above cost implemented comparison and evaluation of various processes and regeneration felling work performed during development of forest areas. Article [8] is performed comparing biometric parameters of Scots pine wood cultures of different ages creates rows and biogroups and exemplary construction of mathematical models in the process of development of reforestation clumps (biogroups); including the use of polynomial approximation.

The paper [9] the regularities of natural regeneration in various site conditions for the coniferous-broad-leaved forests Average Volga region, scoring the technique of quantitative and qualitative characteristics for the objective purpose regret regeneration methods for taxation allocated. Using STATISTICA environment in the module "Classification and Regression Trees" cluster analysis of basic silvicultural factors influencing the presence and density of undergrowth. Interpolated developed a scale forest inventory of indicators to assess the prospects of methods of reforestation. Each scale correction coefficient adjusted force effect of this factor on the occurrence of regrowth. Scoping silvicultural factors accumulate the analyzed taxation allocated. Based on the amount of points. The paper [10] is search for the approaches in assessing the success of reforestation. A system of six figures and equations for their calculation; final grade is calculated on the basis of the generalized integral index. Article [11] carried out a cumulative analysis of the dynamics of reforestation in the Russian Federation. The study employed methods of analysis and synthesis methods for calculating statistical absolute, relative, average values, methods of construction and study time series, spreadsheet, graphics, monographic, statistical grouping method.

Thus, the analysis of literature shows that for evaluating the effectiveness reforestation currently widely used statistical whose coefficient, point patterns, and construction of standard integrated indicators. As the experience of the application of these methods and models, they have a rather low flexibility, do not allow a significant variation of the complex of the studied parameters do not take into account expert assessments. In this paper we propose a new method, based on the method of fuzzy sets theory [12], to evaluate the effectiveness of reforestation in three groups: the dynamics of indicators of the reforestation, the dynamics of the implementation of measures for thinning and information about patrolling the territory of the forest fund.

2   Experimental Part

2.1   General Principles of Methodology for Assessing the Effectiveness of Reforestation

To develop a methodology for assessing the effectiveness of reforestation, we used the basic principles of the author's assessment methodology, created for the effectiveness of agricultural production in the region [13, 14]. The methodology is based on fuzzy-multiple data aggregation tools, which is a modification of the standard five-level fuzzy [0, 1] - classifiers.

Stage 1. Formation of a list of significant indicators, on the basis of which the effectiveness of reforestation is assessed: 1) group of dynamics of reforestation; 2) a group of dynamics of the implementation of measures for thinning; 3) a group patrolling the territory of the forest fund.

Stage 2. The ranking of the studied indicators according to their importance for assessing the effectiveness of reforestation, the calculation of their weight coefficients based on expert estimates.

Stage 3. Calculation of normalized (that is, belonging to the interval [0, 1]) numerical values of the studied indicators for the considered period n years on the basis of formulas determined by the meaning of the problem.

Stage 4. Setting linguistic variables. To the indicators selected in Stage 3, we compared linguistic variables with term sets of five terms: "very low level of the indicator"; "Low rate"; "Average level indicator"; "High level indicator"; "A very high level indicator." The membership functions of linguistic variables are determined using standard trapezoidal functions.

We introduce the linguistic variables: γ = 'a comprehensive assessment of the effectiveness of reforestation'; γ_1 = 'assessment of the dynamics of reforestation'; γ_2 = 'evaluation of dynamics of implementation of measures to thinning'; γ_3 = 'rating patrolling forest areas'. Universal set for each linguistic variable is numeric interval [0, 1], and the set values of the four variables γ, γ_1, γ_2, γ_3 – term set G = {G1, G2, G3, G4, G5}, where G1 – 'steady trend towards reduction in growth'; G2 –'tendency to decrease growth'; G3 –'tendency to stagnation'; G4 – 'upward trend'; G5 – 'steady upward trend'. The accessory functions also have a standard trapezoidal shape.

Stage 5. The transition from the numerical values of indicators to the numerical values of estimates based on the general algorithm of the standard five-level fuzzy [0, 1] - classifiers.

Stage 6. Linguistic recognition of the obtained numerical estimates in accordance with the definition of the term set G = {G1, G2, G3, G4, G5}; analysis of the assessments of the effectiveness of reforestation, recommendations for the correction of the current situation.

2.2   Data

Estimation of the dynamics of reforestation performed based on a set of 15 indicators taking into account the positive and negative of the dynamics of their changes (data are taken from the annual publication 'Ecological Bulletin Don' [15] (Table 2):. 1) reforestation, total, ha; 2) including burns; 3) the survival rate of coniferous trees, %; 4) the survival rate of deciduous trees, %; 4) cultural care of forest plantations, ha; 5) soil cultivation under forest plantations, ha; 6) cultivation standard planting material thousand. Pcs. 7) procurement and purchase of seeds, kg; 8) putting young trees in the category of plants; 9) write-offs of forest cultures, all ha; 10) including decommissioned forest crops the first year of establishment; 11) incorporated valuable forest plantations - pine; 12) - locust acacia; 13) - oak.

Data are presented for 10 year, from 2009 to 2018.

Assessment of the dynamics of the implementation of measures for thinning carried out based on a set of five indicators: 1) clarifying and cleaning; 2) thinning; 3) accretion cuttings; 4) clear sanitary cuttings; 5) selective sanitary felling.

Evaluation patrol forest areas carried out based on a set of three indicators for 10 years, from 2009 to 2018: 1) number of patrol routes (units); 2) the total length of patrols (km); 3) the number of raids carried out on the territory of the forest fund of the Rostov region (units).

Data are presented for 10 years, from 2009 to 2018. Assessment of the dynamics of the implementation of measures for thinning carried out based on a set of five indicators in the 10 years from 2009 to 2018: 1) Clarifying and cleaning; 2) thinning; 3) Accretion cuttings; 4) Clear sanitary cuttings; 5) Selective sanitary felling.

Evaluation patrol forest areas carried out based on a set of three indicators for 10 years, from 2009 to 2018: 1) Number of patrol routes (units); 2) The total length of patrols (km); 3) The number of raids carried out on the territory of the forest fund of the Rostov region (units).

Thus, the effectiveness of reforestation in the Rostov region is necessary to assess 23 indicators based on available statistical data within 10-years old.

Therefore, the calculation of aggregate values x_i (i = 1, 2, ..., m1) studied performance regeneration during the period N-s performed based on the scheme, an integrating time-series data for each of the indicators, and taking into account the importance of the different time periods due to the weight coefficients:

$$x_i = \sum_{i=1}^{N-1} k_i I_i; \quad k_i = \frac{2(N - i)}{(N - 1)N} \qquad (2)$$

where ki - weighting coefficients determined on the basis of rules Fishburn, and the numbering of the time periods in the reverse order (i.e., in this example, the first period - in 2018, and the last, 10 the, 2009). Ii - normalized, i.e. divided by the highest value, the value of the indicator data. The growth of most of the indicators corresponds to an increase of efficiency of regeneration. At the same time, growth indicators 11 and 12 of Table 1 corresponds to a decrease of efficiency of regeneration, so the aggregated values are calculated as one minus the value calculated by formula (2). For calculations based on the above-described circuits designed software package [16]. Understanding sustainability also includes structural balance.

3   Results and Discussion

The calculation of estimates γ_1 = "assessment of the dynamics of reforestation";
   γ_2 = "evaluation of dynamics of implementation of measures to thinning";
   γ_3 = "rating patrolling forest areas".
   and γ = "a comprehensive assessment of the effectiveness of reforestation".
   The highest aggregated value of the indicator corresponds to the best situation, and
the smallest to the worst. As follows from Table 1, the highest values in time dynamics
correspond to such indicators as the reforestation fund (thousand ha), the survival rate
of coniferous and deciduous annuals. The worst situation is with indicators such as
reforestation in burned areas (ha), the laying of valuable forest crops - oak and pine.
Thus, the proposed technique allows for the quantitative comparison and ranking of the
dynamics of heterogeneous indicators, incomparable on the basis of conventional
models. The obtained aggregate assessment shows the dynamics of reforestation in the
Rostov Region corresponds to the term "good" rather than to "satisfactory" term.
   Analogical, indicators for assessing the dynamics of the implementation of measures for thinning can be ranked, in order of dynamics deterioration, as follows:

Table 1. Calculation of the assessment of the dynamics of reforestation, Rostov region: the aggregated value of the index

		Index	Weight indicator	Terms 1	2	3	4	5
1.	Reforestation fund (1000 ha)	0.933	1/15	0	0	0	0	1
2.	Reforestation, all (ha)	0.615	1/15	0	0	0.3	0.7	0
3.	including burns	0.158	1/15	0.9	0.1	0	0	0
4.	Survivalannual coniferous, %	0.843	1/15	0	0	0	0.1	0.9
5.	Survival annual deciduous%	0.874	1/15	0	0	0	0	1
6.	Cultural care of forest plantations, ha	0.797	1/15	0	0	0	0.5	0.5
7.	Soil cultivation under forest crops, ha	0.701	1/15	0	0	0	1	0
8.	Cultivation of standard planting material, 1000 Pcs	0,719	1/15	0	0	0	1	0
9.	Procurement and purchase of seeds, kg	0.480	1/15	0	0	1	0	0
10.	Putting young trees in category of plants	0.456	1/15	0	0	1	0	0
11.	Write-offs of forest crops, only, ha	0.726	1/15	0	0	0	1	0
12.	Including decommissioned forest crops the first year of creation, ha	0.709	1/15	0	0	0	1	0
	Incorporated valuable forest crops, ha							
13.	- Pine	0.427	1/15	0	0.2	0.8	0	0
14.	- locust acacia	0,792	1/15	0	0	0	0.6	0.4
15.	- oak	0.228	1/15	0.2	0.8	0	0	0
	Weight terms			11/150	11/150	31/150	59/150	38/150

γ_1 = 11/150 * 0,125 + 11/150 * 0.3 + 31/150 * 0.5 + 59/150 * 0.7 + 38/150 * 0.885 = 0.634
("More good than satisfactory")

selective sanitary cutting, through cutting, clarification and cleaning, thinning, clear sanitary cutting. The resulting aggregate assessment shows that the assessment of the dynamics of the implementation of measures on thinning in the Rostov Region corresponds, rather, to the term "good" than "satisfactory".

Finally, the indicators for assessing the patrol of the forest fund territory can be ranked in the following order: the total length of the patrol routes; the number of raids conducted on the territory of the forest fund; number of patrol routes. The obtained aggregate assessment shows that the assessment of patrolling the territory of the forest fund of the Rostov region corresponds to the term "excellent".

Table 2 aggregated the three assessments considered into the final assessment of the effectiveness of reforestation in the Rostov Region. At the same time, based on Table 2, it is possible to analyze the effectiveness of reforestation by groups. As the values of the estimates indicate, the best situation is in the area of patrolling the territory of the forest fund ("excellent"). Rather, "good" than "satisfactory" can be assessed the dynamics of indicators for reforestation. Finally, the dynamics of the implementation of thinning operations can be assessed as "satisfactory" rather than "good". The final assessment of the effectiveness of reforestation in the Rostov region is "good".

Thus, the analysis shows that in order to improve the situation with reforestation in the Rostov Region, it is necessary to pay attention to the implementation of measures for logging, primarily thinning and clear cutting. In the field of general indicators of reforestation, attention should be paid to reforestation in burned areas (ha), as well as the laying of valuable forest crops - oak and pine.

Table 2. Calculation of final assessment the effectiveness of regeneration in Rostov region

	Evaluation	Numerical evaluation value	Weight indicator	Terms				
				1	2	3	4	5
1.	Evaluation of the dynamics of reforestation, G1	0.63	0.5	0	0	0.2	0.8	0
2.	Assessment of the dynamics of the implementation of measures for thinning, G2	0.59	0.3	0	0	0.6	0.4	0
3.	Evaluation patrolling forest, G3	0.87	0.2	0	0	0	0	1
Weight terms				0	0	0.28	0.52	0.2

$\gamma = 0 * 0,125 + 0 * 0.3 + 0.28 * 0.5 + 0.52 * 0.7 + 0.2 * 0.885 = 0.87$ ("good")

4 Conclusion

The methodology for a comprehensive assessment of the effectiveness of forestry regeneration in the region on the basis of standard 5-point classifiers develops approaches to modeling sustainability. The practical significance of the methodology is that it makes it possible to form a comprehensive assessment of the effectiveness of reforestation from time series of three groups: the dynamics of reforestation indicators, the dynamics of the implementation of thinning activities and information on patrolling, the territory of the forest fund on the basis of available reliable data. The contribution of each of the considered parameters is estimated using a weighting factor that reflects its importance, which increases the objectivity of the result.

In comparison with similar assessment methods, the proposed approach has a number of advantages, such as: 1) simple calculation scheme; 2) the significance of indicators that can vary depending on the characteristics; 3) the opportunity to take into account the influence of each parameter in the overall result is presented; 4) the possibility of ranking regions according to the effectiveness of reforestation has been taken into account; 5) ensures that optimal corrective action is taken by stakeholders.

References

1. Prudsky, V.G., Neznakina, K.V.: Methods of assessing the effectiveness of the mechanism for the integrated use of forest resources in the region. PSU Bull. Ser.: Econ. **4**(19), 60–65 (2013)
2. Runova, E.M., Grebenuk, A.L.: Reproduction of forests based on sustainable forest management criteria. Actual Probl. Forestry Complex **15**, 47–50 (2006). (in Russian)
3. Ivanov, V.A., Brezinskaya, L.V., Ivanov, A.V., Friedrich, I.E.: Ecological and economic evaluation of stem wood quality losses in the plantation after the impact of forest fires and insect. Herald KrasGAU **7**, 101–122 (2017). (in Russian)
4. Zekunova, A.I.: Evaluating the effectiveness of reforestation. GIABA **10** (2008). (in Russian)
5. Rodionov, A.V.: Resource conservation as a criterion of development of forest areas. Invest. Russia **8**, 278–283 (2005). (in Russian)
6. Rodionov, A.V.: Evaluation of forest technologies based on resource conservation. Herald MSFU. Forest Gazette **6**, 11–18 (2005). (in Russian)
7. Pisareva, S.V.: About statistical analysis of mathematical models of crop growth forest. Sci. Statements BSU. Ser.: Math. Phys. **27**(276), 104–108 (2017). (in Russian)
8. Black, L.V., Chernykh, D.V., Denisov, S.A.: Forestry statistical approach for assigning reforestation methods. The universities. Forest Mag. **4**(358), 9–23 (2017). https://doi.org/10.17238/issn0536-1036. (in Russian)
9. Rozhkov, L.N., Kuzmenko, M.V., Kulagin, A.P., Homets, V.N.: Evaluation of the structure and productivity of forests with reforestation and afforestation. In: Proceedings BSTU. Series 1: Forestry, Natural Resources and Recycling of Renewable Resources, vol. 1, pp. 115–117 (2012). (in Russian)
10. Salimov, G.A., Hasrudikov, R.M.: Forest resources assessment of reproduction in the Russian Federation. News OGAU **2**(70), 90–94 (2018). (in Russian)
11. Nedosekin, A.: Fuzzy financial management. AFA Library, Moscow (2003). (in Russian)

12. Alekseychik, T.V., Bogachev, T.V., Karasev, D.N., Sakharova, L.V., Stryukov, M.B.: Fuzzy method of assessing the intensity of agricultural production on a set of criteria of the level of intensification and the level of economic efficiency of intensification. In: Advances in Intelligent Systems and Computing, vol. 896, pp. 790–798 (2019). https://doi.org/10.1007/978-3-030-04164-9_83

13. Vovchenko, N.G., Stryukov, M.B., Sakharova, L.V., Domokur, O.V.: Fuzzy-logic analysis of the state of the atmosphere in large cities of the industrial region on the example of Rostov region. In: Advances in Intelligent Systems and Computing, vol. 896, pp. 709–715 (2019). https://doi.org/10.1007/978-3-030-04164-9_93

14. Ecological Don Gazette. On the state of the environment and natural resources of the Rostov region in 2018. The government of the Rostov region, Rostov-on-Don (2018)

15. Albekov, A.U., Arapova, E.A., Karasev, D.N., Strukov, M.B., Sakharova, L.V.: A program for assessing the intensity of agricultural production through a fuzzy 5-point classifier. Federal Service for Intellectual Property, Computer Registration Certificate No. 2018613875 (2018)

16. Kuzminov, A.N., Korostieva, N.G., Dzhukha, V.M., Ternovsky, O.A.: Economic coenosis stability: methodology and findings. Contemp. Issues Bus. Finan. Manage. Eastern Eur. **100**, 61–70 (2018)

Weighted Assessment of the Microcredit Borrower Solvency Using a Fuzzy Analysis of Personal Data

Elchin Aliyev[✉] and Zaur Gaziyev

Institute of Control Systems of ANAS, Vahabzadeh str. 9,
1141 Baku, Azerbaijan
elchin.aliyev@sinam.net, Z.gaziyev@gmail.com

Abstract. In practice, the opinions of different analysts or those responsible for making credit decisions often differ, especially if controversial situations are considered that have many acceptable alternative solutions. As a result, in assessing the solvency of potential microloan borrowers, the subjective opinion of the expert and the incompetent or deliberate interpretation of the information resulting in the adoption of decisions that are detrimental to the microfinance organization are overweight. To increase the degree of objectivity, the paper discusses an approach to assessing the responsibility and solvency of microloan borrowers, based on the use of the fuzzy maxmin convolution method of multi-criteria choice of alternatives under uncertainty. Taking into account the weakly structured personal data of applicants this approach allows them to be flexibly and promptly assessed for microcredits. The applied qualitative criteria of assessment are weighed on the base of agreed expert opinions relative to priority of each of them. An important advantage of the applied model is that it is simple, convenient to use and able to adapt to the requirements of various commercial banks and micro-financial organizations. The model has been tested on the example of ten hypothetical microloan borrowers.

Keywords: Microcredit · Borrower solvency · Credit scoring · Fuzzy set · Membership function · Maxmin convolution

1 Introduction

In the context of the dynamic development of the financial services market, micro crediting, as a relatively new banking service, began to gain popularity throughout the post-Soviet space. According to data of the international organization Microfinance Information Exchange [1], at the end of 2017 Azerbaijan occupied the 2nd place in the microcredit market among the CIS countries with a specific weight of 24% and a credit portfolio of 337.6 Mio. US dollars. At the scale of Azerbaijan, this means that a fairly large part of its adult population at least once in their life obtained microcredits. Nevertheless, for the microfinance sector of the economy of Azerbaijan, system risk management still remains an unsolved problem, especially since over the past years there has been a volatile increase of microcredit value, its depth and geography. There are still great opportunities for further development in the republic, since the microcredit market is still "underserved". Along with this, there are changes in the markets for goods and services, in the technologies for their delivery. All this becomes the reason that new requirements for the accuracy of assessing the responsibility and solvency of microloan borrowers are increasing, and financial resources are being reduced. New effective and progressive methods of operational comprehensive assessment are needed in combination with the borrower's credit history, which is a stable, effective and inexpensive source for analyzing the responsibility and solvency of microloan borrowers.

In the process of micro crediting, the assessment of the borrower's credit history is usually carried out by an expert who relies on his heuristic knowledge and intuition. But such an assessment still remains the product of the subjective judgments of the expert. Therefore, in real situations, the opinions of different experts often differ, especially in cases where many alternative solutions are considered. Reducing the expert's influence on the decision and increasing the share of objective factors can be achieved by formalizing the borrower's behavior and, in fact, the decision-making procedure for crediting. On the other hand, retrospective credit histories are documents with weakly structured data, i.e. their belonging to a certain type is known. For example, data can be presented in the form of segment $x \in [x_{min}, x_{max}]$, or verbally as "x close to t 13", i.e. in the form of fuzzy sets [2]. Thus, on the basis of these assumptions, it becomes obvious the actuality and necessity of developing a fuzzy method for analyzing the loan portfolio of a microfinance organization for the evaluation of the quality of performance of borrowers' liability to repay their debt based on the construction of membership functions of fuzzy sets describing weakly structured data from corresponding credit history. The application of this approach will allow the loan officer to quickly receive reports with assessments of borrowers' "responsibility", classify their credit histories as, for example, "positive" and/or "negative", as paid or not paid for their liability.

2   Scoring Analysis of the Borrower's Credit History

Scoring analysis, as a tool for assessing the borrower's solvency has been successfully used by banks for many decades. In the field of micro crediting, data on customers are taken directly from their requests, and also, if necessary, from open Internet sources. For each of the following points, borrowers are awarded the corresponding points [3]: x_1 (*age*) – from 0.1 to 0.3 points: the older the person, the more reliable; x_2 (*sex*) – 0 points for men and 0.4 points for women: it means that the weaker sex is considered a more responsible payer and less prone to risks and adventures; x_3 (*settled*) – $0.042 \div 0.42$ points: the value directly depends on the time (number of years) of the borrower's residence at the address provided; x_4 (*work risks*) – 0 points for those working in dangerous undertaking, 0.16 points – for those working at moderate risk to life, 0.55 points – for those working in safe industries; x_5 (*work in a large company*) – 0.21 points: added to the asset, if the borrower works in a large company that is more reliable than small firms; x_6 (*seniority*) – $0.059 \div 0.59$ points: the longer a person works, the more reliable; x_7 (*assets*) – 0.45 points: added for each item separately – for the presence of insurance, property and deposit account. In order to qualify for micro crediting, the borrower must get a grade at least 1.25 points. Moreover, as it is not difficult to calculate, the maximum score on this scale is 3.82. Now, suppose that a bank considers 10 requests from natural persons for the extension of microcredits. As a result of verification and preliminary analysis of personal data for all the above points of graduation, all borrowers were awarded the appropriate points by bank manager, which are summarized in Table 1.

Table 1. Preliminary results of scoring analysis of personal data of microloan applicants.

Borrower	Assessment criterion						
	x_1	x_2	x_3	x_4	x_5	x_6	x_7
a_1	0.161	0.0	0.044	0.00	0.00	0.061	0.45
a_2	0.222	0.4	0.132	0.55	0.21	0.326	1.35
a_3	0.117	0.4	0.231	0.16	0.00	0.350	0.90
a_4	0.185	0.0	0.374	0.55	0.00	0.248	0.45
a_5	0.298	0.4	0.268	0.16	0.00	0.069	0.45
a_6	0.213	0.0	0.412	0.00	0.21	0.473	1.35
a_7	0.109	0.0	0.172	0.55	0.00	0.424	0.45
a_8	0.205	0.4	0.389	0.16	0.21	0.164	0.90
a_9	0.171	0.4	0.303	0.00	0.00	0.504	1.35
a_{10}	0.255	0.0	0.253	0.16	0.21	0.261	0.90

3   Fuzzification of Quality Evaluation Criteria

In essence, evaluation concepts x_i $(i = 1 \div 7)$ of scoring analysis are qualitative criteria, which are the terms of the corresponding linguistic variables. Moreover, the numerical estimates of the considered alternative borrowers are represented by the degrees of compliance with these evaluation concepts. Starting from this, we denote the set of alternative borrowers as $A = \{a_1, a_2, ..., a_{10}\}$, and the set of evaluation criteria as $F = \{F_1, F_1, ..., F_7\}$, where each of the criteria is a fuzzy subset of the discrete universal A in form: $F_i = \{\mu_{Fi}(a_1)/a_1; \mu_{Fi}(a_2)/a_2; ...; \mu_{Fi}(a_{10})/a_{10}\}$ $(i = 1 \div 7)$. As membership functions of these subsets, following Gaussian functions is chosen: $\mu_i(u) = \exp\{-(u - u_i)^2/\sigma_i^2\}$, $u \in [0, u_i]$, $i = 1 \div 7$, where u_i is maximum score in the corresponding gradation of scoring analysis; $\sigma_i^2 = \sum_{k=1}^{41}(u - u_i)^2/41$ is standard deviation. Then the concepts x_i, characterizing the solvency of borrowers, can be described in the form of the following fuzzy sets:

- PREFERABLE (age): $F_1 = \{0.5294/a_1, 0.8185/a_2, 0.3320/a_3, 0.6470/a_4, 0.9999/a_5, 0.7794/a_6, 0.3009/a_7, 0.7430/a_8, 0.5782/a_9, 0.9355/a_{10}\}$;
- PREFERABLE (sex): $F_2 = \{0.0517/a_1, 1.0000/a_2, 1.0000/a_3, 0.0517/a_4, 1.0000/a_5, 0.0517/a_6, 0.0517/a_7, 1.0000/a_8, 1.0000/a_9, 0.0517/a_{10}\}$;
- DESIRED (settled): $F_3 = \{0.0930/a_1, 0.2483/a_2, 0.5488/a_3, 0.9651/a_4, 0.6784/a_5, 0.9989/a_6, 0.3559/a_7, 0.9840/a_8, 0.7946/a_9, 0.6260/a_{10}\}$;

- MINIMAL (work risk): $F_4 = \{0.0517/a_1, 1.0000/a_2, 0.2254/a_3, 1.0000/a_4, 0.2254/a_5, 0.0517/a_6, 1.0000/a_7, 0.2254/a_8, 0.0517/a_9, 0.2254/a_{10}\}$;
- PREFERENTIAL (work): $F_5 = \{0.0517/a_1, 1.0000/a_2, 0.0517/a_3, 0.0517/a_4, 0.0517/a_5, 1.0000/a_6, 0.0517/a_7, 1.0000/a_8, 0.0517/a_9, 1.0000/a_{10}\}$;
- SIGNIFICANT (seniority): $F_6 = \{0.0924/a_1, 0.5525/a_2, 0.6125/a_3, 0.3695/a_4, 0.0992/a_5, 0.8900/a_6, 0.7909/a_7, 0.2134/a_8, 0.9390/a_9, 0.3980/a_{10}\}$;
- SUITABLE (assets): $F_7 = \{0.2680/a_1, 1.0000/a_2, 0.7195/a_3, 0.2680/a_4, 0.2680/a_5, 1.0000/a_6, 0.2680/a_7, 0.7195/a_8, 1.0000/a_9, 0.7195/a_{10}\}$.

It is obvious that the criteria F_i $(i = 1 \div 7)$ have different significance in assessing the responsibility and solvency of borrowers. It is necessary to identify the generalized weights α_i $(i = 1 \div 7)$ of these criteria. Therefore, we use the method of expert estimates of the weighted coefficients of the estimated indicators from personal data.

4 Expert Identification of Evaluation Criteria Weights

Expert identification of criteria weights for assessing the responsibility of borrowers implies: 1) ranking of criteria x_i for their priority; 2) cluster estimate of the normalized values of the generalized weights of the criteria x_i based on their relative influence on the solvency level of borrower. Based on this scheme, suppose that through an independent survey of 15 experts, rank estimates of the priority of criteria x_i were obtained. At the same time, each expert was proposed to arrange x_i relative to the principle: the most important criterion should be indicated by the number "1", the next less important criterion – by the number "2" and, further, in decreasing order of expert preference. The obtained rank estimates for the all criteria are summarized in Table 2.

To establish the degree of consistency of expert opinions, the Kendall's concordance coefficient W is used, which demonstrates the multiple rank correlation of expert opinions. According to [4], this coefficient is calculated by formula: $W = 12 \cdot S/[m^2(n^3 - n)]$, where m is the number of experts; n is the number of evaluation criteria; S is the deviation of expert opinions from the average value of the ranking of criteria x_i, which is calculated as [4]: $S = \sum_{i=1}^{n} [\sum_{j=1}^{m} r_{ij} - m(n+1)/2]^2$, where r_{ij} is the rank of the i-th criterion established by the j-th expert. In our case (see Table 2), $r_{ij} \in \{1, 2, ..., 7\}$, and the deviation is calculated as $S = \sum_{i=1}^{7} [\sum_{j=1}^{15} r_{ij} - 15(7+1)/2]^2 = 4984$. Then the Kendall's concordance coefficient will be the following number: $W = 12 \cdot 4984/[15^2(7^3-7)] = 0.7911$, which indicates a fairly strong consistency of expert conclusions relative to importance degrees of the criteria x_i $(i = 1 \div 7)$. Further, suppose that at the preliminary stage the values of normalized estimates of the criteria weights are established by experts and also are summarized in Table 2.

According to [6], preliminary calculations for the subsequent identification of the criteria weights are carried out by group averaging of normalized estimates by formula

$$\alpha_i(t+1) = \sum_{j=1}^{m} w_j(t)\alpha_{ij} \tag{1}$$

where $w_j(t)$ is the degree of competence of the j-th expert $(j = 1 \div m)$ at time t, which is calculated by following equalities:

$$\begin{cases} w_j(t) = [1/\eta(t)] \sum_{i=1}^{n} \alpha_i(t) \cdot \alpha_{ij} \ (j = \overline{1, m-1}), \\ w_m(t) = 1 - \sum_{j=1}^{m-1} w_j(t), \ \sum_{j=1}^{m} w_j(t) = 1. \end{cases} \tag{2}$$

where $\eta(t)$ is the normalizing factor calculated as $\eta(t) = \sum_{i=1}^{n} \sum_{j=1}^{m} \alpha_i(t)\alpha_{ij}$.

Table 2. Expert judgments of criteria ranks and normalized values of criteria weights.

Expert	Evaluation criteria, their ranks and normalized estimates													
	x_1		x_2		x_3		x_4		x_5		x_6		x_7	
	r_{1j}	α_{1j}	r_{2j}	α_{2j}	r_{3j}	α_{3j}	r_{4j}	α_{4j}	r_{5j}	α_{5j}	r_{6j}	α_{6j}	r_{7j}	α_{7j}
e_1	5	0.075	7	0.025	3	0.200	6	0.050	2	0.250	4	0.100	1	0.300
e_2	6	0.050	7	0.025	4	0.100	5	0.075	1	0.300	3	0.200	2	0.250
e_3	6	0.050	5	0.075	2	0.250	7	0.025	3	0.200	4	0.100	1	0.300
e_4	5	0.075	6	0.050	2	0.250	7	0.025	1	0.300	4	0.100	3	0.200
e_5	4	0.100	7	0.025	1	0.300	6	0.050	3	0.200	5	0.075	2	0.250
e_6	5	0.075	7	0.025	4	0.100	6	0.050	2	0.250	3	0.200	1	0.300
e_7	7	0.025	5	0.075	3	0.200	4	0.100	2	0.250	6	0.050	1	0.300
e_8	6	0.050	7	0.025	3	0.200	5	0.075	1	0.300	4	0.100	2	0.250
e_9	4	0.100	6	0.050	3	0.200	7	0.025	2	0.250	5	0.075	1	0.300
e_{10}	5	0.075	7	0.025	4	0.100	6	0.050	1	0.300	3	0.200	2	0.250
e_{11}	6	0.050	7	0.025	4	0.100	5	0.075	2	0.250	3	0.200	1	0.300
e_{12}	5	0.075	7	0.025	2	0.250	6	0.050	3	0.200	4	0.100	1	0.300
e_{13}	4	0.100	7	0.025	5	0.075	6	0.050	2	0.250	3	0.200	1	0.300
e_{14}	4	0.100	6	0.050	3	0.200	7	0.025	2	0.250	5	0.075	1	0.300
e_{15}	7	0.025	6	0.050	1	0.300	5	0.075	4	0.100	2	0.250	3	0.200
Σ	79	1.025	97	0.575	44	2.825	88	0.800	31	3.650	58	2.025	23	4.100

The group averaging process is completed under the following condition [6]

$$\max_{i=1\div 7} \{|\alpha_i(t+1) - \alpha_i(t)|\} \leq \varepsilon \tag{3}$$

where ε is the accepted miscalculation. Then, assuming that $\varepsilon = 0.001$ and at the initial stage $t = 0$ the experts' competency indicators are identical as $w_j(0) = 1/15$, according to (1) the average values for the groups of normalized estimates of the criteria weights in a first approximation are obtained as $\alpha_i(1) = \sum_{j=1}^{15} w_j(0)\alpha_{ij} = \sum_{j=1}^{15} \alpha_{ij}/15$ $(i = 1 \div 7)$. In this case, we have: $\alpha_1(1) = 0.0683$; $\alpha_2(1) = 0.0383$; $\alpha_3(1) = 0.1883$; $\alpha_4(1) = 0.0533$; $\alpha_5(1) = 0.2433$; $\alpha_6(1) = 0.1350$; $\alpha_7(1) = 0.2733$. Obviously, condition (3) does not satisfy for this approximations. Therefore, to go to the next stage, it is necessary to calculate the corresponding normalizing factor η (1):

$$\eta(1) = \sum_{i=1}^{7} \sum_{j=1}^{15} \alpha_i(1)\alpha_{ij} = 1.025 \cdot 0.0683 + 0.575 \cdot 0.0383 + 2.825 \cdot 0.1883 + 0.800 \cdot 0.0533$$
$$+ 3.650 \cdot 0.2433 + 2.025 \cdot 0.1350 + 4.100 \cdot 0.2733 = 2.9490.$$

Then, according to (2) or, more specifically, from the equalities:

$$\begin{cases} w_j(1) = [1/\eta(1)] \sum_{j=1}^{14} \alpha_i(1) \cdot \alpha_{ij}(j = \overline{1, 14}), \\ w_{15}(1) = 1 - \sum_{j=1}^{14} w_j(1), \sum_{j=1}^{15} w_j(1) = 1, \end{cases}$$

new indicators of expert competence are obtained in the form of the following numbers:
$w_1(1) = 0.0726$; $w_2(1) = 0.0700$; $w_3(1) = 0.0709$; $w_4(1) = 0.0703$; $w_5(1) = 0.0691$;
$w_6(1) = 0.0706$; $w_7(1) = 0.0706$; $w_8(1) = 0.0719$; $w_9(1) = 0.0719$; $w_{10}(1) = 0.0702$;
$w_{11}(1) = 0.0705$; $w_{12}(1) = 0.0713$; $w_{13}(1) = 0.0696$; $w_{14}(1) = 0.0719$; $w_{15}(1) = 0.0086$.

Further, assuming $\alpha_i(2) = \sum_{j=1}^{15} w_j(1)\alpha_{ij}$, in the 2nd approximation the average values for the groups of normalized estimates of criteria weights are obtained in the form of the following numbers: $\alpha_1(2) = 0.0708$; $\alpha_2(2) = 0.0377$; $\alpha_3(2) = 0.1823$; $\alpha_4(2) = 0.0520$; $\alpha_5(2) = 0.2514$; $\alpha_6(2) = 0.1282$; $\alpha_7(2) = 0.2776$. Checking these values for the fulfillment of condition (3): $\max\{|\alpha_i(2)-\alpha_i(1)|\} = \max\{|0.0708-0.0683|$; $|0.0377-0.0383|$; $|0.1823-0.1883|$; $|0.0520-0.0533|$; $|0.2514-0.2433|$; $|0.1282-0.135|$; $|0.2776-0.2733|\} = 0.0081 > \varepsilon$,
we make sure that it is not satisfied again. Therefore, it is necessary to calculate the following normalizing factor:

$$\eta(2) = \sum_{i=1}^{7} \sum_{j=1}^{15} \alpha_i(2)\alpha_{ij} = 1.025 \cdot 0.0708 + 0.575 \cdot 0.0377 + 2.825 \cdot 0.1823$$
$$+ 0.800 \cdot 0.0520 + 3.650 \cdot 0.2514 + 2.025 \cdot 0.1282 + 4.100 \cdot 0.2776 = 2.9662.$$

In this case, the relevant indicators of expert competence will be the numbers:
$w_1(2) = 0.0726$; $w_2(2) = 0.0700$; $w_3(2) = 0.0709$; $w_4(2) = 0.0703$; $w_5(2) = 0.0691$;
$w_6(2) = 0.0706$; $w_7(2) = 0.0706$; $w_8(2) = 0.0719$; $w_9(2) = 0.0719$; $w_{10}(2) = 0.0702$;
$w_{11}(2) = 0.0705$; $w_{12}(2) = 0.0713$; $w_{13}(2) = 0.0696$; $w_{14}(2) = 0.0719$; $w_{15}(2) = 0.0086$. Then, assuming for the 3rd approximations $\alpha_i(3) = \sum_{j=1}^{15} w_j(2)\alpha_{ij}$, the group averaging for normalized estimates of criteria weights are obtained in the form of the following numbers: $\alpha_1(3) = 0.0710$; $\alpha_2(3) = 0.0376$; $\alpha_3(3) = 0.1816$; $\alpha_4(3) = 0.0519$; $\alpha_5(3) = 0.2523$; $\alpha_6(3) = 0.1275$; $\alpha_7(3) = 0.2781$. As seen from:

$$\max\{|\alpha_i(3) - \alpha_i(2)|\} = \max\{|0.0710 - 0.0708|; |0.0376 - 0.0377|; |0.1816 - 0.1823|; |0.0519 - 0.0520|; |0.2523 - 0.2514|; |0.1275 - 0.1282|; |0.2781 - 0.2776|\} = 0.0009 < \epsilon,$$

condition (3) is already fulfilled, which is the basis for the interruption of calculations. This means that $\alpha_1(3)$, $\alpha_2(3)$,..., $\alpha_7(3)$ are the final generalized weights of the corresponding criteria x_i.

5 Assessment and Ranking of Borrowers' Reliability and Solvency Using the Fuzzy Maxmin Convolution Method

So, taking into account the various degrees of importance of the evaluation criteria, let us form the set of optimal alternatives A by intersection of the above obtained fuzzy sets $F_i(a)$ in the form: $A = F_1^{\alpha_1} \cap F_2^{\alpha_2} \cap ... \cap F_7^{\alpha_7}$, where, according to [5], the j-th alternative borrower $(j = 1 \div 10)$ is the best if the following equation

$$\mu_A(a_j) = \max\{\mu_A(a_1), \mu_A(a_2), \ldots, \mu_A(a_{10})\}$$

is satisfied. In this case, the set of optimal alternatives is formed as follows:
$A = \{\min\{0.5294^{0.0710}; \quad 0.0517^{0.0376}; \quad 0.0930^{0.1816}; \quad 0.0517^{0.0519}; \quad 0.0517^{0.2523}; 0.0924^{0.1275}; 0.2680^{0.2781}\}; \min\{0.8185^{0.0710}; 1.0000^{0.0376}; 0.2483^{0.1816}; 1.0000^{0.0519}; 1.0000^{0.2523}; 0.5525^{0.1275}; 1.0000^{0.2781}\}; \min\{0.3320^{0.0710}; 1.0000^{0.0376}; 0.5488^{0.1816}; 0.2254^{0.0519}; 0.0517^{0.2523}; 0.6125^{0.1275}; 0.7195^{0.2781}\}; \min\{0.6470^{0.0710}; 0.0517^{0.0376}; 0.9651^{0.1816}; 1.0000^{0.0519}; 0.0517^{0.2523}; 0.3695^{0.1275}; 0.2680^{0.2781}\}; \min\{0.9999^{0.0710}; 1.0000^{0.0376}; 0.6784^{0.1816}; 0.2254^{0.0519}; 0.0517^{0.2523}; 0.0992^{0.1275}; 0.2680^{0.2781}\}; \min\{0.7794^{0.0710}; 0.0517^{0.0376}; 0.9989^{0.1816}; 0.0517^{0.0519}; 1.0000^{0.2523}; 0.8900^{0.1275}; 1.0000^{0.2781}\}; \min\{0.3009^{0.0710}; 0.0517^{0.0376}; 0.3559^{0.1816}; 1.0000^{0.0519}; 0.0517^{0.2523}; 0.7909^{0.1275}; 0.2680^{0.2781}\}; \min\{0.7430^{0.0710}; 1.0000^{0.0376}; 0.9840^{0.1816}; 0.2254^{0.0519}; 1.0000^{0.2523}; 0.2134^{0.1275}; 0.7195^{0.2781}\}; \min\{0.5782^{0.0710}; 1.0000^{0.0376}; 0.7946^{0.1816}; 0.0517^{0.0519}; 0.0517^{0.2523}; 0.9390^{0.1275}; 1.0000^{0.2781}\}; \min\{0.9355^{0.0710}; 0.0517^{0.0376}; 0.6260^{0.1816}; 0.2254^{0.0519}; 1.0000^{0.2523}; 0.3980^{0.1275}; 0.7195^{0.2781}\}\} = \{0.4735, 0.7765, 0.4735, 0.4735, 0.4735, 0.8575, 0.4735, 0.8212, 0.4735, 0.8892\}$.

As a result, decisions relative to estimates of borrowers' responsibility and solvency are found from the equality $\max\{\mu_A(a_j)\} = \max\{0.4735, 0.7765, 0.4735, 0.4735, 0.4735, 0.8575, 0.4735, 0.8212, 0.4735, 0.8892\}$, which shows, that the most responsible is the borrower a_{10} with a total rating 0.8892. The others are ranked in descending order of their own estimates: a_6 (0.8575), a_8 (0.8212), a_2 (0.7765), etc.

6 Conclusion

To estimate the responsibility and solvency of potential microcredit borrowers the following weighted evaluation criterion [6]

$$C_k = \frac{\sum_{i=1}^{7} \alpha_i x_{ki}}{\sum_{i=1}^{7} \alpha_i m_i} \cdot 100\% \tag{4}$$

can be also applied, where α_i is the weight of the i-th evaluation criterion; x_{ki} is the score of the k-th borrower for the i-th point of scoring analysis; m_i is maximal score for the i-th point of scoring analysis. The results of estimates using the criterion C_k, scoring estimates, and, in fact, the estimates obtained using the fuzzy maxmin convolution method are presented in Table 3.

As can be seen from Table 3, the proposed fuzzy approach to ranking of microcredit borrowers by consolidating of expert judgements relative to priority of criteria for all points of scoring analysis differs significantly from scoring evaluation and from the weighted evaluation by formula (4). But we, in fact, did not set the such problem. The important here is that the fuzzy method of maxmin convolution more "flexibly" evaluates the responsibility and solvency of potential microloan borrowers. Moreover, this approach was tested on the example of arbitrarily selected ten potential borrowers, which predetermined the choice of the discrete universe $\{a_1, a_2, ..., a_{10}\}$, on the basis of which qualitative evaluation criteria were described by appropriate fuzzy subsets. In the case of a larger number of microloan applicants, the quality of the description of the criteria for assessing the indicators F_i $(i = 1 \div 7)$ by fuzzy sets will noticeably improve, which will inevitably positively affect the adequacy of the ranking of borrowers.

Table 3. Microcredit borrower ranking results.

Borrower	Scoring		Weighted estimate		Maxmin convolution	
	Estimate	Order	Estimate	Order	Estimate	Order
a_1	0.716	10	23.63	10	0.4735	5
a_2	3.190	1	85.81	2	0.7765	4
a_3	2.158	5	57.15	6	0.4735	5
a_4	1.807	7	41.31	7	0.4735	5
a_5	1.645	9	35.22	9	0.4735	5
a_6	2.658	3	89.74	1	0.8575	2
a_7	1.705	8	38.26	8	0.4735	5
a_8	2.428	4	67.11	4	0.8212	3
a_9	2.728	2	80.94	3	0.4735	5
a_{10}	2.039	6	63.42	5	0.8892	1

References

1. Microfinance Information Exchange. https://web.archive.org/web/20140816205120/https://www.themix.org/about
2. Zadeh, L.A.: The concept of a linguistic variable and its application to approximate reasoning. Inf. Sci. **8**(3), 199–249 (1965). https://doi.org/10.1016/j.ins.2006.06.003
3. Molchanov, K.: We Increase the Probability of a Credit – Instructions. Credit Online. https://www.liga.net/creditonline/uvelichivaem-veroyatnost-vydachi-kredita-instrukciya-ot-liga-kreditonlajn. (in Russian)

4. Lin, A.S., Wu, W.: Statistical Tools for Measuring Agreement. Springer, New York (2012)
5. Andreichenkov, A.V., Andreichenkova, O.N.: Analysis, synthesis, planning decisions in the economy. Finance and Statistics, Moscow (in Russian) (2000)
6. Mardanov, M.J., Rzayev, R.R.: One approach to multi-criteria evaluation of alternatives in the logic basis of neural networks. Adv. Intell. Syst. Comput. **896**, 279–287 (2018). https://doi.org/10.4018/978-1-4666-7248-2

Artificial Intelligence for Lassa Fever Diagnosis System

Elbrus Imanov[1](✉) and Fidelis Jumare Asengi[2]

[1] Department of Computer Engineering, Near East University,
North Cyprus via Mersin 10, Nicosia, Turkey
elbrus.imanov@neu.edu.tr
[2] Department of English, Acme Business School Education, Ojodu, Nigeria
fjumare4@gmail.com

Abstract. Currently, knowledge base intelligent system is an important part of Artificial Intelligence that is comprehensively utilized in the field of medical science for identification, medical analysis, and remedy for various kinds of illnesses. The vast majority of the outcomes acquired from different sorts of diagnosis expert systems display reliable and close results to human decision. A comprehensive framework for the analysis and treatment of Lassa fever is yet insufficient.

Study is conducted to develop a system that will diagnose Lassa fever. The information obtaining technique in improvement of that framework was achieved across talking with the therapeutic experts, and expertise was represented in the ruling-based plan of action. These rules decide if an individual is infected with Lassa fever and will go further to predict the level of the infection i.e. either mild Lassa fever, severe Lassa fever or critical Lassa fever by the virtue of the test results, signs and symptoms entered by the user, and answers to the questions by the intelligent system. VPES programming software was utilized for the structure of this process and the process was tested with the data of a few suspected patients with great accuracy, and the results were confirmed and supported by a specialist doctor based on his diagnosis.

The created framework can be utilized proficiently for analysis of Lassa fever where the number of suspected patients is much. Consequently, it will assist the medicinal experts with quick and accurate determinations, and can be used in the lack of experts as well.

Keywords: Artificial intelligent · Expert system · Knowledge base · Virtual program · Expert · Lassa fever · Lassa fever diagnosis system

1 Introduction

Artificial Intelligence (AI) is machine simulation of the processes of human intelligence, particularly computer systems. Such operations tend to involve reasoning and self-correcting learning. Computer programming associated with AI concept is used by computers to perform and mimic the tasks of human experts [1]. AI is sparkling by looking at how person cerebrum reasons, and how individuals settle on a choice and work regardless of the reality endeavoring to report a hazardous task, after that audition

R. A. Aliev et al. (Eds.): ICAFS 2020, AISC 1306, pp. 540–547, 2021.
https://doi.org/10.1007/978-3-030-64058-3_67

the comes to fruition of that examination as a foundation of making astutely PC program and frameworks. Some methods such as structural knowledge and procedural metadata are adopted in AI concept to enable acting and thinking like a human being [2]. An ES is a branch of AI expends to answer a determined problem in a specific state. Artificial intelligence is a field in computer science that includes study and development of computer systems to replace the human intelligence [3]. Computer system that is created becomes intelligent to mimic the actions of a human specialist. An expert system is a computer system that tries to mimic a person specialist. The terms of Expert System or Knowledge base intelligence system is nearly new to be mentioned for the computer process that has the matching as a person specialist in its expertise foundation. For the modeling of a human problem-solving process and an adaptive behavior, we can effectively use the rule base system [4]. Knowledge base intelligent system is most common implementations of the AI. Inside an expert system the field to be examined by human being intellectuals is known as task domain [5]. A specialist is a human being that has a reliable skill in a specific domain. Accordingly, a computer system is constructed to replace a specialist to give a solution to a specific problem, which system is called Knowledge Base Intelligent System. Rule base system has a broad range of decision-making application areas including business, technological, planning, and medication issues [6]. In the light of all those, it would be highly recommended to have a computerized system that would act as a parallel healthcare facility, such as the diagnosis of infectious diseases, to provide in the lack of the proper diagnosis system where number of available experts is not adequate in the congested health facilities, and in the situation of long distances until patients can reach the healthcare service [7]. The aim and objective of this paper tend to experience a user-friendly system algorithm that will be utilized for evaluation and diagnosis of Lassa fever. Lassa fever is an African predominantly infection, which is found in some of the West African countries. For this reason, the research work will be concentrated in Africa and precisely Nigeria because most of the clinical data are collected from Nigeria. According to researchers, Lassa disease continues to be a perpetually important health issue in Nigeria. If doctors become more experienced with the disease and as medical services grow in rural communities, under awareness and under-reporting will be decreased. This can be accomplished by continuing training of health professionals [8]. Lassa virus is categorized as a pathogen by the National Institute of Allergy and Infectious Diseases that restricts the access to laboratory testing. Precautions for bio-safety are proposed for handling samples who are potential patients. In 2014 the World Health Organization made a request for Lassa fever early diagnostic tests. This study, along with their strengths and limitations, offers a brief overview of the difficulties of detecting Lassa fever and the numerous diagnostic tests offered for it [9]. As a result of late diagnosis or lack of access to adequate medical treatment, many patients have died unconsciously in both the villages and urban areas, as a result of the fact that fully equipped medical centers to deal with Lassa infected patients are located in faraway places. The proposed system can be helpful for health professionals as a technique that will help to shorten the length of waits and to offer correct diagnosis for Lassa fever [7].

2   Knowledge Base Intelligent System and Method

Once the proper expert system design tool is chosen and the needed knowledge is acquired, we may begin to develop the ES. At the very beginning, to enable the build and comprehension of knowledge we choose to develop a systematic flow diagram, matrix decision tree or may be other plans. As a result of these guides, knowledge is translated into "If-Then" rules. When we have achieved the basic outline, we can start making a prototype of one of the system parts. We can improve the system to achieve a complete system, once the performance of the system is recognized as satisfactory [10]. Knowledge base system, which is based on task domain and knowledge engineer, can indicate the information in a reasonable way [11]. The field specialist is the human that has the essential skill to answer the difficulty and suggest solutions. The skill designer receives skill from the expert and converts it into an arrangement acceptable for the system to answer the patient. Transition methods consisting of representing one element with another depend on the intelligent system which includes concepts of knowledge [12]. Specialist process domain knowledge is stored within Knowledge base and the given unit is extremely critical because, the efficient implementation of the process counts on quality and dependence of the skill held within [13]. Part of rule base intelligent system process is presented in Fig. 1.

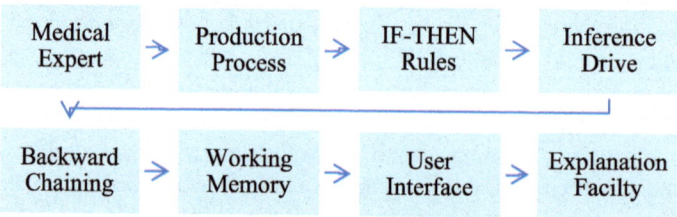

Fig. 1. Knowledge base intelligent system.

The development of ES involves the merging of several components resulting in the process of controlling with main objective idea, actual, ruling, deduction drive, etc. [14]. Therefore, we define the system as an expert system rather than an ordinary software program. In order to solve the problem, inference engine selects the appropriate type of search for operation. Automated advisiory program or ES imitates the reasoning process and learining of the specialists to solve a specific problem [15]. Actually, deduction driver operates the expert system, explain what ruling is functional, and then completes the ruling and interprets when a satisfactory suspension is achieved. For conclusion part, different rules are used by the inference engine, and the justification is provided by the explanation facility [16]. There are forward and backward chaining technique mainly used for ruling based process. The forward chaining is a data driven search. The system searches the condition area going from idea condition to the initial condition at applying the opposite drive. Backward chaining is an idea controller or progressive explorer. Testing hypothesis for human-based problem solving is like backward chaining. For instance, a specialist doctor may be skeptical about few issues related to a patient [16].

The developed system framework uses data gathered from an expert specifically through direct interview with a doctor, besides other useful sources such as books, papers, and journals to generate the rules to improve the VP-Expert system for diagnosing LF patients, by using expert system approach. We can classify the designed system development process into two relevant groups, which are knowledge acquisitions and knowledge representation. Knowledge acquisition involves data gathered. Knowledge representation identifies the form of knowledge represented and it involves coding of IF-THEN statement. Prior to program loading for the purpose of consultation, the system is run on VP expert software. Data was collected for the knowledge acquisition part of the system development process via direct interview with the Lassa fever healthcare specialists, besides studying the related medicine books, articles and scientific publications. The designed system is a rule-based expert system, where IF suggests the conditions and THEN offers the solutions. In order to transfer the knowledge of experts to these rules basic stages must be dealt with, through diagram and the decision tables. There are three modes of diagnosis: Simple Lassa fever, Severe Lassa fever, and Critical Lassa fever. The treatment advices comprise six condition questions along with several of those condition questions such as input of diagnosis process, which has 8 inputs, and includes 5 laboratory analysis, and body temperature, also signs, symptoms, duration of signs and systems. When a user answers all the questions via the options displayed on the user interface window, the inference engine goes through all the rules in accordance with the response of the user to give out the result of the diagnosis. Worktable of sign and symptom is presented in Table 1.

Table 1. Decision making worktable of sign and symptom.

Signs and Symptoms	Mild Lassa	Severe Lassa	Critical Lassa
Symptom1			
Slight fever	**yes**		
Headache	**yes**		
Weakness	**yes**		
Symptom2			
Bleeding gum		**yes**	
Bleeding eyes		**yes**	
Bleeding nose		**yes**	
Nausea and vomiting		**yes**	
Facial, neck swollen		**yes**	
Diarrhea		**yes**	
Chest pain		**yes**	
Sore throat		**yes**	
Back ache		**yes**	

(*continued*)

Table 1. (*continued*)

Signs and Symptoms	Mild Lassa	Severe Lassa	Critical Lassa
Symptom3			
Deafness			**yes**
Convulsion			**yes**
Tremor			**yes**
Stroke			**yes**
Thirst			**yes**
Drowsiness			**yes**

Tests decision making set of facts include the obligatory analysis report consists of reverse transcriptase polymerase chains reactions RTPCR, Serology SRLG, Rapid diagnostic lateral flow assay RDFLA, Rapid diagnostic test for malaria RDTM, and Blood culture for typhoid BCT and finally the decision making for medical treatment's state. Worktable of test is presented in Table 2.

Table 2. Worktable of the test.

Patients	RDTM	BCT	RTPCR	SRLG	RDLFA	DIAGNOSIS
Facts	Negat.	Negat.	Negat.	Negat.	Negat.	Negat.
Facts	Posit.	Posit.	Posit.	Posit.	Posit.	Posit.
Facts	Negat.	Posit.	Posit.	Posit.	Posit.	Posit.
Facts	Posit.	Negat.	Posit.	Posit.	Posit.	Posit.
Facts	Negat.	Negat.	Posit.	Posit.	Posit.	Posit.
Facts	Posit.	Negat.	Negat.	Negat.	Negat.	Negat.
Facts	Posit.	Negat.	Negat.	Negat.	Negat.	Negat.
Facts	Negat.	Posit.	Negat.	Negat.	Negat.	Negat.
Facts	No	No	No	No	No	No

Diagnosing expert system for Lassa fever has been coded using a VP-ES shell, which is an accurate method for developing expert systems, so it is solely known to expert system developers. The framework has an inference engine to search the knowledge base for providing feedback, an editor for coding knowledge base rules, and the user interfaces to manage questions, so, it asks the related questions to clients, and offers recommendations and justifications where it is suitable. This expert system's production rules contain 8 queries about the relevant attributes to be served as an input of the developed system.

```
PLEASE ENTER YOUR RDTM RESULT?
PLEASE ENTER YOUR BCFT RESULT?
PLEASE ENTER YOUR RTPCR RESULT?
PLEASE ENTER YOUR SRLG RESULT?
PLEASE ENTER YOUR RDLFA RESULT?
PLEASE ENTER YOUR BODY TEMPERATURE?
WHAT ARE SIGNS - SYMPTOM YOU are SEEING - FEELING?
WHAT IS DURATION of SIGN - SYMPTOM YOU are SEEING
FEELING?
```

3   Design and Presentation of the Developed System

All the rules, pathways and associations between the attributes were checked with appropriate adjustment in the process of building the system. The VP expert operating system is used for this system design and, ultimately, the system designed is introduced. During operation, a user contacts with the system via the user interface. The "User Interface" possesses three windows that are questions, rules, and facts window. The question window is where users meet the system questions, and their alternatives. Secondly, rules window displays the rules for how the user will respond. The user's answers are displayed in fact window, that may provide the final decision of the given system. The user can see vivid clarification for the system's results. After the system has received the facts, the system presents the final decision. The user can view the compelling explanation of the decision-making process of the system that provides main advantage of the facts and rules windows. Once the facts received by the system, it forwards a final answer or a decision. The Fig. 2 is presented exhibit the application of the VP-Expert i.e. user interface.

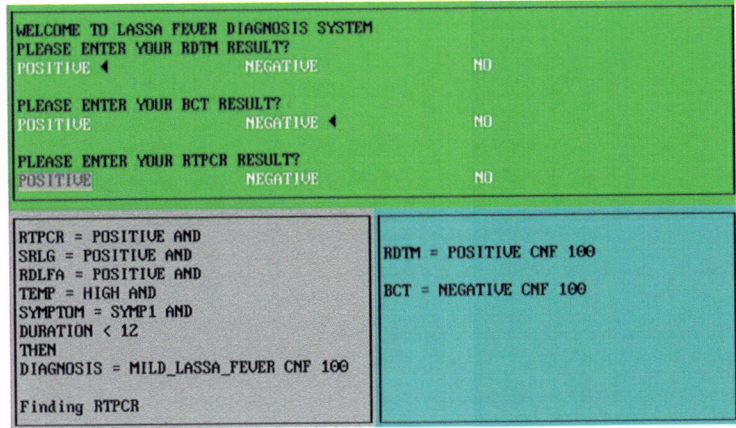

Fig. 2. Depiction of VP expert reflection about result user and system.

The LFDS program performed well as validated by domain experts. Specialists also verified that this system will be used by Lassa fever specialists and will be readily available to all, which will alleviate the difficulty of finding a professional doctor if they are in limited numbers. Accurate and earlier Lassa fever diagnosis is vital for the human life, and therefore the system evolved with the intention of being the best solution to this challenge. Figure 3 is presented for identification result.

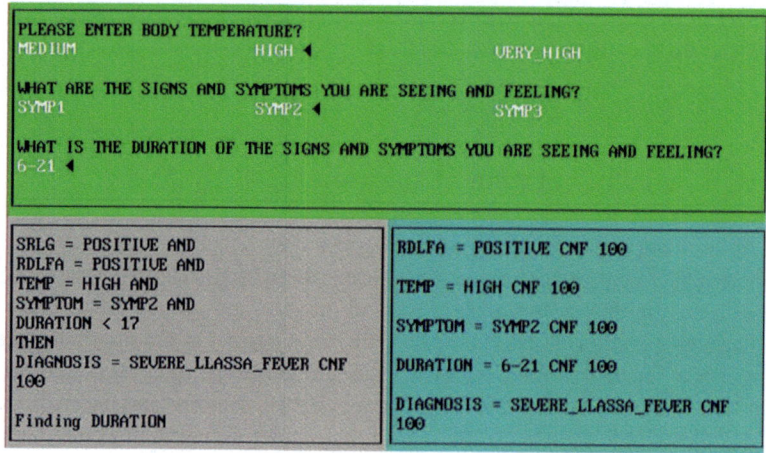

Fig. 3. Depiction of VP expert reflection about identification result.

When a doctor has the ability to manually assess 100 LF patients per day, then LFDS now has a potential ability to do diagnosis for 300 LF patients a day. As a result, system makes it simpler, faster and also more accurate to diagnose this fatal virus. That suggests the process's expected reliability after all to building a fast and correct system option instead of a thinking person.

4   Conclusion

Presently, knowledge base intelligent system is used in many areas, beginning from medicine department, military system, law section, economics, and so on. They are being used to monitor robot in engineering and industrial sectors, where they are interconnected to vision systems. All the studies that involve decision making process are the possible implementation areas of an expert system. Knowledge based intelligence system is the Artificial Intelligence program that realize special degree ability within creating practical solutions for some tasks by transmitting knowledge about certain tasks. The LFDS has been established and the extent of knowledge accession and depiction is clearly represented. The rule base system that forms the IF and THEN rules facts is based on edition from doctor experts and the information quality, 175 condition-action pair rules were developed. The condition-action pairs entail 8 inputs condition rules and have 4 outputs action rules for process. The Lassa fever diagnostic system was extended to apply VP-Expert process shell and tested for disease control with great accuracy using few data of suspected infected patients from the Nigeria Center. Every procured result from the system was seen and confirmed by a medical expert dealing with Lassa fever cases and it content the regulation of Lassa fever identification knowledge base intelligent system.

References

1. Aliev, R., Huseynov, O.H.: Decision Theory with Imperfect Information. Word Scientific, River Edge (2005)
2. Studer, R., Benjamins, R.V., Fensel, D.: Knowledge engineering, principles and methods. Data Knowl. Eng. **25**(1–2), 161–197 (1998). https://doi.org/10.1016/S0169-023X(97)00056-6
3. Patel, M., Virparia, P., Patel, D.: Web based fuzzy expert system and its application- a survey. Int. J. Appl. Inf. Syst. **1**(7), 11–15 (2012)
4. Alcala, R., Cordon, O., Casillas, J., Herrera, F., Zwir, S.J.: Techniques for learning and tuning fuzzy rule-based systems for linguistic modeling. In: Knowledge-Based Systems, pp. 889–941. Academic Press (1999)
5. Mishkoff, H.: Understanding Artificial Intelligence. Dallas, Texas, Instrument learning Centre (1985)
6. Aliev, R., Aminzadeh, F., Jamshidi, M.: Fuzzy Expert Systems, Soft Computing: Fuzzy Logic Neural Networks and Distributed Artificial Intelligence, pp. 99–108. NJ: PTR Printing Prentice Hall (1994)
7. Abiola, H.M., Alaba, A.A., Luka Joy, D.: Expert system for lassa fever diagnosis using rule based approach. Ann. Comput. Sci. Ser. **15**(2), 68–74 (2017)
8. Denue, B.A., Stephen, M., Dauvoux: The unending threat of Lassa fever in Nigeria, Port Harcourt Med. J. **11**(3), 113–121 (2017)
9. Raabe, V., Koehler, J.: Laboratory diagnosis of lassa fever. J. Clin. Microbiol. **55**(6), 1629–1637 (2017). https://doi.org/10.1128/JCM.00170-17
10. Nilsson, N.J.: Principles of Artificial Intelligence. Narosa Publishing House, New Delhi (1998)
11. Kapoor, N., Bahl, N.: Comparative study of forward and backward chaining in artificial intelligence. Int. J. Eng. Comput. Sci. **5**(4), 16239–16242 (2016). https://doi.org/10.18535/IJECS%2FV5I4.32
12. Zadeh, L.A.: Computing with words and perceptions a paradigm shift. In: Proceedings of the IEEE International CONFERENCE 2009, on Information Reuse and Integration, pp. 450–452. IEEE Press (2009). https://doi.org/10.1109/IRI.2009.5211627
13. Tabibi, S.T., Zaki, T.S., Ataeepoor, Y.: Developing an expert system for diabetics' treatment Advices. Int. J. Hospital Res. **2**(3), 155–162 (2013)
14. Merritt, D.: Rubik's cube. In: Building Expert Systems in Prolog, pp. 207–218 (1989)
15. Bobillo, F., Delgado, M., Gómez-Romero, J., López, E.: A semantic fuzzy expert system for a fuzzy balanced scorecard. Expert. Syst. Appl. **36**(1), 423–433 (2009). https://doi.org/10.1016/j.eswa.2007.09.020
16. Roseline, P., Tauro, C.J., Ganesan, N.: Design and development of fuzzy expert system for integrated disease management in finger millets. Int. J. Comput. Appl. **56**(1), 31–35 (2012)
17. Sagheb-Tehrani, M.: Expert system development: some issues of design process. ACM SIGSOFT. **30**(2), 1–5 (2005). https://doi.org/10.1145/1050849.1050864

Methodological Foundations of Fuzzy-Multiple Assessment of the State of Economic Systems Based on Aggregation of Indicators

I. Akperov Gurru[1]([⊠]) [iD] and Sakharova Lyudmila[2] [iD]

[1] Private Educational Institution of Higher Education «SOUTHERN UNIVERSITY (IMBL)», Rostov-on-Don, Russia
pr@iubip.ru
[2] Rostov State University of Economics, Rostov-on-Don, Russia
l_sakharova@mail.ru

Abstract. We presented a general model for a comprehensive assessment of the state of economic systems based on aggregation of a hierarchy of heterogeneous indicators. The model is based on the use of systems of fuzzy-logical conclusions - standard fuzzy multi-level [0, 1] - classifiers previously used in the economy to assess the risk of bankruptcy of a company. The result of the application of the method is a "balance sheet of estimates", in which normalized values of indicators are summarized. Linguistic recognition of the final comprehensive assessment makes it possible to evaluate the dynamics of the system not only quantitatively (the numerical value of the corresponding fuzzy variable), but also qualitatively ("how bad" or "how good"). Analysis of the estimates of indicators allows us to identify those that lead to a decrease in the value of a comprehensive assessment; therefore, to determine in which areas work should be done to improve the situation. The information obtained can serve as the basis for a targeted analysis of production in detail, as well as for making management decisions. The technique is concretized for the case of a comprehensive assessment of the effectiveness of economic activity, which occupies an important place in management analysis.

Keywords: Mathematical model · Methods for assessing the state of economic systems · Integrated assessment · Scorecard · Fuzzy logic

1 Introduction

Assessing the state of economic systems by a set of indicators is a complex multidimensional problem of mathematical modeling. Currently, there are many classical single- and multi-parameter estimation models for the totality of indicators, carried out, as a rule, by calculating a variety of integral indices. However, almost all of these models, despite their practical importance, have a number of significant shortcomings from the point of view of the theory of mathematical modeling. Consider in more detail the problem on the example of a comprehensive assessment of the effectiveness of economic activity.

© The Author(s), under exclusive license to Springer Nature Switzerland AG 2021
R. A. Aliev et al. (Eds.): ICAFS 2020, AISC 1306, pp. 548–554, 2021.
https://doi.org/10.1007/978-3-030-64058-3_68

The methodology for a comprehensive assessment of the effectiveness of economic activity occupies an important place in management analysis [1]. Its use allows you to get:

1. An objective assessment of past activities, to find reserves to improve business efficiency.
2. Feasibility study on the transition to new forms of ownership and management.
3. A comparative assessment of producers in the competition and the choice of partners.

The efficiency of economic activity is based on the intensification of production. At present, the general principles of methods for assessing the intensification of production are developed in detail. The standard algorithm of the assessment methodology includes the following steps [2]: 1) determination of qualitative indicators of resource use; 2) the establishment of the ratio of resource growth per 1% increase in output; 3) calculation of the share of the influence of intensity on the increase in the volume of production; 4) determination of the relative saving of resources; 5) calculation of comprehensive estimates of the extension and intensification of production. However, its implementation in practice is often difficult due to obstacles of both methodological and organizational nature.

The essence of the problem is as follows [2, 3]. To calculate a comprehensive assessment in practice, it is necessary to analyze a number of indicators that reflect the extension and intensification of production. The indicators of the extensiveness of development are quantitative indicators of the use of resources: the number of employees, the value of expended objects of labor, the amount of depreciation, the volume of fixed assets and advanced working capital. Indicators of development intensity are qualitative indicators of the use of resources, i.e. labor productivity (or labor intensity), material productivity (or material intensity), capital productivity (or capital intensity), the number of circulating assets turnover (or the working capital fixing ratio). To formulate private estimates for each of the indicators, as a rule, indicators of growth rates are used, which are the ratio of the value of the indicator in the study period to its value in the previous period percent ($E = P_2/P_1$). The obtained rates are then substituted into standard formulas characterizing the degree of prevalence of the intensive type of management over the extensive type. If this comprehensive assessment is built for a small (or relatively homogeneous) enterprise for a small number of periods, the existing evaluation algorithms allow you to get a satisfactory result. However, if the enterprise being evaluated (or even the manufacturing industry) includes many divisions (workshops, sub-sectors, etc.), the complexity of the task increases by an order of magnitude and goes beyond the scope of standard economic analysis. The problem arises from integrating the obtained private estimates by units and periods under consideration. Mathematical methods in this area are not yet sufficiently developed. Two groups are distinguished: deterministic integrated estimation methods and stochastic integrated estimation methods [4].

Among the group of deterministic integrated estimates, the most common is the method of sums, in which the estimate is calculated by simply summing the actual values. A necessary condition for a correct assessment in this case is the unidirectionality of the studied indicators. That is, the system of indicators should be organized in such a way that an increase in the value of any of the indicators corresponds to an improvement in the results of the evaluated economic activity, and a decrease in its value corresponds to a deterioration in the results. The unidirectionality of private indicators allows us to rank production facilities by increasing or decreasing values of the integral indicator.

Evaluation of the results of economic activity by the method of amounts can be built on various private indicators and not only in comparison with the plan, but also in previous periods (dynamics assessment) and with reference values of indicators for a group of production facilities.

The disadvantage of the sum method is the mutual leveling of indicators, that is, a sufficiently high estimate of the integral indicator can be obtained in this case due to overlapping indicators with low growth rates with high growth rates. In some cases, this deficiency can be eliminated if, instead of the total integral indicator, two particular indicators are calculated, the sum of positive and the sum of negative deviations of the values of particular indicators from the comparison base. However, there are cases when the specified technique does not guarantee the correct result.

When using the geometric mean method, we calculate the normalized values of the estimated indicators, that is, prisoners in the interval from 0 to 1. At the same time, we take the value corresponding to the lowest possible value of the indicator as zero, and the unit value corresponding to its largest value. The geometric mean method is advisable to apply if the number of evaluated indicators is relatively small, and most of their values are close to unity. The use of the geometric mean method is also possible only in the case of unidirectional influence of all the estimated parameters on a comprehensive assessment. Otherwise, we take indicators that are inverse to the values of the initial indicators.

The basis of the distance method is to consider the proximity of objects according to the compared indicators to the standard object. The main problem in this case is the choice of a standard. A conditional object with maximum elements for all indicators can be taken as a standard. In some cases, a typical object is considered one whose values are equal to the arithmetic mean levels of indicators in the study population. Sometimes the authors propose to use as a standard the full correspondence of the values of real production indicators to those that were planned; at the same time, they note that deviations from the planned indicators both in smaller and larger directions are undesirable. However, in the aggregate of economic objects where asymmetric distributions predominate, the arithmetic mean as a characteristic of a typical, standard object loses its meaning.

Methods of stochastic integrated assessment include a variety of expert statistical methods, as well as methods of component analysis, which are rather cumbersome and complicated in practical application.

We have proposed a general model of a comprehensive assessment of the state of a system of systems based on aggregation of a hierarchy of indicators. The model is based on the use of fuzzy-logical inference systems, fuzzy multi-level [0, 1] - classifiers, allowing aggregation of normalized values of indicators within the framework of two-dimensional convolution [5, 6]. The main idea is that all measurements are made on a qualitative basis, using a fuzzy classification of quantitative factors. Preference systems in the hierarchy of indicators are implemented due to the system of weights. In a two-dimensional convolution of one of the systems of weights, the weights of factors appear, and the other as a system of weights are the nodal points of standard fuzzy multilevel [0, 1] - classifiers. When developing the model, we used and summarized our methodologies for assessing the efficiency of agricultural production [1–3], as well as evaluating the regions of the region for compliance with the principles of environmental nature management [7, 8].

2   General Methodology for a Comprehensive Assessment of Systems Based on Aggregates of Indicators

Let the system under consideration consist of several subsystems, each of which can be estimated by a complex of heterogeneous indicators defined by time series.

1. *Evaluation of each individual subsystem* can be made on the basis of data aggregation by means of fuzzy-logical inference systems - fuzzy multi-level [0, 1] - classifiers.The following algorithm is used to form the estimate.

Stage 1. Formation of a list of indicators relevant to the formation of a comprehensive assessment of the system. The values of indicators in this case should be represented by time series for N years.

Stage 2. Ranking the importance of the studied indicators for evaluation, calculation of their weight coefficients based on expert estimates. It is required to enter weights for each indicator r_i, i = 1, 2,..., M (a prerequisite $\sum_{i=1}^{M} r_i = 1$). The following options are possible:

1)   indicators are equivalent, therefore, have the same weight: $r_i = 1/M$;
2)   indicators are ranked in descending order of their importance and weights are determined according to the Fishburn rule:

$$r_i = \frac{2(M - i + 1)}{(M + 1)M};$$

3)   the weights are determined on the basis of the share contribution of the direction: for example, if the intensification is determined by three groups, then the weight of each group is 1/3; if in this case two directions are considered within the group, then the weight of each of them will be $\frac{1}{3} \cdot \frac{1}{2} = \frac{1}{6}$, etc.
4)   weights are determined based on estimates of expert economists (based on pairwise comparisons of alternatives, etc.)

Stage 3. Calculation of normalized numerical characteristics of indicators belonging to the segment [0, 1]. Formulas are drawn up based on which the aggregation of time series is made; these formulas are determined by the nature of the indicators and the meaning of the problem being solved. Suppose we need to obtain normalized values of indicators x_i (i = 1, 2,..., M) taking into account their values in different years; however, different years have different significance for the assessment. In this case, we can use the formula:

$$x_i = \sum_{i=1}^{N-1} k_i I_i, \quad k_i = \frac{2\,(N - i)}{(N - 1)N}$$

where k_i are the weighting coefficients determined on the basis of the Fishburn rule; for the numbering of time periods, a countdown is used; I_i is the value of the indicator for years. Moreover, if the calculations are made to rank identical subsystems, then the indicators are normalized as follows: the largest in the I_{max} regions is selected, all indicators are divided into it.

If the study is carried out only with the aim of assessing the dynamics of a given parameter, then aggregation can be performed on the basis of the following scheme:

$$x_i = 0,5\left(1 + \sum_{i=1}^{N-1} k_i I_i\right), \quad k_i = \frac{2\,(N - i)}{(N - 1)N}$$

where ki are weights determined based on the Fishburn rule; Ii is an integer function that takes three values $(-1, 1, 0)$ in the case of positive, negative and zero dynamics.

Stage 4. We carry out the task of linguistic variables. Moreover, the normalized values of indicators (Stage 3) are the numerical characteristics of fuzzy variables with a universal set in the form of a segment [0, 1]. We use them to compare linguistic variables with term sets of five terms reflecting the level of the indicator: "very low"; "low"; "middle"; "high"; "very high". The membership functions of linguistic variables are determined using trapezoidal functions.

In addition, we introduce a linguistic variable: γ = "a comprehensive assessment of the state of the system. "The universal set for the linguistic variable is the numerical segment [0, 1], and the set of values γ is the term set $G = \{G_1, G_2, G_3, G_4, G_5\}$, where terms reflect tendencies: G_1 - "resistant to growth decrease"; G_2 - "to decrease growth"; G_3 - "towards stagnation"; G_4 - "to increase growth"; G_5 - "resistant to growth increase". Membership functions also have a trapezoidal shape (Table 1).

Stage 5. The transition from the numerical characteristics of indicators to the numerical characteristics of estimates based on the general algorithm of the standard five-level [0, 1] - classifiers. The rule of transition from the values of the indicators x_i (i = 1, 2,..., M) to the weights pi of the terms of the linguistic variable γ has the form:

$$p_l = \sum_{i=1}^{M} r_i \cdot \mu_{il}(x_i), \quad l = 1, ..., 5.$$

Here $\mu_{il}(x_i)$ are the values of the membership function of the values of the indicators x_i. The value of the variable γ itself is determined by the formula:

$$\gamma = \sum_{k=1}^{5} p_k \cdot \bar{g}_k,$$

where $\bar{g}_k$ — are the nodal points of the classifier, that is, the centers of gravity of its terms (0.125; 0.3; 0.5; 0.7; 0.885).

Stage 6. We carry out linguistic recognition of the obtained numerical estimates, guided by the definition of the term set $G = \{G_1, G_2, G_3, G_4, G_5\}$. In addition, we analyze the resulting assessment based on normalized values of indicators and formulate recommendations for correcting the current situation.

As follows from the definition of terms, the value $\gamma \leq 0,5$ indicates, in general, a decrease in the intensification of production and the need to analyze the situation as a whole. If $\gamma \geq 0,5$ and linguistic recognition indicates a "tendency towards stagnation", then it is required to analyze individual indicators and select those for which the normalized values are minimal. Obviously, it is these indicators that lead to a decrease in the value of a comprehensive assessment, and, therefore, it is in these areas that work should be done to improve the situation.

Table 1. Membership functions of subsets of term-sets of linguistic variables

Terms B_{ij}, C_{ij}(level of the indicator), G_i, (tendencies), l=1,2,3,4,5	Membership function of fuzzy sets
B_{i1}, C_{i1} – "very low"; G_1 – "resistant to growth decrease";	$\mu_1(x) = \begin{cases} 1, & \text{if } 0 \leq x < 0,15 \\ 10(0,25 - x), & \text{if } 0,15 \leq x < 0,25 \\ 0, & \text{if } 0,25 \leq x \leq 1 \end{cases}$
B_{i2}, C_{i2} – "low"; G_2 – "to decrease growth";	$\mu_2(x) = \begin{cases} 0, & \text{if } 0 \leq x < 0,15 \\ 10(x - 0,15), & \text{if } 0,15 \leq x < 0,25 \\ 1, & \text{if } 0,25 \leq x < 0,35 \\ 10(0,45 - x), & \text{if } 0,35 \leq x < 0,45 \\ 0, & \text{if } 0.45 \leq x \leq 1 \end{cases}$
B_{i3}, C_{i3} – "middle"; G_3 – "towards stagnation";	$\mu_3(x) = \begin{cases} 0, & \text{if } 0 \leq x < 0,35 \\ 10(x - 0,35), & \text{if } 0,35 \leq x < 0,45 \\ 1, & \text{if } 0,45 \leq x < 0,55 \\ 10(0,65 - x), & \text{if } 0,55 \leq x < 0,65 \\ 0, & \text{if } 0,65 \leq x \leq 1 \end{cases}$
B_{i4}, C_{i4} – "high"; G_4 – "to increase growth";	$\mu_4(x) = \begin{cases} 0, & \text{if } 0 \leq x < 0,55 \\ 10(x - 0,55), & \text{if } 0,55 \leq x < 0,65 \\ 1, & \text{if } 0,65 \leq x < 0,75 \\ 10(0,85 - x), & \text{if } 0,75 \leq x < 0,85 \\ 0, & \text{if } 0,85 \leq x \leq 1 \end{cases}$
B_{i5}, C_{i5} – "very high"; G_5 – "resistant to growth increase"	$\mu_5(x) = \begin{cases} 0, & \text{if } 0 \leq x < 0,75 \\ 10(x - 0,75), & \text{if } 0,75 \leq x < 0,85 \\ 1, & \text{if } 0,85 \leq x \leq 1 \end{cases}$

As follows from the description, the estimate obtained for the subsystem is normalized. The set of assessments of all subsystems of the system can be aggregated into a comprehensive assessment of the entire system. In addition, assessments of individual systems can serve as input for aggregation into integrated assessments of higher-level systems.

Among the advantages of the proposed assessment method, we note the following: 1) universality due to the presence of a single work algorithm for economic and social systems of various types; 2) variability due to the ability to change the set of parameters, as well as their weight coefficients without complicating the model; 3) the ability to take into account expert opinions by changing the weights of the parameters; 4) simplicity of software implementation; 5) the ability to aggregate assessments of simpler systems into assessments of more complex systems.

3 Conclusion

The proposed method for constructing a comprehensive assessment of the intensification of production and its pace allows us to assess the state of the enterprise, industry, etc. based on a set of series of intensification indicators for an arbitrary number of time periods. The method is simple to use, easy to formalize in the form of software systems. The result of the application of the method is a kind of "balance sheet of estimates", in which unified data on indicators and periods are summarized. Linguistic recognition of the final comprehensive assessment allows us to judge the dynamics of the intensification process as a whole, and the numerical value of the corresponding fuzzy variable allows us to give it a quantitative assessment ("how bad" or "how good"). The result obtained is easily analyzed based on the "balance sheet of grades" and allows you to select indicators that lower the final grade. In turn, the analysis of evaluations of indicators with negative dynamics helps to track the development of the process and highlight the periods in which the dynamics were the worst. The information obtained can serve as the basis for a targeted detailed analysis of the investigated production and the adoption of appropriate management decisions.

References

1. Bakanov, M.I., Sheremet, A.D.: Theory of Economic Analysis: A Textbook, 4th edn., 416 p. Finance and Statistics (2001). (In Russian)
2. Alekseeva, A.I., Vasiliev, Yu.V., Maleeva, A.V., Ushvitsky, L.I.: Comprehensive Economic Analysis of Economic Activity: Training Manual, 672 p. (2006). Finance and Statistics (In Russian)
3. Lysenko, D.V.: Comprehensive Economic Analysis of Economic Activity: Textbook for Universities. INFRA-M, 420 p. (2008). (In Russian)
4. Sheremet, A.D.: Integrated Business Analysis. INFRA-M, 415 p. (2006). (In Russian)
5. Konysheva, L.K.: Fundamentals of the Theory of Fuzzy Sets: A Training Manual. St. Petersburg: Peter, 192 p. (2011). (In Russian)
6. Nedosekin, A.O.: Financial Management on Fuzzy Sets: Monograph. Audit and Finance. Analysis, Moscow, 162 p. (2003). (In Russian)
7. Alekseychik, T.V., Bogachev, T.V., Karasev, D.N., Sakharova, L.V., Stryukov, M.B.: Fuzzy method of assessing the intensity of agricultural production on a set of criteria of the level of intensification and the level of economic efficiency of intensification. Adv. Intel. Syst. Comput. **896**, 790–798 (2019). https://doi.org/10.1007/978-3-030-04164-9_83
8. Arapova, E.A., Lukyanova, G.V., Sakharova, L.V., Akperov, G.I.: Fuzzy-Logic analysis of the level of comfort and environmental well-being of the urban environment on the example of large cities of Rostov region. Adv. Intel. Syst. Comput. **896**, 643–650 (2018). https://doi.org/10.1007/978-3-030-04164-9_84

Problem Solving Based on Z-Numbers

K. A. Mammadova$^{(\boxtimes)}$ ⓘ, A. B. Sultanovaⓘ, E. N. Aliyevaⓘ,
and A. N. Huseynovaⓘ

Candidate of Technical Sciences, Lecturer at the Department of Computer
Engineering, Azerbaijan State Oil and Industry University, Baku, Azerbaijan
ka.mamedova@yandex.ru, saxira@mail.ru,
yegane.aliyeva.1969@mail.ru, hasanova_a@inbox.ru

Abstract. In most cases, initial information in making real decisions is partially reliable. This can be explained by data source insecurity, wrong interpretations, inexperience, etc. Formation of information (Z-information) based on Z-numbers represents the justified value of the variable in the natural language according to natural language (NL) based reliability. There is one point that needs attention. One Z-information represents imperfect, i.e. incomplete information inherent in reality. At the same time, compared to a fuzzy number a Z-information has a stronger image in terms of human perception. This approach allows us to find the significance of unknown probabilities with a certain reliability, using the expected paradigm of usefulness.

Keywords: Z-numbers · Fuzzy number · Discrete Z-numbers · Utility value · Defuzzification · Decision making · Expected utility

1 Introduction

In decision-making theory some information related to decision making may be popularized. The first popularization is used for precise numbers; the second one is based on the use of intervals. In fact, the decision-making theory is based on two popularizations. The decision-making theory can be classified as follows: theory of expected utility and perspectives using in numerical information; theory based on the set of initial information as maxmin expected utility using information estimated leg in interval. The third popularization is fuzzy sets. Within the third popularization framework, there exist several scientific-practical works devoted to linguistics terms, fuzzy membership function, fuzzy multicriterial decision-making and others. Unfortunately, all approaches of decision-making based on three popularizations do not possess real reliability.

In majority of information based on decision-making existence of uncertainty requires to the more careful. As a result, it becomes difficult to form the rational decision-making ability for giving inexact, fuzzy or incomplete information [1].

At present, on the most efficient research fields is the multicriterial decision-making problem. As long as the uncertainty and complexity exist, the fuzzy sets [2] will be considered as key problems widely used on decision-making process. Z-numbers based calculation have the generalized forms of calculation with random numbers, clear and

R. A. Aliev et al. (Eds.): ICAFS 2020, AISC 1306, pp. 555–564, 2021.
https://doi.org/10.1007/978-3-030-64058-3_69

fuzzy numbers, intervals based on this sequence, we can distinguish the level of generalization as follows: computation in numbers (zero level); calculation in intervals (first level); calculation in fuzzy numbers (second level); [2]; calculation in random numbers (third level); calculation in Z-numbers (fourth level). The potential for building models increases due to increase in the autonomy level economy, risk assessment, solution analysis, planning in some fields as an analysis of cause and effect relationships [1].

In Z-numbers, sufficient reliability of real information in life must be taken into account [3–10]. Zadeh, offered a new Z-numbers more appropriate to describe uncertainty. The Z-numbers requires both restraint and reliability. Unlike the elastic fuzzy numbers, the ability of description of real information is higher in Z-numbers [10]. Z-numbers is used to offer the solution to conduct calculation with not completely reliable numbers. In the following example Z-numbers convincing [1]. Azerbaijan's population (about 10 million, very sure). Petrol's price in the next 2 years (below 50 $, likely). The goal of the present paper to study decision-making methods generalizing the existing approach to the expected utility based on Z-information. Compared to other works related to Z-information based decision-making, this method are based on direct calculation on Z-numbers without converting to fuzzy ones. In conducting direct calculations on Z-numbers, no loss of information during conversion. In this paper, using the expected utility program, operative approach to the solution of problems by means of Z-information was considered this approach is based on the calculations of original Z-numbers (i.e. not converting to fuzzy numbers) offered by L.A.Zadeh. In the paper, issue on application of the offered suggestion related to the solution of a business-decision making problems.

2 Preliminaries

During investigation, an approach based on expected utility for solving decision-making problem by Z-information is offered. This approach is based on calculation on Z-numbers with for the operations offered in [1–10].

Definition 1. A fuzzy set which is defined on a universe set X is given as: $A = \{(x, \mu_A(x)/x \in X)\}$, were $\mu_A : X \rightarrow [0, 1]$ is the membership function of A. $\mu_A(x)$ is a membership degree of every x contained in A.

Definition 2. The trapezoidal fuzzy number $\tilde{A}$ can be written as $\tilde{A} = (a_1, a_2, a_3, a_4)$ this case, the trapezoidal fuzzy number $\tilde{A}$ will be commented is follows:

$$\mu_{\tilde{A}}(x) = \begin{cases} \frac{x-a_1}{a_2-a_1}, & a_1 \leq x \leq a_2 \\ 1, & a_2 \leq x \leq a_3 \\ \frac{x-a_4}{a_3-a_4}, & a_3 \leq x \leq a_4 \\ 0, & otherwise \end{cases} \quad (1)$$

The a_1, a_2, a_3 and a_4 are real numbers. The condition $a_1 \leq a_2 \leq a_3 \leq a_4$ is satisfied, where .... If $a_2 = a_3$, then $\tilde{A} = \{x, \mu_{\tilde{A}}(x) | x \in [0, 1]\}$ is converted into a triangular fuzzy number. Thus, the triangular fuzzy number can be considered as a particular case of a trapezoidal fuzzy number.

Assume that $\tilde{R} = \{x, \mu_{\tilde{R}}(x) | x \in [0, 1]\}$ is a fuzzy set. Similar to expression (2) the membership function of the fuzzy number $\tilde{R} = (b_1, b_2, b_3, b_4)$ is determined in the following form:

$$\mu_{\tilde{R}}(x) = \begin{cases} \frac{x-b_1}{b_2-b_1}, & b_1 \leq x \leq b_2 \\ 1, & b_2 \leq x \leq b_3 \\ \frac{x-b_4}{b_3-b_4}, & b_3 \leq x \leq b_4 \\ 0, & otherwise \end{cases} \tag{2}$$

Here b_1, b_2, b_3 and b_4 are real numbers. The condition $b_1 \leq b_2 \leq b_3 \leq b_4$ is satisfied.

Definition 3. As was explained in [6], Z-numbers were determined to make conduct calculations with not completely reliable numbers. Z-numbers is the pair of ordered fuzzy numbers expressed as $Z = (\tilde{A}, \tilde{R})$. Calculation by Z-numbers is based on determination of accuracy of information. For simplicity assume that, $\tilde{A}$ and $\tilde{R}$ are trapezodial fuzzy numbers the first component, $\tilde{A}$ is a limitation imposed on real value of variable X. The second component R expresses of realibility or other notions close to it as trust, reliability and others. These components were described in Fig. 1.

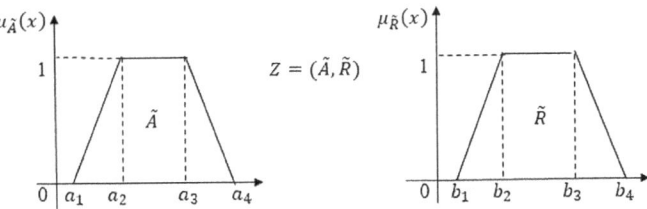

Fig. 1. Description of Z-numbers.

Examples on Z numbers: "Kamil is an excellent student" (very high, it is assumed). The first part, "very high" is the limitation of assesment of Kamil's knowledge, i.e. the value of A. The second part "it is assumed" is R and is a probability size indicating accuracy of this information. "A very comfortable car" (very high, very high).

Definition 4. (a fuzzy number described in here discrete form [7–10]). The Fuzzy subset A of R with a membership function $\mu_A : \mathcal{R} \to [0, 1]$ becomes a fuzzy number if this number possesses finite carriers, i.e. there exists the numbers $x_1, x_2, \ldots, x_n \in \mathcal{R}$ satisfying the condition $x_1 < x_2 < \ldots < x_n$. At the same time, there exist the numbers s, t given by $1 \leq s \leq t \leq n$ and satisfying the following conditions:

For any natural numbers i satisfying the condition $s \leq i \leq t - \mu_A(x_i) = 1$.

For any natural numbers i, j satisfying the condition

$$1 \leq i \leq j \leq s - \mu_A(x_i) \leq \mu_A(x_j).$$

For any natural numbers satisfying the condition

$$t \leq i \leq j \leq n - \mu_A(x_i) \geq \mu_A(x_j).$$

Definition 5. (Probability site of a discrete Fuzzy numbers [7]). Assume that A is a discrete fuzzy number, $P(A)$ is a probability siza of A. The probability size can be defined as follows:

$$P(A) = \sum_{i=1}^{n} \mu_A(x_i) p(x_i) = \mu_A(x_{j1}) p_j(x_{j1}) + \mu_A(x_{j2}) p_j(x_{j2}) + \ldots + \mu_A(x_{jn_j}) p_j(x_{jn_j})$$

Below, the definition of Fuzzy numbers are considered [6–9]. Here, Fuzzy numbers are considered in non interactive form.

Definition 6. (Summation of discrete Fuzzy numbers [1–5]). Summation of discrete Fuzzy numbers

$$A_{12}^{\alpha} = \left\{ x \in \left\{ supp(\tilde{A}_1) + supp(\tilde{A}_2) \right\} \middle| \min\{A_1^{\alpha} + A_2^{\alpha}\} \leq x \leq \max\{A_1^{\alpha} + A_2^{\alpha}\} \right\} \qquad (3)$$

where,

$$supp(\tilde{A}_1) + supp(\tilde{A}_2) = \left\{ x_1 + x_2 \middle| x_j \in supp(\tilde{A}_j), j = 1, 2 \right\}$$

$$min\{A_1^{\alpha} + A_2^{\alpha}\} = min\left\{ x_1 + x_2 \middle| x_j \in supp\left(\tilde{A}_j^{\alpha}\right), j = 1, 2 \right\}$$

$$max\{A_1^{\alpha} + A_2^{\alpha}\} = max\left\{ x_1 + x_2 \middle| x_j \in supp\left(\tilde{A}_j^{\alpha}\right), j = 1, 2 \right\}$$

$$\mu_{\tilde{A}_1 + \tilde{A}_2}(x) = \sup\{\alpha \in [0, 1] x \in \tilde{A}_1^{\alpha} + \tilde{A}_2^{\alpha}\}.$$

Definition 7. (Multiplication of discrete Fuzzy numbers [10]). Multiplication of discrete Fuzzy numbers $\tilde{A}_{12} = \tilde{A}_1 \cdot \tilde{A}_2$ is described by the cuts a of discrete Fuzzy numbers as was given:

$$A_{12}^{\alpha} = \left\{ x \in \left\{ supp(\tilde{A}_1) \cdot supp(\tilde{A}_2) \right\} \middle| \min\{A_1^{\alpha} \cdot A_2^{\alpha}\} \leq x \leq max\{A_1^{\alpha} \cdot A_2^{\alpha}\} \right\}$$

where,

$$supp(\tilde{A}_1) \cdot supp(\tilde{A}_2) = \left\{ x_1 \cdot x_2 \middle| x_j \in supp(\tilde{A}_j), j = 1, 2 \right\}$$

$$min\{A_1^{\alpha} \cdot A_2^{\alpha}\} = min\left\{ x_1 \cdot x_2 \middle| x_j \in supp\left(\tilde{A}_j^{\alpha}\right), j = 1, 2 \right\}$$

$$max\{A_1^\alpha \cdot A_2^\alpha\} = max\{x_1 \cdot x_2 | x_j \in supp\left(\tilde{A}_j^\alpha\right), j = 1, 2\}$$

$$\mu_{\tilde{A}_1 \cdot \tilde{A}_2}(x) = \sup\{\alpha \in [0, 1] x \in \tilde{A}_1^\alpha \cdot \tilde{A}_2^\alpha\}$$

Definition 8. (discrete distribution of probability). Discrete distribution of probability is determined as the function P. Here it is assumed that the discrete random number X have K numbers different values determined by the probability $P(X = x_i) = p(x_i)$. The probability $P(x)$ should satisfy the condition $0 \leq p(x_i) \leq 1$ for every and $\sum_{i=1}^{n} p(x_i) = 1$.

Definition 9. (converfence of discrete probability distribution). Assume that the distribution functions X_1 and X_2 discrete random variables p_1 and p_2. The distribution function $X_1 * X_2$ is given in the following form [3–5]:

$$p_{12}(x) = \sum_{x=x_1 * x_2} p_1(x_1) p_2(x_2)$$

Definition 10. (Z-numbers [4]). The discrete Z-numbers is expressed in the form of the ordered pairs $Z = \left(\tilde{A}, \tilde{B}\right)$. Here $\tilde{A}$ and $\tilde{B}$ are discrete fuzzy numbers consisting of the values of the random variable X. $\tilde{A}$ is a simitation imposed in the values of the variable X. $\tilde{B}$ is a fuzzy limitation imposed on $\tilde{A}$.

$$P\left(\tilde{A}\right) = \tilde{B}$$

The notion is limitation is more approriate to the notion of generalization than to the notion of restriction. Restriction can be showen as a generalized limitation. Probability distribution is a generalized restriction, lent it its known that this is not limitation [9, 10]. The notion of Z^+ is connected with the discrete Z-numbers, i.e. the Z^+-numbers consists of the pair of fuzzy number A and random number R and is defined as follows:

$$Z^+ = \left(\tilde{A}, R\right)$$

Here as in the discrete Z-numbers, $\tilde{A}$ is a fuzzy limitation value of the random variable X. R plays the role of distrbution of probabilities P [3–5].

$$P(A) = \sum_{i=1}^{n} \mu_A(x_i) p(x_i) = \mu_A(x_{j1}) p_j(x_{j1}) + \mu_A(x_{j2}) p_j(x_{j2}) + \ldots + \mu_A(x_{jn_j}) p_j(x_{jn_j})$$

3   Statement of the Problem

In this section, decision-making methods generalizing the expected utility approach based on Z information. This method is based on direct calculation on Z-numbers without converting to fuzzy numbers. Even the problem statement differs from other research works. Direct calculation with Z-numbers without conversion with great probability excepts information los.

Build an algorithm for assessing student knowledge. $Z = (A, B)$. A is the assessment threshold of a student's knowledge. $A_1 = \{0.1, 0.3, 0.5, 1.0, 0.7\}$ and $A_2 = \{0.1, 0.6, 1.0, 0.8, 0.3\}$. B is a measure of the probability that the information is true, i.e. $B_1 = \{$satisfactory = 0.2, sufficient = 0.5, excellent = 1, good = 0.7, very good = 0.3,$\}$ and $B_2 = \{$satisfactory = 0.1, sufficient = 0.3, excellent = 0.7, very good = 1.0, good = 0.5,$\}$. Assume, that the pair of fuzzy (A_1, B_1) and (A_2, B_2) numbers was given in the following form:

$$A1 = 0/1 + 0.3/2 + 0.5/3 + 1.0/4 + 0.7/5$$
$$B1 = 0.2/0, 3 + 0.5/0., 5 + 1.0/0.7 + 0.7/0.9 + 0.3/1$$
$$A2 = 0/1 + 0.6/2 + 1.0/3 + 0.8/4 + 0.3/5$$
$$B2 = 0.1/0.3 + 0.3/0.5 + 0.7/0.7 + 1.0/0.9 + 0.5/1$$

4   Solution of the Problem

Step 1. In the first step for calculating Z_{12} number we convert it to the discrete Z^+ number. Accept $Z_1^+ = (A_1 + A_2), Z_2^+ = (R_1 + R_2)$. R_1 and R_2 expresses the discrete probabilities p_1 and p_2. Applying the method for solving linear programming problem, the discrete distribution. For points $b_{11} = b_{21} = 0.3$ discrete distributions P_1 and P_2 of probability is calculated (Fig. 2.):

ck1	0,000	0,300	0,500	1,000	0,700		0,301
vk1	0,500	0,170	0,100	0,130	0,100		1,000
ck2	0,000	0,600	1,000	0,800	0,300		0,300
vk1	0,423	0,060	0,100	0,080	0,337		1,000

Fig. 2. Discrete disributions P_1 and P_2 of probability

ck_1 - described A_1 is fuzzy numbers, $vk1$ - A_1 is the probability dimension of fuzzy numbers, ck_2- described A_2 is fuzzy numbers, $vk_1 - A_2$ is the probability dimension of fuzzy numbers Thus, the distributed discrete probabilities have the following values at the point 1, 2, ..., 5 of x:

$$p_1 = 0.5/1 + 0.17/2 + 0.1/3 + 0.13/4 + 0.1/5$$
$$p_2 = 0.42/1 + 0.06/2 + 0.1/3 + 0.08/4 + 0.34/5$$

The sum of probabilities should satisfy the condition $\sum_{i=1}^{5} p_1(x_{1i}) = 1$.

The discretely distributed probabilities p_1 and p_2 were calculated for the points b_{12}, b_{22}, b_{13}, b_{23}, b_{14}, b_{24} and b_{15}, b_{25} in the same way.

Step 2. In this step we will determine the discrete number $Z_{12}^{+} = (A_1 + A_2, R_1 + R_2)$. Here, at first the number $A_{12} = A_1 + A_2$ is calculated. We distinguish the fuzzy numbers A_1 and A_2 from expression (1) in levels A_1^{α} and A_2^{α} by degrees $\alpha = 0, 0.3, 0.5, 0.6$, $0.7, 0.8, 1$ and apply the summation will $A_{12} = A_1 + A_2$.

$$A_{12} = 0/1 + 0/2 + 0/3 + 0.3/4 + 0.5/5 + 0.6/6 + 1/7 + 0.8/8 + 0.7/9 + 0.3/10$$

Step 3. In the next step, for calculating $R_1 + R_2$ we use the formula

$$p_{12}(x) = \sum_{x = x_{1i} + x_{2i}} p_1(x_{1i}) p_2(x_{2j})$$

In the step 1, using discretely distributed the probability values p_1 and p_2, for every value of x we get 5 (for our example) $p_{12}(x)$. For example, calculate the probability $p_{12}(x)$ as the point $x = 5$.

$$x = x_{11} + x_{24} = 1 + 4 = 5; \quad x = x_{12} + x_{23} = 2 + 3 = 5; \quad x = x_{13} + x_{22} = 3 + 2 = 5;$$
$$x = x_{14} + x_{21} = 4 + 1 = 5;$$

$$p_{12}(5) = p_1(1) \cdot p_2(4) + p_1(2) \cdot p_2(3) + p_1(3) \cdot p_2(2) + p_1(4) \cdot p_2(1)$$
$$= 0.5 \cdot 0.08 + 0.17 \cdot 0.1 + 0.1 \cdot 0.06 + 0.13 \cdot 0.42 = 0$$

Similarly we calculate $p_{12}(x)$ at the points 1, 2, 3, 4, ..., 10 of x. So, we get $Z_{12} = (A_1 + A_2, R_1 + R_2) = (A_1 + A_2, p_{12})$. Here, according to the formula (3) at the points $b_{11} = b_{21} = 0, 3$ we calculate p_{12} and get the following result:

$$p_{12} = 0/1 + 0.21/2 + 0.1/3 + 0.1/4 + 0.12/5 + 0.24/6 + 0.09/7$$
$$+ 0.06/8 + 0.05/9 + 0.03/10$$

Step 4. In this step we understand that "real" probability are not exact. Only p_1 and p_2 created by B_1 and B_2 have fuzzy limitations μ_{p1} and μ_{p2}. When solving a linear programming problem we calculate membership degree of $\mu_{pj}(x_j), j = 1, 2$. How let's consider finding of membership degree of μ_{p1} and μ_{p2} of distributions p_1 and p_2. Using the known values A_1 and p_1, they are calculated according to the following formula

$$\mu_{p1}(p_1) = \mu_{B1}\left(\sum_{k=1}^{n1} \mu_{A1}(x_{1k})p_1(x_{1k})\right)$$

$$\mu_{p1}(p_1) = 0 \cdot 0.5 + 0.3 \cdot 0.17 + 0.5 \cdot 0.1 + 1 \cdot 0.13 + 0.7 \cdot 0.1$$
$$= 0 + 0.051 + 0.05 + 0.13 + 0.07 = 0.301 \approx 0.3.$$

From expression (1) we determine $\mu_{p1}(0.3) = \mu_{B1}(0.3) = 0.2$. We calculate $\mu_{p2}(p_2)$ in the same way:

$$\mu_{p2}(p_2) = 0 \cdot 0.42 + 0.6 \cdot 0.06 + 1 \cdot 0.1 + 0.8 \cdot 0.08 + 0.3 \cdot 0.34$$
$$= 0 + 0.036 + 0.1 + 0.064 + 0.101 = 0.301 \approx 0.3$$

We determine $\mu_{p2}(0.3) = \mu_{B2}(0.3) = 0.1$. So we get membership degree of p_1 and p_2.
Step 5. At this step we should determine the fuzzy limitation μ_{p12}.

$$\mu_{p12}(p_{12}) = max_{p1,p2}\left[\mu_{p1}(p_1) \wedge \mu_{p2}(p_2)\right]$$

Here, $\wedge$ - is a minimum operator. We apply the $\wedge$ condition between the known values of fuzzy membership degrees of p_{12}.

$$\mu_{p12}(p_{12}) = \mu_{p1}(p_1) \wedge \mu_{p2}(p_2) = 0.1 \wedge 0.2 = 0.1$$

All p_{12} are constructed according to steps 1–5, in the similar. We determine the probability degree $P(A_{12})$ taking into account the condition

$$b_{12s} = \sum p_{12s}(x_k)\mu_{A_{12}}(x_k), \quad \mu_{B12}(b_{12s}) = sup(\mu_{p12s}(p_{12s}))$$

$$P(A_{12}) = \sum \mu_{A_{12}}(x_{12k})p_{12s}(x_{12k}) = 0 \cdot 0.21 + 0 \cdot 0.1 + 0.3 \cdot 0.1 + 0.5 \cdot 0.12$$
$$+ 0.6 \cdot 0.21 + 1 \cdot 0.09 + 0.8 \cdot 0.06 + 0.6 \cdot 0.05 + 0.3 \cdot 0.03 = 0.41$$

The calculated value $P(\ P(A_{12}) = b_{12} = 0.4$ is one of the possible values of probability size B_{12}. Now remind that

$$\mu_{B_{12}}\left(b_{12} = \sum_k \mu_{A_{12}}(x_{12k})p_{12}(x_{12k})\right) = \mu_{p12}(p_{12}),$$

Now remind that, $\mu_{p12}(p_{12}) = 0.1$. We get $b_{12} = \sum \mu_{A_{12}}(x_{12k})p_{12s}(x_{12k})$, $\mu_{B_{12}}(b_{12} = 0.41) = 0.1$. Conducting similar calculations, for our example B_{12} is calculated at 25 points. B_{12} with its membership degree is described as follows:

$$B_{12} = 0.1/0.41 + 0.2/0.44 + 0.2/0.51 + 0.2/0.49 + 0.2/0.51 + 0.1 + /0.49$$
$$+ 0.3/0.56 + 0.3/0.62 + 0.5/0.62 + 0.5/0.64 + 0.1/0.52 + 0.3/0.59$$
$$+ 0.7/0.69 + 1/0.75 + 0.5/0.79 + 0.1/0.59 + 0.3/0.65 + 0.7/0.74$$
$$+ 0.7/0.89 + 0.5/0.93 + 0.1/0.61 + 0.3/0.68 + 0.3/0.77 + + 0.3/0.94 + 0.3/1.$$

So the summation $Z_{12} = Z_1 + Z_2$ is given graphically in Fig. 3.

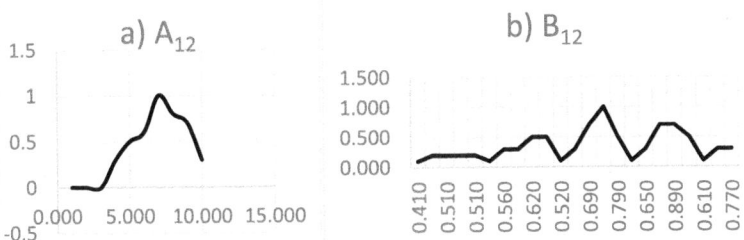

Fig. 3. The results of summation of discrete Z-numbers: a) A_{12}, b) B_{12}

Figure 3 shows the threshold for assessing knowledge (A) on a 10-point scale and a measure of probability (B), which indicates that it is true.

5 Conclusion

In this paper, an approach to make Z-information based decision based on exact calculation on Z-numbers is offered. This approach uses the expected utility paradigm and is applied to solving management problems in the economy.

References

1. Zadeh, L.A.: Fuzzy sets. Inform. Comput. **8**(3), 338–353 (1965)
2. Zadeh, L.A.: A note on Z-numbers. Inform. Sci. **181**(14), 2923–2932 (2011)
3. Aliev, R.R., Bodur, E.K., Mraiziq, D.A.T.: Z-number based decision making for economic problem analysis. In: Proceedings 7th International Conference on Soft Computing, Computing with Words and Perceptions in System Analysis, Decision and Control (ICSCCW 2013), Izmir, Turkey, pp. 251–257 (2013)
4. Aliev, R.A., Alizadeh, A.V., Huseynov, O.H.: The arithmetic of discrete Z-numbers. Inform. Sci. **290**, 134–155 (2015)
5. Kang, B., Wei, D., Li, Y., Deng, Y.: A method of converting Z-number to classical fuzzy number. J. Inform. Comput. Sci. **9**(3), 703–709 (2012)
6. Chen, S.-M.: Evaluating weapon systems using fuzzy arithmetic operations. Fuzzy Sets Syst. **77**(3), 265–276 (1996)

7. Wang, G., Wu, C., Zhao, C.: Representation and operations of discrete fuzzy numbers. South. Asian Bul. Math. **29**(5), 1003–1010 (2005)
8. Zadeh, L.A.: Calculus of fuzzy restrictions. In: Sets and Their Applications to Cognitive and Decision Processes, pp. 1–39. Academic Press, New York (1975)
9. Zadeh, L.A: Generalized theory of uncertainty (GTU)—principal concepts and ideas. Comput. Stat. Data Anal. 15–46 (2006)
10. Aliev, R.A., Zeinalova, L.M.: Decision making under Z-information. In: Human-Centric Decision-Making Models for Social Sciences, vol. 502, pp. 233–252. Springer, Germany (2014)

Trends in Azerbaijan's Electricity Security for Short-Term Periods

Nurali Yusifbayli[1(✉)] [iD] and Valeh Nasibov[2] [iD]

[1] Azerbaijan State University of Oil and Industry,
16/21 Azadliq, Baku, Azerbaijan
yusifbayli.n@gmail.com
[2] Azerbaijan Research and Design-Prospecting Institute of Energetics,
94, Zardabi, Baku, Azerbaijan
nvaleh@mail.ru

Abstract. Electricity security for short-term tasks is presented by the combination of four subsystems, namely: energy resource supply subsystem, energy generation subsystem, transmission and distribution of electricity subsystem, import of electricity. To determine energy security for each subsystem, linguistic variables, rules and membership functions were used. The resultant level of Azerbaijan Republic's short-term electricity security is determined by using the security of subsystems according to developed table.

Keywords: Energy system · Short-term periods · Supply of gas · Energy resource supply · Energy generation · Transmission and distribution of electricity · Subsystem

1 Introduction

Energy security acts as a component of the energy trilemma, both in assessing the effectiveness of energy sector performance and in assessing its sustainability [1–4]. Electricity energy security, in turn, is an essential component of energy security [5]. The short-term trends in Azerbaijan's electricity security using the fuzzy-set theory are considered in this paper. As shown in [5], when considering short-term electricity security, indicators such as environmental impact, rapid growth of demand, depletion of natural resources, availability and volatility of prices related to long-term energy security can be excluded from consideration, and in such approach, such important factors of energy security as management, institutional and investment factors can only be taken into account indirectly. At such approach, the power industry is presented in the form of four interconnected subsystems: energy resource supply subsystem, energy generation subsystem, transmission and distribution of electricity subsystem, import of electricity. For each subsystem the most characteristic indicators are selected and external and internal risks and resiliencies are separately grouped.

R. A. Aliev et al. (Eds.): ICAFS 2020, AISC 1306, pp. 565–571, 2021.
https://doi.org/10.1007/978-3-030-64058-3_70

2 Electricity Security for Short-Term Periods

As shown in [6], when studying the problems of electricity security for short-term periods, one can use letter designations from A to E, as shown in Fig. 1, where A corresponds to the smallest risks and maximum resilience, and E corresponds to the greatest risks and the least resilience. When applying linguistic variables for the subsystem security classification, one can get the following equivalences: A– "excellent", B– "normal", C– "good", D– "bad" and E– "very bad". The selected indicators of each subsystem accept one of three values: low, medium and high.

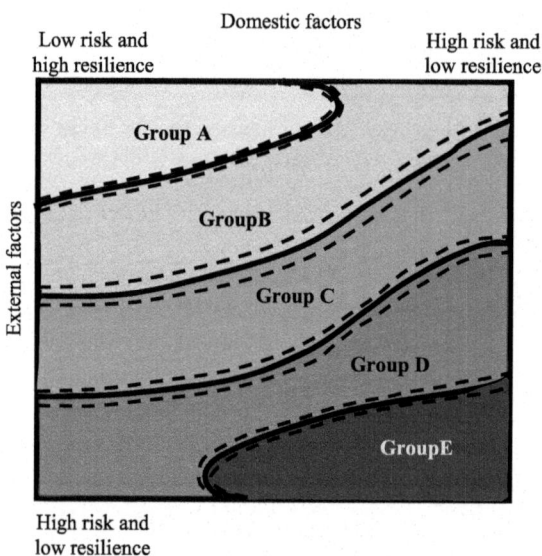

Fig. 1. Energy security profiles

When determining the security on Fig. 1, the obtained result proves to be rather qualitative, than quantitative. For example, receiving a value of C - "not bad", it is difficult to suppose, whether this result is closer to the border of B - "normal" or to the border of D - "bad", or it applies strictly to C. Reduction of area of each state can be achieved by increasing the number of states with "almost perfect", "almost normal", "not so bad" and so on. But in this case by reducing the inaccuracy of the value, the scheme of adequate response becomes complicated many times, and subjectivity increases when making a decision about corresponding of the indicators' values to one or another range [6].

To solve the problem of fuzziness of indicators' values, consider their dynamics' changes and obtain the quantitative value of security on the basis of linguistic information, the expressions of fuzzy-set theory and fuzzy logic can be used.

To solve the problem under consideration, Mamdani type fuzzy model is used.

Defuzzification of a fuzzy set is carried out as usual by the method of the center of gravity [7–11].

As is obvious from Fig. 2, the electricity security is determined on the basis of a three-layer model. At first, the energy resource supply level (natural gas supply) of the country is estimated, the output of which, along with two other inputs, is one of the inputs of the Supply of gas subsystem, the output of which is one of 4 inputs of electricity security [12, 13]. Using the above, the model of fuzzy inference for evaluation of each subsystem's security of electricity separately is drawn. It should be noted that in thermal power plants, where more than 90% of electricity is generated, natural gas is mainly used (almost 100%), therefore, only Supply of gas is considered in the energy resource supply subsystem. Below are the results of calculations for the subsystems' security and electricity security for 2013.

2.1 Supply of Gas

Data-in of the subsystem "Supply of gas" is given in Table 1.

Table 1. Data-in of "Supply of gas" subsystem

Supply of gas -*SNGS*			
Data-in	Terms		
	L-low	M-medium	H-high
DI-dependence on gas imports	< 9.98%	29.97–39.87%	>69.80%
II-infrastructure of gas imports	>59.95%	29.97–59.95%	<29.97%
RP-supplier diversity	>59.95%	29.97–59.95%	<29.97%
PQ-gas storage capacity	<49.96%	49.96–100%	>100%

The corresponding membership functions are selected for each subsystem. Securities of all subsystems are shown in Table 2.

Table 2. Security of all subsystems

Output	A	B	C	D	E
%	84.99–100	62.97–84.99	38.89–62.97	17.90–38.89	0–17.90

To estimate the security level of each subsystem it needs to use the fuzzy knowledge base, drawing up in the form of table of rules.

The "Supply of gas" subsystem's security for the short-term period is 92.5% [14].

2.2 Energy Resource Supply

The inputs to the energy resource supply are "Supply of gas", "Diversity of fuel", "Diversity of gas", as it is shown in Table 3.

Table 3. Data-in of "Energy resource supply" subsystem

Energy resource supply - *PFE*			
Data-in	Terms		
	L	M	H
SNGS -output of "Supply of gas" subsystem	59.97–100%	39.87–60%	0–39.87%
VF-diversity of fuel	>63.92%	32.97–63.92%	<32.97%
DPD- diversity of gas	>63.92%	32.97–63.92%	<32.97%

The security of this subsystem for the Azerbaijan will be 74%.

2.3   Energy Generation

Data-in of the subsystem "Energy generation" is shown in Table 4.

Table 4. Data-in of "Energy generation" subsystem

Energy generation-*EP*			
Data-in	Terms		
	L	M	H
G-domestic energy generation	<79.93%	79.93–89.89%	>89.89%
R-power reserve	<14.91%	14.91–24.92%	>39.87%
CI-equipment wear	<14.91%	14.91–29.95%	>39.87%
MP-level of distributed generation	<14.91%	14.91–29.95%	>39.87%

After defuzzification of output parameter, and with taking the Data-in *G*-100%, *R*-20%, *CI*-25%, *MP*–25% [14], this subsystem's security for Azerbaijan turns out to be 72.8%, which also corresponds to "normal" level.

2.4   Transmission and Distribution of Electricity

Data-in of the subsystem "Transmission and distribution of electricity" are given in Table 5.

Table 5. Data-in of "Transmission and distribution of electricity" subsystem

Transmission and distribution of electricity-*TDE*			
Data-in	Terms		
	L	M	H
WS-level of substation wear	<24.89%	29.87–49.92%	>59.94%
WT-level of transformer wear	<24.89%	29.87–49.92%	>59.94%
WL-level of air line wear	<24.89%	29.87–49.92%	>59.94%
SBR-balancing of region	<39.85%	39.85–69.93%	>68.98%

Calculating the security of "Transmission and distribution of electricity" subsystem at Data-in of WS-67%, WT-62%, WL-60%, SBR-60% [14] we shall receive 28.5%, which corresponds to D-"poor" security level.

2.5 Import of Electricity

The most important Data-in and their ranges for evaluation of "Import of electricity" subsystem's security are LI-electricity import, II-composition of imports, RMC -inter-system connection reserve.

Security of "Import of electricity" subsystem with Data-in LI-0.5% [13] turns out to be 92.5%, which corresponds to A -"excellent" security level.

3 Electricity Security

As shown in Fig. 2, the Electricity security is defined using four subsystems.
Electricity security data values are given in the Table 6.

Table 6. Data-in of "Electricity security"

Electricity security			
Data-in	Terms' meanings		
	L	M	H
PFE-energy recourse supply	0–38.92%	38.92–62.89%	62.89–100%
EP-energy generation	0–38.92%	38.92–62.89%	62.89–100%
TDE-transmission and distribution of electricity	0–38.92%	38.92–62.89%	62.89–100%
CEI-import of electricity	0–38.92%	38.92–62.89%	62.89–100%

The part of fuzzy knowledge base to estimate the electricity security is shown in Table 7.

Table 7. The part of fuzzy knowledge base to estimate the electricity security

№	Energy recourse supply	Energy generation	Transmission and distribution of electricity	Import of electricity	Result (output)
1	AB	AB	AB	AB	A
2	AB	AB	AB	C	B
3	AB	AB	AB	DE	B
4	AB	AB	DE	C	C
5	AB	AB	DE	DE	D

The obtained estimated values of electricity security of subsystems for 2013: energy resource supply-74%, energy generation-72.8%, transmission and distribution of electricity- 28.5%, import of electricity- 92.5%, the electricity security of Azerbaijan will constitute 74% and corresponds to the value "normal".

Similar studies for the electricity security of Azerbaijan have been conducted for 2015, 2018 and 2019, the results of which are presented in Table 8.

Table 8. Electricity security of Azerbaijan over the past few years

	Energy recourse supply, %	Energy generation, %	Transmission and distribution of electricity, %	Import of electricity, %	Output, %
2013	74	72.8	28.5	92.5	74
2015	74	65.3	27	92.5	72.5
2018	74	64	26	92.5	68.1
2019	74	65.2	28	92.5	72.1

As is obvious from the Table 9, the value of the electricity security of Azerbaijan over recent years has been changing in accordance with the changes in the security values of the Energy generation and Transmission and distribution of electricity subsystems. The state of the other two subsystems remains almost unchanged. The electricity security in recent years is shown in Fig. 2.

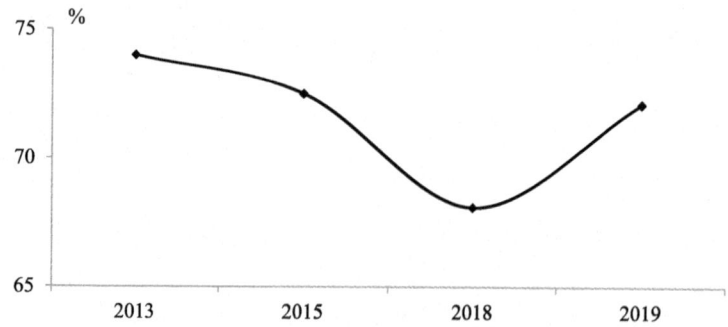

Fig. 2. Trends in electricity security of Azerbaijan

As is obvious from the Fig. 2, the electric power security state since 2013 was gradually deteriorating, mainly due to increased deterioration of generating and electric grid equipment, which was reflected in the system accident in the summer of 2018. As a result of large-scale rehabilitation works carried out, the electricity security in 2019 has slightly restored.

Assessment of the scope of rehabilitation works makes it possible to speak that by the end of 2020, Azerbaijan's electricity security will reach the value of 2013 - 74%.

4   Conclusions

Electricity security for short-term periods can be investigated using 4interconnected subsystems.

Considering the fuzziness and imperfection of the indicators, and the dynamics of their change, the electricity security can be identified with the security of subsystems using the fuzzy-set and fuzzy logic theories.

The following values were determined (as of 2013) for the electricity security and subsystem components of Azerbaijan, using the fuzzy-set theory: energy resource supply of power industry – 74%; energy generation – 72.8%Transmission and distribution of electricity – 28.5%; import of electricity – 92.5%; Azerbaijan's electricity security will be 74%.

According to the developed method, the electricity security of Azerbaijan in recent years (2015–2019) was determined, which showed a gradual deterioration in electricity security until 2018 and some recovery in 2019.

References

1. World Energy Trilemma, 2013 Energy Sustainability Index
2. World Energy Trilemma Index 2019
3. The Global Energy Architecture Performance Index Report 2013
4. Fostering Effective Energy Transition A Fact-Based Framework to Support Decision-Making
5. Yusifbayli, N., Nasibov, V.: Models of Azerbaijan energy security research. Energy pol. (3), 50–59 (2013)
6. Jessica, J.: The IEA Model of Short-term Energy Security (MOSES). Primary EnergySources and Secondary Fuels. International Energy Agency (2011)
7. Zadeh, L.A.: The concept of a linguistic variable and its application to approximate reasoning—I (1975)
8. Aliev, R.A., Aliev, P.P.: Sof Computing Fuzzy sets and systems. Baku (1996)
9. Aliev, R.A., Pedrycz, W., Fazlollahi, B., Huseynov, O.H., Alizadeh, A.V., Guirimov, B.G.: Fuzzy Logic-Based Generalized Decision Theory with Imperfect Information. Inf. Sci. **189**, 18–42 (2011)
10. Shtovba, S.D.: Designing fuzzy systems using MATLAB (2007)
11. Mamdani, E.H., Assilian, S.: Experiment in linguistic synthesis with a fuzzy logic controller. Int. J. Man-Mach. Stud. **7**(1), 1–13 (1975)
12. Nasibov, V.Kh.: Application of the fuzzy-set theory to the tasks of Azerbaijan electroenergetics security for short-term periods. Journal is registered in the library of the US congress, vol. 9, №. 4 (35), pp. 37–50, San Diego (2014)
13. Nasibov, V.Kh.: Determination of Azerbaijan electric power industry security for longtermperiods on the basis of fuzzy deduction. J. Multidisc. Eng.Sci. Stud. (JMESS) **2** (3), 363–373 (2016)
14. Materials of State Statistical Committee of Azerbaijan Republic. https://www.stat.gov.az/

Tiger Detection Using Faster R-CNN for Wildlife Conservation

Mohamad Ziad Altobel$^{(\boxtimes)}$ ⓘ and Melike Sah ⓘ

Department of Computer Engineering, Near East University, via Mersin 10, Nicosia, North Cyprus, Turkey
{mohamadziad.altobel,melike.sah}@neu.edu.tr

Abstract. The world population of tigers has been steadily declining over the years. Three of the nine major subspecies of tigers has extinct, and now tigers are declared as an endangered species. The tiger population does not actually need human aid to live, but it is important that they can be monitored and protected from poachers. For this purpose, artificial intelligence methods can be used to remotely monitor tigers in their habitat. This is the aim of this work. In this study, we use Faster R-CNN for tiger detection. Our software is implemented using Tensor Flow in Python and it can be easily integrated to motion sensor cameras, which can be used for remote monitoring of tigers in their habitat. In this way, the captured tiger images can be analyzed by conservation centers. We evaluated the efficiency of our tiger detection approach both quantitatively and qualitatively. We use ATRW tiger detection dataset for quantitative evaluations. In particular, this is the first time faster R-CNN has been applied for tiger detection in ARTW dataset. Results show that the faster R-CNN performs better than other popular deep learning based detectors.

Keywords: Deep learning · Faster R-CNN · Tiger detection · Python · Wildlife conservation

1 Introduction

Wildlife conservation has become a very important topic in our society as wildlife has come under threat from climate change and especially threats from human agents such as poachers. According to the UN Environment program, scientists estimate that between 150 to 200 species of plants, animals, and insects go extinct everyday as a result of climate change and human agents. This number is expected to grow if measures are not put in place to put a stop to it. One of such animals that have become endangered is the tiger. According to the World Wide Fund for Nature (WWF) [1], there are approximately 3,900 tigers left in the world, with the Balinese Tiger going extinct in the 1940s, the Caspian Tigers in the 1970s, the Javan Tiger in the 1980s, and the South China Tiger being reportedly extinct in the wild sometime in the 1990s. Figure 1 shows a graphical representation of the declining of tiger population between 1965 to 2010 [2].

In a bid to slow the decline of the tiger population and save the species, wildlife conservation centers have been working for years. However, there is still a lot of work

R. A. Aliev et al. (Eds.): ICAFS 2020, AISC 1306, pp. 572–579, 2021.
https://doi.org/10.1007/978-3-030-64058-3_71

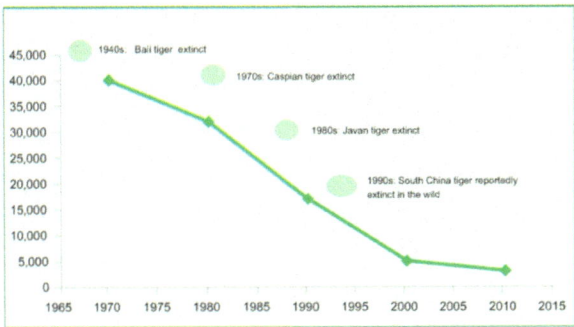

Fig. 1. World tiger population

left to be done, as even though these measures have slowed down the decline in the tiger population, the population continues to decline. The work that is done by these wildlife conservation centers is very important because these mass extinctions of species cause imbalances in the ecosystem that has a slow but dangerous effect on the planet. As an example, the extinction of apex predators in an ecosystem can cause an unchecked growth of the population of their prey, which would consequently put a major strain on the availability of resources such as food for their population.

One of the major steps in the work to reduce the decline of the tiger population is to monitor the tiger population and all threats to them and their habitat. To be able to do this, methods that use artificial intelligence (AI) can be employed. As we well know, it would be difficult, costly and dangerous to have always people live in the wild and monitor the population. Therefore, there is a need for intelligent monitoring using AI. This incorporation of machine learning into wildlife conservation is already happening in places like India for example. Computer vision is already being used to monitor animal population and habitat, and monitor poaching incidents [3, 4]. Some more advanced systems are even able to identify, count and even specify the activities of the animals in an image [5]. In addition, deep learning is also applied for wildlife conservation. For example, in the works of [15, 16], deep learning methods are utilized for the detection and classification of sea turtles.

The aim of this research is to use an accurate deep learning model that would be able to identify tigers in still images that would be captured from motion sensor cameras. This would go a long way in helping scientists capture as many pictures as possible from the wild, analyze the pictures, and return only pictures containing the tigers therefore helping them to monitor properly the population and any possible threats that may come to them. For this purpose, we train a faster R-CNN [6] object detector using Amur Tiger Re-Identification in the Wild (ATRW) dataset [7]. This dataset is one the largest dataset for wildlife conservation that contains images of Amur tigers taken from zoos in Chin. In the wild, less than 600 Amur tigers left, therefore conservation is crucial. Our results show that on the ATRW dataset, we can achieve the top detection accuracy among the other tiger detection methods.

The rest of the paper is organized as follows: Sect. 2 discusses related work. Section 3 explains the Faster R-CNN architecture as well as settings and training of

ATRW dataset. Section 4 discusses evaluations, and finally Sect. 5 is conclusions and future work.

2 Related Work

In this section, we discuss the related works that use computer vision for monitoring and detection of tigers.

There are a few AI systems already being used for wildlife conservation of tigers, such as PAWS (Protection Assistant for Wildlife Security) and M-STrIPES (Monitoring Systems for Tigers Intensive Protection and Ecological Status). PAWS is a system designed at the university of Southern California that uses machine learning to predict the behavior of poachers and the routes they are most possible to take [8]. This system helps to optimize the use of resources in patrols, as patrolling is still the most efficient way to prevent poaching. It was first tested in Uganda in 2014 and is now in regular use in Indonesia.

M-STrIPES is a system that is commissioned for use in India's tiger reserves by the NTCA (National Tiger Conservation Authority) [9]. It is a software that consists of protocols for recording patrol routes, law enforcement, recording trespassing and wildlife crimes and so forth.

In the literature, there are two works that use deep learning like the one used in this study for tiger detection [6, 10]. In the work of [10], Kupyn and Pranchuk use a model that is depend on Feature Pyramid Network (FPN). To improve the accuracy, they use Depthwise Separable Convolutions and lightweight FD-MobileNet model. Furthermore, they also modify RetinaNet architecture in order to decrease complexity.

Other very popular deep neural networks for object detection are Yolo [11], SSD [12], SSD MobileNet [13, 14] and Faster R-CNN [6]. They are proven to be efficient compared to other deep learning-based detectors. YOLO (You Look Only One) is an effective object detection algorithm based on regression in lieu of choosing interesting sections of the image, in running the algorithm for one time [11]. Yolo predicts bounding boxes for each class of the object using center of the class, height, width, a value to correspond the class of the object and the probability for expecting an object in this box. Yolo algorithm divides the image into cells using 19×19 grid each one predicting 5 bounding boxes and this results in 1805 bounding boxes for every image. But most of them will not contain an object and that why it give the predicting value PC which helping to remove the useless bounding boxes for the image and the bounding boxes with the high pc value will serve us to find the objects. Li et al., apply Yolo V3 for tiger detection on the ATRW dataset [7].

SSD (Single Shot Multibox Detector) is utilized for real-time detection tasks [12]. It consist of two parts; the first one for feature map extraction and the second part applies the CNN network to detect the objects. For the feature extraction pre trained networks such as VGG16, ImageNet can be used. SSD use the class score and the location in a simple way using convolutional filters. SSD MobileNet is a lightweight version of the object detector that is suitable when the computational resources are limited. MobileNets apply depth wise convolution in order to decrease the complexity of the network architecture. SSD MobileNet V2 [13] and V3 [14] are utilized by Li et al. [7] for tiger detection on the ATRW dataset. They use ImageNet pre-trained network as backbone architecture for the SSD MobileNet models [7].

Lastly Faster R-CNN is another popular object detector. In this study, we use faster R-CNN for tiger detection since among the other popular deep networks, faster R-CNN generally has better accuracy. In terms of time complexity, Yolo is the fastest, followed by SSD MobileNet, SSD, and then Faster R-CNN. Since, detection accuracy is also very important, we evaluate faster R-CNN in our work. In order to allow deployment on sensor cameras, we have implemented the tiger detector using Python Tensor Flow. In the next section, we explain the faster R-CNN.

3 Tiger Detection Using Faster R-CNN in Tensor Flow

In this section, first we explain the faster R-CNN architecture. Then, we discuss the ATRW dataset, and training and configurations for tiger detection.

3.1 Faster R-CNN

Faster R-CNN is an object detection architecture that was designed by Ren et al. [6]. It is created as a solution to the complexity required for accurate localization in object identification problems and as an improvement on the initial R-CNN and Fast R-CNN. The Faster R-CNN is made up of three parts: Feature extraction, Region Proposal Network (RPN) and detection network (Fig. 2).

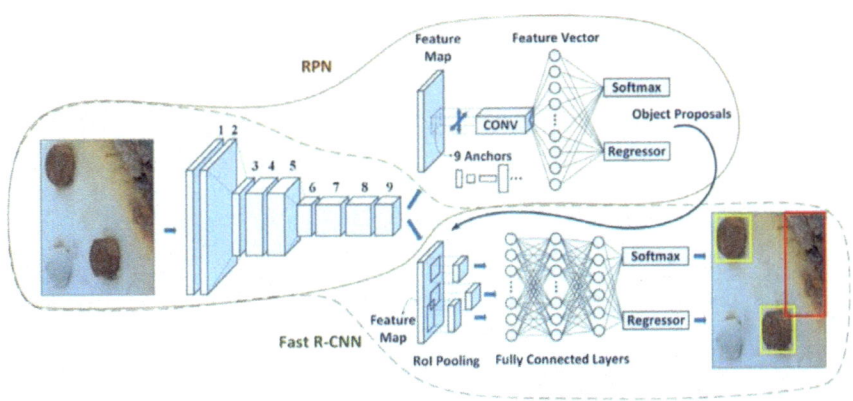

Fig. 2. Faster R-CNN architecture [6]

First, feature extraction is applied using a pre-trained network. Then, the extracted features are feed into a CNN architecture called Region Proposal Network (RPN) that offers hypothetical locations of the object to be identified. The RPN returns rectangular objects which are its proposals as the locations of the object. It also returns a value that represents its possibility of membership to a particular class as opposed to the rest of the image. Here, selective search algorithm is used to choose sections that have similar textures and colors and puts them in separate boxes. Softmax activation function is used to classify these selections. A linear regressor is also used to create bounding box localizations for each of the layers. Finally, in the last part, the Fast R-CNN object detector is applied. It takes the feature maps that are the output of the initial CNN and performs ROI pooling on them. These ROI are then passed on to a fully connected layer where they are flattened and passed on to the output layer to be classified and given bounding boxes.

3.2   Amur Tiger Re-identification in the Wild (ATRW) Dataset

The dataset we used in was provided under the CC BY-NC-SA 4.0 license - which is for the purpose of non-commercial research [7]. They processed sample video frames from 8,000 video clips of 92 Amur tigers taken from 10 zoos in china. This dataset is created with a focus on aiding re-Identification for wildlife and is available with bounding boxes, pose key points, and tiger identity annotations. The dataset contains 2,760 images of size 1920 × 1080. For training and testing, we divided the dataset into two groups of 2,060 images for training and 700 images for testing. This makes a 3-to-1 ratio or 75% to 25% division. The dataset is also available with bounding boxes (annotations). A sample annotated image from the ATRW is shown in Fig. 3.

Fig. 3. A tiger image from ATRW dataset with the annotated bounding boxes [7]

3.3   Faster R-CNN Implementation Using Tensor Flow

We run our work on Google Colab because it allows us to execute the code on Google Clouds server, and take advantage of Google hardware, such as GPUs and TPUs. This way, we could use TensorFlow GPU, rather than the regular TensorFlow. TensorFlow GPU provides more processing power and reduces the training time of the network by a factor of eight. First, we downloaded the ATRW dataset [7] consisting of the image dataset and its corresponding annotation files in XML format. In order to be used in TensorFlow, XML data is converted to CVS using TFRecords. Next, we split the dataset into two parts for training and testing of 75:25 ratio. We use Faster R-CNN Inception V2 model from TensorFlow's model zoo. As a backbone architecture we utilize Coco pre-trained model.

The training process lasted around 7 h with Google Colab TensorFlow GPU. The loss value started at 3.0 and went as low as 0.106 over 250,000 steps (2500 step for every epoch it makes 100 epochs). Stochastic gradient descent is utilized as solver, learning rate is 0.001 and anchor box confidence is 0.5.

4   Evaluations

First, we compare the performance of Faster R-CNN with other popular detectors in the ATRW dataset. Then, we qualitatively examine the performance of the tiger detection using faster R-CNN in ATRW dataset and tiger images from the Web.

4.1   Comparison of the Results on ATRW Dataset

Mean average Precision (mAP) with an overlap ratio of IOU = 0.5:0.05:0.95 with the annotated bounding boxes are calculated. Results in Table 1 shows that Faster R-CNN performs better than SSD MobileNetV2 and V3 and YoloV3. SSD MobileNet and Yolo models are well known for real-time detection performances, but their accuracy is below the accuracy of the Faster R-CNN. If we do not want to compromise the accuracy, rather than performing the detection in the motion sensor camera using a faster detector, processing of images can be transferred to the cloud. In this case, Raspberry Pi can be used together with Python for the detection and remote monitoring of tigers using Faster R-CNN. Since, accurate detection is crucial.

Table 1. Comparison of performances using Mean Average Precision (mAP) among IoU = 0.50:0.05:0.95 for Faster R-CNN, SSD, SSD MobileNetV2, SSD MobileNetV3, YoloV3.

Method	mAP (IoU = 0.50:0.05:0.95)
SSD MobileNetV2 [7]	0.419
SSD MobileNetV3 [7]	0.473
YoloV3 [7]	0.237
Faster R-CNN (this study)	**0.608**

4.2   Visual Examination of Tiger Detection Using ATRW and Web Images

In this section, we illustrate sample tiger detection performances using faster R-CNN. In Fig. 4a shows detection results from ATRW dataset and Fig. 4b shows results from randomly downloaded images from the Web. These illustrations confirm that faster R-CNN can accurately detect tigers even if it is trained from a different dataset.

Fig. 4. Sample tiger detection performances: (a) from the ATRW dataset, and (b) from randomly downloaded Web images.

5 Conclusions and Future Work

Tigers are endangered species and there is an urgent need to monitor and analyze their behaviors in their habitats for wildlife conservation. This can be achieved by artificial intelligence (AI) methods, such as the one used in this study. We presented a deep learning method for tiger detection using Faster R-CNN. Results show that faster R-

CNN provides very accurate detection compared to other popular deep learning detectors. Visual examinations also confirm this. In future, we plan to apply detectors for other wild animals in the habitat of tigers, such as dears, wild pigs, etc. This is particularly important for wildlife conservation. Using AI integrated motion cameras, animal movement, migration patterns and changing behaviors can be monitored remotely without interrupting wild animals. One of the challenges to achieve this aim is to obtain image datasets for the training of detectors.

References

1. World Wide Fund. https://www.worldwildlife.org/species/tiger
2. World tiger population. https://tigerpopulation.weebly.com/background.html
3. Thangarasu, R., Kaliappan, V.K., Surendran, R., Sellamuthu, K., Palanisamy, J.: Recognition of animal species on camera trap images using machine learning and deep learning models. Int. J. Sci. Technol. Res. **8**(10), 2613–2622 (2019)
4. Chen, G., Han, T.X., He, Z., Kays, R., Forrester, T.: Deep convolutional neural network based species recognition for wild animal monitoring. In: 2014 IEEE International Conference on Image Processing (ICIP), pp. 858–862 (2014)
5. Norouzzadeh, M.S., Nguyen, A., Kosmala, M., Swanson, A., Packer, C., Clune, J.: Automatically identifying wild animals in camera trap images with deep learning. ArXiv, abs/1703.05830 (2017)
6. Ren, S., He, K., Girshick, R., Sun, J.: Faster R-CNN: towards real-time object detection with region proposal networks. IEEE Trans. Pattern Anal. Mach. Intell. **39**(6), 1137–1149 (2017)
7. Li, S., Li, J., Lin, W., Tang, H.: Amur tiger re-identification in the wild. arXiv:1906.05586 (2019)
8. Fang, F., Ford, B., Yang, R., Tambe, M., Lemieux, A.M.: PAWS: game theory based protection assistant for wildlife security. In: Gore, M. (ed.) Conservation Criminology. Wiley-Blackwell, Hoboken (2017)
9. M-STrIPES. https://en.wikipedia.org/wiki/M-STrIPES
10. Kupyn, O., Pranchuk, D.: Fast and efficient model for real-time tiger detection in the wild. In: The IEEE International Conference on Computer Vision (ICCV) (2019)
11. Redmon, J., Divvala, S., Girshick, R., Farhadi, A.: You only look once: unified, real-time object detection. In: CVPR, pp. 779–788 (2016)
12. Liu, W., Anguelov, D., Erhan, D., Szegedy, C., Reed, S., Fu, C.Y., Berg, A.: SSD: single shot multibox detector. In: ECCV, pp. 21–37 (2016)
13. Howard, A.G., Zhu, M., Chen, B., et al.: MobileNets: efficient convolutional neural networks for mobile vision applications. arXiv:1704.04861 (2017)
14. Sandler, M., Howard, A., Zhu, M., Zhmoginov, A., Chen, L.C.: MobileNetV2: inverted residuals and linear bottlenecks. In: CVPR, pp. 4510–4520 (2018)
15. Badawy, M., Direkoglu, C.: Sea turtle detection using faster R-CNN for conservation purpose. In: ICSCCW. LNCS (2019)
16. Attal, Z., Direkoglu, C.: Sea turtle species classification for environmental research and conservation. In: ICSCCW. LNCS (2019)

Properties of Join and Meet Operations Over Z-Numbers

Akif V. Alizadeh$^{(\boxtimes)}$ (ID)

Department of Control and Systems Engineering, Azerbaijan State
Oil and Industry University, 20 Azadlig Avenue, Baku AZ1010, Azerbaijan
akifoder@yahoo.com, a.alizadeh@asoiu.edu.az

Abstract. In this study we initiate research on join and meet operations over Z-numbers. We formalize set theoretic operations as complement, join and meet of Z-numbers. Some important properties of these operations are studied.

Keywords: Z-arithmetic · Probabilistic arithmetic · Associativity law · Commutativity law · Z-numbers

1 Introduction

Several studies exist on computation with Z-numbers, which utilize the classical fuzzy arithmetic and probabilistic arithmetic. Zadeh suggests a general framework of computation of a Z-number-valued function based on Zadeh's extension principle [1]. However, the proposed approach is characterized by high complexity, it requires to deal with variational problems.

In [2] they consider Z-number as a pair of fuzzy numbers following the Zadeh's interpretation and propose a new way to formalization. Z-numbers are viewed in the frameworks of possibility theory, imprecise probabilities and belief functions. The authors show that it is easier to use random numbers than convex numbers of probabilities for Z-numbers. In general, the authors propose a new fundamental study of Z-numbers.

Kang et al. [3] proposed to deal with Z-numbers which naturally arise in the areas of control, decision making, modeling and others. The approach is based on converting a Z-number to a fuzzy number based on an expectation of a fuzzy set. However, converting Z-numbers to fuzzy numbers leads to loss of original information reducing the benefit of using original Z-number-based information. The work of Zadeh [4] is devoted to computation over continuous Z-numbers and several important practical problems. The suggested investigation is based on the use of normal probability density functions for modeling random variables. Aliev and colleagues [2, 5, 6] suggested a general and computationally effective approach to computation with Z-numbers. The approach is applied to computation of arithmetic and algebraic operations, t-norms and s-norms, and construction of typical functions. However, this approach is also based on classical fuzzy arithmetic.

The main disadvantage of classical interval arithmetic and fuzzy arithmetic is that fundamental properties of arithmetic operations over real numbers are lost [7–12]. This

© The Author(s), under exclusive license to Springer Nature Switzerland AG 2021
R. A. Aliev et al. (Eds.): ICAFS 2020, AISC 1306, pp. 580–590, 2021.
https://doi.org/10.1007/978-3-030-64058-3_72

creates problems with solving fuzzy equations, defining derivatives of fuzzy functions etc. In order to resolve this problem, in [9–11] they introduced basics of a new approach to fuzzy arithmetic which relies on the use of so-called horizontal membership functions (HMFs).

Thus, there is a need for investigation of properties of lattice operations over Z-numbers. This would provide a strong basis for the theory of Z-numbers and its further development. In this paper, we study properties of Z-numbers under additive arithmetic operations. Validity of the study is illustrated in examples.

In this study, we initiate research on join and meet operations over Z-numbers. We formalize set theoretic operations as complement, join and meet of Z-numbers. Some important properties of these operations are studied. Examples are included to illustrate the proposed approach.

The paper is organized as follows: Sect. 2 includes basic concepts used in the paper. In Sect. 3 we define join and meet operations over Z-numbers. Section 4 concludes properties of join and meet operations under Z-Numbers. Section 5 conclusion are given.

2 Preliminaries

Definition 1. Negation of Probability Distribution. Yager proposed a method for negation of probability distribution (N-PD), which provides a new idea for information representation.

Supposing a probability distribution (PD) $P = (p_1, p_2, p_3, \ldots, p_n)$ defined on the set $X = (x_1, x_2, x_3, \ldots, x_n)$, where $0 \leq p_i \leq 1$, $\sum_{i=1}^{n} p_i = 1$.

Then the NPD was defined as [13]

$$\bar{p}_i = \frac{1 - p_i}{n - 1} \qquad (1)$$

so, the NPD is denoted by P $\bar{P}$ as follows $\bar{P} = (\bar{p}_1, \bar{p}_2, \bar{p}_3, \ldots, \bar{p}_n)$.

Since the mutual exclusivity of probabilities and their negation are also normalized, so the NPD also satisfies $0 \leq \bar{p}_i \leq 1$, $\sum_{i=1}^{n} \bar{p}_i = 1$.

Definition 2. A Z-number. A Z-number is an ordered pair $Z = (A, B)$, where A is a fuzzy number playing a role of a fuzzy constraint on values that a random variable X may take: X *is A*, and B is a fuzzy number with a membership function $\mu_B : [0, 1] \rightarrow [0, 1]$, playing a role of a fuzzy constraint on the probability measure of A: $P(A)$ *is B* [14].

$P(A)$, induced by a set of distributions G:

$$G = \left\{ p_Z(x) : \int_X p_Z(x)dx = 1, \int_X p_Z(x)A(x)dx \text{ is } B, x \in X \right\}. \qquad (2)$$

3   Negation, Join and Meet Operations Over Z-Numbers

We consider three main operations of Z-numbers $Z_i = (A_i, B_i)$ which are defined as follows.

3.1   Negation of Z-Numbers

Definition 3. Negation of Z-number. Considering a discrete Z-number $Z = (A, B)$ where A is a discrete fuzzy number with a membership function $\mu_A : \{a_1, a_2, \ldots, a_n\} \rightarrow [0, 1]$, playing a role of a fuzzy constraint on a random variable X that may take: X is A, and B is also a discrete fuzzy number with a membership function $\mu_B : \{b_1, b_2, \ldots, b_n\} \rightarrow [0, 1]$, playing a role of fuzzy constraint on the probability measure of A: $P(A)$ is B.

The concept of discrete Z^+-number is closely related to a discrete Z-number.

Given a discrete Z-number $Z = (A, B)$, then Z^+-number is a pair consisting of a fuzzy number, A and a random number R: $Z^+ = (A, R)$, where A is the same role as in Z-number and R is the probability distribution p, such that

$$P(A) = \sum_{i=1}^{n} \mu_A(x_i)p(x_i) \tag{3}$$

Similarly, the negation of Z-number can be defined as: $\bar{Z} = (\bar{A}, \bar{B})$, where $\bar{A}$ is the negation of discrete fuzzy number A with a membership function $\mu_{\bar{A}} : \{a_1, a_2, \ldots, a_n\} \rightarrow [0, 1]$, satisfying

$$\mu_{\bar{A}} = 1 - \mu_A \tag{4}$$

and random variables X do not change, because they cover all possible outcomes of the event. Where $\bar{B}$ is also a discrete fuzzy number with a membership function $\mu_{\bar{B}} : \{b_1, b_2, \ldots, b_n\} \rightarrow [0, 1]$, we maintain the value of μ_B and $P(\bar{A})$ is $\bar{B}$.

Then

$$P(\bar{A}) = \sum_{i=1}^{n} \bar{\mu}_A(x_i)\bar{p}(x_i) \tag{5}$$

where $\bar{p}(x_i)$ is the negation of probability distribution calculated by Eq. (1).

3.2   Join of Z-Numbers

Join of Z-numbers $Z_i = (A_i, B_i)$, $i = 1, 2$ is defined as

$$Z_{12} = Z_1 \sqcup Z_2 = \max(Z_1, Z_2) = (A_{12}, B_{12}).$$

At first, let us consider the case of continuous Z-numbers.

The join $A_{12} = MAX(A_1, A_2)$ of the fuzzy numbers A_1 and A_2 is defined as follows:

$$A_{12}(z) = MAX(A_1, A_2)(z) = \sup_{z=\max(x,y)} \min [A_1(x), A_2(y)]. \tag{6}$$

The set G_{12} of resulting distributions p_{12} is defined as $G_{12} = G_1 \sqcup G_2$, where

$$G_i = \left\{ p_{Z_i}(x) : \int_X p_{Z_i}(x)dx = 1, \int_X p_{Z_i}(x)A_i(x)dx \text{ is } B_i, x \in X \right\} \tag{7}$$

for $i = 1, 2$.

Convolution $p_{12} = p_1 \circ_{\max} p_2$ of probability distributions is defined as

$$p_{12}(x) = p_1(x)F_1(x) + p_2(x)F_2(x)$$

where F_1 and F_2 are cumulative distribution functions:

$$F_1(x) = \int_{-\infty}^{x} p_1(x)dx, \quad F_2(x) = \int_{-\infty}^{x} p_2(x)dx.$$

Then

$$G_{12} = \{p_{12}(x) : p_{12}(x) = p_1(x)F_1(x) + p_2(x)F_2(x), p_i(x) \in G_i\},$$
$$B_{12} = \{(\mu_{p_{12}}(p_{12}), \mu_{A_{12}} \cdot p_{12}) : p_{12} \in G_{12}\}.$$

3.3   Meet of Z-Numbers

Meet of Z-numbers $Z_i = (A_i, B_i)$, $i = 1, 2$ is defined as

$$Z_{12} = Z_1 \sqcap Z_2 = \min(Z_1, Z_2) = (A_{12}, B_{12}).$$

The meet $MIN(A_1, A_2)$ of the fuzzy numbers A_1 and A_2 is defined as follows:

$$A_{12}(z) = MIN(A_1, A_2)(z) = \sup_{z=\min(x,y)} \min [A_1(x), A_2(y)]. \tag{8}$$

The set G_{12} of resulting distributions p_{12} is defined as $G_{12} = G_1 \cap G_2$, where G_i defined as (7).

Convolution $p_{12} = p_1 \circ_{\min} p_2$ of probability distributions is defined as

$$p_{12}(x) = p_1(x) + p_2(x) - p_1(x)F_1(x) - p_2(x)F_2(x),$$

where F_1 and F_2 are cumulative distribution functions:

$$F_1(x) = \int\limits_{-\infty}^{x} p_1(x)dx, \quad F_2(x) = \int\limits_{-\infty}^{x} p_2(x)dx.$$

Then

$$G_{12} = \{p_{12}(x) : p_{12}(x) = p_1(x) + p_2(x) - p_1(x)F_1(x) - p_2(x)F_2(x), p_i(x) \in G_i\}.$$

Thus,

$$B_{12} = \{(\mu_{p_{12}}(p_{12}), \mu_{A_{12}} \cdot p_{12}) : p_{12} \in G_{12}\}.$$

4   Properties of Lattice Operations on Z-Numbers

Theorem 1. Commutative law for meet of Z-numbers is satisfied:

$$Z_{12} = Z_1 \sqcap Z_2 = Z_2 \sqcap Z_1 = Z_{21}.$$

Proof. For A_{12} one has:

$$
\begin{aligned}
A_{12}(z) = MIN(A_1, A_2)(z) &= \sup_{z=\min(x,y)} \min\,[A_1(x), A_2(y)] \\
&= \sup_{z=\min(y,x)} \min\,[A_2(y), A_1(x)] = MIN(A_2, A_1)(z) = A_{21}(z).
\end{aligned}
$$

Because

$$
\begin{aligned}
A_{12} &= MIN(A_1, A_2) = \min(A_1(\mu, \alpha_1), A_2(\mu, \alpha_2)) = \\
&\min([a_{1L}(\mu, \alpha_1), a_{1R}(\mu, \alpha_1)], [a_{2L}(\mu, \alpha_{21}), a_{2R}(\mu, \alpha_2)]) = \\
&\min(\{a_1(\mu, \alpha_1) : a_1(\mu, \alpha_1) = a_{1L}(\mu) + \alpha_1 \cdot (a_{1R}(\mu) - a_{1L}(\mu))\}, \\
&\{a_2(\mu, \alpha_2) : a_2(\mu, \alpha_2) = a_{2L}(\mu) + \alpha_2 \cdot (a_{2R}(\mu) - a_{2L}(\mu))\}) = \\
&\min(\{a_2(\mu, \alpha_2) : a_2(\mu, \alpha_2) = a_{2L}(\mu) + \alpha_2 \cdot (a_{2R}(\mu) - a_{2L}(\mu))\}, \\
&\{a_1(\mu, \alpha_1) : a_1(\mu, \alpha_1) = a_{1L}(\mu) + \alpha_1 \cdot (a_{1R}(\mu) - a_{1L}(\mu))\}) = \\
&\min([a_{2L}(\mu, \alpha_{21}), a_{2R}(\mu, \alpha_2)], [a_{1L}(\mu, \alpha_1), a_{1R}(\mu, \alpha_1)]) = \min(A_2(\mu, \alpha_2), A_1(\mu, \alpha_1)) = \\
&MIN(A_2, A_1) = A_{21}
\end{aligned}
$$

The set of distributions of intersection Z_{12} can be described as follows.
$G_{12} = \{p_{12}(x) : p_{12}(x) = p_1(x) + p_2(x) - p_1(x)F_1(x) - p_2(x)F_2(x), p_i(x) \in G_i\}$,
where

$$G_i = \{p_{Z_i}(x) :$$
$$\int\limits_{-\infty}^{+\infty} p_{Z_i}(x)dx = 1, \int\limits_{-\infty}^{+\infty} p_{Z_i}(x)\mu_{A_i}(x) \text{ is } B_i,$$
$$\int\limits_{-\infty}^{+\infty} xp_{Z_i}(x) = \int\limits_{-\infty}^{+\infty} x\mu_{A_i}(x)dx / \int\limits_{-\infty}^{+\infty} \mu_{A_i}(x))dx\}$$

for $i = 1, 2$.

Thus, G_{12} is a set of convolutions $p_{12}(x)$. $p_{12}(x)$ satisfies commutativity property [15]:

$$p_{12}(x) = p_1(x) + p_2(x) - p_1(x)F_1(x) - p_2(x)F_2(x)$$
$$= p_2(x) + p_1(x) - p_2(x)F_2(x) - p_1(x)F_1(x) = p_{21}(x).$$

Therefore, G_{12} also satisfies commutativity property:

$$G_{12} = \{p_{12}(x) : p_{12}(x) = p_1(x) + p_2(x) - p_1(x)F_1(x) - p_2(x)F_2(x), p_i(x) \in G_i\} =$$
$$\{p_{21}(x) : p_{21}(x) = p_2(x) + p_1(x) - p_2(x)F_2(x) - p_1(x)F_1(x), p_i(x) \in G_i\} = G_{21},$$
$$G_{12} = G_1 \sqcap G_2 = G_2 \sqcap G_1 = G_{21}.$$

Then

$$B_{12} = \{(\mu_{p_{12}}(p_{12}), \mu_{A_{12}} \cdot p_{12}) : p_{12} \in G_{12}\} = (\mu_{p_{21}}(p_{21}), \mu_{A_{21}} \cdot p_{21}) : p_{21} \in G_{21}\}$$
$$= B_{21}.$$

Thus, $Z_{12} = Z_{21}$. The proof is completed.

Theorem 2. Commutative law for join of Z-numbers holds:

$$Z_{12} = Z_1 \sqcup Z_2 = Z_2 \sqcup Z_1 = Z_{21}.$$

Proof. For A_{12} one has:

$$A_{12}(z) = MAX(A_1, A_2)(z) = \sup_{z=\max(x,y)} \min [A_1(x), A_2(y)].$$
$$A_{21}(z) = \max(A_2, A_1)(z) = \sup_{z=\max(y,x)} \min [A_2(y), A_1(x)]$$
$$= \sup_{z=\max(x,y)} \min [A_1(x), A_2(y)] = A_{12}.$$

Because

$$A_{12} = MAX(A_1, A_2) = \max(A_1(\mu, \alpha_1), A_2(\mu, \alpha_2)) =$$
$$\max([a_{1L}(\mu, \alpha_1), a_{1R}(\mu, \alpha_1)], [a_{2L}(\mu, \alpha_{21}), a_{2R}(\mu, \alpha_2)]) =$$
$$\max(\{a_1(\mu, \alpha_1) : a_1(\mu, \alpha_1) = a_{1L}(\mu) + \alpha_1 \cdot (a_{1R}(\mu) - a_{1L}(\mu))\},$$
$$\{a_2(\mu, \alpha_2) : a_2(\mu, \alpha_2) = a_{2L}(\mu) + \alpha_2 \cdot (a_{2R}(\mu) - a_{2L}(\mu))\}) =$$
$$\max(\{a_2(\mu, \alpha_2) : a_2(\mu, \alpha_2) = a_{2L}(\mu) + \alpha_2 \cdot (a_{2R}(\mu) - a_{2L}(\mu))\},$$
$$\{a_1(\mu, \alpha_1) : a_1(\mu, \alpha_1) = a_{1L}(\mu) + \alpha_1 \cdot (a_{1R}(\mu) - a_{1L}(\mu))\}) =$$
$$\max([a_{2L}(\mu, \alpha_{21}), a_{2R}(\mu, \alpha_2)], [a_{1L}(\mu, \alpha_1), a_{1R}(\mu, \alpha_1)]) =$$
$$\max(A_2(\mu, \alpha_2), A_1(\mu, \alpha_1)) = MAX(A_2, A_1) = A_{21}$$

The set of distributions of join Z_{12} can be described as follows.

$$G_{12} = \{p_{12}(x) : p_{12}(x) = p_1(x)F_1(x) + p_2(x)F_2(x), p_i(x) \in G_i\},$$
$$G_i = \{p_{Z_i}(x) : \int_{-\infty}^{+\infty} p_{Z_i}(x)dx = 1, \int_{-\infty}^{+\infty} p_{Z_i}(x)\mu_{A_i}(x) \text{ is } B_i,$$
$$\int_{-\infty}^{+\infty} xp_{Z_i}(x) = \int_{-\infty}^{+\infty} x\mu_{A_i}(x)dx / \int_{-\infty}^{+\infty} \mu_{A_i}(x))dx\}$$

for $i = 1, 2$.

$p_{12}(x)$ satisfies commutativity property [15]:

$$p_{12}(x) = p_1(x)F_1(x) + p_2(x)F_2(x) = p_2(x)F_2(x) + p_1(x)F_1(x) = p_{21}$$

Therefore, G_{12} also satisfies commutativity property:

$$G_{12} = \{p_{12}(x) : p_{12}(x) = p_1(x)F_1(x) + p_2(x)F_2(x), p_i(x) \in G_i\} =$$
$$\{p_{21}(x) : p_2(x)F_2(x) + p_1(x)F_1(x) = p_{21}, p_i(x) \in G_i\} = G_{21},$$
$$G_{12} = G_1 \sqcup G_2 = G_2 \sqcup G_1 = G_{21}.$$

Then

$$B_{12} = \{(\mu_{p_{12}}(p_{12}), \mu_{A_{12}} \cdot p_{12}) : p_{12} \in G_{12}\} = (\mu_{p_{21}}(p_{21}), \mu_{A_{21}} \cdot p_{21}) : p_{21} \in G_{21}\}$$
$$= B_{21}.$$

Thus, $Z_{12} = Z_{21}$. The proof is completed.

Theorem 3. Associative law holds for meet of Z-numbers:

$$(Z_1 \sqcap Z_2) \sqcap Z_3 = Z_1 \sqcap (Z_2 \sqcap Z_3).$$

Proof. For A parts of Z-sets one has:

$$A_{(12)3}(z) = MIN(A_{12}, A_3)(z) = \sup_{z=\min(x,y)} \min[A_{12}(x), A_3(y)]$$
$$= \sup_{z=\min(x,y)} \min[\sup_{x=\min(x_1,x_2)} \min[A_1(x_1), A_2(x_2)], A_3(y)].$$

As min operation satisfies associativity condition, then

$$\sup_{z=\min(x,y)} \min [\sup_{x=\min(x_2,x_3)} \min [A_2(x_2),A_3(x_3)],A_1(y)]$$
$$= \sup_{z=\min(x,y)} \min [A_1(y),A_{23}(x)] = A_{1(23)}(z).$$

Then, for any fuzzy numbers A_1, A_2, A_3 the following property hold

$$MIN(MIN(A_1,A_2),A_3)) = MIN(A_1,MIN(A_2,A_3)), \text{ so } A_{(12)3} = A_{1(23)}.$$

Let us now consider G sets:

$$G_{(12)3} = \{p_{(12)3}(x) : p_{(12)3}(x) = (p_1 \circ_{\min} p_2) \circ_{\min} p_3, p_i(x) \in G_i, i = 1,\dots,3\},$$
$$G_{1(23)} = \{p_{1(23)}(x) : p_{1(23)}(x) = p_1 \circ_{\min} (p_2 \circ_{\min} p_3), p_i(x) \in G_i, i = 1,\dots,3\}.$$

As min operation for random variables satisfies associativity property $p_{(12)3}(x) = p_{1(23)}(x)$ [15], we have $G_{(12)3} = G_{1(23)}$.
As, $A_{(12)3} = A_{1(23)}$, and $G_{(12)3} = G_{1(23)}$, then $B_{(12)3} = B_{1(23)}$.
Therefore, $Z_{(12)3} = Z_{1(23)}$. The proof is completed.

Theorem 4. Associative law holds for join of Z-numbers:

$$(Z_1 \sqcup Z_2) \sqcup Z_3 = Z_1 \sqcup (Z_2 \sqcup Z_3)$$

Proof. For A parts of Z-sets one has:

$$A_{(12)3}(z) = \max(A_{12},A_3)(z) = \sup_{z=\max(x,y)} \min [A_{12}(x),A_3(y)]$$
$$= \sup_{z=\max(x,y)} \min [\sup_{x=\max(x_1,x_2)} \min [A_1(x_1),A_2(x_2)],A_3(y)].$$

As max operation satisfies associativity condition, then

$$\sup_{z=\max(x,y)} \min [\sup_{x=\max(x_2,x_3)} \min [A_2(x_2),A_3(x_3)],A_1(y)]$$
$$= \sup_{z=\max(x,y)} \min [A_1(y),A_{23}(x)] = A_{1(23)}(z).$$

Then, for any fuzzy numbers A_1, A_2, A_1 the following property hold

$$MAX(MAX(A_1,A_2),A_3)) = MAX(A_1,MAX(A_2,A_3)).$$

Let us now consider G sets:

$$G_{(12)3} = \{p_{(12)3}(x) : p_{(12)3}(x) = (p_1 \circ_{\max} p_2) \circ_{\max} p_3, p_i(x) \in G_i, i = 1,\dots,3\},$$
$$G_{1(23)} = \{p_{1(23)}(x) : p_{1(23)}(x) = p_1 \circ_{\max} (p_2 \circ_{\max} p_3), p_i(x) \in G_i, i = 1,\dots,3\}.$$

As max operation for random variables satisfies associativity property $p_{(12)3}(x) = p_{1(23)}(x)$ [15], we have $G_{(12)3} = G_{1(23)}$.

As $A_{(12)3} = A_{1(23)}$, and $G_{(12)3} = G_{1(23)}$, then $B_{(12)3} = B_{1(23)}$. Therefore, $Z_{(12)3} = Z_{1(23)}$. The proof is completed.

Theorem 5. Distribute law holds for Z-numbers:

$$Z_1 \sqcap (Z_2 \sqcup Z_3) = (Z_1 \sqcap Z_2) \sqcup (Z_1 \sqcap Z_3),$$
$$Z_1 \sqcup (Z_2 \sqcap Z_3) = (Z_1 \sqcup Z_2) \sqcap (Z_1 \sqcup Z_3).$$

Proof. Let us consider $Z_1 \sqcap (Z_2 \sqcup Z_3) = (Z_1 \sqcap Z_2) \sqcup (Z_1 \sqcap Z_3)$. The proof of $Z_1 \sqcup (Z_2 \sqcap Z_3) = (Z_1 \sqcup Z_2) \sqcap (Z_1 \sqcup Z_3)$ is analogous.

In opened notation:

$$(A_1, B_1) \sqcap ((A_2, B_2) \sqcup (A_3, B_3)) = ((A_1, B_1) \sqcap (A_2, B_2)) \sqcup ((A_1, B_1) \sqcap A_3, B_3)).$$

Let us show that in HMF-based fuzzy arithmetic the distributive law holds for any fuzzy sets A_1, A_2, A_3 [8–12]:

$$A_{1(23)} = MIN(A_1, MAX(A_2, A_3)) = MAX(MIN(A_1, A_2), MIN(A_1, A_3)) = A_{(12)(13)}.$$

In HMF-based representation, fuzzy sets A_1, A_2, A_3 are described as follows.

$$A_1(\mu, \alpha_1) = [a_{1L}(\mu, \alpha_1), a_{1R}(\mu, \alpha_1)] =$$
$$\{a_1(\mu, \alpha_1) : a_1(\mu, \alpha_1) = a_{1L}(\mu) + \alpha_1 \cdot (a_{1R}(\mu) - a_{1L}(\mu))\},$$
$$A_2(\mu, \alpha_2) = [a_{2L}(\mu, \alpha_2), a_{2R}(\mu, \alpha_2)] =$$
$$\{a_2(\mu, \alpha_2) : a_2(\mu, \alpha_2) = a_{2L}(\mu) + \alpha_2 \cdot (a_{2R}(\mu) - a_{2L}(\mu))\},$$
$$A_3(\mu, \alpha_3) = [a_{3L}(\mu, \alpha_3), a_{3R}(\mu, \alpha_3)] =$$
$$\{a_3(\mu, \alpha_3) : a_3(\mu, \alpha_3) = a_{3L}(\mu) + \alpha_3 \cdot (a_{3R}(\mu) - a_{3L}(\mu))\}.$$

We need to prove

$$\min(A_1(\mu, \alpha_1), \max(A_2(\mu, \alpha_2), A_3(\mu, \alpha_3)))$$
$$= \max(\min(A_1(\mu, \alpha_1), A_2(\mu, \alpha_2)), \min(A_1(\mu, \alpha_1), A_3(\mu, \alpha_3)))$$

The left part can be expressed as follows:

$$\min(A_1(\mu, \alpha_1), \max(A_2(\mu, \alpha_2), A_3(\mu, \alpha_3))) =$$
$$\min(a_{1L}(\mu) + \alpha_1 \cdot (a_{1R}(\mu) - a_{1L}(\mu)),$$
$$\max((a_{2L}(\mu) + \alpha_2 \cdot (a_{2R}(\mu) - a_{2L}(\mu))), (a_{3L}(\mu) + \alpha_3 \cdot (a_{3R}(\mu) - a_{3L}(\mu)))))$$

One can see that $A_1(\mu, \alpha_1), A_2(\mu, \alpha_2), A_3(\mu, \alpha_3)$ are real numbers for any values of $\mu, \alpha_1, \alpha_2, \alpha_3$. Then one has

$$\min(a_{1L}(\mu) + \alpha_1 \cdot (a_{1R}(\mu) - a_{1L}(\mu)),$$
$$\max((a_{2L}(\mu) + \alpha_2 \cdot (a_{2R}(\mu) - a_{2L}(\mu))), (a_{3L}(\mu) + \alpha_3 \cdot (a_{3R}(\mu) - a_{3L}(\mu)))) =$$
$$\max(\min(a_{1L}(\mu) + \alpha_1 \cdot (a_{1R}(\mu) - a_{1L}(\mu)), (a_{2L}(\mu) + \alpha_2 \cdot (a_{2R}(\mu) - a_{2L}(\mu))),$$
$$\min(a_{1L}(\mu) + \alpha_1 \cdot (a_{1R}(\mu) - a_{1L}(\mu)), (a_{3L}(\mu) + \alpha_3 \cdot (a_{3R}(\mu) - a_{3L}(\mu))) =$$
$$\max(\min(A_1(\mu, \alpha_1), A_2(\mu, \alpha_2)), \min(A_1(\mu, \alpha_1), A_3(\mu, \alpha_3))).$$

Thus, $A_{1(23)} = A_{(12)(13)}$.
Let us now consider G sets:

$$G_i =$$
$$\{p_{Z_i}(x) : \int_{-\infty}^{+\infty} p_{Z_i}(x)dx = 1, \int_{-\infty}^{+\infty} p_{Z_i}(x)\mu_{A_i}(x) \text{ is } B_i, \int_{-\infty}^{+\infty} xp_{Z_i}(x) = \int_{-\infty}^{+\infty} x\mu_{A_i}(x)dx / \int_{-\infty}^{+\infty} \mu_{A_i}(x))dx,$$
$$i = 1, \ldots, 3\}.$$

Let us prove that

$$p_{1(23)}(x) = p_1 \circ_{\min} (p_2 \circ_{\max} p_3) = (p_1 \circ_{\min} p_2) \circ_{\max} (p_1 \circ_{\min} p_3) = p_{(12)(13)}(x).$$

We recall that

$$p_{1(23)}(x) = \int_V p(x_1, x_2, x_3)dv = \int_V p(x_1, x_2, x_3)dx_1dx_2dx_3,$$

where $V = \{v = (x_1, x_2, x_3) : x = \min(x_1, \max(x_2, x_3))\}$ and $p(x_1, x_2, x_3)$ is a joint distribution. As independent random variables are considered, we have:

$$p(x_1, x_2, x_3) = p_1(x_1)p_2(x_2)p_3(x_3).$$

At the same time, $p_{(12)(13)}(x) = \int_W p_1(x_1)p_2(x_2)p_3(x_3)dw,$

where $W = \{w = (x_1, x_2, x_3) : x = \max(\min(x_1, x_2), \min(x_1, x_3))\}$.
As $\min(x_1, \max(x_2, x_3)) = \max(\min(x_1, x_2), \min(x_1, x_3))$, one has $V = W$. Thus,

$$p_{1(23)}(x) = p_{(12)(13)}(x).$$

Therefore, one has $G_{1(23)} = G_{(12)(13)}$, where

$$G_{1(23)} = \{p_{1(23)}(x) : p_{1(23)}(x) = p_1 \circ_{\min} (p_2 \circ_{\max} p_3), p_i(x) \in G_i, i = 1, \ldots, 3\},$$
$$G_{(12)(13)} = \{p_{(12)(13)}(x) : p_{(12)(13)}(x) = (p_1 \circ_{\min} p_2) \circ_{\max} (p_1 \circ_{\min} p_3), p_i(x) \in G_i, i = 1, \ldots, 3\}.$$

As $A_{(12)3} = A_{(12)(13)}$, and $G_{1(23)} = G_{(12)(13)}$, then $B_{1(23)} = B_{(12)(13)}$.
Thus, $Z_{1(23)} = Z_{(12)(13)}$. The proof is completed.

5 Conclusion

It is proved that the basic laws of Join and Meet Operations holds for Z-numbers. The proofs are based on the analogous properties of fuzzy numbers and probabilistic arithmetic. The obtained results are necessary for strong formulation of such important concepts as ordering of Z-numbers and other concepts.

References

1. Zadeh, L.A.: A note on Z-numbers. Inf. Sci. **181**, 2923–2932 (2011)
2. Aliev, R.A., Alizadeh, A.V., Huseynov, O.H.: The arithmetic of continuous Z-numbers. Inf. Sci. **373**, 441–460 (2016)
3. Kang, B., Wei, D., Li, Y., Deng, Y.: A method of converting Z-number to classical fuzzy number. J. Inf. Comput. Sci. **9**, 703–709 (2012)
4. Zadeh, L.A.: Methods and systems for applications with Z-numbers. United States Patent, Patent No.: US 8,311,973 B1, Date of Patent: Nov 13 (2012)
5. Aliev, R.A., Alizadeh, A.V., Huseynov, O.H.: The arithmetic of discrete Z-numbers. Inf. Sci. **290**, 134–155 (2015)
6. Aliev, R.A., Alizadeh, A.V., Huseynov, O.H., Jabbarova, K.I.: Z-number based linear programming. Int. J. Intell. Syst. **30**, 563–589 (2015)
7. Piegat, A., Plucinski, M.: Computing with words with the use of inverse RDM models of membership functions. Appl. Math. Comput. Sci. **25**(3), 675–688 (2015)
8. Piegat, A., Plucinski, M.: Fuzzy number addition with the application of horizontal membership functions. Sci. World J. **2015**, 16p (2015). Article ID 367214
9. Piegat, A., Landowski, M.: Is the conventional interval-arithmetic correct? J. Theoret. Appl. Comput. Sci. **6**(2), 27–44 (2012)
10. Piegat, A., Landowski, M.: Multidimensional approach to interval uncertainty calculations. In: Atanassov, K.T., et al. (eds.) New Trends in Fuzzy Numbers, Intuitionistic: Fuzzy Numbers, Generalized Nets and Related Topics, Volume II: Applications, pp. 137–151. IBS PAN - SRI PAS, Warsaw (2013)
11. Piegat, A., Landowski, M.: Two interpretations of multidimensional RDM interval arithmetic - multiplication and division. Int. J. Fuzzy Syst. **15**, 488–496 (2013)
12. Piegat, A., Plucinski, M.: Some advantages of the RDM-arithmetic of intervally-precisiated values. Int. J. Comput. Intell. Syst. **8**(6), 1192–1209 (2015)
13. Yager, R.R.: On the maximum entropy negation of a probability distribution. IEEE Trans. Fuzzy Syst. **23**(5), 1899–1902 (2015)
14. Zadeh, L.A.: Probability measures of fuzzy events. J. Math. Anal. Appl. **23**(2), 421–427 (1968)
15. Williamson, R.C., Downs, T.: Probabilistic arithmetic. I. Numerical methods for calculating convolutions and dependency bounds. Int. J. Approx. Reason. **4**(2), 89–158 (1990)

Z-Number Based Approach to Strategic Analysis in Tourism

A. M. Nuriyev[✉] ⓘ

Azerbaijan State Oil and Industry University,
Azadlig Avenue, 20, Baku AZ1010, Azerbaijan
aznouriyev@gmail.com

Abstract. In this study we are analyzing the possibility of utilizing Z-numbers for improving the quality of quantitative strategic analysis. The existing quantitative methods of strategic analysis do not consider the reliability of expert's estimates. Data used in calculations (value of factors, their importance) of Internal and External Factors Evaluation (IFE and EFE), Strategic Position and Action Evaluation (SPACE) and Quantitative Strategic Planning (QSPM) Matrices are considered with exact numerical values, although they were obtained from various sources with different level of confidence and, usually, these data have non-numerical nature. In order to take into account, the reliability of information in strategic analysis, it is proposed to use suggested by L. Zadeh a bi-component Z-number $Z = (A, B)$ [1] for assessing factors and their relevance. Parts A and B, described by fuzzy numbers, most accurately express expert's estimates. Z-value based quantitative strategic analysis framework is proposed. A general and computationally effective approach suggested (Z-QSPM, Z-SPACE) for computations with Z-numbers, can be utilized for quantitative strategic analysis. The efficiency of the proposed approach is illustrated by an example from tourism sector–determination and evaluation of tourism development strategy for the region.

Keywords: Quantitative strategic analysis · Z-number · Fuzzy number · Z-QSPM · Z-SPACE · Tourism development · Z-IDE · Z-EFE

1 Introduction

At present, the utilizing of strategic analysis in all spheres of human activity is a fact of life. Strategic analysis is a starting point for strategy development of the organization (enterprise, region, etc.) and it requires a realistic assessment of the resources and capabilities. Future orientation of the strategic management requires carrying out both internal and external analysis of the environment for exploring opportunities for the development and evaluating possible threats. Multiplicity of the important factors to be considered, poorly structured and unstructured data and incomplete baseline information, are key features of the strategic decision-making. Uncertainty arises from the processes outside and within the organization. The external environment is characterized by the variability of factors, affecting the organization. The uncertainty associated with internal factors stems from the interconnectedness and interdependence of

© The Author(s), under exclusive license to Springer Nature Switzerland AG 2021
R. A. Aliev et al. (Eds.): ICAFS 2020, AISC 1306, pp. 591–600, 2021.
https://doi.org/10.1007/978-3-030-64058-3_73

the components and subsystems of the organization. Generally, strategic decision makers rely on information with a high degree of uncertainty. At the same time, in strategic planning, it is important to identify and analyze as many alternatives as possible at the early stages of the decision-making process, in order to reduce the possibility of the further expensive errors and wrong decisions.

The probability theory, theory of possibilities and fuzzy approach are widely used for the processing of uncertain information in the strategic decision-making process. However, the methods used (analysis of statistical data or expert evaluations) do not fully consider the degree of reliability (certainty or confidence) of information.

Utilizing of Z-numbers allows to take into consideration uncertainty and reliability of the available information for decision-making. In our study we consider the application of Z-numbers for solving strategic management problems in tourism sector. The tourism sector is characterized by inseparability, heterogeneity, intangibility, perishability. A multi-stage Z-value based strategic decision-making structure has been presented. Initial stage consists of the strategic situation analysis by preparing SWOT-matrix, Z-value based IFE and EFE matrices, and Z-value based SPACE matrix. On the next stage, by using Z-value based QSPM, the relative attractiveness of the alternative strategies is assessed and justification for the strategy selection is provided.

2 Preliminary

***Definition 1. Properties of discrete Z-numbers* [2].** For any $Z_1, Z_2, Z_3 \in Z$, where Z is set of discrete Z-numbers, the following properties are holding:

(1) $Z_1 + Z_2 = Z_2 + Z_1$; (2) $(Z_1 + Z_2) + Z_3 = Z_1 + (Z_2 + Z_3)$
(3) $Z + O^z = Z$ for all $Z \in Z$, where O^z is the zero Z-number.

***Definition 2. Comparison of Z-numbers on the base of fuzzy Pareto optimality (FPO) principle* [2, 3].** According to this principle, two Z-numbers $Z_1 = (A_1, B_1)$ and $Z_2 = (A_2, B_2)$ are compared by calculate the functions n_b, n_e, n_w, which evaluate how much one of the Z-numbers is better, equivalent and worse than the other with respect to the first and the second component A and B. Degree of optimality (*do*) of Z-numbers are calculated based on computed $n_b(Z_i, Z_j)$ $n_e(Z_i, Z_j)$ $n_w(Z_i, Z_j)$.

If $do(Z_1) > do(Z_2)$ then $Z_1 > Z_2$, if $do(Z_1) < do(Z_2)$ then $Z_1 < Z_2$ and $Z_1 = Z_2$ otherwise.

***Definition 3. Z-valued weighted arithmetic mean* [2, 3].** A weighted arithmetic mean operator *WA()* assigns the unique Z-number Z_w to real-valued weighting vector $W = (W_1, W_2, \dots W_n)$ and Z-valued vector $Z = (Z_1, Z_2, \dots, Z_n)$. Components A_w, B_w are determined as follows:

$A_W^\alpha = [W_1 A_1 + W_2 A_2 + \dots + W_n A_n]^\alpha$ for each $\alpha \in [0, 1]$, $R_W = W_1 R_1 + W_2 R_2 + \dots + W_n R_n$. B_W is computed in accordance with methodology given in [2].

3 Method

3.1 Quantitative Strategic Analysis

Methods of strategic analysis are widely described in the scientific literature and, therefore, we shall not dwell on them in detail, but briefly state their essence. Strategic analysis a priory supposes teamwork (brainstorming, form-filling, interview, and review of literature etc.) and organization of expert discussions. SWOT analysis is mainly applied at the initial stage of assessment of the situation and allows its structured definition. For further analysis, as well as a description of the influencing factors, there is a need to rank them and determine their priority.

The outcome of these evaluations led to the formation of IFE and EFE matrices which are containing key internal (strengths and weaknesses) and external (opportunities and threats) factors, weights indicating their importance and ratings about the activity of the organization with respect to this factor. SPACE matrix [4] is used to define the type of strategy to be selected based on an assessment of IDE/EFE factors.

The final stage is compiling of the Quantitative Strategic Planning Matrix [5], which allows to choose the most appropriate strategy. Key factors with their weights and alternatives (proposed strategies) are pointed in the matrix. The strategy is determined by computing the sum of all Weighted Attractiveness Scores (WAS) for each of the alternatives and comparing them.

The basic idea is that a required type of strategy is in a space given by two internal and two external dimensions. It allows to search the type of strategy in the created space of external and internal factors. i.e. the space of strategies is generated by SWOT analysis. Applications of these methods in the tourism sector and the compilation of the appropriate matrices with both crisp and fuzzy numbers are shown in [6–9].

3.2 Importance of Z-Numbers for Strategic Analysis

Working with experts and using their knowledge and expertise is a key factor of strategic analysis. It should be noted that for above mentioned strategic matrices the exact values for factor weights and estimates are used [4–6, 9], but how these values are determined (through teamwork) is not exactly specified. As we know, the strategic analysis involves a group of people and the assessment made by each participant may vary. At present, methods for assessment of expert opinions consistency and averaging the expert opinions are widely used for the mathematical processing of expert evaluations. But what about conflicting expert estimates? If the experts do not agree to reassessment and desired degree of consistency is not reached? Moreover, expert's estimates are seldom expressed in crisp numbers, on the contrary, experts usually express their judgments in the form, for example, *«very likely that the price of the product X will increase»*. Moreover, the information used by the experts have not 100% degree of confidence. Experts often have to rely not on their own observations and measurements, but on information obtained from other people, media, financial reports, etc. How can a quantitative strategic analysis be conducted in such sphere as the tourism? Because, the mathematical nature of the data used in tourism management can be very different - numerical (numbers, vectors, functions) and non-numerical (vectors of heterogeneous features, binary relations, fuzzy sets).

From this point of view, the use of Z-numbers allows for an adequate reflection of expert knowledge and assessment in strategic analysis. Z-numbers allow to consider both the values of the evaluated parameters and the degree of confidence in these values. The creation in recent years of efficient methods of computation with Z-numbers [2], including the multiplication of crisp numbers and Z-numbers, makes it possible to process individual estimates, converting them into collective decisions, rank the elements, as well as to ensure a further transition to strategic decision-making.

4   Results

We are considering the use of Z-numbers for strategic analysis based on example from the tourism sector. Let's suppose, we should define a strategy for the development of the tourism sector in some region, with favorable geographical location, climate and ecology and certain historical sites.

Step 1. By performing the SWOT analysis, a group of experts identified the factors that affect the development of the tourism in the region. Various factors [6–9] of tourism development were considered and the SWOT-matrix was compiled (Table 1).

Table 1. SWOT-factors of tourism sector development

Strengths	Weaknesses
S_1 - Geographical location	W_1 - lack of accessible accommodation places
S_2 - Historical places	W_2 - Poor management and low skills in tourism and hospitality enterprises
S_3 – Weather	W_3 - Low transportation service
S_4 – Ecology	W_4 - Image and reputation in the tourist market
S_5 - Potential labor force	
S_6 - Experience in event organization	
Threats	Opportunities
T_1 - Possibility of natural disasters (rainfalls, foods, earthquake)	O_1 - Possibility of branding cultural, natural, historical features
T_2 - Destruction of historical places	O_2 - Ability of development of tourist infrastructure
T_3 - Environmental degradation	O_3- Favorable investment environment
T_4 - Increasing tariff rates	O_4 - Opportunity of personnel training
T_5 - Increasing number of competitors	O_5 - Increasing awareness of the potential tourists and strengthening marketing

In the next stage, matrices of IFE and EFE are compiled, the weights of each factor and its rating are determined by experts. In the classical form of this matrix [5], the weights are crisp numbers from 0.0 to 1.0 and ratings are crisp numbers from 1 to 4 (below average, average, above average, superior). However, as noted above, it is difficult to fully reflect the views of the experts with crisp numbers. Moreover, after processing of the expert opinion by the averaging or reassessment, the estimates may not be entirely adequate.

In our example, we use Z-numbers with parts A and B expressed by fuzzy triangular numbers (FTN). As a core of FTN, expressing the rating and weight, the numbers from 1 to 4 and from 0 to 1 are used only for presentation purposes. If expert's opinion is represented as range of values, then fuzzy trapezoidal numbers can be used in parts A and B. It should be noted that according given task, other ranges for ratings and

Table 2. The encoded linguistic terms for A and B components of Z-numbers

Variable	A (level)	A (TMF value)	B (level)	B (TMF value)
Rating	About 1	0/1, 1/1, 0/0	Not sure (VS)	1/0.05, 1/0.05, 0/0.25
	About 2	0/1, 1/2, 0/3	Not very sure (NVS)	0/0.05, 1/0.25, 0/0.5
	About 3	0/2, 1/3, 0/4	Sure (S)	0/0.25, 1/0.5, 0/0.75
	About 4	0/3, 1/4, 0/5	Very sure (VS)	0/0.5, 1/0.75, 0/1
	About 5	0/4, 1/5, 1/5	Extremely sure (ES)	0/0.75, 1/1, 1/1
Weight	About 0.03	0/0.02, 1/0.03, 0/0.04	Not sure	1/0.05, 1/0.05, 0/0.25
	About 0.05	0/0.04, 1/0.05, 0/0.06	Not very sure	0/0.05, 1/0.25, 0/0.5
	About 0.07	0/0.06, 1/0.07, 0/0.08	Sure	0/0.25, 1/0.5, 0/0.75
	About 0.1	0/0.09, 1/0.1, 0/0.11	Very sure	0/0.5, 1/0.75, 0/1
	About 0.13	0/0.12, 1/0.13, 0/0.14	Extremely sure	0/0.75, 1/1, 1/1
	About 0.15	0/0.14, 1/0.15, 0/0.16		
	About 0.17	0/0.16, 1/0.17, 0/0.18		
Attractiveness	Not very atr	0/1, 1/1, 0/0	Not sure	1/0.05, 1/0.05, 0/0.25
	Attract	0/1, 1/2, 0/3	Not very sure	0/0.05, 1/0.25, 0/0.5
	Very attract	0/2, 1/3, 0/4	Sure	0/0.25, 1/0.5, 0/0.75
	Ext. attract	0/3, 1/4, 0/5	Very sure	0/0.5, 1/0.75, 0/1
			Extremely sure	0/0.75, 1/1, 1/1

weights might be used because the search of a strategy is conducted in the space of SWOT factors. For normalization of the Z-weights of the factors can be used standard technique, as for crisp numbers. Differences in expert's judgments are expressed in part B of Z-number - degree of confidence, i.e. if there is a strong difference of opinion, this can be expressed by a correspondingly lower value of part B.

The linguistic terms of A and B components of Z-numbers and appropriate triangle membership function (TMF) values are shown in Table 2.

Step 2. After working with the experts and processing their opinions, weights and factor ratings are determined and the resulting IFE matrix is shown in Table 3.

Table 3. IFE matrix

Factor	Z-weight	Z-rating	Z-weighted score
S_1	About 0.17, VS	About 4, ES	(0.48, 0.68, 0.0), (0.43, 0.75, 0.95)
S_2	About 0.15, VS	About 4, VS	(0.42, 0.6, 0.8), (0.3, 0.63, 1)
S_3	About 0.1, VS	About 3, S	(0.18, 0.3, 0.44), (0.19, 0.43, 1)
S_4	About 0.17, ES	About 4, VS	(0.48, 0.68, 0.9), (0.43, 0.74, 1)
S_5	About 0.13, ES	About 3, VS	(0.24, 0.39, 0.56), (0.43, 0.75, 1)
S_6	About 0.07, VS	About 3, S	(0.12, 0.21, 0.32), (0.17, 0.43, 0.75)
W_1	About 0.1, ES	About 1, ES	(0, 0.1, 0.22) (0.63, 0.98, 1)
W_2	About 0.03, S	About 2, VS	(0.02, 0.06, 0.12) (0.31, 0.48, 0.75)
W_3	About 0.05, S	About 1, ES	(0, 0.05, 0.12) (0.24, 0.5, 0.75)
W_4	About 0.03, S	About 2, VS	(0.02, 0.06, 0.12) (0.31, 0.48, 0.75)

For calculation of Z-Weighted Score of Internal Factors we use

$$Z_{int} = \sum_{k=1}^{6} ZW_k^s \cdot ZR_k^s + \sum_{n=1}^{4} ZW_n^W \cdot ZR_n^w \tag{1}$$

Here, ZW_k^s – Z-weight and ZR_k^s – Z-rating of Strengths,
ZW_n^w – Z-weight and ZR_n^w – Z-rating of Weaknesses.

Z-Weighted Score for each factor is calculated by multiplication the Z-values of weights and ratings of appropriate factor.

Depending on features of analysis, the weights or ratings can also be expressed by scalar - i.e. instead of a Z-number (About 0.1, extremely sure) can be used suitable crisp number, for example 0.11. It should be noted that, commonly, crisp number 0.11 and Z-number (About 0.1, extremely sure) are not interchangeable because they reflect different information. As result of calculation the Total Z-weighted score Z_{int} = (1.96, 3.13, 4.5), (0, 0.04, 0.43). For calculation of Z-Weighted Score of External Factors we use

$$Z_{ext} = \sum\nolimits_{k=1}^{5} ZW_k^t \cdot ZR_k^t + \sum\nolimits_{n=1}^{5} ZW_n^o \cdot ZR_n^o \qquad (2)$$

Here, ZW_k^t – Z-weight and ZR_k^t – Z-rating of Threats, ZW_n^0 – Z-weight and ZR_n^0 – Z-rating of Opportunities. Total Z-weighted score $Z_{ext} = (1.66, 2.89, 4.27), (0.01, 0.21, 0.82)$ (Table 4).

Table 4. EFE matrix

Factor	Z-weight	Z-rating	Z-weighted score
T_1	About 0.08, VS	About 4, ES	(0.21, 0.32, 0.45), (0.46, 0.74, 1)
T_2	About 0.04, VS	About 2, VS	(0.03, 0.08, 0.15), (0.32, 0.63, 1)
T_3	About 0.04, VS	About 2, VS	(0.03, 0.08, 0.15), (0.32, 0.62, 1)
T_4	About 0.04, ES	About 3, ES	(0.06, 0.12, 0.2), (0.63, 0.98, 1)
T_5	About 0.1, ES	About 4, ES	(0.27, 0.4, 0.55), (0.63, 0.98, 1)
O_1	About 0.15, VS	About 1, VS	(0, 0.15, 0.32), (0.3, 0.63, 1)
O_2	About 0.19, ES	About 4, ES	(0.54, 0.76, 1), (0.63, 0.98, 1)
O_3	About 0.13, ES	About 4, ES	(0.36, 0.52, 0.7), (0.63, 0.98, 1)
O_4	About 0.1, VS	About 2, VS	(0.04, 0.2, 0.33), (0.3, 0.63, 1)
O_5	About 0.13, VS	About 2, VS	(0.12, 0.26, 0.42), (0.3, 0.62, 1)

Step 3. The results are used to determine the type of strategies in Z-SPACE matrix. In classical method [4] factors are evaluated by numbers from 0 to 6. In other cases, the numbers from 1 to 4 are used [6, 8]. All of this is aimed at determining the size of the respective strategy quadrants. In cases of Z-evaluations the maximal values $Z_{max\text{-}int}$ и $Z_{max\text{-}ext}$ of internal and external factors are used. To set the origin of coordinates, we can use the lowest estimate in Z-Rating or can set own value, for example $Z_{min} = (0, 1, 2), (0.25, 0.5, 0.75)$. To define quadrants (Fig. 1), the average values of factors $Z_{aver\text{-}int}$ and $Z_{aver\text{-}ext}$ are calculated or, we can define other Z-numbers as average, for example
$Z_{aver\text{-}int} = (1.5, 2.5, 3.5), (0.25, 0.5, 0.75)$ and $Z_{aver\text{-}ext} = (1.5, 2.5, 3.5), (0.25, 0.5, 0.75)$

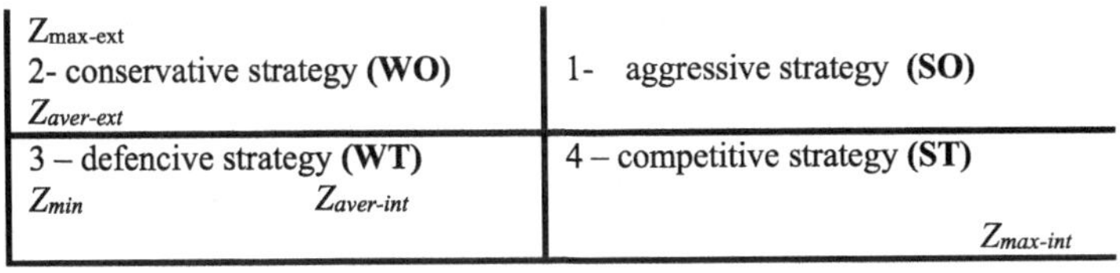

$Z_{max\text{-}ext}$ 2- conservative strategy **(WO)** $Z_{aver\text{-}ext}$	1- aggressive strategy **(SO)**
3 – defencive strategy **(WT)** Z_{min} $\qquad$ $Z_{aver\text{-}int}$	4 – competitive strategy **(ST)** <div align="right">$Z_{max\text{-}int}$</div>

Fig. 1. Z-SPACE-matrix plot

$Z_{int} > Z_{aver-int}$ and $Z_{ext} > Z_{aver-ext}$ means the aggressive strategy; $Z_{int} < Z_{aver-int}$ and $Z_{ext} > Z_{aver-ext}$ - conservative strategy; $Z_{int} < Z_{aver-int}$ и $Z_{ext} < Z_{aver-ext}$ – defencive strategy; $Z_{int} > Z_{aver-int}$ и $Z_{ext} < Z_{aver-ext}$ – competitive strategy. Then calculated Z_{ext}, Z_{int} are compared with $Z_{aver-ext}$, $Z_{aver-int}$

$$do\,(Z_{ext}, Z_{aver-ext}) = 0.82, \; do\,(Z_{aver-ext}, Z_{ext}) = 1 \; and \; do\,(Z_{int}, Z_{aver-int})$$
$$= 0.6, \; do\,(Z_{aver-int}, Z_{int}) = 1$$

The results fall into the 3rd quadrants and preferable is a defencive strategy-WT.

However, if to use the strengths and opportunities, then utilizing the concept of the pessimism coefficient $\beta > 0.5$ in formula below [10], Z-numbers can be re-compared

$$r(Z_i, Z_j) = \beta\,do(Z_j) + (1 - \beta)\,do(Z_i) \tag{3}$$

If $r(Z_i,\, Z_j) > 0.5 \cdot (do(Z_j) + (1 - \beta)\,do(Z_i))$ then $Z_i > Z_j$, if $r\,(Z_i,\, Z_j) < 0.5 \cdot (do(Z_j) + (1 - \beta)\,do(Z_i))$, then $Z_i < Z_j$, and $Z_i = Z_j$, otherwise.

In our case $Z_{ext} > Z_{aver-ext}$ and $Z_{int} > Z_{aver-int}$ and aggressive strategy (SO) is possible.

Table 5. Z-QSPM

Factor	Weight	(Strategy N1)	(Strategy N2)
		Z-AS	Z-AS
S_1	0.17	VA, ES	EA, VS
S_2	0.15	VA, S	EA, VS
S_3	0.1	EA, VS	NVA, VS
S_4	0.17	VA, ES	EA, ES
S_5	0.13	VA, ES	EA, ES
S_6	0.07	—	—
W_1	0.1	NVA, VS	A, VS
W_2	0.03	—	—
W_3	0.05	A, ES	VA, VS
W_4	0.03	A, VS	VA, ES
T_1	0.08	—	—
T_2	0.04	—	—
T_3	0.02	—	—
T_4	0.04	VA, VS	A, VS
T_5	0.1	EA, ES	A, VS
O_1	0.15	A, VS	EA, VS
O_2	0.19	EA, ES	VA, VS
O_3	0.13	EA, ES	VA, VS
O_4	0.1	—	—
O_5	0.13	A, VS	VA, VS

Step 4. We build QSPM with two SO strategies *(1 - developing tourism entertainment infrastructure, 2 - creating positive image as recreational place).* Experts evaluate for each factor Z-value of attractiveness score (AS). The AS should be defined for each strategy. If factor does not affect both strategies, then value of AS replaced by a dash. To demonstrate the mentioned above possibility of the joint using Z-numbers and crisp numbers in strategic analysis, we take the crisp numbers for weights (the values of the core of membership function of part A expressing the Z-value of weight) (Table 5).

Total attractiveness of each strategy is calculated according the expression

$$Z = \sum_{s=1}^{n1} W_s \cdot Sas_s + \sum_{w=1}^{n2} W_w \cdot Sas_w + \sum_{t=1}^{n3} W_t \cdot Sas_t + \sum_{o=1}^{n4} W_o \cdot Sas_o$$

Here $W_{s,w,t,o}$ – weights of factors, $Sas_{s,w,t,o}$ – Z-values of AS,
n_1, n_2, n_3, n_4 – number of factors with defined AS.

Total attractiveness for strategy N1 is expressed by $Z_1 = (3.24, 4.88, 6.52), (0.03, 0.29, 1)$ and for strategy N2 by $Z_2 = (3.61, 5.25, 6.79), (0.02, 0.27, 0.93)$.

According to the comparison of Z_1 and Z_2 $do(Z_1, Z_2) = 0.91$, $do(Z_2, Z_1) = 1$, and therefore the strategy N2 is preferable.

5 Conclusion

In our study we examine the application of Z-value based quantitative strategic analysis. Suggested approach used for strategic analysis in tourism sector. Uncertainty is a priori presented in strategic analysis and traditional quantitative models have a limited capacity of description and processing. In strategic analysis process researchers deal with both quantitative data and linguistic information and assess factors based on the intuition and experience. The possibility of constructing strategic analysis matrices (IFE, EFE, SPACE, QSPM) using Z-numbers is shown. By using the Z-numbers strategic analyst can evaluate factors of strategy, as well as their importance weights and attractiveness, then Z-numbers arithmetic technique can be used to calculate the overall attractiveness for each alternative. Recent advantages in theory of Z-numbers allow to process expert's information that takes into account both value of uncertain variable and reliability of value. Suggested approach is applicable not only to the tourism sector, but also to other spheres of activities.

References

1. Zadeh, L.A.: A note on Z-numbers. Inf. Sci. **181**(14), 2923–2932 (2011)
2. Aliev, R.A., Huseynov, O.H., Aliev, R.R., Alizadeh, A.V.: The Arithmetic of Z-Numbers: Theory and Applications. World Scientific, Singapore (2015)
3. Babanli, M.B., Huseynov, V.M.: Z-number-based alloy selection problem. Proc. Comput. Sci. **102**, 183–189 (2016). ISSN 1877-0509
4. Louw, L., Radder, L.: SPACE Matrix: a tool for calibrating competition. Soc. Sci. **4**, 549–559 (1998)

5. David, M., David, R., David, F.: The quantitative strategic planning matrix (QSPM) applied to a retail computer store. Coast. Bus. J. **8**, 42–52 (2009)
6. Simovic, S., Milinković, Z., Ljubojevic, A., Jovanović, J., Babić, K.P.: Selection process of sport tourism development strategy in Banja Vrućica Spa Resort: a quantitative analysis (2020)
7. Feili, H., Qomi, M., Sheibani, S., Azmoun, G.: SWOT analysis for sustainable tourism development strategies using fuzzy logic. In: 3rd International Conference of Science & Engineering in the Technology Era, Denmark, vol. 30 (2017)
8. Delis, A., Hodijah, S.: Development of local community-based tourism potential as an alternative for increasing sources of community income in Kerinci Regency, Indonesia. Russ. J. Agric. Soc.-Econ. Sci. **93**, 167–173 (2019)
9. Sadin, H., Zanganeh, A., Neshat, A., Talkhabi, H.R.: Strategic planning for development of rural tourism of North Astar Abad District of Gorgan with SWOT and fuzzy QSPM method (2015)
10. Aliev, R., Huseynov, O., Serdaroglu, R.: Ranking of Z-numbers and its application in decision making. Int. J. Inf. Tech. Decis. Making **15**, 1–17 (2016)

Estimation of Prediction Intervals for Artificial Neural Network-Based Rainfall-Runoff Modeling

Vahid Nourani[1] , Fahreddin Sadikoglu[2] ,
Nardin Jabbarian Paknezhad[1(✉)] , and Elnaz Sharghi[1]

[1] Department of Water Resources Engineering, Faculty of Civil Engineering,
University of Tabriz, 5166616471 Tabriz, Iran
{nourani,n.jabbarian,sharghi}@tabrizu.ac.ir,
jabbarian.nardin@gmail.com, elnaz_sharghi@yahoo.com
[2] Department of Electrical and Electronic Engineering, Faculty of Engineering,
Near East University, POBOX: 99138 Mersin 10, Nicosia, TRNC, Turkey
fahreddin.sadikoglu@neu.edu.tr

Abstract. In this study, the uncertainty associated with artificial neural network (ANN) for rainfall-runoff modeling on daily and monthly scales was evaluated by Prediction Intervals (PIs) for two watersheds, West Nishnabotna River basin in the United States and Lighvanchai River basin in Iran. Upper Lower Bound Estimation (LUBE) method was applied to construct the PIs. Furthermore, the Bootstrap technique, as a benchmark model, was applied to evaluate the uncertainty of ANN. In the LUBE method, the ANN is trained by minimizing the objective function via the genetic algorithm optimization method, and the objective function contains the width and coverage criteria of PIs evaluation. PIs coverage probability and PIs width values, respectively were up to 20% higher and 30% lower in the LUBE method compared to the Bootstrap method. Moreover, the CWC measure was considerably lower for LUBE method than the Bootstrap method. Also, the Lighvanchai basin modeling showed more accurate results than the West Nishnabotna River basin, which is due to four well defined regular seasons of the Lighvanchai basin.

Keywords: Rainfall-runoff · Prediction Interval · Artificial neural network · Bootstrap · Upper Lower Bound Estimation

1 Introduction

Accurate rainfall-runoff modeling is one of the main challenges in water and watershed management. True estimation of runoff is efficient in many cases like urban planning, land use, reduction in the impact of flood, and damage to the environment. The relationship between the rainfall and runoff values has been a complicated issue and hot topic in hydrology, due to spatial and temporal changes of factors that influence runoff. Therefore, many complicated hydrological models were applied for simulation of this stochastic process. As regards to the uncertainty and complication of considering the physical parameters, it is obvious that the application of black-box modeling can be an effective alternative because, in black-box modeling, the relation between inputs and output was obtained, independent of any information of the procedure. Recently, the Artificial Neural Networks (ANNs) are applied in different hydrological modeling to identify the complicated non-linear relation between input and output. The ANN is capable of modeling rainfall-runoff and it has been expressed in some researches (e.g., see, [1]).

Despite the appropriate ability of ANNs in hydrological modeling, the point predictions of the ANN models do not express the error and uncertainty of modeling, which is involved in the system being modeled [2]. With this regard, quantifying the ANN-based modeling reliability will be helpful in hydrological studies. Prediction Intervals (PIs) are appropriate tools to quantify the uncertainty of modeling. Decision makers should decide on the management of the critical issues associated to the rainfall. Therefore, by applying the PIs, uncertainties are considered and the achieved results present the important information to properly manage, decide and design the equipment's associated to the model. PIs express that if the issue is critical so pessimistic decision making is used and decision is made according to upper bound but it may not be economical; however, optimistically decision and consideration of the lower bound leads to high risk. There are several methods that calculate the PIs like delta [3], Bayesian [4], and Bootstrap [5]. The Bootstrap method has been applied in some previous studies about the ANN-based hydrological modeling (e.g., [2]). In some previous studies Lower Upper Bound Estimation (LUBE) method was applied to evaluate the uncertainty of ANN modeling (e.g. see [6]).

In this paper, the LUBE technique was used to construct the PIs. This technique is independent of any knowledge about the bounds of PIs or data distribution and it has a good coverage of observed values [7]. Quantitatively comparing the constructed PIs states that the LUBE method has higher speed and lower uncertainty. The ANN-based LUBE method includes outputs correspond to PIs in contrast to the classic ANNs, which consider one output as point prediction.

2   Methods

2.1   Study Area and Data Sets

The data from the West Nishnabotna River (a sub-basin of Missouri River in United States) and the Lighvanchai River (a sub-basin in Iran) which have various geomorphology were considered in this study. The first set of data used in this paper is for Lighvanchai watershed, placed in northwest Iran at Azerbaijan province. The time series data obtained from Iran Water & Power Resources Development Co. (IWPC). This basin is placed at the latitude of $37° 43'$ and $37°50'$ and longitude of $46°22'$ and $46°28'$, in the northern slope of Sahand Mountain (northwestern Iran). The area of the basin is 142 km^2 (Fig. 1). Its height is in the range of 1263 m and 3679 m. The length of the Lighvanchai River is 28.5 km. Lighvanchai River is a perennial flow as the regime of the river is snowy and the watershed has a permanent snow cover in different altitudes. Thus the snowmelt has a significant influence on the river flow and consequently, its drainage basin density is low. This basin weather is mostly wet and humid and has four well defined seasons. The topography is steep with an average slope of 11%. Consequently, the soils are disposed to erosion to some extent. The second basin, the West Nishnabotna River watershed is the main sub-basin of the Missouri River watershed that is placed in southwestern Iowa. The river is located in the longitude 95° 67′ W and latitude 40°51′ N. Also, the area of the watershed is 7600 km². This watershed consists of the wide highlands and valleys. The data for this basin were extracted from waterwatch.usgs.gov web site.

2.2   Evaluation Criteria

Point Prediction Evaluation. One of the frequent measures to assess the performance of the modeling is the Correlation Coefficient (CC), which reveals the linear relativity between two variables and can range from −1 to 1. The values more than 0 present a positive dependence and values lower than 0 express a negative relevant, while the value of 0 means there isn't any correlation between the two variables. CC is calculated as Eq. 1.

$$CC = \frac{\sum (R_i - \bar{R}) - \sum (Z_i - \bar{Z})}{\sqrt{\sum (R_i - \bar{R})^2 \sum (Z_i - \bar{Z})^2}} \tag{1}$$

As well, to assess the exactness of modeling, the Determination Coefficient (DC) is applied. DC explains the agreement of the observed value and the output of the modeling and it is between 0 to 1. The higher DC, presents the adaptation between the output and observed values. DC is measured as Eq. 2.

$$DC = 1 - \frac{\sum_{i=1}^{N} (R_i - Z_i)^2}{\sum_{i=1}^{N} (R_i - \bar{R})^2} \tag{2}$$

Root Mean Square Error (RMSE) is the criterion of the efficiency of model which is as Eq. 3.

$$RMSE = \sqrt{\frac{\sum_{i=1}^{N} (R_i - Z_i)^2}{N}} \tag{3}$$

PIs Assessment Measures. In order to quantify the PIs, the criteria of PI coverage probability (PICP) and mean PI width (MPIW) were used. The first criterion corresponds to the encompassments of the intervals. The wider PIs led to higher values of PICP [7]. PICP is calculated as Eq. 4.

$$PICP = \frac{1}{N} \sum_{i=1}^{N} c_i \tag{4}$$

$$if\ L(X_i) < x_i < U(X_i) \rightarrow c_i = 1;\ else \rightarrow c_i = 0$$

where $U(X_i)$ and $L(X_i)$ present respectively the upper and lower bounds of PIs corresponding to the ith sample.

Another criterion for quantification of PIs is their width. Equation (5) shows the normalized width [7].

$$NMPIW = \frac{1}{nR} \sum_{i=1}^{n} L(X_i) - U(X_i) \qquad (5)$$

where R is span of maximum and minimum of observed values. NMPIW is a dimensionless measure presenting the average width of PIs and the targets range.

Each of these criteria separately can't lead to a clear conclusion due to their inverse relationship. Therefore, the combinational coverage width-based criterion (CWC) is introduced, which contains both above criteria to evaluate the PIs [7] as follow:

$$CWC = NMPIW(1 + \gamma(PICP)e^{-\eta(PICP-\mu)}$$

$$\gamma = \begin{cases} 0, & PICP \geq \mu \\ 1, & PICP < \mu \end{cases}$$

The constants η and μ are two hyperparameters determine the penalty associated with lower PICP. μ presents the nominal confidence level associated with PIs and can be set to $1 - \alpha$. η magnifies the low variation between PICP and μ. The value of η is considered 10 in this study.

2.3   PIs Construction by LUBE Method

The procedure of proposed method is to obtain two outputs, which indicated the upper and lower bounds. The first step was driving the data into subsets of train and verify. The assigned weights were considered as initial weights. Then Genetic Algorithm (GA) starts generating new sets of weights by using the elitism, cross over, and mutation operators. Elitism determines number of the individuals with the lowest CWC. A selection operator was used for selecting the parents for the next generation. The mutation mechanism randomly generates new populations. The GA optimization ended when the number of iterations achieved the intended number or when the successive iterations didn't improve.

2.4   PIs Construction by Bootstrap Method

In the Bootstrap method, different neural networks (B networks) are created with randomly selected sub-sets among the main set of N samples [5]. This technique presumes that ensemble of some ANNs would lead to lower estimation errors. The superiority of this technique is about its independence from any complicated calculation of non-linear function. The randomly selected samples from total data are applied to train each of the ANNs. N is the number of main datasets and B is the number of the trained networks. A model as $f_{ANN}(x)$ is fitted to each of the generated bootstrap subset and the bootstrapping estimate is calculated as the average and variance of each model as:

$$\hat{y}_{boot}(x) = \frac{1}{B} \sum_{b=1}^{B} f_{ANN}^{b}(x) \tag{6}$$

$$\hat{\sigma}_{boot}^{2}(x) = \frac{1}{B-1} \sum_{b=1}^{B} \left(f_{ANN}^{b}(x) - \hat{y}_{boot}(x) \right)^{2} \tag{7}$$

For constructing the PIs [$l.u$], the value of observation X with normal distribution probability of P and according to Eqs. 6 and 7, $P(l < X < u)$ is as follow:

$$P(l < X < u) = P\left(\frac{l - \hat{y}_{boot}}{\hat{\sigma}_{boot}} < \frac{X - \hat{y}_{boot}}{\hat{\sigma}_{boot}} < \frac{u - \hat{y}_{boot}}{\hat{\sigma}_{boot}} \right) \tag{8}$$

where $z = \frac{X - \hat{y}_{boot}}{\hat{\sigma}_{boot}}$, the standard score of X (Table 1 represents the corresponding values of Z and PIs) is distributed as standard normal [8] hence:

$$\frac{l - \hat{y}_{boot}}{\hat{\sigma}_{boot}} = -z; \quad \frac{u - \hat{y}_{boot}}{\hat{\sigma}_{boot}} = z, \tag{9}$$

or

$$l = \hat{y}_{boot} - z\hat{\sigma}_{boot}; \quad u = \hat{y}_{boot} + z\hat{\sigma}_{boot}$$

Table 1. Values of Z for different PIs [8].

PI	z
75%	1.15
90%	1.64
95%	1.96
99%	2.58

3 Results

Since the performance of any data-driven method depends on the inputs of the model, selecting dominant inputs is critical in any ANN-based modeling. For selecting effective inputs of ANN, Mutual Information (MI) between target and time series of rainfall and runoff up to 20 and 13 lags, respectively for daily and monthly modeling were calculated and time series, which led to higher MI were chosen as inputs. For the West Nishnabotna River, on daily scale, the ANN inputs were only runoff from 1 to 5 time steps lag times. Also on monthly scale, the inputs of the ANN were combination of rainfall with one step lag time and runoff time series up to two steps lag times. For the Lighvanchai on daily scale runoff time series up to three lag times were used as inputs of the ANN. On monthly scale combination of rainfall with 1 step lag time with time series of runoff up to three time steps and 12th and 13th lag times for monthly scales were used as inputs. Therefore, four ANNs with different inputs were trained in this study on monthly and daily scales; the used time series as inputs of ANN were tabulated in Table 2. The 75% and 25% of time series were splitted to train and verification subsets, respectively.

Table 2. Point prediction results for two watersheds in both daily and monthly scales.

Scale	Basin	ANN	Input	DC train	DC verification	CC train	CC verification	RMSE* train (normalized)	RMSE* Verification (normalized)
Daily	Lighvanchai	1	$Q_{t-1}, Q_{t-2}, Q_{t-3}$	0.93	0.9	0.96	0.95	0.02	0.03
	West Nishnabotna River	2	$Q_{t-1}, Q_{t-2}, Q_{t-3}, Q_{t-4}, Q_{t-5}$	0.7	0.7	0.84	0.83	0.02	0.03
Monthly	Lighvanchai	3	$I_{t-1}, Q_{t1}, Q_{t2}, Q_{t-3}, Q_{t-12}, Q_{t-13}$	0.71	0.64	0.84	0.8	0.07	0.08
	West Nishnabotna River	4	$I_{t-1}, Q_{t-1}, Q_{t-2}$	0.55	0.5	0.67	0.69	0.08	0.1

Therefore, the data in periods 1989–2007 and 1978–2005 were used for the training, respectively for the Lighvanchai and the West Nishnabotna River and in periods 2008–2015 and 2006–2015 for the validation purpose. Since the point prediction of ANN didn't give any description about the accuracy of modeling, so, PI construction methods as the LUBE and the Bootstrap methods were used. The values of the PIs accuracy measurements were tabulated in Table 3.

Table 3. Criteria of PIs quantification via LUBE and Bootstrap methods.

Scale	ANN model	LUBE			Bootstrap		
		PICP	NMPIW	CWC	PICP	NMPIW	CWC
Daily	1	0.93	0.25	0.25	0.89	0.11	0.229765
	2	0.88	0.05	0.1	0.77	0.03	0.19484
Monthly	3	0.69	0.17	1.12	0.6	0.19	2.47
	4	0.7	0.17	1.04	0.51	0.11	2.81

4 Discussion

A three-layer feed-forward ANN with back propagation training algorithm of levenberg-Marquardt and the tangent sigmoid as the activation function were used in the models. To distinguish the appropriate ANN model, the epoch and hidden neuron numbers were determined by trial and error procedure. RMSE, DC, and CC measures were used to evaluate the performance of methods. In the Table 2, the results of ANN modeling were shown. As tabulated in Table 2, modeling on daily scale led to better performance and DC of validation sets were 10% and 15% higher than monthly scale respectively for the Lighvanchai and the West Nishnabotna River watersheds. This may be due to the consideration of the basin soil moisture, which causes the lower water infiltration, so the estimation of runoff values may be more accurate. In addition, stronger autoregressive time series of daily scale modeling compared to monthly scale modeling, which causes a better performance of modeling on daily scale. Moreover, more samples in the daily data with regard to the monthly date cause better model calibration on daily scale. For the Lighvanchai and the West Nishnabotna River watersheds modeling on the daily scale leading to appropriate results. The Lighvanchai on both monthly and daily scales showed higher performance with validation DC of about 20% and 26% higher than the West Nishnabotna River, respectively for monthly and daily scales modeling. Low efficiency of the ANN in West Nishnabotna River may be due to highly variable weather and rainfall patterns for this watershed and also vast area of this basin compared to the Lighvanchai basin. Moreover, the construction of dams in this watershed might impact on the river regime.

As shown in Table 3, the values of PICP for the daily estimation were 29% and 18% higher than monthly estimation for the Lighvanchai, and the West Nishnabotna River basins, respectively and the values for NMPIW were 20% and 17% narrower for daily modeling in the Lighvanchai and the West Nishnabotna River basins, respectively. Considering the values of CWC indicated that on daily scale it was considerably lower than monthly scale, which showed the higher accuracy on daily scale modeling. The overall outcome for PIs construction corresponded to the results of the point prediction. The length of the dataset in training may have important role in the performance of the PIs. On daily scale, the value of PICP for the Lighvanchai basin was up to 9% higher than the West Nishnabotna River; but on monthly scale, both watersheds approximately showed similar modeling performance. In order to compare the obtained results by LUBE technique, the Bootstrap technique was applied to calculate the PIs. Hence, 80 ANN models developed with each bootstrap dataset of 7440, 5114, 242,166 samples among 14811, 10227, 485, 335 samples for the West Nishnabotna River and the Lighvanchai basins, for the daily and monthly rainfall-runoff modeling, respectively. Then, total samples were trained by obtained weights of subsets training to achieve the outputs and to calculate the PIs (as Eqs. 6, 7 and 9); the obtained results with confidence interval of 90% are presented in Table 3. It is concluded from Table 3, in most cases modeling of the Lighvanchai got higher values of PICP up to 17% compared to the West Nishnabotna basin modeling. Moreover, the Bootstrap method for both cases on daily scale modeling led to respectively 38% higher and 11% lower, PICP and NMPIW values than on the monthly modeling, which denoted the more accurate results of the modeling on daily scale.

As shown in Table 3, considering CWC measure indicated that modeling of the Lighvanchai watershed got better performance on both daily and monthly scales for the Bootstrap method but for the LUBE method yielded almost similar performance on monthly scale for both basins but on daily scale, again the Lighvanchai modeling showed better performance. Comparison of two methods indicated that the PI obtained by the LUBE method was narrower than that obtained via the Bootstrap method. PICP values for the LUBE method for all models were higher than those from Bootstrap method, but NMPIW values from Bootstrap method were higher than most of those from the LUBE method, moreover the criterion of CWC was lower for constructed PIs via LUBE method. In most of the cases the Bootstrap method could lead to better performance in catching the peak flows, but with higher PIs width in comparison to the LUBE method. Moreover, it presented overestimated runoff values for most samples. This may be due to randomly selecting the samples for training the Bootstrap subsets, therefore subsets may not consist the feature of the main training set. As tabulated in Table 3, PICP values were higher for models 1 and 2 (i.e. for daily models) and NMPIW values were lower for models 1 and 2 which indicated more precise outcomes of daily modeling compared to the monthly modeling.

5 Conclusions

In this paper, ANN-based modeling, for rainfall-runoff modeling for two watersheds of Lighvanchai and West Nishnabotna River, was carried out on monthly and daily time scales. The obtained PIs via LUBE and Bootstrap methods were compared with each other. Analysis indicated that daily scale modeling led to more reliable results in both PIs construction methods. Values of PICP, respectively were up to 29% and 21%, higher on daily scale compared to monthly scale for constructed PIs of Lighvanchai and West Nishnabotna River basin modeling. Moreover, the value of CWC was significantly lower on daily scale modeling. Results indicated that LUBE method constructed PIs with high quality, values of PICP for constructed PIs via LUBE method were up to 19% higher than those constructed via Bootstrap method, which denoting the superiority of LUBE method with regard to the Bootstrap method. Comparison between the performance of the two basins, indicated that the Lighvanchai showed better performance compared to the West Nishnabotna River basin, PICP values for constructed PIs of the Lighvanchai basin modeling were up to 21% higher than those for the West Nishnabotna River basin. The regular four seasons for the Lighvanchai basin cause better performance of its modeling.

References

1. Tokar, A.S., Johnson, P.A.: Rainfall-runoff modeling using artificial neural networks. J. Hydrol. Eng. 4(3), 232–239 (1999). https://doi.org/10.1061/(ASCE)1084-0699(1999)4:3 (232)
2. Kasiviswanathan, K.S., Sudheer, K.P.: Quantification of the predictive uncertainty of artificial neural network based river flow forecast models. Stoch. Environ. Res. Risk Assess. 27(1), 137–146 (2013). https://doi.org/10.1007/s00477-012-0600-2
3. Chryssolouris, G., Lee, M., Ramsey, A.: Confidence interval prediction for neural network models. IEEE Trans. Neural Netw. 7(1), 229–232 (1996). https://doi.org/10.1109/72.478409
4. MacKay, D.J.C.: A practical Bayesian framework for backpropagation networks. Neural Comput. 4(3), 448–472 (1992)
5. Efron, B., Tibshirani, R.J.: An Introduction to the Bootstrap. CRC Press, Boca Raton (1994)

6. Nourani, V., Paknezhad, N.J., Sharghi, E., Khosravi, A.: Estimation of prediction interval in ANN-based multi-GCMs downscaling of hydro-climatologic parameters. J. Hydrol. **579**, 124226 (2019). https://doi.org/10.1016/j.jhydrol.2019.124226

7. Khosravi, A., Nahavandi, S., Creighton, D., Atiya, A.F.: Lower upper bound estimation method for construction of neural network-based prediction intervals. IEEE Trans. Neural Netw. **22**(3), 337–346 (2011). https://doi.org/10.1109/TNN.2010.2096824

8. Leavenworth, R.S., Grant, E.L.: Statistical Quality Control. Tata McGraw-Hill Education, New York (2000)

Study of Quality of Plastic Details Without Mechanical Destroying by Using Fuzzy Approach

Djahid Kerimov$^{(\boxtimes)}$ (iD)

Azerbaijan State Oil and Industry University, Azadlig 34, Nasimi, Baku,
Azerbaijan
haciyevanaila64@gmail.com

Abstract. Using the theory of experiment planning, the relationship is determined between mechanical indicators (strength, hardness, surface quality, etc.).
Groups of mechanical indicators (hardness, cleanliness of surface, etc.) were
experimentally studied, and the effect and nature of changes in mechanical
uncontrolled parameters can be predicted. The use of fuzzy logic is considered to
account for imprecision of information on relation between these indicators.

Keywords: Plastic details · Mechanical indicators · Strength · Hardness ·
Cleanliness of surface · Empirical equation · Imprecise information · Fuzzy
logic

1 Introduction

The joint action of temperature and relative humidity of the ambient air is a real
condition for operational products of plastic.

It should be noted that most of the work devoted to the study of the influence of
these parameters on the stability of the properties of polymer materials was carried out
on polyamides. Systematic research in this direction is not conducted. Unfortunately,
we do not have the opportunity to compare the obtained results, as the methodology for
conducting such studies is different [1].

Tables 1, 2, and 3 show the data on changing the value of thermoplastics quality
indicators based on the results of climatic tests.

An analysis of the data shows that the properties of thermoplastics deteriorate with
changes in ambient temperature and humidity. The shrinkage value is exposed the
largest change, especially for details made of ABS-plastic. This is due to the ability of
ABS-plastic to absorb moisture, where the quality of surface the details of this material
has improved.

As can be seen from the presented tables, details from ABS-plastic have best
stability of the values of quality indicators in climatic conditions.

Thus, under operating conditions, it is necessary to take into account the possibility
of structural changes of thermoplastics. The latter depend heavily on the initial degree
of orientation and crystallinity. From this it follows the conclusion about the influence
of the technology of manufacturing plastic details on the stability of quality indicators,

R. A. Aliev et al. (Eds.): ICAFS 2020, AISC 1306, pp. 610–615, 2021.
https://doi.org/10.1007/978-3-030-64058-3_75

Table 1. Values of quality indicators of details made of polystyrene the emulsion grades of (ABS)-plastic before (I) and after (II) climatic tests

Quality indicators	Sb,%	ρ,kg/m3	Ra,mkm	HB,MPa	σc,MPa
I	0,74	1046,85	1,24	112,98	44,16
II	0,80	1044,62	1,11	106,50	41,67
Change, %	8,1	0,4	10,4	5,7	5,6

Table 2. Values of quality indicators of details made polyamide tar of 68 (PA) before (I) and after (II) climatic tests

Quality indicators	Sb,%	ρ,kg/m3	Ra,mkm	HB,MPa	σc,MPa
I	0,74	1046,85	1,24	112,98	44,16
II	0,80	1044,62	1,11	106,50	41,67
Change, %	8,1	0,4	10,4	5,7	5,6

Table 3. Values of quality indicators of details made polypropylene (PP) tar of 68 (PA) before (I) and after (II) climatic tests

Quality indicators	Sb,%	ρ,kg/m3	Ra,mkm	HB,MPa	σc,MPa
I	2,41	911,46	1,13	72,95	34,46
II	2,62	908,02	0,99	69,07	32,02
Change, %	8,9	0,4	12,3	5,3	7,1

as technological parameters prove to be very important in the formation of a particular structure. Let us also mention that the information on the values of the mentioned indicators is imprecise. Thus, we consider using fuzzy logic to model the relationship between these indicators.

2 The Method of Determining the Strength of Products Without Any Disruption of Its Structure

As is known, the determination of the strength of the finished product is practically impossible without its direct destruction, which in turn leads, to the impossibility of its operation.

The proposed method of determining the strength of products without any disruption of its structure and properties takes into account the inhomogeneity of distribution the properties of material, as well as makes it possible to assess the quality of the article as a whole.

This method is especially appropriate for the manufacture of plastic details in the established technological process, where the finished details are assembled after their manufacture.

This stage the results of a study to establish a multicomponent link of the quality indicators of plastic details are presented. It should also be noted that the quality indicators chosen by us are characterized by relatively low laboriousness intensity of their definition directly on the detail itself, possibility of using serial equipment, simple method of determination, high sensitivity to change in quality of the detail and low error of their evaluation [2].

Preliminary results of the studies showed that the relationship of the studied indicators under imprecise information can be described by the following fuzzy equation:

$$\tilde{\sigma}_{calc} = \tilde{S}_b^{n_1} \tilde{R}_a^{n_2} \widetilde{HB}^{n_3} \tilde{K}^{n_4}, \tag{1}$$

where $\tilde{K}$ – fuzzy-valued coefficient of volume, $\tilde{K} = \frac{\tilde{V}_g}{\tilde{V}_f} = \frac{\tilde{M}_g}{\tilde{\rho}_g \tilde{V}_f}$; $\tilde{V}_g, \tilde{M}_g, \tilde{\rho}_g$ – respectively the fuzzy-valued variables of volume, mass and density of the detail; $\tilde{V}_f$ – the fuzzy-valued volume of the form cavity.

The problem is to determine the coefficients n_1, n_2, n_3, n_4 of Eq. (1).

Find logarithms for both sides of the Eq. (1), after that we get

$$\ln \sigma_{calc} = n_1 \ln S_b + n_2 \ln R_a + n_3 \ln HB + n_4 \ln K. \tag{2}$$

Let us present the logarithms in the following form:

$$\ln \sigma_{calc} = \sigma'_{calc}; \quad \ln S_b = S'_b; \quad \ln R_a = R'_a; \quad \ln HB = HB'; \quad \ln K = K',$$

Then Eq. (2) takes the form

$$\sigma'_{calc} = n_1 S'_b + n_2 R'_a + n_3 HB' + n_4 K'. \tag{3}$$

To simplify Eq. (3), all variables are converted to a standardized form:

$$\sigma''_{calc} = \frac{\sigma''_{calc_1} - \sigma'_{calc}}{\sqrt{D_{\sigma'_c}}}; \quad S''_b = \frac{S''_{b_1} - S'_b}{\sqrt{D_{S'_b}}}; \quad R''_a = \frac{R''_{a_1} - R'_a}{\sqrt{D_{R'_a}}};$$

$$HB'' = \frac{HB''_1 - HB'}{\sqrt{D_{HB'}}}; \quad K'' = \frac{K''_1 - K'}{\sqrt{D_{K'}}},$$

where $\bar{\sigma}'_{calc}, \bar{S}'_b, \bar{R}'_a, \overline{HB'}, \bar{K}'$ and $\sqrt{D_{\sigma'_{calc}}}, \sqrt{D_{S'_b}}, \sqrt{D_{R'_a}}, \sqrt{D_{HB'}}, \sqrt{D_{K'}}$ – respectively mathematical expectations and mean quadratic deviations of corresponding variables determined by known method [3]. Now Eq. (3) in a standardized form can be written in the following form:

$$\sigma''_{calc} = n_1 S''_b + n_2 R''_a + n_3 HB'' + n_4 K''. \tag{4}$$

The coefficients n_1, n_2, n_3, n_4 are found by the least square method from the condition

$$A = \sum \left[\sigma''_{calc} - \left(n_1 S''_b + n_2 R''_a + n_3 HB'' + n_4 K'' \right) \right]^2 = \min. \tag{5}$$

The minimum of condition for expression (5) will be

$$\frac{\partial A}{\partial n_1} = 2 \sum \left[\sigma''_{calc} - \left(n_1 S''_b + n_2 R''_a + n_3 HB'' + n_4 K'' \right) \right] \cdot S''_b = 0 ;$$

$$\frac{\partial A}{\partial n_2} = 2 \sum \left[\sigma''_{calc} - \left(n_1 S''_b + n_2 R''_a + n_3 HB'' + n_4 K'' \right) \right] \cdot R''_a = 0 ;$$

$$\frac{\partial A}{\partial n_3} = 2 \sum \left[\sigma''_{calc} - \left(n_1 S''_b + n_2 R''_a + n_3 HB'' + n_4 K'' \right) \right] \cdot HB'' = 0 ; \tag{6}$$

$$\frac{\partial A}{\partial n_4} = 2 \sum \left[\sigma''_{calc} - \left(n_1 S''_b + n_2 R''_a + n_3 HB'' + n_4 K'' \right) \right] \cdot K'' = 0 .$$

By performing permutations and simplifications, one may write down a system of linear Eqs. (6) with respect to the coefficients

$$\sum \sigma''_{calc} S''_b = n_1 \sum S''_b S''_b + n_2 \sum R''_a S''_b + n_3 \sum HB'' S''_b + n_4 \sum K'' S''_b ;$$

$$\sum \sigma''_{calc} R''_a = n_1 \sum S''_b R''_a + n_2 \sum R''_a R''_a + n_3 \sum HB'' R''_a + n_4 \sum K'' R''_a ;$$

$$\sum \sigma''_{calc} HB'' = n_1 \sum S''_b HB'' + n_2 \sum R''_a HB'' + n_3 \sum HB'' HB'' + n_4 \sum K'' HB'' ; \tag{7}$$

$$\sum \sigma''_{calc} K'' = n_1 \sum S''_b K'' + n_2 \sum R''_a K'' + n_3 \sum HB'' K'' + n_4 \sum K'' K'' .$$

The solution of the system of Eqs. (7) was carried out on a computer.
The calculation results for a specially designed program are given in Table 4.

Table 4. The value of the coefficients of Eqs. (4) for details from (ABS)-plastic, polyamide tar (PA), polypropylene tar (PP).

Material for details	Coefficients of Eq. (4)			
	n1	n2	n3	n4
ABS	0,0738835	−0,1437933	0,0593783	0,4139313
PA	0,0313837	−0,0914700	0,4013026	0,2029645
PP	0,0531273	−0,0762046	0,4762118	0,9572402

Substituting the numerical values of the coefficients n_1, n_2, n_3, n_4 in Eq. (4), we obtain for ABS

$$\sigma''_{calc} = 0,07388S''_b - 0,14379R''_a + 0,05937HB'' + 0,41393K'', \tag{8}$$

for PA

$$\sigma''_{calc} = 0,03138S''_b - 0,09147R''_a + 0,40130HB'' + 0,20296K'', \tag{9}$$

for PP

$$\sigma''_{calc} = 0,05312S''_b - 0,07620R''_a + 0,47621HB'' + 0,95724K''. \tag{10}$$

3 Discussion and Conclusions

To be used in practice in solving Eqs. (8)–(10), we move to the natural form of mathematical dependence and, by determining the coefficients sought, we obtain equations for determining the durability of details:

for ABS

$$\sigma_{calc} = S_b^{0,0399} R_a^{-0,0257} HB^{0,0612} K^{1,5991} \exp(3,5557), \tag{11}$$

for PA

$$\sigma_{calc} = S_b^{0,0433} R_a^{-0,0418} HB^{2,8953} K^{2,0897} \exp(1,5755), \tag{12}$$

for PP

$$\sigma_{calc} = S_b^{0,0191} R_a^{-0,0138} HB^{0,3337} K^{1,1888} \exp(2,1409). \tag{13}$$

In order to determine the adequacy of the obtained dependence, mean quadratic deviations of observed values σ_{obs} from the calculated calculations σ_{calc} predicted by Eqs. (11)–(13) $D_{\sigma calc}$ and relative to mean 3 value σ_{obs}, D_0, were calculated, given in Table 5.

Table 5. The value of $D_{\sigma calc}$ and D_0 for details from (ABS)-plastic, polyamide tar (PA), polypropylene tar (PP).

Material for details	Dσcalc	D0
ABS	1,734321	1,544924
PA	1,483756	1,023296
PP	1,152331	0,635057

Since the second values are smaller than the first ones, therefore, the obtained Eqs. (11)–(13) adequately reproduce the investigated dependence.

The results obtained show that the strength of the plastic detail is directly proportional to the shrinkage, hardness and density values and inversely proportional to the surface roughness value. Changes in volume characteristics (surface features) have the greatest influence on the strength of the detail [4].

The obtained dependence is, in our opinion, of great practical importance for quality control of plastic details of oil equipment. The obtained empirical equation is not only a mathematical connection between individual parameters, but also one of the main evaluations of the quality of the finished plastic detail. We consider using fuzzy logic to model relationship between these parameters under imprecise information.

References

1. Kerimov, D.A.: Scientific bases and practical methods of optimization of quality indicators of plastic details of oil field equipment. Dissertion Doctor of Technical Sciences. Baku. Azerbaijan Institute of Oil and Chemistry named after M. Azizbekov (1985)
2. Mitropolsky, A.K.: Technique of statistical calculations. Moscow, Statistics (1971)
3. Nalimov, V.V., Chernova, N.A.: Statistical methods of extreme experiments planning. Moscow, Statistics (1965)
4. Kerimov, D.A., Gasanova, N.A.: Determination of quality of plastic details without disruptions. In: 13th International Conference on Theory and Application of Fuzzy Systems and Soft Computing — ICAFS-2018, Warsaw, Poland, August 27–28, 2018 Springer Nature Switzerland AG 2019, Advances in Intelligent Systems and Computing (AISC), Vol. 896, pp. 848–851. Springer, Cham (2019). The conference proceedings will be published in the Springer Nature (indexed in Web of Science, SCOPUS and Elsevier)

Developing a Software Product for Neuro-Fuzzy Risk Prediction to Ensure University's Information Security

E. V. Zhilina$^{(\boxtimes)}$ ⓘ, E. V. Efimova ⓘ, N. A. Rutta ⓘ,
and A. R. Savskaya ⓘ

Rostov State University of Economics,
B. Sadovaya St., 69, 344002 Rostov-on-Don, Russia
`black-2@mail.ru`, `efim19732008@yandex.ru`,
`rutic79@mail.ru`, `sarl@list.ru`

Abstract. Based on the neuro-fuzzy models and integration of Matlab components the authors have developed desktop – a software product for risk prediction related to university's information security. The developed UML model displays the interaction of the main functions and components of the software project. The developed ER model allows to automate the writing of sql scripts for the database.

Keywords: Risk · Information security · Hybrid models · Software product

1 Introduction

Creating the conditions necessary for ensuring information security constitutes a big concern in educational institutions (universities) [1]. The use of information technologies in educational organizations is an effective tool for increasing productivity and providing educational services. Since information systems in this sector are often not structured, they are accompanied by increased risks for violations of information security [2]. Despite the interest of scientists and specialists in this problem, there are still no unified comprehensive methods for predicting information risks in universities. Little attention has been paid to improving the accuracy of forecasts without the use of traditional expert surveys [3]. The use of neuro-fuzzy approach for estimating the information security risks enables to obtain reliable forecasts of output linguistic variables, both quantitative and qualitative [4].

The purpose of the study is to develop software based on the algorithms of neuro-fuzzy modeling, which allows predicting monthly possible information risks for educational institutions based on "infections" statistics.

For the purpose of estimating IT security risks in the university (using the example of statistical data from the IT department of Rostov State Economic University (RINH)) a fuzzy inference system has been developed, which enables to evaluate the magnitude of each risk for the whole list uncovered by experts [5, 6]:

- Unauthorized access to confidential information (risk 1).
- Unauthorized access to personal information (risk 2).

- Password attack (risk 3).
- DDOS-attack (risk 4).
- Virus or malware infection (risk 5).

To achieve efficiency of IT departments of educational institutions, there is need to develop a software product, which enables reliable information risk prediction. Neuro-fuzzy prediction algorithms will allow to make such calculations and minimize costs when implementing them. The development of the visual UML-model and the ER-model of the software project will significantly reduce the cost of programming and configuring it.

The following describes in more detail the stages of software project development.

2   Main Part

2.1   Fuzzy Modelling

A MATLAB package, an ANFIS system and a fuzzification block have been used to model neuro-fuzzy information security risk assessment system. The steps for neuro-fuzzy risk prediction modeling in the university are as follows:

1. Preparing baseline data. Collecting files with the extension *.dat. File naming according to the risks list.
2. Loading training data into the ANFIS. Generation of fuzzy inference system (FIS-structure) based on algorithm Sugeno, which is a hybrid network model.

   In the 'Properties' section of ANFIS you can set model parameters:

- the number of input/output variables;
- the type of membership function (in the experiment triangular functions have been used) for certain linguistic terms (in our case, 3 linguistic terms for each variable);
- viewing information about number of rows in the samples.

The «Structure» section of ANFIS enables to visualize fuzzy inference system organized as neuro-fuzzy network [7]. Figure 1 as an example shows trained neuro-fuzzy network «Virus or malware infection», consisting of two input variables (year, month of infection) and one output variable (the number of infections) (see Fig. 1).

3. Setting parameters of training neuro-fuzzy (hybrid) network:

- choosing a method for training hybrid network: the back-propagation method or the hybrid method, combining the method of least squares and the gradient descent method;
- specifying Error Tolerance, the default values equal zero;
- specifying the number of Epochs, the default values equal 3 Epochs (it is generally recommended to increase the number of Epochs; in the experiment under consideration the default values equalled 40-50).

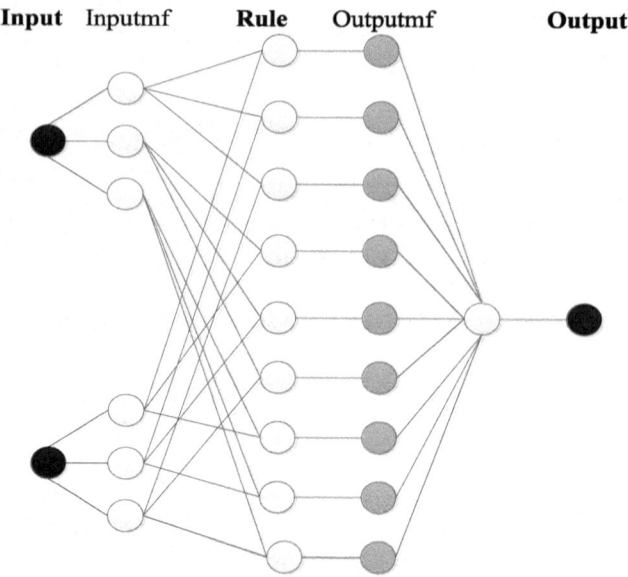

Fig. 1. Trained neuro-fuzzy network «Virus or malware infection».

4. Training hybrid network (Train now).
5. Storing a fuzzy inference system in an external file with the extension *.fis.
6. Visualizing variables of a fuzzy inference system. The following figures represent triangular membership functions of one of the input variables (year of infection, Input1) of the model «Virus or malware infection» (see Fig. 2).

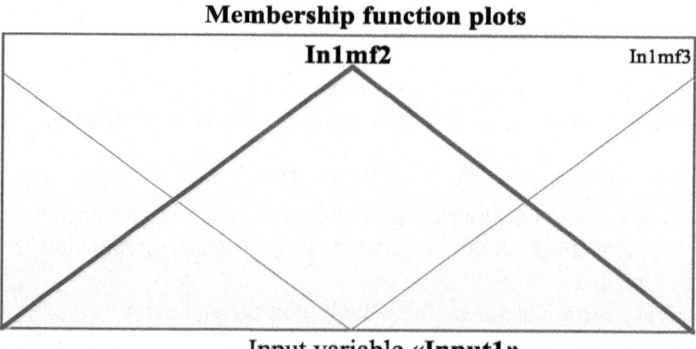

Fig. 2. Membership functions of the fuzzy inference system «Virus or malware infection».

7. Verification of adequacy of the created hybrid fuzzy network model, calculation of the predicted virus infection value for the short-term period, for one of the subsequent months in the year:

- loading into the MATLAB Workspace the necessary fis-file of the analysed fuzzy inference system through a command line: fis = readfis();
- predicting the analysed information security risk. For instance, the forecast for June 2020: evalfis([2020 6], fis).

Hybrid models related to other information security risks for the IT department of Rostov State Economic University (RINH)) have been developed in a similar way.

2.2   Developing UML-Model of the Software Product

Visual modeling methods can reduce costs and the time required to develop software, which allows to considerably improve the software quality. Using UML methodology enables: component-based technologies; RAD-tools - visual programming; working with design patterns; visual modeling, CASE-tools.

Figure 3 shows the main diagram Use Case of the project under development. The major precedents are: operations with database tables, system set up, operations with reports [8].

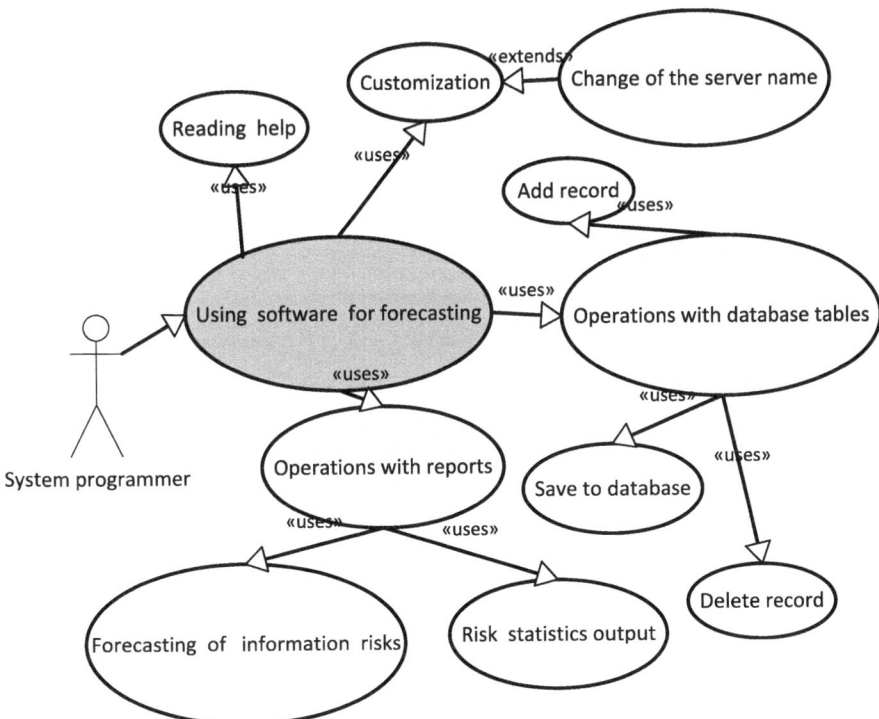

Fig. 3. Use case diagram of the project.

The project intends to use the following external components: MS SQL Server (for database source), Cristal Reports (for reports and diagrams), and dll compiled using Matlab (to handle volumes of data of the fis-model).

2.3   Developing ER-Model

The logical type of ER-model of the project related to information risk prediction in universities (see Fig. 4) has been developed, where entities (risks, combating risks, incident, vulnerabilities, object, intruder and risks statistical data) and their attributes are specified [9].

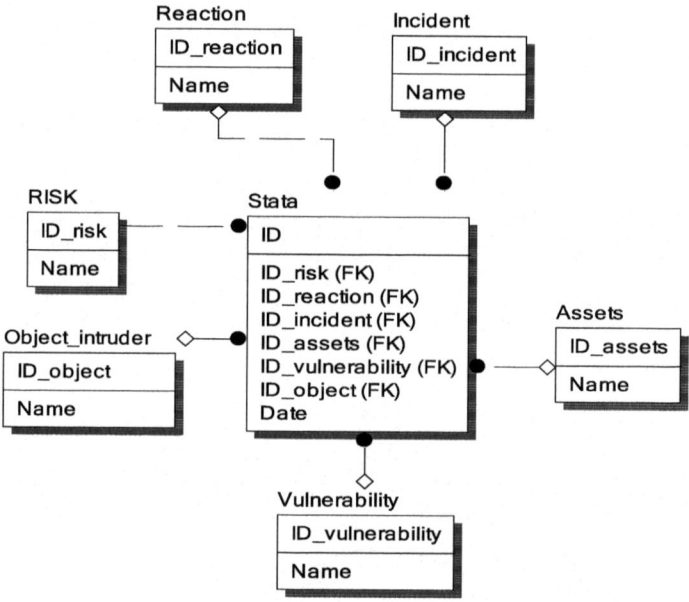

Fig. 4. ER-model of the database project (logical type).

In the next phase of the project implementation the physical type of ER-model has been developed, where types of attributes, primary keys and the links between entities are specified. Data types have been established with regard to the chosen Database Management System (DBMS) for further physical implementation of database (MS SQL Server). The main data types of the model are varchar, integer and datetime. At this stage fields parameters are specified, they act as a counter unit: you can define the starting number (IDENTITY column) and the increment value for each subsequent number. Here is an example of script for the entity «Risk»:

```
CREATE TABLE RISK (ID_risk integer IDENTITY (1,1), Name
varchar(250)NOT NULL)
```

For quick access to statistical data VIEW (V_1) was created, which will enable to replace text values of corresponding entity names, not their codes (ID), the fragment of a script:

```
CREATE VIEW V_1(RISK, Date, Object_intruder, Incident,
Reaction, Assets, Vulmerability) AS SELECT RISK.Name, Stata.
Date, ... FROM RISK, Stata, ...
```

2.4 Software Development of the Project

The software product created using the C# programming language for the purpose of estimating IT security risks in universities enables to obtain predicted values based on violation statistical data and prediction period option (month, year, date range), which significantly reduces the costs of approbation procedure of neuro-fuzzy models (design pattern Windows Forms). The product includes the integration of COM- component from MATLAB.

The software design included the following steps:

- Filling of primary directories of the database.
- Filling the database of information security risks based on university IT-department statistics.
- Development of screen forms.
- Programming of events arising during implementation of algorithms of interaction with the database.
- Development of a third-party library in Matlab to compile neuro-fuzzy modeling algorithms.
- Programming of events arising during implementation of algorithms for forecasting information risks.
- Development of reporting documentation interfaces to visualize the results of violation statistics and predicted values of information risks at the university.

The project is launched via startup parent form, containing the top main application menu: «Tables», «Reports», «Service» and «Help». The fragment of the initialization code of the application system parameters:

```
public static string file = @"Source\Init.sys";
private void Form1_Load(object sender, EventArgs e)
{try{StreamReader str = new StreamReader(file);
string server = str.ReadLine(); str.Close(); }}
```

The «Tables» section is composed of subsections to access all project tables (Risk, Intruder, Risks statistical data etc.). The code fragment of child forms of the project:

```
private Stata OpenStata()
{foreach (Form frm in Application.OpenForms)
if (frm is Stata){frm.Activate(); return frm as Stata; }
Stata form = new Stata();
form.MdiParent = this; form.Show(); return form; }
```

The «Reports» section enables to move on to specifying criteria for detailed report on statistical data related to violations identified and for creating predicted values of information security risks based on results of neuro-fuzzy modeling, on the basis of which the reports and corresponding diagrams are created. The fragment of class code for obtaining predicted values (namespace System, System.Data, System.Data. SqlClient, System.Windows.Forms, MathWorks.MATLAB.NET.Arrays, IB are connected):

```
private void button4_Click(object sender, EventArgs e)
{... INsert(YearFirst, MonthFirst, file[i], risk[i]); }
private void INsert(int year, int month, string file, string
ugroz)
{... str = Fis(file, year, month).ToString();
double prognoz = double.Parse(correctStr.ToString()); ... }
private MWNumericArray Fis(string file, int x, int y)
{MWArray[] res = null;
MWNumericArray descriptor = null;
MyFisClass Myobject = new MyFisClass();
res = Myobject.IB(1, file, x, y);
descriptor = (MWNumericArray)res[0];return descriptor;}
```

3 Conclusion

The neuro-fuzzy information security risk prediction models (including models – «Unauthorized access to confidential information», «Unauthorized access to personal information», «Password attack», «DDOS-attack», «Virus or malware infection») based on Sugeno adaptive fuzzy logic and developed by using the Matlab software tools and Anfis system give the possibility to obtain short-term predicted values after training initial statistical data network.

The developed UML-model of the project, differs in the description of the functional purpose of the system, contains information processes for conducting forecasting, displays the structure of objects and links in the system, which allows you to significantly reduce the cost of developing and configuring a developed software product.

The developed software product for estimating the forecast values of information risks arising in the field of education is distinguished by the ability to build a forecast based on available statistics of violations and choosing a forecast period (month, year, date range), which allows significantly reducing the cost of testing models of neuro-fuzzy prediction.

The introduction of the developed software will allow educational institutions to reduce the appearance of security risks and to facilitate the smooth functioning of universities' computing power.

References

1. Tishchenko, E., Serpeninov, O., Zhilina, E.: Comparative analysis of the computerized testing systems in education. In: Proceedings of the International Scientific Conference Innovative Approaches to the Application of Digital Technologies in Education and Research SLET 2019. CEUR Workshop Proceedings **2494** (2019). ISSN: 161300732
2. Zegzhda, P.D., Zegzhda, D.P., Pechenkin, A.I., Poltavtseva, M.A.: Modeling of information systems to solve the problem of security management (in Russian). Problems Inform. Secur. Comput. Syst. **3**, 7–16 (2016)
3. Tishchenko, E.N.: Analysis of The Security of Economic Information Systems. RSUE, Rostov-on-Don (2003). (in Russian)
4. Zhilina, E.V., Popova, L.K., Rutta, N.A., Sheydakov, N.E.: Fuzzy model of functioning of educational-laboratory and production capacities of the educational cluster in the information security field. In: Aliev, R.A., Kacprzyk, J., Pedrycz, W., Jamshidi, M., Babanli, M.B., Sadikoglu, F.M. (eds.) ICSCCW 2019. AISC, vol. 1095, pp. 704–711. Springer, Cham (2020). https://doi.org/10.1007/978-3-030-35249-3_91
5. Zhilina, E.V., Efimova, E.V., Tishina, A.R.: Expert risk analysis of the introduction of an enterprise information confidential information accounting system (in Russian). In: International scientific and practical conf. of Russia and the EU: Development and Prospects, pp. 816–821. RSUE, Rostov-on-Don (2016)
6. Kozachok, V.I., Vlasova, S.A.: Factors determining the information security of the corporation (in Russian). Central Russ. Bull. Soc. Sci. **5**(35), 30–34 (2014). ISSN: 2071-2367
7. Sklyarov, A.V., Tishchenko, E.N., Efimova, E.V., Zhilina, E.V.: Evaluation of the effectiveness of information security measures on protected economic systems using artificial neural networks (in Russian). Econ. Sci. **177**, 77–81 (2019). https://doi.org/10.14451/1.177.77
8. Zhilina, E.V., Ediev, T.A., Savskaya, A.R.: UML - modeling software for accounting and forecasting information risks in educational institutions (in Russian). In: International Scientific and Practical Conference of Problems, Prospects and Directions of the Innovative Development of Science, pp. 32–37. RSUE, Rostov-on-Don (2018)
9. Zhilina, E.V., Ediev, T.A., Suchkov, O.S.: Development of ER - model of a software project for accounting for information threats in universities (in Russian). In: International Scientific and Practical Conference of Innovative Mechanisms for Solving Problems of Scientific Development, pp. 28–33. RSUE, Rostov-on-Don (2018)

Identification of a Fuzzy Model of the Coking Process

K. R. Aliyeva[(✉)] [iD]

Department of Instrument-Making Engineering, Azerbaijan State
Oil and Industry University, 20 Azadlig Ave., AZ1010 Baku, Azerbaijan
kamalann64@gmail.com

Abstract. Coke process in oil refinery plant is complex dynamic process that characterized by deep uncertainty of its regime. In order to optimize techno-logical regime (rephrase) of the process, it is essential to construct the dynamic model of this unit. This paper will consider the identification of the dynamic model of coke process as dynamic regression model with fuzzy coefficients. Identification process is based on the use of gathering uncertain data of this process and fuzzy mean square error minimization procedure. As a result, we construct numerical dynamic model the considered unit.

Keywords: İdentification · Fuzzy numbers · Fuzzy dynamic model · Coking process

1 Introduction

Coke processing is a complex physicochemical process. According to the given ratios, the average temperature in all chambers is the most important parameter in the oil refining process and, therefore, it is a key to obtain high-quality coke [1, 2]. Basic options for design of the coking process in industry are periodic coking, continuous coking and semi-continuous (delayed) coking. Periodic coking is interesting method as it is expected to coke to be quenched in cubes, thereby obtaining the most valuable product - lumpy coke used in the production of electrodes. Identification of this pro-cesses have great practical importance [4]. Gary et al. [3] gave good descriptions of the petroleum coke and the delayed coking operation, coking reactions, product separation and technology economics. Basically, delayed coking is severe thermal cracking operation that used to upgrade and convert petroleum residual materials (e.g., bottoms from atmospheric and vacuum distillation of crude oil) to liquid and gas product streams (fuel gas, naphtha, and coke gas oil), where it leaves behind a large amount of concentrated solid carbon material, petroleum coke.

The coker gas oils may be a further process in the hydrotreaters and used as feedstock for other cracking processes. Many ways have been proposed to improve the control performance of a fuzzy control system through the identification of fuzzy model. The fuzzy logic controller adjusts the input and output universes in a real-time manner [6, 7], rather than producing fuzzy rules [8, 9]. This paper presents identifi-cation of a fuzzy model of the coking process and structured as follows. Section 2

R. A. Aliev et al. (Eds.): ICAFS 2020, AISC 1306, pp. 624–630, 2021.
https://doi.org/10.1007/978-3-030-64058-3_77

includes description of process of the complex that produce coke with defined parameters, which is achieved by preparing raw materials and selecting the necessary coking conditions. Section 3 consists necessary definitions used for the identification of a fuzzy model of the coking process. Section 4 contains statement of the problem. The solution approach for the considered problems is given in Sect. 5. Section 6 is the conclusion.

2 Description of Coke Process

In the process of delayed coking, petroleum coke is obtained from heavy oil residues. The essence of the technological process lies in the fact that the raw material pre-heated to a high temperature - a mixture of tar and a mixture of asphalt deasphalting, selective cleaning extract and cracking residue is pumped into coke ovens where coking is carried out due to heat accumulated by raw materials.

The technological scheme of the reactor block of coke oven chambers is made in the form of two parallel streams having one tube furnace and two coke chambers for each molasses. The duration of each chamber is 30 h. The coking process is carried out in two sequentially operating reaction chambers and proceeds as follows. Raw materials from the furnace enter the distillation column, then flowing down towards the gaseous coking product coming from the coke chambers, the raw material is heated. Continuous coking process is aimed at maximum yield of distillate fractions. In semi-continuous (delayed) coking method, raw materials heated to high temperatures (470–510 °C) are fed into large coke oven chambers where thermal reactions occur cracking to form coke and gaseous products. From the bottom distillation columns raw materials and recirculated coking products enter the furnace, where the secondary raw materials are heated and then fed to the reaction chambers. To maintain a certain temperature regime in the coking chambers, in order to obtain better coke, heated light coking gas oil is fed into them. At this time, the second chamber freed from coke is being prepared for work. In the process of feeding into the chamber a vaporous light gas oil of secondary raw materials heated in furnaces, a coke formation reaction takes place, which accumulates in the chambers, and pairs of coking products - gasoline, light and heavy gas oil and gas are sent to the distillation column. The coker gas oil is suitable feed to catalytic cracking unit or to a hydrocracking unit. The feed to delayed coker is typically a vacuum residue. It combines with a recycle oil from the coker fractionation bottom and with steam stream and is heated in a charge furnace to reach thermal cracking temperature of 485–505 °C. With a short residence in a furnace tube coking of the feed material is delated until it is reaching the bottom of a large coking drum downstream of the furnace [5]. The coker drum has a long residence time, allowing the heavy liquids to polymerize and dehydrogenate to form petroleum coke and to deposit the coke on the sides of the drum. The coke builds on the drums from bottom up. When the accumulated coke reaches a certain height, the drum is taken off-line, and the coke deposit is removed by a hydraulic drill. The removal of coke from a drum consumes a large amount of water, and contaminated water requires treatment before water reuse [2]. The steam and oil vapors that do not coke leave the top of the coker drum and quenched with cold gas oil to stop the reactions before fractionation column, where they are washed with heavy gas oil. The washed zone liquid falls to the bottom of the fractionation column, where it mixes with fresh residual charge [5]. The main fractionator separates the vapors from the wash zone into fuel gas, naphtha or unstabilized gasoline, light gas oil, heavy gas oil, and a bottom residue stream. The residue stream is recycled to the coke inlet as a coker feed.

The coke-filled chamber is disconnected from the feed stream. The maximum coke level should be less than 21.5 m. From the top of the distillation column, gas, gasoline vapors, and water vapor are discharged to a gas separator, where gas, gasoline, and water are separated.

Light and heavy gas oils from the distillation column enter the stripping columns. The output and quality of the coking products are coke, gasoline, gas, light and heavy gas oil. The mode parameters of the delayed coking process have a noticeable effect—the recirculation coefficient (u_1), the temperature of the secondary raw materials (u_2), the temperature of the light gas oil (u_3), and the top temperature of the reaction chamber (u_4). An increase in the temperature of secondary raw materials, light gas oil and the recirculation coefficient leads to an increase in the yield of coke, gasoline, light gas oil and a recirculation coefficient leads to an increase in the output of coke, gasoline, light gas oil and gas, but to an increase in the yield of heavy gas oil.

The main indicator of the quality of raw materials for the coking process is its density (z_1). The target product of the coking process is coke (y). According to the fractional composition of the coke, the process of delayed coking is divided into fractions depending on the size of the pieces: coke breeze for the production of abrasives and other products with the size of pieces over 0 to 8 mm, coke with the size of pieces 8 mm to 250 mm. The disturbing influences affecting the process are the density and consumption of primary raw materials (z_2). The control actions are the recirculation coefficient, the top temperature of the reaction chamber.

3 Preliminaries

Definition 1. Fuzzy Numbers. [11, 12]. A fuzzy number is a fuzzy set A on R and have next features: a) A is a normal fuzzy set; b) A is a bell shape fuzzy set; c) α-cut of A, A^α is a enclosed interval for each $\alpha \in (0, 1]$; d) the support of A, supp(A) is bounded where

$$\underline{\mu}_A(\alpha) \leq \bar{\mu}_A(\alpha), \ 0 \leq \alpha \leq 1.$$

Definition 2. Operations on Fuzzy Numbers. [11, 12]. The fuzzy extension principle is the basic tool for fuzzy calculus; it extends functions of real numbers to functions of non-interactive fuzzy quantities and it allows the extension of arithmetic operations and calculus to fuzzy arguments. In terms of the α-cuts, the four arithmetic operations and the scalar multiplication for $k \in R$ are obtained by the well-known interval arithmetic:

Addition:

$$\begin{cases} u+v = (u^- +v^-, u^+ +v^+) \\ \alpha \in [0,1], [u+v]_\alpha = \left[u_\alpha^- + v_\alpha^-, u_\alpha^+ + v_\alpha^+ \right] \end{cases} \tag{1}$$

Scalar multiplication:

$$\begin{cases} \begin{cases} ku = (ku^-, ku^+) & if \quad k > 0 \\ ku = (ku^+, ku^-) & if \quad k < 0 \end{cases} \\ \alpha \in [0,1], [ku]_\alpha = \left[\min\{ku_\alpha^-, ku_\alpha^+\}, \ \max\{ku_\alpha^-, ku_\alpha^+\} \right] \end{cases} \tag{2}$$

Subtraction:

$$\begin{cases} u-v = (u^- - v^-, u^+ - v^+) \\ \alpha \in [0,1], [u-v]_\alpha = \left[u_\alpha^- - v_\alpha^-, u_\alpha^+ - v_\alpha^+ \right] \end{cases} \tag{3}$$

Multiplication:

$$\begin{cases} & u \times v = \left((uv)^-, (uv)^+ \right) \\ & (uv)_\alpha^- = \min\{ u_\alpha^- v_\alpha^-, u_\alpha^- v_\alpha^+, u_\alpha^+ v_\alpha^-, u_\alpha^+ v_\alpha^+ \} \\ \alpha \in [0,1], & \begin{cases} (uv)_\alpha^- = \min\{ u_\alpha^- v_\alpha^-, u_\alpha^- v_\alpha^+, u_\alpha^+ v_\alpha^-, u_\alpha^+ v_\alpha^+, \} \\ (uv)_\alpha^+ = \max\{ u_\alpha^- v_\alpha^-, u_\alpha^- v_\alpha^+, u_\alpha^+ v_\alpha^-, u_\alpha^+ v_\alpha^+, \} \end{cases} \end{cases} \tag{4}$$

Division:

$$\begin{cases} \frac{u}{v} = \left(\left(\frac{u}{v}\right)^-, \left(\frac{u}{v}\right)^+ \right) & if \ 0 \notin \left[v_0^-, v_0^+ \right] \\ \alpha \in [0,1], \begin{cases} \left(\frac{u}{v}\right)_\alpha^- = \min\{ \frac{u_\alpha^-}{v_\alpha^-}, \frac{u_\alpha^-}{v_\alpha^+}, \frac{u_\alpha^+}{v_\alpha^-}, \frac{u_\alpha^+}{v_\alpha^+}, \} \\ \left(\frac{u}{v}\right)_\alpha^+ = \max\{ \frac{u_\alpha^-}{v_\alpha^-}, \frac{u_\alpha^-}{v_\alpha^+}, \frac{u_\alpha^+}{v_\alpha^-}, \frac{u_\alpha^+}{v_\alpha^+}, \} \end{cases} \end{cases} \tag{5}$$

4 Fuzzy Identification of Dynamic Objects

Let, $X = (X_1, X_2, \ldots, X_n)$, $U = (U_1, U_2, \ldots, U_m)$ are fuzzy vectors of the perturbing and control parameters and Y is the fuzzy output parameter of the object.

Suppose that the object is expressed by the fuzzy difference equation [10, 13].

$$\begin{aligned} ..Y_{k+1} &= A_0 + A_1 * x_{1k} + A_2 * x_{2k} + \cdots + A_n * x_{nk} + A_{n+1} * y_k + .. \\ &+ B_1 * u_{1k} + B_2 * u_{2k} + \cdots + B_m * u_{mk} \end{aligned} \tag{6}$$

$A = (A_0, A_1, \ldots, A_{n+1})$ and $B = (B_0, B_1, \ldots, B_m)$ - are vectors of fuzzy coefficients. The problem of identification of the object expressed by Eq. (6) aims to estimate the coefficients $A_i (i = \overline{0, n+1})$ and $B_j (j = \overline{1, m})$. To evaluate the quality of the model, we will compare the current value of the output Y with the value (y) obtained from Eq. (6). As a result, we obtain the following system of fuzzy equations.

$$\begin{cases} A_0 + A_1 * x_{10} + \cdots + A_n * x_{n0} + A_{n+1} * y_0 + B_1 * u_{10} + \cdots + B_m * u_{m0} = \tilde{Y}_1 \\ A_0 + A_1 * x_{11} + \cdots + A_n * x_{n1} + A_{n+1} * y_1 + B_1 * u_{11} + \cdots + B_m * u_{m1} = \tilde{Y}_2 \\ \cdots\cdots\cdots\cdots\cdots\cdots\cdots\cdots\cdots\cdots\cdots\cdots\cdots\cdots\cdots\cdots\cdots\cdots \\ A_0 + A_1 * x_{1,N-1} + \cdots + A_n * x_{n,N-1} + A_{n+1} * y_{N-1} + B_1 * u_{1,N-1} + \cdots + B_m * u_{m,N-1} = \tilde{Y}_N \end{cases} \quad (7)$$

The main task is to determine such values of the coefficients $A_i(i = \overline{0, n+1})$ and $B_j(j = \overline{1,m})$ that Eq. (7) is ensured. In practice, it is not possible to satisfy Eq. (7) completely, so it is necessary to determine such values $A_i(i = \overline{0, n+1})$ and $B_j(j = \overline{1,m})$, that the difference between the left and right sides of Eq. (6) is minimal, in other words,

$$\begin{aligned} J = \{ &(A_0 + A_1 * x_{10} + \cdots + A_n * x_{n0} + A_{n+1} * y_0 + B_1 * u_{10} + \cdots + B_m * u_{m0} - \tilde{Y})^2 \\ &+ (A_0 + A_1 * x_{11} + \cdots + A_n * x_{n1} + A_{n+1} * y_1 + B_1 * u_{11} + \cdots + B_m * u_{m1} - \tilde{Y}_2)^2 \\ &+ (A_0 + A_1 * x_{1,N-1} + \cdots + A_n * x_{n,N-1} + A_{n+1} * y_{N-1} + B_1 * u_{1,N-1} + \cdots + B_m * u_{m,N-1} - \tilde{Y}_N) \}/N \end{aligned} \quad (8)$$

The function (7) is a function of ordinary fuzzy variables and should be minimal. To minimize it, it is used α - level approach, and then aggregation corresponding α level sets.

5   Identification of Coke Processing

We consider that $X = \{X_1, X_2, \cdots, X_n\}$ and $U = \{U_1, U_2, \cdots, U_m\}$ are the vectors of the excitation and control parameters, and Y is the fuzzy output parameter of the object. As a result of the observation of the object, in Table 1 we show the following N fuzzy expressions of the excitation, control and output variables.

When determining the fuzzy model of the coking process, the required number of observations are made in the coking process to identify the object. The inability to measure the coke collected in the chambers leads to a subjective assessment of the coke product obtained at the output by the technologist-operator.

X_1 - raw material density, X_2 - primary consumption, u_{11} - secondary consumption (I stream), u_{12} - secondary consumption (II stream), u_{13} - secondary consumption (II stream), u_{14} - secondary consumption (IV stream), u_2 - secondary consumption temperature, u_3 - light gas oil temperature, u_4 - top temperature reaction chamber, $\tilde{Y}$ - coke output.

Given the physical characteristics of the coking process, a fuzzy linear difference equation was used to describe the situation.

$$Y_{k+1} = a_0 + Y_k * A_1 + z_{1k} * A_2 + \ldots + B_1 * U_{1k} + \ldots + B_4 * U_{4k}$$

Thus, the solution of the problem of identification of the coking process is based on the determination of the coefficients $A_i(i = \overline{0,2})$ and $B_j(j = \overline{1,4})$ by the method of identification of fuzzy dynamic models given in session 4. As a result, the following values of the coefficients $A_i(i = \overline{0,2})$ and $B_j(j = \overline{1,4})$ are obtained.

Table 1. Fuzzy expressions of the excitation, control and output variables.

N	X_1	X_2	u_{11}	u_{12}	u_{13}	u_{14}	u_2	u_3	u_4	$\tilde{Y}$
1	0.9453	70	26.8	26.8	25.6	25.2	490	515	420	0
2	0.9453	80	26.0	26.0	24.8	25.6	492	515	425	23.1
3	0.9453	80	26.6	26.6	25.6	25.2	485	515	425	48.2
4	0.9453	76	26.4	25.0	25.0	26.4	486	512	425	74.0
5	0.9453	76	26.4	25.0	25.0	26.4	498	520	425	99.6
6	0.9453	70	26.6	25.6	25.6	25.2	485	515	425	48.2
7	0.9456	70	27.0	25.0	25.0	25.6	498	520	425	147.4
8	0.9456	70	27.0	25.0	25.0	28.4	498	520	425	169.5
9	0.9456	70	27.0	25.0	25.0	26.4	498	520	425	192.9
10	0.9456	70	26.4	25.0	25.0	25.6	498	520	425	216.0
11	0.9456	76	26.4	25.4	25.6	24.4	490	510	424	239.2
12	0.9456	76	26.4	26.4	25.6	24.0	495	510	423	264.0
13	0.9436	76	26.4	26.4	25.6	24.0	492	510	420	268.6
14	0.9436	76	26.4	26.4	25.6	24.0	490	510	420	313.4
15	0.9436	76	26.4	26.4	25.6	24.0	490	520	410	338.3
16	0.9436	76	26.8	26.4	25.4	24.8	490	520	410	363.3
17	0.9436	76	26.8	26.8	24.0	24.8	490	500	430	385.9
18	0.9436	76	26.8	26.8	24.2	24.8	490	515	430	408.5
19	0.9436	76	26.8	26.8	24.0	24.8	492	510	430	431.5

$A_0 = 0.00005$
$A_1 = 0.5/13.327 + 0.8/16.5935 + 1.0/24.763 + 0.8/26.4177 + 0.5/19.9155$
$A_2 = 0.5/0.0035 + 0.8/0.0152 + 1.0/-0.0236 + 0.8/-0.0232 + 0.5/-0.0065$
$B_1 = 0.5/-0.8245 + 0.8/-0.0465 + 1.0/-0.8241 + 0.8/-0.8205 + 0.5/-0.6164$
$B_2 = 0.5/0.0052 + 0.8/0.004 + 1.0/0.0038 + 0.8/0.0041 + 0.5/0.0077$
$B_3 = 0.5/0.005 + 0.8/0.0051 + 1.0/0.005 + 0.8/0.0051 + 0.5/0.007$
$B_4 = 0.5/0.0041 + 0.8/0.006 + 1.0/0.0053 + 0.8/0.0045 + 0.5/0.0092$

The Y output of coke unit product may be calculated on the following model:

$Y_{k+1} = 0.00005Y$
$+ (0.5/13.327 + 0.8/16.593 + 1.0/24.76 + 0.8/26.4177 + 0.5/19.915)z_{1k}$
$+ (0.5/.0035 + 0.8/0.015 + 1.0/-0.0236 + 0.8/-0.023 + 0.5/-0.0065)z_{2k}$
$+ (0.5/-0.8245 + 0.8/-0.0465 + 1.0/-0.824 + 0.8/-0.820 + 0.5/-0.616) + U_{1k}$ (9)
$+ (0.5/0.0052 + 0.8/0.004 + 1.0/0.0038 + 0.8/0.0041 + 0.5/0.0077)U_{2k}$
$+ (0.5/0.005 + 0.8/0.0051 + 1.0/0.005 + 0.8/0.0051 + 0.5/0.007)U_{3k}$
$+ (0.5/0.0041 + 0.8/0.006 + 1.0/0.0053 + 0.8/0.0045 + 0.5/0.0092)U_{4k}$

When $J = \sum\limits_{i=1} \left| \tilde{\tilde{Y}} - \tilde{Y} \right|$ and $N = 19$ function $J = U_{\alpha} J^{\alpha}$ receives the following values:

$$J = 0.5/1.26 + 0.8/1.25 + 1.0/1.22 + 0.8/0.95 + 0.5/1.16$$

6 Conclusion

Fuzzy dynamic numerical model for multistage optimization of coke process in oil refinery plant is considered. The model is presented as dynamic regression model with the fuzzy coefficients. Computer simulation shows effectiveness of the proposed model.

References

1. Berkutov, N.K., Stepanov, Y.V., Popova, N.K.: The relation between coke quality and blast-furnace performance. Steel Transl. **37**(5), 438–441 (2007)
2. Yurin, N.I., Morozov, O.S., Likhacheva, O.L.: Influence of coke quality on blast-furnace performance. Steel Transl. **41**(11), 924–927 (2011)
3. Gary, J.H., Handwerk, G.E.: Petroleum Refining: Technology and Economics. Taylor & Francis, Boca Raton (2005)
4. Uddin, M.N., Rebeiro, R.S.: Online efficiency optimization of a fuzzy-logic-controller-based IPMSM drive. IEEE Trans. Ind. Appl. **47**(2), 1043–1050 (2011)
5. Ellis, P.J., Paul, C.A.: Tutorial: delayed coking fundamentals. Presented at the 5–9 March 2000, AIChE 2000 Spring National Meeting, Atlanta, GA (2000)
6. Li, H.X., Miao, Z.H., Wang, J.Y.: Variable universe adaptive fuzzy control on the quadruple inverted pendulum. Sci. China (Ser. E) **5**(2), 213–224 (2002)
7. Cao, D.Y., Zeng, S.P., Li, J.H.: Variable universe fuzzy expert system for aluminum electrolysis. Trans. Nonferrous Metals Soc. China **21**(2), 429–436 (2011)
8. Pedrycz, W.: Fuzzy Control and Fuzzy Systems. Research Studies Press, New York (1989)
9. Pedrycz, W., Gudwin, R.R., Gomide, F.A.C.: Nonlinear context adaptation in the calibration of fuzzy sets. Fuzzy Sets Syst. **88**(1), 91–97 (1997)
10. Aliev, R.A., Aliyev, R.R.: Soft Computing and Its Applications. World Scientific Publishing Company, Singapore (2001)
11. Zadeh, L.A.: Similarity relations and fuzzy orderings. Inform. Sci. **3**(2), 177–200 (1971). https://doi.org/10.1016/S0020-0255(71)80005-1
12. Zadeh, L.A.: Fuzzy sets as a basis for a theory of possibility. Fuzzy Sets Syst. **1**, 3–28 (1978)
13. Aliev, R.A., Mamedova, G.A., Aliyev, R.R.: Fuzzy Sets Theory and Its Application. Tabriz University Press, Tabriz (1993)

Automated Malaria Parasite Detection Using Artificial Neural Network

Emre Özbilge[1(✉)] ⓘ, Emrah Güler[2,3] ⓘ, Meryem Güvenir[4] ⓘ,
Tamer Şanlıdağ[3,7] ⓘ, Ahmet Özbilgin[5] ⓘ, and Kaya Süer[6] ⓘ

[1] GenBiomics R&D, Gazi Mağusa Teknoloji Geliştirme Bölgesi (GMTGB),
Famagusta, Northern Cyprus, Mersin 10, Turkey
emreozbilge@gmail.com
[2] Faculty of Medicine, Department of Medical Microbiology and Clinical
Microbiology, Near East University, Nicosia, Cyprus
emrah.guler@neu.edu.tr
[3] Near East University, DESAM Institute, Nicosia, Cyprus
tamer.sanlidag@neu.edu.tr
[4] Vocational School of Health Services, Near East University, Nicosia, Cyprus
meryemguvenir@hotmail.com
[5] Faculty of Medicine, Department of Parasitology, Manisa Celal Bayar
University, Manisa, Turkey
a.ozbiligin@yahoo.com
[6] Faculty of Medicine, Department of Infectious Diseases and Clinical
Microbiology, Near East University, Nicosia, Cyprus
kaya.suer@neu.edu.tr
[7] Faculty of Medicine, Department of Medical Microbiology, Manisa Celal
Bayar University, Manisa, Turkey

Abstract. Malaria is still an infectious disease that causes high mortality in
endemic regions. It is thought that it will maintain importance in the future,
especially due to people travelling from African countries where malaria is
endemic to its eradicated regions. Therefore, rapid and accurate diagnosis is a
critical step in the effective treatment of malaria and reducing mortality rates.
This paper provides a malaria diagnosis system using an artificial neural network approach with SURF (Speeded Up Robust Features) method that helps the
clinicians to predict and locate infected cell with malaria on the sample thin
blood smear image. The performance of the proposed neural network and local
image feature extraction technique SURF were analyzed statistically and presented in this paper. The network was trained using only 45 infected thin blood
smear images and was then tested with 200 (100 infected and 100 non-infected)
unseen images. The experimental results showed that the proposed system
identified the malaria parasite with 93% accuracy, 86% sensitivity and 100%
specificity.

Keywords: Image processing · Neural network · Machine learning · Malaria ·
Plasmodium

R. A. Aliev et al. (Eds.): ICAFS 2020, AISC 1306, pp. 631–640, 2021.
https://doi.org/10.1007/978-3-030-64058-3_78

1 Introduction

Malaria is an infectious disease caused by *Plasmodium* parasites that infect red blood cells in human. There are five *Plasmodium* species that cause human malaria: *Plasmodium falciparum*, *Plasmodium vivax*, *Plasmodium ovale*, *Plasmodium malariae* and *Plasmodium knowlesi* [1]. The disease is transmitted by the bite of the infected female *Anopheles* mosquito and can be life threatening [2]. According to World Health Organization (WHO) data 228 million malaria cases were reported in 2018 and 405,000 people died from malaria infection. Approximately 93% of all cases and 94% of malaria-related deads were found in African continent. Moreover, 67% of the deaths (272,000) were children under five years old [3].

Immediate and accurate diagnosis of malaria is critical in the effective treatment of the diseases. Although the high sensitivity in diagnosis is important in any case, it is even more important in sensitive groups such as young children and people who are not immune to malaria, where the disease can progress rapidly and cause death [4]. In addition, the necessity of early diagnosis and correct treatment is indisputable, since a certain level of resistance is observed against almost all available antimalarial drugs recently [5]. The main purpose of malaria treatment and therefore early diagnosis is to reduce mortality, morbidity and socioeconomic losses, to prevent spread and to minimize the risk of developing resistance [6].

Microscopy based on the search for parasites in peripheral blood is still accepted as the gold standard method in diagnosis despite developing diagnostic tests [6, 7]. Hundreds of millions of blood smears are screened each year to detect malaria parasites and infected red blood cells. Determining the parasitemia in the blood is important for the diagnostic purposes, beside that detecting drug resistance, measuring the effectiveness of antimalarial drug and determining the severity of the disease are also related to parasitemia [8]. Microscopic diagnosis, a burdensome and time-consuming procedure, is largely based on the experience of the microscopist [2, 9]. When parasitemia is low and mixed *Plasmodium* species are present, all area of the blood smears must be examined. However, this requires a lot of time and a physically exhausting process, especially in the eyes, is experienced for large amounts of smears. In addition, there are problems in diagnosis and treatment due to the large number of cases and limited number of experienced technicians in areas where malaria is endemic [1]. In conclusion, we believe that a new, faster, more practical and reliable diagnosis method is required because the false negative and false positive results increased the mortality, morbidity and resistant *Plasmodium* species.

Recent study by [10], authors first segmented the images using thresholding method described in [11], after applying several processes to the cell images, the skewness, distance colour and angle colour features of the cell images were obtained. Then, obtained features are presented to support vector machine for learning to classify the red blood cell images as infected or healthy cells. Consequently, the proposed system demonstrated 94.6% accuracy. In this paper, an alternative method using artificial neural network with SURF (Speeded Up Robust Features) method is proposed for predicting and locating the infected cells with malaria on the thin blood smear images. The SURF method provides the automated feature extraction by detecting infected red blood cells with *Plasmodium* parasites on the image and then the proposed neural network is capable of model the non-linear relationship between the extracted features and its corresponding outcome whether infected or healthy.

In this study, thin blood smears were prepared by using the blood of patients who diagnosed with malaria at the Near East University Hospital, Microbiology Laboratory. Blood smears were stained with Giemsa solution as described in [12].

The aim of this paper is to implement automated malaria detection system to help the clinicians to diagnose the thousand number of patients' blood smear samples fast and reliable. The structure of the paper is as follows: Sect. 2 presents image processing method and neural network approach used in this study. Afterwards, the experimental results of the proposed malaria detection system is given in Sect. 3. Finally, Sect. 4 points out some conclusion and suggests some extension for the proposed system.

2 Malaria Diagnosis System

An overview of the malaria parasite detection system is shown in Fig. 1. The system receives thin blood smear image, then the local image features from the input image are extracted using SURF (Speeded Up Robust Features) detector. After having obtained a set of SURF descriptors for the current input image, they are presented to a neural network in order to identify whether the presented SURF descriptor is classified as malaria parasite or not. By this way, the proposed system is able to identify the location of malaria parasite on the corresponding image. This is because, each of the SURF descriptor extracted from the image is associated with a location of infected red blood cell on the image and all the descriptors are evaluated on the neural network individually.

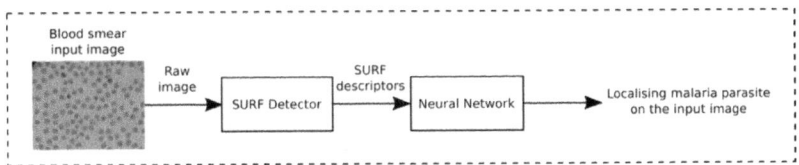

Fig. 1. Automated malaria parasite detection system.

The following sections will be described each block of the malaria parasite detection system given in Fig. 1 in detail.

2.1 SURF Detector

Figure 2a shows the sample thin blood smear image which is clearly seen that there is uniform colour distribution on the image expect for the regions which are stained with Giemsa stain. Instead of training entire image on the neural network, because using entire image can causes slow down the convergence of the neural network learning, due to the fact that the stainless pixels on the image provide irrelevant information about malaria parasite and are dominating over the image. To select the smeared regions on the input image, the SURF method was used [13]. This method is a fast algorithm to detect the region of interest on the image and also provides invariant local features on the detected smeared areas on the image.

The invariant feature property of the SURF method is very useful for producing invariant similar feature representation of the detected malaria parasite on the thin blood smear image even the shape of the parasite on the image has different rotation or scale. Figure 2b shows the extracted SURF descriptors on the corresponding image. The length of each SURF descriptor comprises 64 long feature vectors. In addition, the different number of SURF descriptors could be obtained from each thin blood smear images which depends on the dense of the smeared regions on the corresponding images.

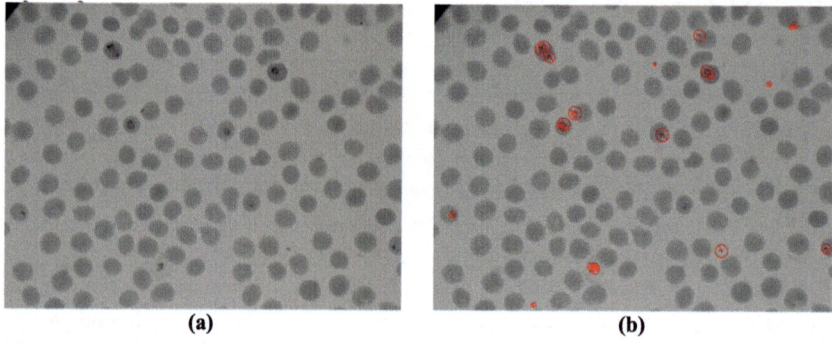

(a)   (b)

Fig. 2. (a) Blood smear image. (b) Extracted SURF descriptors, each red circle indicates the extracted SURF descriptor.

2.2   Artificial Neural Network

To classify the extracted each SURF descriptor from the thin blood smear image whether is one of the malaria forms, a feedforward neural network was used. The network was receiving 64-dimensional SURF descriptor from each individual location extracted from the image, and then the network makes prediction whether the presented SURF descriptor is positive or negative for the malaria parasite. Figure 3 shows the configuration of the network for the malaria parasite detection system. The network constituted of inputs to the network, hidden layer and output layer, for the hidden layer, 50 nodes and for the output layer, 2 output nodes were used in total. In order to find the optimal number of the hidden nodes on the network, cross-validation test was applied with various number of hidden nodes and the results showed that the lowest cross-validation error was obtained with 50 hidden nodes. In fact, by using a smaller number of hidden nodes leaded underfitting, on the other hand using large number of the hidden nodes caused overfitting problem of the network [14, 15].

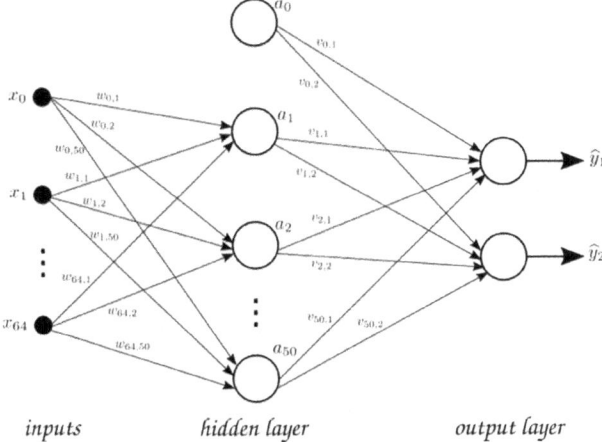

Fig. 3. The architecture of 2-layer feedforward neural network for classification.

The activation output of each hidden nodes a_i can be computed as follows:

$$a_i = f\left(\sum_{j=0}^{64} w_{j,i} x_j\right), \tag{1}$$

$$f(t) = \beta \, tanh(\gamma t), \tag{2}$$

where x_j indicates the inputs of the network, x_0 is a bias term and set to be 1, $w_{j,i}$ is the connection weight from the input j to the hidden node i. The function $f(\cdot)$ is a hyperbolic tangent activation function where β and γ are recommended to set 1.7159 and $\frac{2}{3}$ respectively [16].

After obtaining hidden outputs of the network, the network prediction outputs $\hat{\mathbf{y}}$ for the current SURF input vector are computed as given:

$$\hat{y}_k = g\left(\sum_{i=0}^{50} v_{i,k} a_i\right), \tag{3}$$

$$g(t) = \frac{1}{1+e^{-t}}, \tag{4}$$

where k indicates the index of the corresponding network output ($k \in \{1, 2\}$), a_0 is the bias term and set to be 1, $v_{i,k}$ is the connection weight from the hidden node i to the network output k and the function $g(\cdot)$ is the sigmoid activation function. Here the sigmoid function clamps the output value in the range between 0 and 1, and this value indicates the estimated probability of the corresponding output. The network prediction output $\hat{y}_1$ indicates the estimated probability of malaria parasite negative, on the other hand, the prediction output $\hat{y}_2$ is the estimated probability of malaria parasite positive. This is the multi-class classification problem so that in order to make final prediction whether the current SURF input vector $\mathbf{x}$ can be classified as positive or negative of being malaria parasite, the maximum predicted probability value among the network outputs is selected ($max\{\hat{y}_1, \hat{y}_2\}$) to determine the predicted class.

In order to train the network, the extracted SURF vectors and their corresponding desired outputs of being either positive or negative for malaria parasite from the images are arranged as the following vector form:

$$\mathbf{x}^{(i)} = [x_1 x_2 x_3 \cdots x_{64}], \quad \mathbf{y}^{(i)} = [y_1 y_2] \tag{5}$$

where i indicates the index of the current inputs-outputs vector of the network. It is also important to note that the desired outputs have to satisfy $y_1 y_2 \in \{0, 1\}$.

Data Preparation for Training. In order to determine the desired outputs of the extracted each SURF input vector $\mathbf{x}^{(i)}$ from the 45 infected thin blood smear images, an expert identified each stained region on the images whether is negative or positive for the malaria parasite. As a result, whenever the corresponding region was decided to be negative by the expert, the desired output y_1 is set to be 1 and the y_2 is set to be 0, on the other hand, whenever the region is positive, in this time y_1 is set be 0 and y_2 is set to be 1 while constituting outputs vector for the corresponding extracted SURF vectors.

During the preparation process, it is important to note that in each image there is only few SURF descriptors which were identified as a positive for malaria parasite by the manual examination. On the other hand, each image has more than a half of the SURF descriptors are negative. As a result, while constituting training dataset for the network, the negative inputs-outputs data vectors become dominant in the entire training dataset. This situation could create a problem for the network training because a neural network usually requires balance representation of the both classes (*i.e.* positive, negative) during the training, otherwise the class which contains minor data cannot be learned by the network because those minor data could be identified as an outliers by the network. To overcome this problem, it is important to use enough number of sample images during the training process.

Network Training. The network connection weights $\mathbf{W}, \mathbf{V}$ are learned by optimizing the following logistic regression cost function defined in (6). The cost function was optimized using the conjugate gradient method as developed in Matlab by Carl Edward Rasmussen.

$$J(\mathbf{V}, \mathbf{W}) = -\frac{1}{m}\left[\sum_{i=1}^{m}\sum_{j=1}^{2} y_j^{(i)} ln\left(\hat{y}_j^{(i)}\right) + \left(1 - y_j^{(i)}\right) ln\left(1 - \hat{y}_j^{(i)}\right)\right] \\ + \frac{\lambda}{2m}\left[\sum_{i=1}^{64}\sum_{j=1}^{50} (w_{i,j})^2 + \sum_{i=1}^{50}\sum_{j=1}^{2} (v_{i,j})^2\right] \tag{6}$$

where m indicates the total number of training data, λ is a regularisation parameter and during the network training various λ values were tested in order to obtain optimal network connection weights to eliminate the overfitting problem.

The steps of the training algorithm of the neural network for the malaria parasite classification is given in Algorithm 1.

Algorithm 1: The steps of the neural network training with cross-validation.

Input: Provide training dataset $< \mathbf{X}_{train}, \mathbf{Y}_{train} >$, validation dataset $< \mathbf{X}_{valid}, \mathbf{Y}_{valid} >$
Output: Obtain the optimal connection weights $< \mathbf{W}, \mathbf{V} >$

1   Define set of regularization parameters $\boldsymbol{\lambda} = \{\lambda_1, \lambda_2, \lambda_3, \cdots, \lambda_n\}$
2   Initialize network connection weights $< \mathbf{W}_{init}, \mathbf{V}_{init} >$
3   **for** i = 1; *each* λ_i ; i = i + 1 **do**
4   Set $< \mathbf{W}_{\lambda_i}, \mathbf{V}_{\lambda_i} > = < \mathbf{W}_{init}, \mathbf{V}_{init} >$
5   **for** j = 1; j < Max_Repetition; j = j + 1 **do**
6   Compute the activation of the hidden nodes for all training input data $\mathbf{X}_{train}$ using (1)
7   Compute the network prediction outputs using (3)
8   Compute the cost of the network $J(\mathbf{W}_{\lambda_i}, \mathbf{V}_{\lambda_i})$ for all training dataset $< \mathbf{X}_{train}, \mathbf{Y}_{train} >$ using (6)
9   Use conjugate gradient method to update all connection weights $< \mathbf{W}_{\lambda_i}, \mathbf{V}_{\lambda_i} >$
10   **end**
11   Compute the cost of the network $J(\mathbf{W}_{\lambda_i}, \mathbf{V}_{\lambda_i})$ for all validation dataset $< \mathbf{X}_{valid}, \mathbf{Y}_{valid} >$ using (6) by setting $\lambda = 0$
12   **end**
13   Select the best connection weights of the network: $< \mathbf{W}, \mathbf{V} > = min(J(\mathbf{W}_{\lambda_i}, \mathbf{V}_{\lambda_i}))$

3 Experimental Results

Firstly, each image was analysed whether the trained neural network was able to correctly identify all the positive malaria regions on the images. In other words, the performance of the network was verified for the classification of the extracted SURF descriptors. The network predicted outputs (positive or negative) for each detected region on the corresponding image were compared against the expert's decision on the same images. As a result, Fig. 4a shows the results of the identified malaria parasites on the image as shown with the original image and extracted SURF descriptors on the same image side by side. It is clearly seen that both the network predictions and the expert's decision were agree with each other. However, it is also possible that some SURF vectors on the image can be identified as negative even though it is positive by the expert, this is shown in Fig. 4b at the right top of the image marked with 'A'. Basically, at that location on the image there is 2 different SURF vectors extracted (*i.e.* one shown with red circle and other is white circle), the first SURF vector was classified as positive (shown with white circle) that is same as the expert's manual identification as well, but the second one was not detected as positive (shown with red circle). This is because, extracted SURF descriptor does not carry enough information about the stained shape, in other words, it was not the strongest descriptor for that region. In fact, it is not clear by human eyes both stained shapes belong to each other as shown in Fig. 4c. To overcome this problem, one way to do it is that; whenever the SURF detector provides more than one descriptor for the same stained shape on the image, those SURF descriptors could be combined as a one before presented to the network. Therefore, this process will prevent to misclassify the input descriptor by the network.

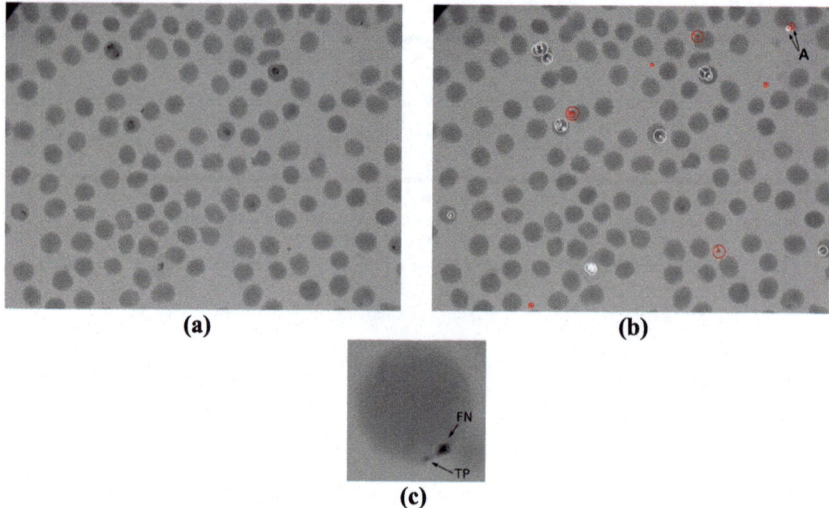

Fig. 4. (a) Original thin blood smear image, (b) extracted SURF descriptors. White circles indicate that the network predicts those SURF descriptors as positive for malaria parasite and the red circle indicates where the network predicts as negative, (c) detail view for region 'A', TP implies True Positive and FN implies False Negative.

In order to evaluate the performance of the network, several statistical analyses were carried out. To do this firstly, 200 (100 negative - 100 positive) unseen thin blood smear images were presented to the network. Then, instead of examining all individual SURF descriptors in an image whether or not the trained network predictions and the expert's decisions match each other, we check if the network predicts any positive SURF descriptor in the image in order to classify the corresponding image as a positive. Otherwise, if all the SURF descriptors in an image are predicted as negative the image will be classified as a negative. The network predictions and the expert's decisions were compared in 2×2 confusion matrix. As a result, Table 1 shows the resultant network performance over 200 unseen images. The accuracy of the network was found 93%. The sensitivity value was 86% where the sensitivity shows the performance of the network that detects the malaria parasites. Finally, the specificity value was 100% that indicates the network perfectly rejects all the stained shapes on the image which are all negative for malaria parasite. This was expected results because there was imbalance between the negative and positive classes of the training dataset. There were 1420 and 187 SURF descriptors which were obtained from 45 training images, used for negative and positive classes respectively during the network training, hence possibly some form of the malaria parasite was not learned because of not existing in the training dataset.

Table 1. Confusion table for quantitative assessment of the neural network classification for malaria parasite.

	Positive by expert	Negative by expert
Predicted positive	86	0
Predicted negative	14	100

4 Conclusion

This paper demonstrated local feature selection method and neural network approach to detect malaria parasites on the thin blood smear images. The SURF method was used in this paper as a local feature extraction algorithm to identify infected red blood cells on the images. This method provided robust, fast and good results to select the infected red blood cells on the image, as a result, the proposed neural network was forced to focus only the selected regions on the image during the training. As a result, the network performed to classify all negative malaria parasite images perfectly. This is very important in order not to treat the patients who are not infected with malaria wrongly. On the other hand, the network was failed to detect only 14% of the infected patients, this situation requires for further investigation to find out the reason of the failure by the network. However, some idea can be; using fewer positive data during the training might lead the network not to predict those images strong enough as an infected, or some failed images could represent different malaria form than the learned ring form.

Another issue was also found out that the SURF feature extraction method sometimes provides multiple SURF descriptors for the same stained region (see Fig. 4c) on the corresponding image, this was because of the structure of the SURF algorithm itself. Actually, this was not an issue because the network predicted at least one of the positive SURF descriptors at the detected region correctly. Therefore, an extension will be implemented as a post-decision mechanism for eliminating the unwanted faulty prediction in order to improve the overall performance of the network as a future work.

References

1. Vijayalakshmi, A., et al.: Deep learning approach to detect malaria from microscopic images. Multimedia Tools Appl. 1–21 (2019)
2. Rajaraman, S., Jaeger, S., Antani, S.K.: Performance evaluation of deep neural ensembles toward malaria parasite detection in thin-blood smear images. PeerJ **7**, 69–77 (2019)
3. World Health Organization: Malaria, 14 January 2020. https://www.who.int/news-room/fact-sheets/detail/malaria. Accessed 19 June 2020
4. Republic of Turkey: Malaria Case Management Guide. General Directorate of Public Health, Department of Zoonotic and Vector Diseases (2019)
5. Salman, Ö., Erbaydar, T.: Drug resistant malaria. TAF Prev. Med. Bull. **15**(4) (2016)
6. Ardıç, N., Koru, Ö.: Rapid diagnostic tests in malaria. Nobel Medicus J. **8**(2) (2012)

7. Çapın Özmen, B.B., Sönmezer, M.Ç., Tortop, S., Ünalan, T., Bölek, H., Altıntop, S.E., İnkaya, A.Ç., Metan, G., Ergüven, S.: The importance of awareness for malaria regarding prophylaxis and early diagnosis: two imported malaria cases in Turkey. Mikrobiyoloji bulteni **53**(4), 472–479 (2019)

8. Poostchi, M., Silamut, K., Maude, R.J., Jaeger, S., Thoma, G.: Image analysis and machine learning for detecting malaria. Transl. Res. **194**, 36–55 (2018)

9. Sorgedrager, R.: Automated malaria diagnosis using convolutional neural networks in an on-field setting. Master's thesis, Faculty of Mechanical, Maritime and Materials Engineering, Delft University of Technology (2018)

10. Sunarko, B., Bottema, M., Iksan, N., Hudaya, K.A.N., Hanif, M.S., et al.: Red blood cell classification on thin blood smear images for malaria diagnosis. J. Phys. Conf. Ser. **1444**, 012036 (2020). IOP Publishing

11. Otsu, N.: A threshold selection method from gray-level histograms. IEEE Trans. Syst. Man Cybern. **9**(1), 62–66 (1979)

12. Korkmaz, M., Ok, Ü.Z.: Parazitolojide laboratuvar. Türkiye Parazitoloji Dernegi (2011)

13. Bay, H., Tuytelaars, T., Van Gool, L.: SURF: speeded up robust features. In: Leonardis, A., Bischof, H., Pinz, A. (eds.) ECCV 2006. LNCS, vol. 3951, pp. 404–417. Springer, Heidelberg (2006). https://doi.org/10.1007/11744023_32

14. Bishop, C.M.: Pattern Recognition and Machine Learning. Springer, New York (2006)

15. Haykin, S.: Neural Networks and Learning Machines. Prentice Hall, Upper Saddle River (2009)

16. LeCun, Y., et al.: Generalization and network design strategies. Connect. Perspect. **19**, 143–155 (1989)

Integral Estimate of the Tourist Potential by Knowledge Compilation Based on Expert and Consumer Preferences

Ramin Rzayev[1]($\boxtimes$) (iD) and Inara Rzayeva[2] (iD)

[1] Institute of Control Systems of ANAS, Vahabzadeh Street 9,
AZ1141 Baku, Azerbaijan
`raminrza@yahoo.com`
[2] Azerbaijan State University of Economics, Istiglaliyat Street 6,
AZ1101 Baku, Azerbaijan
`ina3r@mail.ru`

Abstract. The concept of a combined system for assessing and ranking of the regions relative to the levels of development of their tourism potentials is proposed. This concept is based on the compilation of knowledge presented in the form of expert and consumer assessments of heterogeneous natural, cultural and historic, and socio-economic indicators. Within the framework of this system, it is proposed to compile the expert-consumer assessments in the fuzzy information environment using the fuzzy inference system and fuzzy maxmin convolution method. The assessment results obtained for various scenarios of the tourist potential of hypothetical regions in the form of acquired knowledge can be compiled using a three-layer feedforward neural network. After the necessary training on various scenarios of the tourist potential, the feedforward neural network turns into a universal assessment mechanism and it is able to induce an integral estimation of the tourism potential for any region given by itself indicators.

Keywords: Tourism potential · Knowledge compilation · Expert judgement · Fuzzy set · Neural network

1 Introduction

To make and plan the economic decisions, intelligent systems are currently being developed, since these systems are able to synthesize decisions more efficiently than humans. The methods used here extensively take into account the opinions of specialists from the subject area (experts), whose heuristic knowledge and preferences are replenished and modified due to changes in the environment. Naturally, this requires the permanent participation of experts in the decision-making process in the interests of the company and consumers of its goods (services). As a rule, consumers of travel services are well aware of their preferences and they intuitively can assess the degree of their satisfaction (utility) from the tourist solutions offered by travel agency. Their assessments are also able to ensure the targeted activities of tourism companies and the adequacy of their tourism decisions. In this case, the dissatisfaction of the tourist

R. A. Aliev et al. (Eds.): ICAFS 2020, AISC 1306, pp. 641–648, 2021.
https://doi.org/10.1007/978-3-030-64058-3_79

services consumer (TSC) with one or another indicator of tourist potential becomes an additional important reason for adjusting the tourist policy and/or revising the vision relative to tourist potential of the region as a whole. For example, new components of tourism potential can appear that require the introduction of a new criteria. Therefore, the process of multi-criteria assessment of the tourist potential of the region is dynamic, i.e. it implies multiple (iterative) assessment for short time intervals, and, as a result, it requires the use of flexible information support systems.

2   The System Structure for Assessing the Tourism Potential Using Expert Knowledge and Assessments of TSC

Of the many methods and approaches to decision-making, the most interesting are those that take into account multi-criteria and uncertainty, and also allow the selection of solutions from a variety of alternatives of different nature in the presence of criteria having different types of measurement scales. Currently, these are the methods of utility theory, hierarchy analysis and fuzzy set theory, which are widely embodied in computer and information support systems. This choice is determined by the fact that these methods most satisfy the requirements of universality, allowance for the multi criteria of the choice under uncertainty from the discrete or continuous set of alternatives, and the ease of preparation and processing of expert information.

The expected utility method is widely used among a group of axiomatic decision-making methods under risk and uncertainty. The main idea of this approach is to obtain quantitative estimates of the utility from the possible outcomes that are the consequences of the decision-making process. Further, it is possible to choose the best outcome on the base of these estimates.

At the same time, traditional approaches to decision making are mainly based on the preferences of the group of people responsible for developing the decision. However, in the context of dynamically developing economies and the competitive environment, the opinion of the ultimate users of goods and services becomes dominant. Therefore, to obtain estimates of the region tourist potential, it is proposed to use information about the preferences of TSCs, which can provide public regulation of the tourist decision-making process. At the same time, it is necessary to mean that, firstly, potential TSCs have all the information necessary for making a consumer decision, and secondly, TSCs are able to rank all conceivable outcomes based on the ability of the tourist offer to ensure their satisfaction for one or another indicator of tourism potential. Every TSC always tries to maximize the level of its satisfaction or, as economists call it, its utility, which is defined as an individual perception of itself satisfaction from the tourist offers.

In assessing the tourist potential in each concrete case every TSC relies on its own subjective judgments and mentally form its own preference scale. Even if he evaluates the quantitative characteristics of the tourist offer, his assessment still has a "fuzzy color", because he has to use quality categories all the time. Therefore, considering the assessment of the tourist potential as the composite outcome, each of its components could be represented as the term of the linguistic variable described by the appropriate fuzzy set [1]. Moreover, the level of tourist potential for every TSC is the qualitative category rather than the quantitative category, and therefore, the linguistic variable with the fuzzy values can be used as a measure of utility. In such cases, for determining the region tourist potential taking into account the opinions (satisfaction) of TSCs the corresponding functional dependence can be established by constructing the suitable fuzzy inference system consisting of fuzzy logical rules of the form [1]:

$$\text{If } x_1 \text{ is } F_{1k} \text{ and } x_2 \text{ is } F_{2k} \text{ and } \ldots \text{ and } x_n \text{ is } F_{nk}, \text{ then } y \text{ is } Y_j, \tag{1}$$

where F_{ik} ($i = 1 \div n$, $k = 1, 2, \ldots$) is the fuzzy set that describes the term of the linguistic variable reflecting the k-th level of satisfaction on the i-th indicator of the tourism potential; Y_j ($j = 1, 2, \ldots$) is the fuzzy set reflecting the quality level of the region tourist potential. Starting from a fuzzy paradigm, to assess the region tourist potentials the information system is proposed, which combines both expert knowledge and TSC assessments and some fuzzy models of multi-criteria assessment of alternatives. The structure of this system is shown in Fig. 1.

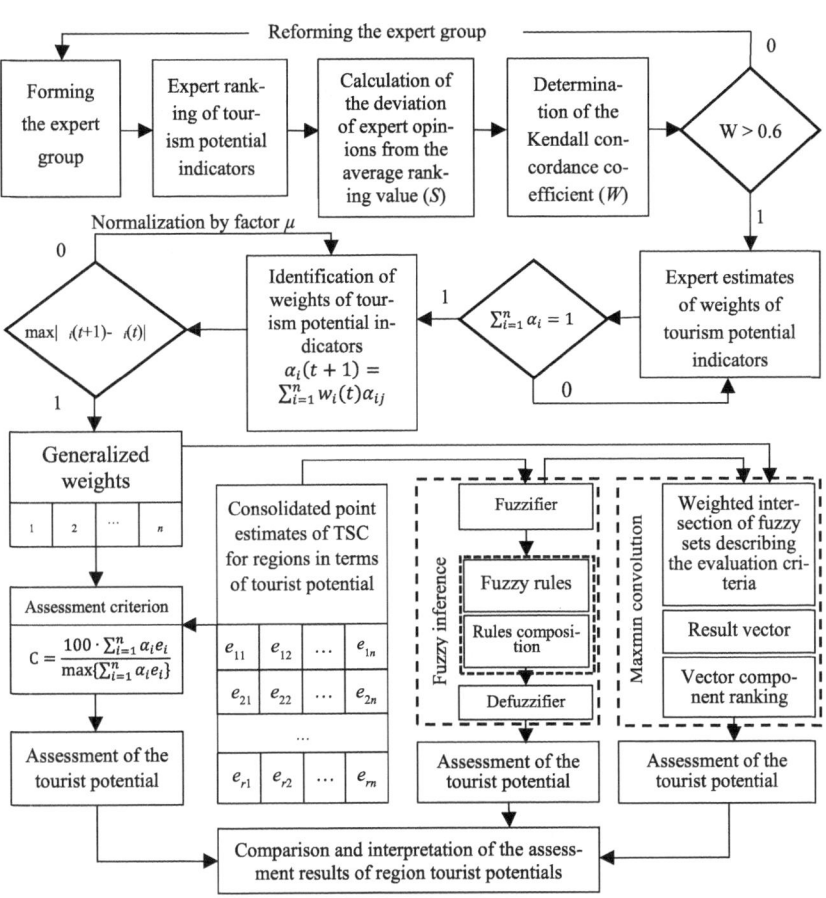

Fig. 1. The system structure for assessing the region tourist potential.

The adequacy of the procedures proposed in the framework of this system for assessing tourism potentials depends on many factors, including: identification of the weights of tourism potential indicators based on initial expert judgements; the proper organization, collection and storage of consolidated estimates of TSCs relative to satisfactory of tourism potential for each of the indicators separately; the appropriate choice of implicative rules (1); identification of membership functions to restore the fuzzy sets as qualitative evaluation criteria.

3 Identification of Weights of Tourism Potential Indicators Based on Expert Judgements

Tourism potential is a multifactor category, which is characterized by a system of natural, cultural-historic, and socio-economic indicators x_i ($i = 1 \div n$), which have different influences on the competitiveness of the region in the market of tourism service activities. Identification of weights for these indicators as criteria for assessing the competitive advantages of regions in the field of tourism service activities implies, firstly, ranking the indicators for their priority, and secondly, a group assessment of the normalized values of the generalized indicator weights based on their relative influence on the level of competitiveness of regions. To implement this scheme, profile experts are involved, whose knowledge and judgements in the subject area become the basis for further calculations.

So, at the initial stage of the study, a group of m experts is invited to rank indicators of tourist potential. Using an independent questionnaire of experts, the ranking estimates of the priority of indicators x_i ($i = 1 \div n$) are established according to the principle: the most important indicator is denoted by the number "1", the next less important indicator is denoted by the number "2" and then in descending order of expert preferences. The rank estimates of indicators obtained in this way are summarized in the corresponding base and studied for their consistency. To establish the degree of consistency of expert rank assessments, the Kendall's concordance coefficient is usually used. This coefficient demonstrates the multiple rank correlation of expert judgements and according to [2] it can be calculated by the formula:

$$W = 12 \cdot S / \left[m^2 \left(n^3 - n \right) \right],$$

where m is the number of invited experts; n is the number of indicators of the tourist potential, and S is the deviation of expert judgements from the average ranking of indicators x_i, which is calculated by the formula [2]:

$$S = \sum_{i=1}^{n} \left[\sum_{j=1}^{m} [r_{ij} - m(n+1)/2]^2 \right],$$

where $r_{ij} \in \{1, 2, \ldots, n\}$ is the rank of the i-th indicator established by the j-th expert.

At the preliminary stage of independent questioning, each j-th expert is also invited to establish the value of the normalized weight estimate for the i-th indicator in the form α_{ij} under $\sum_{i=1}^{n} \alpha_{ij} = 1$, $\alpha_{ij} \geq 0$. The results of this survey are also summarized in the appropriate base.

Further, starting from the data obtained at the preliminary stage of independent questioning of the expert community, the iterative process of identifying the generalized weights of the indicators x_i is activated in the form of stepwise averaging over groups of normalized expert judgements. In particular, averaging over the i-th group of normalized estimates of the weights of the indicators x_i is carried out according to [3] as

$$\alpha_i(t+1) = \sum_{j=1}^{m} w_j(t)\alpha_{ij},$$

where $w_j(t)$ is the competency degree of the j-th expert in the t-th approximation. At the initial moment, i.e. at $t = 0$, each expert competency degree is equal to $1/m$. At the next steps of the iteration, the competency degrees of experts are calculated by the following equalities [3]:

$$\begin{cases} w_j(t) = [1/\eta(t)] \sum_{i=1}^{n} \alpha_i(t) \cdot \alpha_{ij} \ (j = \overline{1, m-1}), \\ w_m(t) = 1 - \sum_{j=1}^{m-1} w_j(t), \ \sum_{j=1}^{m} w_j(t) = 1, \end{cases}$$

where $\eta(t)$ is the normalizing factor, which is calculated by the formula [3].

$$\eta(t) = \sum_{i=1}^{n} \sum_{j=1}^{m} \alpha_i(t)\alpha_{ij}.$$

The averaging process is finished under $\max_{i} \{|\alpha_i(t+1) - \alpha_i(t)|\} \leq \varepsilon$, where ε is the accuracy of the calculations, which is acceptable to the user.

If the numbers $\alpha_i(\tau)$ $(i = 1 \div n)$ obtained at the τ-th step are satisfied the conditions $\max_{i} \{|\alpha_i(\tau) - \alpha_i(\tau - 1)|\} \leq \varepsilon$ and $\sum_{i=1}^{n} \alpha_i(\tau) = 1$, then they are considered as generalized weights of the corresponding indicators x_i $(i = 1 \div n)$ of the tourist potential.

4   Weighted Evaluation of the Region Tourism Potential

The method of weighted evaluation involves a discussion of influences on the level of tourist potential by a group of specially invited experts from among of TSCs. All experts are provided with a full set of indicators x_i $(i = 1 \div n)$ and they are offered to obtain an independent qualitative assessment of the region tourist potential for each indicator individually, for example, according to the following five-point rating system: 5 – TOO HIGH; 4 – HIGH; 3 – AVERAGE; 2 – LOW; 1 – TOO LOW; 0 – ABSENT. Further, expert evaluations are analyzed for their consistency (or inconsistency) according to the rule: the maximum allowable difference between two expert estimates relative to tourism potential by any indicator x_i $(i = 1 \div n)$ should not exceed 3 units. This rule allows to filter unacceptable deviations in expert evaluations of tourism potential by each indicator. The total tourist potential index is derived using the following evaluation criterion [3]:

$$C = \frac{\sum_{i=1}^{n} \alpha_i e_i}{\max \left\{ \sum_{i=1}^{n} \alpha_i e_i \right\}} \times 100,$$

where $C \in [0, 100]$; α_i is the weight of i-th indicator, which is established by above technique; e_i is consolidated assessment of tourism potential by the i-th indicator, which is calculated as the arithmetic average of the corresponding expert (or TSC) estimates on the five-point scale. The minimum value of the index C means the minimum level of tourist potential in the region, and vice versa, the closer the obtained value of the index C to 100, the better the development of tourist potential in the given region.

5 Multifactor Evaluation of Tourist Potential by the Fuzzy Maxmin Convolution Method

In fuzzy applications, qualitative evaluation criteria are understood as some evaluation concepts (terms of linguistic variables), and numerical estimates of the considered alternatives are represented by the degrees of compliance with these concepts or, another words, by the values of the corresponding membership functions, for example, by the values of membership functions corresponding to the concepts: "PREFERRED NATURAL-CLIMATE CONDITIONS", "DESIRED CULTURAL-HISTORIC SIGHTS" OR "SUITABLE SOCIO-ECONOMIC ENVIRONMENT" [1]. Starting from this, suppose that $A = \{a_1, a_2, \ldots, a_r\}$ is the set of regions, where tourist clusters are formed, and $\{x_i\}_{i=1 \div n}$ is the set of indicators, which are considered as the quality criteria (or evaluation concepts) F_i for assessing the tourism potential. To describe these criteria, appropriate membership functions $\mu_{Fi}(a)$ are chosen that restore the corresponding fuzzy subsets of the universe A in the form:

$$F_i = \{m_{Fi}(a_1)/a_1, m_{Fi}(a_2)/a_2, \ldots, m_{Fi}(a_r)/a_r\}.$$

For example, the following Gaussian function $\mu_{F_i}(a_k) = \exp\{-[e_i(a_k) - 5]^2/\sigma_i^2\}$ can be chosen as the membership function, which determines the relation degree of the tourism potential of the region a_k ($k = 1 \div r$) to the evaluation criterion F_i. Here, $e_i(a_k)$ is a consolidated expert assessment of the tourism potential for the k-th region relative to the criterion F_i by the five-point scale; σ_i^2 is the parameter (density) that determines the width of the function profile.

In the case of using the fuzzy *maxmin* convolution method, the choice of the region with the greatest tourist potential is carried out by intersection of the fuzzy sets F_i [4]:

$$D = F_1 \cap F_2 \cap \ldots \cap F_n, \tag{2}$$

---

[1] In this case, we intentionally do not specify the indicators of tourism potential, as this will be the subject of the next study.

where, according to [1, 4], the intersection operation is realized by the formula: $\mu_D(a_k) = \min\{\mu_{F_1}(a_k), \mu_{F_2}(a_k), \ldots, \mu_{F_n}(a_k)\}$, $k = 1 \div r$. In this case, the region a^* has the greatest tourist potential, if its corresponding membership function has the greatest value among the components of the vector $[\mu_D(a_1), \mu_D(a_2), \ldots, \mu_D(a_r)]$, i.e. provided by equality $\mu_D(a^*) = \max\{\mu_D(a_1), \mu_D(a_2), \ldots, \mu_D(a_n)\}$. The other regions are ranked by the level of tourist potential in descending order of values of their membership functions.

As noted above, indicators of tourism potential, and, hence, the evaluation criteria F_i have various degrees of importance. According to the technique presented in Sect. 3, these degrees are identified in the form of weights α_i $(i = 1 \div n)$. Then, to choice the region with the greatest tourist potential and to rank the others it is necessary to replace intersection (2) by the following weighted intersection: $D = F_1^{\alpha_1} \cap F_2^{\alpha_2} \cap \ldots \cap F_n^{\alpha_n}$.

6 Assessment of Tourist Potential using a Fuzzy Inference

This approach is based on the verbal proofs of experts relative to the level of tourist potential of a hypothetical region. After fuzzification of the terms of input and output linguistic variables, these verbal proofs are ultimately reflected in the form of implicative rules of the form (1). For each of the such rules the intersections of the fuzzy sets from the left-hand sides are determined: $(x_1 = F_{1k}) \cap (x_2 = F_{2k}) \cap \ldots \cap (x_n = F_{nk})$, and only after that an implication operation is used to represent these rules, for which there are different fuzzy implementation methods [3]. As a result, the fuzzy relations with membership functions $\mu_{Ri}(x, u)$ are transformed into the corresponding matrices R_i, where according to the principle $\mu_R(x, u) = \min\{\mu_{Ri}(x, u)\}$ the intersection of R_i forms a common functional solution: $R = R_1 \cap R_2 \cap \ldots \cap R_n$. In this case, the fuzzy conclusion relative to the tourist potential is described for the region by the fuzzy set F and it is determined by the composition rule $G = F \circ R$, where G is the fuzzy subset of the discrete universe $U = \{0, 0.1, 0.2, \ldots, 1\}$; "$\circ$" denotes the operation of rules composition, for example, in the form of $\mu_G(u) = \max\{\min[\mu_F(x), \mu_R(x, u)]\}$.

Evaluation and comparison of regions is carried out on the basis of point estimates of fuzzy conclusions (fuzzy subsets $C \subset U$) relative to their tourism potentials. For this purpose, first, for the corresponding $C \subset U$ α-level sets ($\alpha \in [0, 1]$) are established in the form of $C_\alpha = \{i | \mu_C(u) \geq \alpha, u \in U\}$. Then, for each of them, the corresponding cardinal number $M(C_\alpha)$ are determined by the formula $M(C_\alpha) = \sum_{j=1}^{p}(u_j/p)$. As a result, a numerical estimate of the fuzzy set C, reflecting the level of tourist potential of the corresponding region, is obtained from equality: $F(C) = (1/\alpha_{\max}) \int_0^{\alpha_{\max}} M(C_\alpha) d\alpha$.

7 Conclusion

Figure 2 shows the scheme of a three-layer neural network, which after training can be a universal signal converter. The network converts the input signals taking into account of its parameters (weights of connections and thresholds of non-linear neurons). During the training process, these weights are adjusted so that the neural network induces the desired outputs. The process begins with an arbitrary choice of weights and, therefore, at the beginning the neural network outputs will be significantly different from the desired ones. Comparing its current outputs with the corresponding desired ones, the network adjusts the weights of its connections in order to reduce this difference. Actually, this is the basis of any supervisor algorithm for training the neural network. After training on the sufficient number of examples of estimates of tourist potential, the neural network at its output reproduces the signal, which identifies the integral assessment of the tourist potential y for the particular region as follows

$$y = \sum_{i=1}^{p} w_i \phi \left[\sum_{k=1}^{n} w_{ik} e_k - \theta_i \right],$$

where e_k is the point expert estimate of the region tourist potential relative to k-th indicator; w_{ik} and w_i are the weights of the input and output connections, respectively; θ_i is the threshold of i-th nonlinear neuron from the hidden layer; $\varphi(\cdot)$ is the activation function of the sigmoid type $\varphi(x) = 1/[1 + \exp(-(x-\theta))]$.

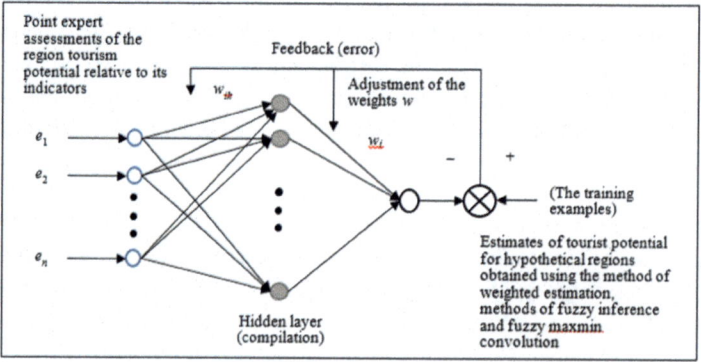

Fig. 2. Neural network as a tool for compiling the integral estimate of tourism potential.

References

1. Zadeh, L.: The concept of a linguistic variable and its application to approximate reasoning-I. Inform. Sci. **8**(3), 199–249 (1975). https://doi.org/10.1016/0020-0255(75)90036-5
2. Lin, A.S., Wu, W.: Statistical tools for measuring agreement. Springer, New York (2012)
3. Telnov, G.V.: Approach to the formation of the final assessment of the level of mastering the material of the academic discipline during the intermediate certification of trainees on the basis of weighted coefficients of the evaluated features. The Bulletin of the Adyghe State University, the series "Natural-Mathematical and Technical Sciences. **1**(154), 119–127 (2015) (in Russian)
4. Andreichenkov, A.V., Andreichenkova, O.N.: Analysis, synthesis, planning decisions in the economy. Finance and Statistics, Moscow (2000). (in Russian)

Fuzzy-Multiple Methodology for Assessing the Sustainability of Agricultural Production in the Regions of the Region and their Ranking

Elizabeth A. Arapova$^{(\boxtimes)}$ ⓘ, Natalia D. Rodionova ⓘ,
Artamonova Anna ⓘ, and Rakhmetova Lilia ⓘ

Rostov State University of Economics, Rostov-on-Don, Russia
dist_edu@ntti.ru, ndrodionova@mail.ru,
l.dov2010@mail.ru, anya10043@gmail.com

Abstract. We have presented a model for assessing the sustainability of agricultural production based on the aggregation of many significant indicators. The model is based on the use of fuzzy-logical inference systems—fuzzy five-level [0,1]—classifiers used by Nedosekin to assess the state of economic systems, including the risk of bankruptcy of an enterprise. To assess the sustainability of agriculture in each of the districts of the agricultural region, we propose to aggregate the assessments of its three subsystems: economic, social and environmental. We construct each of the estimates of the subsystems by aggregating the normalized values of significant indicators selected in accordance with generally accepted assessment methods. The fuzzy multiple method proposed by us has such advantages over existing analogues as universality, taking into account the significance of indicators by varying the weight coefficients. The methodology was tested by statistical data on the indicators of the Rostov region, taken from Federal State Statistic Service of Russia. It has been established that all districts of the Rostov region can be attributed to the term "development close to sustainable", and the last two districts can equally well be attributed to the fifth term "high level of sustainability".

Keywords: Mathematical model · Methods for assessing the state of economic systems · Integrated assessment · Scorecard · Fuzzy logic

1 Introduction

The problem of assessing the sustainability of production in a region, including agricultural production, is, from a mathematical point of view, the problem of qualimetry of the economic system based on a set of significant indicators of different origins. There are various approaches both to determining the economic sustainability of a region [1–3], and to assessing the level of sustainable development of economic systems. Nevertheless, for them it is possible to identify general principles. The mathematical implementation of the assessment consists of two important stages: 1) selection and justification of a system of indicators characterizing the stability of production; 2) development on their basis of a system of qualitative and quantitative indices characterizing the degree of production stability. If the first stage is currently fully developed, then the second stage requires additional research, including involving the fuzzy logic apparatus.

The concept of sustainable agricultural development is inextricably linked with the growth of food production, the efficient use of economic and intellectual resources, improving the welfare and quality of life of the villagers, stable and balanced nature management [4]. The authors indicate that for a quantitative characteristic of the stability of the agricultural sector, it is advisable to use a system of indicators that includes three groups: economic, social and environmental development.

Currently, a number of models have been developed to assess the degree of sustainability of the development of the agricultural region [5–7]. So, in [5], it is noted that a generalization of world experience in developing indicators of sustainable development made it possible to distinguish two approaches: 1) the construction of a system of indicators, each of which reflects individual aspects of sustainable development: environmental, economic, social, institutional; 2) the construction of an integral indicator, on the basis of which one can judge the degree of sustainability of socio-economic development. An integrated indicator is usually carried out on the basis of the following groups of indicators: environmental and economic; environmental, social and economic; environmental. At the same time, indicators of the sustainability of the development of the agricultural sector of the region are not unchanged, once and for all established. They are determined on the basis of internal factors characterizing the economic, social and environmental development of the agricultural sector of the region. The estimation generation algorithm consists of five stages: 1) forming a list of indicators; 2) an assessment of the level of development of the agricultural sector in the region for each indicator; 3) calculation of a comprehensive measure of sustainability by groups; 4) the formation of an integral indicator, for example, as the root of the cubic from the product of complex stability indicators for each of the three groups; 5) interpretation of the integrated assessment of the sustainability of the development of the regional agricultural sector. For these purposes, it is necessary to establish threshold values of the sustainability index, which can range from 0 to 1. For example, five levels of sustainability of the regional agrarian sector can be distinguished (Table 1).

Table 1. Interpretation of threshold values of the integral index of sustainable development of the regional agricultural sector

Sustainability level	Index interval boundaries	System stability
1	$0,8 < I < 1,0$	High level of stability
2	$0,6 < I < 0,8$	Sustainable development
3	$0,4 < I < 0$	Close to sustainable development
4	$0,2 < I < 0$	Signs of volatility
5	$0 < I < 0,2$	State of crisis

The developed technique allows you to get a relative assessment of the sustainability of the development of the agricultural sector of the region, to reflect its place in the aggregate of regions. It is noted that with a small amount of information necessary for calculation, the integral index has a certain sensitivity and informativeness. However, in the case of a large amount of data, the model becomes uninformative.

Thus, the analysis shows that existing models have such disadvantages as: 1) locality, attachment to a specific problem being solved; 2) low universality, lack of unified algorithms for constructing estimates; 3) the impossibility of ranking indicators, as well as the impossibility of taking into account expert opinions when building estimates.

From the point of view of mathematical modeling, the problem can be interpreted as follows. To assess sustainability, we can consider the system of indicators used to measure effectiveness, by groups (economic, social, environmental). Sustainable development of the system (district, region, country or a separate sub-industry) corresponds to: 1) finding the indicators on the basis of which integrated estimates are built in the regulatory framework determined from the meaning of the task, which, in turn, corresponds to the known intervals of the integrated estimates; 2) the positive dynamics of indicators, which is numerically reflected in the integral estimates of the respective groups.

In this work, we proposed a methodology for assessing the sustainability of agricultural production in the regions of the region, based on the above interpretation. To calculate the assessment, we used the elements of methods developed by us earlier and applied to assess the efficiency of agricultural production in the region based on time series of data throughout the region [5, 6]. Then, we used the methods developed by us to assess the ecological state of the region based on the aggregation of assessments of individual regions [7–10]. To construct estimates, we used the universal apparatus of the theory of fuzzy sets—a system of fuzzy five-level [0,1]—classifiers.

2 The Mathematical Tools of the Methodology for Assessing the Sustainability of Production: A System of Fuzzy Multi-Level [0,1]—Classifiers

A quantitative assessment of the sustainability of agriculture in the regions of the Rostov region was made on the basis of fuzzy-multiple aggregation of estimates of three subsystems: economic, social and environmental.

Evaluation of each individual subsystem can be made on the basis of data aggregation through systems of fuzzy five-level classifiers. The algorithm includes six stages.

Stage 1. Compilation of a list of indicators that have a significant impact on the formed assessment.

Stage 2. Ranking the importance of the studied indicators for evaluation, calculation of their weight coefficients, reflecting their significance for the final assessment.

Stage 3. Calculation of the normalized values of indicators belonging to a unit segment. The formulas used for this are determined by the meaning of the task and the nature of the indicator (static or dynamic).

Stage 4. Setting linguistic variables. The normalized values of the indicators calculated in Step 3, we associate linguistic variables with term sets consisting of five terms: "very low level of the indicator"; "Low rate"; "Average level of an indicator";

"High level indicator"; "A very high level indicator." The membership functions of linguistic variables are determined using standard trapezoidal functions [5, 6]. In addition, we introduce a linguistic variable: γ = "comprehensive assessment of the state of the system." A linguistic variable has a unit segment as a universal set. Its term set consists of five terms: G = {G1, G2, G3, G4, G5}, where G1 is the "steady tendency to decrease in growth" (for dynamic indicators) or "very low level" (for static); G2 is "downward trend" or "low rate"; G3 is "tendency to stagnation" or "average indicator level"; G4 is "upward trend" or "high level indicator"; G5 is "steady growth trend" or "very high level of indicator".

Stage 5. Calculation of numerical values of estimates based on normalized values of indicators, using standard five-level classifiers [5, 6].

Stage 6. Recognition of the obtained numerical estimates based on the classification of terms of the term set G. Analysis of the obtained estimates based on the normalized values of indicators; selection of indicators leading to a decrease in the assessment; formulation of practical recommendations to change the current situation.

3 The System of Indicators for Assessing the Sustainability of Agricultural Production in the Region

In this study, the time series of indicators influencing the sustainability of agricultural production are used as experimental material. The data was obtained from open Internet sources, including the Federal State Statistic Service of Russia (Rosstat) database.

3.1 Assessment of the Economic Subsystem

Indicators of the level of intensification of production (by district) (indicators for the year 2008–2018): 1) the availability of agricultural machinery in agricultural organizations at the end of the year, pieces (tractors, combine harvesters, corn combines, forage harvesters, potato harvesters, beet harvesters—separately); 2) sown area of crops, hectare (farms of all categories, agricultural organizations, households); 3) livestock and poultry in households of rural settlements, head (cattle, cows, pigs, sheep, goats, poultry, horses, rabbits, bee colonies—separately); 4) introduced mineral fertilizers (in terms of 100% nutrients) for crops of agricultural crops in agricultural organizations, centner. The list of indicators can be significantly expanded.

Indicators of the level of economic efficiency of intensification (by district) (indicators for the year, 2008–2018): 1) the volume of agricultural production (in actual prices), thousand rubles, farms of all categories; 2) the volume of crop production (in actual prices), thousand rubles; 3) the volume of livestock production (in actual prices), thousand rubles; 4) agricultural production index (in comparable prices; as a percentage of the previous year), percentage; 5) the index of crop production (in comparable prices; as a percentage of the previous year), percentage; 6) the index of livestock production (in comparable prices; as a percentage of the previous year), percentage.

3.2 Assessment of the Social Sphere

The set of indicators for the formation of an assessment of the social efficiency of agriculture (by region) has the form: 1) the average monthly salary of employees of organizations, the ruble (2016–2019); 2) the volume of social payments to the population and taxable cash incomes to the population, thousand rubles, the value of the indicator for the year (2011–2016); 3) total fertility rate, per mille (2010–2018); 4) general mortality rate, per mille (2010–2018); 5) the number of doctors of all specialties (2008–2013); 6) the number of nursing staff (2008–2013); 7) the number of medical beds (2008–2013); 8) the commissioning of individual residential buildings on the territory of the municipality, sq.m. of total area, square meter of total area.

3.3 Assessment of the Ecological System

The set of indicators for the formation of an assessment of compliance with the principles of environmental management has the following form: air pollution—1) the amount of pollutants emanating from all stationary sources of emission, thousand tons (by years); 2) the proportion of untreated emissions into the atmosphere; water resources—3) total water withdrawal; 4) water intake from underground sources; 5) transportation losses; 6) discharge of contaminated water without treatment; 7) the volume of circulating and sequentially used water; management of production and consumption waste—8) the share of waste disposal facilities included in the state register of waste disposal facilities (units) in the total quantity in the Rostov Region (authorized landfills); 9) the share of landfills for reclamation (pcs) in the total amount in the Rostov region (unauthorized landfills); 10) specially protected natural territories (SPNA)—the share of specially protected territories; 11) the costs of financing environmental programs.

4 The Study of the Sustainability of Agricultural Production in the Regions of the Rostov Region

For each of the 43 districts of the Rostov Region, according to the method described above, an assessment was made of the sustainability of its economic, social and environmental subsystems, Table 1. Based on the estimates constructed, using the same system of fuzzy multi-level [0,1]—classifiers, we built final estimates of the sustainability of agricultural production in each district, Table 2.

Table 2. Calculation of the final assessment of the sustainability of agricultural production for the regions of the Rostov region: (1)—assessment of the effectiveness of agricultural production (weight 0.4); (2)—Assessment of the social efficiency of agriculture in the regions of the Russian Federation (weight 0.3); (3)—Assessment of compliance with environmental management principles (weight 0.3); (4)—Final sustainability score

Municipal district	(1)	(2)	(3)	(4)
Azov	0.6154	0.5502	0.505	0.5524
Aksay	0.5000	0.5821	0.564	0.5276
Bagaevsky	0.6074	0.6122	0.539	0.5832
Belokalitvinsky	0.5000	0.5070	0.579	0.5174
Bokovsky	0.6000	0.6634	0.539	0.6000
Verkhnedonskoy	0.5000	0.5904	0.597	0.5524
Veselovsky	0.5156	0.5540	0.523	0.5024
Volgodonsk	0.5240	0.5540	0.504	0.5024
Dubovsky	0.5509	0.5816	0.508	0.5196
Egorlyksky	0.5266	0.5972	0.501	0.5283
Zavetinsky	0.4935	0.5845	0.504	0.5207
Zernogradsky	0.5692	0.6053	0.539	0.5485
Zimovnikovsky	0.6499	0.5754	0.504	0.5952
Kagalnitsky	0.4394	0.6286	0.524	0.5387
Kamensky	0,5704	0,5617	0,501	0,5233
Kashar	0,5738	0,5823	0,494	0,5384
Konstantinovsky	0,5000	0,5012	0,559	0,5054
Krasnosulinsky	0,5185	0,4946	0,459	0,5000
Kuibyshevsky	0,5998	0,5073	0,515	0,5398
Martynovsky	0,5572	0,4911	0,515	0,5058
Matveevo-Kurgan	0,5140	0,5193	0,654	0,5600
Millerovsky	0,5000	0,5587	0,579	0,5226
Milyutinka	0,5000	0,5795	0,464	0,5177
Morozovsky	0,556	0,5587	0,513	0,5100
Myasnikovsky	0,5000	0,5254	0,573	0,5138
Neklinovsky	0,6000	0.5004	0.539	0.5400
Oblivsky	0.5000	0.5795	0.508	0.5177
October	0.5228	0.5399	0.543	0.5000
Oryol	0.5072	0.5067	0.564	0.5084
Peschanokopskiy	0.5000	0.624	0.511	0.5444
Proletarian	0.5756	0.5957	0.564	0.5563
Remontnensky	0.5399	0.5656	0.365	0.4584
Rodionovo–Nesvetaysky	0.5406	0.5740	0.504	0.5144
Salsky	0.5043	0.5360	0.499	0.5000
Semikarakorsky	0.5000	0.4862	0.504	0.5000
Soviet	0.5024	0.5214	0.504	0.5000
Tarasovsky	0.5094	0.4794	0.476	0.5000

(continued)

Table 2. (*continued*)

Municipal district	(1)	(2)	(3)	(4)
Tatsinsky	0.5746	0.5034	0.582	0.5389
Ust-Donetsk	0.6024	0.5913	0.348	0.5067
Tselinsky	0.5101	0.5377	0.575	0.5150
Tsimlyansky	0.4622	0.5029	0.374	0.4544
Chertkovsky	0.5000	0.4889	0.506	0.5000
Sholokhovskiy	0.6157	0.5160	0.551	0.5532

Table 3 ranks the Rostov Region in terms of agricultural sustainability.

Table 3. Ranking of districts of the Rostov region with linguistic recognition of assessment

G_3—"Close to sustainable development"

1.	Tsimlyansky	0.4544	23.	Dubovsky	0.5196
2.	Remontnensky	0.4584	24.	Zavetinsky	0.5207
3.	Krasnosulinsky	0.5000	25.	Millerovsky	0.5226
4.	October	0.5000	26.	Kamensky	0.5233
5.	Salsky	0.5000	27.	Aksay	0.5276
6.	Semikarakorsky	0.5000	28.	Egorlyksky	0.5283
7.	Советский	0.5000	29.	Kashar	0.5384
8.	Tarasovsky	0.5000	30.	Kagalnitsky	0.5387
9.	Chertkovsky	0.5000	31.	Tatsinsky	0.5389
10.	Veselovsky	0.5024	32.	Kuibyshevsky	0.5398
11.	Volgodonsk	0.5024	33.	Neklinovsky	0.5400
12.	Konstantinovsky	0.5054	34.	Peschanokopskiy	0.5444
13.	Martynovsky	0.5058	35.	Zernogradsky	0.5485
14.	Ust-Donetsk	0.5067	36.	Azov	0.55
15.	Oryol	0.5084	37.	Verkhnedonskoy	0.5524
16.	Morozovsky	0.5100	38.	Sholokhovskiy	0.5532
17.	Myasnikovsky	0.5138	39.	Proletarian	0.5563
18.	Rodionovo–Nesvetaysky	0.5144	40.	Matveevo–Kurgan	0.56
19.	Tselinsky	0.5150	41.	Bagaevsky	0.5832
20.	Belokalitvinsky	0.5174	42.	Zimovnikovsky	0.5952
21.	Milyutinka	0.5177	43.	Bokovsky	0.6000
22.	Oblivsky	0.5177			

It is seen that all areas of the Rostov region can be attributed to the term "development close to sustainable", and the last two regions can equally well be assigned to the fifth term "high level of sustainability". Thus, the developed general methodology has shown its applicability for assessing and ranking the sustainability of agricultural development of the regions of the region on the example of the Rostov region.

References

1. Antonova, M.A.: Theoretical and methodological foundations of the study of sustainable development of regions (In Russian). Soc. Polit. Econ. Law **4**, 113–119 (2013)
2. Zhurova, L.I., Toporkov, A.M.: A comparative analysis of approaches to assessing the sustainable development of economic systems (In Russian). Vestnik of VUiT. **4**, 1–13 (2017)
3. Tretyakova, E.A., Alferova, T.V., Pukhova, Yu.I.: Analysis of methodological tools for assessing the sustainable development of industrial enterprises (In Russian). Perm Univ. Herald. Econ. **4**(27), 133–139 (2015)
4. Ivanov, V.A., Ponomareva, A.S.: Methodological foundations of sustainable development of the agricultural sector (In Russian). Econ. Soc. Changes: Facts, Trends, Forecast. **4**, 109–131 (2011)
5. Martynov, K.P.: Methodology for assessing the sustainability of the development of the regional agricultural sector (In Russian). Theory Pract. Soc. Dev. **8**, 316–318 (2013)
6. Uskova, T.V.: Sustainable development management of a region: monograph (In Russian). ISERT RAS, Vologda (2009)
7. Bor, V.N.: The relationship of efficiency and sustainability in the development of small business (In Russian). Prod. Organ. **1**, 12–16 (2009)
8. Alekseychik, T.V., Bogachev, T.V., Karasev, D.N., Sakharova, L.V., Stryukov, M.B.: Fuzzy method of assessing the intensity of agricultural production on a set of criteria of the level of intensification and the level of economic efficiency of intensification. Adv. Intell. Syst. Comput. **896**, 790–798 (2019). https://doi.org/10.1007/978-3-030-04164-9_83
9. Vovchenko, N.G., Stryukov, M.B., Sakharova, L.V., Domakur, O.V.: Fuzzy-logic analysis of the state of the atmosphere in large cities of the industrial region on the example of Rostov Region. Adv. Intell. Syst. Comput. **896**, 709–715 (2019). https://doi.org/10.1007/978-3-030-04164-9_93
10. Albekov, A.U., Arapova, E.A., Karasev, D.N., Strukov, M.B., Sakharova, L.V.: A program for assessing the intensity of agricultural production through a fuzzy 5-point classifier. Federal Service for Intellectual Property. Computer Registration Certificate No. 2018613875
11. Arapova, E.A., Lukyanova, G.V., Sakharova, L.V., Akperov, G.I.: Fuzzy-logic analysis of the level of comfort and environmental well-being of the urban environment on the example of large cities of Rostov Region. Adv.Intell. Syst. Comput. **896**, 643–650 (2019). https://doi.org/10.1007/978-3-030-04164-9_84

A Technique for Aggregate Assessment of Agricultural Production Efficiency in a Region Based on the Theory of Fuzzy Sets

Galina A. Batishcheva[1]($\boxtimes$) (ID), Evgeniy N. Tishchenko[1] (ID),
Natalia S. Grigorieva[2] (ID), and Zhanna Ya. Kolycheva[2] (ID)

[1] Rostov State University of Economics, Rostov-on-Don, Russia
gbati@mail.ru, celt@inbox.ru
[2] Private Educational Institution of Higher Education «Southern University
(IMBL)», Rostov-on-Don, Russia
grigorievans@bk.ru, ruhnjak@mail.ru

Abstract. A methodology has been developed for a comprehensive assessment and ranking of the efficiency level of agricultural regions of the region based on indicators of two groups: the level of production intensification and the level of economic efficiency of intensification. As a mathematical apparatus for forming the assessment, systems of odd-logical conclusions—standard five-level [0.1] - classifiers—were used. The technique allows you to rank areas based on the estimates. The contribution of each of the considered indicators is estimated using a weight coefficient reflecting its significance. The technique was tested in 43 districts of the Rostov region.

Keywords: Agricultural production efficiency · A set of indicators · Fuzzy-logical conclusion

1 Introduction

As noted in [1], "efficiency" is the relative effect, the effectiveness of the process, operation, project; the ratio of the result to the costs, expenses that caused or ensured its receipt. It is indicated that it is advisable to consider the effectiveness at different levels of the organization of farming: at the level of an enterprise, region and country as a whole. At the same time, the article notes that at present there are no unified criteria for determining the efficiency of agriculture.

In the work [2] it is indicated that one should distinguish between the level of intensity or, in other words, the investments of means of production, and the economic efficiency of intensification. In the work [3] it is indicated that in modern conditions, the main criterion of effectiveness is the ability of self-financing of an agricultural organization to ensure expanded reproduction based on innovative processes. It is proposed to use a system of economic indicators, including achieved level of labor productivity, size of profit, and level of profitability in the whole economy, industries and types of products, the return on capital investment because of innovation ionic activity, the cost of production. In the articles [4, 5] it is argued that industry performance should reflect two components of a market economy: satisfied demand, providing efficient supply, when through an optimal supply option, while satisfying demand, highly efficient production is ensured. The resulting performance indicators of the agricultural industry are technological and economic characteristics.

Researcher [6] also considers agricultural production efficiency as a complex multi-purpose, open system consisting of functional and organizational subsystems. Functional subsystems (technological, social, environmental, and economic) express the content of agricultural production, while organizational ones express the form of their manifestation. To evaluate each subsystem, its own set of indicators is considered. In [7, 8], the choice of a system of criteria and indicators of agricultural production efficiency is substantiated in detail. In the work [9], net income, profit and profitability are considered as the main indicators of production efficiency.

In the work [10] it is noted that, in accordance with the classical theory, all indicators characterizing the level of effectiveness are divided into two groups. The first group - resource indicators of production efficiency - the ratio of the result of production to a specific resource or their combination (resource productivity, land productivity, capital productivity, labor productivity, etc.). They characterize the efficiency of the use of applied resources. The second group - costly indicators of production efficiency - the ratio of the result of production-to-production costs (consumed resources). This cost, material consumption, labor, profitability, etc. Thus, as the analysis shows, to assess the effectiveness of agricultural production, a system of disparate indicators is introduced, each of which characterizes the agricultural production of the region in a given aspect. There are practically no models in which integral estimates of the effectiveness of agricultural production or its subjects would be used.

We have developed a methodology for a comprehensive quantitative assessment of the intensity of agricultural production based on a combination of criteria of two groups: the level of intensification of production and the level of economic efficiency of intensification of production in agriculture [11, 12]. The scores for each group are calculated using five-level fuzzy classifiers. Manufacturers aggregate the values of all indicators, pay attention to each of the groups. The technique is implemented in the form of software that has passed state registration [13]. The above methodology is applied to the entire region (the Rostov Region) as a whole. It does not take into account the heterogeneous development of agricultural production in various regions of the region, the level and dynamics of production intensification and the economic efficiency of production intensification. However, the ranking of districts by the level and dynamics of agricultural production is key in the formation of agricultural management policies at the regional level.

Therefore, in this work, the presented methodology is developed and aimed at assessing and ranking agricultural production in the regions of the region. The technique was tested on statistical materials of the Rostov region taken from Federal State Statistics Service (Rosstat) [14].

2   Mathematical Apparatus and Experimental Material

The operation algorithm can be described as follows.

Stage 1. Selecting a scorecard for each of the two subgroups; indicators should refer to generally accepted systems of indicators used to assess each of the subgroups. In addition, they must meet the requirement of data completeness, that is, be without gaps and unified for all areas of the region. In addition, it is necessary to rank them in order of importance based on expert assessments. Indicators of the level of intensification of production (by district)

1. The presence of agricultural machinery in agricultural organizations at the end of the year, unit, value of the indicator for the year (weight 1/4, 2008-2018): 1.1. Tractors (without tractors on which earthmoving, reclamation, and other machines are mounted) (weight 1/6 inside the group - total $1/6 * 1/3 = 1/18$); 1.2. Combine harvesters; 1.3. Combine harvesters; 1.4. Forage harvesters; 1.5. Potato harvesters;1.6. Beet harvesting machines (without toppers).
2. Sown area of crops, hectare, the value of the indicator for the year (weight 1/4).
3. Livestock and poultry in households of rural settlements, head, indicator value for the year (weight 1/4): 3.1. Cattle; 3.2. Cows; 3.3. Pigs; 3.4. Sheeps; 3.5. Goats; 3.6. Bird; 3.7. Horses; 3.8. Rabbits; 3.9. Bee families.
4. Mineral fertilizers were applied (in terms of 100% of nutrients) for crops of agricultural crops in agricultural organizations, cwt (weight 1/4).

Indicators of the level of economic efficiency of intensification (by district) (weights 1/6, 2008–2018).

1. The volume of agricultural production (in actual prices), thousand rubles, the value of the indicator for the year, households of all categories.
2. The volume of crop production (in actual prices), thousand rubles, the value of the indicator for the year.
3. The volume of livestock production (in actual prices), thousand rubles, the value of the indicator for the year.
4. Index of agricultural production (in comparable prices; as a percentage of the previous year), percentage, value of the indicator for the year.
5. The index of crop production (in comparable prices; as a percentage of the previous year), percentage, value of the indicator for the year.
6. Livestock production index (in comparable prices; as a percentage of the previous year), percentage, value of the indicator for the year.

Stage 2. Calculation of normalized values of indicators of two subgroups. The values must be uniform, that is, they must belong to a unit segment. In this case, the formulas used are determined by the meaning of the indicators and the objectives of the study. For example, let the values of the indicator for N years be given: x_i $(i = 1,2, ..., m_1)$. At the same time, we assume that the values of the nearest time periods are more important for the study. Then, to form an estimate, we use the formula:

$$x_i = 0,5\left(1 + \sum_{i=1}^{N-1} k_i I_i\right) \qquad k_i = \frac{2(N-i)}{(N-1)N},$$

Designations here: k_i is Fishburne weight coefficients, numbering of time periods is carried out in reverse order, that is, 2017–2018 are taken for the first period, and the last, 11th, 2007–2008. The integer function I_i takes on the value "−1" if there is a negative increase in the i-th indicator; value "1" m if there is a positive increase; value "0" if there is stagnation, zero growth.

Stage 3. Setting linguistic variables. In Stage 2, we have defined the normalized values for the indicators. Let us compare them linguistic variables with a universal set in the form of a unit segment and term sets of five terms that have the following meaning: "Very low level of an indicator"; "Low rate"; "Average level indicator"; "High level indicator"; "A very high level indicator". The membership functions of linguistic variables are defined using standard trapezoidal functions [6].

We introduce the linguistic variables: g = "a comprehensive assessment of the intensity of agricultural production"; g_1 = "assessment of the level of intensification of production in agriculture"; g_2 = "assessment of the economic efficiency of the intensification of production in agriculture." The universal set of each linguistic variable is a unit segment. The set of values of all three variables g, g_1, g_2 consists of five terms: G = {G_1, G_2, G_3, G_4, G_5}, where G_1 is the "stable trend" to reduce growth"; G_2 – "growth tendency"; G_3 - "stagnation tendency"; G_4 - "upward trend"; G_5 – "steady growth trend". The accessory functions have a standard trapezoidal shape.

Stage 5. Aggregation of indicator values into the required estimates based on fuzzy five-level classifiers, in accordance with the standard algorithm.

Stage 6. Definition of terms of linguistic variables, analysis of the results obtained. The algorithm is organized in such a way that a decrease in the normalized values of indicators leads to a decrease in the final grade and vice versa. Therefore, it is quite easy to trace the indicators, the values of which are the lowest, which means they play a negative role in the final assessment. Consequently, to improve the situation, a set of measures is needed, leading to an increase in the corresponding values.

3 Computational Experiment

3.1 Assessment of the Level of Intensification of Production (by District)

An assessment of the level of intensification of agricultural production in the region is based on indicators for 11 years. Table 1 shows the results of the estimates for 43 districts of the Rostov Region, as well as their ranking with linguistic recognition of the estimate. 34 districts can be assigned to term G_3 – "tendency to stagnation", and the remaining 9 to term G_4 - "tendency to growth". There are no areas that could be attributed to the terms G_1 – "steady tendency to decrease growth", G_2 – "tendency to decrease growth", and G_5 - "steady tendency to growth".

Next, aggregation of estimates is carried out by municipal districts of the Rostov region, where the proportion of farmland of municipalities is used as weighting factors. As a result of the calculations, we obtain a comprehensive assessment of the level of intensification of agricultural production in the region based on an analysis of its municipalities. Thus, for the Rostov region $g_1 = 0.545$, which corresponds to the term G3 – "the tendency to stagnation."

Table 1. Ranking of districts of the Rostov region with linguistic recognition of assessment

G_1 – "stable trend to reduce growth"; noDistricts				
G_2 – "growth tendency"; noDistricts				
G_3 – "stagnation tendency"				
1.	Aksay	g = 0.4839	18. Myasnikovsky	g = 0.5202
2.	Belokalitvinsky	g = 0.5074	19. Oblivsky	g = 0.4963
3.	Bokovsky	g = 0.5035	20. October	g = 0.5297
4.	Verkhnedonskoy	g = 0.5226	21. Oryol	g = 0.4783
5.	Veselovsky	g = 0.5656	22. Peschanokopskiy	g = 0.5322
6.	Volgodonsk	g = 0.574	23. Proletarian	g = 0.5904
7.	Egorlyksky	g = 0.5766	24. Remontnensky	g = 0.5899
8.	Zavetinsky	g = 0.4705	25. Rodionovo-Nesvetaysky	g = 0.5348
9.	Zernogradsky	g = 0.5858	26. Salsky	g = 0.5220
10.	Kagalnitsky	g = 0.5034	27. Semikarakorsky	g = 0.4697
11.	Kamensky	g = 0.5355	28. Soviet	g = 0.4166
12.	Konstantinovsky	g = 0.4919	29. Tarasovsky	g = 0.5594
13.	Krasnosulinsky	g = 0.5345	30. Tatsinsky	g = 0.4999
14.	Matveevo-Kurgan	g = 0.5242	31. Ust-Donetsk	g = 0.5548
15.	Millerovsky	g = 0.4761	32. Tselinsky	g = 0.5601
16.	Milyutinka	g = 0.4817	33. Tsimlyansky	g = 0.5507
17.	Morozovsky	g = 0.4710	34. Chertkovsky	g = 0.5435
G_4 – "upward trend"				
35.	Azov	g = 0.6248	40. Kuibyshevsky	g = 0.6470
36.	Bagaevsky	g = 0.6037	41. Martynovsky	g = 0.6016
37.	Dubovsky	g = 0.6009	42. Neklinovsky	g = 0.6545
38.	Zimovnikovsky	g = 0.6453	43. Sholokhovskiy	g = 0.6157
39.	Kashar	g = 0.6164		
G_5 - "steady growth trend": no Districts				

3.2   Assessment of the Level of Intensification of Production (by District)

Table 2 shows the results of ranking the districts; it was found that one region can be attributed to the term G_2 = "a tendency to decrease in growth"; 35 regions to the term G_3 = "tendency to stagnate" and 7 regions to the term G_4 = "tendency to growth". No regions were found that could be attributed to the terms G_1 = "steady tendency to decrease growth" and G_5 = "steady tendency to growth".

A comprehensive assessment of the level of economic efficiency of intensification of agricultural production in the Rostov region, calculated on the basis of aggregation of the above estimates, is $g_2 = 0.533$, which corresponds to the term G_3 – "tendency to stagnation". The weight ratios here also include the share of agricultural land of municipalities.

Table 2. Ranking of districts of the Rostov region with linguistic recognition of assessment

G_1 – "stable trend to reduce growth"; no Districts				
G_2 – "growth tendency"				
1.	Kagalnitsky	$g = 0,3894$		
G_3 –"stagnation tendency"				
2.	Azov	$g = 0,5906$	20.	Myasnikovsky	$g = 0,5466$
3.	Aksay	$g = 0,4967$	21.	Oblivsky	$g = 0,4963$
4.	Belokalitvinsky	$g = 0,4995$	22.	October	$g = 0,5728$
5.	Verkhnedonskoy	$g = 0,5472$	23.	Oryol	$g = 0,5572$
6.	Veselovsky	$g = 0,4709$	24.	Peschanokopskiy	$g = 0,5327$
7.	Volgodonsk	$g = 0,5000$	25.	Proletarian	$g = 0,5852$
8.	Dubovsky	$g = 0,5108$	26.	Remontnensky	$g = 0,4894$
9.	Egorlyksky	$g = 0,5108$	27.	Rodionovo-Nesvetaysky	$g = 0,5906$
10.	Zavetinsky	$g = 0,4435$	28.	Salsky	$g = 0,5543$
11.	Zernogradsky	$g = 0,5834$	29.	Semikarakorsky	$g = 0,5251$
12.	Kashar	$g = 0,5574$	30.	Soviet	$g = 0,5858$
13.	Konstantinovsky	$g = 0,5311$	31.	Tarasovsky	$g = 0,5454$
14.	Krasnosulinsky	$g = 0,5685$	32.	Tselinsky	$g = 0,4772$
15.	Kuibyshevsky	$g = 0,5528$	33.	Tsimlyansky	$g = 0,4115$
16.	Martynovsky	$g = 0,5556$	34.	Chertkovsky	$g = 0,5288$
17.	Matveevo-Kurgan	$g = 0,5640$	35.	Neklinovsky	$g = 0,5486$
18.	Millerovsky	$g = 0,5484$	36.	Sholokhovskiy	$g = 0,6000$
19.	Milyutinka	$g = 0,4848$			
G_4 – "upward trend"				
37.	Bagaevsky	$g = 0,6037$	41.	Morozovsky	$g = 0,6060$
38.	Bokovsky	$g = 0,6842$	42.	Tatsinsky	$g = 0,6246$
39.	Zimovnikovsky	$g = 0,6046$	43.	Ust-Donetsk	$g = 0,6476$
40.	Kamensky	$g = 0,6204$			
G_5 - "steady growth trend": no Districts				

3.3 Assessment of the Level of Intensification of Production (by District)

The final assessment of the efficiency of agricultural production g = "a comprehensive assessment of the intensity of agricultural production" in the region was calculated based on the aggregation of the previously obtained estimates g_1 = "assessment of the level of intensification of production in agriculture"; g_2 = "assessment of the economic efficiency of intensification of production in agriculture" (Fig. 1).

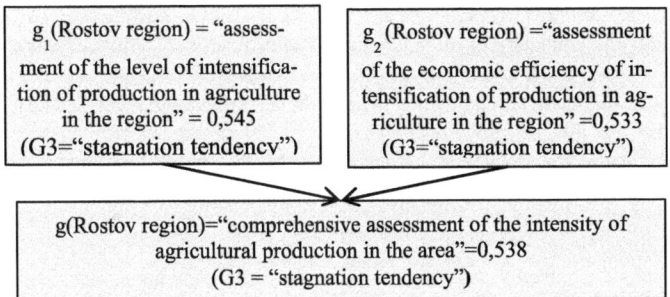

Fig. 1. Calculation of the final assessment of the effectiveness of agricultural production in the region.

It was found that g (Rostov region) = "a comprehensive assessment of the intensity of agricultural production in the region" = 0.538, which corresponds to the term G3 = "tendency to stagnation").

4   Conclusion

Thus, a technique has been developed that allows one to calculate quantitative and qualitative assessments of the intensity of agricultural production in the regions of the region on the basis of a set of significant indicators. At the same time, indicators can have different dimensions and characterize different areas of agricultural production. As a result of rationing based on the proposed formulas, the indicators are unified. First, their values belong to the unit segment; secondly, they are unidirectional, that is, an increase in their value leads to an increase in the final grade and vice versa. The values are aggregated into the final score using the standard fuzzy five-level classifier algorithm. For the assessment, linguistic recognition is performed in accordance with the initially defined term-set. The methodology allows ranking the regions of the region and analyzing which of them are outsiders and for what reason. Consequently, the technique allows us to formulate a set of measures aimed at leveling the level of intensity of agricultural production in the region.

References

1. Teterkina, A.M.: The essence of the efficiency of agricultural production. Econ. Probl. **5**, 35–142 (2005)
2. Tsatkhlanova, T.T.: Methodological aspects of assessing and improving the efficiency of agricultural production. UEkS **33**, 48–63 (2011)
3. Novichkova, O.V.: Actual problems of the efficiency of agricultural production. AGAU Bull. **11**, 101–125 (2012)
4. Volkova, E.A., Churilova, K.S., Shchegorets, A.A.: Methods and indicators of the economic efficiency of agricultural production. TSUE Bull. **1**(81), 24–38 (2017)

5. Volkova, E.A., Muratov, A.A., Tuaeva, E.V., Churilova, K.S., Ryzhkov, V.A.: Complex assessment of the efficiency of production and use of grain silage from cereal crops in dairy farming. Far Eastern Agrarian Bull. **3**(47), 74–85 (2018)
6. Umavov, Yu.D.: Economic efficiency of agricultural production: theoretical aspect. RPE **10** (48) (2014)
7. Sandu, I.S., Svobodina, V.A.: The effectiveness of agricultural production (guidelines). M.: FGBNU "Rosinformagrotech", 228 p. (2013)
8. Ermolenko, O.D.: Factors of increasing the efficiency of agricultural enterprises in Russia. Acc. Stat. **2**(42), 26–42 (2016)
9. Starchenko, I.V., Chabanny, A.A.: Theoretical foundations and indicators of the efficiency of agricultural production. In: Problems of the Modern Economy: Materials of the IV International Science Conference, Chelyabinsk: Two Komsomol Members, pp. 101–104 (2015)
10. Lobova, S.V., Ponkina, E.V.: The essence of efficiency in the context of current paradigms of economics and management. Econ. Anal. Theory Pract. **4**(355), 81–92 (2014)
11. Arapova, E.A., Lukyanova, G.V., Sakharova, L.V., Akperov. G.I.: Fuzzy-logic analysis of the level of comfort and environmental well-being of the urban environment on the example of large cities of Rostov region. In: ICAFS 2018, Advances in Intelligent Systems and Computing, vol. 896, pp. 643–650 (2018). https://doi.org/10.1007/978-3-030-04164-9_84
12. Vovchenko, N.G., Stryukov, M.B., Sakharova, L.V., Domokur, O.V.: Fuzzy-logic analysis of the state of the atmosphere in large cities of the industrial region on the example of Rostov region. Advances in Intelligent Systems and Computing, vol. 896, pp. 709–715 (2019). https://doi.org/10.1007/978-3-030-04164-9_93
13. Albekov, A.U., Arapova, E.A., Karasev, D.N., Strukov, M.B., Sakharova, L.V.: A program for assessing the intensity of agricultural production through a fuzzy 5-point classifier. Federal Service for Intellectual Property, Registration of Computer Program Certify, No. 2018613875 (2018)
14. Federal State Statistics Service (Rosstat). www.gks.ru/dbscripts/munst/munst60/DBInet.cgi. Accessed 03 Mar 2020

An Application of the Fuzzy VIKOR Method to an Investment Problem

Aynur I. Jabbarova[(✉)] [iD]

Azerbaijan State Economic University, Baku, Azerbaijan
stat_aynur@mail.ru

Abstract. In the information age, there is a growing need in application of efficient decision-making methods. In particular, it is related to investment decisions characterized by fuzzy information on decision criteria. In this paper we consider application of fuzzy VIKOR method to ranking investment alternatives under imprecise information.

Keywords: Investment problem · MCDM · Fuzzy VIKOR method

1 Introduction

It is difficult to assess the place and role of investments in the economy of any country. Only the inflow of investment resources allows the economic system to reach a new level of development [1].

Many problems in the field of investment have been studied mainly on the basis of optimization. This approach is not always the best, as social and economic conditions have a greater impact on the volume of investment. This is due to the fact that the problem must be clearly formulated in the context of impacts. It is more appropriate to use fuzzy programming to solve the problem [2].

The degree of determination is to rank the alternatives of preference in order. This is based on the criteria defined for each alternative involved. This practice is known as multiple-criteria decision-making (MCDM). There are many approaches to MCDM, one of which is known as VIKOR [3].

The VIKOR method has been developed to deal with MCDM issues with non-conflicting and incomparable criteria. It is noted that a compromise is possible to resolve conflicts. On the other hand, objective weights factors based on the Shannon concept of entropy can be used to regulate the subjective weights factors determined by decision makers [4].

2 Preliminaries

Definition 1. **Fuzzy sets [5].** Let X be a classical set of objects, called the universe, whose generic elements are denoted x. Membership in a classical subset A of X is often viewed as a characteristic function μA from X to $\{0,1\}$ such that

R. A. Aliev et al. (Eds.): ICAFS 2020, AISC 1306, pp. 665–671, 2021.
https://doi.org/10.1007/978-3-030-64058-3_82

$$\mu_A(x) = \begin{cases} 1 & \text{iff } x \in A \\ 0 & \text{iff } x \notin A \end{cases},$$

where $\{0,1\}$ is called a valuation set; 1 indicates membership while 0 non-membership.

If the valuation set is allowed to be in the real interval $[0,1]$, then A is called a fuzzy set [6–11]. $\mu_A(x)$ is the grade of membership of x in A

$$\mu_A : X \to [0, 1].$$

As closer the value of $\mu_A(x)$ is to 1, so much x belongs to A.

A is completely characterized by the set of pairs:

$$A = \{(x, \mu_A(x), \quad x \in X\}.$$

Definition 2 [5]. Arithmetic operations on triangular fuzzy numbers (TFNs) with linear LR-representation. TFN is completely represented by a triplet:

$$A = (a_1, a_2, a_3). \tag{1}$$

This can be expressed analytically as

$$\mu_A(x) = \begin{cases} 0 & \text{if } x \leq a_1 \\ (x - a_1)/(a_2 - a_1) & \text{if } a_1 \leq x \leq a_2 \\ (x - a_1)/(a_2 - a_1) & \text{if } a_3 \leq x \leq a_2 \\ 0 & \text{if } x \geq a_3 \end{cases} \tag{2}$$

Let two triangular fuzzy numbers A and B be given:

$$A = (a_1, a_2, a_3); \quad B = (b_1, b_2, b_3),$$

Then we define

$$A + B = (a_1 + b_1, a_2 + b_2, a_3 + b_3),$$
$$A - B = (a_1 - b_3, a_2 - b_2, a_3 - b_1),$$
$$A \cdot B \approx (a_1 \cdot b_1, a_2 \cdot b_2, a_3 \cdot b_3), \quad a_1, b_1 \geq 0,$$
$$k \cdot A = \begin{cases} (ka_1, ka_2, ka_3) & k > 0 \\ (ka_3, ka_2, ka_1) & k > 0, \\ (0, 0, 0) & k = 0 \end{cases}$$
$$A : B \approx (a_1/b_1, a_2/b_2, a_3/b_3), \quad a_1 \geq 0, b_1 > 0,$$
$$0 \notin \text{supp} A, \quad 0 \notin \text{supp} B.$$

3 An Investment Problem and Its Solution

A company plans to invest in various areas. Assume five areas are considered: agriculture, small business development, construction, tourism sector and transport. Any alternative is evaluated by four criteria: product quality, C_1, degree of risk, C_2, the amount of income generated C_3, and the level of environmental impact, C_4. Fuzzy decision matrix is shown in Table 1. The importance weights of criteria are $w_1 = (0.2, 0.25, 0.3)$, $w_2 = (0.15, 0.2, 0.25)$, $w_3 = (0.3, 0.35, 0.4)$, $w_4 = (0.05, 0.2, 0.35)$. We use the fuzzy VIKOR method [12] to solve the problem.

Table 1. Fuzzy decision matrix.

	Quality of product	Degree of risk	Volume of income	Environmental impact
Agriculture (f_1)	(7, 9, 9)	(7, 9, 9)	(5, 7, 9)	(3, 5, 7)
Development of small businesses (f_2)	(5, 7, 9)	(7, 9, 9)	(3, 5, 7)	(5, 7, 9)
Construction (f_3)	(5, 7, 9)	(3, 5, 7)	(5, 7, 9)	(5, 7, 9)
Tourism sector (f_4)	(3, 5, 7)	(3, 5, 7)	(7, 9, 9)	(5, 7, 9)
Transportation (f_5)	(5, 7, 9)	(5, 7, 9)	(5, 7, 9)	(5, 7, 9)

At the first step, we determine the best fuzzy value $f_i^* = (a_i^*, b_i^*, c_i^*)$ and the worst fuzzy value $f_i^o = (a_i^o, b_i^o, c_i^o)$ using the following equations [12]:

$$\tilde{f}_i^* = \max_j \tilde{f}_{ij}, \quad \tilde{f}_i^o = \min_j \tilde{f}_{ij}, \quad \text{for i} \in \text{B},$$

$$\tilde{f}_i^* = \min_j \tilde{f}_{ij}, \quad \tilde{f}_i^o = \max_j \tilde{f}_{ij}, \quad \text{for i} \in \text{C}.$$

where B is the set of benefit criteria and C is the set of cost criteria. The results obtained are given in Table 2.

Table 2. The best fuzzy value f_i^* and the worst fuzzy value f_i^o.

	Quality of product	Degree of risk	Volume of income	Environmental impact
f_i^*	(7, 9, 9)	(7, 9, 9)	(7, 9, 9)	(5, 7, 9)
f_i^o	(3, 5, 7)	(3, 5, 7)	(3, 5, 7)	(3, 5, 7)

At the second step, the fuzzy difference $\tilde{d}_{ij}$ is normalized using the equations given below:

$$\tilde{d}_{ij} = \left(\tilde{f}_i^* - x_{ij}\right) \Big/ \left(c_i^* - a_i^o\right) \quad \text{for } i \in B,$$

$$\tilde{d}_{ij} = \left(\tilde{x}_{ij} - f_j^*\right) \Big/ \left(c_i^o - a_i^*\right) \quad \text{for } i \in C.$$

The results obtained are given in Table 3.

Table 3. Normalization of fuzzy $\tilde{d}_{ij}$ difference

	Quality of product	Degree of risk	Volume of income	Environmental impact
(f_1)	(−0.33, 0, 0.33)	(−0.33, 0, 0.33)	(−0.33, 0.33, 0.67)	(−0.33, 0.33, 1)
(f_2)	(−0.33, 0.33, 0.67)	(−0.33, 0, 0.33)	(0, 0.67, 1)	(−0.67, 0, 0.67)
(f_3)	(−0.33, 0.33, 0.67)	(−0.33, 0.33, 0.67)	(−0.33, 0.33, 0.67)	(−0.67, 0, 0.67)
(f_4)	(0, 0.67, 1)	(0, 0.67, 1)	(−0.33, 0, 0.33)	(−0.67, 0, 0.67)
(f_5)	(−0.33, 0.33, 0.67)	(−0.33, 0.33, 0.67)	(−0.33, 0.33, 0.67)	(−0.67, 0, 0.67)

At the third step, separation measures calculations are carried out. The following equations are used in the calculation:

$$\tilde{S}_j = \sum_{i=1}^{5} (\tilde{w}_j \cdot \tilde{d}_{ij}),$$

$$\tilde{R}_j = \max_i \sum_{i=1}^{5} (\tilde{w}_j \cdot \tilde{d}_{ij}),$$

where $\tilde{S}_j = \left(S_j^a, S_j^b, S_j^c\right)$ is a fuzzy weighted sum of the separation measure of f_j from the best value f_i^*, $R_j = \left(R_j^a, R_j^b, R_j^c\right)$ refers to the separation measure of f_j from the best value f_i^o. The results obtained are given in Table 4.

Table 4. $\tilde{S}_j$ and $\tilde{R}_j$ separation measures

	$\tilde{S}_j$	$\tilde{R}_j$
(f_1)	(−0.23333, 0.183333, 0.8)	(−0.01667, 0.116667, 0.35)
(f_2)	(−0.15, 0.316667, 0.916667)	(0, 0.233333, 0.4)
(f_3)	(−0.2, 0.333333, 0.95)	(0, 0.133333, 0.266667)
(f_4)	(−0.13333, 0.3, 0.916667)	(0, 0.166667, 0.3)
(f_5)	(0.25, 0.266667, 0.866667)	(−0.03333, 0.116667, 0.266667)

At the fourth stage, $\tilde{Q}_j = \left(Q_j^a, Q_j^b, Q_j^c\right)$ the value is calculated by the following equation [12]:

$$\tilde{Q}_j = v\left(\tilde{S}_j - \tilde{S}^*\right) \Big/ \left(S^{oc} - S^{*a}\right) + v\left(\tilde{R}_j - \tilde{R}^*\right) \Big/ \left(R^{oc} - R^{*a}\right),$$

$\tilde{S}^* = \text{MIN}_j \tilde{S}_j = (-0.25, 0.183333, 0.8),$ $S^{oc} = \text{MAX}_j S_j^c = 0.95,$ $\tilde{R}^* = \text{MIN}_j \tilde{R}_j = (0.03333, 0.11667, 0.266667),$ $\tilde{R}^{oc} = \text{MAX}_j R_j^c = 0.4,$ v is a weight for the strategy of "majority criteria" (or "maximum utility"), where $1 - v$ represents the weight of the individual regret [13]. The obtained results are shown in Table 5.

Table 5. $\tilde{Q}_j$ value

	$\tilde{Q}_j$
(f_1)	(−0.75747863, 0, 0.879808)
(f_2)	(−0.703526, 0.190171, 0.986111)
(f_3)	(−0.72436, 0.254808, 0.846154)
(f_4)	(−0.69658, 0.106303, 0.870726)
(f_5)	(−0.78365, 0.034722, 0.811432)

At the fifth stage, we use the following well-known diffusion method to convert fuzzy numbers to crisp scores:

$$\text{Crisp}\left(\tilde{N}\right) = (2b + a + c)/4$$

The results are given in Table 6.

Table 6. Defuzzied values of $\tilde{S}_j, \tilde{R}_j, \tilde{Q}_j$

	$\tilde{S}_j$	$\tilde{R}_j$	$\tilde{Q}_j$
(f_1)	0,233333333	0,141667	0,030582
(f_2)	0,35	0,216667	0,165732
(f_3)	0,354166667	0,133333	0,157853
(f_4)	0,345833333	0,158333	0,096688
(f_5)	0,2875	0,116667	0,024306

At the sixth step, alternatives are sorted based on $\tilde{S}_j$, $\tilde{R}_j$, $\tilde{Q}_j$ values (Table 7).

Table 7. Sorting all alternatives

	$\tilde{S}_j$	$\tilde{R}_j$	$\tilde{Q}_j$
(f_1)	f_1	f_5	f_5
(f_2)	f_5	f_3	f_1
(f_3)	f_4	f_1	f_4
(f_4)	f_2	f_4	f_3
(f_5)	f_3	f_2	f_2

At the seventh step, the conditions C1 and C2 of fuzzy VIKOR method are verified:

Condition C1 is not satisfied. $f^1, f^2, \ldots, f^5$ are compromise solutions. Therefore, we calculate like this: $Q(f^{(5)} - f^{(1)}) = 0, 14 < \frac{1}{4} = 0.25$.

Condition C2 is satisfied.

4   Conclusion

In this paper we considered multi-criteria decision-making problem of investment under fuzzy information. The fuzzy VIKOR method has been used to solve the problem. The obtained results confirm the accuracy and correctness of the applied approach.

References

1. Polyakova, A.A.: Investment and its role in economic growth. Omsk Univ. Bull. Econ. Ser. **3**, 14–16 (2017). (in Russian)
2. Tsuda, H., Saito, S.: Application of fuzzy theory to the investment decision process. In: Lodwick, W.A., Kacprzyk, J. (eds). Fuzzy Optimization, STUDFUZZ, vol. 254, pp. 365–387
3. Musani, S., Jemain; A.A. Ranking schools' academic performance using a fuzzy VIKOR, J. Phys. Conf. Ser. **622**(1), 012036. https://doi.org/10.1088/1742-6596/622/1/012036
4. Shemshadi, A., Shirazi, H., Toreihi, M., Tarokh, M.J.: A fuzzy VIKOR method for supplier selection based on entropy measure for objective weighting. Expert Syst. Appl. **38**(10), 12160–12167 (2011). https://doi.org/10.1016/j.eswa.2011.03.027
5. Aliev, R.A., Aliev, R.R.: Soft Computing and Its Application. World Scientific, River Edge (2001)
6. Zadeh, L.A.: Fuzzy sets. Inf. Control **8**, 338–353 (1965)
7. Aliev, R.A., Mamedova, G.A., Aliev, R.R.: Fuzzy Sets Theory and Its Application. Tabriz University Press, Tabriz (1993)
8. Klir, G.J., Yuan, B.: Fuzzy Sets and Fuzzy Logic: Theory and Applications. PRT Prentice Hall, Upper Saddle River (1995)

9. Zimmermann, H.J.: Fuzzy Set Theory and Its Applications, 2nd revised edn. Kluwer Academic Publishers, Boston (1993)

10. Klir, G.J., Clair, U.S., Yuan, B.: Fuzzy Set Theory. Foundations and Applications. PTR Prentice Hall, Upper Saddle River (1997)

11. Aliev, R.A., Aliev, F.T., Babaev, M.D.: Fuzzy Process Control and Knowledge Engineering. Verlag TUV Rheinland, Koln (1991)

12. Afful-Dadzie, E., Nabareseh, S., Oplatková, Z. K.: Fuzzy VIKOR approach: evaluating quality of internet health information. In: Proceedings of the Federated Conference on Computer Science and Information Systems, vol. 2, pp. 183–190 (2014). https://doi.org/10.15439/2014F203

13. Opricovic, S.: Fuzzy VIKOR with an application to water resources planning. Expert Syst. Appl. **38**(10), 12983–12990 (2011). https://doi.org/10.1016/j.eswa.2011.04.097

Research of Optimal Production Modes of Plastic Details by Fuzzy Logic-Based Modelling

Djahid A. Kerimov[✉] [iD]

Azerbaijan State Oil and Industry University, Azadlig 34,
Nasimi, Baku, Azerbaijan
haciyevanaila64@gmail.com

Abstract. The determination of optimal manufacturing modes of some products requires experimental and theoretical studies, the volume of which increases significantly with the growth of the number of shapes and standard sizes. The organization of such studies with a view to development recommendations for the production and operation of these products can lead to a significant consumption of material and labor resources. We consider the use of fuzzy logic for dealing with imprecision of information related to product parameters.

Keywords: Optimal regimes · Plastic details · Modeling · Quality indicators · Reliability · Longevity imprecise information · Fuzzy logic

1 Introduction

In various industries the need arises for the manufacture many nomenclature products with a large complex of shapes and sizes. Each of these articles is characterized by a set of quality indicators which are formed in the manufacturing process and determine the initial quality of the product. Establishment of initial optimal level of product quality increases its reliability and durability during long-term operation. The initial quality level of the products mainly depends on the regimes of their manufacture. Therefore, often the establishment of the optimum quality level of the article is reduced to the determination of the optimal regime of its manufacture.

One effective way to reduce specified costs is to select and study the most typical representatives from a given set the many of products. However, with a large number of products and more than two compromising indicators of quality, this choice is not only difficult, but sometimes even impossible.

The choice of typical representatives, based on experience and intuition, in complex situations can lead to erroneous results in further studies. In connection with this, in this work proposes a method of formally analyzing and selecting the most suitable of a large number details of machine building in general and details of petroleum machine building in particular. Since this method operates on quantitative relationships and uses the basic characteristic of the system of quality indicators of the studied detail objects, then the results of the research will sufficiently reflect the quality of uniform groups of objects.

© The Author(s), under exclusive license to Springer Nature Switzerland AG 2021
R. A. Aliev et al. (Eds.): ICAFS 2020, AISC 1306, pp. 672–677, 2021.
https://doi.org/10.1007/978-3-030-64058-3_83

The setting of the task can be formulated as follows.

The many objects have been given $X = (X_1, X_2, \ldots, X_m)$. The m properties of each object $X_i = (X_{i1}, X_{i2}, \ldots, X_{im})$ are known. It is required to divide the many objects into some number of groups so that each group includes the objects with the similar properties, and determine the representatives of groups that characterize the remaining objects of each homogeneous group. We consider using fuzzy logic to account for imprecision of information related to the properties.

2   The Study of the Optimal Manufacturing Regimes

In order to solve this task, we will use quantitative characteristics in the form of a measure of proximity, criterion of optimality of division and criterion of selection of a typical representative. When grouping a certain type of products with several quality indicators corresponding measure of proximity is the distance between any pair of objects of the original the many [1]. It is obvious that if a distance function is given $\rho = (X_1, X_i)$ for the many X, then objects close in the sense of this metric are uniform and belong to the same group.

As a measure of the distance between the details under imprecise information, it is proposed to use a fuzzy weighted Euclidean distance

$$\tilde{\rho}_{i1} = \sqrt{\sum_{k=1}^{m} \tilde{\omega}_k^2 (\tilde{X}_{ik} - \tilde{X}_{jk})^2}, \tag{1}$$

where $\tilde{X}_{ik}$ – is the fuzzy value of the k - th quality indicator of the i - th object; $\tilde{\omega}_k$ – fuzzy weighting coefficient (the degree of importance of the k - th quality indicator). The basic ω_k values within $\tilde{\omega}_k$ satisfy:

$$\sum_{k=1}^{m} \omega_k = 1, \qquad \omega_k > 0. \tag{2}$$

The determination of weighting coefficients $\tilde{\omega}_k$ by information only contained in the source data $(\tilde{X}_{ik})$ does not always produce the desired effect. Therefore, expert evaluation methods can be used to solve the task.

When ranking quality indicators by experts to calculate weighting coefficients, the following formula is most convenient [2]:

$$k = v_1 + \frac{Y_k - Y_1}{Y_m - Y_1}(v_m - v_1). \tag{3}$$

in which – the total rank of the k - the feature; Y_1 – the total rank of the least important feature; Y_m – the total rank of the most important feature; v_1 – the assigned weight of the least important feature; v_m – the assigned weight of the most important feature.

However, the weights calculated by formula (3) do not satisfy condition (2), therefore their it is necessary normalize by the formula

$$\omega_k = \frac{\upsilon_k}{\sum\limits_{k=1}^{m} -\upsilon_k}. \tag{4}$$

As a criterion for the optimality of grouping of objects, such criterion should be used, that is based on measuring the distance (1) and ensures consistent optimality of the grouping procedure, since the number of groups is not specified. Based on that it is possible to use average distance between elements of uniform group - the intragroup distance or average grouped of distance between objects of different groups, i.e. intergroup distance [3]. Given simplicity, we use the criterion of dividing by minimum the average grouped of distance

$$\rho = m_v in \left\{ \vec{\rho}_v = \sum_{S=1}^{\rho} \bar{\rho}_{vS} \right\}. \tag{5}$$

Here $\bar{\rho}_{vS}$ – the average distance between the objects of the F_S group, determined by the formula

$$\bar{\rho}_{vS} = \frac{2}{l(l-1)} \sum_{\substack{i,j=1 \\ i \subset j}}^{t} \rho_{tji}; \quad i,j \in F_S, \ S = \overline{1,\rho}, \tag{6}$$

here v – level number of the partitioning into groups; S – groups number of received at the v - the level; ρ – groups number of received at the v - the level; l – the objects number in the S - the group.

The use of the criterion (5) enables to implement a hierarchical grouping procedure, which is characterized by clarity of results and gives a complete analysis of the structure of the studied the many of objects. The hierarchical grouping procedure considered here is based on the search of distance matrix elements and consists of sequentially combining the groups of elements first closest and then increasingly distant from each other. Although some hierarchical procedures differ in the cumbersomeness of their computational implementation, at the objects number $n < 150$ the amount of machine memory and implementation time required by these algorithms are not high and with a small number of objects $(n \leq 30)$ they can be realized even manually.

The algorithm of the hierarchical sequential combining of objects is as follows.

In the first level of the hierarchical tree, each X_1 object is considered as a separate group. Next, at each level, the two closest groups merge. Work of an algorithm comes to an end when all initial objects are united in one group [1, 3].

In connection with the use in this work of the proximity measure in the form of distance, we construct matrix by similarity of objects as follows

$$R = \|\rho_{ij}\|j \quad (i,j = 1,2,\ldots,n). \tag{7}$$

Since $\rho_{ij} = \rho_{ji}$ at $i \neq j$ and $\rho_{ij} = 0$ at $i = j$, then matrix (7) will be triangular with zero elements on the main diagonal. The minimum element of matrix R is denoted by C_1, the second minimal element after C_1 – through C_2, etc. Let's make the increasing row [3] out of C_1 elements

$$c_1 < c_2 < \ldots < c_t \tag{8}$$

Where

$$t = \frac{1}{2}n(n-1) - 1.$$

The many pairs of objects corresponding to the members of a row (8) is denoted by

$$A_{ij}^1, A_{ij}^2, \ldots, A_{ij}^*. \tag{9}$$

Consider two pairs of objects linked if:

a) these pairs have a common object;
b) there is a pair A_{ij}^* with which both the first and second pairs have a common object.

Considering a pair A_{ij}^1 at the first step, we determine the composition of the first group formed at the first level of the hierarchical tree, i.e. the group G_n. Next we consider the pair A_{ij}^2. If the pair A_{ij}^2 turns out to be associated with A_{ij}^1, then we include the objects A_{ij}^2 in the group G_n and instead get a G_{21} with a large number of elements. If the pair A_{ij}^2 turns out to be unrelated to the pair A_{ij}^1, then from the objects of the pair A_{ij}^2 we'll form a new group G_{22} and accordingly we'll get the groups of the second level $G_{21} = G_n$ and G_{22}.

Continuing, thus, at some level $v < n$ of hierarchical division, we get merging all the objects into a single group $G_{v1}(X_1, X_2,\ldots, X_n)$. Usually this number of such levels is small compared to the number of objects n. After obtaining the final number of partitioning, starting with the second, for each partitioning, we calculate the values $\bar{\rho}_{vS}, \rho_v$ and determine the optimal partitioning.

Let some number of groups $G_{v1}, G_{v2}, \ldots, G_{v\rho}$ defined. From each group, it is necessary to choose an object that is most characteristic for analyzing the quality of all objects in a given group. If the representatives of the obtained groups are investigating for design the optimum manufacturing mode of each set of products, it is necessary the main quality indicator to be improved should be used, and it is possible to choose the representative with the worst quality indicator.

The fact is that while ensuring the optimum quality of such a product, the remaining homogeneous products in the same mode will all the more satisfy the optimality condition. The procedure for selecting a typical representative can be formalized as follows:

$$n = \max_i \{n_i\}, \quad i \in \Gamma_{vS}; \quad S \in \{1, \rho\}, \tag{10}$$

if the main indicator subject to minimized;

$$n = \min_i \{n_i\}, \quad i \in F_{vS}; \quad S \in \{1, \rho\}, \tag{11}$$

if the main indicator subject to maximized.

If n > 10, than expediently solve the task on a computer.

3 Discussion and Conclusions

The above-stated technique has been applied to the decision of a problem of grouping and a choice of typical representatives of details of oil mechanical engineering at considerable quantity of forms and the sizes $n = 1, 2, ..., v$, that has allowed to design optimum modes of pressing and casting under pressure at relatively lower costs.

The task was modeled in an algorithmic language and implemented on a computer.

The obtained results confirmed the reliability and effectiveness of the proposed method.

When solving a specific problem, the averaged relative values of the following details quality indicators were taken as initial data: X_{i1}; X_{i2}; X_{i3}; X_{i4}.

On the basis of expert ranking and analysis of these indicators, the following values of weighting coefficients were obtained:

$$\omega_1 = \frac{11}{S_2}; \quad \omega_2 = \frac{4}{13}; \quad \omega_3 = \frac{4}{13}; \quad \omega_1 = \frac{5}{13}; \quad \omega_5 = \frac{11}{13}.$$

For the convenience of presentation of the results as an example we will give of grouping data for ten details.

The minimum value of the partition criterion is obtained at the sixth step of the grouping procedure

$$\min \rho_v = \rho_{6.1} + \rho_{6.2} = 0,156 + 0,099 = 0,255.$$

Consequently, at this step, the best split is obtained with the following compositions of groups:

$$G_1 = \{3, 4, 5, 6, 7, 8, 10\}; \quad G_2 = \{6, 9\}; \quad G_3 = \{1\}; \quad G_4 = \{2\}.$$

Such result is in good accord with the data of physical analysis of ten details and the results of numerous researches of groups of authors [4, 5].

Since the G_3 и G_4 groups consist of only one element, the selection criteria the representative is considered for the G_1 and G_2 groups. In this example, shrinkage of parts plays a dominant role and has the largest value of weight coefficient; therefore, when determining the optimal mode, it is advisable to minimize shrinkage of parts with a restriction of other quality indicators. Consequently, to select representatives, we use criterion (10) and select parts with maximum shrinkage $i = 7$ from group G_1 and $i = 9$ from group G_2.

As it can be seen from the given example, instead of geometry of ten details it is enough to study details 1 and 2 which made up separate groups and geometry of details 7 and 9 from groups G1 and G2, i.e. just four details. Consequently, the number of details to be further investigated has decreased by 2,5 times.

Obviously, with the rise in the source number of details, the effectiveness of the proposed approach becomes more significant.

References

1. Beshelev, S.D., Guvich, F.G.: Mathematic-Static Methods of Expert Estimations. Statistics, Moscow (1974)
2. Babanly, A.Yu.: Grouping correlated data. In: Reports of a Scientific and Technical Conference on the Results of Scientific and Research Works for 1968–1969. Moscow Power Engineering Institute (1970)
3. Kerimov, D.A.: Scientific bases and practical methods of optimization of quality indicators of plastic details of oil field equipment. Abstract of the dissertation of Doctor of Technical Sciences. Azerbaijan Institute of Oil and Chemistry named after M. Azizbekov, Baku (1985)
4. Kerimov, D.A., Mustafayev, A.D.: Grouping of plastic parts according to the criteria of quality indicators. In: Abstracts of a Scientific and Technical Conference. The Development of Oil Mechanical Engineering in Azerbaijan in the X five-year plan – Baku (1978)
5. Kerimov, D.A., Gasanova, N.A.: Development of optimal regimes of cooling for plastic parts of thermosets. SAEQ Sci. Appl. Eng. Q. 17–19 (2017)

Mathematical Model of Supply and Demand Management

Alexander V. Bratishchev[1], Galina A. Batishcheva[2(✉)] [ID],
Maria I. Zhuravleva[2], and Guzenko Natalia[2]

[1] Don State Technical University, Rostov-on-Don, Russia
avbratishchev@spark-mail.ru
[2] Rostov State University of Economics, Rostov-on-Don, Russia
gbati@mail.ru, zhurmari@mail.ru, musamav@mail.ru

Abstract. The article provides a complete bifurcation analysis of the mathematical model of the dynamic system "Supply and demand" proposed by V.P. Milovanov. The behavior of trajectories at infinity is studied using the Poincare transform. All phase portraits of the system were obtained using theoretical analysis and numerical experiment in the Matlab+Simulink package. The system was rough in the open first quarter of the phase plane. A system of additive control of both monetary and commodity flows is constructed by analytical design of aggregated regulators to achieve a given dynamic equilibrium from an arbitrary initial state. A class of acceptable achievable States is highlighted. The numerical experiment shows the stability of this state in General. This model allows you to predict the development of the process for any pre-set initial state of the system, as well as manage the system parameters for the design of a pre-set dynamic equilibrium.

Keywords: Cash and commodity flows · Autonomous system · Equilibrium state · Phase portrait · Stability · Aggregated variable · Invariant variety · Synergetic regulator

1 Introduction

In the monograph [1] two models of supply and demand dynamics are proposed in the case when demand adjusts to supply, and supply (at constant prices) tends to provide effective demand. This article considers a model of the type

$$\begin{cases} x_t' = -a_1 x + a_2 y - a_3 x^2 - a_4 xy =: f_1(x, y) \\ y_t' = b_1 x - b_2 xy =: f_2(x, y) \end{cases} \tag{1}$$

Here x(t) - is the quantity of the product at the time t the manufacturer has (the manufacturer's offer), y(t) - the amount of money available to the consumer for purchasing the product x. Four factors influence the rate of change in the offer in different directions: the moral and physical wear of the product $-a_1 x$, the availability of money from the consumer $a_2 y$, the deterioration of the quality of the product released over time $-a_3 x^2$, and the product's sell-out rate $-a_4 xy$. The speed of demand changes is influenced by two factors: the quantity of the product from the manufacturer $b_1 x$ and the sales of the product $-b_2 xy$. Parameters are assumed to be arbitrary positive numbers.

We study the model using the methods of the theory of bifurcation analysis [2].

2 Analysis of Final Equilibrium States

In economic terms, we will be interested in the first quarter of the phase plane. It is easy to see that the system has three States of equilibrium

$$
S_1 = (0,0), \quad S_{2,3} = \left(\frac{-(a_1 b_2 + a_4 b_3) \pm \sqrt{(a_1 b_2 + a_4 b_3)^2 + 4a_2 b_1 a_3 b_2}}{2a_3 b_2}, \frac{b_1}{b_2} \right)
$$

The state S_2 is in the first, and the state S_3 – is in the fourth quarter of the phase plane. Using Lyapunov's first approximation stability theorem [3], it is established that regardless of the parameter values, the state S_1 is always a saddle and a rough equilibrium state, the state S_2 is a rough stable node, and the state S_3 – is a rough unstable node. This result is confirmed numerically by constructing a phase portrait in the vicinity of these States for specific parameter values using the Matlab+Simulink mathematical package (Fig. 1).

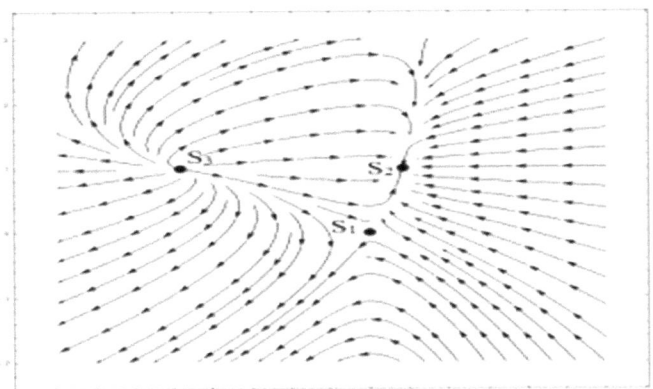

Fig. 1. Fragment of the phase plane of the system (1).

Conclusion. It is known [4] that a quadratic system (1) can only have limit cycles surrounding a single singular point that is a focus, as well as no more than two foci in the finite part of the plane. Therefore, the system in question has no cycles.

3 Analysis of Equilibrium States at Infinity

The construction of the scheme [2] of a polynomial Autonomous system (1) implies the study of equilibrium States at infinity. The latter is performed using the Poincare transform.

To determine the nature of possible equilibrium States ($\pm\infty$, 0), such a Poincare transform is used [2] $x = \frac{1}{u}$, $y = \frac{v}{u}$. The system (1) then switches to the system (2):

$$\begin{cases} u_t = a_3 u + a_1 u^2 + a_4 uv - a_2 u^2 v \\ v_t' = b_1 u + (a_3 - b_2)v + a_1 uv + a_4 v^2 - a_2 uv^2 \end{cases} \qquad (2)$$

Equating the right parts to zero, we find two equilibrium States $S_1^0 = (0,0)$ and $S_2^0 = \left(0, \frac{b_2-b_3}{a_4}\right)$, and they merge when $b_2 = a_3$.

It follows from Lyapunov's first approximation stability theorem that in the case $b_2 < a_3$ of the state $S_1^0 = (0,0)$ is an unstable node, in the case $b_2 > a_3$ it will be a saddle. Similarly, in the case $b_2 < a_3$ of the second state $S_2^0 = \left(0, \frac{b_2-a_3}{a_4}\right)$ will be a saddle, and in the case $b_2 > a_3$ of it will be an unstable node.

Separately, we consider the case of a multiple state of equilibrium when $b_2 = a_3$. In this case, the system (2) must be reduced to a canonical form and use the multiple equilibrium position theorem [2]. This condition turns out to be an unstable saddle node.

To determine the nature of possible equilibrium States $(0, \pm\infty)$, the Poincare transform [2] of the form is used $x = \frac{v}{u}$, $y = \frac{1}{u}$. The system (1) then switches to the system (3):

$$\begin{cases} u_t' = b_2 uv - b_1 u^2 v \\ v_t' = a_2 u - a_4 v - a_1 uv + (b_2 - a_3)v^2 - b_1 uv^2 \end{cases} \qquad (3)$$

Equating the right parts to zero, we find 2 equilibrium States $S_3^0 = (0,0)$ and $S_4^0 = \left(0, \frac{a_4}{b_2-a_3}\right)$.

Conclusion. The equilibrium States S_2^0 and S_4^0 define and are defined by the same equilibrium state at infinity of the Autonomous system (1). The equilibrium State $S_3^0 = (0,0)$ is a multiple. The same method establishes that it will be a stable saddle node.

4 Complete Bifurcation Analysis of the System

Let's go to the full description of the phase portraits of the system (1). Recall that the scheme is called information about the nature of equilibrium States, limit cycles, and the course of separatris of the phase portrait of the Autonomous system [2]. The diagram allows you to get a schematic image of all phase portraits. For system (1), there are 3 of them, depending on the values of the parameters a_3 and b_2. The use of the Matlab+Simulink package was necessary in order to use numerical experiments to understand which States connect separatrises. As a result, we obtained such a schematic partition of phase spaces into elementary cells that are invariant sets (Figs. 2, 3 and 4).

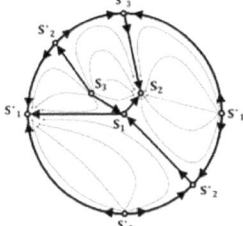 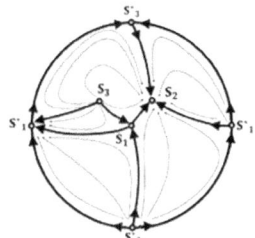 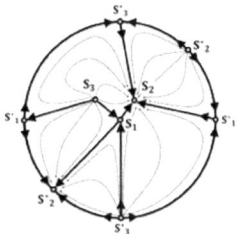

Fig. 2. Phase portrait, $a_3 > b_2$.

Fig. 3. Phase portrait, $a_3 = b_2$.

Fig. 4. Phase portrait, $a_3 < b_2$.

Conclusion. All trajectories of each such cell behave in the same way asymptotically. This knowledge allows us to qualitatively describe the development of the economic two-dimensional process in question in time according to its initial state.

5 Building a Synergetic Regulator of the System (1)

Synergetic control of interrelated processes of supply and demand is [5, 7, 8] to make this two-dimensional process change according to a certain law, for example, to stabilize in a predetermined state, regardless of its initial state. This is achieved by using additive control of the rate of change of any of these processes:

$$\begin{cases} x'_t = -a_1 x + a_2 y - a_3 x^2 - a_4 xy + u_1(x, y) \\ y'_t = b_1 x - b_2 xy \end{cases}, \quad \begin{cases} x'_t = -a_1 x + a_2 y - a_3 x^2 - a_4 xy \\ y'_t = b_1 x - b_2 xy + u_2(x, y) \end{cases}$$

(4)

The control function $u_i(x, y)$ is defined by the function (aggregated variable) $\psi_i(x, y)$, which sets the invariant set $\psi_i(x, y) = 0$, of the projected controlled system, also called a synergetic regulator. This set is assumed to be attractive in the sense that the aggregate variable $\psi_i(x(t), y(t))$, tends to zero on the trajectories of the projected controller. Therefore, the equilibrium States of the regulator must lie at the intersection of the curves $\psi_i(x, y) = 0$ and $f_j(x, y) = 0$, $i \neq j$. That is the choice of an aggregated variable is subject to the requirement of transversality.

Let the condition of attraction be represented as a differential equation with a stable General equilibrium state 0, for example

$$(\psi(x(t), y(t)))'_t = -\frac{1}{T} \psi(x(t), y(t)),$$

where the value $T > 0$ determines the rate of approach $\psi(x(t), y(t))$ to zero [5]. Then differentiating this equation, we get the missing equation of the regulator.

Since the required state must also lie on the curve defined by the unchangeable right side of the source system, only the points of the last curve can be potential terminal States. This means that to solve the problem, you must also vary the parameters of the right part of the a_i (or b_i) of the source system. It is often possible to change one or more parameters to cover the entire first quarter with graphs of the corresponding curves (the points of which make economic sense). If it is possible to prove that any terminal state is stable, then we can assume that the problem of synergetic control of the system is solved. is imposed on the selection of an aggregated variable. Let's apply these considerations to our system.

First, we will control the rate of change x(t). The Terminal States must lie on the curve $b_1x - b_2xy = 0$. Limited to non-zero States, we have a horizontal semi-direct $y = b_1/b_2$ in the first quarter. You need to select an aggregated variable with a parameter so that when the latter changes, all the points of the semidirect intersect with the corresponding curves, and so that the resulting terminal States of the controller are stable. The simplest family has the form $\psi = y - ax$. On the trajectories of the projected controller, it must satisfy the equation $(\psi(x(t), y(t)))'_t = -\frac{1}{T}\psi(x(t), y(t))$, where the value $T > 0$ determines the speed of convergence of the trajectories of the controller with the manifold $y - ax = 0$ Differentiate this expression, replace y'_t with $f_2(x, y)$, and resolve it relatively x'_t. Receive $x'_t = \frac{-1}{\psi'_t}\left(\psi'_y f_2(x, y) + \frac{1}{T}\psi\right)$. Combining this equation with the first equation of the system (1), we obtain the following equation:

$$\begin{cases} x'_t = \frac{1}{\alpha}\left(b_1x - b_2xy + \frac{1}{T}(y - \alpha x)\right) \\ y'_t = b_1x - b_2xy \end{cases}$$

By resolving the system

$$\begin{cases} b_1x - b_2xy = 0 \\ y - ax = 0 \end{cases},$$

we obtain the following non-zero equilibrium state $S = \left(\frac{b_1}{ab_2}, \frac{b_1}{b_2}\right)$. Obviously, when $a > 0$ all the points of the semi-direct $y = \frac{b_1}{b_2}$ from the first quarter are run. Let's check the stability condition of this state [6].

$$f'_{2y} - \frac{\psi'_y}{\psi'_x}f'_{2x} = -b_2x + \frac{1}{\alpha}(b_1 - b_2y)\Big|_S = -\frac{b_1}{\alpha} < 0$$

It is performed when $a > 0$.

We verified the last result by designing the S-model of the regulator (2) and constructing $b_1 = b_2 = a = 1$ a phase portrait of the regulator in the vicinity of its equilibrium state (Fig. 5). It can be seen that all the trajectories of the phase portrait shrink to the terminal state (1, 1).

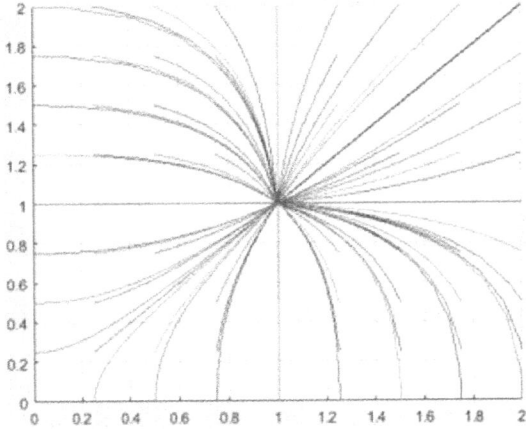

Fig. 5. Fragment of a phase portrait in the vicinity of the terminal state.

Note: So, any valid terminal state has the form $= \left(\frac{b_1}{ab_2}, \frac{b_1}{b_2} \right), a > 0.$

Therefore, if we want the pre-set point (x_0, y_0) of the first quarter to be the terminal state of the controller with an aggregated variable $\psi = y - ax$, we must first select b_1, b_2 such that $\frac{b_1}{b_2} = y_0$, and then select the parameter $a := \frac{b_1}{b_2 x_0}$.

Let's now find an additive control of the variable's rate of change $y(t)$. The hyperbola graph $-a_1 x + a_2 y - a_3 x^2 - a_4 xy = 0$ suggests this choice of an aggregated variable $\psi := a_3 x + a_4 y - a, a > 0$, since the family of corresponding curves is transversal to this curve. Similarly, we obtain the following equation of the regulator

$$\begin{cases} x'_t = -a_1 x + a_2 y - a_3 x^2 - a_4 xy \\ y'_t = \frac{-1}{a_4} \left(a_3(-a_1 x + a_2 y - a_3 x^2 - a_4 xy) + \frac{1}{T}(a_3 x + a_4 y - \alpha) \right) \end{cases}$$

The equilibrium state of the latter is equal to

$$S = \left(\frac{\alpha a_2}{\alpha a_4 + a_1 a_4 + a_2 a_3}, \frac{\alpha(\alpha + a_1)}{\alpha a_4 + a_1 a_4 + a_2 a_3} \right)$$

Let's check the stability condition of this state.

$$f'_{1x} - \frac{\psi'_x}{\psi'_y} f'_{1y} \bigg|_S = -a_1 - \frac{a_2 a_3}{a_4} - \alpha < 0$$

6 Conclusion

1) Thus, the terminal state of the regulator is stable at $a > -a_1 - \frac{a_2 a_3}{a_4}$. In terms of economic meaning, we are interested in values $a > 0$.

2) If we want the pre-set point of the first quarter (x_0, y_0) to be the terminal state of the controller, we first select such positive parameter values a_i that the equality is fulfilled $a_2 y_0 = a_1 x_0 + a_3 x_0^2 - a_4 x_0 y_0$. And then we create an aggregated variable $\psi := a_3 x + a_4 y - a_0$, where $a_0 := -a_1 x_0 + a_2 y_0$ (or from the previous equality follows $a_0 := a_3 x_0^2 + a_4 x_0 y_0$).

The stability of this state is also checked by numerical experiments.

References

1. Milovanov, V.P.: Non-Equilibrium socio-economic systems: synergetics and self-organization, 265 pp. Editorial URSS, Moscow (2001)
2. Bautin, N.N., Leontovich, E.A.: Methods and techniques of qualitative research of dynamical systems on the plane, 496 pp. Nauka, Moscow (1976)
3. Malkin, I.G.: Theory of motion stability. Moscow: Nauka, 532 pp. (1966)
4. YeYangian: Transl. AMS. Providence. RhodeIsland **66** (1986)
5. Kolesnikov, A.A.: Synergetic methods of managing complex systems. The theory of system analysis. M.: Komkniga, 240 p. (2006)
6. Bratshchev, A.: On the characteristic polynomial of balance an autonomous system that has an attractive invariant manifold. Dif. Eqn. Control Process. **2**, 15–23 (2017)
7. Bratishchev, A.V., Batishcheva, G.A., Zhuravleva, M.I.: Bifurcation analysis and synergetic management of the dynamic system "Intermediary activity". Advances in Intelligent Systems and Computing, vol. 896, pp. 659–667. In: 13th International Conference on Application of Fuzzy Systems Soft Computing, ICAFS 2018 (2019)
8. Bratishchev, A.V., Batishcheva, G.A., Denisov, M.Y., Zhuravleva, M.I.: Bifurcation analysis and synergetic control of a dynamic system with several parameters. Advances in Intelligent Systems and Computing, vol. 1095, pp. 639–646. In: 10th International Conference on Theory Application of Soft Computing, Computing with Words and Perceptions - ICSCCW-2019; Prague, Czech Republic (2020)

Investment Projects Portfolio Analyses Using Fuzzy Evaluation Methods

Adila Ali$^{(\boxtimes)}$

Institute of Control Systems of ANAS, Vahabzadeh Str. 9,
AZ1141 Baku, Azerbaijan
aliadelae@gmail.com

Abstract. The paper considers approaches to solving the problem of multi-criteria assessment of an investment proposals portfolio in the presence of information on financial indicators of projects from reliable sources. Within the framework of the proposed approaches and the formed set of financial indicators, a comprehensive methodology for comparative analysis and selection of investment decisions is proposed, which is based on the application of the principles of Pareto and Board, as well as the fuzzy maximin convolution method under equal (in important) evaluation criteria, and the fuzzy inference system. When using fuzzy methods of analysis, each of the financial indicators is considered as a qualitative criterion for evaluating the effectiveness of an investment project, which is interpreted by suitable fuzzy sets. Based on the results of the calculations, the ranking of hypothetical alternative investment projects was carried out using the considered multi-criteria evaluation methods and their comparative analysis was carried out, which ultimately allows financing the most effective projects.

Keywords: Investment project · Financial evaluation indicator · Pairwise comparison of alternatives · Fuzzy set · Membership function · Fuzzy reference

1 Introduction

Evaluation and ranking of investment decisions are important stages both in the selection and modeling of investment portfolio. If alternative investment projects pursue similar goals, then each project must be analyzed separately to determine the effect of its implementation regardless of other investment projects. Nevertheless, a comparative analysis of alternative investment projects allows to provide financing for the most effective of them. In [1] a comprehensive approach to such analysis and selection of investment projects was considered. This approach involves the following procedures: (1) forming of a metrics for multicriteria evaluation and comparative analysis of alternative projects; (2) justification for the selection of projects using analysis techniques; (3) establishing a criterion in order to form an optimal investment projects portfolio.

In most cases, the effectiveness of investment projects is assessed by the following financial indicators: x_1 – Net Present Value (NPV); x_2 – Profitability Index (PI); x_3 – Internal Rate of Return (IRR); x_4 – Payback Period (PP); x_5 – Return On Investment (ROI). At that, the method selection for evaluating the investment projects and forming an investment portfolio is determined by concrete aim of the investor. However, existing methods for evaluating and selecting of investment projects do not always adequately reflect the values of separate indicators, which in a sense are qualitative (weakly structured) evaluation criteria. The best way to describe them is appropriate verbal reproduction, or better, fuzzy set. Actually, this paradigm formed the basis of this paper, which allows the use the fuzzy approaches to multi-criteria evaluation of projects in the presence of weakly structured data of indicators x_i ($i = 1 \div 5$).

2 Problem Definition

Suppose that ten investment projects a_k ($k = 1 \div 10$) are considered by investor for the selection of the best and/or selection of the best of them. In the framework of metrics of the comparative estimation of efficiency these projects are presented in Table 1.

Table 1. Investment projects and their indicators.

Project	Indicators of the comparative estimation of efficiency				
	x_1	x_2	x_3	x_4	x_5
	NPV ($)	PI (%)	IRR (%)	PP (%)	ROI (%)
a_1	1100000	1.14	28	1.80	28
a_2	900000	1.25	25	1.90	32
a_3	700000	1.35	40	1.60	36
a_4	850000	1.05	13	1.20	24
a_5	550000	1.45	20	1.40	25
a_6	1250000	1.15	35	1.65	22
a_7	300000	1.20	24	1.50	26
a_8	1300000	1.13	18	1.70	30
a_9	1500000	1.24	15	2.00	23
a_{10}	630000	1.18	30	1.30	27

Presenting the indicators x_i ($i = 1 \div 5$) in the system of comparative estimation of investment efficiency in the form of weakly structured terms of linguistic variables, it is necessary to adapt the fuzzy methods of multi-criteria evaluation to the solution of the problem of assessing, ranking and choosing the best investment project under limited financial resources.

3 Choice of an Investment Decision by Pareto and Board Principles

The justification of the choice according to the Pareto principle suggests that the best is solution, which in all indicators is not worse than the first, but at least in one indicator is better than it. In support of projects comparing, so-called preference tables are compiled that demonstrate the advantages of various investment decisions. As a result, Pareto principle provides several and/or more solutions than provided by limited financial resources. Therefore, in addition to this, Board's principle of selection is applied, according to which investment projects are ranked in descending order of the corresponding indicators. As a result, the project with the maximum value of the total rank is recognized as the best decision.

The Pareto principle provides for the selection a few of the best from the investment projects portfolio. Thus, considering the investor's capabilities the total amount of financing of selected projects is established by their amount of financing [2]. At the first stage, within the framework of the metrics x_i $(i = 1 \div 5)$ projects are ranked. For the considered projects from Table 1 this ranking presented in Table 2.

Table 2. Project ranking.

Order	Indicators of project comparative appraisal					Order	Indicators of project comparative appraisal				
	x_1	x_2	x_3	x_4	x_5		x_1	x_2	x_3	x_4	x_5
1	a_9	a_5	a_3	a_9	a_3	6	a_4	a_{10}	a_7	a_3	a_7
2	a_8	a_3	a_6	a_2	a_2	7	a_3	a_6	a_5	a_7	a_5
3	a_6	a_2	a_{10}	a_1	a_8	8	a_{10}	a_1	a_8	a_5	a_4
4	a_1	a_9	a_1	a_8	a_1	9	a_5	a_8	a_9	a_{10}	a_9
5	a_2	a_7	a_2	a_6	a_{10}	10	a_7	a_4	a_4	a_4	a_6

At the next step of the Pareto principle, a comparative analysis of projects is carried out according to indicators x_i $(i = 1 \div 5)$ by establishing pairwise preferences. In Table 3, these preferences are established according to the following principle: for example, for project a_1, the square of intersection of row PI and column a_2 is denoted as "−", because PI value for project a_1 is less than for project a_2, and the sign "+" is displayed at the intersection with column a_4, because PI value for project a_1 is greater than for project a_4. If the values of indicators for alternative projects are equal, then the sign "0" is set. According to the Pareto principle, projects with columns that do not contain the symbol "−" are preferred. For example, for project a_1, column a_4 contains only symbols "+", which means that project a_1 is superior to project a_4. Columns a_4 and a_7 contain the symbols "+" in segment a_2; columns a_7 and a_{10} contain the symbols "+" in segment a_3; and column a_4 contains the symbols "+" in segment a_8. Therefore, only relative to a_4, a_7 and a_{10} there are projects that have an advantage. For pairwise comparison of the remaining projects the Pareto principle is again applied by trivial algorithm, which is easily implemented on a computer.

The Pareto principle produces more investment decisions than necessary. Therefore, to complete the comparative analysis of investment projects, the Board selection rule is applied, according to which projects are ranked for each indicator in descending order with the assignment of the corresponding rank values (see Table 4) and the total

Table 3. Comparative analysis of projects by pairwise preferences of their indicators.

a_1	a_2	a_3	a_4	a_5	a_6	a_7	a_8	a_9	a_{10}
NPV	+	+	+	+	−	+	−	−	+
PI	−	−	+	−	−	−	+	−	−
IRR	+	−	+	+	−	+	+	+	−
PP	−	+	+	+	+	+	+	−	+
ROI	−	−	+	+	+	+	−	+	+

a_2	a_1	a_3	a_4	a_5	a_6	a_7	a_8	a_9	a_{10}
NPV	−	+	+	+	−	+	−	−	+
PI	+	−	+	−	+	+	+	+	+
IRR	−	−	+	+	−	+	+	+	−
PP	+	+	+	+	+	+	+	−	+
ROI	+	−	+	+	+	+	+	+	+

a_3	a_1	a_2	a_4	a_5	a_6	a_7	a_8	a_9	a_{10}
NPV	−	−	−	+	−	+	−	−	+
PI	+	+	+	−	+	+	+	+	+
IRR	+	+	+	+	+	+	+	+	+
PP	−	−	+	+	−	+	−	−	+
ROI	+	+	+	+	+	+	+	+	+

... ...

a_8	a_1	a_2	a_3	a_4	a_5	a_6	a_7	a_9	a_{10}
NPV	+	+	+	+	+	+	+	−	+
PI	−	−	−	+	−	−	−	−	−
IRR	−	−	−	+	−	−	−	+	−
PP	−	−	−	+	+	+	+	−	+
ROI	+	−	−	+	+	+	+	+	+

... ...

a_{10}	a_1	a_2	a_3	a_4	a_5	a_6	a_7	a_8	a_9
NPV	−	−	−	−	+	−	+	−	−
PI	+	−	−	+	−	+	−	+	−
IRR	+	+	−	+	+	−	+	+	+
PP	−	−	−	+	−	−	−	−	−
ROI	−	−	−	+	+	+	+	−	+

rank is calculated for each decision (see Table 5). As a result, the project with the highest value of the total rank is considered the best. As can be seen from Table 5, projects a_2 and a_3 have the highest ratings and their choice is the best investment decision.

Table 4. Project ranking by the Board principle.

Rank	Indicators of project comparative appraisal				
	x_1 (NPV)	x_2 (PI)	x_3 (IRR)	x_4 (PP)	x_5 (ROI)
10	a_9	a_5	a_3	a_9	a_3
9	a_8	a_3	a_6	a_2	a_2
8	a_6	a_2	a_{10}	a_1	a_8
7	a_1	a_9	a_1	a_8	a_1
6	a_2	a_7	a_2	a_6	a_{10}
5	a_4	a_{10}	a_7	a_3	a_7
4	a_3	a_6	a_5	a_7	a_5
3	a_{10}	a_1	a_8	a_5	a_4
2	a_5	a_8	a_9	a_{10}	a_9
1	a_7	a_4	a_4	a_4	a_6

Table 5. Ranks of the compared projects.

Project	Indicators of project comparative appraisal					Total	Order
	x_1	x_2	x_3	x_4	x_5		
a_1	7	3	7	8	7	32	3
a_2	6	8	6	9	9	38	1
a_3	4	9	10	5	10	38	2
a_4	5	1	1	1	3	11	10
a_5	2	10	4	3	4	23	8
a_6	8	4	9	6	1	28	6
a_7	1	6	5	4	5	21	9
a_8	9	2	3	7	8	29	5
a_9	10	7	3	10	2	32	4
a_{10}	3	5	8	2	6	24	7

4   The Choice of the Investment Decision by the Fuzzy Approaches

According to the above reasoning, the indicators x_i ($i = 1 \div 5$) are qualitative criteria for evaluating investment projects, which can be described by appropriate fuzzy subsets of the universe $U = \{a_1, a_2, ..., a_{10}\}$, where a_j denotes the j-th project. To restore these subsets the Gaussian membership functions $\mu_i(u) = \exp\{-(u - u_i)^2/\sigma_i^2\}$ are used, where $u \in [0, u_i]$; u_i is the maximum value of the x_i; $\sigma_i^2 = \sum_{k=1}^{51} (u - u_i)^2/51$ is standard deviation. Then, the evaluation concepts can be described by the following fuzzy sets:

- ENOUGH (NPV): $F_1 = \{0.8096/a_1;\ 0.6217/a_2;\ 0.4296/a;\ 0.5725/a_4;\ 0.3038/a_5;\ 0.9208/a_6;\ 0.1494/a_7;\ 0.9486/a_8;\ 1.0000/a_9;\ 0.3682/a_{10}\}$;

- HIGH (PI): $F_2 = \{0.8730/a_1, 0.9451/a_2, 0.9860/a_3, 0.7977/a_4, 1.0000/a_5, 0.8806/a_6, 0.9155/a_7, 0.8653/a_8, 0.9396/a_9, 0.9021/a_{10}\}$;
- SUITABLE (IRR): $F_3 = \{0.7654/a_1, 0.6586/a_2, 1.0000/a_3, 0.2584/a_4, 0.4759/a_5, 0.9546/a_6, 0.6217/a_7, 0.4072/a_8, 0.3134/a_9, 0.8306/a_{10}\}$;
- ACCEPTABLE (PP): $F_4 = \{0.9707/a_1, 0.9926/a_2, 0.8880/a_3, 0.6217/a_4, 0.7654/a_5, 0.9130/a_6, 0.8306/a_7, 0.9354/a_8, 1.0000/a_9, 0.6950/a_{10}\}$;
- HIGH (ROI): $F_5 = \{0.8636/a_1, 0.9640/a_2, 1.0000/a_3, 0.7189/a_4, 0.7578/a_5, 0.6381/a_6, 0.7952/a_7, 0.9208/a_8, 0.6789/a_9, 0.8306/a_{10}\}$;

To identify the best investment decision, let us carry out a convolution of relevant information about each alternative project a_k. To do this, it is necessary to determine the totality of optimal alternatives by meet of fuzzy sets F_i, including project estimates for the given evaluation criteria. In this case, these criteria are considered as equivalent. Then the selection rule is determined by intersection $C = F_1 \cap ... \cap F_5$, where the optimal decision is the choice of the project for which the value of the corresponding membership function is the largest, and meet of fuzzy sets F_i is realized as $\mu_C(a_k) = \min\{\mu_{Fi}(a_k)\}$ [3, 4]. In this case, the totality of optimal alternatives are formed as:

$A = \{\min\{0.8096; 0.8730; 0.7654; 0.9707; 0.8636\}; \min\{0.6217; 0.9451; 0.6586; 0.9926; 0.9640\}; \min\{0.4296; 0.9860; 1.0000; 0.8880; 1.0000\}; \min\{0.5725; 0.7977; 0.2584; 0.6217; 0.7189\}; \min\{0.3038; 1.0000; 0.4759; 0.7654; 0.7578\}; \min\{0.9208; 0.8806; 0.9546; 0.9130; 0.6381\}; \min\{0.1494; 0.9155; 0.6217; 0.8306; 0.7952\}; \min\{0.9486; 0.8653; 0.4072; 0.9354; 0.9208\}; \min\{1.0000; 0.9396; 0.3134; 1.0000; 0.6789\}; \min\{0.3682; 0.9021; 0.8306; 0.6950; 0.8306\}\}$.

Then, the resultant vector with components reflecting the priority of investment projects has the following form: $\max\{\mu_C(a_k)\} = \max\{0.7654; 0.6217; 0.4296; 0.2584; 0.3038; 0.6381; 0.1494; 0.4072; 0.3134; 0.3682\}$. It follows that the best investment decision is the choice of the project a_1 with the highest value 0.7654. Further, the projects are ranked in descending order: $a_6 - 0.6381$, $a_2 - 0.6217$, $a_3 - 0.4296$, $a_8 - 0.4072$, $a_{10} - 0.3682$, $a_9 - 0.3134$, $a_5 - 0.3038$, $a_4 - 0.2584$, $a_7 - 0.1494$.

To choose the best investment decision using fuzzy inference system a basis verbal model is formulated by the following consistent judgments:

e_1: "If net present value of project is enough and its payback period is acceptable, then the project is satisfactory";
e_2: "If in addition the project has a high profitability index, then it is more than satisfactory";
e_3: "If, in addition to all the above requirements, the internal rate of return of the project is suitable and the return on investment is high, then the project is perfect";
e_4: "If net present value of project is enough and its profitability index is high, project internal rate of return is suitable and its return on investment is high, then the project is very satisfactory";
e_5: "If net present value of project is enough, profitability index and return on investment are high, but project payback period is unacceptable, then it is satisfactory";
e_6: "If profitability index of project and its return on investment are not high, then the project is unsatisfactory".

Analysis of these judgments as verbal model detects the inputs in the form of fuzzy sets F_i ($i = 1 \div 5$) and the output terms of linguistic variable y. To describe the output terms by appropriate fuzzy sets let us choose the discrete universe set as $U = \{0, 0.1, \ldots, 1\}$. Then, according to [3] $\forall u \in U$ we have: S = SATISFACTORY: $\mu_S(u) = u$; MS = MORE THAN SATISFACTORY: $\mu_{MS}(u) = u^{1/2}$; VS = VERY SATISFACTORY: $\mu_{VS}(u) = u^2$; P = PERFECT: $\mu_P(u) = 1$, if $u = 1$ and $\mu_P(u) = 0$, if $u < 1$; US = UNSATISFACTORY: $\mu_{US}(u) = 1 - u$.

Thereby, the judgments $e_1 \div e_6$ can be rewritten in symbolic form as follows:

e_1: $(x_1 = F_1)$ & $(x_4 = F_4) \Rightarrow (y = S)$; e_2: $(x_1 = F_1)$ & $(x_2 = F_2)$ & $(x_4 = F_4) \Rightarrow (y = MS)$;

e_3: $(x_1 = F_1)$ & $(x_2 = F_2)$ & $(x_3 = F_3)$ & $(x_4 = F_4)$ & $(x_5 = F_5) \Rightarrow (y = P)$;

e_4: $(x_1 = F_1)$ & $(x_2 = F_2)$ & $(x_3 = F_3)$ & $(x_5 = F_5) \Rightarrow (y = VS)$;

e_5: $(x_1 = F_1)$ & $(x_2 = F_2)$ & $(x_4 = \neg F_4)$ & $(x_5 = F_5) \Rightarrow (y = S)$;

e_6: $(x_2 = \neg F_2)$ & $(x_5 = \neg F_5) \Rightarrow (y = US)$.

According to the rule of meet of fuzzy sets [4], for the left sides of the rules $e_1 \div e_6$ the following takes place:

$\mu_{M1}(a) = \min\{\mu_{A1}(a), \mu_{A4}(a)\}$, $M_1 = \{0.8096/a_1; 0.6217/a_2; 0.4296/a_3; 0.5725/a_4; 0.3038/a_5; 0.9130/a_6; 0.1494/a_7; 0.9354/a_8; 1.0000/a_9; 0.3682/a_{10}\}$;

$\mu_{M1}(a) = \min\{\mu_{A1}(a), \mu_{A2}(a), \mu_{A4}(a)\}$, $M_2 = \{0.8096/a_1; 0.6217/a_2; 0.4296/a_3; 0.5725/a_4; 0.3038/a_5; 0.8806/a_6; 0.1494/a_7; 0.8653/a_8; 0.9396/a_9; 0.3682/a_{10}\}$;

$\mu_{M3}(a) = \min\{\mu_{A1}(a), \mu_{A2}(a), \ldots, \mu_{A5}(a)\}$, $M_3 = \{0.7654/a_1; 0.6217/a_2; 0.4296/a_3; 0.2584/a_4; 0.3038/a_5; 0.6381/a_6; 0.1494/a_7; 0.4072/a_8; 0.3134/a_9; 0.3682/a_{10}\}$;

$\mu_{M4}(a) = \min\{\mu_{A1}(a), \mu_{A2}(a), \mu_{A3}(a), \mu_{A5}(a)\}$, $M_4 = \{0.7654/a_1; 0.6217/a_2; 0.4296/a_3; 0.2584/a_4; 0.3038/a_5; 0.6381/a_6; 0.1494/a_7; 0.4072/a_8; 0.3134/a_9; 0.3682/a_{10}\}$;

$\mu_{M5}(a) = \min\{\mu_{A1}(a), \mu_{A2}(a), 1 - \mu_{A4}(a), \mu_{A5}(a)\}$, $M_5 = \{0.0293/a_1; 0.0074/a_2; 0.1120/a_3; 0.3783/a_4; 0.2346/a_5; 0.0870/a_6; 0.1494/a_7; 0.0646/a_8; 0/a_9; 0.3050/a_{10}\}$;

$\mu_{M6}(a) = \min\{1 - \mu_{A2}(a), 1 - \mu_{A5}(a)\}$, $M_6 = \{0.1270/a_1; 0.0360/a_2; 0/a_3; 0.2023/a_4; 0/a_5; 0.1194/a_6; 0.0845/a_7; 0.0792/a_8; 0.0604/a_9; 0.0979/a_{10}\}$.

As a result, the rules $e_1 \div e_6$ can be rewritten in more compact form as follows:

e_1: $(x = M_1) \Rightarrow (y = S)$; e_2: $(x = M_2) \Rightarrow (y = MS)$; e_3: $(x = M_3) \Rightarrow (y = P)$; e_4: $(x = M_4) \Rightarrow (y = VS)$; e_5: $(x = M_5) \Rightarrow (y = S)$; e_6: $(x = M_6) \Rightarrow (y = US)$.

Transformations of these rules using the Lukasiewicz implication $\mu(a, u) = \min\{1, 1 - \mu(a) + \mu(u)\}$ forms the fuzzy relations in the form of corresponding matrices, the meet of which forms a decision in the form of the next matrix R.

$$R = \begin{array}{c|ccccccccccc}
 & 0 & 0.1 & 0.2 & 0.3 & 0.4 & 0.5 & 0.6 & 0.7 & 0.8 & 0.9 & 1 \\
\hline
a_1 & 0.1904 & 0.2346 & 0.2346 & 0.2346 & 0.2346 & 0.2346 & 0.2346 & 0.2346 & 0.2346 & 0.2346 & 0.8730 \\
a_2 & 0.3783 & 0.3783 & 0.3783 & 0.3783 & 0.3783 & 0.3783 & 0.3783 & 0.3783 & 0.3783 & 0.3783 & 0.9640 \\
a_3 & 0.5704 & 0.5704 & 0.5704 & 0.5704 & 0.5704 & 0.5704 & 0.5704 & 0.5704 & 0.5704 & 0.5704 & 1.0000 \\
a_4 & 0.4275 & 0.5275 & 0.6275 & 0.7275 & 0.7416 & 0.7416 & 0.7416 & 0.7416 & 0.7416 & 0.7416 & 0.7977 \\
a_5 & 0.6962 & 0.6962 & 0.6962 & 0.6962 & 0.6962 & 0.6962 & 0.6962 & 0.6962 & 0.6962 & 0.6962 & 1.0000 \\
a_6 & 0.0870 & 0.1870 & 0.2870 & 0.3619 & 0.3619 & 0.3619 & 0.3619 & 0.3619 & 0.3619 & 0.3619 & 0.8806 \\
a_7 & 0.8506 & 0.8506 & 0.8506 & 0.8506 & 0.8506 & 0.8506 & 0.8506 & 0.8506 & 0.8506 & 0.8506 & 0.9155 \\
a_8 & 0.0646 & 0.1646 & 0.2646 & 0.3646 & 0.4646 & 0.5646 & 0.5928 & 0.5928 & 0.5928 & 0.5928 & 0.9208 \\
a_9 & 0.0000 & 0.1000 & 0.2000 & 0.3000 & 0.4000 & 0.5000 & 0.6000 & 0.6866 & 0.6866 & 0.6866 & 0.9396 \\
a_{10} & 0.6318 & 0.6318 & 0.6318 & 0.6318 & 0.6318 & 0.6318 & 0.6318 & 0.6318 & 0.6318 & 0.6318 & 0.9021
\end{array}$$

According to [3], the fuzzy conclusion concerning the k-th project ($k = 1 \div 10$) is presented by fuzzy subset E_k of the universe U. The values of the corresponding membership function are from the k-th row of the matrix R. To point estimate these conclusions the defuzzification procedure is used. In particular, for defuzzification of fuzzy conclusion concerning the assessment of the 8-th project: $E_8 = \{0.0646/0;$ $0.1646/0.1;$ $0.2646/0.2;$ $0.3646/0,3;$ $0.4646/0.4;$ $0.5646/0.5;$ $0.5928/0.6;$ $0.5928/0.7;$ $0.5928/0.8;$ $0.5928/0.9;$ $0.9208/1\}$, it is necessary to establish the α-level sets $E_{8\alpha}$ and calculate its cardinal number as $M(E_{8\alpha}) = \sum_{j=1}^{n} r_j/n$ ($r \in E_\alpha$). In this case, we have:

- $0 < \alpha < 0.0646$: $\Delta\alpha = 0.0646$, $E_{8\alpha} = \{0; 0.1; 0.2; ...; 0.9; 1\}$, $M(E_{8\alpha}) = 0.50$;
- for $0.0646 < \alpha < 0.1646$: $\Delta\alpha = 0.1$, $E_{8\alpha} = \{0.1; 0.2; ...; 0.9; 1\}$, $M(E_{8\alpha}) = 0.55$;
- for $0.1646 < \alpha < 0.2646$: $\Delta\alpha = 0.1$, $E_{8\alpha} = \{0.2; 0.3; ...; 0.9; 1\}$, $M(E_{8\alpha}) = 0.60$;
- for $0.2646 < \alpha < 0.3646$: $\Delta\alpha = 0.1$, $E_{8\alpha} = \{0.3; 0.4; ...; 0.9; 1\}$, $M(E_{8\alpha}) = 0.65$;
- for $0.3646 < \alpha < 0.4646$: $\Delta\alpha = 0.1$, $E_{8\alpha} = \{0.4; 0.5; ...; 0.9; 1\}$, $M(E_{8\alpha}) = 0.70$;
- for $0.4646 < \alpha < 0.5646$: $\Delta\alpha = 0.1$, $E_{8\alpha} = \{0.5; 0.6; 0.7; 0.8; 0.9; 1\}$, $M(E_{8\alpha}) = 0.75$;
- for $0.5646 < \alpha < 0.5928$: $\Delta\alpha = 0.0282$, $E_{8\alpha} = \{0.6; 0.7; 0.8; 0.9; 1\}$, $M(E_{8\alpha}) = 0.80$;
- for $0.5928 < \alpha < 0.9208$: $\Delta\alpha = 0.3280$, $E_{8\alpha} = \{1\}$, $M(E_{8\alpha}) = 1$.

Then, the numerical estimation of the investment project a_8 is obtained as follows:

$$F(E_8) = \frac{1}{\alpha_{max}} \int_0^{\alpha_{max}} M(E_{8\alpha})d\alpha = \frac{1}{0.9208}[0.0646 \cdot 0.5 + 0.1 \cdot 0.55 + ... + 0.3280 \cdot 1]$$
$$= 0.7687.$$

The following numerical estimates for other projects are obtained by similar actions: $F(E_1) = 0.8682$; $F(E_2) = 0.8038$; $F(E_3) = 0.7148$; $F(E_4) = 0.5763$; $F(E_5) = 0.6519$; $F(E_6) = 0.8243$; $F(E_7) = 0.5355$; $F(E_9) = 0.7786$; $F(E_{10}) = 0.6498$.

Table 6. Project ranking by various methods.

Project	Pareto principle	Board principle		Maximin convolution		Fuzzy inference	
	Order	Estimate	Order	Estimate	Order	Estimate	Order
a_1	3	32	3	0.7654	1	0.8682	1
a_2	1	38	1	0.6217	3	0.8038	3
a_3	2	38	2	0.4296	4	0.7148	6
a_4	10	11	10	0.2584	9	0.5763	9
a_5	7	23	8	0.3038	8	0.6519	7
a_6	6	28	6	0.6381	2	0.8243	2
a_7	8	21	9	0.1494	10	0.5355	10
a_8	4	29	5	0.4072	5	0.7687	5
a_9	5	32	4	0.3134	7	0.7786	4
a_{10}	9	24	7	0.3682	6	0.6498	8

5   Conclusion

Table 6 shows the results of project ranking obtained by different methods. Despite the fact that all the above approaches to the assessment and ranking of investment projects are based on the same consistent information about alternatives, the results are still noticeably different. First of all, this is explained by different ways of interpreting the source information and different approaches to making investment decisions. So, the Pareto and Board principles are based on the rational and partly balanced approach based on pairwise comparisons of alternative projects. The fuzzy maxmin convolution method, which was applied without taking into account the factors weights of the evaluation, reflects a pessimistic approach that ignores the "positive" sides of the alternatives. It identifies the best alternative as having minimal weaknesses in all criteria. The method of fuzzy inference implements a heuristic approach and, thus, is able to satisfy the basic requirements of the investor. Being "flexible" relative to the initial data the fuzzy inference system has a significantly greater potential relative to the presented methods, because it able to involve managers in the computing process and generate more rational investment decisions by the possibility of its structural and parametric optimization, for example, using the appropriate neural network.

References

1. Bystrov, O.F., et al.: Management of Investment Activities in the Regions of the Russian Federation. INFRA-M, Moscow (2008). (in Russian)
2. Application of the Pareto and Board Methods for Choosing of Investment Projects. https:// afdanalyse.ru/publ/investicionnyj_analiz/teorija/primenenie_metodov_pareto_i_borda_pri_ vybore_investicionnykh_proektov/27-1-0-330. (in Russian)
3. Andreichenkov, A.V., Andreichenkova, O.N.: Analysis, Synthesis, Planning Decisions in the Economy. Finance and Statistics, Moscow (2000). (in Russian)
4. Zadeh, L.: The concept of a linguistic variable and its application to approximate reasoning-I. Inf. Sci. 8(3), 199–249 (1975). https://doi.org/10.1016/0020-0255(75)90036-5

Phenomenon of Information and Informational Ecology: Interaction and Definitions on the Language of Soft Computing

Popov Oleg[1]([⊠]) [iD] and Martynov Boris[2] [iD]

[1] Educational-Scientific Centre "Internauka" Rostov Region,
Rostov-on-Don, Russia
orion777@bmail.ru
[2] Southern University (Institute of Management Business and Law) Rostov
Region, Rostov-on-Don, Russia
martynov@iubip.ru

Abstract. The multidimensional nature of the concept of "information" as a phenomenon of the modern digital world is revealed. The basic properties and attributes of information are considered, and appropriate mathematical measures of information are presented. In the context of the formation of a secure geoinformation environment, within the framework of the "Digital Earth" project, the critical importance of the formation of information ecology as a scientific direction at the intersection of social ecology, synergetic and systems theory is indicated.

Keywords: Information and its measure · Geoinformation space · Security · Ecology · Social ecology · Information ecology · System of systems

1 Introduction

Theoretical justification of physical reality, which is undoubtedly important in the scientific picture of the world, is reflected in the philosophical and natural science ideas of A. Einstein. The term "physical reality" appears in different roles for Einstein and includes different content. The philosopher D. Gribanov, analyzing the legacy of Einstein, highlights the following aspects of it [1]:

(1) as a concept that covers the objects of the external world studied by physics, primarily matter and field (or, according to Gribanov, physical reality of the first order);

(2) as a concept that covers "the relation of objects of the external world to each other, as well as the physical properties of objects that are manifested in these relations, such as time, space, mass, energy, inertia, speed, acceleration, etc." (second-order physical reality);

(3) as representing the reflected physical reality in scientific concepts, principles, theories and in general in physics - an integral essence of substance, ideal in nature or information (third-order reality).

Based on the research of A. Einstein and D. Gribanov, the Russian philosopher A. Lisin concludes that "theoretically" complete "physical reality, capable of displaying

R. A. Aliev et al. (Eds.): ICAFS 2020, AISC 1306, pp. 694–701, 2021.
https://doi.org/10.1007/978-3-030-64058-3_86

the entire integrity of the world (being of the universe, substance), contains at least three orders of realities, which are naturally constructed from the same three fundamental concepts: substance, field and information" [2]. And further, criticizing the incompleteness of the physicalist picture of the world, the philosopher argues that "if we recognize the objectivity of the existence of information and if we consider it one of the three attributes of substance (along with matter and energy), the scientific picture of the world should contain not just a mention of information, this alone is not enough; information as an attribute should become a full component of this scientific picture» [2, c. 330].

Natural science understanding of the nature and phenomenon of information, in which its main substantive feature is determined by the absence of material (or field) origin, is given in the concise definition of the founder of cybernetics N. Wiener, according to which "information is information, not matter and not energy" [3].

2 Multidimensional Concept "Information". Basic Features and Attributes

The set of information measures (Kulbak I_C, Fischer I_F, Shannon H_S, Kolmogorov K_f, Lovtsov – Knyazev $H_{ЛК}$, Shileiko – Kochnev $I_{ШК}$, Petrov $\Delta_П$, Sukhov H_C (D, D_o), Kharkevich I_X, Moiseev Δ_M, Shrader $H_{Ш}$, Hartley H_H), used in geoinformational environment (GIE), allowing to execute the informational principle, must correspond to the following main requirements: adequacy, consistency, efficiency, additivity, clarity [4].

Information attributes – fundamental properties of information (common to all types of information):

meaningfulness (meaning or concept is an invariant of information processing and transformation);

"memorability" (information is a memorized choice of one option from a variety of possible options)

hierarchy (information at the top management levels is determined by the presence of General, system-wide, individual, and other thesauri necessary for its reception or generation);

value (information has consumers and therefore has a certain value and quality in the sense of a thesaurus or goal);

"structurization" (allows us to perceive the phenomena of the world as signals that have structural identifiable informative parameters);

connectedness with a certain self-organizing object (system).

One of the most important characteristics of information is its adequacy.

The adequacy of information is the level of conformity of the image created with the help of information to a real object, process, or phenomenon. The correctness of decision-making depends on the degree of adequacy of information. The adequacy of information can be expressed in three forms: syntactic, semantic, and pragmatic.

Syntactic adequacy reflects the formal and structural characteristics of information, without affecting its semantic content. The syntactic level takes into account the type of media and the way information is presented, the speed of its transmission and processing, the size of information representation codes, the reliability and accuracy of converting these codes, and so on. Information viewed in this way is usually called data.

Semantic adequacy determines the degree to which an object's image corresponds to the object itself. The semantic content of the information is taken into account here. At this level, the information reflected by the information is analyzed and semantic connections are considered. Thus, semantic adequacy is manifested in the presence of unity of information and the user. This form is used for forming concepts and representations, identifying the meaning, content of information and its generalization.

Pragmatic adequacy reflects the compliance of information with the management goal implemented on its basis. The pragmatic properties of information are manifested when there is a commonality of user information and management goals. At this level, the consumer properties of information related to the practical use of information are analyzed, with the correspondence of its target function to the system's activity. Each form of adequacy corresponds to its own measure of the amount of information.

The syntactic measure of information operates with impersonal information that does not express a semantic relationship to the object. At this level, the amount of data in a message is measured by the number of characters in that message. In the case of the syntactic form of adequacy the most common measures of R. Hartley H_H и K. Shannon - H. Weener H_S [5].

A *semantic measure of information* is used to measure the semantic content of information. The most common here is a thesaurus measure that links semantic properties of information with the user's ability to accept the received message. A thesaurus is a collection of information that a user or system has. The consumer receives the maximum amount of semantic information when its semantic content is coordinated with its thesaurus, when the incoming information is clear to the user and carries previously unknown information. The semantic measure of the amount of information is related to the content coefficient, which is defined as the ratio of the amount of semantic information to the total volume of data.

When taking into account the semantic properties (quality) of information in the form of a variety of corresponding subsets of the thesaurus, you can use a combinatorial measure of the diversity of the system-wide thesaurus (knowledge stock) of Yu. Shrader [5]. In addition, in the case of a semantic form of adequacy, methods and models of soft mathematics and fuzzy logic are appropriate [6, 7].

The pragmatic measure of information determines its usefulness and value for the management process. Usually, the value of information is measured in the same units as the target system management function.

We believe that in the case of solving the problem of determining the value of information in the decision-making and coordination subsystems, we can use the probabilistic measure of the expediency of A. Harkevich management [5]. In addition, in this case, it is also advisable to use fuzzy logic methods [6, 7].

The rational distribution between the above forms of adequacy of appropriate mathematical measures of information representation is given in Table 1.

Extremely complex in its variety of manifestations, the concept of "information" becomes quite simple in its General definition, from which it is possible to establish the following. Any real object has the property of diversity, but it is infinitely large in

Table 1. Distribution of mathematical measures in informational system

Form	Mathematical measure of information
Syntactical	H_S (p_m, N) – Shannon
	H_H (N) – Hartley
Semantic	H_{III} (m, T) – Shrader
Pragmatic	I_X (P_G) – Kharkevich

quantitative terms. For material objects this variety of as material and energy components, which allows to transfer part of the diversity using the radiation from a single object (object, source) to another (the carrier, the receiver - information), where this partial diversity can be maintained if the receiver has the properties of (interoperable) media non-native variety, which is perceived, under certain conditions, information.

3   Information and Geoinformational Space

One of the most important concepts used in the process of transforming socio-economic systems within the framework of the Digital Earth project is the unified geographic information space (UGIS). Given that there is currently no unambiguous interpretation of some terms, we will fix the variant of their use in this paper. The concept of space itself is interpreted "as an objective reality, a form of existence of matter characterized by length and volume" [12]. Among the many known spaces, we note the information and geographical spaces. Moreover, the latter occupies a central place in the Earth Sciences. It is understood as "the form of existence of geographical objects and phenomena within a geographical envelope; a set of relations between geographical objects located on a specific territory and developing over time" [12]. It is assumed [12] that space, like any complex system (more precisely, System of Systems), includes three components – matter, energy and information, the description of each of which reflects its own characteristics of the entire system.

For a well-coordinated operation of the UGIS model, all its components must have interoperability, which is interpreted as "the ability of two or more information systems or their components to exchange information and use the information obtained as a result of exchange" [12]. And taking into account the fact that information systems of the model should not be interoperable, and "work according to independent algorithms, do not have a single control point, and the overall management is determined only by a single set of standards - the interoperability profile" [12], we have to deal with System of Systems. Accordingly, the problem arises of determining the requirements for these standards, criteria for their adequacy to solve problems, on the one hand, and the adequacy of the functional relationships of the real UGIS and its model at the level of semantic layers.

To develop a detailed methodology for taking into account the mutual information influence of natural and artificial objects in different regions when assessing the socio-economic properties of the territory, it is necessary to form a developed knowledge base. The authors considered these issues when developing the principles of separate semantic layers on the example of the so-called digital plan schemes [12].

The main function of the information environment for a person is the transmission of specific signals that are creatively restored to knowledge by the addressees [8]. The process of transforming information into knowledge, the organization of knowledge, its management, and the organization and management of information transfer processes are the main tools for creating an information environment.

4   Human Security in Geoinformational Environment: Cyber Security and Environmental Approach

Human interaction with the information environment, which is part of the GIS, creates problems that may have social and environmental meaning. Socio-ecological meaning in the final form can be expressed in the survival, vital activity, and viability of a person.

According to A. Lisin, "all physical processes without exception are accompanied by one or another — and often global-influence of not only creative, but also destructive information, which is encoded and mostly escapes the perception of the human consciousness, insofar as we have to admit that, interfering in fundamental physical processes, humanity itself does not know what it is creating" [2]. One of the founders of quantum computing S. Lloyd gives this example: "A one-kilogram piece of matter that is completely converted to energy is the scientific definition of a 20-megaton hydrogen bomb. Exploding nuclear weapons process a huge amount of information, the initial composition of which is set by the initial configuration; the result of processing is encoded in the emitted radiation" [2].

In this regard, information security becomes an actual and integral part of the culture of modern society.

Despite the fact that the concept combination "information security" sufficiently characterizes its content, specialists in various fields of scientific knowledge interpret it in their own way. From the point of view of the main functions performed by the

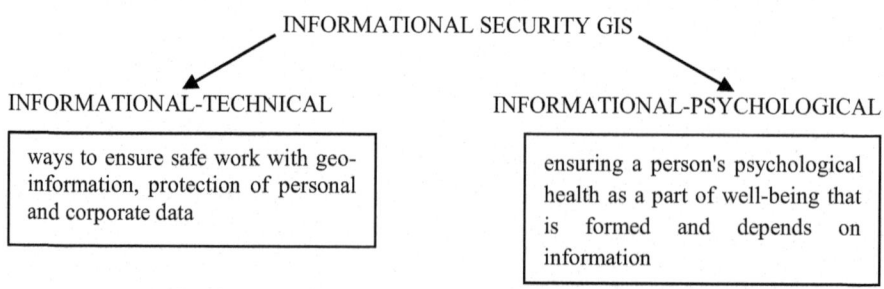

Fig. 1. Basic trends of informational security GIS

geographic information system (GIS) in society, we can offer the following scheme (Fig. 1).

Information technology security is associated with measures to protect information from unauthorized access, destruction, modification, disclosure, and delays in access. This activity is usually based on special information security systems and includes a set of measures to protect the processes of processing, storing and providing information [9].

The goals of the information and psychological direction are, in interpretation given by Eremin, "prevention of the negative impact of information on the mental, physical and social well-being of an individual, social groups, and the population as a whole, prevention of diseases of the population associated with information, improvement of the information environment" [10].

Thus, in the process of human interaction with the information environment, technical, social, medical and psychological aspects of human existence and development arise. Such a variety of impacts undoubtedly requires a systematic approach to mastering the culture of information security. With this approach, information security is transformed into an integral part of a more General system.

The basic principles of synergetics [11], as well as the theory of systems that allow for accurate research of complex systems, through merging them into a more functional and productive "metasystem" or SoS (system of systems) [12], open the way to solving these questions.

5   Fuzzy Information and Means of Calculating

Methods of probability theory, decision theory, information theory, or control theory are usually used to handle insufficiently defined quantities. However, such problems, which cannot be strictly formalized, can be solved by a person using subjective and vague ideas. The realization of this fact led to the emergence of a mathematical discipline-fuzzy logic, which eliminated the contradiction between the rigor of mathematics and the uncertainties of the real world.

The device of blurred (fuzzy) logic allowed to formalize fuzzy concepts and knowledge (NON-factors), operate with them and, accordingly, make fuzzy conclusions.

The method of measuring fuzzy information includes "forming a conditional scale of fuzzy reference objects-terms on the base set by assigning standard representatives. Subsequently, all values of the base set are fuzzified by using membership functions with the semantics of the original incomplete measurement information. Then, on the basis of this conditional scale, incomplete measurement information is fuzzified, which is used for preliminary measurement of fuzzy information by determining its fuzzy inclusion in standard terms" [15]. In this case, the term with the highest inclusion is considered a preliminary result of measuring fuzzy information. Further, the preliminary result is corrected using its defuzzification operations, taking into account the values of the membership functions in all terms of the scale and all its typical values. The secondary correction is introduced by repeated fuzzification of the updated source information and its subsequent defuzzification.

With the evolution of "not the biosphere, but its connections with one of its inseparable but extremely destructive parts — with humanity, the role of information increases even more. And it is becoming more and more important not just to inform people about the state of the biosphere, but to know its information and management networks" [13].

Therefore, despite the fact that the three-dimensional field studies in environmental and, in particular, geo-environmental Informatics not sufficiently elaborated information and the ecological approach is associated with a logical dominant information in the totality of environmental components of the biosphere, which postulate environmentalists, opens up new possibilities for developing the methodological principles of research of geoinformation space.

The scientific direction of information ecology is currently being developed in the works of a number of authors [10, 14].

You can pay attention to the following definition, which integrates the goals and subject content of the emerging science. "Information ecology is a science that studies the regularities of the influence of information on the formation and functioning of a person, human communities and humanity as a whole, on individual and social relationships with the surrounding information environment, as well as interpersonal and intergroup information interactions" [10].

6 Conclusion

Thus, in the context of the formation of the subject of information ecology, in relation to the concept of a safe geo-information space for humans, the following issues are most relevant:

- mechanisms for informing the human society about its current needs and safe ways of development, ways to quickly check public information that is allegedly unreliable;
- prevention of negative impact of information on the mental, physical and social well-being of an individual, social groups, and the population as a whole;
- elimination of the discrepancy between the high level of technology development and the low level of human culture (implementation of computer and information ethics);
- research of consequences of influence of information and technical means on change of the state of the surrounding geoinformation environment;
- ways to preserve and restore the system of the surrounding geoinformation environment, in relation to information ecology;
- further development of the theory of information ecology as a doctrine of human survival in an unstable geoinformation space.

  The practical purpose of this work is to use research results to solve actual problems of monitoring geoinformation systems [12] with the introduction of elements of social and technological forecasting.

References

1. Gribanov, D.P.: Einstein's Philosophical Views And the Development of Relativity Theory. Moscow (2010)
2. Lisin, A.I.: Ideal: the General Theory of the Ideality of Matter. Publishing house Ikar, Moscow (2012)
3. Weener, N.: Cybernetics or Control and Communication in An Animal and A Machine. Nauka, Moscow (1983)
4. Lovtsov, D.A.: The information theory ergosystem basics. Leg. Inform. **3**, 4–20 (2019). https://doi.org/10.21681/1994-1404-2019-3-04-20
5. Khramov, V.V.: Theory of Information Processes and Systems: Educational and Methodological Guide. Rostov-on-Don (2011)
6. Zade, L.A: The role of soft computing and fuzzy logic in understanding, designing, and developing intelligent information systems. In: Proceedings of Artificial Intelligence, Vol. 2 (3), pp. 7–11. (2001)
7. Khramov, V.V.: Method for Aggregating Multiple Sources of Fuzzy Information. Taganrog (2001)
8. Bazuluk O.A.: Space Travels – Traveling Mentality: Course of Lectures. Kiev (2012)
9. Tsvetkov, V.Y.: Information security and geo-Informatics (the state of the problem). In: Proceedings of the Universities. Geodesy and Aerial Photography, vol. 1, pp. 106–121 (2001)
10. Eremin, A.L.: Noogenesis and Theory of Intelligence. Soviet Kuban, Krasnodar (2005)
11. Haken, G.: Information and Self-Organization. Macroscopic Approach To Complex Systems. Moscow (1991)
12. Kramarov, S.O., Khramov, V.V.: System engineering approach to research of complex multidimensional systems based on soft models. In: Intellectual Resources-Regional Development, vol. 4(1), pp. 222–228 (2018)
13. Ramers, N.F.: Ecology: Theories, Laws, Rules, Principles, and Hypotheses. Moscow (1994)
14. Makalsky, L.M., Makarov, A.K., Tsekhanovich, O.M.: Information Ecology. Ontoprint, Moscow (2017)
15. Ryzhakov, M.V., Ryzhakov, V.V.: Method for Measuring Fuzzy Information. Patent of the Russian Federation No. 2565494 (2015)

Analysis the Image Classification Problem Based on Transfer Learning

Eldar Zeynallı[(✉)] [ID]

Azerbaijan State Oil and Industry University, 20 Azadlyg Ave.,
Baku AZ1010, Azerbaijan
eldar.zeynalli@gmail.com

Abstract. To solve the main problem of recognition accuracy, many image classification models have been implemented. A lot of attention was paid to Machine Learning. In this work, we will examine the problem of image classification related on transmission training to study whether it will work better in point of accuracy and efficiency with new sets of image data through Transfer Learning. Transfer Learning is a method of using the knowledge of a pre-trained model in another task. In this article, we will compare the image classification results of Logistic Regression (LR), Linear SVM and Random Forest Classifiers (RFC) using the pre – trained VGG-16 model. Image classification problem is implemented using Caltech - 101 and Flowers - 17 datasets.

Keywords: Transfer learning · Feature extraction · VGG-16 model · Logistic regression · SVM and random forest classifiers · Image classification

1 Introduction

The classification of images with a large number of object categories has a different application value. Problems of this kind arise in various fields: for example, image searching on the keywords, object recognition for robotics, etc. Until recently, a few algorithms has been done to solve this problem, including due to computational complexity. In few years deep learning allured the attention of researchers and also used over a large range of real applications. Unlike regular machine learning methods, deep learning algorithms are used to learn advanced features from large volumes of information. Extraction of features from data and classification are stages of machine learning.

Information dependence is one of the main problem of deep learning. Thus, unlike regular machine learning methods, it needs a large dataset to understand the information more deeply. Developing a large – scale dataset is a very complex and expensive process. In this case, Transfer Learning can significantly simplify the process. Using Transfer Learning can significantly improve the performance of the learning process. Thus, the use of labeled data from similar or related fields ensures that the selected machine learning algorithm achieves more accurate results.

Authors of the paper [1] discussed how transferring features even from distant tasks can get accurate result than using random features. They demonstrated an approach

R. A. Aliev et al. (Eds.): ICAFS 2020, AISC 1306, pp. 702–708, 2021.
https://doi.org/10.1007/978-3-030-64058-3_87

how transferability is affected to optimization problem. Particle swarm optimization done for optimizing the fuzzy model and the accuracy of work was 93.11% [2].

In literature many works are devoted to define defect detection of the fruit [3] and from this we can say agriculture field is the place where machine learning [4] and deep learning [5–9] are used.

In the paper [10] mainly discussed choosing method for being fresh or old papaya fruits and how they got 100% of accuracy with VGG19 related on transfer learning approach. Transfer learning helps to find the faults in the physical process with the diagnostic knowledge contained in the computer simulation [11]. Transfer learning helped to get maximum recognition accuracy in biometric identification using VGG19 with non – ideal images [12]. Inception – v3 accuracy and efficiency works perfect on new image dataset via Transfer Learning [13]. Attentive feature distillation and selection helps to get more accurate result with Transfer Learning [14].

Random forest is one of the supervised learning algorithm. The basic idea of the method is combining models rises overall result. Random forest constructs some decision trees and integrates them together to get an precise and steady prediction. Importance feature of the random forest algorithm is that it is to calculate the relations vital affect on the prediction. One of the advantages of random forest is its versatility. In this article we will analyze the image classification problem related to transfer learning.

The structure of the paper is written as follows. Section 2 gives the preliminaries. A statement of the problem and its solution is given in Sect. 3. Finally, Sect. 4 sums up the paper.

2 Preliminaries

Logistic Regression [15]. Logistic regression estimates the relation among the dependent and independent variables by calculating probabilities with a logistic function. The logistic function as a sigmoid function is given below:

$$\sigma(t) = \frac{e^t}{e^t + 1} = \frac{1}{1 + e^{-t}},$$

If t linear function then it express as $t = kx + b$ and $p : \mathbb{R} \to (0, 1)$, then general logistic function is defined as

$$p(x) = \sigma(t) = \frac{1}{1 + e^{-(kx + b)}}$$

The parameter β of logistic regression is defined by using the following expression:

$$y = \begin{cases} 1, & \beta_0 + \beta_1 x + \varepsilon > 0 \\ 0, & else \end{cases}$$

Where ε is an error demonstrated by the standard logistic distribution.

Support Vector Machine [16]. In machine learning support-vector machines analyze data used for regression analysis and classification.

A training dataset consisted of n points are represented the following form

$$(\vec{x}_1, y_1), \ldots, (\vec{x}_n, y_n),$$

If y_i are either 1 or -1, it shows that the point belongs to $\vec{x}_i$. Where $\vec{x}_i$ indicate a p-dimensional vector. Our aim is to define the "maximum-margin hyper plane" which splits the group of points $\vec{x}_i$ for every $\vec{y}_i = 1$ from the group of points for which $\vec{y}_i = -1$, as it defined, so the distance between the hyperplane and the nearest point $\vec{x}_i$ from either group is maximized.

3   Statement of the Problem

The aim is to analyze the image classification problem based on transfer learning.

There are two datasets: Caltech-101 [17] and Flowers-17 [18]. The CALTECH – 101 dataset introduced by Fei – Fei et al. in 2004. It is a popular benchmark dataset for object detection. The dataset consists of 8677 images belonging to 101 different categories including soccer balls, bicycles, elephants, and even human brains. A fragment of the CALTECH – 101 dataset is given in Fig. 1.

airplanes

celing fan

ketch

joshua tree

water lily

Fig. 1. A fragment of five classes in the CALTECH – 101 dataset.

The Flowers – 17 dataset introduced by Nilsback and et al. [18]. A fragment of the Flowers-17 dataset is given in Fig. 2. Generally it is preferable to use 1000–5000 instance for each class while training a deep neural network [19]. But in this work classification improved using feature extraction - transfer learning method.

4   Solution of the Problem

In the first step, each of Logistic Regression, Linear SVM, and Random Forest classifiers was trained and tested separately with two datasets.

For both of datasets 75% of the data will be used for the train and 25% for the testing purposes separately. In the second step, the classifiers will be trained and tested in the same way. However, this time the separately extracted features for each dataset with pretrained VGG – 16 model used for training and testing purposes. VGG16 is a convolutional neural network model.

Fig. 2. A fragment of five classes in the FLOWERS – 17 dataset.

Transfer Learning with Pre-trained Models. ImageNet is a computer vision research project aimed at labeling and categorizing images almost 22000 separate object categories. The process of merging small networks and using them as initializations for the larger, deeper networks is called *pre-training*. Pretrained models are already trained on ImageNet.

Transfer learning proposes a different training paradigm. In this work power of transfer learning used for pre – trained model VGG – 16 as an effective feature extractor.

Extraction Process is based on the following steps:

1. Input an image to the network
2. Propagate through the network
3. Calculate the classification probabilities

For the feature extraction, the propagation process can be stopped at any layer. We remove fully connected layers (head).

In max pooling layer have 512 filter with each of size 7 x 7. After forward propagate single image through this network we can use these 7 x 7 x 512 = 25088 values as a feature vector. We get vector for N images as design matrix if we repeat this process for entire dataset. It works for datasets even VGG – 16 was not trained on.

Analysis. As mentioned earlier, first, each of our classifiers is trained with datasets itself, and the classification results is reviewed. This process performed separately for each dataset. Then we extract features and classify our datasets with the pretrained VGG-16 model using ImageNet weights. Finally, all results will be compared on various indicators.

Training with Dataset Itself. After training each of the Linear SVM, Random Forest, and Logistic Regression Classifiers separately with datasets, we see which dataset has the highest classification accuracy for each classifier in Fig. 3.

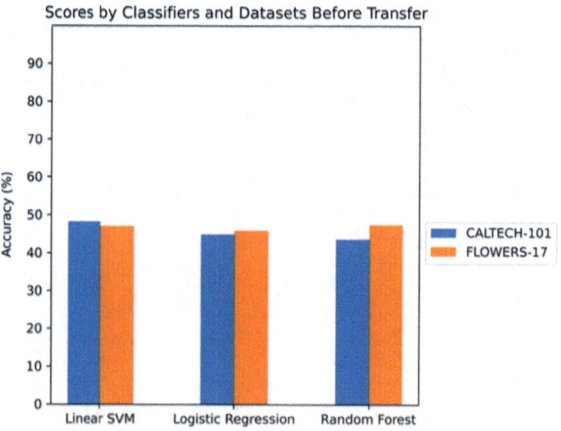

Fig. 3. Scores by classifiers and datasets after training with *dataset itself.*

Training with Extracted Features. After training each of the Linear SVM, Random Forest, and Logistic Regression Classifiers separately with extracted feature by VGG – 16, we see which dataset has the highest classification accuracy for each classifier in Fig. 4.

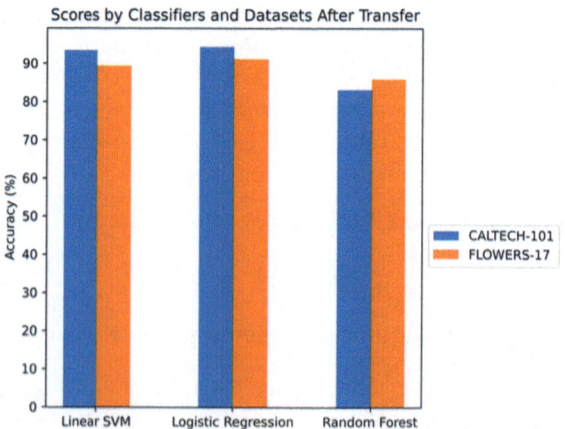

Fig. 4. Scores by classifiers and datasets after training with *extracted features.*

5 Conclusion

All calculation were made in Python environment, using Tensorflow, Keras and scikit-learn libraries, which are widely used in machine learning. As we can see in Table 1, the model trained with ***extracted features*** for each dataset was able to get almost 2 times more accurate results than the model trained with dataset itself.

Table 1. Comparision of the results

Dataset	LR trained with (accuracy, %)		SVMC trained w.ith (accuracy, %)		RFC trained with (accuracy, %)	
	Dataset itself	Extracted features	Dataset itself	Extracted features	Dataset itself	Extracted features
CALTECH-101	45.023	94.654	48.387	93.594	43.687	83.272
FLOWERS- 17	45.882	91.176	47.058	89.418	46.471	85.882

The following results is defined: The Logistic Regression and Linear SVM Classifiers classified the dataset CALTECH – 101 more accurately than the FLOWERS – 17 and the Random Forest Classifier classified the dataset FLOWERS – 17 more accurately than the CALTECH – 101. Visual representation of the results is given in Fig. 5.

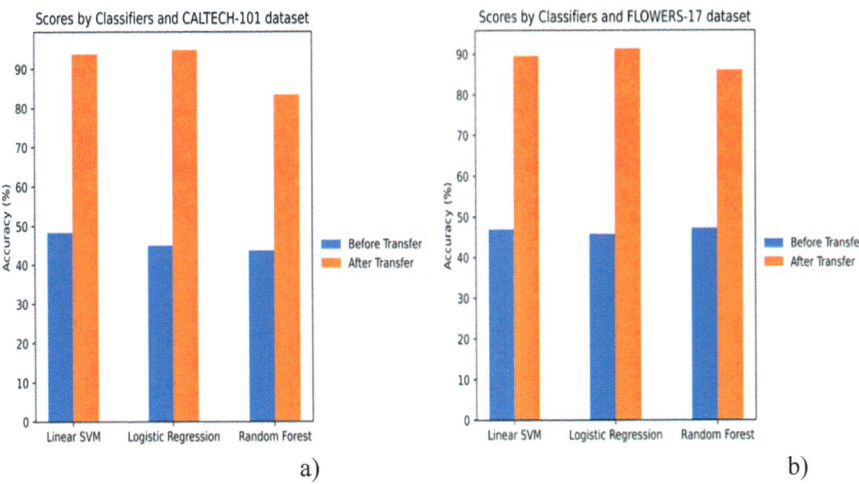

Fig. 5. Comparison (a) CALTECH – 101 dataset, (b) FLOWERS – 17 dataset.

References

1. Jason, Y., Jeff, C., Yoshua B., Hod L.: How transferable are features in deep neural networks? (2014). arXiv:1411.1792[cs.LG]
2. Marimuthu, S., Roomi, S.M.: Particle swarm optimized fuzzy model for the classification of banana ripeness. IEEE Sens. J. **17**(15), 4903–4915 (2017)
3. Devi, P.L., Varadarajan, S.: Defect fruit image analysis using advanced bacterial foraging optimizing algorithm. IOSR J. Comput. Eng. **14**(1), 22–26 (2013)
4. Huang, T., Yang, R., Huang, W., Huang, Y., Qiao, X.: Detecting sugarcane borer diseases using support vector machine. Inform. Process. Agric. **5**(1), 74–82 (2018)
5. Sharma, P., Berwal, YPS., Ghai W.: Performance analysis of deep learning CNN models for disease detection in plants using image segmentation. Inform. Process. Agric. 1–9 (2019). https://doi.org/10.1016/j.inpa.2019.11.001. (in press)
6. Ji, M., Zhang, L., Wu, Q.: Automatic grape leaf diseases identification via united model based on multiple convolutional neural networks. Inform. Process. Agric. (2019). https://doi.org/10.1016/j.inpa.2019.10.003
7. Guo, X., Zhao, X., Liu, Y., Li, D.: Underwater sea cucumber identification via deep residual networks. Inform. Process. Agric. **6**(3), 307–315 (2019)
8. Muhammad, H.A., Abdul, B.: Weed detection in canola fields using maximum likelihood classification and deep convolutional neural network. Inform. Process. Agric. (2019). https://doi.org/10.1016/j.inpa.2019.12.002
9. Gardashova, L.A., Gahramanli, Y., Babanli, M.: Fuzzy neural network based analysis of the process of oil product sorption with foam polystyrene. Int. J. Eng. Res. Appl. **7**(9), 85–90 (2017)
10. Behera, S.K., Rath, A.K., Sethy, P.K.: Maturity status classification of papaya fruits based on machine learning and transfer learning approach. Inform. Process. Agric. (2020). https://doi.org/10.1016/j.inpa.2020.05.003
11. Li, W., Gu, S., Zhang, X., Chen, T.: Transfer learning for process fault diagnosis: Knowledge transfer from simulation to physical processes. https://doi.org/10.1016/j.compchemeng.2020.106904
12. Kumari, P., Seeja, K.R.: Periocular biometrics for non-ideal images: with off-the-shelf Deep CNN & Transfer Learning Approach. https://doi.org/10.1016/j.procs.2020.03.234
13. Hussain, M., Bird, J., Faria, D.: A Study on CNN Transfer Learning for Image Classification. Advances in Computational Intelligence System (2020). (in press)
14. Kafeng, W., Xitong, G., Yiren, Z., Li, X., Dejing, D., Cheng-Zhong, X.: Pay attention to features, transfer learn faster CNNs. In: International Conference on Learning Representations. https://openreview.net/forum?id=ryxyCeHtPB
15. Rodríguez, G.: Lecture Notes on Generalized Linear Models. http://data.princeton.edu/wws509/notes/
16. Cortes, C., Vapnik, V.N.: Support-vector networks. Mach. Learn. **20**(3), 273–297 (1995). https://doi.org/10.1007/BF00994018. S2CID: 206787478
17. Li, F.-F., Fergus, R., Pietro, P.: Learning generative visual models from few training examples: an incremental bayesian approach tested on 101 object categories. Comput. Vis. Image Underst. **106**, 59–70 (2007)
18. Nilsback, M.-E., Andrew, Z.A.: Visual vocabulary for flower classification. In: CVPR (2), pp. 1447–1454. IEEE Computer Society (2006). http://dblp.uni-trier.de/db/conf/cvpr/cvpr2006-2.html#NilsbackZ06
19. Goodfellow, I., Yoshua B., Aaron,C.: Deep Learning. MIT Press (2016). http://www.deeplearningbook.org

Breast Cancer Classification Using Deep Learning

Umit Ilhan$^{(\boxtimes)}$ 🆔, Kaan Uyar 🆔, and Erkut Inan Iseri 🆔

Near East University, TRNC via Mersin 10, Nicosia, Turkey
`umit.ilhan@neu.edu.tr, kaan@uyar.com, erkut@iseri.eu`

Abstract. Cancer in any form is one of the most deadly illnesses in the world. Scientists are investigating into this disease and developing methods and treatments to fight it. The recent surveys show that breast cancer is also one of the major causes of mortality rate among female population around the world. Breast cancer's definition may be explained as some old cells that aggressively grow out of control to form a population of a harmful mass in the breast tissue. Eventually, as a result they lead to the formation a malignant tumor. Deep learning (DL) that is the subfield of machine learning algorithms provides a powerful tool to help experts to analyze, model and make sense of complex clinical data across a broad range of medical applications. The aim of this study is to develop an efficient system to classify breast tumors as malignant and benign. This system is divided in two stages. The first stage is the normalization of the data. The second stage is the classification of tumors. The accuracy of the approach is 98.42%. The overall result showed that the DL outperformed the previous studies where the same data set was used.

Keywords: Deep learning · Classification · Breast cancer · Wisconsin (Diagnostic) data set

1 Introduction

Breast cancer cases not detected in early phase can't be cured and mostly result with mortality because of the aggressive nature with very high probability of spreading to other parts of the body. It is a worldwide cancer type. According to Cancer Research Fund, nearly 12.7 million women are affected by the disease and each year 7.6 million of them die [1]. The future projections are giving no hope but instead the cancer research centers are estimating that by the year 2030 around 26 million people will suffer from the breast cancer disease.

Mostly women between the ages 50–70 are the vulnerable ones to have breast cancer. Basically the disease starts in the cells of the breast tissue or in the glands that produce milk. Unless it is detected in an early stage, very slowly they populate to be a growing tumor by attacking the surrounding cells. Eventually over time they make their wat to armpit lymph nodules. When the cancer reaches to these nodules the whole body becomes vulnerable to the threat because the lymph nodules open the way to all organs of the body. The studies in 2014 in USA showed that there were 2.8 million women with diagnoses of breast cancer [2] including those being treated and have their

© The Author(s), under exclusive license to Springer Nature Switzerland AG 2021
R. A. Aliev et al. (Eds.): ICAFS 2020, AISC 1306, pp. 709–714, 2021.
https://doi.org/10.1007/978-3-030-64058-3_88

treatment completed. The statistics showed that those women whose first degree relatives (mother, daughter or sister) had history of diagnoses of breast cancer, regardless of the end result, would have a doubled possibility of being diagnosed with breast cancer. Fifteen percent women diagnosed with breast cancer had a first degree family member with the same disease. On the other hand 5-10% of breast cancer cases shown to have connection with the gene mutations inherited from the parents [3].

Against all the fatality and the mortality of the breast cancer, there is a very good chance to stop the spread of the tumorous cells with an early detection of the disease. Among other cancer types the breast cancer is the most curable one when diagnosed in an early stage. Early diagnosis is vital in cancer diagnosis, treatment planning, and evaluation of treatment outcome. If a patient with a tumor is not treated correctly and early, the patient's chances of survival may decrease and result in death [4]. Experts in the diagnosis of the tumors may show a tendency to error. Increasing the diagnostic capabilities of experts and reducing the time spent for correct diagnosis is possible with computer-aided systems.

Several different machine learning techniques have been used for the prediction and classification of breast cancer using different data sets from early detection and diagnosis (early, clinical, imaging and pathological) stages [8–11]. Genetic algorithm (GA) based trained recurrent fuzzy neural network (RFNN) and adaptive neuro-fuzzy inference system (ANFIS) used by Uyar et al. [8] for the Breast Cancer Coimbra Dataset (BCCD) [5]. Dalwinder et al. [9] proposed wrapper method that uses Ant Lion optimization for searching feature weights and optimal parametric values of the neural networks simultaneously for the classification of the Wisconsin original Breast Cancer (WBC) [5], Wisconsin Diagnostic Breast Cancer (WDBC) [5], and BCCD [5] datasets. Shahnaz et al. [10] used K- Nearest Neighbor (KNN), Support Vector Machine (SVM), Naïve Bayes (NB), Random Forests, Logistic Regression, Multilayer Perceptron (MLP) and Convolutional Neural Network (CNN) for the classification of the WDBC data set. Agarap [11] performed z-score normalization for WDBC data set and used rectified linear units (ReLU) as the classification function in a DNN.

In this work deep learning (DL) was used for classification of normalized breast cancer data. The aim of this study is to find out the impact of DL method on the classification problem and see whether or not it results with a more accurate prediction model for WDBC data set cases that contains 32 features for each of 569 patients. The remaining part of the paper is organized as follows: Sect. 2 describes the method used, Sect. 3 gives the results and discussions, and the conclusion is given in the last section.

2   Method

2.1   Dataset

The Wisconsin Diagnostic Breast Cancer (WDBC) data set that has been taken from the UCI Machine Learning Repository [5] is used in this study. There are 569 patient data (contains 357 benign and 212 malignant type tumors) with 32 attributes (id number, diagnosis and ten real valued features that computed for each cell nucleus: radius, texture, perimeter, area, smoothness, compactness, concavity, concave points, symmetry and fractal dimension) in the WDBC dataset. Features computed from image of a fine needle aspirate (FNA) of a breast mass. They describe characteristics of the cell nuclei present in the image.

In neural networks there is a need to normalize the data which is the most important step in pre-processing. The data is scaled into values in the interval of 0 and 1 before feeding them into the network. This process will improve the performance of the multilayer neural network. The scaling should be repeated for each attribute [6]. In this study, data is normalized using the following formula.

$$New\ value(after\ normalization) = \frac{Current\ value - Minimum\ value}{Maximum\ value - Minimum\ value}$$

In this study, the dataset is divided into two sub-sets. One sub-set is used for training (512 instances) and second one is used for testing (57 instances). The testing set corresponded to 10% of the data set. The training set is deliberately set to 10% of the data set to be in consistence with the works to be compared with. The training set is determined by selecting data randomly from each class with the same ratio. This process is repeated 10 times.

2.2   Deep Neural Networks

Deep neural networks (DNNs) are members of commonly known as the machine learning tools. They allow the system to learn complex nonlinear functions of a given input leading to a minimized error cost. The feed forward DNN can be used for classification tasks with a structure that has an input layer feeding the data as a vector, some hidden layers exceeding 2 levels where a transformation is applied to the output of the previous layer obtaining a higher degree of representation, and an a final layer giving the computed output of the DNN. To compute the cost, the output and the reference label are compared and the error criteria applied to calculate the cost [7]. The designed DNN network in this study is trained using backpropagation algorithm with maximum 5000 epchs and 0.09 learning rate and contains 3 hidden layers with 150 neurons each as shown in Fig. 1.

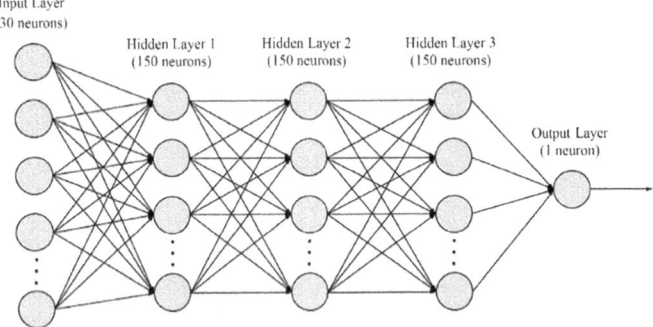

Fig. 1. The DNN Structure.

712   U. Ilhan et al.

2.3   Evaluation Criteria

The performance of the study is evaluated for sensitivity, specificity and accuracy of the testing set. As for the testing set and overall performance, the analysis is carried out to show TN, FN, TP and FP values. Table 1 gives the definition of each case.

Table 1. The definition of TN, FN, TP and FP values.

Abbreviation	Full form	Definition
TN	True Negative	Benign tumor detected as benign
FN	False Negative	Malignant tumor detected as benign
TP	True Positive	Malignant tumor detected as malignant
FP	False Positive	Benign tumor detected as malignant

3   Results

The DL method is applied on normalized WDBC data set to find out whether or not some improvement can be achieved on the classification problem of breast cancer. The proposed methodology is implemented using a system having i7 microprocessor and 16 GB RAM, Windows OS and MATLAB. Table 2 outlines the results obtained with DL method.

Table 2. The performance of the proposed system.

TP	FN	TN	FP	Sensitivity	Specificity	Accuracy
20	1	36	2	95.24%	100%	98.25%
20	1	36	0	95.24%	100%	98.25%
21	0	34	2	100%	94.44%	96.49%
21	0	36	0	100%	100%	100%
21	0	36	0	100%	100%	100%
20	1	36	0	95.24%	100%	98.25%
20	1	36	0	95.24%	100%	98.25%
19	2	36	0	90.48%	100%	96.49%
21	0	35	1	100%	97.22%	98.25%
21	0	36	0	100%	100%	100%

4   Discussion

Dalwinder et al. [9] compared their classification performance with previous 22 works that have used different features and classifier such as SVM, neuro-fuzzy system, type-2 fuzzy inference system, and DNN [10, 11] based on 10-fold cross validation in the case of WDBC data set, and achieved better mean accuracy. Table 3 shows the comparison of the proposed and 3 previous works [9–11] based on 10-fold cross-validation.

Table 3. Comparison of the proposed work with previous works based on 10-fold cross-validation.

	Data Normalization	Sensitivity (%)			Specificity (%)			Accuracy (%)		
		Min	Max	Mean	Min	Max	Mean	Min	Max	Mean
Dalwinder et al. [9]	Yes	96.69	96.75	96.43	99.17	99.72	99.52	98.24	98.61	98.37
Shahnaz (DNN) [10]	No									98.06
Agarap [11]	Yes									87.96
Proposed work	Yes	90.48	100	97.14	94.44	100	99.17	96.49	100	98.42

It can be seen that the proposed work attains 98.42% mean accuracy showing improvement as compared to the [9–11].

5 Conclusion

DL classification technique is used in this study to evaluate Wisconsin Diagnostic Breast Cancer (WDBC) dataset to identify the benign and malignant tumor types. There are 569 patient data in the dataset which contains 357 benign and 212 malignant type breast tumors. The system achieved mean success rate in the classification of malignant and benign tumors 97.14% and 99.17% respectively. The overall success rate of the system is 98.42%. The DL approach provides higher degree of classification and hence better mean accuracy compared with previous studies that used 10-fold cross validation applied on the WDBC data set.

References

1. Gao, Q., Wang, X.Y., Qiu, S.J., Yamato, I., Sho, M., Nakajima, Y., Xu, Y.: Overexpression of PD-L1 significantly associates with tumor aggressiveness and postoperative recurrence in human hepatocellular carcinoma. Clin. Cancer Res. 15(3), 971–979 (2009). https://doi.org/10.1158/1078-0432.CCR-08-1608
2. Sengupta, D., Bhargava, D.K., Dixit, A., Sahoo, B.S., Biswas, S., Biswas, G., Mishra, S.K.: ERR β signalling through FST and BCAS2 inhibits cellular proliferation in breast cancer cells. Br. J. Cancer 110(8), 2144–2158 (2014)
3. Acharya, A., Cutts, M., Dean, J., Haahr, P., Henzinger, M., Hoelzle, U., Lawrence, S., Pfleger, K., Sercinoglu, O., Tong, S.: Information retrieval based on historical data. U.S. Patent No. 7,346,839. Washington, DC: U.S. Patent and Trademark Office (2008)

4. Fritz, A., Percy, C., Jack, A., Health, W., Organization, W.H.: International Classification of Diseases For Oncology (ICD-O), 3rd edn. World Health Organization, Geneva, Switzerland (2013)
5. UCI Machine Learning Repository, Breast Cancer Wisconsin (Diagnostic) Data Set. https://archive.ics.uci.edu/ml/datasets/Breast+Cancer+Wisconsin+(Diagnostic)
6. Thein, H.T.T., Tun, K.M.M.: An approach for breast cancer diagnosis classification using neural network. Adv. Comput. Int. J. (ACIJ) 6(1), 1–11 (2015). https://doi.org/10.5121/acij.2015.6101
7. Lozano-Diez, A., Zazo, R., Toledano, D.T., Gonzalez-Rodriguez, J.: An analysis of the influence of deep neural network (DNN) topology in bottleneck feature based language recognition. Plos One. 12(8), Article Number: e0182580 (2017). https://doi.org/10.1371/journal.pone.0182580
8. Uyar, K., Ilhan, U., Ilhan, A., Iseri, E.I.: Breast cancer prediction using neuro-fuzzy systems. In: 2020 7th International Conference Electrical, Electronics Engineering (ICEEE), pp. 328–332. IEEE Press (2020). https://doi.org/10.1109/iceee49618.2020.9102476
9. Dalwinder, S., Birmohan, S., Manpreet, K.: Simultaneous feature weighting and parameter determination of Neural Networks using Ant Lion Optimization for the classification of breast cancer. Biocybern. Biomed. Eng. 40(1), 337–351 (2020). https://doi.org/10.1016/j.bbe.2019.12.004
10. Shahnaz, C., Hossain, J., Fattah, S.A., Ghosh, S., Khan, A.I.: Efficient approaches for accuracy improvement of breast cancer classification using Wisconsin database. In: 2017 IEEE Region 10 Humanitarian Technology Conference (R10-HTC), pp. 792–797. IEEE Press (2017). https://doi.org/10.1109/r10-htc.2017.8289075
11. Agarap, A.F.: Deep Learning Using Rectified Linear Units (RELU). ArXiv Prepr ArXiv180308375 (2018). https://arxiv.org/pdf/1803.08375.pdf

Application of Fuzzy Logic Model for Daylight Evaluation in Computer Aided Interior Design Areas

Ahmed Valiyev, Rahib Imamguluyev$^{(\boxtimes)}$ ⬡, and Gahramanov Ilkin

Odlar Yurdu University, AZ1072 Baku, Azerbaijan
oyu-asp@mail.ru, rahib.aydinoglu@gmail.com,
ilkin.gahramanov@wbc.com

Abstract. Today, the consumption of electricity in indoor and outdoor lighting systems is constantly increasing. It can be used as a concept of sustainable development as the main source of daylight for a nation, because the sunny area has very promising energy saving opportunities. The daylight system consists of daylight such as windows and glasses and various shading devices to control sunlight and glare. At a time when computer technology is evolving, architects and designers need to pay attention to window sizes and sunlight factors in their CAD (Computer Aided Design) projects. Fuzzy Logic allows you to talk about a certain percentage of daylight factors, both a bright set and a member of the middle set, and perhaps a dark set. In designs developed with CAD programs, it is possible to change the values of the sun with a fuzzy approach and get more accurate results. Calculations of the size and condition of solar radiation and windows for enough daylight have been developed and presented graphically with classic and fuzzy models. It is recommended that the application of this technique be effectively applied in the field of internal assessment of daylight. Based on the obtained values, calculations were made in Matlab Fuzzy Toolbox section and the results were developed by 3DsMax in the design of the Flour Products facility at Sumgayit station in Absheron region. Other possible design applications are recommended.

Keywords: Fuzzy logic · Daylighting · Daylighting calculations · Matlab · Interior design · Interior daylight evaluation · CAD · 3DsMax

1 Introduction

Proper distribution of daylight in interior design is one of the most important parts of architecture. At a time when computer technology was developing, it is possible to visualize this work through CAD programs before the building was built. So, designers determine the area where the building will be built on Google Earth. Learns the directions of the sun and determines the direction of the building. He then models the building in 3Dsmax based on the results obtained. Once the model is ready, determines the windows of the building based on the direction of the sun. In 3DsMax, it is advisable to use fuzzy logic to determine the brightness values of the sun. Thus, designers can determine the degree of brightness based on fuzzy logic. Fuzzy Logic is an extension of

© The Author(s), under exclusive license to Springer Nature Switzerland AG 2021
R. A. Aliev et al. (Eds.): ICAFS 2020, AISC 1306, pp. 715–722, 2021.
https://doi.org/10.1007/978-3-030-64058-3_89

the fuzzy set theory developed about 55 years ago. The purpose of fuzzy logic was to eliminate the problems associated with traditional binary logic, where the conditions are completely true or false [1–3]. Fuzzy Logic allows something to be partly true and partly false. To give a simple example, if 4% of the room is lit by sunlight, the room is dark. or if 60% is lit by sunlight, the room is very light. However, with fuzzy logic, more accurate results can be obtained by taking the illumination factor in the range of 1–100%. To apply the results to the model developed in 3DsMax, we first select create-system in the command panel, select Daylight for daylight from the drop-down section and place it in the desired direction on the stage. In the Control Parameters section, enable Date, Time and Location, and in the time section, enter time and date (see Fig. 1).

Fig. 1. 3DS MAX sunlight and daylight systems

To change the brightness of the Sun in 3DsMax, select the sun, select modify, and change the "Multiplier" value based on fuzzy values in the General Parameters section (see Fig. 2).

Fig. 2. 3DsMax sunlight general parameters

2 Fuzzy Logic

Fuzzy Logic is a mathematical theory that was founded for the first time in 1961 by the Azerbaijani mathematician Lutfi Alasker Zadeh. As mentioned in its name, it is a kind of 'logic'. In fact, we are not very foreign to the subject of logic from high school mathematics lessons. Especially for those who do not like mathematics, I think its place is more separate. Perhaps, I do not know, because mathematics is one of the rare verbally weighted topics. But let's remember the concepts that take place in our memory such as "and, or, if, and only if." We would try to understand whether the facts mentioned in the sentences we suggest suggesting whether they overlap with the facts in these lessons. When we said (1), we were saying 'false' (0) to those who said otherwise, the concepts we just mentioned were meant to make the relations between these propositions, namely 1 and 0, meaningful. Although we did not understand enough that he was trying to explain, the infrastructure of today's computer and electronic technologies actually emerged from the logic issue we saw in high school, because logic in the simplest sense, whispers to our ear how we can transfer it to the world of numbers with certain thinking methods in order to solve any real life problem. uses the possibilities of mathematics and we also use the methods that logic offers us. Fuzzy Logic works exactly for the same purpose. In fuzzy logic, one element can be elements of more than one set. The characteristic function in classical logic for the relationship of an element u with a set A: If $\mu A(u) = 1$ is the element of Set A, and if $\mu A(u) = 0$, u is not the element of Set A, so an object belongs to a group or not [3–5, 7–11].

In fuzzy sets, an object can be a partial member of a group. The degree of membership is defined by a generalized characteristic function called the membership function:

$$\mu A (u) : U \rightarrow [0,1]$$

The U general cluster is the fuzzy subset of AU [10–12].

In many sub-branches of engineering sciences such as control systems, automation, robotics and cybernetics, we find traces of fuzzy logic. [11]. Thus, Architects and Designers can achieve more successful results by applying a fuzzy logic model in interior design.

3 Daylight Criteria in İnterior Design

Daylight is not a new concept in design. The use of daylight in architecture as a design principle dates to ancient Rome. In the buildings of the period, the principle of drawing light from the wall openings into the interior was taken as a basis, and the location of the building was determined by daylight. Before the spaces were illuminated with electricity, the connection of the space with daylight was of great importance to the architects. With the shortage of oil and energy in the 1970s, the importance of daylight was understood, scientific research was conducted on this topic, and it became an important topic that is still being studied due to the continuing energy problem.

Daylight is an important spatial design introduction that improves the quality of space and ensures the integration of man with nature. In recent years, with the concept of sustainability, which is often discussed, the improvement of spatial comfort has become more demanding of designers. The fact that energy efficiency is more talked about in design has led designers to consider these topics earlier in the design phase.

The lighting effect of daylight attracts more people's attention than artificial lighting and is more distinguished with increasing satisfaction. Brightness levels chosen by residents may vary depending on a person's sensitivity to light, sleep patterns, age, sense of comfort, vision, and type of movement. Daylight brightness is the most important factor in design. It is expedient to determine this brightness value based on fuzzy logic. The high brightness level chosen for visual comfort can provide better visual performance in the space but can also cause discomfort. According to the standards, depending on the average age of users, the brightness values in office buildings vary from 300 lx to 500 lx. The value of the brightness level set at 500 lx for users over 55 is set at 300 lx for users under 55 (see Table 1) [4–6].

Table 1. Lighting levels (in Lux) required by various public spaces.

Place	General	Special
	Lux	Lux
Homes		
Living rooms	50	500
Cuisines	125	250
Bedrooms	50	250
Entrance hall, stairs, roof snow, warehouse, garages	50	250
Offices		
Painting offices, cadaster, map	2500	
Projects, technical drawing, architecture	750	
Decorative pictures and sketches	500	
Accounting related devices	500	
Calculations	400	
Typewriter room	500	
Hall with documents	100	
Management offices	250	
Waiting rooms	150	
Conference rooms	200	
Malls		
Showcases	1000	
Purchasing centers in large cities (general lighting)	1000	
Additional lighting with spotlight	5000	
Other places (general lighting)	500	
Additional lighting with spotlight	2500	
Store Interior		
Department stores	500	
Trade centers of major cities	500	
Other places	250	
Location of very small parts	200	

3.1 Daylight Brightness Factor

On a surface, the division of the luminous flux per unit area by the surface area is defined as "Light level (E; lm/m2)". The symbol is 'E' and the unit is lx. Calculated by the following formula [5].

$$E(LuminousLevel) = \Phi(Bright\,Flux)/A(Area) \tag{1}$$

When the intensity of light falling on a certain surface does not change, the level of brightness in that area does not change. However, the linear structure, structure and distribution of light within the spectral structure may change. However, even if the light intensity remains the same, the brightness level decreases with distance.

4 Daylight and Brightness Effects in 3DsMAX

The designer must know enough CAD programs to illuminate any interior space. First, the designer must determine the window dimensions of the object and the direction of the sun in the Computer-Aided Design programs (see Fig. 3).

Fig. 3. Window dimensions and sun direction

3DsMax uses V-Ray and Corona plugins to get a better visual image. Each of these plugins has separate sunlight. Unlike the 3DsMAX program's default sunlight, V-ray and Corona lights allow for more realistic results. Sunlight in the V-Ray plugin is called V-Ray Sun, and sunlight in the Corona plugin is called CoronaSun.

When lighting, attention should be paid to the colors and brightness effects used in the interior. To change the brightness value of the material given to the object in 3DsMax, first select the M-key on the keyboard and select the required material cell in the window that opens. Then select the Reflect section below and enter the brightness value in the Value section of the window that opens. Here, in the traditional way, there is 0- not brightness, 255-full brightness [6–8]. However, based on fuzzy logic, designers can get more accurate results by taking this value in the range 0–255 (see Fig. 4).

Fig. 4. Changing the brightness value of the material

5   Practical Application of Fuzzy Technologies

The first step for applying fuzzy logic to a system is to determine the inputs and outputs of the system. Considering the proper lighting of the interior design, the most important expectation is that the daylight will be shared properly. In order to meet these expectations, certain parameters related to brightness, which form the outputs of the fuzzy logic model, gain importance. Fuzzy logic model's inputs and outputs are determined as shown in (see Fig. 5). With its fuzzy logic model of day brightness, brightness is aimed. Membership function numbers, names, lower and upper limits of all parameters are determined according to the effects of input and output parameters on the problem to be modeled. Figure 4 shows the membership functions of the input parameters and Fig. 5 shows the lower and upper limit values of the membership functions of the output parameters [6–11].

After determining the membership functions, lower and upper limit values of the parameters required to establish the model, 27 rules were created to establish the necessary relationships between the parameters affecting the system.

The value of the current surface reflectance factor is categorized into one of four fuzzy sets: dark, medium, light and very light. Similarly, the value of the opening is described as closed, medium, open and very open [8–13].

This value is in the range [0–100]. When the process is complete, a list is created and can be used visually.

Every linguistic variable has four fuzzy values with triangular or trapezoid membership functions, as follows:

- For input variables – Fig. 5: D-dark; M – Medium; L – Light; VL – Very Light
- For output variables – Fig. 6: VL – very low; L – low; M – medium; H – high; VH – very high.

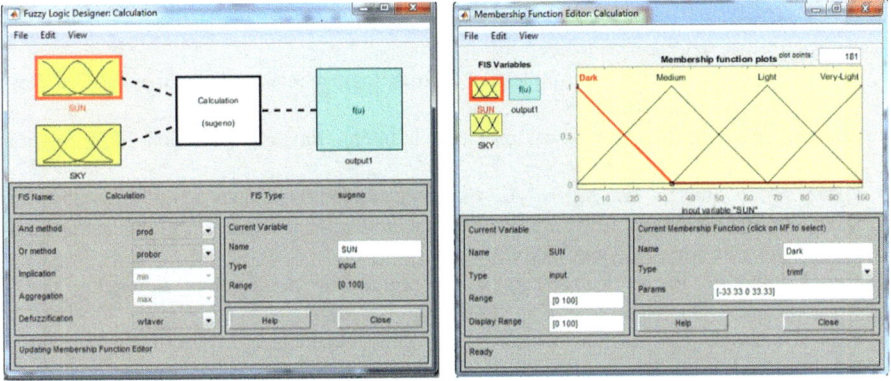

Fig. 5. Input variables

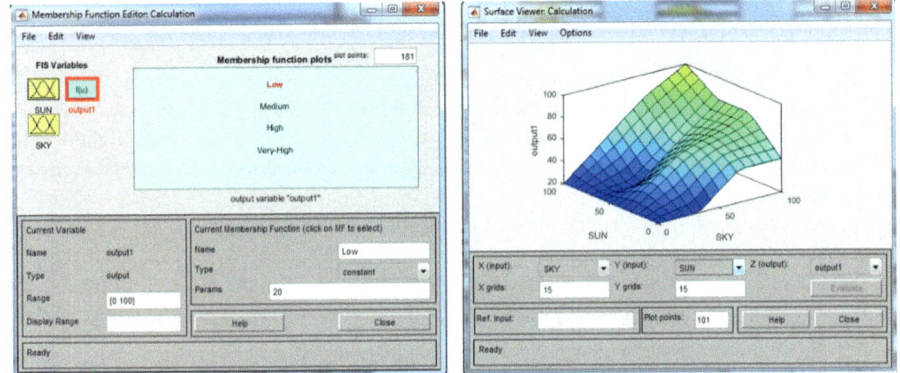

Fig. 6. Output variables.

6 Conclusion

The evaluation of the alternatives in Fig. 1 in the MATLAB\ Fuzzy Change Systems note described above allowed the following results to be obtained: Alternatively, in 3dsMax, the brightness value of daylight was set to dark, medium, light and very light instead of dark and light.

The results are more rhetorical, showing the effectiveness of the assessment of the brightness factor of daylight based on fuzzy additional theory. On the other hand, given other simple but vague and difficult formation criteria, it is useful to use fuzzy set theory when determining lighting with a simple objective method based on a perfect interior design. and decisions based on them are reasonable and economically feasible.

References

1. Zemmouri, N., Schiller, M.E.: Application of fuzzy logic in interior daylight estimation. Rev. Energ. Ren. **8**, 8–62 (2005)
2. Praharaj, Mayarani: Application of fuzzy logic in interior daylight evaluation. Elixir Sustain. Arc. **80**, 31339–31344 (2015)
3. Cziker, A., Chindris, M., Miron, A.: Fuzzy controller for indoor lighting system with daylighting contribution. In: ELECO 2007 5th International Conference on Electrical and Electronics Engineering, Turkey (2007)
4. Cao, C., Zhu, X.: Energy management using optimal fuzzy logic control in wireless sensor network. Int. J. Online Biomed. Eng. **14**(9), 35–52 (2018)
5. Ümit Arpacıoğlu, Cemal İrfan Çalışkan, Bahar Şahin, Nazlı Ödevci: Tasarim Kuram **16**(29), 53–78 (2020). https://doi.org/10.14744/tasarimkuram.2020.70783
6. Rea, M.S.: IESNA Lighting Handbook. 9th edn. Illuminating Engineering (2000)
7. Zhang, L., Yu, Y., Ma, H., Zhang, Y., Cao, P.: Design of photovoltaic power supply MPPT circuit for WSN node based on current observation. Int. J. Online Eng. **14**(7), 45–61 (2018)
8. Cziker, A., Chindris, M., Miron, A: Implementation of fuzzy logic in daylighting control. In: 11th International Conference on Intelligent Engineering Systems, pp. 195–200. INES (2007)
9. Görgülü, S., Ekren, N.: Energy saving in lighting system with fuzzy logic controller which uses light-pipe and dimmable ballast. Energy Build. **61**, 172–176 (2013)
10. Lighting Controls Association: Introduction to Lighting Automation. http://lightingcont rolsassociation.org/2005/02/14/introduction-to-lighting-automation-for-the-home/. Accessed 30 Apr 2020
11. Zadeh, L.A.: Fuzzy logic, neural networks, and soft computing. Commun. ACM **37**, 77–84 (1994)
12. IEA Task 21, Application Guide for Daylight Responsive Lighting Control. https://www.iea-shc.org/Data/Sites/1/publications/8-8-1%20Application%20Guide.pdf. Accessed 30 Apr 2020
13. Daylighting Control, Design and Application Guide. https://www.legrand.us/-/media/brands/wattstopper/resources/application-guide/ws-appguide-daylightingcontrol.ashx. Accessed 30 Apr 2020

Synthesis Algorithms for Neural Network Regulator of Dynamic System Control

A. N. Yusupbekov[1]([✉]) [ID], J. U. Sevinov[1] [ID], U. F. Mamirov[1] [ID],
and T. V. Botirov[2] [ID]

[1] Tashkent State Technical University, Tashkent, Uzbekistan
abek71@mail.ru, sevinovjasur@gmail.com,
uktammamirov@gmail.com
[2] Navoi State Mining Institute, Navoi, Uzbekistan
btvl979@mail.ru

Abstract. The main purpose of this research is to investigate stable synthesis algorithms of a neural network regulator for controlling dynamic systems. In this research, the synthesis of neural network regulator and analysis of the modern condition of intelligent control systems has been conducted. This paper presents stable synthesis algorithms of a multi-mode neural network controller on the basis of methods for solving variational inequalities, which can ensure the consistency of desired estimations and high accuracy of the intelligent control system. Based on estimation theory, the regular algorithms have been proposed in combination with adaptive identification algorithms. These derived algorithms can provide consistency of desired estimations with convergence properties and can be used in solving different kinds of problems related to system synthesis for controlling dynamic objects used for several functional purposes.

Keywords: Neural network regulator · Theory of artificial intelligence · Neural networks · Dynamic system · Control object · Adaptive identification · Estimation theory · Incorrectly posed problem

1 Introduction

The neural network system (NNS) plays a crucial role in the process of creating automatic control systems for complex dynamic objects. There are a number of ways to utilize NNS in control [1–10].

When designing a neural network controller (NNC), the problem arises of determining the dimension of a multilayer direct distribution neural network (MDDNN), which provides the required accuracy of functioning. It is impossible to divide small dimension NNC systems into classes with the desired accuracy. The redundancy of the dimension significantly affects the performance of the NNC during its software modeling. Therefore, when choosing the dimension of the MDDNN, it is required to determine its sufficient value to avoid a situation in which it is impossible to achieve the desired partition after continuous training.

© The Author(s), under exclusive license to Springer Nature Switzerland AG 2021
R. A. Aliev et al. (Eds.): ICAFS 2020, AISC 1306, pp. 723–730, 2021.
https://doi.org/10.1007/978-3-030-64058-3_90

2 Problem Definition

Let us consider the formalization of the problem of NNC synthesis based on the criteria of minimum complexity, while meeting the specified requirements for the accuracy, stability and quality of transient processes on a fixed set of operating modes of the object [9, 10]. Suppose that the dynamics of the control object (CO) is described by the differential equation "input – output", given implicitly [9]:

$$\varphi(y^{(n)}, y^{(n-1)}, \ldots, y; u^{(m)}, u^{(m-1)}, \ldots, u) = 0, \tag{1}$$

where $u = u(t)$ and $y = y(t)$ – the input and output of the investigated object, respectively; m and n - maximum orders of derivatives $u^{(i)}$, $y^{(j)}$ for input and output variables $u(t)$ and $y(t)$, $(m \le n)$; $\phi(\cdot)$ – is some non-linear function.

As a neural network, you can use a recurrent neural network. The dynamics of this network is described by the difference equation

$$u(k) = F(u(k-1), u(k-q), v(k), \ldots, v(k-p)), \tag{2}$$

where $F(\cdot)$ – a nonlinear function with respect to the indicated $(p+q+1)$ arguments, the specific form of which depends both on the selected activation functions of neurons and on the values of the weights of synaptic connections $W_{\alpha\beta}$, W_{β}.

The output signal NS $u(k)$ at the current time instant k depends both on the current value of the input signal $v(k)$ and on the past values of the input and output signals delayed by the corresponding number of clock cycles (sampling periods).

Assuming that a set of basic modes of operation of the CO is set, characterized by a set of coordinates $M_r = (u_0^{(r)} y_0^{(r)})$, $(r = 1, 2, \ldots, R)$, where $u_0^{(r)} y_0^{(r)}$ – fixed (steady-state) values of the input and output of the object for the r-rd basic mode; R – the number of such modes. In order to maintain the preset value $y = y_0^{(r)}$ at the output of the object, it is enough to apply a constant reference action (setpoint) $g_0^{(r)}$ to the input of the control system. An indispensable requirement for the functioning of the automatic control system (ACS) is the stability of the closed-loop control system at each of the above-mentioned basic modes of operation of the object.

3 Formalized Problem Statement

First, assuming that the deviations of coordinates

$$\Delta y = y^{(r)} - y_0^{(r)}, \ \Delta u^{(r)} = u^{(r)} - u_0^{(r)}, \ \Delta v = v^{(r)} - v_0^{(r)}, \ \Delta g^{(r)} = g^{(r)} - g_0^{(r)},$$

are small relative to their base values $v_0^{(r)}$, $u_0^{(r)}$, $v_0^{(r)}$, $g_0^{(r)}$ in the r-th mode, we write down the discrete transfer function of the linearized object $W_{CO}^{(r)}(z)$ for the r-th mode of operation (M_r), which, taking into account expression (1), will have the form

$$W_{CO}^{(r)}(z) = \frac{\Delta Y^{(r)}(z)}{\Delta U^{(r)}(z)} = \frac{a_0^r z^{-m} + \ldots + a_{m-1}^{(r)} z^{-1} + a_m^{(r)}}{b_0^r z^{-n} + \ldots + b_{n-1}^{(r)} z^{-1} + b_n^{(r)}}, \tag{3}$$

where $\Delta Y^{(r)}(z)$ and $\Delta U^{(r)}(z)$ – discrete Laplace images for deviations $\Delta y^{(r)}(k)$ and $\Delta u^{(r)}(k)$, and the coefficients $a_p^{(r)}$, $(\rho = 0, 1, 2, \ldots, m)$ and $b_y^{(r)}$, $(\gamma = 0, 1, 2, \ldots, n)$ depend on the form of the nonlinear function $\varphi(\cdot)$ of the coordinates of the base mode $u_0^{(r)}$, $y_0^{(r)}$, and the selected sampling period T_0.

Similarly, from the difference Eq. (2) you can go to the discrete transfer function of the linearized neural network $W_{NN}^{(r)}(z)$ for the r-nd mode of operation of object

$$W_{NN}^{(r)}(z) = \frac{\Delta U^{(r)}(z)}{\Delta V^{(r)}(z)} = \frac{c_0^r z^{-m} + \ldots + c_{p-1}^{(r)} z^{-1} + c_p^{(r)}}{d_0^r z^{-n} + \ldots + d_{q-1}^{(r)} z^{-1} + d_p^{(r)}}, \tag{4}$$

where $\Delta V^{(r)}(z)$ is a discrete Laplace image for a deviation of $\Delta v^{(r)}(k)$, and the coefficients $c_s^{(r)}$, $(s = 0, 1, 2.., p)$ and $d_t^{(r)}$, $(t = 0, 1, 2.., q)$ depend on several factors at once: numbers neurons σ in the hidden layer; type of activation function of neurons; weights of synaptic connections $W_{\alpha\beta}$, $W_{\alpha\beta}, (\alpha = 1, 2, \ldots, p + q + 1; \beta = 1, 2, \ldots, \sigma)$ as well as from the value of the network input $v_0^{(r)}$ on the r-st basic mode of ACS operation.

Considering the above, the characteristic equation of the linearized ACS obtained for the r-th mode of operation M_r takes the form:

$$1 + T_0 / (1 - z^{-1}) W_{CO}^{(r)}(z) W_{NN}^{(r)}(z) = 0, \tag{5}$$

where the discrete transfer functions $W_{CO}^{(r)}(z)$ and $W_{NN}^{(r)}(z)$ are determined by expressions (3) and (4).

4   Polynomial Notation

Turning to the polynomial form of notation, we can rewrite Eq. (5) as follows:

$$H^{(r)}(z) = h_0^{(r)} z^{-L} + h_1^{(r)} z^{-(L-1)} + \ldots + h_{L-1}^{(r)} z^{-1} + h_L^{(r)} = 0 \tag{6}$$

where $H^{(r)}(z)$ – the characteristic polynomial of the closed ACS for the r-rd operating mode of the CO; $L_{max}\{n + q + 1; m + p\}$ - the order of the characteristic polynomial $H^{(r)}(z); h_0^{(r)}; h_1^{(r)}, \ldots, h_L^{(r)}$ - the numerical coefficients of this polynomial, depending on $(p + q + 1)$ unknown coefficients c_s, d_t of the discrete transfer function $W_{NN}^{(r)}(z)$, which, in turn, are functions of the tunable weights of the synaptic connections $W_{\alpha\beta}$, W_β, the neural network. For definiteness, in (3), (4), (6) we can take: $b_0^{(r)} = d_0^{(r)} = h_0^{(i)} = 1$.

To ensure stability and a given quality of transient processes in the r-th basic operating mode of the ACS, we require the fulfillment of condition

$$H^{(r)}(z) = H_d^{(r)}(z), \tag{7}$$

where $H_d^{(r)}(z) = z^{-L} + h_{1,d}^{(r)} z^{-(L-1)} + \ldots + h_{L-1,d}^{(r)} z^{-1} + h_{L,d}^{(r)}$ – is the desired characteristic polynomial of a closed ACS with a given distribution of roots $(|z_l^*| < 1, l = 1, 2, \ldots, L)$.

Considering a set of R basic operating modes of system $M_r, (r = 1, 2, \ldots, L)$ for each of which condition (7) must be satisfied, we arrive at a system of R equations

$$\begin{cases} H^{(1)}(z) = H_d^{(1)}(z), \\ \cdots\cdots\cdots\cdots\cdots\cdots, \\ H^{(R)}(z) = H_d^{(R)}(z); \end{cases} \tag{8}$$

where the choice of polynomials $H_d^{(1)}(z), \ldots, H_d^{(R)}(z)$ is carried out taking into account the possible difference in the requirements for quality indicators (regulation time, overshoot) at each of the specified modes of operation of the object.

Considering that the coefficients of the polynomials $H^{(r)}(z)$ on the left side of Eq. (8) depend on the tunable parameters of the NS-controller - the weights of synaptic connections $W_{\alpha\beta}, W_{\alpha\beta}, (\alpha = 1, 2, \ldots, p+q+1; \beta = 1, 2, \ldots, \sigma)$ one can rewrite Eq. (8) as follows

$$\left\{ \begin{array}{c} \text{mode'} \\ M_1 : h_1^{(1)}(W_{11}, \ldots, W_{p+q+1,v}; W_1, \ldots, W_\sigma) = h_{1,d}^{(1)}; \\ h_2^{(1)}(W_{11}, \ldots, W_{p+q+1,v}; W_1, \ldots, W_\sigma) = h_{2,d}^{(1)}; \\ \cdots\cdots\cdots\cdots\cdots\cdots\cdots\cdots\cdots\cdots\cdots\cdots\cdots; \\ h_L^{(1)}(W_{11}, \ldots, W_{p+q+1,v}; W_1, \ldots, W_\sigma) = h_{1,d}^{(1)}; \\ \text{-----------------------} \\ \vdots \\ \text{-----------------------} \\ \text{mode'} \\ M_R : h_1^{(R)}(W_{11}, \ldots, W_{p+q+1,v}; W_1, \ldots, W_\sigma) = h_{1,d}^{(1)}; \\ h_2^{(R)}(W_{11}, \ldots, W_{p+q+1,v}; W_1, \ldots, W_\sigma) = h_{2,d}^{(R)}; \\ \cdots\cdots\cdots\cdots\cdots\cdots\cdots\cdots\cdots\cdots\cdots\cdots\cdots; \\ h_L^{(R)}(W_{11}, \ldots, W_{p+q+1,v}; W_1, \ldots, W_\sigma) = h_{L,d}^{(R)}. \end{array} \right. \tag{9}$$

It is easy to see that the total number of nonlinear algebraic equations forming system (9) is

$$(NE) = RL = R\max\{n + q + 1; m + p\}, \tag{10}$$

while the number of unknown parameters included in it $\{W_{\alpha\beta}, W_{\alpha\beta}\}$ will be:

$$(NP) = (p + q + 1)\sigma + \sigma = (p + q + 2)\sigma. \tag{11}$$

We require the fulfillment of the condition of certainty of the system of Eqs. (9), which assumes that the number of unknown parameters (NP) included in it should be no less than the number of equations (NE) connecting them.

Substituting expressions (10) and (11) in (12), we obtain inequality $(p + q + 1)\sigma \geq R\max\{n + q + 1; m + p\}$ which, in turn, breaks down into two conditions:

$$(p + q + 2)\sigma \geq R\{n + q + 1\}; \ (p + q + 2)\sigma \geq R\{m + p\}; \tag{12}$$

or, after performing elementary transformations:

$$(p + q + 2)\sigma - Rq \geq R(n + 1); (p + q + 2)\sigma - Rq \geq Rm. \tag{13}$$

Using relations (13), it is possible to determine the required values p and q, i.e. the number of delay elements at the input and output of the network, as well as the number of neurons in the hidden layer a, which guarantee the presence of at least one solution to the system of Eq. (9). In this case, the orders m and n of the differential equation of the object (1) and the number R of the basic operating modes of the CO, in relation to which the requirements for the stability and quality of control processes are formulated, are used as the initial data.

5   Solution

We write Eq. (9) in the form:

$$H(W) = h, \quad W \in D(H) \tag{14}$$

defined by the operator $H : D(H) \rightarrow G$, $D(H) \subset G$, where G - a real Hilbert space.

Under the above assumptions, problem (14) is generally incorrect [11–20]. The incorrectness of problem (14) means that solution (14) may be absent or unstable to small variations in the initial data. One of the classical methods of studying nonlinear equations consists in regularizing these equations with a small term [21], which improves their qualitative characteristics, with the subsequent derivation of a priori estimates of solutions and the tendency of the regularization parameter to zero. In this way, various results were established related to the solvability and smoothness of solutions of various classes of nonlinear equations [21]. One of the main approaches to the design of such algorithms is the principle of iterative regularization [17].

The vector $\hat{W}$ will be determined based on the minimization of a functional of the form:

$$J(W) = \sum_{j=1}^{l} (H_j(W) - h_j)^2 \tag{15}$$

i.e. the required estimate of the vector $\hat{W}$ is $\hat{W} = \arg\inf_{W \in R^k} J(W)$, where $l = R\max$ $(n+q+1, m+p)$, $k = (p+q+2)\sigma$.

Functional $J(W)$ in (15) is a convex differentiable functional defined on a closed convex set $Q \subset G$ of a Hilbert space G. It is well known [17] that the problem of minimizing a convex functional on a closed convex set $Q \subset G$ of a Hilbert space can be reduced to the problem of solving a variational inequality with respect to W:

$$(J'(W), W - d_j) \leq 0, \quad \forall d_j \in Q \tag{16}$$

Here $J'(W)$ is the usual gradient in the smooth case and the element of the subdifferential $J(W)$ in the nonsmoothed case. If $J(W)$ – a convex functional, then $J'(W)$ – a monotone operator [17]. The monotonicity of operator $J'(W)$ means that condition $W, d_j \in G$ is satisfied for all elements

$$(J'(W_2) - J'(W_1), W_2 - W_1) \geq 0, \quad \forall W_1, W_2 \in Q \tag{17}$$

where the bracket $(\cdot , \cdot)$ on the left side of (17) means the dot product.

Thus, it is necessary to determine such W, for which

$$(F(W), W - d_j) \leq 0, \quad \forall d_j \in Q \tag{18}$$

where $F(W)$ – a monotone operator.

We will assume that instead of the exact operator F, its approximation $F_\delta : G \to G$, $\rho(F_\delta - F) \leq \delta$, $\delta \geq 0$, $F_\delta \in F$, is available, where F – a certain class of approximate operators, the parameter δ characterizes the level of error in setting F and is assumed to be known. Let $M(W)$ – an operator with the strong monotonicity property on $Q \subset G$, that is, $(M(W_1) - M(W_2), W_1 - W_2) \geq C_M \|W_1 - W_2\|^2$, $C_M > 0$. Instead of the main inequality (18), it is advisable to use the family of auxiliary inequalities

$$(F_\delta(W) + \varepsilon M(W), W - d_j) \leq 0, \quad \forall d_j \in Q, \ \varepsilon > 0 \tag{19}$$

Variational inequalities (19) have significantly better properties, since the operator $F_\delta(W) + \varepsilon M(W)$ is strongly monotone for a fixed $\varepsilon > 0$, and standard iterative methods are applicable to their solution [16–18]. On the other hand, the variational inequalities (19) have, for $\varepsilon \to 0$, the Browder-Tikhonov approximating properties [17] for problem (18).

In the case under consideration, one can, for example, accept $M(W) = W$ and $C_M = 1$. Inequality (19) can be called a regularized inequality. Additive εM is similar to the stabilizing additive of AN Tikhonov in the theory of ill-posed extremal problems [17, 18]. If we assume only the existence of a solution to problem (16), then under broad assumptions about F and M, we can assert that for any $\varepsilon > 0$ there exists W_ε - the only solution to (19) and, moreover, there is a limit relation

$$\lim_{\varepsilon \to 0} \| W_\varepsilon - W^* \| = 0 \tag{20}$$

where $W^* \in A$ – the only solution of the variational inequality $(M(W), W - d_j) \leq 0$, $\forall d_j \in A$, A – set of solutions (17).

Following [17–22], it can be shown that under the conditions considered above, the iterative sequence $\hat{W}_r$ can be written in the following form:

$$\hat{W}_{r+1} = P_Q(\hat{W}_r - \alpha_{r,j}(F_\delta(\hat{W}_r) + \varepsilon_{r,j} M(\hat{W}_r))), \quad r = 0, 1 \tag{21}$$

where P_Q – a metric projector; $\alpha_{r,j} > 0$, $\varepsilon_{r,j} > 0$ – regularization parameters, r – iteration number. To ensure the convergence of the iterative process (20–21), parameters $\alpha_{r,j}$ and $\varepsilon_{r,j}$ must satisfy the following conditions

$$\lim_{r \to \infty} \frac{\alpha_{r,j}}{\varepsilon_{r,j}} = 0, \quad \lim_{r \to \infty} \varepsilon_{r,j} = 0, \quad \sum_{n=1}^{\infty} \alpha_{r,j} \varepsilon_{r,j} = \infty, \quad \lim_{r \to \infty} \frac{|\varepsilon_{r,j} - \varepsilon_{r+1,j}|}{\alpha_{r,j} \varepsilon_{r,j}^2} = 0.$$

As an example of sequences $\alpha_{r,j}$ and $\varepsilon_{r,j}$ satisfying the required conditions, we can take sequences $\alpha_{r,j} = (1+r)^{-1/2}$, $\varepsilon_{r,j} = (1+r)^{-p}$, $0 < p < 1/2$.

Stopping the considered iterative process can be carried out on the basis of relations of the form [16–18]: $\lim_{\delta \to 0} \delta/\varepsilon_{r,j(\delta)} = 0$, $\lim_{\delta \to 0} \delta^{1/2}/\varepsilon_{r,j(\delta)}^2 = 0$.

6   Conclusion

Above mentioned derived algorithms (with adaptive identification algorithms based on estimation theory) can provide consistency of desired estimations with convergence properties, and can be employed to solve different kinds of problems regarding system synthesis for controlling dynamic objects used for several functional purposes.

References

1. Sigeru, O., Khalid, M.B., Rubiyah, Y.: Neuro-Control and its Applications. Springer, London (1996)
2. Rutkovskaya, D., Pilinsky, M., Rutkovsky, L.: Neural Networks, Genetic Algorithms and Fuzzy Systems. Hotline-Telecom, Moscow (2004)
3. Terekhov, V.A., et al.: Neural Network Control Systems. Book 8. (Neurocomputers and Their Application). IPRZHR, Moscow (2002)
4. Vasiliev, V.I., Panteleev, S.V.: Neurocontrol - a new section of the theory of complex systems control. Neurocomput. Dev. Appl. 5, 33–45 (2005)

5. Uskov, A.A., Kuzmin, A.V.: Intelligent Control Technologies. Artificial Neural Networks and Fuzzy Logic. Hotline-Telecom, Moscow (2004)
6. Kocijan, J.: Modelling and Control of Dynamic Systems Using Gaussian Process Models. Springer, New York (2016)
7. Yusupbekov, N.R., Marahimov, A.R., Gulyamov, Sh.M., Igamberdiev, H.Z.: APC fuzzy model of estimation of cost of switches at designing and modernizations of data-computing networks. In: AICT 2010, pp. 244–249 (2010). https://doi.org/10.1109/icaict.2010.5612015
8. Xakimovich, S.I., Maxamadjonovna, U.D., Askarxodjaevna, B.H.: Adaptive system of fuzzy-logical regulation by temperature mode of a drum boiler. IIUM Engin. J. **21**(1), 182–192 (2020). https://doi.org/10.31436/iiumej.v21i1.1220
9. Vasiliev, V.I., Ilyasov, B.G.: Intelligent Control Systems. Theory and Practice. Radio Engineering, Moscow (2009)
10. Egupov, N.D.: Methods of Robust, Neuro-Fuzzy and Adaptive Control. Bauman Mosc. State Tech. Univ., Moscow (2002)
11. Verlan, A.F., Sizikov, V.S.: Integral Equations: Methods, Algorithms, Programs. Naukovadumka, Kiev (1986)
12. Kolomeitseva, M.B., Ho, D.L.: Adaptive control systems for dynamic objects based on fuzzy controllers. Sputnik+Company, Moscow (2002)
13. Yusupbekov, N.R., Igamberdiev, H.Z., Sevinov, J.U.: Formalization of identification procedures of control objects as a process in the closed dynamic system and synthesis of adaptive regulators. J. Advan. Res. Dyn. Cont. Sys. **12**(06), 77–88 (2020). https://doi.org/10.5373/JARDCS/V12SP6/SP20201009
14. Yusupbekov, N.R., Igamberdiev, H.Z., Mamirov. U.F.: Algorithms of sustainable estimation of unknown input signals in control systems. J. Mult. Val. Log. Soft Comput. **33**(1–2), 1–10 (2019). https://www.oldcitypublishing.com/pdf/9291
15. Sevinov, J.U., Zaripov, O.O., Zaripova, Sh.O.: The algorithm of adaptive estimation in the synthesis of the dynamic objects control systems. Inter. J. Adv. Sci. Tech., **29**(5), 1096–1100 (2020). http://sersc.org/journals/index.php/IJAST/article/view/7887
16. Alifanov, O.M., Artyukhin, E.A., Rumyantsev, S.V.: Extremal Methods for Solving Ill-Posed Problems. Science, Moscow (1988)
17. Bakushinsky, A.B., Goncharsky, A.V.: Iterative Solution Methods for Ill-Posed Problems. Science, Moscow, (1989). (in Russian)
18. Vainikko, G.M., Veretennikov, A.Yu.: Iterative Procedures in Ill-Posed Problems. Science, Moscow (1986)
19. Igamberdiev, H.Z., Sevinov, J.U., Yusupbekov, A.N.: Regular algorithms for identifying the parameters of an object and a controller in a closed-loop control system. J. Chem. Tech. Cont. Manag. **6**, 50–54 (2017)
20. Zaripov, O.O., Shukurova, O.P., Sevinov, J.U.: Algorithms for identification of linear dynamic control objects based on the pseudo-concept concept. Inter. J. Psy. Rehab. **24**(3), 261–267 (2020). https://doi.org/10.37200/IJPR/V24I3/PR200778
21. Bakushinsky, A.B., Kokurin, M.Yu.: Iterative methods for solving irregular equations. Lenand, Moscow (2006)
22. Igamberdiyev, H.Z., Yusupbekov, A.N., Zaripov, O.O., Sevinov, J.U.: Algorithms of adaptive identification of uncertain operated objects in dynamical models. Procedia Comput. Sci. **120**, 854–861 (2017). https://doi.org/10.1016/j.procs.2017.11.318

Modeling of Oil and Gas Production for an Enterprise by Using Fuzzy Cobb-Douglas Function

V. J. Axundov$^{(\boxtimes)}$ (iD)

Research Laboratory of Intelligent Control and Decision Making Systems in Industry and Economics, Azerbaijan State Oil and Industry University, Azadlig ave., 20, 1010 Baku, Azerbaijan
azeri46@mail.ru

Abstract. Classical modeling methods are not developed to account for fuzziness or imprecision. Real data on production volumes of enterprise may be characterized by fuzziness. In this paper, we applied the theory of fuzzy sets to modeling of production at oil and gas enterprises under imprecise information.

Keywords: Fuzzy number · Cobb-Douglas function · Reduced value of product

1 Introduction

As a rule, when considering economic problems, we have to deal with imprecision of source data and a lack of information on influential factors and their relationship. In such cases, modeling problems can be solved by using interbal or fuzzy models [1]. In [2] interval analysis is used to apply Leontief "input-output" model under imprecise information on production. The fuzzy set theory-based model is used to build binomial option pricing model in [3]. Analysis of existing literature showed that fuzzy set theory is widely used for economic modeling problems. One kind of such problems systematic forecasting of production at oil and gas production enterprises under fuzzy information.

In this paper we use fuzzy set theory to apply Cobb-Douglas function for modeling of production in a large enterprise under imprecision of information related to production volume, capital, labor and other quantities.

The paper is organized as follows. The second section of the article classifies the factors affecting the volume of final products at oil and gas production enterprises. A procedure for calculating the reduced value of the product produced in this area is used. In the third section of the article, fuzzy Cobb-Douglas model is used to calculate the fuzzy values of the end product at the oil and gas enterprise under study.

R. A. Aliev et al. (Eds.): ICAFS 2020, AISC 1306, pp. 731–737, 2021.
https://doi.org/10.1007/978-3-030-64058-3_91

2   Calculation of the Reduced Value of the End Product Manufactured at Oil and Gas Production Enterprises

In practice, when constructing a production function for parametric classes, the argument list is expanded to include new parameter vectors:

$$Q = F(K, L; w) \tag{1}$$

where Q - is the nominal value of the product produced.

For the practical application of the problem, it is necessary to choose a class of functions and parameters (w = β, α, A) in such a way as to ensure good equality of indicators. It should be noted that due to the uncertainty of the different origin of the data, it is impossible to completely ensure this equality for all values of Yt, Kt, Lt. To take into account the uncertainty of modeling and statistics at oil and gas production enterprises, it is necessary to link the model to a regression system of equations that takes into account the additive error. In order to get a more accurate forecast in this area of production, it is necessary to take into account the uncertainty of measurement evaluation [3]:

$$Q = F(K, L; w) + \varepsilon_t \quad t = \overline{0, T}, \tag{2}$$

where ε_t - represents the uncertainty in the measurement evaluation. Symbolically, the problem of uncertainty in data measurements is called the problem of evaluating data measurements.

However, possibilistic uncertainty related to real-world production data does not allow to use classical regression models. This must be taken into account when establishing the production function in oil and gas production enterprises, since they have a significant impact on the volume of the end product produced.

In the process of forecasting oil and gas production, it is important to divide the factors affecting the volume of the end product into 2 groups. The first group includes the evaluation of data measurements (ε_t). Examples of this group of factors are changes in the volume of production in accordance with international agreements, significant fluctuations in fuel prices, changes in the exchange rate of the national currency. These types of factors are assessed by adjusting statistical indicators. The second group of factors can often be data uncertainty, their inaccuracy, statistical errors in the assessment of economic events and characteristics. These types of factors can be estimated using fuzzy set theory.

First, let's look at the measurement evaluation (ε_t) factors in the oil and gas production forecasting process.

As noted above, the expected uncertainty in the measurement evaluation is characterized by the first group of influencing factors, which is determined by the formula that we propose below:

$$\varepsilon_t = (H \pm P) \cdot N \cdot K \tag{3}$$

where H is the natural volume of oil and gas exported in the current year; P-due to international agreements, geological factors, etc. the expected changes in the export of oil (gas); N is -average annual export price of 1 barrel of oil (or 1000 m³ of gas); K-average annual exchange rate of the national currency to 1 US dollar. It should be noted that if the indicators of the current year or previous years are analyzed, then P = 0.

Using Eqs. (2) and (3), we can express the nominal value of the end product as

$$Q = Y + \varepsilon_t \quad \text{or,} \quad Y = Q - \varepsilon_t \tag{4}$$

Y - the reduced value of the product produced.

The study was conducted at the level of an oil and gas company to show practical solutions to the problem. Table 1 shows the technical and economic indicators of the oil and gas production facility used to build the model.

Table 1. Technical and economic indicators of oil and gas production enterprise

Years	Amount of oil produced	Average annual price of barrel of oil	Amount of gas produced (1000 m³)	Average annual cost of 1000 m³ of gas	Total cost of oil and gas produced (USD)	Average annual cost of fixed assets	Average number of employees
2007	8800.9	72.7	387239.6	205.5	1055766	5162385	64145
2008	8651.3	97.7	380657.2	284.3	1099439	6974067	65083
2009	8543.3	61.9	375905.2	126.2	1200078	7946556	71635
2010	8459.7	79.6	372226.8	127.2	1245439	8244627	75552
2011	8400.9	111	369639.6	129	1291819	9065173	78844
2012	8289.8	111.4	364751.2	99.1	1230998	10776896	70901
2013	8314.9	108.8	365855.6	143.2	1182075	11665000	65568
2014	8320.6	98.9	366106.4	141.04	1081111	12134000	56460
2015	8160.6	52.4	359066.4	82.6	1087576	17236000	52576
2016	7522.4	44	330985.6	85.5	1083597	20116000	50933
2017	7427.1	54.53	326792.4	89.6	1100498	25669000	50122
2018	7542.3	72.93	331861.2	85.75	1114527	28259000	50332

Y (the reduced value of the product) was calculated using the indicators in Table 1 and formulas (4), (3) and is given in Table 2.

The definition of the reduced value of the product allows you to more accurately estimate the real volume of production in value terms. This economic indicator was used in the next section to predict the end product under conditions of uncertain information.

Table 2. The reduced value of the product of oil and gas produced at the oil and gas production facility

Value	2007	2008	...	2016	2017	2018
The reduced value of the product (Oil) Y'_n	387239.6	380657.2	...	330985.6	326792.4	331861.2
The reduced value of the product (Gaz) Y'_q	495393.5	640364.8	...	517637.7	502959.7	539014.6
TOTAL (Y)	882633.1	1021022	...	848623.3	829752.1	870875.8

3 Forecasting the End Product at Oil and Gas Enterprises Under Imprecise Information

To obtain more adequate results in the process of forecasting oil and gas production, it is necessary to take into account the uncertainty of the data, their inaccuracy, statistical errors, etc.

In the calculations during the study, a fuzzy Cobb-Douglas model was used to predict the end product in oil and gas production:

$$\tilde{Y}' = \tilde{A} \cdot \tilde{K}^{\tilde{\alpha}} \cdot \tilde{L}^{\tilde{\beta}} \tag{5}$$

where the values of $\tilde{Y}'$, $\tilde{K}$, $\tilde{L}$, and parameters $\tilde{A}, \tilde{\alpha}, \tilde{\beta}$ are described by fuzzy numbers. Using fuzzy data, we will calculate fuzzy prices for the end product of this industry based on fuzzy estimates of the number of fixed assets and the average number of employees.

Table 3 shows fuzzy estimates (in form of triangular fuzzy numbers, TFN) of the average annual cost of fixed assets and the average number of employees of an oil and gas production company, taking into account the degree of probable difference in 2007–2018.

Using this table, construction of Cobb-Douglas production function under fuzzy information (described by Eq. (5)) is based on a solution to the following optimization problem:

$$\sum_{years} d\left(\tilde{Y}, \tilde{Y}'\right) \rightarrow \min \tag{6}$$

s.t.

$$\underline{\tilde{A}} \leq \tilde{A} \leq \overline{\tilde{A}}, \tag{7}$$

$$\underline{\tilde{\alpha}} \leq \tilde{\alpha} \leq \overline{\tilde{\alpha}}, \tag{8}$$

Table 3. Fuzzy estimates of Y, K, L for oil and gas production (2003–2018).

	Average annual cost of fixed assets, (thousand dollars) $\widetilde{K}$	Average number of employees (thousand people) $\widetilde{L}$	Cost of extracted oil and gas (thousands of dollars) (the reduced value of the product) $\widetilde{Y}'$
1	2	3	4
2007	(5007.5; 5162.4; 5317.3)	(62.2; 64.1; 66.1)	(848.7; 882.6;917.9)
2008	(6764.8; 6974.0; 7183.3)	(63.1; 65.1; 67.0)	(981.8; 1021; 1061.9)
2009	(7549.2; 7946.5; 8343.9)	(68.1; 71.6; 75.2)	(890.0; 946.1; 1005.6)
2010	(7832.4; 8244.6; 8656.9)	(71.8; 75.6; 79.3)	(908.0; 965.2;1026.0)
2011	(8611.9; 9065.1; 9518.4)	(74.9; 78.8; 82.8)	(898.2; 954.8; 1014.9)
2012	(1(0238.1; 10776.9; 11315.7)	(67.4; 70.9; 74.4)	(881.2; 936.7; 995.7)
2013	(1(1081.7; 11665.0; 12248.3)	(62.3; 65.6; 68.8)	(899.0; 955.6; 1015.8)
2014	(1(1527.3; 12134.0; 12740.7)	(53.6; 56.5; 59.3)	(905.7; 962.7; 1023.4)
2015	(1(6029.5; 17236.0; 18442.5)	(48.9; 52.6; 56.3)	(847.8; 926.7; 1012.8)
2016	(1(8707.9; 20116.0; 21524.1)	(47.4; 50.9; 54.5)	(776.4;848.6; 927.6)
2017	(2(3872.2; 25669.0; 27465.8)	(46.6; 50.1; 53.6)	(759.2; 829.8; 906.9)
2018	(2(6280.9; 28259.0; 30237.1)	(46.8; 50.3; 53.9)	(796.8; 70.9; 951.9)

$\widetilde{Y}$ denotes fuzzy reduced values of end product (Table 3), $\widetilde{Y}'$ is a fuzzy value of function (5), $\widetilde{A}$, $\widetilde{\alpha}$ and $\widetilde{\beta}$ are the fuzzy parameters sought for the fuzzy Cobb-Douglas production function, $\widetilde{\overline{A}}$, $\underline{\widetilde{A}}$ and $\underline{\widetilde{\alpha}}$, $\widetilde{\overline{\alpha}}$, $\underline{\widetilde{\beta}}$, $\widetilde{\overline{\beta}}$ are the boundary conditions of the ranges of these parameters. Distance between fuzzy numbers d is adopted from [4]. Comparison of fuzzy numbers $\leq$ is considered as formulated in [5]. The problem is to find such values of $\widetilde{A}$, $\widetilde{\alpha}$ and $\widetilde{\beta}$ that minimize d between fuzzy data (Table 3) and fuzzy values of Cobb-Douglas production function. $\widetilde{A}$, $\widetilde{\alpha}$ and $\widetilde{\beta}$ parameters are taken for the studied oil and gas production enterprise as follows:

$$\underline{\tilde{A}} = (0.98; 1; 1.02), \tilde{\bar{A}} = (294.12; 300; 306),$$
$$\underline{\tilde{\alpha}} = (0.098; 0.1; 0.102), \tilde{\bar{\alpha}} = (0.882; 0.9; 0.918),$$
$$\underline{\tilde{\beta}} = (0.098; 0.1; 0.102), \tilde{\bar{\beta}} = (0.882; 0.9; 0.918).$$

By solving problem (6)–(8), the fuzzy parameters $\tilde{A}$, $\tilde{\alpha}$ and $\tilde{\beta}$ for the investigated oil and gas production enterprise were found as follows:

$$\tilde{A} = (256.2109; 261.4397; 266.6685).$$
$$\tilde{\alpha} = (0.580053; 0.591891; 0.603729).$$
$$\tilde{\beta} = (0.098; 0.1; 0.102).$$

The obtained value of (6) of the solution was 57776.45 thousand manats, which is of approx. 6% accuracy. Thus, the obtained results can be used for modeling of production in the considered enterprise with acceptable accuracy. The corresponding values of Cobb-Douglas production function $\widetilde{Y}'$ are shown in Table 4.

Table 4. Fuzzy Cobb-Douglas function values

Years	TFN-based values of production function, $\widetilde{Y}'$
2007	(848685.67; 882633.10; 917938.42)
2008	(981751.88; 1021021.96; 1061862.84)
2009	(890021.64; 946093.00; 1005696.86)
2010	(907999.94; 965203.94; 1026011.79)
2011	(898207.45; 954794.52; 1014946.57)
2012	(881207.11; 936723.16; 995736.72)
2013	(898991.40; 955627.86; 1015832.42)
2014	(905653.51; 962709.68; 1023360.39)
2015	(847806.31; 926652.30; 1012830.96)
2016	(776416.54; 848623.28; 927545.25)
2017	(759151.02; 829752.06; 906919.00)
2018	(796775.63; 870875.76; 951867.21)

The fuzzy values of end product and those of Cobb-Douglas production function $\widetilde{Y}'$ are shown graphically below (Fig. 1, the lower bound, the core and the upper bound of the TFNs are shown).

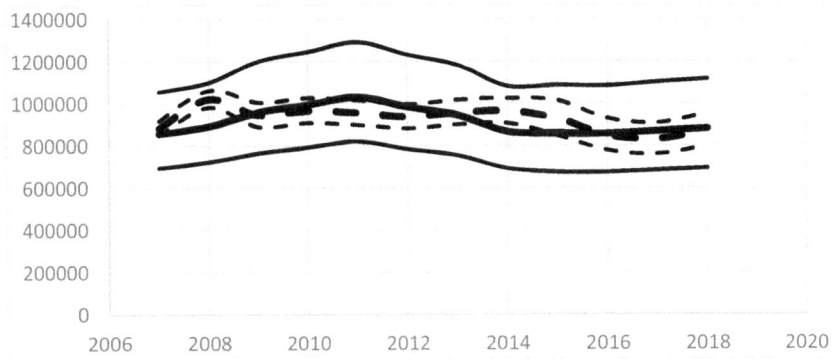

Fig. 1. Fuzzy Cobb-Douglas function graph (solid curve) and imprecise data (dashed curve).

4 Conclusion

In order to get a more adequate forecast of end product of oil and gas production it is necessary to take into account imprecise information. In this paper we used fuzzy Cobb-Douglas model to account for fuzzy values of end product, capital, labor and production parameters. The obtained result show adequacy of the used approach.

References

1. Aliev, R.A., Fazlollahi, B., Aliev, R.R.: Soft Computing and Its Applications in Business and Economics. Springer, Berlin (2004)
2. Akhundov, V.J., Mammadova, S.K., Aliyev, A.M.: Investigation of the "Input-Output" model of the Azerbaijani economy in interval information conditions. In: Aliev, R.A., Kacprzyk, J., Pedrycz, W., Jamshidi, M., Sadikoglu, F.M. (eds.) 13th International Conference on Theory and Application of Fuzzy Systems and Soft Computing—ICAFS-2018. Advances in Intelligent Systems and Computing, vol. 896. Springer, Cham (2019). https://doi.org/10.1007/978-3-030-04164-9_100
3. Lee, C.F., Tzeng, G., Wang, S.: A fuzzy set approach for generalized CRR model: an empirical analysis of S&P 500 index options. Rev. Quant. Finan. Acc. **25**, 255–275 (2005). https://doi.org/10.1007/s11156-005-4767-1
4. Zhang, X., Chen, W.M.: New similarity of triangular fuzzy number and its application. Adv. Inform. Tech. p. 7 (2014). Article ID 215047. https://doi.org/10.1155/2014/215047
5. Zhang, G.-Q.: Fuzzy Number-Valued Measure Theory. Tsinghua University Press, Beijing (1998)

Neural Networks - Based Ecological Forecasting

A. B. Sultanova$^{(\boxtimes)}$ ⓘD and M. Y. Abdullayeva ⓘD

Azerbaijan State University of Oil and Industry, Baki 1010, Azerbaijan
saxira@mail.ru, mayaabdullayeva@hotmail.com

Abstract. The article is devoted to the use of neural networks to predict changes in the structure of waste depending on the level of well-being of the population. To predict the dependence of gross national income on municipal solid waste, a neural network was built based on the MultiLayer Perceptron (MLP), which predicts the time series. MLP is based on sigmoidal neurons. It was found that, based on a fuzzy model, in countries with a middle high-income population, the share of organic waste in MSW decreases, while the share of recyclable fractions increases. Based on statistical data, a fuzzy model was compiled for the classification of municipal solid waste depending on the level of well-being of the population and a methodology for teaching the fuzzy model algorithm was developed.

Keywords: Municipal solid waste · Neural networks · Organic waste · Income level

1 Introduction

Since 2008, there has been a steady increase in waste generation indicators in Azerbaijan, which corresponds to the dynamics of increasing the well-being of the population. It is important to note that despite the fact that, according to the World Bank classification, Azerbaijan is classified as a middle-income country, i.e. the potential for the specific growth of municipal solid waste (SHW) in the country is significant, but has not been realized yet. High-income countries generate the most waste per capita, while low-income countries generate the least solid waste per capita. It is possible to expect that the growth of individual indicators of waste formation will be preserved in the near future and will be stabilized in a few years. Growth will be able to increase the population's well-being and change consumer behavior [1].

Within the framework of the forecast of growth of GDP for the near future in the course of 10 years it is possible to expect the achievement of saturation points at the level of up to 400 kg of SHW per person per year. Dependence on the volume of education deviations from the level of well-being has been confirmed by international comparisons. Average allocated volumes of SHW education in countries with a large income are characterized by a large specific volume of waste disposal. According to the data received, countries with a large income are characterized by a large share of expenditures. In addition, according to the size of revenues, the share of organ-Czech expenditures in the SHW is reduced, and the fractions that are suitable for processing are increased [2].

Solid waste management is the only thing that nearly every city government provides for its residents. While service levels, environmental impacts and costs vary widely, solid waste management is arguably the most important municipal service and serves as a prerequisite for other municipal actions. As the world rapidly approaches its urban future, the amount of municipal solid waste (SHW), one of the most important by-products of urban lifestyles, is growing even faster than the rate of urbanization [2].

Changes in the structure of solid waste are influenced by the climate and national characteristics, improved collection, disposal and accounting of waste, as well as an increase in the share of solid waste from small and medium-sized businesses (however, the contribution of small enterprises is insignificant).

In the future, in Azerbaijan, one can expect a further increase in the share of recyclable fractions (paper and cardboard, glass, metal, plastic, etc.), accompanied by a decrease in the share of organic waste [3]. The specific volumes of their education in Azerbaijan are approaching the indicators of European countries. This means that the potential for MSW processing is also approaching the European value, and a further increase in the relative volumes of suitable fractions can be assumed.

The use of traditional forecasting methods does not always provide a satisfactory result. In this regard, information technologies have become widespread, which make it possible to obtain the necessary analytical forecast information. Artificial intelligence technologies are one of the most promising areas. The best forecasting results are shown by technologies based on the use of artificial neural networks.

2 Statement of the Problem

Artificial neural networks (ANN) are a system that resembles biological neural systems and uses the principles of the human brain as a basis. ANN can be applied in various fields for forecasting, classification, optimization, data binding, and so on. In recent years, ANN has been used frequently in financial applications. In this study, ANN is used in forecasting gross domestic product. Gross Domestic Product (GDP) is the market value of all final goods and services produced in a country over a given period.

This is a system consisting of many simple computational elements operating in parallel, the function of which is determined by the structure of the network, the strength of interconnected links, and the calculations are performed in the elements or nodes themselves. A neural network is a set of neurons connected in a certain way.

It has only one hidden layer, since it has been proven that it is sufficient for solving one-dimensional forecasting problems [7–9].

It has just one hidden layer, since it is confirmed to be sufficient enough to solve univariate forecasting problems [9]. In this figure, indices "in," "h," and "o" denote input, hidden, and output layers of the ANN, respectively.

The use of artificial neural networks is recognized as one of the most promising methods for researching sociological, biological, ecological, financial, economic and other complex systems. Such systems are the result of the influence of many factors, including human, therefore, it is considered practically impossible to create a complete mathematical model that would consider all existing restrictions and conditions [4, 5].

3 Solution Method

The problem is solved using neural networks. Simulation of the solution of problems was carried out using the Neural Network Toolbox (NNT) package of the MATLAB system [6–8].

The data to be trained is presented in Table 1, in which the first five numbers columns (X1, X2, X3, X4, X5) will be the input values of the neural network, and the four rows of which represent the output parameters, which will be the desired output value of the neural network. The sixth column (Y) shows the gross national income corresponding to the input parameters of the neural network, and Table 2 shows the changes in the waste structure depending on the level of the population's well-being.

Table 1. Changes in the structure of waste depending on the level of well-being of the population

	X1	X2	X3	X4	X5	Y
1	64	3	0.22	17	5	876
2	59	4	0.29	15	9	3465
3	54	5	0.42	13	14	10725
4	28	7	0.78	17	31	15000

Neural networks were used to predict the dependence of gross national income on municipal solid waste. For this, a database on structural changes between organic waste and national income was created, a training process was carried out and a relationship was obtained. The results obtained make it possible to predict the near future.

The results clearly show that a neural network trained with static and input data can predict the dependence of IRR on SHW (Table 2).

At the same time, depending on the amount of municipal solid waste in accordance with the national income, a neural network can be built. Figures 1, 2, 3, 4 and 5 show the results of computer simulation, changes in the structure of waste depending on the level of well-being of the population, namely, depending on the change in the amount of plastic waste, the amount of organic waste, the amount of paper waste, tons of metal waste and the amount of solid waste. Figure 1 shows the Changes in the amount of organic waste depending on the level of well-being of the population.

The results obtained using neural networks for predicting IRR have many qualities. The size of GNI determines the obligations of membership in this integration, as well as the benefits that can be provided from numerous funds. On the other hand, with the intention of maintaining a sustainable future, the importance of forecasting local traffic in large cities arises for many reasons, such as monitoring the environment and pollution; route guidance in real time; and ITS (Intelligent Transportation System).

Table 2. Changes in the structure of waste depending on the level of well-being of the population using a neural network

	Fraction	Low income countries (<$876 GNI/person)	Low income countries ($876–$3465 GNI/person)	Middle-income countries ($3465–$10725 GNI/person)	High income countries (>$10725 GNI/person)
1	Organic waste, % (X1)	32.9–50.5	50.6–55.2	43.4	45.3–45.3
2	Metal,% (X2)	3.85–4.52	3.7–4.89	5.04	4.81–4.81
3	SHW education t/person/year (X3)	0.531–0.622	0.429–0.52	0.48	0.469–0.51
4	Plastic, % (X4)	13.6–14.1	13.7–14.1	13.2	13
5	Paper and cardboard, % (X5)	9.57–26.4	9.57–24.4	16.8	18.2–18.7

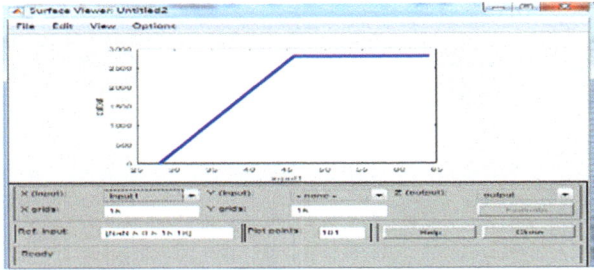

Fig. 1. Changes in the amount of organic waste depending on the level of well-being of the population

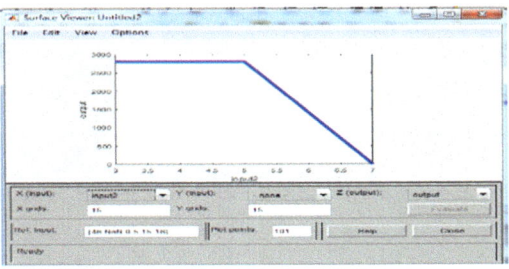

Fig. 2. Changes in the amount of metal waste depending on the level of well-being of the population well-being of the population

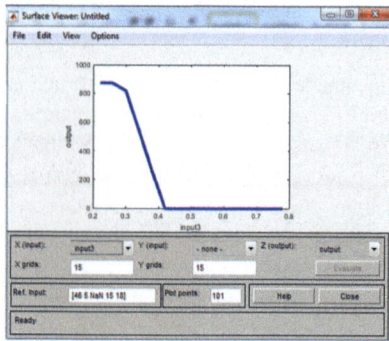

Fig. 3. Changes in the amount of solid waste depending on the level of well-being of the population well-being of the population

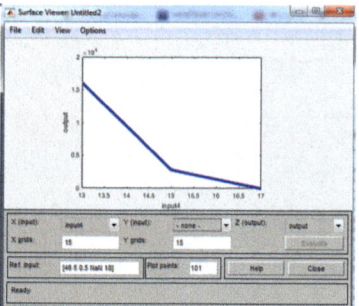

Fig. 4. Changes in the amount of plastic waste depending on the level of well-being of the population

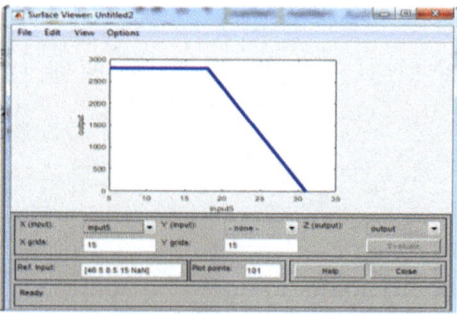

Fig. 5. Changes in the amount of paper waste depending on the level of well-being of the population

4 Conclusions

A fuzzy model for the classification of municipal solid waste, depending on the level of well-being of the population, is proposed, and a methodology for teaching the fuzzy model algorithm based on statistical data has been developed.

As a result of solving the problem, the optimal mode was found:

- number of linear parameters: 243
- number of training data pairs: 4
- number of fuzzy rules: 243
- network error - 0.0485429

The obtained results of computer simulation fully confirmed the reliability of statistical data.

References

1. Aliev, R.A., Fazlollahi, B., Aliev, R.R.: Soft Computing and its Applications in Business and Economics. Springer, Heidelberg (2004)
2. https://www.stat.gov.az/news/index.php?id=4105
3. World Bank Report What a Waste: A Global Review of Solid Waste Management (2012)
4. Kafarov, V.V., Gromov, Y.Y., Matveykin, V.G.: Mathematical modeling of chemical-technological objects in conditions of uncertainty TFKT, vol. 30, pp. 85–90 (1996)
5. Neural Networks Statistica Neural Networks. M.: Telecom (2000)
6. Bishop, C.: Neural Networks for Pattern Recognition. University Press, Oxford (1995)
7. Volkova, A.A., Yakshina, N.V., Shashmurina, E.V.: Use of artificial neural networks for analysis and prediction of environmental risk. In: Environmental Problems of Industrial Regions Materials of the Seventh All-Russian Scientific and Practical Conference—Yekaterinburg, pp. 244–245 (2006)
8. Lin, J., Shen, X., Wang, T., Deyu Li, D., Luo, Y., Wang, L.: Recognition of Fatty Liver Using Hybrid Neural Network. Advances in Neural Networks - ISNN (2006)
9. Ghiassi, M., Saidane, H., Zimbra, D.K.: A dynamic artificial neural network model for forecasting time series events. Int. J. Forecast. 21(2), 341–362 (2005)

Approach to Multi-criteria Fuzzy Optimization Based on Differential Evolution

Babek Guirimov[(✉)] [ID]

State Oil Company of Azerbaijan Republic, SOCAR, SOCAR Tower,
121, H. Aliyev Ave., Baku AZ 1029 Azerbaijan
guirimov@hotmail.com

Abstract. The paper considers a development of original Differential Evolution optimization algorithm to be used for constrained and multi-objective problems. The paper demonstrates how the considered algorithm and implemented software can be used for solving optimization problems.

Keywords: Differential evolution · Multi-objective differential evolution · Multi-objective optimization · Evolutionary algorithm

1 Introduction

Differential Evolution (DE) Optimization is a very efficient and effective population based evolutionary optimization method [1]. Like in other methods of this class, a set of individuals with different properties in artificial population, with every generation undergo crossover and mutation operations to progress towards a predefined desired property (fitness). After a number of generations, one or more individuals representing potential solutions are selected from the population based on the value of their property.

Unlike traditional genetic algorithm DE does not use coding and decoding procedures. The DE's Mutation and Crossover operators manipulate real-valued problem parameters (decision variables) of potential solutions to generate new potential solutions. Mutation consists in performing certain vector operations on a few randomly selected individuals to produce a trial solution. A usual strategy for Mutation is to randomly select three different individuals and add the weighted vector difference of the second and third to the first. There exist more sophisticated strategies as well. Crossover determines which vector components (problem parameters) to inherit from the parent solutions to the child. Subject to the Selection criterion, the trial solution may replace an existing solution to become a new candidate solution. The Selection criterion, based on objective function(s), matches the properties of competing individuals to determine which of the two is better.

The objective of DE Algorithm is solving an optimization problem in conditions where other traditional optimization methods fail [1]. The failure may be due to various factors: complex objective function(s) with multiple global minima, non-smooth, non-continuous, non-differentiable objective functions, additional conditions constraining the input variables with some functional criteria and other.

© The Author(s), under exclusive license to Springer Nature Switzerland AG 2021
R. A. Aliev et al. (Eds.): ICAFS 2020, AISC 1306, pp. 744–750, 2021.
https://doi.org/10.1007/978-3-030-64058-3_93

The original version of algorithm [1–4] is based on numerical data and fits the purpose of single-objective numerical function optimization. To make DE suitable for processing with extended objective and data type concepts including fuzzy, Z-number and others, a special adaptation of the algorithm is needed.

Some improvements and extensions of the original DE algorithm have been suggested by researchers to adopt it for multi-objective optimization. A good survey of algorithms based on DE is provided in [2, 4–6, 8, 9].

In the presented paper, a new version of DE algorithm is suggested – DE with Constraint Satisfaction Function – DEC. The algorithm not only provides good ability of Multi-Objective DE optimization, but also allows account of possible parameter constraints with supplying an Error Function (in addition to several objective functions). The algorithm is simpler than many algorithms suggested by researchers, does not use a separate population for each objective function like in Cooperative MODE [4]. In MODEA [7] version of algorithm, two separate populations are used: current and advanced. The performance of the algorithm is tested on a number of non-trivial single and multi-objective benchmark optimization problems with constraints [10, 11]. The implemented software library can also be effectively used in processing application problems described by fuzzy and Z-number-valued data in decision making, optimization and control [12–14].

2   Multi-objective DE with Constraints

The DE individuals are vectors made of parameters the values of which need to be optimized. The DE parameters usually are numerical (real) values. Therefore, for some problems that use non-numerical (e.g. fuzzy-number valued) parameters or parameters constrained in some other way, some procedures may be required for mapping from the space of problem to the space of DE vectors (RN, there N is the dimensionality of vectors).

According to the standard DE optimization strategy, the population progresses with each generation as follows [1] (OF stands for Objective Function):

1. Choose a next vector V_i ($i = 1, \ldots, ps$);
2. Choose randomly different 3 vectors from P: V_{r1}, V_{r2}, V_{r3} each of which is different from current V_i
3. Generate trial vector $V_t = V_{r1} + f \times (V_{r2} - V_{r3})$;
4. Generate new vector from trial vector V_t. Individual vector parameters of V_t are inherited with probability cr into the new vector V_{new}. If the property of V_{new} is better than the property of V_i, (i.e. $OF(V_{new}) < OF(V_i)$) in population P the vector V_i is replaced by V_{new}.

In advanced versions of DE (one of such versions is proposed in this paper), the current value of desired property for any individual is evaluated by more than one special functions (including e.g. fitness/objective/error/feasibility/constraint satisfaction).

In the suggested version of the algorithm, the state property of each vector is evaluated by two groups of objective functions (OFs):

Cost Functions (CFs) – express the state of the solution in terms of quality (traditionally, the lower the better) and

Error (degree of constraint dissatisfaction) Function (EF) – expresses the state of vector in terms of meeting all numerical restrictions over its components (i.e. individual parameters). In this particular implementation, the value meeting all requirements is assumed to produce a non-positive value of the Error Function, the lower the better.

In the suggested version of DE, the preference of solution V_1 over solution V_2 ($V_1 \succ V_2$) is determined in single-objective version as follows:

IF $CF(V_1) < CF(V_2)$
  THEN IF $EF(V_1) \leq 0$ OR $EF(V_1) \leq EF(V_2)$ THEN V_1 IS BETTER THAN V_2
  ELSE V_1 IS NOT BETTER THAN V_2
ELSE IF $CF(V_2) < CF(V_1)$
   THEN IF $EF(V_2) \leq 0$ OR $EF(V_2) \leq EF(V_1)$ THEN V_1 IS NOT BETTER THAN V_2
  ELSE V_1 IS BETTER THAN V_2
ELSE
  IF $EF(V_1) < EF(V_2)$ THEN V_1 IS BETTER THAN V_2
ELSE V_1 IS NOT BETTER THAN V_2

In the multi-objective version, the preference of solution V_1 over solution V_2 ($V_1 \succ V_2$) is determined as follows:

IF V_1 DOMINATES V_2 $\square CF_k(.)$, k=1,...,N
  THEN IF $EF(V_1) \leq 0$ OR $EF(V_1) \leq EF(V_2)$ THEN V_1 IS BETTER THAN V_2
  ELSE V_1 IS NOT BETTER THAN V_2
ELSE IF V_2 DOMINATES V_1 $\square CF_k(.)$, k=1,...,N
   THEN IF $EF(V_2) \leq 0$ OR $EF(V_2) \leq EF(V_1)$ THEN V_1 IS NOT BETTER THAN V_2
  ELSE V_1 IS BETTER THAN V_2
ELSE
  IF $EF(V_1) < EF(V_2)$ THEN V_1 IS BETTER THAN V_2
ELSE V_1 IS NOT BETTER THAN V_2

The domination of solution V_1 over solution V_2 is decided with help of the following simple algorithm:

```
SET STATUS = 0;
FOR ALL CFk, k=0,...,N {
        IF(CFk(V1)< CFk(V2)) THEN SET STATUS = 1;
        ELSE IF CFk(V1)> CFk(V2) THEN {
                SET STATUS = -1;
                LEAVE FOR CYCLE;
        }
}
IF STATUS = 1, THEN the solution V1 dominates over V2,
IF STATUS = 0, THEN V1 and V2 are equivalent,
IF STATUS = -1, THEN V1 is dominated by V2.
```

We would like to emphasize it that to account for the constraint $C(V) \leq 0$, the error (constraint satisfaction) function $EF(V) = C(V)$ cannot be replaced by additional objective function $CF_{N+1}(V) = C(V)$, because the error function has higher preference than the cost function (all cost functions are of the same preference), and positive and negative values of cost function are treated differently when performing selection for the better solution.

DE algorithm has a few control parameters, proper setting of which may significantly shorten the time required for the population to converge to the desired solution. The recommended adaptive method for setting the DE control parameters *PopSize*, *F*, and *Cr* is as follows:

PopSize is set to NF × NP × 10,
F = Rand(), if Rand() < 0.1, or 0.5, otherwise,
Cr = 0.5 + Rand()/2, if Rand() < 0.1, or 0.9, otherwise,

where N_F is the number of cost functions (CF_k, $k = 1,..., N_F$), N_P is the number of optimization parameters (dimensionality of solution), Rand() is a generated uniform pseudo-random number in [0, 1].

3 Test Problems

3.1 Example 1. Single Objective Optimization Problem with Constraint

Minimize the function [10] (Fig. 1):

$$f(x, y) = \sin(y) \exp((1 - \cos(x))^2) + \cos(x) \exp((1 - \sin(y))^2) + (x - y)^2,$$

subject to constraint:

$$(x+5)^2 + (y+5)^2 \leq 25.$$

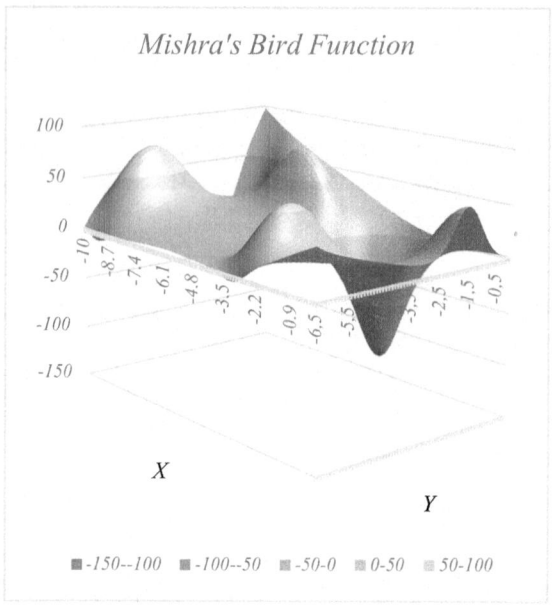

Fig. 1. Mishra's Bird Function graph

For applying the suggested algorithm, we set:

$$V = (x, y),$$
$$CF(V) = f(x, y),$$
$$EF(V) = (x+5)^2 + (y+5)^2 - 25.$$

The DE population size (*PopSize*) is set to 20. The other DE controls (*F* and *Cr*) are chosen adaptively by using the adaptive method for setting the DE control parameters *PopSize*, *F*, and *Cr* (Sect. 2). Population vectors are randomly set in ($x \in [-10, 0], y \in [-6.5, 0]$). The suggested algorithm instantly found the correct solution: $(x, y) = (-3.1302468, -1.5821422)$.

3.2 Example 2. Viennet Function

The problem [11] is described by three objective functions:

$$f_1(x, y) = 0.5(x^2 + y^2) + \sin(x^2 + y^2)$$
$$f_2(x, y) = \frac{(3x - 2y + 4)^2}{8} + \frac{(x - y + 1)^2}{27} + 15$$
$$f_3(x, y) = \frac{1}{x^2 + y^2 + 1} - 1.1 \exp(-(x^2 + y^2))$$

For this rather complex multi-objective problem, *PopSize* was set to 1000, $F = 0.5$, $Cr = 1$.

The problem has an infinite number of Pareto-Optimal solutions. The DE populations converge to these solutions. The solutions found by the proposed algorithm is demonstrated in Fig. 2. For convenience, the solutions, significantly separate by ranges, are divided into two sub-populations. Both groups of solutions are extracted inside a single converged DE population.

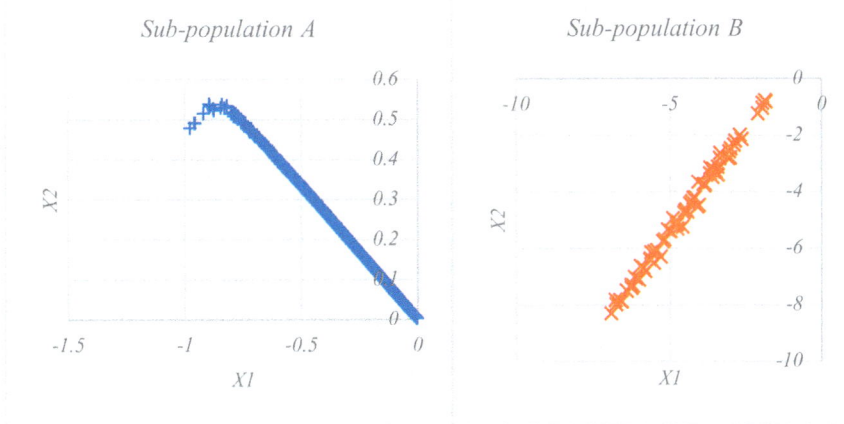

Fig. 2. Two group of solutions found by DEC algorithm

4   Conclusion

The work considers a simple and efficient algorithm based on DE for constrained and multi-objective optimization. The algorithm has been successfully tested on benchmark and application problems. The software implemented on the basis of the suggested method has demonstrated excellent performance, which allows its use for complex processing problems involving practical applications described with fuzzy and Z-numbers.

References

1. Storn, R., Price, K.: Differential evolution – a simple and efficient heuristic for global optimization over continuous spaces. J. Global Optim. **11**, 341–359 (1997). https://doi.org/10.1023/A:1008202821328
2. Sreedhar, D., Binu Rajan, M.R.: Differential evolution based multiobjective optimization – a review. Int. J. Comput. Appl. **63**(15), 14–19 (2013). https://doi.org/10.5120/10541-5019
3. Robič, T., Filipič, B.: DEMO: differential evolution for multiobjective optimization. In: Coello Coello, C.A., Hernández Aguirre, A., Zitzler, E. (eds.) EMO 2005. LNCS, vol. 3410, pp. 520–533. Springer, Heidelberg (2005). https://doi.org/10.1007/978-3-540-31880-4_36

4. Wang, J., Zhang, W., Zhang, J.: Cooperative differential evolution with multiple populations for multiobjective optimization. IEEE Trans. Cybern. **46**(12), 2848–2861 (2016). https://doi.org/10.1109/TCYB.2015.2490669

5. Nakayama, H.: Multi-objective optimization and its engineering applications. In: Dagstuhl Seminar Proceedings 04461. Practical Approaches to Multi-objective Optimization, http://drops.dagstuhl.de/opus/volltexte/2005/234

6. Huband, S., Hingston, P., Barone, L., While, L.: A review of multiobjective test problems and a scalable test problem toolkit. IEEE Trans. Evol. Comput. **10**(5), 477–506 (2006)

7. Ali, M., Siarry, P., Pant, M.: An efficient differential evolution based algorithm for solving multi-objective optimization problems. Eur. J. Oper. Res. **217**, 404–416 (2012). https://doi.org/10.1016/j.ejor.2011.09.025

8. Emmerich, M.T.M., Deutz, A.H.: A tutorial on multiobjective optimization: fundamentals and evolutionary methods. Nat. Comput. **17**(3), 585–609 (2018). https://doi.org/10.1007/s11047-018-9685-y

9. Horn, D., Wagner, T., Biermann, D., Weihs, C., Bischl, B.: Model-based multi-objective optimization: taxonomy, multi-point proposal, toolbox and benchmark. In: Gaspar-Cunha, A., Henggeler Antunes, C., Coello, C.C. (eds.) EMO 2015. LNCS, vol. 9018, pp. 64–78. Springer, Cham (2015). https://doi.org/10.1007/978-3-319-15934-8_5

10. Mishra, S.K.: Some New Test Functions for Global Optimization and Performance of Repulsive Particle Swarm Method, 23 August 2006. Available at SSRN: https://ssrn.com/abstract=926132. http://dx.doi.org/10.2139/ssrn.926132

11. Okabe, T., Jin, Y., Olhofer, M., Sendhoff, B.: On test functions for evolutionary multi-objective optimization. In: Yao, X., et al. (eds.) PPSN 2004. LNCS, vol. 3242, pp. 792–802. Springer, Heidelberg (2004). https://doi.org/10.1007/978-3-540-30217-9_80

12. Aliev, R.A., Pedrycz, W.: Fundamentals of a fuzzy-logic-based generalized theory of stability. IEEE Trans. Syst. Man Cybern. Part B (Cybern.) **39**(4), 971–988 (2009)

13. Aliev, R.A., Huseynov, O.H., Zeinalova, L.M.: The arithmetic of continuous Z-numbers. Inf. Sci. **373**, 441–460 (2016)

14. Aliev, R.A., Alizadeh, A.V., Huseynov, O.H., Jabbarova, K.I.: Z-number-based linear programming. Int. J. Intell. Syst. **30**(5), 563–589 (2015)

Fuzzy Evaluation of Impact of the Professional Female Tourist Guides Jobs on Their Family Life

Saide Sadikoglu[(⊠)] [iD]

Faculty of Tourism and Hotel Management, Near East University, Mersin 10,
North Cyprus, Turkey
saide.sadikoglu@neu.edu.tr

Abstract. Professional tourist guides have become one of the essential elements of today's tourism sector. A significant part of the service activities provided by tourism enterprises to tourists is provided by tourist guides. The number of female employees takes an important place in the professional tourist guide profession, which is a crucial service group. In addition to the difficulties of their jobs, female tourist guides also carry out their family-related responsibilities. Tourist guides are handy in providing satisfaction to tourists coming to their countries with the holiday, coming to the country again and ensuring customer loyalty. Tourist guides do not only represent their own country but also act as intermediaries who serve their country's people and culture in the best way. At this stage, female guides are the faces of the country, and the first people tourists see. Female guides have become one of the most critical elements of the tourism industry. This paper is devoted to the analysis of the perceptions and attitudes of professional tourist guides regarding the work and family relationships using fuzzy-logic approach. As the case study was chosen professional female guides in the Turkish Republic of Northern Cyprus (TRNC).

Keywords: Professional tourist guide · Work-Family conflict · Fuzzy evaluation

1 Introduction

The increasing economic situation of the countries and respectively of the leisure time of the people positively effect on the tourist receipt. Tourist guides are very effective in ensuring the satisfaction of tourists coming to their countries with the holiday, and ensuring customer loyalty [1]. Tourist guides do not only represent their own country in the best way but also act as individuals who reflect their country's people. Especially guides are the first people with whom tourists contact face-to-face and get an image of the country.

Guides have significant duties and contributions, as well as responsibilities at an equal rate in the tourism industry. What makes this duty important is that tourist guides represent their countries to foreigners with their personalities [2]. Being a guide requires a great accumulation of knowledge, and beyond that, multifaceted education.

© The Author(s), under exclusive license to Springer Nature Switzerland AG 2021
R. A. Aliev et al. (Eds.): ICAFS 2020, AISC 1306, pp. 751–759, 2021.
https://doi.org/10.1007/978-3-030-64058-3_94

Guides should be a good manager, a good culture person, a good psychologist, a good sociologist, in addition to being a small giant of knowledge [3].

A significant part of the service activities provided by tourism enterprises to tourists are carried out through tourist guides [4]. A tourist guide must take responsibility for the correctness of given information about the country or region to visited tourists. They should be an ambassador who makes the presentation and show hospitality to tourists in a way that ensures that tourists come back to a country or region, and a host who creates an environment in which tourists feel comfortable [5].

According to Aristoteles, free-time is serious work, and Aristoteles expressed this by saying "We work for having free time" [6]. Therefore, it can be said that the aim is having free time while working is a tool in this sense because when the concept of time is examined, "free time" in today's society is no longer an element that human beings can freely possess. The great part of our lives passes with "working time" [7]. Free-time activities are in an active position to increase life satisfaction of individuals [8]. The economic forces, movements, adaptation levels, goals, incidents in the life and mental events are important determinants of happiness. In its primary sense, the happiness level of an individual can be qualified as the extent to which an individual generally assesses the total quality of his life positively [9]. The concept of happiness is related to achieving the level of importance given by an individual to the sources of joy [10].

At this point, it is necessary to mention that the quality of life is a concept that cannot be considered separately from the concept of life satisfaction and happiness, and these are interrelated concepts [11]. It is possible to talk about the interaction between the concepts of work and family. There is a mutual relationship between working life, and both are important [12, 13]. In this sense, work-family conflict is a form of conflict between roles in the fulfilment of familial responsibilities resulting from the characteristics of work [14]. Therefore, work and family-related roles are the most comprehensive roles that are simultaneous in the long term in human life. There may not always be harmony between the tasks that the individual has undertaken in these two areas [15].

Literature analysis shows that many publications related with tourism researches are based on statistical approaches. There are some studies related with application of fuzzy logic approach in tourism industry. But in existing publications are considered application of fuzzy-logic for job satisfactions of hotel employees, hotel selection, consumer satisfaction etc. In [16] is considered the measurement of job satisfaction using fuzzy set. The integration of fuzzy sets and conjoint analysis for the evaluation of job satisfaction of hotel employees is given in [17]. Hotel selection using fuzzy expert system is examined in [18]. In [19] is described the Fuzzy logic approach in the modeling of sustainable tourism development management.

Thus, this is the first study related to the application of fuzzy-logic approach to evaluate the impact of the tourist guide jobs on their family life using fuzzy-logic.

2 Population of the Study

The study was conducted to determine the positive and negative factors that female guides experience when practising their profession and leading their family lives at the same time. Female tourist guides actively working as professional guides in the TRNC were addressed as the population of the study. As of 2018, the number of active professional female tourist guides of the KITREB, the official tourist guides association of the TRNC, is 116. All these female guides were reached, and the sample number was completed as 116 and the confidence interval as 99%. The test value for the t-test was assigned as 2.

3 Methodology

3.1 Data Collection

In the research was used survey technique as a data collection tool. For this purpose, were used the questionnaires given in [7, 20]. A total of 20 questions were asked from survey participants. The data were measured by using a 5-point Likert-type scale ("Strongly Agree"-SA, "Agree-A", "Undecided"-U, "Disagree"-D, "Strongly Disagree"- SD) (Table 1).

Table 1. Survey questions to female tourist guides

№	Questions	SD	D	U	A	SA
X_1	I cannot spare enough time for myself in this profession					
X_2	I cannot make long-term plans because of my profession					
X_3	My job keeps me away from family activities more than necessary					
X_4	My working hours in the profession prevent me from spending time with my children					
X_5	There is a lot of discussion with family members because of my profession					
X_6	I do not think my profession is an obstacle to happy family life					
X_7	The time I spare for my family often overlaps with my working hours					
X_8	It is hard to sustain this profession for married individuals					
X_9	I do not think my profession is an obstacle to a happy marriage					
X_{10}	The fact that the profession is backbreaking due to working hours					

(continued)

Table 1. (*continued*)

№	Questions	SD	D	U	A	SA
X_{11}	The fact that there are unethical elements in tourism					
X_{12}	The fact that female tourist guides are worried about their life safety during tours					
X_{13}	It is hard to work during pregnancy					
X_{14}	As female guides, we are exposed to tourists' inappropriate offers					
X_{15}	Being a woman is a disadvantage in our profession					
X_{16}	The fact that being a tourist guide is not regarded as a profession					
X_{17}	Not having enough income to provide family subsistence					
X_{18}	There is an inadequacy of facilities in tours to fulfil the toilet and food needs					
X_{19}	The fact that the legislation for guides does not adequately protect the rights of tourist guides					
X_{20}	Is professional guidance training suitable for women?					

3.2   Linguistic Interpretation

The trapezoidal membership functions (MF) are used for fuzzification of input data. Analogical to the Likert scales were used the following linguistic Degree of acceptance: *"Strongly Agree"*, *"Agree"*, *"Undecided"*, *"Disagree"*, *"Strongly Disagree"*. Then data processing is performed according to the following steps:

Table 2.   The encoded linguistic terms for Degree of Acceptance

Scale	Level	Trapezoid MF value
1	Strongly disagree	1/0, 0.5/0.05, 0/0.1
2	Disagree	0/0.05, 1/0.15, 1/0.25, 0/0.35
3	Undecided	0/0.3, 1/0.4, 1/0.5, 0/0.6
4	Agree	0/0.55, 1/0.65, 1/0.75, 0/0.85
5	Strongly Agree	0/0.8 1/0.9, 1/1, 1/1

Step 1. The codebook for linguistic variable Degree of Acceptance are compiled. Graphically it can be shown in diagram below (Fig. 1):

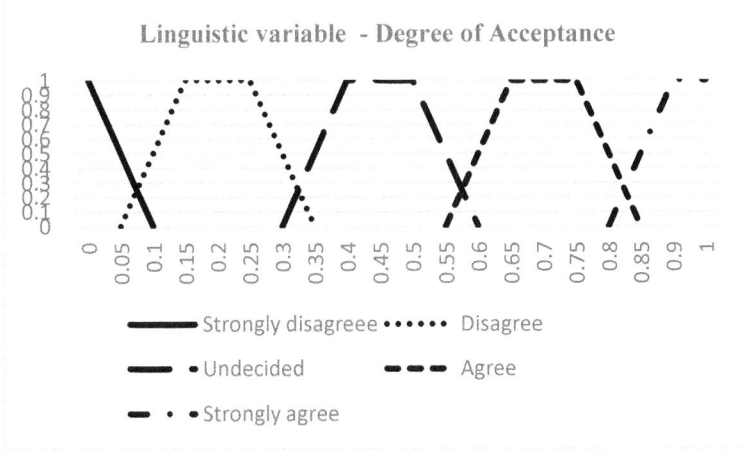

Fig. 1. Graphically representation of linguistic variables

Step 2. The Degree of Acceptance values are calculated based on the survey results
For each question the Crisp Value of Acceptance Degree *(Ad)* are calculated according below rule:

$$\text{If } X_{i\text{-}yes} \neq X_{i\text{-}no} \quad \text{then } Ad_i = X_{i\text{-}yes} - X_{i\text{-}no}, \text{ otherwise } 0$$

Then obtained values of Acceptance degree are normalized according this formula

$$Ad_{i\_norm} = \frac{Ad_i}{|\max\{Ad_i, \ldots, Ad_n\} - \min\{Ad_i, \ldots, Ad_n\}|}$$

where n = 1, 20
and the normalized value of Acceptance degree in the scale [0, 1]

$$Ad_{i\_final} = Ad_{i\_norm} + |\min\{Ad_{1\_norm}, \ldots, Ad_{n\_norm}\}|$$

Step 3. Then according to the obtained results on step 2 (Ad$_{final}$) the linguistic values of Acceptance degree for each question are defined and below table are compiled:

756   S. Sadikoglu

Table 3. The linguistic values of Degree of Acceptance

№	Questions	Ad_{i_final}	Degree of acceptance
X_1	I cannot spare enough time for myself in this profession	0,00	Strongly disagree
X_2	I cannot make long-term plans because of my profession	0,61	Agree
X_3	My job keeps me away from family activities more than necessary	0,32	Disagree
X_4	My working hours in the profession prevent me from spending time with my children	0,36	Undecided
X_5	There is a lot of discussion with family members because of my profession	0,00	Strongly disagree
X_6	I do not think my profession is an obstacle to happy family life	0,61	Agree
X_7	The time I spare for my family often overlaps with my working hours	0,74	Agree
X_8	It is hard to sustain this profession for married individuals	0,57	Undecided
X_9	I do not think my profession is an obstacle to a happy marriage	0,72	Agree
X_{10}	The fact that the profession is backbreaking due to working hours	0,48	Undecided
X_{11}	The fact that there are unethical elements in tourism	0,93	Strongly agree
X_{12}	The fact that female tourist guides are worried about their life safety during tours	0,67	Agree
X_{13}	It is hard to work during pregnancy	1,00	Strongly agree
X_{14}	As female guides, we are exposed to tourists' inappropriate offers	0,78	Agree
X_{15}	Being a woman is a disadvantage in our profession	0,86	Strongly agree
X_{16}	The fact that being a tourist guide is not regarded as a profession	0,56	Undecided
X_{17}	Not having enough income to provide family subsistence	0,87	Strongly agree
X_{18}	There is an inadequacy of facilities in tours to fulfil the toilet and food needs	1,00	Strongly agree
X_{19}	The fact that the legislation for guides does not adequately protect the rights of tourist guides	1,00	Strongly agree
X_{20}	Is professional guidance training suitable for women?	0,81	Agree

It should be noticed that some values of Ad_{i_final} can be assigned simultaneously to two fuzzy sets (with different values of membership functions), for example Ad_{3_final} can belong both to "Undecided" and to "Disagree", Ad_{8_final} - both to "Undecided" and to "Disagree", Ad $_{16\_final}$ - both "Undecided" and to "Disagree", Ad $_{20\_final}$ - both to "Undecided" and to "Disagree". Ultimately, belonging to a fuzzy set is determined by the higher value of the membership function.

Step 4. According to the [21] the total aggregated Degree of Acceptance of the survey questions can be computed by calculating the mean of the fuzzy numbers expressing the values of linguistic variable:

For n trapezoidal fuzzy numbers (a_i, b_i, c_i, d_i), $i = 1, ..., n$ the average calculating

$$X_{avg} = \left(\frac{1}{n} \sum_1^n a_i, \sum_1^n b_i, \sum_1^n c_i, \sum_1^n d_i \right)$$

Calculating by this formula we obtain that $X_{avg}=$ (0.5, 0.59, 0.68, 0.75), which corresponds to linguistic value "Agree".

For arithmetic mean $Ad_i = 0.61$, which also refers to the linguistic meaning of "Agree".

3.3 Results

As a result of the variable analysis conducted according to Table 3 the following propositions come to the forefront: Answer 1, the women responded *Strongly disagree* to the statement *I cannot spare enough time for myself in this profession.* According to this result, female tourist guides experience problems with the concept of time since they work with pleasure. Answer 2 - *Agree* to the statement *I cannot make long-term plans because of my profession.* This result shows that female guides do not make long-term plans because of their respect for their profession. Answer 3 - *Disagree* to the statement *My job keeps me away from family activities more than necessary.* These answers show that female tourist guides have problems with participating in activities with their families. Answer 4 - *Undecided to* the statement *My working hours in the profession prevent me from spending time with my children.* The fact that the answers of yes and no are close to one another shows that the working hours are not a problematic issue for spending time with children. In terms of the profession, while the inability to make long-term family plans since it is not definite when a new group will come emerges as a problem, it also means that the working hours of female tourist guides are more flexible compared to other professions. Analogically are interpreted other items in Table 2.

4   Conclusions

It is observed that long and irregular working hours of female tourist guides can prevent their home and social life. This may negatively affect the preference of this profession by female candidates and preferring male guides by tourism agencies etc. For this reason, appropriate conditions should be prepared for female guides, and the necessary measures should be taken primarily by employers, i.e., agencies. Agencies can use female guides in short trips, city tours or museum tours than sending them on long trips.

The application of fuzzy evaluation in the present study, allows to handle subjective measurement errors introduced by survey participants. and to offer the possibility of individual evaluating of the work-family conflict for each female guides.

This type of studies can be used as a source of professional tourist guides' perceptions and attitudes towards marriage and professional relations, as well as the identification and resolution of potential problems they have.

References

1. Yarcan, Ş.: A conceptual evaluation on professional ethics in professional tourist guidance. Anatol. Tourizm Res. J. **18**(1), 33–44 (2007)
2. Cimrin, H.: ABC of Tourism and Tourist Guiding. Akdeniz Publisher, Antalya (1995)
3. Ahipaşaoğlu, S.: The Reasons of Choosing Tour Planning Management and Guidance as a Profession in Travel Businesses (in Turkish). Varol Printing House, Ankara (1997)
4. Güzel, F.Ö.: Role of professional tourist guide in the development of Turkey image. A Study on German tourists (in Turkish). Master's thesis, Balıkesir University Institute of Social Sciences, Tourism and Hotel Management. Department, Balıkesir (2007)
5. Ap, J.K., Wong, K.: Case study on tour guiding: professionalism, issues and problems. Tour. Manag. **22**, 15–63 (2001)
6. Outhwaite, W.: The Blackwell Dictionary of Modern Social Thought. Wiley-Blackwell Publishing, Oxford (2003)
7. Doğan, H., Üngüren, E., Dönmez, K.D.: Relationship between profession and family life: a research on professional tourist guides. Yaşar Univ. J. **20**(5), 3430–3442 (2010)
8. Sop, S.A.: Job pressure, job-free time conflict, job and life satisfaction. Tour. Acad. J. **1**(1), 1–14 (2014)
9. Bülbül, Ş., Giray, S.: Analysis of the relationship between socio-demographic characteristics and happiness perception. Ege Acad. Vis. J. **11**, 113–123 (2011)
10. Yurtseven, R.: The quality of life and happiness sources: The pilot research on managers of 5-star, hotels in Turkey. Management **13**(46), 41–52 (2003)
11. Demirbulat, Ö.G., Avcıkurt, C.: A conceptual assessment on the relationship between tourism and happiness. Balıkesir Univ. J. Soc. Sci. Inst. **34**(18), 77–97 (2015)
12. Whitehead, D.L., Korabik, K., Lero, D.S.: Work-family integration: introduction and overview. In: Whitehead, D.L., Korabik, K., Lero, D.S. (eds.) Theory, and Best Practices, pp. 3–11. Whitehead Academic Press, San Diego (2008)
13. Carlson, D.S., Grzywacz, J.G.: Reflections and future directions on measurement in workfamily research. In: Carlson, D.S., Grzywacz, J.G. (eds.) Handbook of Work-Family Integration. Research, Theory, and Best Practices, pp. 57–73. Whitehead Academic Press, San Diego (2008)
14. Netemeyer, R.G., Boles, J.S., McMurrian, R.: Development and validation of work-family conflict and family-work conflict scales. J. Appl. Psychol. **81**(4), 400–410 (1996)
15. Demirbulat, G.Ö.: A study of the effect of professional tourist guiding profession on family life (in Turkish). Tour. Res. J. **3**(2), 1–21 (2014)
16. Abiyev, R., Saner, T., Eyupoglu, S., Sadikoglu, G.: Measurement of job satisfaction using fuzzy sets. Procedia Comput. Sci. **102**, 294–301 (2016)
17. Abiyev, R., Sadikoglu, G., Abiyeva, E.: Fuzzy evaluation of job satisfaction of hotel employees. In: International Conference on Artificial Intelligence – ICAI 2015, Vegas, Nevada, USA (2015)

18. Ngai, E.W.T., Wat, F.K.T.: Design and development of fuzzy expert system. Omega **31**, 275–286 (2003)
19. Ziyadin, S., Borodin, A., Streltsova, E., Pshembayeva, D.: Fuzzy logic approach in the modeling of sustainable tourism development management. Polish J. Manag. Stud. **19**(1), 492–504 (2019)
20. Eker, I., Özmete, E.: Harmony between work-family life and individual life: the reflection of the demands of working and family lives on individual life. Int. Acad. Soc. Sci. J. AU **3**(4), 1–15 (2012)
21. Škoda, M., Brožova, H.: Weighted fuzzy group decision-making in recruitment. RELAIS **2**(3), 1–15 (2019)

A Pioneer Approach for the Evaluation of Antihypertensive Drug Combinations Through Fuzzy PROMETHEE Method

Günay Kibarer[1] , Şerife Kaba[1] , Omid Mirzaei[1(✉)] ,
and Sedat Köse[2]

[1] Department of Biomedical Engineering, Near East University, Near East
Boulevard, P.O. Box: 99138, Nicosia, TRNC, Mersin 10, Turkey
{aysegunay.kibarer, serife.kaba,
omid.mirzaei}@neu.edu.tr
[2] Liv Hospital, Ankara, Turkey
bilgi@sedatkose.com

Abstract. Hypertension is a major worldwide health problem being a great risk factor for many heart diseases. It is preventable, however, in spite of the existing effective antihypertensive drugs optimal blood pressure control still remains as a problem for most of the patients reducing their life standards. Therefore, a combination therapy is urgently needed to be determined for the most appropriate drug combination and dose frequency. It is really a hard task for medical people to decide which combination is the most effective since so many parameters exist to be taken into consideration. Under these circumstances, fuzzy-PROMETHEE Method is a most powerful tool in multicriteria decision making with regard to the importance and weight of each criteria. Thus, in this study it was aimed to verify the most appropriate drug combination in increasing order of antihypertensive effectiveness. Miscellaneous quatro-combination of angiotensin receptor blocker (ARB), diuretic, calcium channel blocker and betablocker was observed to be the most effective antihypertensive drug combination as the outcome of this study.

Keywords: Hypertension · Antihypertensive drugs · Cardiovascular disease · Combination therapy · Fuzzy-PROMETHEE method · PROMETHEE rainbow

1 Introduction

It is a common fact that the initial prescription for the treatment of hypertension is usually discontinued [1, 2] which is related to several factors, one of them being the low rate of blood pressure control [1] and it reaches a maximum depending on the type of drugs prescribed for the initial treatment like only diuretics, only β-blockers or only angiotensin receptor blockers [1–3] due to their minimum side effects.

Thus, guidelines of Europe has lately proposed that discontinuation may be prevented by starting the treatment with two antihypertensive drugs from different groups instead of only one drug [1–4] which will result in sooner and more effective blood pressure control [5].

R. A. Aliev et al. (Eds.): ICAFS 2020, AISC 1306, pp. 760–765, 2021.
https://doi.org/10.1007/978-3-030-64058-3_95

Antihypertensive combinations involve drugs from various pharmacological classes such as a couple of β-blocker and HCT (diuretic); ACE inhibitors and HCT; angiotensin receptor blocker and HCT; calcium channel blocker and ACE inhibitors; and also special tricombinations of ARB, HCT and calcium channel blocker; and renin inhibitors, HCT and calcium channel blockers which do not interact with each other [1].

Although mortality is reduced to some extent, the National Health and Nutrition Examination Survey (NHANES) reports [2] that monotherapy provides a decrease in blood pressure below 140/90 mm Hg for only 27 percent of antihypertensive patients. Consequently, the target goal may only be achieved by using two or more correct drug combinations for sufficient blood pressure control [1] at a fixed dosage of one or two tablets everyday by the patients with or without any organ failure and having additional diseases like renal failure, liver failure or diabetis in addition to high blood pressure case.

In this research, antihypertensive drug combinations were studied by employing Fuzzy-PROMETHEE technique [6] as a pioneer application of this approach all through the world, with the aim of serving medical people and providing them with a trustable and highlighting information to ease their decisions in prescribing the most effective treatment for their patients living seriously high and uncontrollable blood pressure problems. It was aimed to verify the most appropriate drug combinations in increasing order of antihypertensive effectiveness for the first time in literature which may be classified as "the state of the art" as a useful application of PROMETHEE technique in the field of medicine.

2   Methodology

PROMETHEE method is a multi-criteria decision-making (MCDM) technique and is abbreviated from "Preference ranking organization method for enrichment evaluation" During the year 1965, fuzzy sets as an elongation of the classic set hypothesis were assigned by open sets launched by Zade [6, 7]. PROMETHEE method was developed in 1985. PROMETHEE may be a straightforward concept, which is far too easier to utilize than any other multi-criteria investigation methods which are right now accessible for implementation. In issues with a limited number of options, PROMETHEE is fabulous for ranking with respect to multiple and complex criteria. PROMETHEE needs two sorts of data: (a) The relative significance (weights) of the considered criteria, and (b) data with respect to the preferences of the decision-makers taking interest in participating with the problem under study [6–9].This method is used in so many fields (engineering, education, etc.) for the evaluation of complex systems that have fuzzy parameters and produce effective comparison results. During recent years, researchers have started to employ this method largely in the field of medicine and healthcare as well.

During the year 2018, Maisaini et al. used this method to evaluate lung cancer treatment techniques. The details of this research are as follows: Fuzzy PROMETHEE was utilized to analyze the therapy techniques on the basis of variables like the dose of radiation, treatment cost and period, survival rate, secondary impacts and the expenses of the healing center. Consequently, surgery technique proved to exhibit the best performance compared to other methods for the treatment of lung cancer concerning

762   G. Kibarer et al.

the selected criteria such as importance and weight [10]. This research may be described as "the state of the art" being related to a healthcare application of fuzzy PROMETHEE approach for the topic under study.

Another research about the health field by using fuzzy PROMETHEE was achieved in 2019 by Özsahin et al., who used this method to evaluate lung cancer treatment techniques. Fuzzy PROMETHEE was utilized to analyze and compare the strategies employed in certain pancreatic cancer treatments involving surgery, chemotherapy, radiation, immunotherapy or hormone therapy ending up with the surgery method being more superior to others. Treatment and equipment cost, survival probability and outcoming unfavorable results faced with following surgery were investigated. The outcome of this research provides a valuable information for alternative treatments of pancreatic cancer cases [11].

3   Results and Discussion

In this method (Table 1), 13 antihypertensive drug combinations were selected [1] being most effective in the treatment of patients suffering from high and uncontrolled blood pressure [1]. The importance (positive/negative) and weights of each individually determined 20 criteria involving number of tablet, dose frequency, resistant hypertension with no organ failure, side effects, drug-drug interaction, diabetis, renal failure, heart failure, liver failure, drug resistance, incorrect drug combination/prescription, systolic and diastolic blood pressure, treatment duration, age, hard working conditions, gender, mental disorder, edema, potassium level and mortality were evaluated on the basis of this powerful technique and finally, PROMETHEE rainbow was obtained. Table 2 and Fig. 1 exhibit all the utilized tools during the analysis of input data.

Table 1. Fuzzy linguistic scale for the assessment of analysis.

Linguistic scale for evaluation	Fuzzy number	Rating of criteria
Very High (VH)	(0.75, 1.00, 1.00)	Number of tablet, Dose frequency, Resistant HT with no organ failure, Side effects, Drug-drug interaction, Diabetis, Renal inefficiency, Heart failure, Liver failure, Drug resistance, Incorrect drug combination or prescription, Systolic and diastolic blood pressure, Edema, Potassium level, Mortality
High (H)	(0.50, 0.75, 1.00)	Treatment duration
Moderate (M)	(0.25, 0.50, 0.75)	Hard working conditions
Low (L)	(0, 0.25, 0.50)	
Very Low (VL)	(0, 0, 0.25)	Age, Gender, Mental disorder

Table 2. Antihypertensive drug combinations in increasing order of effectiveness to reduce blood pressure [1].

Drug Combinations	Phi	Phi+	Phi−
Irbesartan + hydrochlorothiazide + lercadip + metoprolol (300 mg/25 mg/10 mg)50 mg)	0.4139	0.5347	0.1208
Telmisartan + hydrochlorothiazide + amlodipine (80 mg/12.5 mg/10 mg) TWYNSTA	0.2931	0.4718	0.1788
Aliskiren + hydrochlorothiazide + amlodipine (150 mg/12.5 mg/5 mg) TEKTURNA	0.2931	0.4496	0.1566
Telmisartan and hydrochlorothiazide (12.5 mg/(80 mg)	0.2513	0.4164	0.1652
Olmesartan and hydrochlorothiazide(12.5 mg/40 mg)	0.2513	0.4164	0.1652
Candesartan and hydrochlorothiazide (/12.5 mg/16 mg)	0.1858	0.3711	0.1853
Irbesartan and hydrochlorothiazide (12.5-25 mg/300 mg	0.0599	0.2905	0.2306
Captopril and hydrochlorothiazide (25 mg/15 mg, 25 mg/25 mg, 50 mg/15 mg, 50 mg/25 mg)	0.1722	0.2271	0.3993
Metoprolol and hydrochlorothiazide (50 mg/25 mg, 100 mg/25 mg, 100 mg/50 mg)	0.2422	0.1866	0.4288
Losartan and hydrochlorothiazide (50 mg/12.5 mg, 100 mg/25 mg)	0.2875	0.1513	0.4388
Valsartan and hydrochlorothiazide (12.5 mg, 160 mg/12.5 mg)	0.2875	0.1513	0.4388
Amlodipine and benazepril (2.5 mg/10 mg, 5 mg/10 mg, 5 mg/20 mg)	0.3258	0.1654	0.4912
Spironolactone and hydrochlorothiazide (25 mg/50 mg, 50 mg/50 mg)	0.4330	0.1221	0.5551

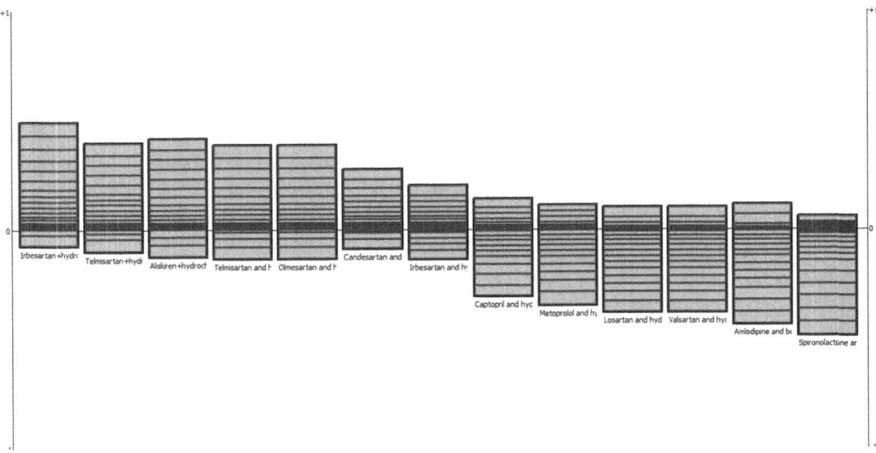

Fig. 1. PROMETHEE rainbow.

The outcome of the input data obtained from PROMETHEE Rainbow (Fig. 1.) with respect to the weight and function of each criterion, with their positive and negative outranking flow, implies that miscellaneous quatro-combination of angiotensin receptor blocker (ARB) (irbesartan), diuretic (hydrochlorothiazide), calcium channel blocker (lercadip) and betablocker (belocZok) is the most effective antihypertensive drug combination followed by the 2 other tri-combinations TWYNSTA and TEKTURNA, respectively, all approved by FDA (Food and Drug Administration) after 12 years of research on patients treated by Mayo Clinique and Cleveland Medical Center [12–14].

4   Conclusion

In this study, a combination therapy was proposed to support a medical doctor in that field which can be trustable in the preparation of a professional prescription for the most appropriate drug combination and dose frequency in cases of uncontrollable blood pressure with systolic BP > 20 mmHg and diastolic BP > 10 mmHg. Since so many parameters exist to be taken into consideration, the most appropriate drug combination in increasing order of antihypertensive effectiveness was determined by employing fuzzy-PROMETHEE Method as a pioneer approach for the first time in literature which may be classified as "the state of the art" in the field of medicine for the evaluation of antihypertensive drug combinations.

References

1. Skolnik, N.S., Beck, J.D., Clark, M.: Combination antihypertensive drugs: recommendations for use. Am. Fam. Phys. **61**(10), 3049 (2000)
2. National Institutes of Health: The sixth report of the Joint National Committee on prevention, detection, evaluation, and treatment of high blood pressure. Arch. Intern. Med. **157**(21), 2413–2446 (1997). https://doi.org/10.1001/archinte.157.21.2413
3. Sica, D.A.: Fixed-dose combination antihypertensive drugs. Drugs **48**(1), 16–24.9 (1994). https://doi.org/10.2165/00003495-199448010-00003
4. Cocke, T.B., Fisch, S., Gilmore, H.R., Okun, R., Tamagna, I.G., Wipplinger, K.: Double-blind comparison of triamterene plus hydrochlorothiazide and spironolactone plus hydrochlorothiazide in the treatment of hypertension. J. Clin. Pharmacol. **17**(5–6), 334–338 (1977). https://doi.org/10.1002/j.1552-4604.1977.tb04613.x
5. Multicenter Diuretic Cooperative Study Group: Multiclinic comparison of amiloride, hydrochlorothiazide, and hydrochlorothiazide plus amiloride in essential hypertension. Arch. Intern. Med. **141**(4), 482–486 (1981). https://doi.org/10.1001/archinte.1981.0034004007802120
6. Badiru, A.B., Cheung, J.: Fuzzy Engineering Expert Systems with Neural Network Applications, vol. 11. John Wiley & Sons, Hoboken (2002)
7. Senvar, O., Tuzkaya, G., Kahraman, C.: Multi criteria supplier selection using fuzzy PROMETHEE method. In: Kahraman, C., Öztayşi, B. (eds.) Supply Chain Management Under Fuzziness. SFSC, vol. 313, pp. 21–34. Springer, Heidelberg (2014). https://doi.org/10.1007/978-3-642-53939-8_2

8. Brans, J., Mareschal, B.: Multiple criteria decision analysis, Chap. 5, Promethee methods (2004)
9. Macharis, C., Springael, J., De Brucker, K., Verbeke, A.: PROMETHEE and AHP: the design of operational synergies in multicriteria analysis: strengthening PROMETHEE with ideas of AHP. Eur. J. Oper. Res. **153**(2), 307–317 (2004)
10. Maisaini, M., Uzun, B., Ozsahin, I., Uzun, D.: Evaluating lung cancer treatment techniques using fuzzy PROMETHEE approach. In: Aliev, R.A., Kacprzyk, J., Pedrycz, W., Jamshidi, M., Sadikoglu, F.M. (eds.) ICAFS 2018. AISC, vol. 896, pp. 209–215. Springer, Cham (2019). https://doi.org/10.1007/978-3-030-04164-9_29
11. Ozsahin, I., Ozsahin, D.U., Nyakuwanikwa, K., Simbanegavi, T.W.: Fuzzy PROMETHEE for ranking pancreatic cancer treatment techniques. In: 2019 Advances in Science and Engineering Technology International Conferences (ASET), pp. 1–5. IEEE (2019). https://doi.org/10.1109/icaset.2019.8714554
12. FDA. https://www.accessdata.fda.gov/drugsatfda_docs/label/2012/021985s023lbl.pdf
13. Mayo Clinic. https://www.mayoclinic.org/diseases-conditions/high-blood-pressure/in-depth/high-blood-pressure-medication/art-20046280)
14. Cleveland Clinic. https://consultqd.clevelandclinic.org/new-hypertension-guidelines-how-to-approach-your-patients/

The Impact of Cultural Values
to the Trustworthiness of New Internet Society

Farida Huseynova[(✉)] 🆔

Azerbaijan State Oil and Industry University, Azadlig Ave., 20,
AZ1010 Baku, Azerbaijan
farida_hus@hotmail.com

Abstract. It is becoming very common to interact with unknown people by Internet in our "global village". In this new context, there is a need for new tools that can help in deciding about the trustworthiness of new societies as 'Trust Networks'. The paper looks at the role of cultural value in determining how much trust on virtual businesses according to countries is expected to have. This paper will describe the impact of cultural variables in determining the level of trust according to Hofstede's value system (power distance, individualism and uncertainty avoidance) in 'Trust Networks, and also help to interact in today's environment of uncertainty.

Keywords: Hofstede's framework · Trust · Value system · Cultural issues · Perceiver · Attitudes · Target · Situation · Time

1 Introduction

Covid-19, first and foremost a human tragedy, affecting hundreds of thousands of people, caused a massive negative impact on world economy. Today in the so called "global village" new societies as Trust Networks appear without delay, which enables us to stay home and make businesses by the Internet. It is becoming common to interact with unknown people. In this new context, there is a need for new tools that can help in deciding about the trustworthiness of other people.

Amid the uncertainty caused by it's crucial that companies adopt an "act now, plan now" strategy. In order to stay ahead in a rapidly changing world it is becoming common to build a new kind of businesses on web and interact with unknown people. Business leaders, affected by social distancing, provide perspectives on the evolving situation and implications for their companies. Industries, companies act now to protect their employees and customers—and prepare now for the short- and long-term online businesses beyond trust for a more sustainable future. In today's world people seek authenticity and integrity from the environment with which they choose to engage. However, there is enormous uncertainty in today's environment with new partners and it creates the gap between them. Experienced businessman usually relying on intuitions provides ways of going on with social interaction without the monitoring of future consequences. This division separates specific relations, situations, and objects into areas which familiarity designates as trust. However, today, even people with higher levels of mental ability are in difficulty today, to process information about new

R. A. Aliev et al. (Eds.): ICAFS 2020, AISC 1306, pp. 766–772, 2021.
https://doi.org/10.1007/978-3-030-64058-3_96

businesses more quickly, solve problems more accurately, and make decisions faster in today's environment of uncertainty.

Azerbaijan is a newly independent and developing country, with its own cultural norms. So people here might be expected also to be less susceptible to common environment, where people are separated by only a small number of acquaintances on the bases of trust.

The study focuses on the modes of trust on new virtual businesses, and particularly on how knowledge, experience, familiarity, and decision-making are combined in the act of trusting on them on the bases of values. Why is it important to know an individual's values? Values often underlie and explain attitudes, behaviors, and perceptions. So knowledge of an individual's value system can provide insight into what makes the person "tick."

This paper will address some cultural issues as power distance, individualism and uncertainty avoidance according to Hofstede's framework, that could contribute to the levels of trust in different regions of the world to be so diametrically opposed.

The paper structured in the following way. In Sect. 2 the challenges and opportunities for finding a sustainable system of Web of trust is analyzed. In Sect. 3 modes of uncertainty of trust according to perceptions are employed to establish a positive expectation. At the same time the role of the cultural values as power distance, individualism, uncertainty avoidance and its affect to decision making on the bases of trust was explored. A number-valued matrix was proposed to describe preferences over multiple criteria in a foreign market selection problem. Complexity of the considered criteria, a DM's preferences is characterized by fuzziness and partial reliability. The solution of the problem-prediction of the relevant outcome of trust is given in Sect. 4.

2 Objectives

It is necessary to glimpse into the future and see what challenges and opportunities lie ahead and find a sustainable system. If we are to enjoy the efficiencies and other benefits of the virtual organization, we will have to rediscover how to run organizations based more on trust than on control" [1].

We should rethink our roles, explore the dangers of a society. This shouldn't surprise since trust is a very human and social concept and it is in the speculations of men. "A lack of trust makes untrustworthy which can affect negatively to future virtual organizations. Simmel [2] claims that: As a hypothesis regarding future behavior, a hypothesis certain enough to serve as a basis for practical conduct, trust is intermediate between knowledge and ignorance about a man. According to Möllering [3] trust, consequently, remains a hypothesis: a positive expectation of the future that somehow overcomes uncertainty despite insufficient knowledge. In what way trust overcomes uncertainty is, however, intensely debated he. In our conceptualization of the domain, we define a trust statement as the explicit opinion expressed by a user about another user regarding the perceived quality of a certain characteristics of the user. For example, on a site where users contribute reviews about products, users could be asked to express a positive trust statement on a user "whose reviews and ratings they have consistently found to be valuable" and a negative trust statement on "authors whose reviews they find consistently offensive, inaccurate, or in general not valuable".

Consequently, there is good reason to suspect that the experience of trusting others is markedly different from the experience of trusting systems in today's environment of uncertainty. Web of trust requires that users describe their beliefs about others. Apparently, there is a need for new tools that can help in deciding about the trustworthiness of businesses. We tried to analyze trust on people in a new way. Our analysis deals with three different aspects of trust:

- factors of perceiver-attitudes
- factors of the situation-time, environment
- factors in the target-novelty.

Finally, the factors are contrasted to other positions in Hofstede's [4] cultural value system, and suggest new ways of theorizing and analyzing trust. Differences in international values according to Hofstede's value dimension on national culture helps us understand why different people act differently.

Managers in different countries make business decisions in different ways such as, at a much slower and more deliberate pace which may bring to distrust? The framework is used to distinguish between different national cultures, and their impact on a business setting.

Referring to the topic of this special issue "the dimensions of culture", we compare the framework and analyze the extremes of culture on society which can help to calculate or guess the future consequences of present decisions. The predominant approach is that it can be helpful in gaining unspecific positive attitude towards the business partners. Trust is a way of relating to others without our considering risk and probability, because trust does not concern the specific intentions of the other, but rather the shared character of human agency [5–7].

3   Research Questions

Trust may be based on a perception of social relations and the consequences of agency that is different from cultural perspectives in decision-making.

In order to investigate the strength of each of the hypotheses, two research questions are pursued in the empirical analysis.

1. What modes of uncertainty of trust according to perceptions are employed to establish a positive expectation
2. How are the cultural values as power distance, individualism, uncertainty avoidance can affect decision making and combine in trust?

Cultural value differences lay the foundation for our understanding of people's behaviors and influence significantly our perceptions. We enter virtual world n with preconceived notions of what "ought" and "ought not" to be. These notions are not value-free; on the contrary, they contain our interpretations of right and wrong and our preference for certain behaviors or outcomes over others. As a result, at times, our own perceptions cloud objectivity and rationality. However, in most instances they can help us in our righteous decision making. We decided to analyze cultural dimensions of values and factors in the situation, factors of perceiver and the factors in the target according to- certain time and unstable environment, attitude of the perceiver and novelty of business. Analyses of the framework; power distance, individualism, and uncertainty avoidance according to our country and its collation with other countries can help us in finding reliable partners in business.

Let us consider extraction of a consistent Z-number-valued matrix to describe preferences over multiple criteria in a foreign market selection problem. We will deal with three criteria that describe a series of economical and institutional characteristics: Institutional Proximity (C_1) (government performance and economic freedom issues), *Economic Proximity* (C_2) (socioeconomic issues) and *Social and Cultural Proximity* (C_3) (cultural characteristics). Due to complexity of the considered criteria, a DM's preferences may be characterized by fuzziness and partial reliability. In view of this, we use partially reliable preference degrees of the Saaty scale to represent comparative importance of criteria (Table 1):

Table 1. Modes of uncertainty of trust according to perceptions and the cultural values (power distance, individualism, uncertainty avoidance)

Time/environment	Factors in the situation	Cultural Dimensions of values
Attitudes	Factors in the perceiver	Individualism/collectivism
Novelty	Factors in the target	Power distance
		Uncertainty avoidance

3.1 Defining the Cultural Dimensions of Values

Hofstede analyzed a large database of employee values scores collected by IBM employees covering more than 70 countries [8]. The studies [9] have upheld the value of his measures. In addition, Hofstede himself claims that commercial airline pilots and students in 23 countries, civil service managers in 14 countries, and up-market consumers in 15 countries have all validated his results [8].

3.2 Power Distance

Hofstede's Power distance Index measures the extent to which the less powerful members of organizations and institutions (like the family) accept and expect that power is distributed unequally. This represents inequality (more versus less), but defined from below, not from above. People of different power levels are trusted differently and it suggests that a society's level of inequality is endorsed by the followers as much as by the leaders.

When we think about trust, we must be sure to think about areas of divergent government structure that operate efficiently. This means that corporations that are able to circumvent the law by legal means would be considered trust by way of bureaucratic inefficiency. We mean avoiding mistrust as an all-inclusive variable comprising of bribes, bureaucratic inefficiency, extortion, fraudulent conversion and embezzlement.

In a low power distance society, superiors and subordinates regard themselves as equal in power and titles and status are less important which leads to harmony and cooperation.

In the countries, because of the lack of equality, superiors are better able to conceal their transactions as they are not required to make their transactions transparent. As subordinates can not dispute the leader's actions, it is argued that there is potentially high violation. It is argued that countries with a high power distance are more likely to accept a lack of equality regarding power and authority; therefore, they say, their individuals are more likely to have corrupt values and to accept corrupt practices [10].

3.3   Individualism/Collectivism

The fundamental issue addressed by this dimension is the degree of interdependence a society maintains among its members. It has to do with whether people's self-image is defined in terms of "I" or "We". In Individualist societies people are supposed to look after themselves and their direct family only. In Collectivist societies people belong to 'in groups' that take care of them in exchange for loyalty.

3.4   Uncertainty Avoidance

Uncertainty avoidance deals with a society's tolerance for uncertainty and ambiguity; it ultimately refers to man's search for Truth. It indicates to what extent a culture programs, its members feel either uncomfortable or comfortable in unstructured situations. Unstructured situations are novel, unknown, surprising, and different from usual. Uncertainty avoiding cultures try to minimize the possibility of such situations by strict laws and rules, safety and security measures, and on the philosophical and religious level by a belief in absolute Truth; 'there can only be one Truth and we have it'.

High Uncertainty avoidance affects high performance, and satisfaction are likely to be higher if their values fit well with the others.

4   Methods

On the bases of surveys, we tried to such strategic compartments create a social topography of relations, situations, and objects excluded from uncertainty and 'surrounded' by areas where the probability [10] of trust is taken for granted.

We decided to learn own cultural values and compare ourselves with other cultures, because values are the rules that tell us what is right and wrong, what we should and should not do, what is more important and less important. These values can be helpful in building trust with other people.

This paper is based on an interview study conducted in Azerbaijan in 2019 and 2020. The 300 people interviewed represent a high level of socio-economic and demographic variation. They were recruited through different organizations. Types of social, cultural and economic differences most closely correlated to different levels of trust within population. By creating an image of social relations characterizing the trustworthy, one can set the person apart from the risky category of strangers.

The assumption here is that variations in social position will also entail variations in the type and quality of social relations as well as in the attitude towards and perceptions of these relations.

Participants come to understand the powerful effects that culture plays in every person's life. It may be used to help participants prepare for living and working in another culture or to learn how to work with people from other departments, disciplines, genders, races, and ages. In most settings, a user has a direct opinion only about a very small portion of users. The remaining users are unknown users. "How much should I trust this unknown user?".

Investigated the relationship of cultural values and a variety of organizational criteria at both the individual and national level of analysis allows us to predict the relevant outcomes of trust. For example, Azerbaijan has a 75 on the cultural scale of Hofstede's analysis. Compared to Arab countries where the power distance is low (80), but comparing to Germany and Austria it is very high (11). On the other hand, the power distance in the United States scores a 40 on the cultural scale. The United States exhibits a more unequal distribution of power in society.

Just focusing on three dimensions we could find out that, when the power distance is of high index and uncertainty avoidances is of low index within a country, it increases the mistrust index. In addition, the paper will also show that as the power distance increases, the uncertainty avoidance decreases, trustworthy also decreases. Finally, the paper will look at the moderating effect that power distance has on the relationship between uncertainty avoidance and trust. These are important issues to address so that businesses will have a better understanding of - the timing, attitudes and novelty - to enter and invest in countries that traditionally have had poor scores on the level of trust but may change based on the changing cultural variables based on value systems.

5   Conclusion

The belief on future, importance of rationality, ability of people to solve problems, and preference for collective decision making for web business can be very helpful nowadays. Unlike neutral or biased standards used to decide what is fair, understanding of - the timing, attitudes and novelty and cultural values as power distance, individualism, and uncertainty avoidance can be helpful in future businesses.

People in our country are not too keen on uncertainty, nevertheless, by planning everything carefully they try to avoid it, reduce the risks to the minimum and proceed with changes step by step. However, sudden environmental changes can be a huge barrier in creation of prosperity of trust in local companies. Therefore it would be advisable to create global Trust Networks for making businesses in a society that relies on rules, laws and regulations, as f.ex. Germany, Austria, Denmark, etc.

In sum, this research suggests that Hofstede's value framework (power distance, individualism, and uncertainty avoidance) may be a valuable way of thinking about differences among cultures, but we should be cautious about assuming all people from a country have the different values.

By our model we try to create trustworthiness; maps of the world for ourselves which we then use to interpret what is going on around us and check the fairness by determining the overall value balance. We hope it can be helpful to predict what will happen and hence keep control of our personal world.

References

1. Handy, C.: Print Book: English, 3rd edn. Penguin, Harmondsworth (1985)
2. Simmel, G.: The secret and the secret society. In: Wolff, K.H. (ed.) The Sociology of Georg Simmel. The Free Press of Glencoe, London (1950)
3. Mollering, G.: The nature of trust: from Georg Simmel to a theory of expectation, interpretation and suspension. Sociology **35**(2), 403–420 (2001)
4. Understanding Cultures & People with Hofstede Dimensions (2015). www.cleverism.com
5. Aliyev, R.A., Fazlollahi, B., Aliyev, R.R.: Soft Computing and Its Applications in Business and Economics. Springer, Heidelberg (2004)
6. Seligman, G., Grøn, A.: Trust, Sociality and Selfhood. Mohr Siebeck, Tübingen (2010). Grøn, A., Welz, C. (eds.)
7. Levinas, E.: The Ethical Demand. University of Notre Dame Press, Notre Dame (2011). Mistzal, B. (ed.)
8. Hill, New York. ITIM International (2011). http://www.geert-hofstede.com/. Accessed 24 May 2011
9. Robertson, C.J., Watson, A.: Corruption and change: the impact of foreign direct investment [Electronic version]. Strateg. Manag. J. **25**(4), 385–401 (2004)
10. Seleim, A., Bontis, N.: The relationship between culture and corruption: a cross-national study. J. Intellect. Cap. **10**(1), 166–184 (2009)

Research and Analysis Indicators
of the Quality of Service Multimedia Traffic
Using Fuzzy Logic

Bayram G. Ibrahimov[1] and Almaz A. Alieva[2][✉]

[1] Azerbaijan Technical University, Huseyn Javid, 25, Yasamal,
Baku, Azerbaijan
i.bayram@mail.ru
[2] Mingechaur State University, Zahid Khalilov, 23, Mingachevir, Azerbaijan
almaz40@gmail.com

Abstract. The paper proposes a solution to an urgent scientific problem related to the development of a streaming network model for research and analysis numerous parameters of the quality of service multiservice traffic using fuzzy logic. The efficiency multiservice communication networks based on the architectural concept subsequent NGN (Next Generation Network) and future networks FN (Future Networks) using the technology building distributed communication networks is analyzed. A new approach to monitoring the indicator quality of service QoS (Quality of Service) and quality of experience QoE (Quality of Experience) differentiated by multimedia traffic classes using the fuzzy logic apparatus is considered. Based on the new approach, a model and fuzzy inference system is proposed, membership functions and fuzzy production rules are determined taking into account the numerous requirements QoS & QoE multimedia traffic and the quality of state of the CS (Current Service) communication channel. In this paper, the formulation of the problem is investigated and approaches to its solution are proposed.

Keywords: Fuzzy logic apparatus · QoS · Hurst coefficient · CS · Membership function · QoE · Communication channel

1 Introduction

The rapid development infrastructure of the digital economy based on a single information space requires new principles and global approaches to building highly efficient multiservice communication networks based on the architectural concept of the following NGN and future networks FN with increased performance [1, 2].

Today, the use technology for building distributed communication networks helps accelerate the launch new multimedia services and applications. These reduces the overall cost their implementation and requires the use innovative technologies SDN (Software Defined Networking), NFV (Network Functions Virtualization) and IMS (Internet Protocol Multimedia Subsystem). Recent data confirm that QoS is multimedia traffic, resource management in networks, sufficient classification management flexibility, as well as QoE using fuzzy logic [2, 3].

R. A. Aliev et al. (Eds.): ICAFS 2020, AISC 1306, pp. 773–780, 2021.
https://doi.org/10.1007/978-3-030-64058-3_97

It was established [2, 4–6] that a special role in multiservice communication networks is played by ensuring the required level QoS and QoE multimedia traffic. For an adequate description of traffic dynamics, as well as for determining the required level QoS and QoE in the network, the necessary creation an adequate system and approaches for assessing the effectiveness of the network in conditions traffic self-similarity is an important task. Currently, research is continuing in the field mathematical modeling multiservice traffic, which allows us to explain the physical causes of the self-similarity phenomenon in traffic and evaluate its effect as on network efficiency. An important parameter self-similar processes is the Hurst parameter. This indicator is used to assess the degree self-similarity of a random process that generates multimedia services and applications [2].

The investigated communication networks based on the new approach using fuzzy logic take into account the following [2, 6]: parameters of self-similar traffic; introduction of new services and applications such as "Triple Play services" & "Bandwidth on Demand"; effective management information and network resources; use of limited network bandwidth resources; QoS service quality indicators; reliability of the network; QoE quality of experience indicators.

Given the above, the task preliminary assessment numerous indicators multimedia traffic using fuzzy logic in multiservice communication networks is relevant.

Based on the study, it was established [2, 7, 8] that an adequate assessment of the performance of multiservice communication networks is crucial for the successful creation of a fuzzy logical inference system.

The paper considers the solution of the problem formulated above - the study and analysis QoS and QoE indicators multimedia self-similar traffic based on the fuzzy control technique that allows you to manage resources in multiservice communication networks using the fuzzy logic.

2 General Statement of the Problem

Given the importance of building multi-service communication networks based on FN [1, 6, 8] with packet switching (ITU-T, Y.3000, …, Y.3499) for QoS and QoE multimedia traffic, special attention should be paid to complex indicators. In the works, a generalized technique for fuzzy control of the QoS classifier [8] and methods for improving the performance communication networks taking into account the self-similarity multimedia traffic [2] using the fuzzy logic are studied.

The indicators of service quality and quality of traffic perception [3], operational distribution information and network resources in multiservice communication networks [1, 5], as well as a method for calculating the Hurst coefficient parameters using fuzzy models to describe [2, 5] and modeling weak formalized multiservice systems and telecommunication processes [8]. The analysis showed [1–3] that the effective load factor of the communication channel is determined by the quality of the QoS & QoE self-similar traffic control.

Based on the study, it was determined [8] that the standard procedure for using the classifier multiservice communication networks for outgoing traffic of a multimedia service different class does not allow organizing the required (Requested Service, RS) quality of service for the telecommunication infrastructure.

Unified telecommunications infrastructure does not have the following:

- sufficient flexibility to control the classifier communication networks;
- limited ability of real communication channels;
- not rational use of the channel resource.

Given these shortcomings, a single telecommunication system can lead to the loss important service packages during the collection information and the management forces and complex hardware multiservice communication networks.

Given the components of the vector, the classifier of multiservice communication networks using a single telecommunications infrastructure is functionally described by the following relationship:

$$Q_{VK}(H, \lambda_i) = G[\rho(H_i), R_{SE}, E_i(\lambda_i)], \quad i = \overline{1, k}, \tag{1}$$

where $E_i(\lambda_i)$ – a function that takes into account the indicators of the quality of state communication channel with when providing i-th multimedia services with intensity λ_i, $i = \overline{1, k}$; R_{SE} – the required indicator of the quality of service and the quality of perception of the telecommunications infrastructure with the uneven nature of the formation of the current state of traffic, $R_{SE} = [R_{QoS}, R_{QoE}]$; $\rho(H_i)$ – coefficient effective loading of the channel multiservice communication networks, taking into account the self-similarity traffic H_i in the provision of i-th multimedia services, $i = \overline{1, k}$. As the degree self-similarity increases, the Hurst parameter takes values from 0.5 to 1. The proposed expression (1) are a comprehensive indicator of the classifier multiservice communication networks that take into account the QoS & QoE multimedia traffic control and indicators of the quality of state communication channels. Thus, in this paper, the formulation of the problem is investigated and new approaches to its solution are proposed.

3   Description of the Fuzzy Inference System

Based on the study, it was established [2, 6] that an important parameter in (1) is the effective load factor communication channels $\rho(\lambda_i, H_i)$ in a fuzzy control system, which changes the control parameter of multiservice communication networks.

Given the self-similarity multimedia traffic $\rho(\lambda_i, H_i)$, the objective functions is to maximize the effective loading communication channels of multiservice communication networks as follows:

$$\rho(\lambda_i, H_i) = [\frac{\lambda_i}{\mu \cdot N_m} \cdot f(H_i)] \xrightarrow{i} \max, \quad i = \overline{1, k}, \tag{2}$$

where λ_i – the flow rate i-th of the multimedia service packet; μ – multimedia service traffic packet stream service speed; N_m – the number of communication channels for the services served at the access point to multimedia services, i.e. the number channel resource units multiservice networks; $f(H_i)$ – a function that takes into account the self-similarity of the incoming load and is equal to $f(H_i) = 2H_i$; H_i – Hurst coefficient for i-th packet flow, $i = \overline{1, k}$.

Expression (2) is the objective function that determines the effective use communication channels using the fuzzy logic apparatus based on the required QoS & QoE parameters, taking into account the dynamics changes in the Hurst coefficient.

However, as the degree of self-similarity increases, the Hurst parameter significantly affects the required QoS & QoE parameters and the quality of the network Q_{CS}.

Based on the proposed approach to monitoring the QoS & QoE indicator multimedia self-similar traffic and the quality of network, we consider the method calculating the Hurst coefficient using the fuzzy logic apparatus.

It should be noted that when using the apparatus in question, the expected value of the coefficient H_i from area H_3 according to the experimental results for various methods can be determined using the fuzzy integral [2]. In this case, it is a priori necessary to determine the distribution of the fuzzy density of the weights of these values $g(H_i)$, $H_i \in H_e$. Here $H_i \in H_e$ means the aggregate of the expected value of the Hurst coefficient in the management QoS & QoE traffic.

Given the processes that occur using fuzzy logic, it is advisable to use the membership function $\mu_W(H_i)$ of an element H_i to a fuzzy set W. The analysis systems and decision-making under fuzzy conditions is the subject of the fundamental work Zade and Bellman [9, 10]. On the basis of [2, 9, 10], which allows changing the QoS & QoE parameters using the fuzzy logic apparatus from the domain magnitude W and H_e there is a set of ordered pairs and is described as follows:

$$W = \{\rho(H_i), \mu_W(H_i)\}, \quad H_i \in H_e \tag{3}$$

where $\mu_W(H_i)$ – function that determines the degree membership of an element H_i to a fuzzy set W. Given the algorithm for calculating the Hurst coefficient using the fuzzy logic apparatus, the membership function is expressed as follows:

$$\mu_W(H) = \sum_i [\mu_W(H_i)/(H_i)], \ H_i \in H_e \tag{4}$$

As a result of the study of a different method for calculating indicators self-similar traffic, the following values of the Hurst coefficient are obtained:

$$H_i = \{(i = 1, \ldots, 4) : H_1 = 0,68, \ \ H_2 = 0,75; \ \ H_3 = 0,86; \ \ H_4 = 0,95\}$$

Now we can determine the value of the function membership function ordered by decreasing degrees as follows:

$$Y(H_i) = \sum_{i=1}^{4} H_i \cdot \mu_W(H_i), \quad i = \overline{1,4} \tag{5}$$

Then, according to the calculated data, numerical values $Y(H_i)$ are found as

$$Y(H_i) : h(H_i) = 0,43, \quad Y(H_2) = 0,52, \quad Y(H_3) = 0,90, \quad Y(H_4) = 0,98$$

Based on the fuzzy control technique, the numerical values of the membership function of the process under study can be determined as follows:

$$\mu_W(H_i) = 0,43H_1^{-1} + 0,52H_2^{-1} + 0,90H_3^{-1} + 0,98H_4^{-1} \tag{6}$$

Given (6) under given conditions, fuzzy measures take the following values:

$$g(F_1) = 0,52, \quad g(F_2) = 0,81, \quad g(F_3) = 0,95, \quad g(F_4) = 1,00$$

Based on the proposed approach and calculation, the QoS & QoE indices are considered both the average service time and the probability packet loss when using the predicted value H_i compared to determining the listed indicators by the average value of the Hurst coefficient $H_i^{av} = 0,670$, where $0 < H_i^{av} \leq 1$.

It is worth noting that when constructing and checking the degree of adequacy of the fuzzy model, the Fuzzy Logic Toolbox software package Matlab was used. The function belonging $\mu_W(H_i)$ to the input variables of the coefficient of effective loading of communication channels in multiservice networks at $H_i^{av} = 0,670$ and $V_i^{ck} = 25$ Mbps. From the graphical dependence it follows that an increase in the effective loading communication channels leads to an increase in the share lost packets, there by reducing the required QoS & QoE parameters.

For various services, its noticeable change begins with the value $\rho(H_i) > 0.55, \ldots, 0.65$ at $H_i^{av} = 0,670$ и $V_i^{ck} = 25$ Mbps. To set membership functions in the Fuzzy Logic Toolbox, it was found that membership functions become non-linear. And this greatly complicates further research. Therefore, it was decided to build mathematical models for assessing the quality of transmitted traffic based on the fuzzy logic apparatus.

In view (1), the formal description of the model is represented by the expression:

$$Q_{VK} = \langle R, \mu_W, T \rangle, \tag{7}$$

where R – many indicators reflecting the required parameters R_{SE} and R_{CS} network, $R = [R_{SE}, R_{CE}]$; T – many indicators describing the probabilistic-time characteristics of the network, $T = TD(\lambda_i, H_i)$; μ_W – membership function. The criterion of the quality of service and perception QoS & QoE of multimedia self-similar traffic using a fuzzy logic apparatus is a probability-time characteristic:

$$T[Mark_{mod}[TD(\lambda_i, H_i)] \xrightarrow{i} \min, \ i = \overline{1,k} \tag{8}$$

where $Mark_{mod}[TD(\lambda_i, H_i)]$ – model assessment of the probability-time characteristics when providing the required multimedia services; $TD(\lambda_i, H_i)$ – delays in transmitting packet streams based on parameters λ_i and H_i, $i = \overline{1,k}$.

Given (7) and (8), the criteria for the quality indicator of the state of a communication channel for representing a variety of services is defined as follows:

$$K_{opt}[Mark_{mod}(R)] \rightarrow K_{opt}[Mark_{treb.}(R)], \quad i = \overline{1,k}, \tag{9}$$

where $K_{opt}[Mark_{treb.}(R)]$ – quality of service and experience R_{SE}, and quality condition of the communication channel R_{CE} when providing the service at the required level. Based on (7), (8) and (9), the criteria or rules for removing the partial uncertainty of the quality state of the communication channel when providing a service that describes the operation of fuzzy logic are represented by the expression:

$$\mu_W = \{0, \ R \notin W, \ \mu_W(R), \ R \in W, \ \mu_W(R) \in (0,1]\} \tag{10}$$

where μ_W – membership function that associates each element of a fuzzy set W with a certain number, which is interpreted as the degree of belonging of the element R to a fuzzy set W. Expressions (7), ..., (10) describes a new approach to constructing a mathematical model for assessing the quality of traffic and the state of the communication channel in communication networks based on the fuzzy logic.

4   Building a Generalized Structural Diagram of a Fuzzy Control System

System-technical analysis [2, 11] showed that having information about the current state of the system and the required services at the access point to multimedia services, it is possible to dynamically control the parameters communication channels.

This mechanism allows you to increase the quality indicators of the communication channel (Current Service, R_{CS}) in the network $Q_i^{CS}(\lambda)$. The indicators of the quality state of the communication channel in communication networks based on a fuzzy control system are functionally described by the following relationship:

$$Q_i^{CS}(\lambda) = F[K_i(\lambda), \ E_i(\lambda)], \ \lambda = \sum_{i=1}^{k} \lambda_i \quad i = \overline{1,k}, \tag{11}$$

where $K_i(\lambda)$ – a coefficient characterizing the intensity multimedia traffic and taking into account the number units of the channel resource N_m multiservice communication networks is determined by the expression:

$$K_i(\lambda) = N_m \cdot [\lambda_i^s / \sum_{i=1}^{N_m} \lambda_i^s] \cdot f(H), \quad i = \overline{1,k}, \tag{12}$$

where λ_i^s – traffic arrival rate i-th service on one of the N_m terminal devices; $E_i(\lambda)$ – total non-delivery rate i-th service packet streams. Given the loss coefficients packets

i-th traffic $K_{i.nn}(\lambda)$ during transmission over communication channels, the total coefficient non-delivery packets i-th traffic is found as follows:

$$E_i(\lambda) = K_{i.nn}(\lambda) + K_{i.BER}(\lambda), \quad i = \overline{1,k}, \tag{13}$$

where $K_{i.BER}(\lambda)$ – the coefficient erroneous reception packet flows i-th traffic when transmitted over communication channels. Expressions (11), (12) and (13) characterize the indicators of the quality state of the communication channel in the network and describe the algorithms of the odd logic system. To implement the algorithms of operation and control of the quality of service classifier systems in dynamics, a fuzzy control system is proposed [2, 8, 12, 13].

A generalized structural diagram fuzzy control system for multiservice communication networks consists of input and output data, a fuzzy control system, a router, a communication channel, and a switch. In this case, the fuzzy control system combines the following functional blocks, as subsystem for assessing the required quality of QoS & QoE, subsystem for assessing the current state of the communication channel R_{CS}, subsystem of fuzzy control of support algorithms R_{SE}.

Thus, the study of the effective loading of communication channels and analysis of the structural diagram shows that when using the fuzzy logic apparatus, the parameters of the required quality of service are clearly higher than when using standard parameters communication network.

5 Conclusion

The system and apparatus of fuzzy logic are investigated, a new approach to the analysis of the QoS & QoE traffic indicator and the qualitative state of the network under the influence of the self-similarity effect taking into account the Hurst coefficient prediction is proposed. Based on the model, the function membership of an element H_i to the set W using the fuzzy logic apparatus is proposed, and the functions graphs $\mu_W(H_i)$ of the input variables of the coefficient of effective loading of communication channels are presented and a generalized block diagram of the fuzzy control systems in multiservice communication networks is constructed.

The proposed approach and the system fuzzy logical control of the classifier multiservice communication networks most fully takes into account the change in the current state of the multimedia service in the telecommunications infrastructure and can improve the quality of traffic transfer with limited channel resources.

References

1. Roslyakov, A.V., Vanyashin, S.V.: Future Networks. PSUTI, Samara (2015)
2. Ibrahimov, B.G., Alieva, A.A.: An approach to analysis of useful quality service indicator and traffic service with fuzzy logic. In: Advances in Intelligent Systems and Computing, vol. 1095, pp. 495–504 (2019). https://doi.org/10.1007/978-3-030-35249-3_63

3. Sokolov, D.A.: Fuzzy quality assessment system. Technol. Means Commun. **4**, 26–28 (2009)
4. Zadeh, L.A.: Fuzzy sets. Inf. Control **8**, 338–353 (1965)
5. Borisov, V.V., Kruglov, V.V., Fedoulov, A.S.: Fuzzy Models and Networks. Hotline – Telecom, Moscow (2012)
6. Ibrahimov, B.G., Humbatov, R.T., Ibrahimov, R.F.: Analysis performance multiservice telecommunication networks with using architectural concept future networks. T-Comm **12** (12), 84–88 (2018)
7. Aliev, R.A., Gurbanov, R.S., Aliev, R.R., Huseynov, O.H.: Investigation of stability of fuzzy dynamical systems. In: Proceedings of the Seventh International Conference on Applications of Fuzzy Systems and Soft Computing. Siegen, Germany, pp. 158–164 (2006)
8. Grebeshkov, A.Yu.: Deciding service provider to manage the connection to the network access based on user rank. T-Comm **7**, 25–27 (2013)
9. Bellman, R., Giertz, M.: On the analytic formalism on the theory of fuzzy sets. Inf. Sci. **5**, 149–157 (1974)
10. Belman, R.E., Zadeh, L.A.: Decision - making in a fuzzy environment. Manag. Sci. **17**, 141–164 (1970)
11. Bianco, B., Fajordo, J.O., Cianonoulakis, I., et al.: Technology pillars in the architecture of future 5G mobile networks: NFV, MEC and SDN. Comput. Stand. Interf. **54**, 216–228 (2017)
12. Zade, L.A.: The role of soft computing and fuzzy logic in the understanding, design and development of information/intelligent systems. Artif. Intell. News. **2–3**, 7–11 (2001)
13. Drutskoy, D., Keller, E., Rexford, J.: Scalable network virtualization in software-defined networks. IEEE Internet Comput. **17**(2), 20–27 (2013). https://doi.org/10.1109/MIC.2012.144

Solution of a Decision Making Problem Under Risk with Z-Information

Konul Jabbarova[1] (ID) and O. H. Huseynov[2(✉)] (ID)

[1] Department of Computer Engineering, Azerbaijan State Oil and Industry University, 20 Azadlig Ave., AZ1010 Baku, Azerbaijan
konul.jabbarova@mail.ru
[2] Research Laboratory of Intelligent Control and Decision Making Systems in Industry and Economics, Azerbaijan State Oil and Industry University, Baku, Azerbaijan
oleg_huseynov@yahoo.com

Abstract. In this paper expected utility under Z-information methodology is surveyed in order to select one of a fixed number of alternative actions in the context of risk, as many real-world solution taking problems are defined by uncertainty and partial reliability. In the final part, we will apply the Fuzzy Pareto Optimality (FPO) principle based on the comparison of Z-numbers.

Keywords: Solution taking · Expected utility · Z-number · Business decision

1 Introduction

A lot of real-world problems of decision making are characterized with partial reliability and uncertainty [1]. Prof. Zadeh introduced the concept of Z-number to describe partial reliability of information. A Z-number is a pair of fuzzy numbers $Z = (A, B)$, where A is a soft constraint on a value of a variable of interest, and B is a soft constraint on a value of a probability measure of A playing a role of reliability of A. In [1] they proposed two different approaches of decision making with Z-information. The first approach is based on reduction of Z-numbers to classic fuzzy numbers and uses Choquet integral with an integrant based on Z-numbers. Based on a given Z-information, a fuzzy measure is calculated. The second approach of decision making with Z-information is based on direct computation with Z-numbers. The numerical examples are used to demonstrate the effectiveness of the approaches of decision making with Z-information.

In [2] an approach to decision making under Z-information based on direct computation over Z-numbers is proposed. This approach uses the expected utility paradigm to solve the benchmark problem in the field of economics.

In [3] two approaches of decision making with Z-information are proposed. The first one is determined based on conversion of the Z-numbers into crisp numbers to find out. The next approach is based on Expected Utility Theory by using Z-numbers. In order to visualize a validity of the aforementioned approaches of decision making with Z-information, numerical examples have been used.

R. A. Aliev et al. (Eds.): ICAFS 2020, AISC 1306, pp. 781–786, 2021.
https://doi.org/10.1007/978-3-030-64058-3_98

In [4] they suggested Expected Utility Theory-based decision-making method under Z-information. An investment problem example is used to illustrate application of the approach proposed. In [5] expected utility's application to decision making in business under information specified by U-numbers is proposed. The attained results conform to decision making characterized as human intuition-based. Six different methodologies [6] are surveyed to choose one of the fixed number of alternative actions in the context of uncertainty.

The paper is structured as follows. In Sect. 2 we include a preliminary material used in the sequel. Section 3 consists of an application of the Expected Utility method under fuzziness and partial reliability of information. Section 5 is conclusion.

2 Preliminaries

Definition 1. A discrete Z-number [7, 8, 9 10]. A discrete Z-number is an ordered pair $Z = (A, B)$ where A is a discrete fuzzy number a fuzzy constraint role on values of a random variable X: X is A. Is a discrete fuzzy number with a membership function $\mu_B : \{b_1, \ldots, b_n\} \to [0, 1]$, $\{b_1, \ldots, b_n\} \subset [0, 1]$, playing a fuzzy constraint role on the probability measure of A: $P(A) = \sum_{i=1}^{n} \mu_A(x_i) p(x_i)$ is B.

Definition 2. Operations over Discrete Z-numbers [7–10]: Assume X_1 and X_2 are discrete Z-numbers and explain information on values of X_1 and X_2. Contemplate referring to $Z_{12} = Z_1 * Z_2$, $* \in \{+, -, \cdot, /\}$. This is a computation of $A_{12} = A_1 * A_2$ as the initial stage.

The second stage includes construction of B_{12}. We put into practice that in Z-numbers Z_1 and Z_2, the 'true' probability distributions p_1, p_2 are not known precisely. Contrary, fuzzy restrictions outlined in terms of the membership functions are available:

$$\mu_{p_1}(p_1) = \mu_{B_1}\left(\sum_{k=1}^{n_1} \mu_{A_1}(x_{1k}) p_1(x_{1k})\right), \mu_{p_2}(p_2) = \mu_{B_2}\left(\sum_{k=1}^{n_2} \mu_{A_2}(x_{2k}) p_2(x_{2k})\right)$$

Probability distributions $p_{jl}(x_{jk})$, $k = 1, \ldots, n$ infer probabilistic uncertainty over $X_{12} = X_1 + X_2$. Given any likely pair p_1, p_2, the convolution $p_{12} = p_1 \circ p_2$ is calculated as

$$p_{12}(x) = \sum_{x_1 + x_2 = x} p_1(x_1) p_2(x_2), \forall x \in X_{12}; x_1 \in X_1, x_2 \in X_2.$$

Given p_{12s}, the value of probability measure of A_{12} is calculated:

$$P(A_{12}) = \sum_{k=1}^{n} \mu_{A_{12}}(x_{12k}) p_{12}(x_{12k}).$$

p_1 and p_2 are described by fuzzy restrictions which infer fuzzy set of convolutions [7, 12]:

$$\mu_{p_{12}}(p_{12}) = \max_{\{p_1,p_2:p_{12}=p_1 \circ p_2\}} \min\{\mu_{p_1}(p_1), \mu_{p_2}(p_2)\}$$

Uncertainty of information on p_{12} infers fuzziness of A_{12} as a discrete fuzzy number B_{12}. The membership function $\mu_{B_{12}}$ is interpreted as

$$\mu_{B_{12}}(b_{12}) = \max \mu_{p_{12}}(p_{12})$$
$$\text{s.t.} \quad b_{12} = \sum_{i=1}^{n} \mu_{A_{12}}(x_i) p_{12}(x_i)$$

As a result, $Z_{12} = Z_1 * Z_2$ is obtained as $Z_{12} = (A_{12}, B_{12})$.

Definition 3. Fuzzy Pareto Optimality (FPO) principle-based comparison of Z-numbers [11]. FPO principle helps us in finding degrees of Pareto Optimality of multiattribute alternatives. This principle is introduced in order to compare Z-numbers as multiattribute alternatives, while one attribute measures value of a variable, the other one measures the related reliability. In conformity with this approach, by directly comparing Z-numbers $Z_1 = (A_1, B_1)$ and $Z_2 = (A_2, B_2)$ one reaches at total degrees of optimality of Z-numbers: $do(Z_1)$ and $do(Z_2)$. These degrees are decided on the basis of a number of components (the minimum is 0, the maximum is 2) with respect to which one Z-numbers dominates another one. Z_1 is considered higher than Z_2 if $do(Z_1) > do(Z_2)$.

3 Statement of the Problem

A construction company is going to construct a dormitory in a specific location and must decide whether to construct a 25-bed, 50-bed, or 100-bed facility. The possible levels of demand are high, medium, or low [6]. The outcomes of the alternatives [6]:

  A1. Construct 25 units and consider low demand;
  A2. Construct 25 units and consider medium demand;
  A3. Construct 25 units and consider high demand;
  B1. Construct 50 units and consider low demand;
  B2. Construct 50 units and consider medium demand;
  B3. Construct 50 units and consider high demand;
  C1. Construct 100 units and consider low demand;
  C2. Construct 100 units and consider medium demand;
  C3. Construct 100 units and consider high demand.

The following conditions of nature are considered: low demand - S_1, medium demand - S_2, high demand - S_3. Based on the uncertainty of future demand, information on possible benefit of the options is defined by fuzziness as well as partial reliability. On the grounds of this, the decision making problem is explained in a form of the following Z-number valued pay-off matrix (Table 1):

Table 1. Z-number valued pay-off matrix.

	S_1	S_2	S_3
Build 25	(Low, Almost Sure)	(Medium, Sure)	(High, Almost Sure)
Build 50	(Low, Almost Sure)	(Medium, Sure)	(High, Almost Sure)
Build 100	(Low, Almost Sure)	(Medium, Sure)	(High, Almost Sure)

Z-number valued probabilities of states of nature: $P(S_1) = (Low, Very Sure)$, $P(S_2) = (Almost\ Medium, Very\ Sure)$, $P(S_2) = (Medium, Very\ Sure)$.

The problem is to calculate Z-number valued EUs of the alternatives $EU(f_i)$ and find the best option as an alternative with the maximal Z-number valued EU:

Find f^* such that $do(EU(f^*)) = \max\limits_{i=1,\dots,3} do(EU(f_i))$.

4   Solution of the Problem

Let us consider the Z-number valued information in Table 1. The codebooks for the used Z-numbers are shown in Tables 2, 3 and 4.

Table 2. The linguistic terms for A parts of Z-numbers for the profit of 25-bed facility

Linguistic value	Fuzzy value
Low (L)	(20, 30, 40)
Medium (M)	(25, 35, 45)
High (H)	(30, 40, 50)

Table 3. The linguistic terms for A parts of Z-numbers for profit of 50-bed facility

Linguistic value	Fuzzy value
Low (L)	(−30, −20, −10)
Medium (M)	(40, 50, 60)
High (H)	(45, 55, 65)

Table 4. The linguistic terms for A parts of Z-numbers for profit of 100-bed facility

Linguistic value	Fuzzy value
Low (L)	(−50, −40, −30)
Medium (M)	(−20, −10, 0)
High (H)	(65, 75, 85)

The information on probabilities of states of nature is also characterized by fuzziness and partial reliability. Therefore, we will consider Z-number valued probabilities of states of nature which are given in Table 5.

Table 5. The linguistic terms for A parts of Z-number valued probabilities

Linguistic value	Fuzzy value
Low (L)	(0, 0.1, 0.2)
Almost Medium (AM)	(0.3, 0.4, 0.5)
Medium (M)	(0.3, 0.5, 0.7)

B parts of the used Z-numbers are given in Table 6.

Table 6. The linguistic terms for B parts of Z-numbers

Linguistic value	Fuzzy value
Almost Sure (AS)	(0.6, 0.7, 0.8)
Sure (S)	(0.7, 0.8, 0.9)
Very Sure (VS)	(0.8, 0.9, 1)

The Z-number valued EUs of the alternatives are found as follows:

$$
\begin{aligned}
EU(f_1) &= P(S_1) \cdot f_1(S_1) + P(S_2) \cdot f_1(S_2) + P(S_3) \cdot f_1(S_3) = (L, AS) \cdot (L, VS) \\
&+ (M, S) \cdot (AM, VS) + (H, AS) \cdot (M, VS) = (0 \quad 0.1 \quad 0.2)(0.8 \quad 0.9 \quad 1) \\
&\cdot (20 \quad 30 \quad 40)(0.6 \quad 0.7 \quad 0.8) + (0.3 \quad 0.4 \quad 0.5)(0.8 \quad 0.9 \quad 1) \\
&\cdot (25 \quad 35 \quad 45)(0.7 \quad 0.8 \quad 0.9) + (0.3 \quad 0.5 \quad 0.7)(0.8 \quad 0.9 \quad 1) \\
&\cdot (30 \quad 40 \quad 50)(0.6 \quad 0.7 \quad 0.8) = (0 \quad 3 \quad 8)(0.53 \quad 0.66 \quad 0.88) \\
&+ (7.5 \quad 14 \quad 22.5)(0.59 \quad 0.73 \quad 0.89) + (9 \quad 20 \quad 35)(0.51 \quad 0.64 \quad 0.78) \\
&= (16.5 \quad 37 \quad 65.5)(0.25 \quad 0.39 \quad 0.6);
\end{aligned}
$$

$$
\begin{aligned}
EU(f_2) &= P(S_1) \cdot f_2(S_1) + P(S_2) \cdot f_2(S_2) + P(S_3) \cdot f_2(S_3) = (L, AS) \cdot (L, VS) \\
&+ (M, S) \cdot (AM, VS) + (H, AS) \cdot (M, VS) = (0 \quad 0.1 \quad 0.2)(0.8 \quad 0.9 \quad 1) \\
&\cdot (-30 \quad -20 \quad -10)(0.6 \quad 0.7 \quad 0.8) + (0.3 \quad 0.4 \quad 0.5)(0.8 \quad 0.9 \quad 1) \\
&\cdot (40 \quad 50 \quad 60)(0.7 \quad 0.8 \quad 0.9) + (0.3 \quad 0.5 \quad 0.7)(0.8 \quad 0.9 \quad 1) \\
&\cdot (45 \quad 55 \quad 65)(0.6 \quad 0.7 \quad 0.8) = (-6 \quad -2 \quad 0)(0.53 \quad 0.66 \quad 0.88) \\
&+ (12 \quad 20 \quad 30)(0.59 \quad 0.73 \quad 0.89) + (13.5 \quad 27.5 \quad 45.5)(0.51 \quad 0.64 \quad 0.78) \\
&= (19.5 \quad 45.5 \quad 75.5)(0.28 \quad 0.41 \quad 0.62);
\end{aligned}
$$

Then, to determine the best option, we will use the FPO principle-based comparison of Z-numbers. The following results are obtained:

Alternative f_1 vs. Alternative f_2: $do(EU(f_1)) = 0$, $do(EU(f_2)) = 1$,
Alternative f_1 vs. Alternative f_3: $do(EU(f_1)) = 1$, $do(EU(f_3)) = 0$,
Alternative f_2 vs. Alternative f_3: $do(EU(f_2)) = 1$, $do(EU(f_3)) = 0$.

Obviously, the best option is f_2.

5   Conclusion

In this paper we considered an application of Z-numbers in a real-world business decision making problem. Probabilities of the states of nature and outcomes of alternatives are described by Z-numbers. Fuzzy Pareto Optimality principle-based comparison is used to compare Z-number valued Expected Utility values. The obtained results are appropriate to human decision making.

References

1. Aliev, R.A, Zeinalova, L.M.: Decision making under Z-information. In: Pedrycz, W., Guo, P. (eds.) Human-Centric Decision-Making Models for Social Sciences, pp. 233–252 Springer (2014)
2. Aliev, R.R., Mraiziq, D.A.T., Huseynov, O.H.: Expected utility based decision making under Z-information and its application. Comput. Intell. Neurosci. Article ID 364512, 11 (2015). https://doi.org/10.1155/2015/364512
3. Gardashova, L.A.: Application of operational approaches to solving decision making problem using Z-numbers. J. Appl. Math. **5**(9), 1323–1334 (2014). https://doi.org/10.4236/am.2014.59125
4. Zeinalova, L.M.: Expected utility based decision making under Z-information. Intell. Autom. Soft Comput. **20**(3), 419–431 (2014). https://doi.org/10.1080/10798587.2014.901650
5. Jabbarova, K.I.: Application of expected utility to business decision making under U-number valued information. In: 13th International Conference on Theory and Application of Fuzzy Systems and Soft Computing, pp. 716–723 (2019). https://doi.org/10.1007/978-3-030-04164-9_94
6. Whalen, T.: Decision making under uncertainty with various assumptions about available information. IEEE Trans. Syst. Man. Cybern. Syst. **14**, 888–900 (1984). https://doi.org/10.1109/nafips.2016.7851588
7. Aliev, R.A., Alizadeh, A.V., Huseynov, O.H.: The arithmetic of discrete Z-numbers. Inf. Sci. **290**, 134–155 (2015). https://doi.org/10.1016/j.ins.2014.08.024
8. Aliev, R.A., Alizadeh, A.V., Huseynov, O.H., Jabbarova, K.I.: Z-number-based linear programming. Int. J. Intell. Syst. **30**(5), 563–589 (2015). https://doi.org/10.1002/int.21709
9. Aliev, R.A., Huseynov, O.H.: Decision Theory with Imperfect Information. World Scientific, Singapore (2014)
10. Aliev, R.A., Huseynov, O.H., Aliyev, R.R., Alizadeh, A.V.: The Arithmetic of Z-Numbers. Theory and Applications. World Scientific, Singapore (2015)
11. Aliev, R.A., Huseynov, O.H., Serdaroglu, R.: Ranking of Z-numbers and its application in decision making. Int. J. Inf. Technol. Decis. Mak. **15**(6), 1503–1519 (2016). https://doi.org/10.1142/S0219622016500310
12. Aliev, R.A., Huseynov, O.H., Zeinalova, L.M.: The arithmetic of continuous Z-numbers. Inf. Sci. **373**, 441–460 (2016)

Identification of Extended Objects of Geoinformation Space by Semantic Triangulation

Gurru I. Akperov(ID) and Vladimir V. Khramov$^{(\boxtimes)}$ (ID)

Private Educational Institution of Higher Education «Southern University (IMBL)», Rostov-on-Don, Russia
pr@iubip.ru, vxpamov@inbox.ru

Abstract. Development refers to management in socio-economic systems and can be used in systems for collecting, converting, processing information and making decisions when the source information is incomplete or unclear.

Keywords: Space monitoring · Identification · Extended object · Semantic triangulation · Geo-information space · Soft models

1 Introduction

Many problems of spatial economy require the creation of digital technologies to support management decisions that provide reliable identification of socio-economic objects on real or virtual (synthesized) images of geographic information space territories [1].

There are no universal methods for automatically identifying objects in images yet. Vast amount of information to identify images is included into their limits [2, 3], in other words in the form of the object outline. Under the outline we understand "a lot of pixels of the investigated objects having at least one neighbor pixel that doesn't correspond to the given object. On the binary two-leveled digital picture every pixel that either don't clearly belong to the investigated object or belong completely, the outline can be considered as determined (mathematical) object" [3]. For the "grey" (multi-gradational) "picture outline is a fragment of an image, where the gradient of the signal function is rapidly changing" [2].

2 Main Part

The task of the automated objects identification of images was considered [4], for example, by the identification of the object's outline and the further comparison of different characteristics of this outline with the model. To such characteristics mostly refer the following ones: the area and length of the outline, rate of the object's outline defined as the relevance of the square of the outline length to its area limited by this outline and etc. According to this approach "first the first photo image with high resolution are scanned. Then the resulted sample matrix is being structured, bringing to the reference scale. The image then is divided into three two-dimensional matrices in the palette of standard colors RGB" [5]. Then outlined images are revealed. The surfaces of objects inside the distinguished outlines Поверхности рельефов объектов внутри выделенных контуров are approximate to the same figures: triangles, squares, hexagons [1]. The area of mosaics in each channel is calculated according to the formula of Heron and/or Peak and compare the resulted areas or coefficients of the object surfaces with their meaning for the models according to the reliability criterion.

The disadvantages of this approach include [5]:

- not all information available for research (signs of the image signal as a whole) is used to identify the object;
- ambiguity of splitting the image into a mosaic of triangles, which leads to a significant error in calculating their area, including using the Heron formula;
- insufficient information content of the only feature - the area of the object, caused by the fuzziness of the original image data, which reduces the reliability of identification.

To solve the task of the given research the authors suggest the following scheme of actions:

- thematic binarization of the given image [4];
- selection of the object's outline based on heuristic dependencies obtained by the authors earlier [4, 6];
- defining the peculiar features of the object's outline and its outlined identification [7];
- calculating the coordinates of the outline (for example, center of gravity of the outline (for example, the center of gravity of a plane figure, restricted by this outline);
- the following fuzzy triangulation [8] of these points with the similar points of neighbor objects;
- fuzzy comparison of the triangulation with the model one (obtained and processed earlier) variant.

Next, the resulting image is decomposed into n two-dimensional matrices, and the contour drawings of the object are selected using spatial differentiation methods in each of the n channels. In this case, the object contour is described [7–9] parametrically: $\{x(s_i), y(s_i)\}$, then these parameters are laid out in rows by orthogonal functions, the coefficients of this decomposition are signs of identification, where, s_i is the length of the i-step along the contour; $\{x(s_i), y(s_i)\}$ are arrays of coordinates of contour points.

If we select the center of gravity of the outline as a characteristic (reference) point in this case its coordinates will be described by fuzzy numbers due to the combination of NON-factors [7]. Similarly, they find the centers of gravity of images neighbor to the given ones, they form the adjacent matrix, or in other words they describe the "context" of the image. Then the final identification is executed by means of the resulted matrix of the image with the reference one.

The method can be implemented in several ways, in particular, when $n = 1$, «the hyperspectral image is processed using a single channel. A significant number of satellites present such images. Alternatively, you can use a monochrome image with several "gray levels"» [8].

At $n = 3$, the initial hyperspectral image is pre-divided (during preprocessing) into three projections R, G, and B. The resulting images are processed using three corresponding channels, and objects are identified by their shape by calculating the features of their contours within the boundaries of the selected channels R, G, and B [5, 10]. Then the calculated features are compared with their reference values and a decision is made to identify them based on a predefined majorization rule (for example, 2 out of 3).

Some satellites, such as LandSat, provides the generated on-board multispectral image. In this case, the received images are processed using n corresponding channels, and objects are identified by their shape (contour) by calculating the features of their contours within the borders of the selected n channels, respectively. Then the calculated features are compared with their standard values and a decision on identification is made based on a predefined majorization rule. In this case, depending on the problem being solved, for example, how dangerous the detected property (attribute value) is, two or three or even one (out of 10–15 channels) with a "dangerous" attribute may be enough to make a decision.

3 Methods and Models

The problem solved by the proposed method of automatic identification of objects in images includes a number of subtasks, in particular:

– subtask for selecting the outline drawing of an item;
– calculation (or selection) of coordinates of characteristic points of this contour, followed by fuzzy Delaunay triangulation [7] of these points with similar contour points of all other objects in the image as a whole.

The technical result is achieved by using the fuzzy triangulation method to identify objects in images:

– decomposing the resulting image into n two-dimensional matrices corresponding to the selected channels and frequency ranges;
– highlight outline drawings of an item in each of the selected channels;
– finding the (indistinct) of the centers of gravity of flat objects in the image, limited to the respective circuits;
– finding the centers of gravity of all objects in a given scene in the image;
– implementation of a fuzzy Delaunay triangulation of the entire irregular set of characteristic points of the scene;
– identification of the resulting multidimensional matrix of adjacency of image objects with the reference one.

Figures 1 and 2 show an example of using automatic object identification in satellite images. For the image (Figs. 1 and 2), the found centers of gravity are shown in Table 1.

Example of contour selection using a heuristic algorithm, discussed in detail in [2] is shown in Fig. 3.

Fig. 1. Source image.

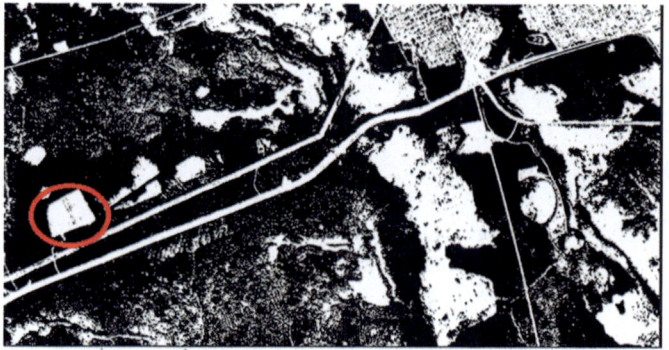

Fig. 2. Initial image after binarization.

Table 1. Centers of gravity of all objects of the specific IPS reflected on the satellite image, for which contour identification was performed [2].

	1	2	3	4	5	6	7	8	9	10	11	12	13	14	15
X	112	134	255	242	410	322	432	455	473	520	555	632	642	789	818
Y	357	538	319	286	183	280	42	100	401	137	404	473	404	396	228

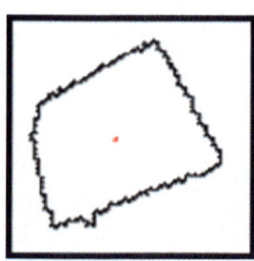

Fig. 3. Example of contour selection using a heuristic algorithm, discussed in detail in [2].

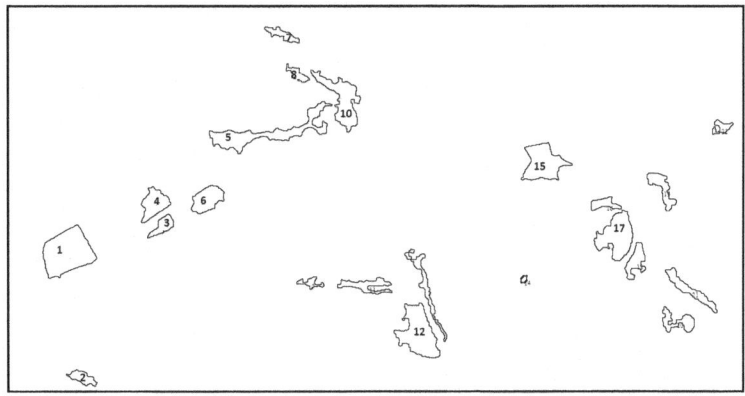

Fig. 4. Selection of item outlines for the entire IPS and their identification.

Then the calculated features are compared with their reference values and a decision is made to identify them based on a predefined majorization rule (for example, two out of three).

Depending on the problem being solved, for example, how dangerous a detected property (attribute value) is, two or three or even one (out of 10–15 channels) with a "dangerous" attribute may be enough to make a decision.

A variant of this method that provides high efficiency and ease of execution. Object identification in relation to a binarized (object-background) image, while contour selection is performed using a special heuristic algorithm [8].

In the proposed method of triangulation, Delaunay is proposed as decoding features that have the useful property that the initial data and, accordingly, the coordinates of the characteristic points of the object are fuzzy, and the possible segments have different lengths. Given the fact that "triangulation, by definition, is a planar graph, all the inner regions of which are triangles, the problem of constructing a triangulation for a given (characteristic) set of two-dimensional points is called the problem of connecting these points by non-intersecting segments so that triangles are formed.

For unambiguity of the triangulation, we can require, for example, the Delaunay condition or its generalization [4, 7]. Moreover, the triangulation satisfies the Delaunay condition if none of the specified triangulation points falls inside the circle described around any constructed triangle. In this case, a triangle of triangulation is considered to satisfy the Delaunay condition if this condition is satisfied by a triangulation composed only of this triangle and its three neighbors [10].

It is obvious that the original IPS images of objects on the earth's surface in satellite technical vision systems can be subjected to Delaunay triangulation.

4   Procedure for Implementing the Method

After scaling the original image (Fig. 1), to increase the accuracy of identification, the original matrix (image) is decomposed into three orthogonal ones, in the standard RGB color palette.

Figure 5 shows a variant of constructing a Delaunay triangulation on a scene fragment of the underlying surface image.

The resulting contour drawing in one of the channels is illustrated in Fig. 4. A Detailed study of the contour selection program is given in [2].

The contour points of each of the "reference" objects have fuzzy coordinates, but the "center of gravity" of the object in the image has significant stability [1, 7], so it can act as a "characteristic point" when performing the Delaunay triangulation scheme. That is why the authors of the article believe that in order to obtain a reliable identification of the image object, in addition to the actual "contour identification", a "contextual", spatial, using Delaunay triangulation is also necessary.

As a result, the technical vision system captures the objects present there. Next, we get a set of "reference points" (centers of gravity) that objectively set the coordinates of these objects at the time of shooting. Sometimes there is a situation when some of the objects in the scene are not identified, Since it is a question of monitoring, that is, regular surveys of this area are made, and preliminary triangulation has already been carried out.

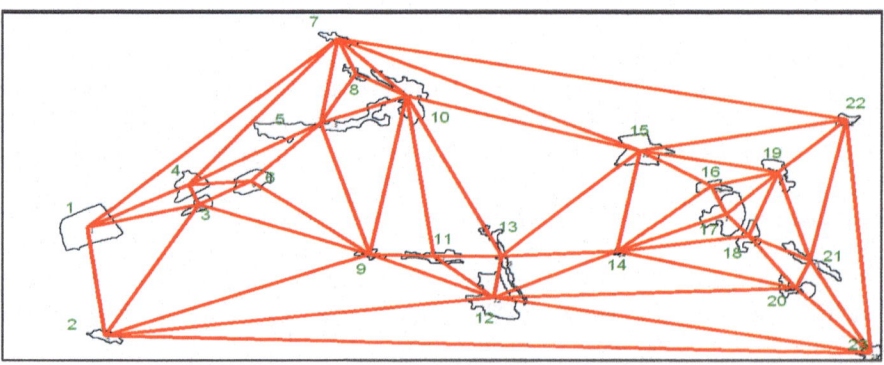

Fig. 5. Delaunay Triangulation in a virtual contour drawing

This allows you to restore information about missing (or unidentifiable at the time of shooting) objects. This approach makes it possible to increase the reliability of the navigation system in a particular area of the territory.

5   Conclusion

Thus, until now, the problem of recognizing objects in the resulting images is usually solved by an operator based on a set of decryption features. At the same time, the largest amount of information contains the shape of the object, and the image contains its contour drawing. The use of the identification method described in this paper can increase the accuracy of identification and, accordingly, improve the quality of management decisions.

References

1. Zadeh, L.: The Concept of a Linguistic Variable and Its Application to Making Approximate Decisions. Mir, Moscow (1976). (in Russian)
2. Khramov, V.V.: About methods of contour coding of models of information objects. Sci. Notes Inst. Manag. Bus. Law Ser.: Inf. Technol. Manag. **1**, 76–81 (2012). (in Russian)
3. Akperov, G.I., Khramov, V.V.: A fuzzy semantic data triangulation method used in the formation of economic clusters in Southern Russia. In: Aliev, R., Kacprzyk, J., Pedrycz, W., Jamshidi, M., Babanli, M., Sadikoglu, F. (eds.) 10th International Conference on Theory and Application of Soft Computing, Computing with Words and Perceptions, ICSCCW 2019. Advances in Intelligent Systems and Computing, vol. 1095, pp. 340–344. Springer, Cham (2020). https://doi.org/10.1007/978-3-030-35249-3_43
4. Akperov, G.I., Khramov, V.V., Gorbacheva, A.A.: Using soft computing methods for the functional benchmarking of an intelligent workplace in an educational establishment. In: Aliev, R., Kacprzyk, J., Pedrycz, W., Jamshidi, M., Babanli, M., Sadikoglu, F. (eds.) 10th International Conference on Theory and Application of Soft Computing, Computing with Words and Perceptions, ICSCCW 2019. Advances in Intelligent Systems and Computing, vol. 1095. Springer, Cham (2020). https://doi.org/10.1007/978-3-030-35249-3_6
5. Khramov, V.V., Gvozdev, D.S.: Intelligent Information Systems: Data Mining. Rostov State University of Railway Transport, Rostov-on-Don (2016). (in Russian)
6. Khramov, V.V.: Concept of ensuring the effectiveness of organizational and technical systems based on a bionic-intellectual approach. Bull. Rostov State Univ. Railway Transp. **2**, 138–141 (2001). (in Russian)
7. Khramov, V.V., Tsarkov, A.N.: Modeling of information processes in ergatic systems based on the principles of self-organization. In: The Collection: Problems of Ensuring the Effectiveness and Stability of Complex Technical Systems. Collection of Works. Ministry of Defense: SMI, pp. 444–447 (2003). (in Russian)
8. Mityasova, O.Yu., Akperov, I.G., Kramarov, S.O., Khramov, V.V.: Certificate of registration of computer programs RUS 2017615097. System of Analysis of Satellite Images (SASI), 13 March 2017 (2017)
9. Akperov, I.G., Kramarov, S.O., Khramov, V.V., Mityasova, O.Yu., Povkh, V.I.: Patent for invention RUS 2640331. Method of identification of extended objects of the earth's surface, 11 December 2015 (2015)
10. Khramov, V.V.: Computer Modeling. Manual for Course and Diploma Design, p. 87 p. Ministry of Defense, Moscow (1992). (in Russian)

Vegetation Index Formation Using Fuzzy Analyses of Object Multispectral Reflection

Elchin Aliyev$^{(\boxtimes)}$⬤ and Fuad Salmanov⬤

Institute of Control Systems of ANAS, Vahabzadeh Str. 9,
AZ1141 Baku, Azerbaijan
{elchin.aliyev,fuad.salmanli}@sinam.net

Abstract. A salient feature of vegetation and its state is spectral reflectivity characterized by large differences in the reflection of radiation of different wavelengths. Knowledge of the relationship between the structure and state of vegetation with its spectrally reflective abilities allows to use of aerospace images to map and identify types of vegetation and their stress state. To work with spectral information, the vegetation indices are created. Based on a combination of brightness values in certain channels, informative for highlighting the object under study, and calculating the object's "spectral index", appropriate image is constructed that corresponds to the index value in each pixel, which allows to select the object under study or evaluate its condition. In order to form a map for assessing the intensity of vegetation and/or recognition of other objects in the absence of vegetation, two fuzzy approaches to the quantitative estimation of photosynthetic activity of biomass are proposed. To identify the function for calculating the vegetation index a fuzzy analysis of multispectral reflection of objects in the red and near-infrared regions of the electromagnetic spectrum is used.

Keywords: Vegetation index · Multispectral reflection of objects · Fuzzy set · Membership function · Fuzzy inference

1 Introduction

The multispectral reflectivity of vegetation, characterized by significant differences in the reflection of radiation of different wavelengths (from 400 nm to 2400 nm), is a characteristic evaluative dimension of vegetation intensity and its condition. Multispectral visualization of the earth's surface provides data collection in both visible and invisible streak of light, which allows the generation of composite color images, as well as vegetation indices. The accumulated knowledge about the cause-effect relations between the structure and vegetation state and its spectrally reflective abilities according to the predicted pattern in the color spectrum made it possible to use aerospace images for mapping and identifying vegetation types and their stress state. Undoubtedly, this is of great applied importance in the field of precision farming and economic planning in the agricultural sector, or, more specifically, for assessing the condition of crop products and, accordingly, forecasting their productivity.

R. A. Aliev et al. (Eds.): ICAFS 2020, AISC 1306, pp. 794–801, 2021.
https://doi.org/10.1007/978-3-030-64058-3_100

To work with multispectral information obtained by remote sensing of the Earth's surface, users usually resort to creating so-called "index" images, which can be collected with a resolution measured in inches per pixel. By combining the brightness values in certain ranges that are informative for recognizing vegetation and next calculating the appropriate "vegetation index", an image is formed that corresponds to the value of this index in each pixel. This allows to recognize the intensity of the vegetation and/or evaluate its current state [1].

Vegetation indices are calculated by performing operations with different spectral ranges of remote sensing data and reflect certain vegetation parameters in a given pixel of the image. Moreover, the existing vegetation indices are formed experimentally (ин empirical analysis), taking into account the features of the reflection of a particular vegetation. Currently, about 160 types of vegetation indices are known, empirically selected on the basis of the known characteristics of the spectral reflectivity curves of vegetation and soils. To optimally determine the reflectivity of vegetation there are in multispectral visualization (see [2–4]) the narrow-band filters based cameras are applied. They provide the information necessary to estimation their current state (Fig. 1).

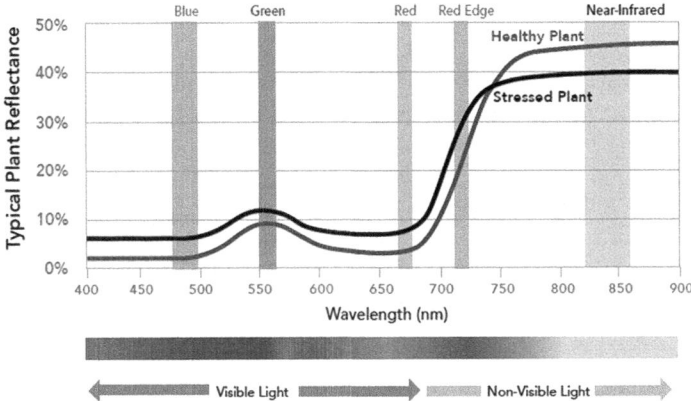

Fig. 1. Spectral reflectance of plants.

2 Problem Definition

Existing approaches to the calculation of vegetation indices, as a rule, are based on two independent parts of the spectrum of the curve shown in Fig. 1, namely, according to [4]: on the reflection in the red region of the spectrum, which accounts for the maximum absorption of solar radiation by chlorophyll of higher vascular plants; and on the reflection in the near infrared region of the spectrum, where the region of maximum reflection of the cellular structures of the leaf is concentrated. The most common vegetation index used to solve problems associated with estimates of vegetation cover is the Normalized Difference Vegetation Index (NDVI), which is calculated by the formula

$$NDVI = (NIR-RED)/(NIR+RED) \qquad (1)$$

where *RED* is the reflection in the red region of the spectrum, *NIR* is the reflection in the near infrared region of the spectrum (750 ÷ 900 nm). The values of the *NDVI* index vary from 0 to 1: the higher its value, the higher the vegetation intensity, and vice versa, the lower the index value, the vegetation is discharged, and the tendency to zero generally indicates open soil.

Reflections of vegetation in the spectral regions are estimated as a percentage, i.e. as shown in Fig. 1, or in decimal units. Moreover, the reflection ranges of the spectral regions, as a rule, does not exceed 50% (or 0.5 units). However, in real situations, the reflection data of the spectral regions must be considered as weakly structured, i.e. those of which their belonging to a certain type is known. In particular, the *RED* and *NIR* values determine the reflection levels in the red and infrared regions, which are estimated by different ranges in different sources, for example, in the form of 620 ÷ 750 nm, 550 ÷ 750 nm or 600 ÷ 700 for red and in the form of 750 ÷ 1300 nm, 750 ÷ 1000 nm or 700 ÷ 1000 for the near infrared region of the spectrum. It is not surprising, since the red and near infrared regions do not have crisp, explicit boundaries. In fact, they are fuzzy sets (or sets with degraded boundaries), i.e. weakly structured areas, which are advisable to describe as fuzzy subsets of the universe, covering a range of wavelengths of the spectrum, for example, from 400 to 900 nm, not taking into account wavelengths in the middle and far infrared ranges of the spectrum. Therefore, it is better to specify the reflections by interval $x \in [x_{min}, x_{max}]$ or in the form of a verbal expression of the type "close to 0.35", i.e. in the form of a fuzzy set. Then, it is necessary to formulate appropriate algorithms for forming the *NDVI* index using fuzzy analysis of data obtained from multispectral sensors based on the results of remote sensing of vegetation.

3 Evaluation the State of Vegetation Using Fuzzy Data Analysis

In [5], to display the Earth's surface a standardized continuous gradient scale of NDVI index values is summarized in the range from −1 to +1 by following Table 1.

Due to the features of reflection in the red and near-infrared regions of the electromagnetic spectrum, natural and artificial (not related to vegetation) objects have fixed NDVI values, which cannot be said about territories covered by vegetation. According to the above considerations, it is advisable to consider the RED and NIR reflection values of vegetation as weakly structured and, therefore, they are best described by fuzzy sets. For example, in the case of heavy vegetation RED and NIR can be reflected by following fuzzy sets: $RED = \{0.1/0.075; 1/0.1; 0.1/0.125\}$, $NIR = \{0.1/0.475; 1/0.5; 0.1/0.525\}$, which describe the evaluation concepts (terms) of the types "*RED* close to 0.1" and "*NIR* close to 0.5", respectively.

Table 1. *NDVI* index representations for different types of objects.

Object type	RED	NIR	NDVI	$NDVI^{\text{def}}$
Heavy vegetation	0.100	0.50	0.700	0.6674
Discharged vegetation	0.100	0.30	0.500	0.5013
Open soil	0.250	0.30	0.025	0.0910
Clouds	0.250	0.25	0.000	0.0000
Snow and/or ice	0.375	0.35	−0.050	−0.0345
Water	0.020	0.01	−0.250	−0.3349
Artificial materials	0.300	0.10	−0.500	−0.4921

The results of calculations of *NDVI* by the formula (1) at the α-levels are follows:

- $\alpha_1 = 0.1$ (at the left), $NDVI_{\alpha 1} = (0.475 - 0.075) / (0.475 + 0.075) = 0.7273$;
- $\alpha_2 = 1$, $NDVI_{\alpha 2} = (0.5 - 0.1) / (0.5 + 0.1) = 0.6667$;
- $\alpha_3 = 0.1$ (at the right), $NDVI_{\alpha 3} = (0.525 - 0.125) / (0.525 + 0.125) = 0.6154$.

In this case, the solution is the following fuzzy set:

$$NDVI_F = NDVI_{a1} \cup NDVI_{a2} \cup NDVI_{a3} = \{0.1/0.7273; \ 1/0.6667; \ 0.1/0.6154\}$$

The corresponding defuzzified value $NDVI^{\text{def}} = 0.6674$ is obtained by follows formula

$$NDVI^{\text{def}} = \left[\sum\nolimits_{i=1}^{3} NDVI_{\alpha_i} \alpha_i\right] / \left[\sum\nolimits_{i=1}^{3} \alpha_i\right].$$

Then the desired value of *NDVI* belongs to the interval $[NDVI^{\text{def}}, NDVI] = [0.6674, 0.7]$.

The index *NDVI* = 0.7 calculated according to formula (1) (see Table 1) is a rounded value. Therefore, the fuzzy set $NDVI_F$ more adequately reflects the heavy vegetation. Its triangular membership function is shown in Fig. 2.

For the others objects the obtained defuzzified values of the corresponding *NDVI* indexes are summarized in Table 1.

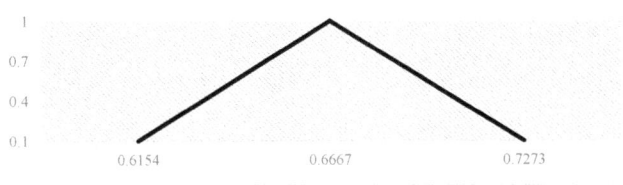

Fig. 2. The triangular membership function of fuzzy set $NDVI_F$.

4 Evaluation the State of Vegetation Using Fuzzy Inference

According to the discrete scale of object gradation presented in Table 1, the following judgments are chosen as a basis:

e_1: "If the reflection in the red region of the spectrum is weak and the reflection in the near infrared region of the spectrum is strong, then the vegetation index is high";

e_2: "If the reflection in the red region of the spectrum is weak and the reflection in the near infrared region is average, then the vegetation index is average";

e_3: "If the reflection in the red region of the spectrum is below average and the reflection in the near infrared region is average, then the vegetation index is low";

e_4: "If the reflection in the red region of the spectrum is below average and the reflection in the near infrared region is below average, then the vegetation index is low";

e_5: "If the reflection in the red region is above average and the reflection in the near infrared region is average, then the vegetation index is more than low";

e_6: "If the reflection in the red region of the spectrum is too weak and the reflection in the near infrared region is too weak, then the vegetation index is very low";

e_7: "If the reflection in the red region of the spectrum is average and the reflection in the near infrared region is weak, then the vegetation index is too low".

Analysis of information fragments $e_1 \div e_7$ in the form of cause-effect relations between the reflection levels of objects in the red and near infrared regions and the types of these objects allows to create a complete set of linguistic variables and implicative rules for forming the appropriate fuzzy inference system. For convenience, all linguistic variables are summarized in Table 2.

Table 2. Variables of the fuzzy inference system.

Type	Denote	Name	Term set	Universe
Input	x_1	Reflection in the red region	{TOO WEAK, WEAK, BELOW AVERAGE, AVERAGE, ABOVE AVERAGE, STRONG}	[0, 0.5]
Input	x_2	Reflection in the near infrared region	{TOO WEAK, WEAK, BELOW AVERAGE, AVERAGE, ABOVE AVERAGE, STRONG}	[0, 0.5]
Output	y	Vegetation index	{TOO LOW, VERY LOW, MORE THAN LOW, LOW, AVERAGE, HIGH}	[−1, 1]

In symbolic form, the fuzzy inference system is presented as follows:

e_1: $(x_1 = $ WEAK$)$ & $(x_2 = $ STRONG$) \Rightarrow (y = $ HIGH$)$;
e_2: $(x_1 = $ WEAK$)$ & $(x_2 = $ AVERAGE$) \Rightarrow (y = $ AVERAGE$)$;
e_3: $(x_1 = $ BELOW AVERAGE$)$ & $(x_2 = $ AVERAGE$) \Rightarrow (y = $ LOW$)$;
e_4: $(x_1 = $ BELOW AVERAGE$)$ & $(x_2 = $ BELOW AVERAGE$) \Rightarrow (y = $ LOW$)$;
e_5: $(x_1 = $ ABOVE AVERAGE$)$ & $(x_2 = $ AVERAGE$) \Rightarrow (y = $ MORE THAN LOW$)$;
e_6: $(x_1 = $ TOO WEAK$)$ & $(x_2 = $ TOO WEAK$) \Rightarrow (y = $ VERY LOW$)$;
e_7: $(x_1 = $ AVERAGE$)$ & $(x_2 = $ WEAK$) \Rightarrow (y = $ TOO LOW$)$.

For the final formation of the fuzzy inference system it is necessary to reflect the introduced terms of linguistic variables to the set of corresponding real numbers (universe) by establishing the membership functions. As the common construction for membership functions of fuzzy sets describing the terms of input and output linguistic variables following Gaussian function is chosen

$$m(u) = \exp[-(u - u_0)/\sigma^2], \qquad (2)$$

where u is the element of the corresponding universe; u_0 is the center; σ^2 is the density. So, following Figs. 3 and 4 show the membership functions for terms of input and output linguistic variables, respectively, which are generated by MATLAB\Fuzzy Inference System editor in accordance with equality (2).

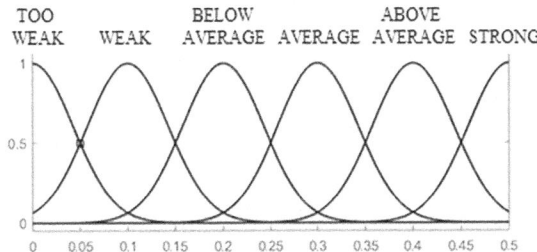

Fig. 3. Membership functions of input fuzzy sets.

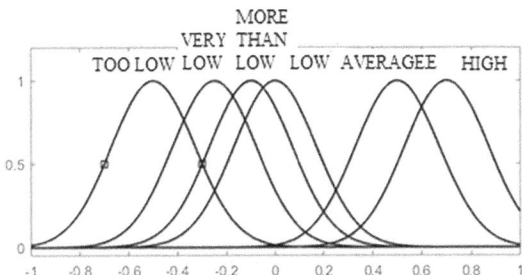

Fig. 4. Membership functions of output fuzzy sets.

After establishing all variables, membership functions and rules of the fuzzy knowledge base one can begin to analyze the operation of the fuzzy inference system. For this, it is convenient to use the graphical interpretation in MATLAB notation in the form of membership surfaces, which are shown in Fig. 5.

It is clearly seen that the quantitative indicator of photosynthetic active biomass does not exceed 0.7, decreases with increasing reflection of the near infrared region of the spectrum, and falls with increasing the reflection of the red region of the spectrum. In particular, after implementation of the rules $e_1 \div e_7$, the MATLAB\FIS editor of the Mamdani type forms an interactive window (Fig. 6), reflecting the fuzzy model of *IDVI* index. For example, when *RED* = 0.1 and *NIR* = 0.5 it generates *IDVI* = 0.687.

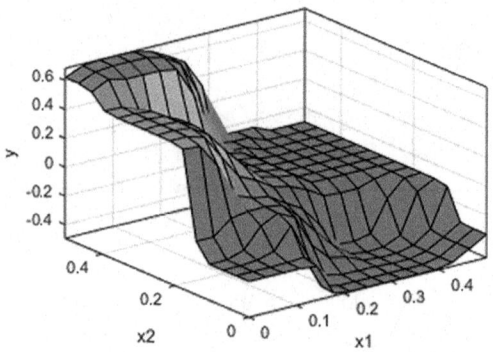

Fig. 5. The dependence of the *IDVI* index on *RED* and *NIR*.

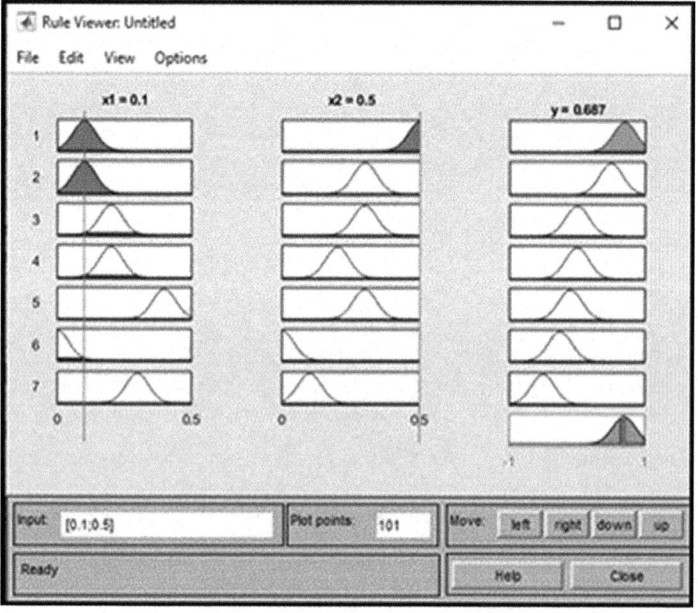

Fig. 6. Fuzzy model of *IDVI*.

5   Conclusion

Table 3 presents the results of the formation of the *NDVI*-index using fuzzy analysis of reflections in the red and near infrared regions of the electromagnetic spectrum.

As can be seen from Table 3, the values of the vegetation index obtained using the two proposed fuzzy approaches differ from the corresponding *NDVI* values calculated by the formula (1). This difference is noticeable in some cases, and in others cases it expressed as a decimal fraction. Nevertheless, in contrast to formula (1), which is

Table 3. Values of vegetation indexes obtained using different methods.

Object type	Quantitative evaluation of photosynthetic active biomass using:		
	Formula (1)	α-levels of fuzzy sets	Fuzzy inference
Heavy vegetation	0.6667	0.6674	0.6870
Discharged vegetation	0.5000	0.5013	0.4500
Open soil	0.0909	0.0910	0.0015
Clouds	0.0000	0.0000	0.0000
Snow and/or ice	−0.0345	−0.0345	−0.0999
Water	−0.3333	−0.3349	−0.2500
Artificial materials	−0.5000	−0.4921	−0.4500

formed empirically, the proposed fuzzy models of the vegetation index are more justified, although they are built on the basis of the discrete gradation scale for the levels of vegetation intensity and other objects under absence of vegetation.

In conclusion, it should be noted that the proposed approaches have a significant resource for improving the adequacy of the problem solution, primarily in the field of optimization of membership functions in both fuzzy approaches. In particular, the approach to the quantitative assessment of photosynthetic active biomass using fuzzy analysis can be significantly improved by introducing a qualitative evaluation criterion with its more adequate fuzzy interpretation. Quantitative evaluation of photosynthetic active biomass using fuzzy inference implies the approximation of the vegetation index, and, according to [6], it is possible by structural and parametric optimization of fuzzy implicative rules.

References

1. Vegetation Indices. Foundations, Formulas, Application. http://mapexpert.com.ua/index_ru. php?id=20&table=news
2. Loupian, E.A., et al.: IKI center for collective use of satellite data archiving, processing and analysis systems aimed at solving the problems of environmental study and monitoring. Sovremennye Problemy Distantsionnogo Zondirovaniya Zemli iz Kosmosa **12**(5), 263–284 (2015). https://doi.org/10.21046/2070-7401-2019-16-5-159-173
3. Abdelrahim, E., Abdelaziem, A.: Monitoring and yield estimation of sugarcane using remote sensing and GIS. American J. Engineering Research. **7**(1) (2018)
4. Official Website of MicaSense. https://micasense.squarespace.com/atlasflight
5. NDVI: Theory and Practice. https://gis-lab.info/qa/ndvi.html
6. Kosko, B.: Fuzzy systems as universal approximators. IEEE Trans. Comput. **43**(11), 1329–1333 (1994). https://doi.org/10.1109/12.324566

The Challenge of Adaptation in Future Networking Environment: Engineering Methodology

N. R. Yusupbekov[1] , Sh. M. Gulyamov[1] ,
and N. B. Usmanova[2(✉)]

[1] Tashkent State Technical University, Tashkent, Uzbekistan
dodabek@mail.ru, shukhrat.gulyamov@mail.ru
[2] Tashkent University of Information Technologies, Tashkent, Uzbekistan
nargizausm@mail.ru

Abstract. The technological strategy towards building the new Internet and efforts of research community in developing the paradigm of Future Network (FN) are devoted mainly to overcome the constraints of today's networks in performance and effectiveness. Through the digital transformation, the future of networking technologies for the coming decades will involve new technologies such as virtual and augmented reality, artificial intelligence, 5G, quantum computing, adaptive and predictive cybersecurity, intelligent Internet of things and many others, opening up for the users unprecedented opportunities of applications and services. However, these next-generation applications will require more sophisticated network functionality to providing much better support for a broad range of requirements. Considering some conceptual features of FN, including the ability to attract the application requirements, this paper describes the issues of adaptation for future networking systems supported by internet infrastructure in different terms. The methodology to engineering the adaptation is proposed to elaborate the behavior of components in complex networking environment for more flexible attraction of features and simple simulation scenario is provided to demonstrate the approach.

Keywords: Future networks · Networking environment · Adaptation · Technology requirements · Agent · Multi-agent system

1 Introduction

The pace of Internet development together with innovations in information and communication technologies designate the digital transformation of society bringing closer to the digital and knowledge economy. Future networks' technologies, including mobile broadband and all optical networks are the key to the revolutionary innovations of tomorrow, enabling to design and build the Future Infrastructure interconnecting and empowering the Digital Society, while digital transformation is evident in almost every aspect of work and personal life today [1, 2].

The research domain in Future Networks is aimed at delivering the next generation of network technologies enabling smart connectivity for all, anywhere, at any time at the highest speed and efficiency so as to meet an overwhelming demand by today's society [3]. Along with tremendous benefits available through technology for both business and society, there are many issues exist on performance, sustainability, security of networking environment. The Future Network (FN) should provide much better support for a broad range of applications, services, and network architectures. In the FN environment, multiple isolated logical networks each with different applications, services, and architectures share the physical infrastructure and resources. Considering some conceptual features of Future networking, including the ability to attract the application requirements, this paper describes the issues of adaptation for future networking systems in different terms.

There are many real-life examples of applying the technologies of FN: techniques of Internet of Things (IoT), Cyber-Physical Systems (CPS), to name a few [4, 5]. As we see from research and development efforts of academia and industry, technologies are being applied to enable intelligent and pervasive systems in infrastructure, software development, cybersecurity, and data. In addition, IoT and artificial intelligence provide foundation for 5G solutions, ubiquitous infrastructure, assistive monitoring, and others with specific usage (e.g. energy management, data mining, block chain, quantum computing, etc.). Wrapping up such areas and application domains, we can note three major characteristics: intelligence, autonomy and real-time behavior [6]. To attract these features in designing and implementing of such systems, the technological paradigm of Multi-Agent Systems (MAS) is used among others.

2 Some Features of Adaptation in FN and Statements for the Research

Different issues are within scientific and practical interest of research community, namely market tendencies and requirements for emerging technologies, the principles of distributed and ubiquitous networks, IoT, autonomous and self-organizing systems, programmable networks and elastic infrastructures, software-defined networks, data management protocols, virtualization and integration methods, information and content-centric networks, the principles of computing and storage (though these are widely represented in publications, e.g. referring to [7–10] as relevant to this paper context), while crucial features can be emphasized when considering the adaptation in network environment.

Adaptation to Enable the Intelligence in Network Architecture. In accordance with the statements of intelligent systems, the implementation of technologies largely depends on the realized principle of system organization, the conceptual basis for the models of objects and processes are represented, as well as on a number of other factors. In this regards, it is important to study the intuitive or associative aspect of information processing (by a human) and its implementation in information technologies: a human is able to flexibly process information, since the brain reflects the distributed representation of information, parallel processing, the ability to learn and self-organization, as well as the ability to integrate information. Technological representation of such ability in technical implementation (intelligent systems), should take into account the following features of information processes, namely:

– *functional*, which is characterized by the integration of uncertain information and the ability to adapt and learn;
– *computing*, which is characterized by highly parallel and distributed processing of multidimensional information with a large number of connections.

These features are developed and integrated in accordance with the development of technologies, implementing of distributed computing resources and data storage, complex intelligent systems, the use of composite applications, etc. and it is obvious that there is a need for various processes of information interaction. Therefore, the nature of future networking is to proceed from primitive to higher levels of cognition: learning, distinguished by the automatic adaptation of connections, thinking, and intelligence. Such transitive conversion would drastically increase the capacity of the network, its intelligence, and expand the possibilities for solving various tasks.

The network should be flexible enough to handle different requirements of various applications [11]. We refer here to 'double side' of network flexibility, i.e. flexible selection of provided functionality is viewed as *otside demands* (from application requirements); there are *inside demands* for flexibility from the network itself (set of protocols, transport technologies, etc.), so far changing network functionality with required level of adaptation is achieved.

Adaptation Engineering for FN Environments. In the concept of adaptation as an active action (control) usually two meanings are feasible to consider: passive adaptation, i.e. adaptation to a fixed environment (in this case, the adaptive system operates in a way that to perform its functions in a given environment in the best way, trying to maximize some performance criteria, e.g. effectiveness of functioning in a given environment) and active adaptation, i.e. search for an environment adequate to the given system (when change in the environment occurs in order to maximize the performance, or active search for an environment is required in which the desired result to be achieved). In both cases, two engineering aspects need to be considered to understand adaptation as a process, i.e. *the purpose of adaptation*, that is, when and how it is defined for the effective functioning of the system; and *the adaptation algorithm*, that is, in what way is the final goal to be achieved. So far, setting the goal and the way to achieve it are to define adaptation as a process. This means that adaptation refers to the management in complex environment defining the process of changing the parameters and structure of the system, and possibly the control actions based on current information with the aim of achieving a certain, usually optimal state of the system with initial uncertainty and changing working conditions [12]. It clearly defines the optimization requirement for a given criterion; however, in most cases the complex systems do not have a single criterion for functioning. Such systems operate in an environment wherein multi criteria define the nature of restrictions, thus selection of adaptation criteria is already an adaptive process itself and should be taken into account when determining adaptation.

3 Methodology and Formal 'Setting the Stage'

From the system management point of view, the adaptation should be considered from the positions of the adaptation object functioning. Taking into account the above-mentioned, herewith the problem of adaptation is described as a way to manage/control the object in the environment of uncertainty both for environment and the object itself. The latter is related primarily with the complexity of the object, which may cause the unclear situation for adequate modeling.

In order to cope with different uncertainty in environment and diverse range of objects' functionalities, the authors suggest to consider the adaptation issues from two points of view for conceptual modeling: 1) when it deals with an environment, the principles of software engineering are applied for modeling the features, and 2) when considering an object the multi agent systems are best suited for enabling individual features. Thus, the first one will be described by the tools of system modeling and the second one – by the agent-based modeling approach. It is interesting to see how formal methods can attract the 'external' requirements of application and technology, while using agents opens to the researcher the broader capabilities each entity within environment can demonstrate.

Logical Spaces for Adaptation Engineering. The first logical space is described using the notions of configuration mechanisms that are able to fit to the context specific requirements (although, other modeling approach can be implemented, depending on the functional requirements of software architecture) [13]. To ensure continuous improvement towards a shorter customization time and reduced cost, controlling of the adaptation process becomes a crucial task [14]. In addition, the flexibility of the customization process has to be determined whether it should allow higher degrees of freedom regarding the model adaptation besides configuration mechanisms whose possible outcome is determined in advance. Thus, the generic models integrating the possible changes and variations is provided because of configurable process and such a model can be configured to a specific solution. Within such considerations we refer to the statements of formal modeling [15] allowing to demonstrate the possibility of configurable model to be guided to a solution that fits to the user's requirements.

Through supporting the different configuration parameters the adaptation process leads to the state that is in relevance with the model, and obviously, to provide such configuration opportunities a configurable model must be able to provide a complete, integrated set of all possible process configurations.

The controlling phase can be structured in a sequence of three consecutive stages: 1) Data Gathering: model data, adaptation data and performance data to be gathered throughout the customization process (model data refers to the final adapted model resulting from the adaptation process; adaptation data – for the information on adaptation process, e.g. time, order, etc.; performance data – for the information on efficiency. 2) Data Merging: to allow comparative analyses in integrating some of the performance data that are directly related to specific model elements. 3) Data Analysis: The data that has been gathered and merged from different customers is being analyzed in order to identify shortcomings and improvements of the reference model and the adaptation mechanisms.

The second logical space is determined by the notions of agents whereas agent is an autonomous entity, interacting within an environment to realize common goals. Being gathered into MAS, collection of autonomous agents can sense the environment they are part of, and act on it in order to realize a purpose.

The management process for MAS will reflect the goal(s) to be realized; such goal then define the a certain specific state of the environment, which is desirable for the application and can not be implemented without outside interference. An agent in the process of communicating with the environment considers those of its parameters, which, on the one hand, determine the state of its intentions (depending on the type of agent) and on the other, can be changed, i.e. agent has the means for such an impact on the environment in which these parameters are changed in the required way, and agent, in defining the goals, responds only to these parameters [16–19]. The parameters of the environment, which determine its needs, but cannot be changed by the agent indirectly affect the behavior during goal setting. Thus, the object perceives the environment as a finite or infinite set of its parameters: $S = (S_1, ..., S_e)$, each of parameters can be of interests for an agent and can be changed. In other words, the situation perceived by the subject is always manageable: $S(U) = (s_1(U), ..., s_e(U))$, where U is the agent management function. The space of situations $\{S\}$, which is formed by the indicated parameters s_i $(i = 1, ..., e)$ is introduced, and each point of this space determines some specific situation that has developed around the agent; through such situational space $\{S\}$ the agent perceives the surrounding environment and various objects. However, the agent formulates its goals not in terms of the environment S: it is more convenient for the agent to operate with other properties called 'target concepts'. Let these target concepts be described by a vector $Z = (z_1, ..., z_k)$, where each target parameter z_i is uniquely determined by the situation S, i.e. $z_i = \psi_i(S)(i = 1, ..., k)$, functions $\psi_i(\bullet)$ determine the relationship between the state of the environment I and the target parameter z_i. In vector form, this relationship is expressed as some definite vector function $Z = \psi(S)$, where $\psi(S) = (\psi_1(S), ..., \psi_k(S))$.

Consider a k- dimensional target space $\{Z\}$, which is convenient for the agent by the fact that for each space point it can express the requirement (goal), and fulfillment of which, according to the agent will lead to the satisfaction of one or more of its requirements. The agent then formulates its goal as a vector target $Z^* = (z^*_1, ..., z^*_k)$, where z^*_i is the ith requirement for the state of the environment S, expressed using the function $\psi_i(S)$. These requirements objectives may have a different character, but their form should be unified. Thus, the process of formulating the objectives of the Z^* of the agent is related firstly with the definition of the vector function $\psi(S)$ and secondly, with the development of requirements imposed on each component of this vector. In general, the target Z^* is reflected in the situation space $\{S\}$, forming the system of target requirements:

$$S* : \begin{cases} \Psi_i(s) = a_i (i = 1, \ldots, s) \\ \Psi_i(s) \geq b_j (j = s+1, \ldots, s+p) \\ \Psi_v(s) \rightarrow \min(v = s+p+1, \ldots, s+p+l) \end{cases}$$

The point or area S^* satisfying these requirements is the state of the environment that the agent is looking for, whether the agent succeeds in achieving this state of the environment depends on its ability to influence the environment, that is, on the type of dependence $S = S(U)$ and from R resources allocated for management: $U \in R$. Obviously, that these resources determine the energy, material, temporal and other management capabilities of U.

Now consider the interaction of the target zone S^* and the trajectory of change of the environment $S(t)$ under the influence of external factors, i.e. the situation drift. If the drift trajectory $S(t)$ passes through the zone, the agent does not need any management/control. It remains to him to wait when external circumstances lead to the fact that $S(t)$ belongs to S^*. The agent prefers to manage the situation, that is, it purposefully affects the environment: $S_t = S\ (U, t)$ and changes it so that to achieve management objectives, i.e. $S(U, t) \in S*$.

4 Simulation of Agent-Based Adaptive Network (Model in AnyLogic)

To demonstrate the features of proposed methodology we implemented the light simulation in AnyLogic: using agent-based modeling whereas individual active components provide system dynamic being the main building blocks (active objects) in the model. Active entities, known as agents, must be identified and their behavior must be defined. Given the environment of agents with potential adopters reflecting the interaction, the global dynamics of the system then emerged from the interactions of the many individual behaviors.

Agent-based model consists of multiple agents and their environment. Agent represents the building block of the agent-based model. Every agent is given a set of rules according to which it interacts with other agents; this interaction then generates the overall system behavior. For convenience the agent in the model is depicted as person. To create agents in AnyLogic, the agent's internal structure is defined using the active object class. The required number of class instances is created, each one representing the individual agent.

The goal of a multi-agent system is to find out how independent processes can interact in a coordinated manner. The agent itself can adapt to changes in the external environment, i.e. it can change behavior when changes in the environment occurred. Simulation model project demonstrated these features and appropriate convergence for adaptation scenario.

5 Conclusion and Further Research Perspectives

The concept of the Future Network is expected to provide functionality and services that go beyond the limitations of modern networking technologies, while at the same time observing several problematic issues related to technical aspects concern the performance. On the research and development agenda of academia and industry, the conceptual vision for developing future network capabilities is being elaborated, together with specific research domain of adaptation possibilities.

In the general setting, based on the context of this research, the task is to determine the conditions of adaptation: in what cases (under what condition) the agent will change its behavior and adapt to changes in the environment and what is the significance of the connection or interaction for such adaptation. On the other hand, it is necessary to find out how different interactions will be formalized, what features and/or functions will be expressed during interaction. In this regard, the study of agents includes such issues as decision making (what decision making mechanism is available to the agent and how the agent decides or evaluates the situation depending on external information and its state); management (including hierarchical interactions between agents); communication (messaging between agents).

Along with proposed in this paper methodology for engineering the adaptation for networking systems, authors state the necessity to define the feedback mechanisms and appropriate model extensions when considering real-time systems. The further research efforts are for conceptual modeling of multi-agent systems for cyber-physical systems, including within technological requirements of Industry 4.0.

References

1. Future Networks & Services: Developing the future of the internet through european research. EU Comm. Inform. Soc. Media (2010). https://doi.org/10.2759/22659
2. Galis, A.: Future networks – design goals and challenges. a viewpoint from ITU-T. In: Seventh International Conference on Autonomic and Autonomous Systems, Keynote (2011). http://www.iaria.org/conferences2011/filesICAS11/ICAS2011_AlexGalis_KeyNote.pdf
3. Subharthi, P., Jianli, P., Raj, J.: Architectures for the Future Networks and the Next Generation Internet. A Survey/Report Number: wucse-2009–69 (2009)
4. Uckelmann, D., Harrison, M., Michahelleset, F. (eds.): Architecting the Internet of Things. Springer-Verlag, Heidelberg (2011). https://doi.org/10.1007/978-3-642-19157-2_1
5. Hamed, H., Olivier, B. (eds.): Recent Advances in Networking, vol. 1. ACM SIGCOMM eBook (2013)
6. Achtaich, A., Souissi, N., Mazo, R., Salinesi, C., Roudies, O.: Designing a framework for smart IoT adaptations. In: International Conference on Emerging Technologies for Developing Countries, pp. 1–10, Marrakech, Morocco. ffhal-01592470 (2017)
7. Yusupbekov, N., Adilov, F., Ergashev, F.: Development and Improvement of Systems of Automation and Management of technological processes and manufactures. J. Autom. Mobile Robot. Intell. Syst. **3**, 53–57 (2017)
8. Zouai, M., Kazar, O., Haba, B., Saouli, H., Benfenati, H.: IoT approach using multi-agent system for ambient intelligence. Int. J. Softw. Eng. Appl. **9**, 15–32 (2017). https://doi.org/10.14257/ijseia.2017.11.9.02
9. Calvaresi, D., Marinoni, M., Sturm, A., Schumacher, M., Buttazzo, G.: The Challenge of real-time multi-agent systems for enabling IoT and CPS. In: Proceedings of the International Conference on Web Intelligence, pp. 356–364. (2017). https://doi.org/10.1145/3106426.3106518

10. Singh, M.P., Chopra, A.K.: The Internet of Things and multi-agent systems: decentralized intelligence in distributed computing. In: IEEE 37th International Conference on Distributed Computing Systems (ICDCS), pp. 1738–1747. Atlanta, GA (2017). https://doi.org/10.1109/icdcs.2017.304
11. Hiramatsu, A., Kawamura, R.: Flexible networking technologies for future networks. NTT Tech. Rev. **8**(10), 1–3 (2012)
12. Schulz, Ph, Wolf, A., Fettweis, G.P., Waswa, A., Soleymani, D.A.: Network architectures for demanding 5G performance requirements: tailored toward specific needs of efficiency and flexibility. IEEE Veh. Technol. Mag. **2**, 33–43 (2019). https://doi.org/10.1109/MVT.2019.2904185
13. Olive, A.: Conceptual Modeling of Information Systems. Springer, Berlin Heidelberg (2007)
14. Padilla, F.A.: Self-adaptation for Internet of things applications. Software Engineering [cs. SE]. Université Rennes 1 (2016). English. ffNNT: 2016REN1S094ff. fftel-01426219v2f
15. Becker, J., Delfman, P. (eds.): Reference Modeling: Efficient Information Systems Design Through Reuse of Information Models. Physica-Verlag A Springer Company, Heidelberg (2007)
16. Rastrigin, L.A.: Adaptation of Complex Systems. Zinatne, Riga (1981). (in Russian)
17. Yusupbekov, N.R., Gulyamov, Sh.M., Kasimov, S.S., Usmanova, N.B.: Knowledge-Based planning for industrial automation systems: the way to support decision making. In: Aliev, R.A., Kacprzyk, J., Pedrycz, W., Jamshidi, M., Sadikoglu, F.M. (eds.) ICAFS 2018. AISC, vol. 896, pp. 873–879. Springer, Cham (2019). https://doi.org/10.1007/978-3-030-04164-9_115
18. Wooldridge, M.: An Introduction to MultiAgent Systems, 2nd edn. Join Wiley & Sons, New Jersey (2009)
19. Pico-Valencia, P., Holgado-Terriza, J.: Agentification of the Internet of Things: a systematic literature review. Int. J. Distrib. Sens. N. **10**, 1–20 (2018). https://doi.org/10.1177/1550147718805945

Machine Learning in Automated Chest Radiographs Classification

Abdulkader Helwan(iD)
and Mohammad Khaleel Sallam Ma'aitah$^{(\boxtimes)}$(iD)

Near East University, Lefkosha, Near East Boulevard,
99138 Nicosia, TRNC, Cyprus
Abdulkader.helwan90@gmail.com,
Mohammad.maaitah@neu.edu.tr

Abstract. The capability of artificial neural networks to use and combine simple learning rules to master different complex tasks is very motivating and has sufficed on lots of challenging pattern recognition problems. Thus, in this paper, we selected a pattern recognition task that can be tough even for humans. The aim is to investigate the capability of neural networks to learn and recognize patterns describing some segmented chest organs in binary segmented images. The geometric descriptions in addition to the entropy of those images have been leveraged on in this work for classification purposes. Two learning algorithms are used for training both feedforward and competitive neural networks which have been considered for the pattern recognition task. Overall, we show that a competitive neural network that falls into unsupervised learning category out-performed the feedforward network which is considered a supervised learning network. This outperformance is in terms of time, error rate, cost, and accuracy.

Keywords: Artificial neural networks · Pattern recognition · Segmented chest organs · Competitive neural networks · Unsupervised learning · Supervised learning

1 Introduction

Chest X-ray radiography images are non-invasive medical scans showing the chest region, non-visible electromagnetic radiations are usually used in these radiography scans [1]. The radiations used are able to penetrate through opaque objects, while some it is absorbed by the object being scanned also, depending on the composition and density of the particular object [2]. In this paper, we consider the use of X-ray radiography for chest scans, which is a very sensitive and important area in view of the body region involved. Generally, a medic may arrange a patient for chest radiography for symptoms such as chest pain, persistent cough, breathing difficulty, and coughing up blood relating to diseases which include pulmonary tuberculosis, lung cancer [2]. The images collected from these scans can help in diagnosing many things about the captured organs such as the condition of the lungs, if cancer is present, and if air or fluid is being trapped in areas around it. Also, the shape and size of the heart can be monitored accordingly (which is related to congestive heart failure); the blood vessels

R. A. Aliev et al. (Eds.): ICAFS 2020, AISC 1306, pp. 810–816, 2021.
https://doi.org/10.1007/978-3-030-64058-3_102

can be examined for anomalies. Furthermore, fractures can also be seen in such scans, and changes that can occur after surgery operations [3, 4].

Instead, we propose an artificial neural network based intelligent system that is capable of accepting the described images or their features, and hence outputs the classes of the chest X-ray segments contained in the images. The capability of neural networks to learn and recognized patterns, especially their geometric descriptions has been leveraged on in this work for classification [4]. The designed systems are trained on processed and segmented images collected from a public medical database [5]. In this work, both supervised and unsupervised learning algorithms for feedforward and competitive neural networks have been considered for the classification task. For supervised learning (feedforward neural network), it is of course required to label all training data for learning, which is a very tedious, costly, and manually intensive process. Alternatively, for an unsupervised learning (using the competitive learning), the process of training data labeling is not required, hence saves time, cost, and the amount of manual input required.

2 Backpropagation Neural Network (BPNN): Feedforward Network

It is a well-known and widely used training rule which is type of supervised learning. It is delta rule generalization which also referred as Least Mean Squares Algorithm (LMS) [6, 7]. In this study, several experiments were carried to evaluate this network over different learning rates and number of neurons.

Figure 1 shows the designed network.

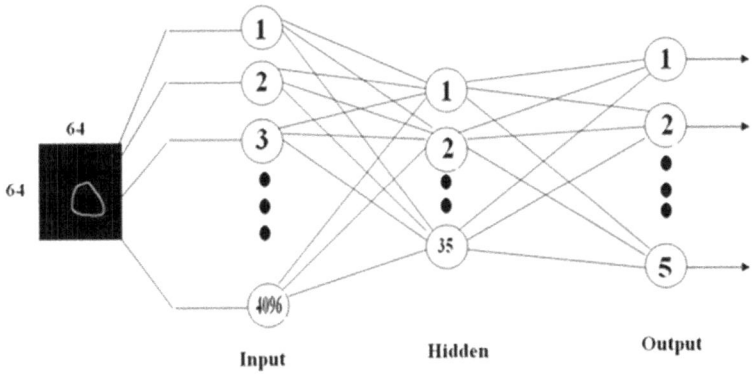

Fig. 1. Backpropagation neural network.

It can be seen in the figure above the number of input neurons equal the total number of input image pixels. i.e. 64×64 input image pixels equals 4096 pixels (number of input neurons).

Table 1 shows the different experiments carried out on the backpropagation networks to classify the five different radiographs.

Table 1. Training parameters for backpropagation networks (64 × 64 input pixels)

Networks	BPPN1	BPNN2	BPNN3	BPNN4
Training samples	620	620	620	620
Hidden neurons	0	35	45	60
Learning rate	0.010	0.0045	0.300	0.15
Momentum rate	0.040	0.0072	0.0504	0.0619
Activation function	Sigmoid	Sigmoid	Sigmoid	Sigmoid
Epochs	1000	1000	1256	1374
Training time (sec)	148	156	184	193
MSE	0.0077	0.0025	0.0056	0.0096

3   Competitive Neural Network (CNN)

Figure 2 shows the designed competitive neural network designed for the classifications of 5 different radiographs. As seen, the network has only two layers since it is a competitive network and needs no labeling for input images. This is because of its unsupervised training algorithm which depends on the winning neurons [8–10]. Therefore, the networks have no error to be reduced since there is no output labeling. Moreover, the output layer has five layers representing the five different organs we are classifying.

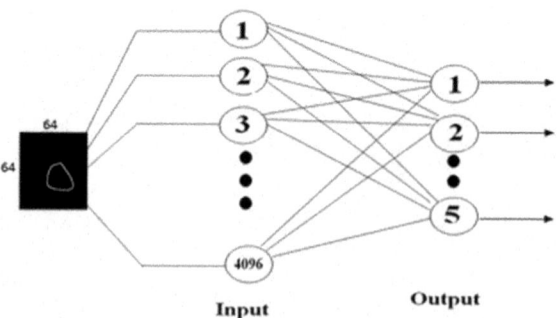

Fig. 2. Competitive neural network.

Table 2. Training parameters for competitive neural network (64 × 64 input pixels)

Networks	CNN1	CNN2	CNN3
Training samples	620	620	620
Learning rate	0.0036	0.05	0.1
Maxim Epochs	100	200	400
Training time (sec)	92	129	158

Table 2 shows parameters adjusted during the training of three different CNNs.

Furthermore, as it was carried out for the backpropagation networks, competitive networks were trained on 128 × 128 pixels images as input, such that the performance of the competitive learning can be compared with when 64 × 64 pixels images were used for training.

The table showing the different trained competitive networks is given below. i.e. Table 3.

Table 3. Training parameters for competitive neural network (128 × 128 input pixels)

Networks	CNN4	CNN5	CNN6
Training samples	620	620	620
Learning rate	0.0062	0.07	0.31
Maximum Epochs	100	200	400
Training time (sec)	97	135	164

4 Results Discussion

The designed backpropagation neural networks and competitive networks are tested using 64 × 64 pixels and 128 × 128 pixels and the results are described below in the following tables. The networks were simulated with the data that were not part of the training data. This has been done such the generalization power of the train networks can be obtained during testing (Fig. 3).

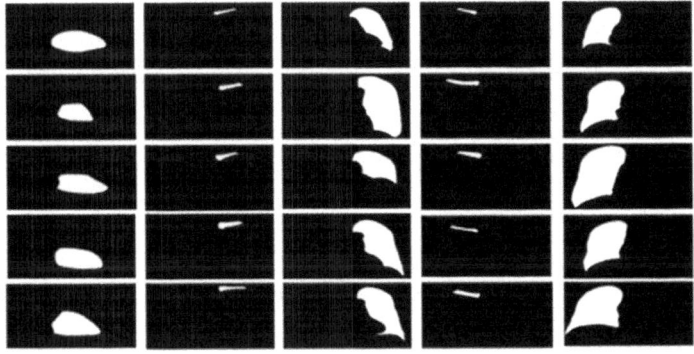

Fig. 3. Samples images for network testing.

The classification rates obtained by the backpropagation networks (BPNNs) trained on different parameters values are shown in Table 4.

Table 4. Recognition rates for BPNNs on training and validation data (64 × 64 pixels)

Network models	Training data (620)	Validation data (350)
BPNN1	92.74%	90.42%
BPNN2	99.19%	98.57%
BPNN3	97.32%	95.36%
BPNN4	98.10%	93.24%

As seen in the Table 4, all networks performed relatively well during training and validation (all above 90%), however one network (BPNN2) outperformed all other networks where it achieved the highest classification rates during training and testing.

Furthermore, to determine the effects of the rescaling the training data, training data images of size 128 × 128 pixels also been used to test the trained backpropagation networks as described previously.

Table 5. Recognition rates for BPNNs on training and validation data (128 × 128 pixels)

Network models	Training data (620)	Validation data (350)
BPNN5	87.43%	85.27%
BPNN6	86.85%	84.14%
BPNN7	88.52%	86.43%
BPNN8	83.49%	88.67%

Table 5 above shows the test results for the designed back propagation networks using 128 × 128 pixels as input image size.

It can be seen from Table 5 that BPPN7 achieved the highest recognition rates on both the training and test data, compared to BPNN5, BPNN6, and BPNN8.

Competitive neural networks which rely on an unsupervised learning algorithm were also trained in this study for the same classification task. These networks are faster to train considering that they have no desired outputs and therefore no error computations and back pass of error gradients for weights update, as it obtains in the backpropagation networks. Table 6 describes the results of using 64 × 64 pixels input image size for training the networks.

Table 6. Recognition rates for CNNs on training and validation data (64 × 64 pixels)

Network models	Training data (620)	Validation data (350)
CNN1	84.21%	81.40%
CNN2	85.23%	84.71%
CNN3	86.57%	76.25%

From the Table 6 above, it can be seen that CNN2 has the highest recognition rates on both the training and test data when the trained networks were simulated. Furthermore, it can be seen that CNN3, though, has a higher recognition rate than CNN2 on the training data, its performance on the test data is lower compared to CNN2. i.e. we can say that CNN3 has lower generalization power as compared to CNN2. The table showing the simulation results for the trained competitive neural networks which were trained on input images of size 128 × 128 pixels is given below as Table 7.

Table 7. Recognition rates for CNNs on training and validation data (128 × 128 pixels)

Network models	Training data (620)	Validation data (350)
CNN4	81.35%	80.04%
CNN5	79.93%	78.65%
CNN6	80.65%	78.97%

It can be seen from Table 7 that CNN4 has the best performance on both the training and test data (highest recognition rate), compared to the other networks CNN5 and CNN6, and trained on the same data.

5 Conclusion

Networks are good at detecting objects which are characterized by shapes and orientations. However, this task gets tougher when the classified objects are human organs such as heart and clavicles. This is because the similarity some organs may have in particular when the images used as input are binary. Moreover, this becomes more and more difficult when networks are supervised in an unsupervised manner. In this paper, supervised and unsupervised approaches were employed to classify 5 different chest organs: the heart, the left clavicle, the left lung, the right clavicle, and the right lung. The classification task is absolute pattern recognition-based learning task as images only differ in shapes. However, some pre-processing techniques were carried out before feeding images to networks in order to extract more accurate and unique features. Experimentally, it was found that an unsupervised learning-based network (competitive neural network) can outperform supervised based network (backpropagation neural network) in such task, as it reached higher accuracy and lower errors.

Conflicts of Interest. No conflict of interest was declared by the authors

References

1. Herrmann, T.L., et al.: Best practices in digital radiography. Radiol. Tech. **1**, 83–89 (2012)
2. Badie, B.M., Mostaan, M., Izadi, M., Alijani, N., Rasoolinejad, M.: Comparing radiological features of pulmonary tuberculosis with and without HIV infection. J. AIDS Clin. Res. **3**(10) (2012). https://doi.org/10.4172/2155-6113.1000188
3. Kim, T.-h.: Pattern recognition using artificial neural network: a review. In: Bandyopadhyay, S.K., Adi, W., Kim, T.-h., Xiao, Y. (eds.) ISA 2010. CCIS, vol. 76, pp. 138–148. Springer, Heidelberg (2010). https://doi.org/10.1007/978-3-642-13365-7_14
4. Nakamori, N., Sabeti, V., MacMahon, H.: Image feature analysis and computer-aided diagnosis in digital radiography: Automated analysis of sizes of heart and lung in chest images. Med. Phys. **3**, 342–350 (1990). https://doi.org/10.1118/1.596513
5. Van, G.B., Stegmann, M.B., Loog, M.: Segmentation of anatomical structures in chest radiographs using supervised methods: a comparative study on a public database. Med. Image Anal. **1**, 19–40 (2006). https://doi.org/10.1016/j.media.2005.02.002
6. Helwan, A., Tantua, D.P.: IKRAI: intelligent knee Rheumatoid arthritis identification. Int. J. Intell. Syst. Appl. **1**, 18–25 (2016). https://doi.org/10.5815/ijisa.2016.01.03
7. Helwan, A., Abiyev, R.: Shape and texture features for the identification of Breast Cancer. In: Proceedings of the World Congress on Engineering and Computer Science, San Francisco, USA, pp. 19–21 (2016)
8. Abiyev, R.H., Ma'aitah, M.K.S.: Deep convolutional neural networks for chest diseases detection. J. Healthc. Eng. (2018). https://doi.org/10.1155/2018/4168538
9. Hanif, M.S., Bilal, M.: Competitive residual neural network for image classification. ICT Express **6**(1), 28–37 (2020). https://doi.org/10.1016/j.icte.2019.06.001
10. Competitive neural networks for image segmentation. In: Tang, H., et al. (eds.) Neural Networks: Computational Models and Applications. Studies in Computational Intelligence, vol 53, pp. 129–144. Springer, Heidelberg (2007). https://doi.org/10.1007/978-3-540-69226-3_9

Machine Learning for Better Understanding of Autistics

Abdulkader Helwan[2] ⓘ, Mustafa Menekay[1] ⓘ,
and Mohammad Khaleel Sallam Ma'aitah[1(✉)] ⓘ

[1] Near East University, Lefkosha, Near East Boulevard,
99138 Nicosia, TRNC, Cyprus
{Mustafa.menekay,Mohammad.maaitah}@neu.edu.tr
[2] Lebanese American University, Byblos, Lebanon
Abedelkader.helwan@lau.edu.lb

Abstract. This paper presents a neural system that assists in understanding the autistics; in particular children who cannot make eye contact with other people and seem aloof. This neural system may help parents and doctors understand their autistics that may have difficulties in understanding and communicating. Thus, such system can identify the emotional and facial status of an autistic who may not capable of showing or saying it.

A backpropagation neural network is used in this work to identify the emotional expressions of autistics by reading their faces. The system is image based work, means that it uses face images in order to recognize their emotional expressions and classify them into seven different emotions: Sad, Happy, Nervous, Surprised, Disgust, Fear, Angry. The classification phase is where the images are classified based on their facial emotional expressions and here a feedforward backpropagation neural network is used. This network is trained using backpropagation learning algorithm, which uses gradient descent algorithm to minimize the error and learn.

Keywords: Autistics · Backpropagation neural network · Gradient descent

1 Introduction

According to the world health organization [1], in 160 children has an autism spectrum disorder (ASD). This neural disorder development can be characterized by some degrees of impaired social communications, interactions, behavior, and language. Autistics, in particular children may not make eye contact with other people and seem aloof. Moreover, some Children with autism may exhibit some difficulties in varying verbal abilities. Autism is dangerous as it may have links to other medical conditions, such as epilepsy and tuberous sclerosis complex [2].

All these challenges make the autistics-doctors communications very hard. Doctors have difficulties to understand, communicate with, and even treat autistics. Hence, there is a need for embedding today's technology into this field in order to facilitate autistics understanding for doctors [3].

As change of expressions on human face is an extreme technique for passing emotions, facial expression acknowledgment will be outstanding amongst other strides

R. A. Aliev et al. (Eds.): ICAFS 2020, AISC 1306, pp. 817–825, 2021.
https://doi.org/10.1007/978-3-030-64058-3_103

for enhancing HMI frameworks. A programmed facial expression acknowledgment framework involves three primary parts: face identification, facial features extraction and facial expression grouping. In the initial step, framework gets input picture and plays out some picture processing strategies with a specific end goal to find the face features [4].

In this paper, we attempt to develop a neural network system that can get the ability of identifying the human emotional status of a patient with autism by reading the faces. The neural network selected to be the heart of the work is backpropagation neural network (BPNN) that showed promising accuracies in classification and face detection tasks [5–7]. This network is first trained on faces of seven different human emotions expressions of different persons. Upon training, the network is tested on different images of faces that have some specific expressions and it is seen that the network was capable of identifying the real emotional expressions of faces [8].

2   Material Methods

In this paper; an emotional facial expression identification system is proposed. This system is image based work, means that it uses face images in order to recognize their emotional expressions and classify them into seven different emotions: Sad, Happy, Nervous, Surprised, Disgust, Fear, Angry. The classification phase is where the images are classified based on their facial emotional expressions and here a feedforward backpropagation neural network is used. This network is trained using backpropagation learning algorithm, which uses gradient descent algorithm to minimize the error and learn [9–11].

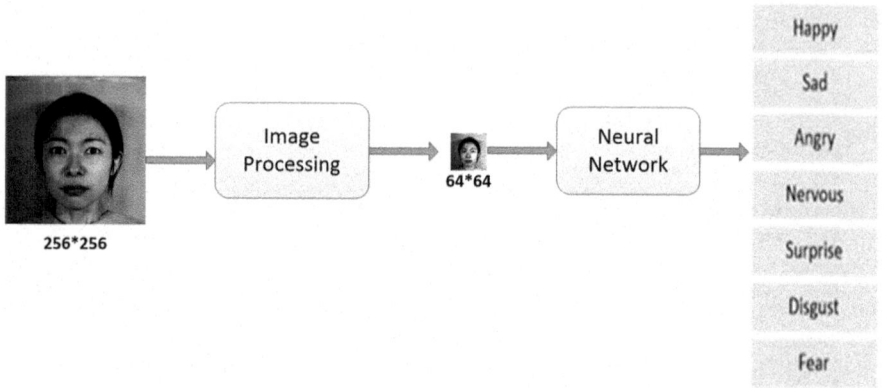

Fig. 1. Flowchart of the proposed emotional expression identification system

Figure 1 shows the flowchart of the emotional facial expressions system. As seen in the Fig. 1, the system is comprised of two main stages. In the first stage, the images are analyzed and processed in which the image intensity adjustment is applied to input images which helps in enhancing the contrast of images that might have some illu-mination effects from the source. Moreover, in this stage image sizes are reduced to 64 * 64 pixels which make them suitable for the neural network input layer.

2.1   Image Database

Images can be described as the "food" of the neural networks. More images mean the smarter and more accurate network will be. Thus, the first step in developing a neural based system is to find a good and public database which will be used for training the network. In this work, face images of different emotional expressions are needed to train our system to be capable of identifying the emotional expressing by checking the human faces. Therefore, the best face expression database was chosen for this task. This database is called The Japanese Female Facial Expression (JAFFE) Database and it is a public database available online for research usage [12].

The database contains of 216 images of 10 females. Among those images there are 7 different facial emotions such as Sad, Happy, Nervous, Surprised, Disgust, Fear, Angry. For each facial emotion there are 3 examples of each female face.

Table 1 shows the description of database and the amount of images it has. As seen, the database contains 216 images of different emotional face expressions.

Table 1. Database description

Facial expression	Number of poses per expression	Number of females
Angry	30	10
Sad	31	10
Happy	31	10
Neutral	31	10
Surprise	31	10
Disgust	31	10
Fear	31	10
Total	216	10

3   Networks Simulation

In this study, a feedforward neural network that uses backpropagation algorithm as a learning algorithm is selected to be used in this work, and it is named as BPNN. Two learning schemes are used for training the neural networks models on the same number of images. The first learning scheme involves images of size 64 * 64 pixels (BPNN1), while the learning scheme uses images of their original size 256 * 256 (BPNN2). The use of two learning schemes of different sizes aims to compare the networks perfor-mances with different image input sizes. The network models are trained on 146 images and tested on 70 different images of the same emotional facial expressions. Note that all networks are simulated using MATLAB software, 2013 version.

3.1   Network Training Scheme 1: 64 * 64 Pixels

The network was trained on 146 emotional expression images obtained from the JAFFE Dataset: 7 images for each different expression. The Table 2 shows the number of training sets which involves different facial expressions. It displays the overall number of database pictures that were trained; the trained images are used in the back propagation learning with adaptive learning and momentum rate to focus the function and to speed up the learning process. In this part, images of size 64 * 64 are used for training the network. During this phase, It layers of the neural network is three which includes; input layer, hidden layer, and output layer. The input layer comprises 4096 neurons since the picture estimate is 64 * 64, however; the hidden layer consists of 100 neurons, which proves significant training while keeping the time expense to a minimum. The output layer contains of 7 neurons; since the emotional expression types are 7 different emotional expressions types. Table 2 shows the training parameters of the network trained using learning scheme 1.

Table 2. Training input parameters of BPNN1 (64 * 64)

Network data	Values
Training images number	146
Input image size	64 * 64 pixels
Hidden neurons number	100
Type of activation function	Sigmoid
Learning rate (η)	0.14
Momentum rate (α)	0.6
Epochs	3000/3000
Training time (sec)	218.13
Reached MSE	0.0006

3.2   Network Training Scheme 2: 256 * 256 Pixels

Same network was also trained on the same images but with their original size which is 256 * 256. Moreover, in this scheme histogram was analyzed for each training image. Figure 2 shows the network architecture that uses learning scheme of input images size 256 * 256. It shows that the number of input neurons changes hence the input images size is changed. The result of the learning of network of the other scheme is shown in Fig. 2.

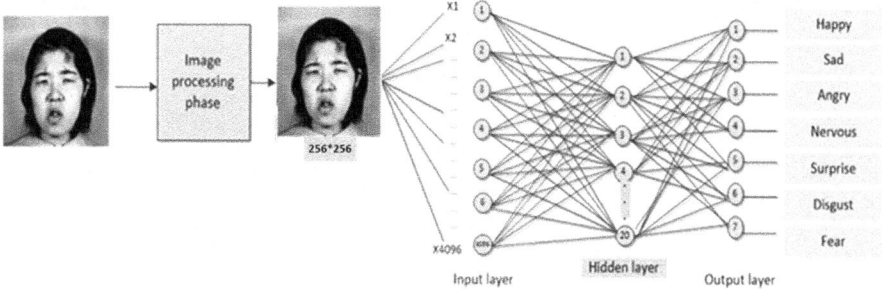

Fig. 2. Network Architecture of BPNN2 (256 * 256)

Table 3 shows the training parameters of the second learning scheme.

Table 3. Training input parameters (256 * 256)

Network data	Values
# of training images	146
Input image size	256 * 256 pixels
# of hidden neurons	100
Type of activation function	Sigmoid
Learning rate (η)	0.14
Momentum rate (α)	0.6
Epochs	759/3000
Training time (sec)	1365.13
Reached MSE	0.0140

3.3 Training Recognition Rates of Learning Schemes 1&2

As can be interpreted from Table 3, the network has shown some differences when trained on different input image sizes. Shows. The network (BPNN1) has achieved a minimum square error of 0.0006 with 3000 as maximum iterations when input images are of size 64 * 64. Note that this was achieved in 218.14 s. On the other hand, Table 4 shows that the network has achieved a higher minimum square error (0.014) than that of learning scheme 1; when input images of size 256 * 256 are used for training the network. Also, this is achieved in very long time of 1365.13 s but with smaller number of iterations 759.

Table 4. Training recognition rates between both learning schemes

	Images size	Total number of images	Recognition rate	Error achieved	Training time (s)
Learning scheme 1	64 * 64	146	99%	0.0006	218.14
Learning scheme 2	256 * 256	146	90%	0.0140	1365.13

Overall, it can be seen that using learning scheme 2 which involves input images of size 256 * 256 results in a higher MSE and required a lower number of iterations then that of learning scheme 1 (64 * 64). However, it is seen that the training of learning scheme 1 requires shorter training time (218.14 s) than that of learning scheme 1 (1365.13 s), and it achieved a lower error compared to that of BPNN2.

Furthermore, it is noticeable that the learning scheme 1 contributes to obtain a better training recognition rate for the network where it achieved 99% which is slightly greater than that obtained when using learning scheme 2 (90%), as shown in Table 4.

As seen in Table 4, the network that was trained using learning scheme 1 outperformed the one that uses learning scheme 2 in terms of training recognition rate, error reached, and training time. This may be due to the large size of input images used in learning scheme 2 where images are of size 256 * 256.

4   Testing the Network Models

Once the network models are trained and achieved a minimum square error rates, they should be tested in order to evaluate their performance in generalizing the recognition of new images, testing means simulating the networks using new images which were not used in training. This aims to investigate the capability of the models of recognizing new images of emotional expressions [13–15].

Note that all network models are tested on the same number of images which are 70 images containing 7 different emotional expressions.

4.1   Testing the Trained Network that Uses Learning Scheme 1&2

After successful iteration of the network, the network is being tested, in which the weights are being checked, in this dissertation new expression images that are non-existent in the literature are being proposed, models like shift and illuminations etc. 70 various emotional expressions pictures were used in testing the evaluation of the trained network that uses learning scheme 1 (BPNN1). The effects of testing and preparing stages are appeared in the accompanying Table 5.

Table 5. Classification rate of the network (learning scheme 1&2)

	Recognition rate of BPNN1 (learning scheme 1)	Recognition rate of BPNN2 (learning scheme 2)
Training	99%	90%
Testing	89%	82%

As seen in Table 5 the network that uses learning scheme 1 has achieved a high recognition rate of 94% which is considered good for such application.

5 Results Discussion

This paper features an overwhelming assignment in machine learning, in like manner in image processing. The backpropagation neural network that is thoroughly trained is then utilized as a part of a non-covering inspecting design to 'identify' images that contain faces of different emotional expressions. Table 5 summarized the results obtained by training and testing both network models. It compares the performance of both models in terms of errors, training time, and accuracies achieved. It is seen that the BPNN1 was capable of achieving a higher recognition rate (94%) than that obtained by BPNN2 (86%). Moreover, BPNN1 has reached a smaller error (0.0006) that achieved by BPNN2 (0.0140). This error is also achieved with shorter time (218.14 s) than the time need for BPNN2 (1365.13 s) to converge.

6 Conclusion

The neural networks have been recently applied in very tough tasks such as classification, detection, and prediction. The success of backpropagation neural networks in classification and identification tasks give positive and optimizing hopes to be used in classifying the human emotional expression into sad, happy, neutral, surprise, fear, angry. Thus, researchers have conducted many researches on how to develop a neural system that has a high accuracy in identifying the human emotional expression by checking the human face. Thus, in this paper, we aim to design a neural system trained using gradient descent to classify different man emotional expressions by only reading their faces in order to understand autistics feeling and needs. The aim is to obtain a high recognition rate with a small error and training time compared to other recent works when the network is tested in new images. This paper aims also to investigate the effects of input sizes in the learning and performance of the neural network.

Experimentally, the network was trained and tested on two input images sizes (64 * 64 pixels) and (256 * 256) and the performance was discussed and compared of each size. Note that the 64 * 64 size was selected as it reduces the computation time of network as well as it preserves the useful features of the image so that the network can learn the different features that distinguish the various expressions.

Finally, the smaller input size can result in a smaller error achieved during training in addition to shorter training time that those obtained when 256 * 256 size is used. It is seen that the input images of 64 * 64 pixels result in a higher performance in terms of recognition rate that obtained when 256 * 256 pixels input images are used.

References

1. World Health Organization, Autism. https://www.who.int/news-room/factsheets/detail/autism-spectrum-disorders
2. Webster, S., Potter, D.: Gaze perception develops atypically in children with autism. Child. Dev. Res. **2011**. https://doi.org/10.1155/2011/462389
3. Lee, H.C., Wu, C.Y., Lin, T.M.: Facial expression recognition using image processing techniques and neural networks. In: Pan, J.S., Yang, C.N., Lin, C.C. (eds.) Advances in Intelligent Systems and Applications - Volume 2. Smart Innovation, Systems and Technologies, vol. 21, pp. 259–267. Springer, Heidelberg (2013). https://doi.org/10.1007/978-3-642-35473-1_26
4. Popović, B., Ostrogonac, S., Delić, V., Janev, M., Stanković, I.: Deep architectures for automatic emotion recognition based on lip shape. In: 12th International Scientific Professional Symposium INFOTEH-JAHORINA, Jahorina, Bosnia and Herzegovina, pp. 939–943 (2013)
5. Schroff, F., Kalenichenko, D., Philbin, J.: FaceNet: a unified embedding for face recognition and clustering. In: Proceedings of the IEEE Conference on Computer Vision and Pattern Recognition, pp. 815–823 (2015). https://doi.org/10.1109/cvpr.2015.7298682
6. Wang, F., et al.: The devil of face recognition is in the noise. In: Ferrari, V., Hebert, M., Sminchisescu, C., Weiss, Y. (eds.) ECCV 2018. LNCS, vol. 11213, pp. 780–795. Springer, Cham (2018). https://doi.org/10.1007/978-3-030-01240-3_47
7. Cao, Q., Shen, L., Xie, W., Parkhi, O.M., Zisserman, A.: A dataset for recognising faces across pose and age. In 13th IEEE International Conference on Automatic Face & Gesture Recognition, FG 2018, pp. 67–74 (2018). https://doi.org/10.1109/fg.2018.00020
8. Helwan, A., Abiye, R.H.: ISIBC: an intelligent system for identification of Breast Cancer. In: Proceedings of International Conference on Advances in Biomedical Engineering, ICABME 2015, Lebanon, pp. 17–20 (2015). https://doi.org/10.1109/icabme.2015.7323240
9. Helwan, A., El-Fakhri, G., Sasani, H., Uzun Ozsahin, D.: Deep networks in identifying CT brain hemorrhage. J. Intell. Fuzzy. Syst. **35**(2), 2215–2228 (2018). https://doi.org/10.1155/2019/4629859
10. Oyedotun, O.K., Olaniyi, E.O., Helwan, A., Khashman, A.: Hybrid auto encoder network for iris nevus diagnosis considering potential malignancy. In: Proceedings of the International Conference on Advances in Biomedical Engineering, ICABME 2015, Lebanon, pp. 274–277 (2015). https://doi.org/10.1155/2019/4629859
11. Helwan, A., Uzun Ozsahin, D.: Sliding window based machine learning system for the left ventricle localization in MR cardiac images. Appl. Comput. Intel. Softw. Comput. (2017). https://doi.org/10.1155/2017/3048181
12. Lyons, M., Akamatsu, S., Kamachi, M., Gyoba, J.: Coding facial expressions with Gabor wavelets. In: Proceedings of the 3rd International Conference on Automatic Face and Gesture Recognition, pp. 200–205. City University of New York, New York (1998)
13. Abiyev, R.H., Helwan, A.: Fuzzy neural networks for identification of Breast Cancer using images' shape and texture features. J. Med. Imaging Health Infor. **8**(4), 817–825 (2018). https://doi.org/10.1007/s11760-010-0177-5

14. Ma'aitah, M.K.S., Abiyev, R., Bush, I.J.: Intelligent classification of liver disorder using fuzzy neural system. Int. J. Adv. Comput. Sci. Appl. **8**(12), 25–31 (2017). https://doi.org/10.14569/ijacsa.2017.081204
15. Abiyev, R.H., Ma'aitah, M.K.S. : Deep convolutional neural networks for chest diseases detection. J. Healthc. Eng. (2018). https://doi.org/10.1155/2018/4168538

Author Index

Printed by Printforce, the Netherlands